Cell Biology of Extracellular Matrix

SECOND EDITION

Cell Biology of Extracellular Matrix

SECOND EDITION

Edited by
Elizabeth D. Hay
Harvard Medical School
Boston, Massachusetts

PLENUM PRESS • NEW YORK AND LONDON

Library of Congress Cataloging-in-Publication Data

Cell biology of extracellular matrix / edited by Elizabeth D. Hay. --
2nd ed.
p. cm.
Includes bibliographical references and index.
ISBN 0-306-43951-4
1. Extracellular matrix. 2. Fibronectins. 3. Collagen. I. Hay,
Elizabeth D.
QP88.23.C43 1991
574.87--dc20 91-34402
CIP

ISBN 0-306-43951-4

A Division of Plenum Publishing Corporation
233 Spring Street, New York, N.Y. 10013

Printed in the United States of America

Contributors

Caroline M. Alexander Laboratory of Radiobiology and Environmental Health, Department of Anatomy, and Programs in Cell and Developmental Biology, University of California, San Francisco, California 94143-0750

David E. Birk Department of Pathology, Robert Wood Johnson Medical School, Piscataway, New Jersey 08854

Marilyn Gist Farquhar Division of Cellular and Molecular Medicine, University of California, San Diego, La Jolla, California 92093-0651

Vincent C. Hascall National Institute of Dental Research, National Institutes of Health, Bethesda, Maryland 20892

Elizabeth D. Hay Department of Anatomy and Cellular Biology, Harvard Medical School, Boston, Massachusetts 02115

Dick K. Heinegård Department of Physiological Chemistry, University of Lund, Lund, Sweden

John E. Heuser Department of Cell Biology and Physiology, Washington University School of Medicine, St. Louis, Missouri 63110

T. F. Linsenmayer Department of Anatomy and Cellular Biology, Tufts University Health Sciences School, Boston, Massachusetts 02111

Robert P. Mecham Department of Cell Biology and Physiology, Washington University School of Medicine, and Department of Medicine, Respiratory and Critical Care Division, Jewish Hospital at Washington University Medical Center, St. Louis, Missouri 63110

Bjorn Reino Olsen Department of Anatomy and Cellular Biology, Harvard Medical School, Boston, Massachusetts 02115

Erkki Ruoslahti Cancer Research Center, La Jolla Cancer Research Foundation, La Jolla, California 92037

Frederick H. Silver Department of Pathology, Robert Wood Johnson Medical School, Piscataway, New Jersey 08854

Bryan P. Toole Department of Anatomy and Cellular Biology, Tufts University Health Science Schools, Boston, Massachusetts 02111

Robert L. Trelstad Department of Pathology, Robert Wood Johnson Medical School, Piscataway, New Jersey 08854

Zena Werb Laboratory of Radiobiology and Environmental Health, Department of Anatomy, and Programs in Cell and Developmental Biology, University of California, San Francisco, California 94143-0750

Thomas N. Wight Department of Pathology, University of Washington, Seattle, Washington 98195

Kenneth M. Yamada Laboratory of Developmental Biology, National Institute of Dental Research, National Institutes of Health, Bethesda, Maryland 20892

Preface

In the ten-year interval since the first edition of this volume went to press, our knowledge of extracellular matrix (ECM) function and structure has enormously increased. Extracellular matrix and cell–matrix interaction are now routine topics in the meetings and annual reviews sponsored by cell biology societies. Research in molecular biology has so advanced the number of known matrix molecules and the topic of gene structure and regulation that we wondered how best to incorporate the new material. For example, we deliberated over the inclusion of chapters on molecular genetics. We decided that with judicious editing we could present the recent findings in molecular biology within the same cell biology framework that was used for the first edition, using three broad headings: what is extracellular matrix, how is it made, and what does it do for cells? Maintaining control over the review of literature on the subject of ECM was not always an easy task, but we felt it was essential to the production of a highly readable volume, one compact enough to serve the graduate student as an introduction and the investigator as a quick update on the important recent discoveries. The first edition of this volume enjoyed considerable success; we hope the reader finds this edition equally useful.

Elizabeth D. Hay

Contents

Chapter 6

Collagen Biosynthesis

Bjorn Reino Olsen

Chapter 7

Matrix Assembly

David E. Birk, Frederick H. Silver, and Robert L. Trelstad

PART III. WHAT DOES MATRIX DO FOR CELLS?

Chapter 9

Proteoglycans and Hyaluronan in Morphogenesis and Differentiation

Bryan P. Toole

Chapter 10

Integrins as Receptors for Extracellular Matrix

Erkki Ruoslahti

Chapter 11

The Glomerular Basement Membrane: A Selective Macromolecular Filter

Marilyn Gist Farquhar

Introductory Remarks

ELIZABETH D. HAY

Cytoskeleton, cell shape, cell migration, control of cell growth and differentiation, these are all subjects that, to be fully understood today, require a consideration of the extracellular matrix (ECM): its composition, role in development, and relationship to the cell surface. The ECM is the relatively stable structural material that lies under epithelia and surrounds connective tissue cells, but the idea that the ECM is an inert supporting material, created by the cells as a mere scaffolding on or in which to reside, is now bygone. Certainly, collagens are a source of strength to the tissues, elastin and proteoglycans are essential to matrix resiliency, and the structural glycoproteins help to create tissue cohesiveness. But the cell, having produced these extracellular macromolecules and influenced their assembly in one way or another, does not then divorce itself of them. The cell continues to interact with its own ECM products, and with the ECM produced by other cells. At the cell surface, matrix receptors link the ECM to the cell interior; the metabolism and fate of the cell, its shape, and many of its other properties are continuously related to and dependent on the composition and organization of the matrix.

The ECM is composed of an even greater variety of molecules than would have been believed possible when the first edition of this volume was produced a decade ago. Excluding all the molecules that must pass through the matrix to reach the cells or are trapped in and/or activated by the ECM, we are left with a basic structural composition of at least four major classes of macromolecules. In Chapter 1, Dr. Linsenmayer brings us up to date on the structure, chemistry, and molecular biology of the different (genetically distinct) types of collagen molecules. Then in Chapter 2, Drs. Wight, Heinegård, and Hascall describe the remarkable progress that has been made in recent years in our knowledge of proteoglycan composition and assembly. In Chapter 3, Drs. Mecham and Heuser describe a unique protein of the ECM, elastin, and give us insights into its structure and its interaction with cells, and into the development of elastic fibers. The final class of ECM molecules, the other structural glycoproteins, consists of relatively large molecules with sugar side chains that are longer than those found in collagen and that are often rich in sialic acid. These glycoproteins are discussed in Chapter 4 by Dr. Yamada, who like many of the authors is a pioneer in his area of expertise.

A volume on the cell biology of ECM would, of course, consider the manner in which cells synthesize ECM, influence its organization, and bring about its degradation. These subjects are covered in the second section of the volume. In Chapter 5, Drs. Hascall, Heinegård, and Wight describe the biosynthesis of glycosaminoglycans (GAG) and the protein cores that, together with GAG, form the proteoglycan (PG) monomers that become assembled into the PG aggregates of the ECM; in both Chapters 2 and 5, the authors take care to relate biochemical aspects of PG organization and synthesis to tissue and cell ultrastructure. In the current edition of this book, Dr. Hascall and his colleagues bring us up to date on PG gene structure and show us how the synthesis of PG involves the Golgi apparatus in essentially every known protein glycosylation pathway. The role of the endoplasmic reticulum and Golgi complex in the synthesis and secretion of collagens is developed in Chapter 6 by Dr. Olsen, who also informs us of recent advances in the area of collagen gene structure and function. In Chapter 7 Drs. Birk, Silver, and Trelstad present a synthesis of our current understanding of collagen assembly in tissue. Finally in Chapter 8, Drs. Alexander and Werb show us what is known and not known about collagenases and the remarkable process of ECM degradation.

The matrix, as indicated earlier, is, in a sense, talking back to the cells that create (and reside on or in) its interstices. In the third section of the book, we consider the phenomenon of cell–ECM communication in more detail. Dr. Toole (Chapter 9) tells us about the role of hyaluronic acid and PG in morphogenesis, cell migration, and cell proliferation, and he presents a model of interaction of hyaluronic acid with the cell surface. Dr. Ruoslahti (Chapter 10) then reviews the recent explosive developments in our understanding of the ECM receptors called integrins. In Chapter 11, Dr. Farquhar argues that the PG component of the glomerular basement membrane (GBM) plays an important role in the filtering function of the glomerulus, and in the last chapter, I give an overview of the effects of collagens and other structural glycoproteins on cell differentiation, cell shape, and cell metabolism, and speculate further on the relationship of the ECM to the cell surface.

We are fortunate in being able to bring together so many investigators who are experts in their fields and who were so willing to make clear to the cell biologist the advances in, and future of, these important subjects. In editing this endeavor, I have not attempted to impose my own views of terminology on these distinguished authors. Thus, the reader will find that some terms are used interchangeably, the best example being the terms *basal lamina* and *basement membrane* (BM). The basal lamina is a zone about 100 nm wide under epithelia and around muscle cells; it consists of a central, compact sheet of collagen, laminin, and other glycoproteins, called the lamina densa, that is separated from the cell by a less electron-dense zone, the lamina lucida externa, and from the underlying connective tissue by a second electron-lucid zone, the lamina lucida interna (lle, lli, Fig. i-1).

Both the laminae lucidae externa and interna contain a layer of PG granules connected by small filaments to the cell, on one hand, and to underlying collagen fibrils, on the other; PG is not visible unless special fixatives are used, hence the terms *rara* or *lucida* are used to refer to the "empty" zone. However, ruthenium red fixation very clearly reveals a layer of PG granules on the inter-

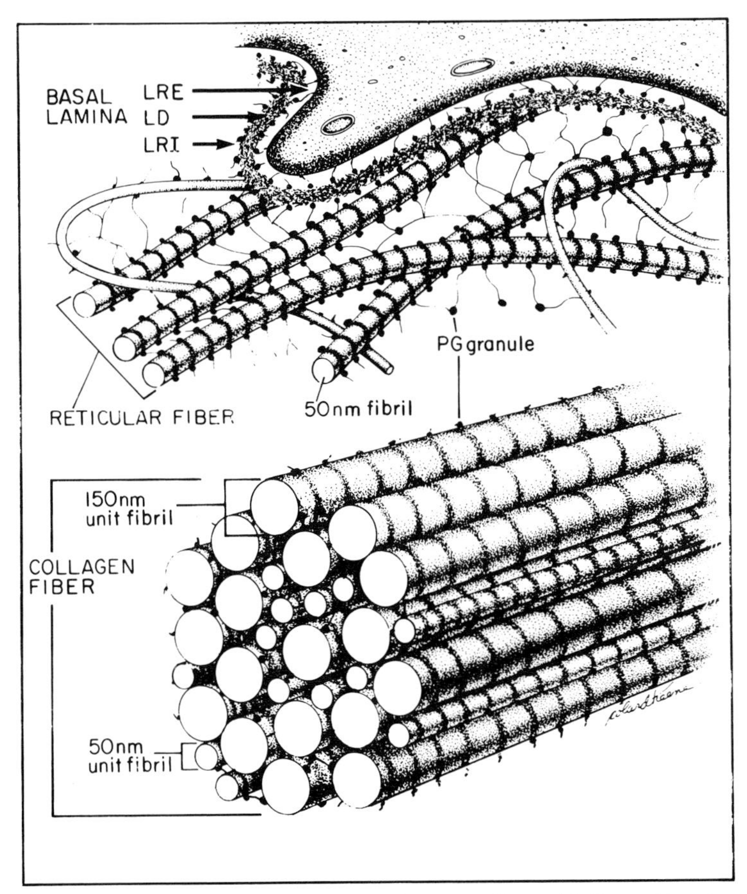

Figure i-1. Diagrams depicting relations of proteoglycans, collagen fibrils, and basal lamina in the tissues. The drawing at the top includes part of the basal cytoplasm of an epithelial cell. The outermost layer of the basal lamina, the lamina rara externa (LRE), is attached to the epithelial cell by small filaments. The LRE and LRI (lamina rara interna) each contain a layer of proteoglycan (PG) granules. The lamina densa (LD) contains collagen and other glycoproteins (see Chapter 11). Collagen fibrils associated with the basal lamina are small (50 nm in diameter) and may form small bundles called reticular fibers that stain with silver salts. The fibrils are covered with proteoglycan granules and are connected by ruthenium red-staining filaments that may consist of hyaluronic acid (see Chapter 2). Collagen fibers are larger than reticular fibers; they contain less proteoglycan and a greater variety of unit collagen fibrils (50–150 nm in diameter).

nal, as well as the external, side of the lamina densa in all tissues examined (Trelstad, Hayashi, and Toole, 1974, *J. Cell Biol.* **62:**815–830).

The term *basement membrane* was originally used by light microscopists to refer to the whole condensation of connective tissue (basal lamina and the associated collagen fibrils, or reticular lamina) under an epithelium. Only in the glomerulus and a few other locations is the basal lamina free of collagen fibrils and thus truly equivalent to the BM visualized previously by light microscopists. Another terminological problem is that in the glomerulus, the basal laminae of the epithelium and endothelium are fused, obliterating the layer we referred to above as the lamina lucida interna; the latter term is given to the PG-rich, juxtaendothelial zone of the GBM. In the interim since the publication of the first edition of this volume, the International Anatomical Nomenclature Committee recommended that the term *lamina lucida interna* only be used to describe the layer next to the endothelium, but this nomenclature fails to recognize the PG-rich inner layer of simple BM described by Trelstad *et al.* and others. This committee recommended the term *lamina fibroreticularis* for the reticular lamina (Laurie and Leblond, 1985, *Nature* **313**:272).

The term *connective tissue* is often used to refer to the collagen fibrils and other ECM molecules that surround cells of the fibroblast family. Some authors, however, include the basal lamina, since this structure is composed of molecules similar to those of the ECM proper and "connects" epithelium, Schwann cells, and muscle to it. By connective tissue "cells," most authors mean cells of the fibroblast family (osteoblasts, chondroblasts, and fibroblasts), but muscle and epithelial cells also secrete ECM and, in this sense, are part of the cellular component of connective tissue. Unlike the connective tissue cells proper (the fibroblast family), however, muscle and epithelial cells are separated from collagen fibrils by basal laminae; the reasons for this are not immediately obvious, but it is likely that these groups of cells have distinct cell surfaces that differ in their interactions with ECM (Chapter 12). Collagen fibrils are 10 nm in diameter or wider (above 25 nm in diameter, they appear striated); the term *filament* is usually used in this book to refer to a fibrous structure less than 10 nm in diameter. Collagen fibrils are often organized into small fibers called reticular fibers, which are argyrophilic as seen under the light microscope, and into larger collagen fibers; the diameters of the fibrils composing these fibers vary in a predictable way (Fig. i-1).

Cell Biology of Extracellular Matrix is a volume written by experts, all of whom have contributed measurably to our understanding of the ECM. They have agreed to present their material succinctly and in a manner understandable and relevant to the cell biologist who has not been working in the field. The chapters are not "review" articles; for full coverage of historical and other details, I have asked the authors to refer the reader to the numerous, more technical reviews that appear in the journals and volumes of the trade and are readily accessible to the interested reader desiring more depth. This volume is written for the cell biologist, then, but it does present the "state of the art" in ample detail to serve as a ready reference for all who wish to think and speak intelligibly about the ECM. Many of the techniques used to study the ECM are presented in footnotes, in figure legends, or in the text. There is much in this volume about disease as well as health. This is, in short, a volume that we believe will convince even the most doubting that understanding of the cell requires understanding of the ECM.

Part I

What Is Extracellular Matrix?

Chapter 1

Collagen

T. F. LINSENMAYER

1. Introduction

Collagenous proteins are a major constituent of all extracellular matrices. Traditionally, the role attributed to collagen has been a structural one. During the past 10 years, however, it has become obvious that collagen comprises a large heterogeneous class of molecules, some with the structural properties classically attributed to "collagen," but numerous others with additional properties. Vertebrates contain at least 15 different types of collagen. These are found in unique tissue-specific patterns, arise during development in defined temporal and spatial patterns, and exhibit different functional properties. Also, collagens have been shown to be involved, either directly or indirectly, in cell attachment and differentiation, as chemotactic agents, as antigens in immunopathological processes, and as the defective component in certain pathological conditions. Thus, in addition to their structural roles, collagens potentially have numerous developmental and physiological functions, many of which remain to be elucidated.

In the present chapter, we will examine the biochemical characteristics of the triple-helical structure present in all collagens. Then we will survey 12 of the various types of collagens with respect to their properties, supramolecular forms, and putative functions. Where appropriate we will also mention and briefly discuss the various methodologies utilized in studying collagens, especially those pertaining to studies of cell and developmental biology.

Due to the breadth of this chapter, only a limited number of citations are made to original publications, and these in general are to work published in the last several years. Instead, when possible, references are made to readily obtainable review articles on each specific topic. More comprehensive citations to

Abbreviations used in this chapter: CM-cellulose, *O*-(carboxymethyl)-cellulose; COL, collagenous domain; NC, noncollagenous domain; P_C-collagen, COOH-terminal peptide extension-collagen; P_N-collagen, NH_2-terminal peptide extension-collagen; SLS, segment-long-spacing.

T. F. LINSENMAYER • Department of Anatomy and Cellular Biology, Tufts University Health Sciences School, Boston, Massachusetts 02111.

Cell Biology of Extracellular Matrix, Second Edition, edited by Elizabeth D. Hay, Plenum Press, New York, 1991.

the "earlier" work can be found in the chapter on collagen in the previous edition of this volume (Linsenmayer, 1981), and extensive reviews of many of the individual collagen types can be found in *Structure and Function of Collagen Types* (Mayne and Burgeson, 1987).

2. The Collagen Molecule

2.1. Triple-Helical Domain(s)

The shape and most of the structural properties of a native collagen molecule are determined by its triple-helical domain(s). In the classic fibril-forming collagens a single triple-helical domain comprises more than 95% of the molecule (Fig. 1-1). Other collagens, however, have multiple triple-helical domains and these may comprise only a fraction of the molecule's mass.

A triple-helical or collagenous domain consists of three separate chains (α chains), each of which contains a characteristic sequence of amino acids twisted in the form of a left-handed helix, termed a polyproline type II helix (Ramachandran and Ramakrishnan, 1976). These three helical chains are then wrapped around one another in a higher-order ropelike fashion to produce the tight, triple-helical structure of the molecule (Traub and Piez, 1971). This conformation is stabilized by interchain hydrogen bonds. These are the bonds that are disrupted during thermal or chemical denaturation, resulting in unfolding of the molecule. The triple-helical conformation is wound such that the peptide bonds linking adjacent amino acids are buried within the interior of the molecule. Thus, the triple-helical region is highly resistant to attack by general proteases such as pepsin. If the molecule is denatured, however, the liberated chains are susceptible to proteolysis and are reduced to small peptides. The only enzymes that are able to efficiently attack the helical portion of native collagen molecules are the collagenases (see chapter on collagenases).

Folding of the component α chains into the proper helical conformation requires that glycine be present as every third amino acid residue. The α chains are therefore composed of a series of triplet Gly-X-Y sequences in which X or Y can be any amino acid (Fig. 1-2). Frequently, X is proline and Y is hydroxyproline. The presence of other amino acids at the X and Y positions does not seem to follow any recognizable or repeating pattern. Having a large number of hydroxyproline residues in the Y position adds stability to the helical structure. Hydroxyproline is formed by the posttranslational enzymatic conversion of proline in the Y position to hydroxyproline (described below). If collagen is synthesized under conditions in which this conversion is blocked, the resulting molecules have a lowered melting temperature that appears to be proportional to the degree of inhibition of proline hydroxylation. At present, the only other amino acid with known functional significance is lysine. Lysine and its enzymatically modified forms participate in covalent cross-linking between chains and molecules, and as sites for sugar attachment.

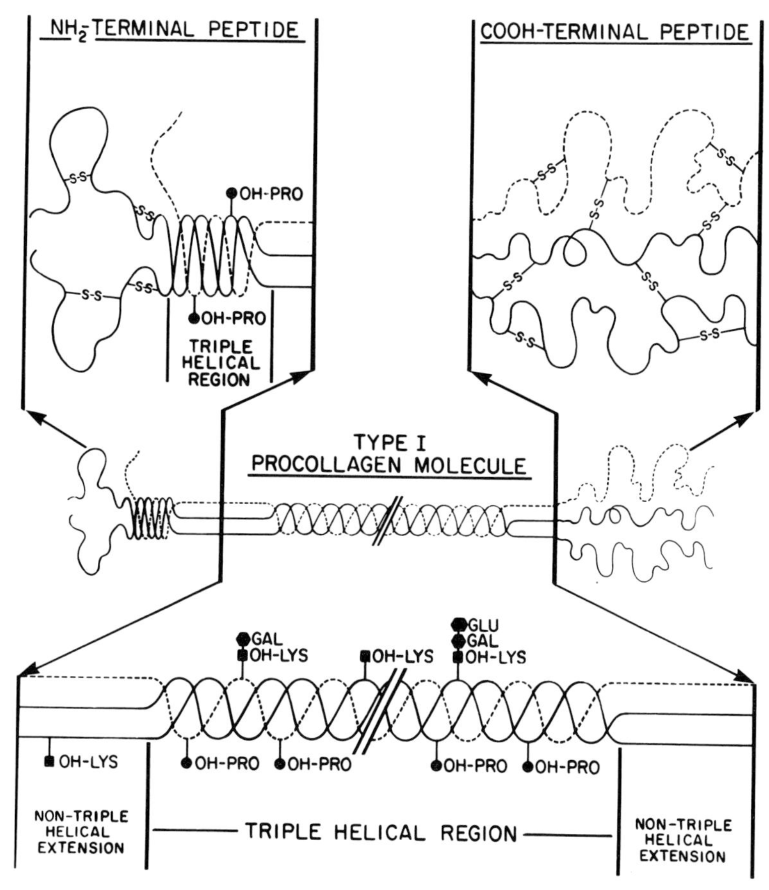

Figure 1-1. Diagram of the major characteristics of the type I collagen molecule and its procollagen form. OH-PRO, hydroxyproline; OH-LYS, hydroxylysine; –S-S–, disulfide bonds; GAL, galactose; GLU, glucose. The extension peptide sugars are discussed in Chapter 6. NH_2-terminal propeptide and COOH-terminal propeptide show the two pieces that are cleaved from procollagen during its processing into a collagen molecule. (Modeled after Prockop *et al.*, 1979.)

2.2. Posttranslational Modifications

2.2.1. Proline

Proline incorporated in the Y position of the Gly-X-Y triplets of newly forming α chains can be enzymatically converted to the 4-isomer of hydroxyproline, 4-hydroxyproline (Grant and Prockop, 1972; Prockop *et al.*, 1979). In a fibrillar collagen, about half the 200 or so prolines undergo this conversion. The enzyme responsible is prolyl hydroxylase. Its substrate speci-

```
        -Gly - X - Y - Gly - X - Y - Gly - X - Y -

NH2-

        -Gly - Pro - Ser - Gly - Pro - Arg - Gly - Leu - Hyp -

        -Gly - Pro - Hyp - Gly - Ala - Hyp - Gly - Pro - Gln -

        -Gly - Phe - Gln - Gly - Pro - Hyp - Gly - Glu - Hyp -

        -Gly - Glu - Hyp - Gly - Ala - Ser - Gly - Pro - Met -

                                                              -COOH
```

Figure 1-2. Amino acid sequence of the cyanogen bromide peptide α1-CD2, derived from the α1 chain of rat skin type I collagen. Note the Gly-X-Y repeating triplets and the presence of hydroxyprolines (Hyp) in the Y position.

ficity is limited to peptide-linked proline residues that precede a glycine residue (i.e., the Y position). In the presence of molecular O_2, α-ketoglutarate, and the cofactors ascorbic acid and ferrous iron, it converts these proline residues into hydroxyprolines. Although prolyl hydroxylase will work on small polypeptides, in general its activity increases with increasing peptide size. In addition to 4-hydroxyproline, different collagens have small, variable amounts (1–7 residues/chain) of 3-hydroxyproline. This second isomer is probably formed by the action of a different enzyme and has no known function.

The formation of 4-hydroxyproline is necessary for helix stability, so it is possible to change the properties of the collagen that is produced. This can be achieved either by modifying the insertion of prolines through the use of amino acid analogues or by deleting one of the substrates or cofactors required for the enzymatic formation of hydroxyproline. Incubating cells in a N_2 environment *in vitro* limits the availability of O_2. The compound αα-dipyridyl, when added to culture medium, chelates available iron. Both of these procedures inhibit the formation of hydroxyproline, which causes the cells to accumulate an unhydroxylated form of collagen called protocollagen. The administration of proline analogues, such as L-azetidine-2-carboxylic acid or *cis*-hydroxyproline, to animals or *in vitro* cultures causes the synthesis of aberrant collagen chains that are inefficiently secreted. These analogues cannot be considered to be specific for collagen, however, because proline is found in other proteins whose synthesis might also be affected. A cofactor that can easily become limiting is ascorbic acid, as it is rapidly destroyed under tissue culture conditions; in long-term cultures fresh ascorbate is added daily.

2.2.2. Lysine

The amino acid lysine, once incorporated into nascent collagen chains, can undergo one or more sequential enzymatic conversions. Lysine residues in the

Y position of Gly-X-Y triplets can be converted to hydroxylysines by the enzyme lysyl hydroxylase (see Fig. 1-1, OH-LYS). The substrate specificity of the enzyme (i.e., lysine preceding a glycine in peptide linkage) and the cofactor requirements, which include molecular O_2, ferrous iron, α-ketoglutarate, and ascorbic acid, are the same as for prolyl hydroxylase. A special case in which the hydroxylation of lysine before glycine rule does not hold is found in the non-triple-helical extensions (discussed later) at both the NH_2 and the COOH terminals. The hydroxylysines in these regions are important in that once they have undergone a further enzymatic modification, they can participate in cross-link formation. In both of these regions, a single lysine residue, not followed by a glycine, is frequently hydroxylated. Possibly these hydroxylysines are formed by the action of a hydroxylase enzyme other than the one that acts on lysines in the helical region. The hydroxylation of lysine is variable in different collagens, and even in the same collagen in different tissues.

Hydroxylysine can undergo further posttranslational enzymatic modifications including glycosylations and oxidations. The glycosylated derivatives are galactosylhydroxylysine (Fig. 1-1, GAL, OH-LYS) and glucosylgalactosylhydroxylysine (Fig. 1-1, GLU, GAL, OH-LYS). It has been suggested that the concentration of these glycosylated derivatives may have some influence in fibril structure but this remains to be critically tested.

From a functional standpoint, the best understood modification of lysine and hydroxylysine is the conversion to their respective aldehyde forms, allysine and hydroxyallysine (Siegel, 1979). These chemically active compounds are necessary for the process of intramolecular and intermolecular cross-link formation (Tanzer, 1973, 1976). The enzyme responsible for this conversion, lysyl oxidase, promotes the conversion of the ε-amino group of lysine and hydroxylysine to highly reactive aldehyde forms. It acts preferentially on these two amino acids when they are found in the NH_2- and COOH-terminal extensions, that is, where cross-linking originates (Fig. 1-1). Unlike the other enzymes that perform posttranslational modifications on collagen, this one is thought to work not intracellularly but extracellularly, and it appears to work preferentially on collagen molecules after they have assembled into fibrillar forms (Siegel, 1979). It seems, then, that one way intermolecular cross-links can be made is for these highly reactive aldehyde compounds, once they themselves have been formed in the terminal extension of one molecule, to react spontaneously with other lysines or hydroxylysines within the triple-helical region of an adjacent molecule.

An example of such a cross-link would be for the aldehyde functional group of a hydroxylysine in the terminal extension of one molecule to form a Schiff's base condensation product with the amino group of a hydroxylysine in the body of an adjacent molecule. It is thought that such a Schiff's base compound, once formed, can undergo an internal oxidation–reduction reaction, called an Amadori rearrangement, giving rise to a stable keto-amine cross-link. While numerous other potential intermolecular and intramolecular cross-link compounds have been isolated and characterized, most are thought to originate with the formation of lysyl oxidase-catalyzed aldehydes.

The enzyme lysyl oxidase itself can be irreversibly blocked by a class of inhibitors called lathyrogens, one of which is a substance termed β-aminopropionitrile. When injected into animals or added to cell or organ cultures, β-aminopropionitrile renders the newly synthesized collagen molecules devoid of cross-links and thus extractable in cold neutral salt solutions (Gross, 1974). The use of this compound and the discovery of its mechanism of action have greatly aided biochemical and metabolic studies on collagen.

3. Procollagen

For certain types of collagens, such as the fibrillar ones (see Section 5.1), the secreted form of the molecules have higher molecular weights than the corresponding form extracted from extracellular matrices (Fessler and Fessler, 1978) (see also Chapter 6). These higher molecular weight forms represent the procollagen (precursor) form(s) of the processed molecules found in the extracellular matrix. In procollagen molecules, each chain has large additional extension peptides, called propeptides, at both NH_2 and COOH termini (Fig. 1-1). As is the case for most proteins destined for extracellular secretion, an even higher molecular weight (pre-pro) form of collagen precursor chain can be identified in cell-free protein-synthesizing systems and in cDNAs encoding the chain.

Procollagen forms are secreted by cells in culture and thus can be isolated from the medium. To isolate procollagen from culture medium, it is necessary to prevent enzymatic removal of the propeptide extensions by inhibiting both specific and nonspecific proteolysis.

For type I collagen, at least two different enzymes (procollagen peptidases) are involved in the processing of procollagen to collagen, one to remove the NH_2-terminal propeptide and a second to remove the COOH-terminal one. The extracellular sequence for removal of the propeptides involves first the removal of all three chains in the NH_2-terminal extension as a single piece. This procollagen intermediate(s) with only the NH_2-terminal propeptide removed is termed P_C-collagen, indicating that it is a procollagen form with an intact COOH-terminal extension but no NH_2-terminal one. Subsequently, the COOH-terminal extensions are also removed en bloc, resulting in a completely processed collagen molecule. In the normal sequence of conversion, molecules with only the NH_2 propeptide intact are usually not found, as this is the first to be removed. Such molecules, called P_N-collagen, are the product of a genetic abnormality such as dermatosparaxis, in which the skin is extremely fragile and easily torn.

The processing of some other procollagens can be more complex (e.g., type V collagen) (Fessler and Fessler, 1987). And other collagens fail to undergo processing, with the intact "procollagen" form being deposited within matrices (discussed later).

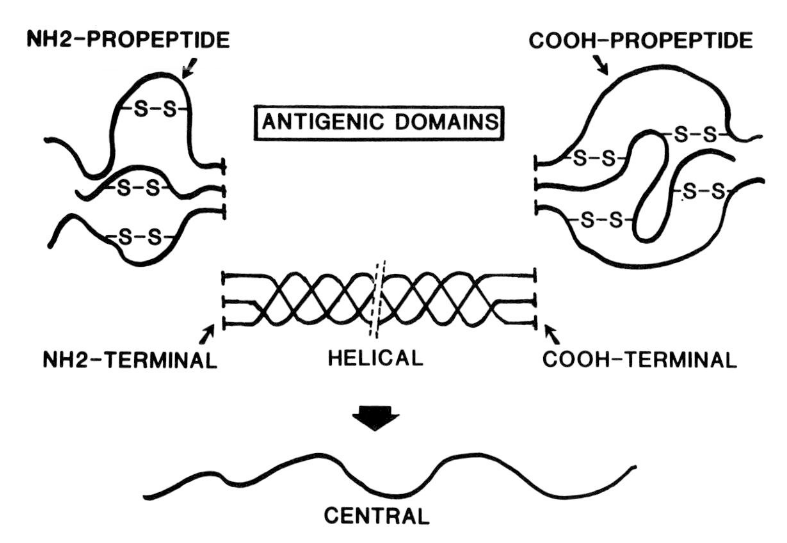

Figure 1-3. Antigenic domains of procollagens and collagens. The individual domains specific for procollagen are the amino (NH_2-propeptide) and carboxy (COOH-propeptide) extensions found only in the procollagen form. Native collagen molecules have determinants in the helical region, as well as in both the amino (NH_2-terminal) and the carboxy (COOH-terminal) nonhelical extensions. Central determinants are found in the chains after denaturation of the molecules. (After Timpl, 1976.)

4. Immunology of Collagens

Immunochemical and immunohistochemical approaches have become of increasing importance to cell and developmental biologists and pathologists. Antibodies can be produced that are specific for the different genetic types of collagens and procollagens, and even for specific domains and sites within these molecules.

Such anticollagen antibodies and domain-specific antibodies can be used as tools for immunocytochemical localizations at both the light and electron microscopic levels, and for molecular and supramolecular reconstructions when immunoelectron microscopy is combined with either rotary shadowing (Fig. 1-7, bottom), isolated fibrils, or thin sections (Figs. 1-5D, 1-10, 1-11C). They can also be used for immunochemical procedures such as enzyme-linked immunoadsorbent assay (ELISA), radioimmunoassay (RIA), and immunoblotting.

Collagen molecules have several different antigenic regions or domains and procollagen molecules have additional ones (Fig. 1-3). In general, the procollagen form is more antigenic. When procollagen is used as an immunogen, most of the antibodies are directed against epitopes within the propeptide extensions at the NH_2- and COOH-terminal ends and not against those within the collagenous body of the molecule. Antibodies against procollagens can also be elicited by immunization with the isolated propeptide extensions themselves, both of which are independently antigenic.

When native, completely processed collagen molecules are used as immunogens, there are still two classes of antibodies that can be elicited: those directed against sites along the triple-helical body of the molecule and those directed against the small NH_2- and COOH-terminal, non-triple-helical extension peptides (sometimes called telopeptides). In the extension peptides, the epitopes are determined largely by the primary amino acid sequences, which are different for each chain. Thus, chain-specific antibodies can be produced. These antibodies react equally well with native or denatured molecules so their epitopes fall into the general category of sequential determinants. Such epitopes depend only on the primary sequence of amino acids within a chain, and do not depend on the conformational state of the chains.

In contrast, antibodies produced against sites along the triple-helical body of native collagen molecules frequently do not react with thermally denatured molecules or isolated α chains. These antibodies are directed against epitopes which are conformation-dependent, and thus exist only when a molecule is in its native form. These are known as helical determinants. They require both the proper amino acid sequences of the different α chains, plus the triple-helical conformation to bring the required amino acids from each chain into proper juxtaposition. In general, these antibodies have the highest degree of specificity for the genetic type of collagen used as an immunogen. They usually do not cross-react with the other genetic types of collagens, but, in certain cases, they do react with the same collagen from another species. It should be realized that *a priori*, no absolute statement can be made about the degree of molecular and species cross-reactivity that will be obtained for an individual antibody.

One final class of epitope is termed the central antigenic determinant. These are linear sequences exposed when the triple-helical portion of the molecule is uncoiled. Antibodies to central determinants are produced when animals are immunized with thermally denatured collagens, isolated α chains, or segments of α chains such as individual CNBr peptides or synthetic peptides. Such antibodies usually do not bind to native molecules; presumably, then, they recognize some portion of the primary amino acid sequence within a single α chain that is exposed upon denaturation and may be either altered or masked when folded into the triple-helical conformation. These antibodies tend to have a good deal of cross-reactivity both with α chains of the same collagen type from different species of animals and with α chains of different collagen types from the same species. This is probably due to different collagens having short regions within their primary sequences that are common to one another.

For the investigator wanting to utilize immunological techniques, the most formidable task is the production of a useful titer of a specific antibody to the desired antigen, with no significant cross-reactivity to other antigens. In the case of collagens, this becomes a formidable task for two reasons. First is the problem of noncollagenous contaminants—as collagen is a poor immunogen, eliciting a useful antibody response requires injecting relatively large amounts of the antigen. So, even when the highest purity collagen preparation is used for immunization, there is the possibility of producing unwanted antibodies to quantitatively minor, yet highly antigenic contaminants that might be present. Second is the problem of producing antibodies that cross-react with several different collagen types. Because the antibodies to a given type of collagen are directed against a heterogeneous group of determinants along the molecule, many of those produced will be against sites that are found only in the type of collagen used for immunization and thus will have the desired specificity. Others, however, may be against sites common to several different types of collagens, and these will have undesirable cross-reactivity.

4.1. Antibody Production

Two general methods are available for producing antibodies with the desired specificity: (1) the conventional method, which involves raising antisera in animals and then using affinity chromatography to remove the unwanted antibodies, and (2) monoclonal antibodies.

4.1.1. Conventional (Polyclonal) Antibodies

Standard procedures exist for the production and purification of conventional antibodies (sometimes referred to as polyclonal antibodies) against collagens (Timpl, 1976). The animal species in which the antibodies are produced will, to some degree, determine the type of the antibody obtained. For example, when native collagen molecules are employed as immunogens in rabbits, the resulting antibodies are largely against the extension peptides (i.e., terminal sites); the same immunogen injected into rodents elicits antibodies predominantly against sites in the triple-helical region (i.e., helical determinants). Subsequent affinity purifications of the antisera on collagen immunoabsorbents is a "must" if one hopes to have antibodies with any degree of "specificity." Such purifications are performed by a series of sequential column chromatographic steps with different genetic types of collagens covalently bound to a solid matrix, such as Sepharose. The first step is to sequentially remove antibodies that cross-react with *each and every* other known collagen type, a truly formidable task as the number of collagens continues to grow (discussed later). Subsequently, the desired antibodies are isolated on a column of the homologous-type collagen. This direct isolation step reduces the relative concentrations of contaminant antibodies that may have been produced if a highly antigenic, minor contaminant in the immunogen collagen had elicited a major antibody response. Each purification step should be monitored by a procedure such as ELISA or RIA, or the inhibition forms of these two assays, and immunoblotting. In this way, it can be determined when the cross-reactive antibodies have been removed, at least within the limits of the assay procedure employed and with respect to the collagens that are available for assay at the time the antibodies are purified.

Recently, synthetic peptides have been employed as immunogens. The sequence information required to produce these is generally derived by sequencing cDNAs. To avoid the problems associated with the collagenous "central" determinants described above (e.g., the epitopes showing cross-reactivity with other collagen types and being masked in native molecules), sequences located in non-triple-helical portions of a molecule are usually employed.

4.1.2. Monoclonal Antibodies

Hybridomas provide an alternative for the production of antibodies, and have become the method of choice for many applications (Linsenmayer and Hendrix, 1982). The monoclonal antibodies produced by these lymphocyte–myeloma hybrids have numerous advantages over conventional antibodies, chief of which is their specificity for a single epitope. This degree of specificity can never be reached with a polyclonal antibody, no matter how exhaustive the purification scheme. In addition, once a monoclonal antibody is fully characterized and found to be one having the desired characteristics, it is available in unlimited quantities with no variation.

Most monoclonals produced against native collagens are directed against helical epitopes. Some, however, have been produced against naturally occurring telopeptides (Summers *et al.*, 1988) and against synthetic peptides (Sugrue *et al.*, 1989).

For the cell biologist, a frequent use of anticollagen monoclonal antibodies is for the localization of the different genetic types of collagens within tissues *in situ* and cells in culture. For light microscopic observations, fluorescence microscopy is the usual method employed to visualize antibody binding to either tissue sections or cells in culture (Fig. 1-4), but the peroxidase–antiperoxidase method is also used. Generally an indirect method is employed, in which the section is first reacted with the monoclonal antibody and subsequently with a fluorescent-tagged secondary antibody directed against the monoclonal itself. The indirect method is frequently employed since the same, labeled secondary antibody (commercially available) can be employed with different primary antibodies. The technique also affords amplification of the signal obtained, which can even be more enhanced simply by performing the reaction sequence more than once (Linsenmayer *et al.*, 1988).

Applications at the electron microscopic level usually employ colloidal gold-labeled antibodies (Figs. 1-5B, 1-10, 1-11C). Double labeling allows for the visualization of two different collagens within the same section. For this, each different primary antibody is tagged with a different size of colloidal gold particle (e.g., 5 and 10 nm) (Birk *et al.*, 1988). For either single or double labeling, the antibodies may be applied to ultrathin sections of frozen tissue or to thick frozen sections of tissue that are subsequently embedded in plastic for thin sectioning.

Recently, other "novel" applications have been devised for monoclonal antibodies. One of these is in elucidating the domain structure of a complex macromolecule, or the structural arrangement of macromolecules within a supramolecular complex. Structural determinations can sometimes be performed

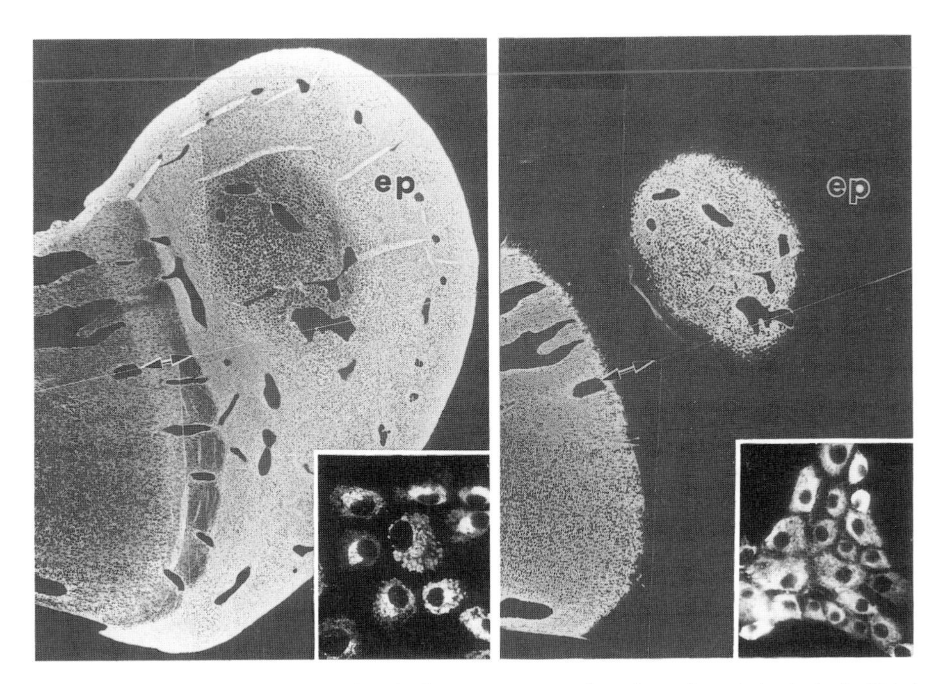

Figure 1-4. Fluorescence micrographs of adjacent tissue sections from the epiphysis (ep) of 21-day chick limb cartilages reacted with monoclonal antibodies against collagen type II (left frame) and collagen type X (right frame), followed by a rhodamine-conjugated antimouse IgG antibody. The double-headed arrows show the growth zone. Only the hypertrophic cartilage matrix is reactive for collagen type X. The inserts show the intracellular distribution of these two collagens in cultured chrondrocytes. The cells were permeabilized before reacting with the antibodies. ×20-insets ×50. (Modified from Schmid and Linsenmayer, 1985.)

by immunoelectron microscopy of sectioned material or on preparations of isolated fibrils. Some analyses, however, such as those which require the visualization of individual molecules, are best performed with rotary shadowed preparations of antibody–antigen complexes (Mayne *et al.*, 1984). A second application can be in evaluating the conformational state of a collagen molecule within a tissue *in situ* employing monoclonal antibodies directed against helical epitopes within a collagen. Since these antibodies bind only to native collagen molecules, and not to denatured ones or to individual α chains, they can serve as probes for whether a collagen molecule has an intact helical structure. When this property of such antibodies is combined with immunohistochemistry, analyses can be performed on tissues *in situ*. Thus far, such analyses have been utilized to determine the thermal stability of intracellular collagens and of those assembled within extracellular matrices (Linsenmayer *et al.*, 1989). Theoretically, applications for this methodology range from the intracellular assembly of collagen molecules to the degradation of extracellular matrices. In the former case, individual α chains, which do not in themselves react with antibody, are assembled into native molecules which do; in the latter the reverse is true, since here native collagen molecules are converted into cleavage products which readily denature.

5. Collagen Types

For a number of years, investigators were impressed with the apparent uniformity of the collagen extracted from different species of animals, and from different tissues within the same animal. It is true that during evolution, the basic size, helical structure, and amino acid composition of the collagens we now classify as "fibrillar" have remained highly conserved. Also, the tissues that were usually examined, adult skin, bone, tendon, and cornea, do indeed contain large amounts of the same genetic type of collagen, now termed collagen type I.

These results, however, were deceiving in that improved extractions and analytical procedures have allowed the identification of more than 14 different types of collagens within vertebrates. A great portion of the research done on collagen today involves analyses of these different collagens with respect to their biochemical characteristics, tissue distributions, regulation of synthesis, and structural and functional roles in developing embryos and adults and in pathological conditions.

5.1. Nomenclature and Classes

In the system of nomenclature that has evolved, each of the 14 known collagen types has been assigned a number generally reflecting the chronological order in which it was discovered. Such numbers are designated by roman numerals (e.g., type I, II, III, etc.) (Bornstein and Sage, 1980). The other parameter which is frequently included in the designation of the molecule is its chain composition. Collagen chains are generally referred to as α chains. In some collagens, all three α chains are identical (e.g., collagen type II), in others all three α chains are different (e.g., collagen type VI), and in still others, two α chains are the same and one is different (e.g., collagen type I). By convention, the different chains of a molecule are designated α1, α2, etc., with the collagen type to which they belong designated by a roman numeral which follows [e.g., the α1 chain of type I collagen is α1(I) and its α2 chain is α2(I)]. The complete designation of a molecule also includes the ratio in which the chains occur. Thus, the type II collagen molecule with its single type of chain is designated $[\alpha1(II)]_3$, the type VI collagen with the three different chains is α1(VI) α2(VI) α3(VI), and the type I collagen molecule with its two different chains is $[\alpha1(I)]_2$ α2(I). For the other collagens see Table I.

For organizational purposes we will subgroup the collagens into six classes, determined either by the supramolecular forms in which they occur within extracellular matrices, or by their molecular size as compared to the prototypical type I collagen. One such organization scheme, shown in Table I,

Table I. Collagens

Class	Type	Chain composition (most common form)	Common distributions
Fibrillar	I	$[\alpha1(I)]_2$ $[\alpha2(I)]$	Fibrous stromal matrices
	II	$[\alpha1(II)]_3$	Cartilage, vitreous, primary corneal stroma, notochord
	III	$[\alpha1(III)]_3$	With type I in heterotypic fibrils
	V	$[\alpha1(V)]_2$ $\alpha2(V)$	With type I in heterotypic fibrils; thin fibrils
	XI	$\alpha1(XI)$ $\alpha2(XI)$ $\alpha3(XI)$	With type II in heterotypic fibrils
Fibril associated	IX	$\alpha1(IX)$ $\alpha2(IX)$ $\alpha3(IX)$	Surface of type II collagen fibrils in cartilage; primary corneal stroma, notochord
	XII	$[\alpha1(XII)]_3$	Certain type I-containing matrices
Network forming	IV	$[\alpha1(IV)]_2$ $\alpha2(IV)$	Basement membranes
Filamentous	VI	$\alpha1(VI)$ $\alpha2(VI)$ $\alpha3(VI)$	100-nm beaded filaments of stromal matrices
Short chain	VIII	?	Descemet's membrane, subendothelial matrices
	X	$[\alpha1(X)]_3$	Hypertrophic cartilage
Long chain	VII	$[\alpha1(VII)]_3$	Anchoring filaments

includes the collagen classes: fibrillar (I–III, V, XI), fibril-associated (IX, XII), network-forming (IV), filamentous (VI), short chain (VIII, X), and long chain (VII). Each class and type of collagen will be discussed in a subsequent section. The most recently discovered collagens (XIII, XIV) are not included (see Summary and Conclusions). In Chapter 6, Olsen subgroups type XIV with the fibril-associated collagens and classifies type XIII with VI and VII in a miscellaneous category.

5.2. The Fibrillar Collagens: Types I, II, III, V, and XI

5.2.1. Supramolecular Arrangements

5.2.1a. Striated (Native) Fibrils. *In vivo* molecules of the fibrillar collagens are found within striated fibrils. *In vitro* they assemble into this form when allowed to aggregate under the proper environmental conditions (see Chapter 7). Depending on whether such native fibrils are positive-stained or negative-stained, they display different morphological patterns, two of which are shown in Fig. 1-5 (A–C). In positive-stained preparations, the stain binds to regions rich in polar amino acids, so the staining pattern reflects the summation of charged residues along the fibril. In negative-stained preparations, the stain permeates the fibrils and becomes trapped in the "hole zones" which are formed by incomplete overlapping of adjacent collagen molecules within the

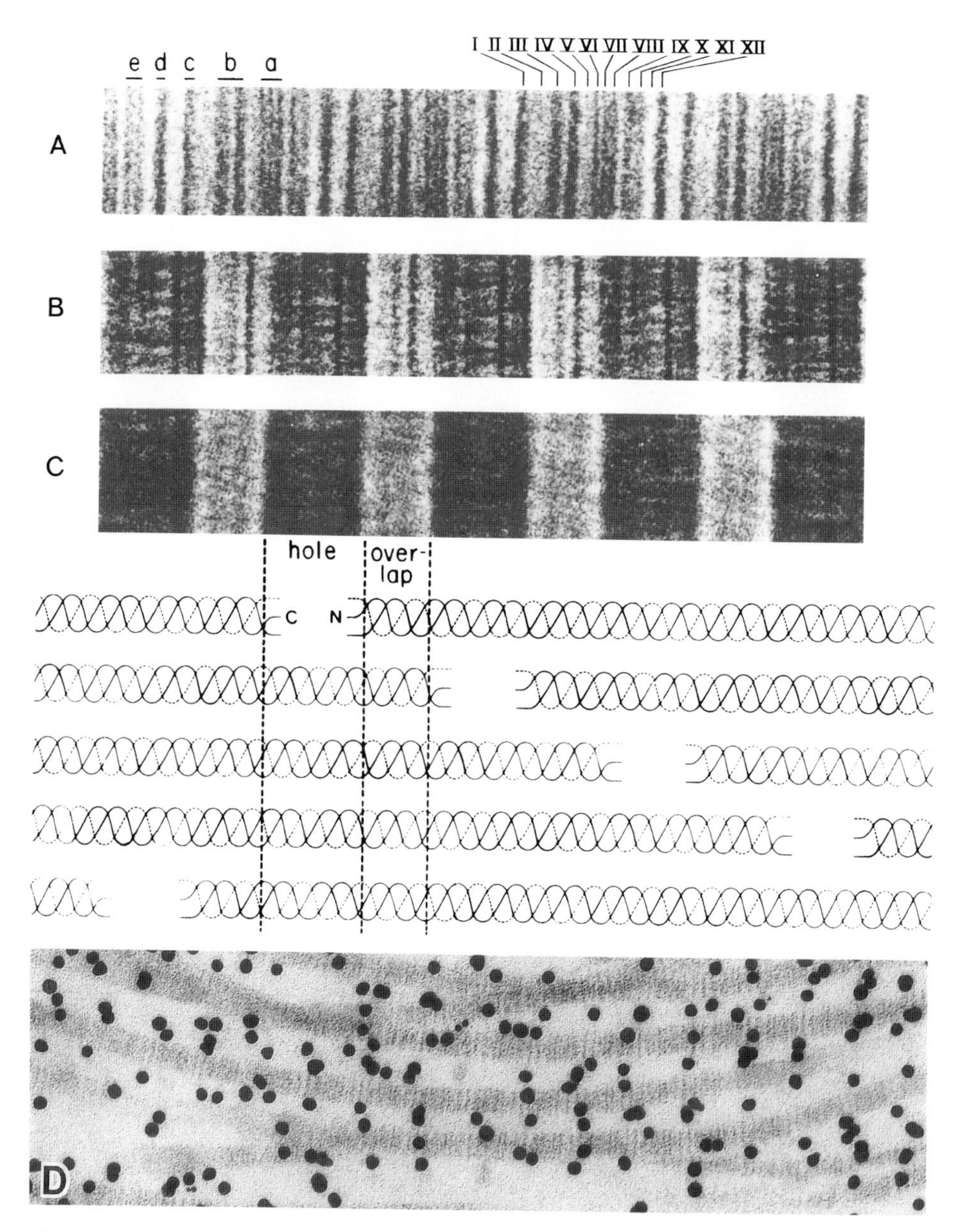

Figure 1-5.(A–C) Electron micrographs of native collagen fibril stained either (A) positive, (B) positive and negative, or (C) negative. (Middle) Schematic diagram showing the quarter-stagger arrangement of collagen molecules within a fibril. The relationships between the "hole" and "overlap" zones are shown by the vertical dashed lines. (D) Section of cornea showing striated fibrils decorated with an anti-type I collagen monoclonal antibody–colloidal gold complex. Many of the gold particles occur in the 64-nm periodicity predicted by the quarter-stagger model. (A–C, courtesy of Dr. R. Bruns; D, courtesy of Dr. D. Birk.)

repeating structure. When fibrils are both positive- and negative-stained, it is possible to see both patterns superimposed. Such preparations provide some of the data that have been used to construct models for the packing of collagen molecules within fibrils.

The most widely accepted model for the packing of molecules in native fibrils (Gross, 1974; Hodge and Petruska, 1963) is diagrammatically shown in Fig. 1-5. The individual molecules are approximately 4.4 times the length of the repeat period (67 nm), called D. They are staggered about one-quarter their length, and a short empty space is present between the NH_2 terminal of one molecule and the COOH terminal of the next. The empty spaces between adjacent molecules are the "hole zones" in which stain accumulates (i.e., the dark bands in negative-stained preparations). The regions in which no stain accumulates (i.e., the light bands) are the regions in which molecular overlapping is complete (the "overlap zone"). In the negative-stained fibrils, then, the basic repeating unit (D) consists of one hole zone, which is about 0.6 D unit long, plus one overlap zone, which is 0.4 D unit long. The 67-nm repeat period can also be identified in the banding pattern of positive-stained fibrils, and the relationship of the two patterns becomes clearly visible in fibrils that are both positive- and negative-stained, as shown in Fig. 1-5. The repeat can also be visualized immunohistochemically with monoclonal antibodies (Fig. 1-5D).

The presence of a hole zone may be necessary for the enzymatic formation of the aldehyde-derived cross-links required for intermolecular cross-link formation. It is known that the required enzyme, lysyl oxidase, acts preferentially to form these compounds in the NH_2- and COOH-terminal, non-triple-helical extension peptides of molecules after they have aggregated into the native fibrillar form (Siegel, 1979). Why the enzyme prefers to act on molecules in native fibrils is still unknown. However, as the terminal extension peptides reside at the borders of each hole zone, the physical presence of the hole zone may be necessary for the enzyme to diffuse into the interstices of fibrils, where it can act on its substrate. It is also thought that only when molecules are in the native fibrillar form are the aldehydes, once formed, brought into juxtaposition with the proper residues in adjacent molecules so that cross-links can then form spontaneously.

Models have been proposed which slightly alter the packing of the molecules. And while these models represent potential refinements and suggest interesting approaches for further experimentation, the basic assumptions of the quarter-stagger model remain intact after more than a quarter of a century.

5.2.1b. Heterotypic Fibrils. Until recently, it was thought that individual collagen fibrils were composed solely of molecules of one fibrillar collagen type (i.e., they were homotypic structures). The tissue-specific differences in fibrils were thought to be derived either from the homotypic assembly of different types of collagens, or from posttranslational modifications of the collagen molecules, such as glycosylation.

Evidence has accumulated, however that the fibrils of many tissues are "heterotypic" structures, composed of different combinations of fibril-forming collagens co-assembled with one another (Hendrix *et al.*, 1982; Linsenmayer *et al.*, 1985, 1989; Birk *et al.*, 1988). It has been hypothesized that such a co-assembly of different collagens could determine certain properties of fibrils. The structural similarities of the fibrillar collagen molecules would allow them to co-assemble; their differences could sterically influence, for example, fibril diameter (Birk *et al.*, 1990; see Chapter 7).

Evidence for such an arrangement came from immunoelectron microscopy (see Chapter 7) and from immunofluorescence observations with monoclonal antibodies against epitopes within the helical domain of the type V collagen molecule (Linsenmayer *et al.*, 1985, 1989). It was observed that when assembled within fibrils, epitopes on type V collagen were antigenically masked (i.e., they were unavailable to antibody), and that unmasking could only be effected by disrupting fibril structure or by removing type I collagen (see Fig. 1-7). This suggested that within fibrils, collagen types I and V were assembled together in such a way that the type I collagen molecules rendered epitopes on the type V molecules unavailable to antibody. Double-labeled immunoelectron microscopic studies have now confirmed this model (Birk *et al.*, 1988; Fig. 7-4).

Other heterotypic combinations of fibrillar collagens have been observed in human skin and tendon (Keene *et al.*, 1987; Fleischmajer *et al.*, 1990) where type III collagen is found co-assembled with type I collagen, in cartilage where type XI collagen is co-assembled with type II (Mendler *et al.*, 1989), and in the early embryonic avian corneal stroma where type II collagen is co-assembled with type I (Hendrix *et al.*, 1982; Linsenmayer *et al.*, 1990).

5.2.1c. Segment-Long-Spacing (SLS) Crystallites. Another form in which fibrillar collagen molecules can assemble, as well as those of some other collagen types (see Figs. 1-11A, 1-12), is the SLS crystallite (for review see Gross, 1974). In such crystallites, the molecules are arranged side by side with the same NH_2-to-COOH-terminal polarity (Fig. 1-11A). The length of the crystallite is the length of the triple-helical domain of the molecule. Since the molecules are in register, the amino acids along their length are also in register, clusters of which can be visualized by staining. For analytical purposes, such crystallites can be formed artificially from solutions of a collagen. These have been used to determine the lengths of molecules, to order peptides in sequencing studies, to identify the large propeptide extension pieces in the precursor forms of collagens (Chapter 6), and to localize the cleavage site of the animal collagenases (Gross, 1974).

Recently, SLS have been found to be the naturally occurring structural form of the collagen within epithelial anchoring fibrils (see Section 5.7).

5.2.2. Type I Collagen

The most abundant type of collagen isolated from many adult connective tissues such as skin, bone, tendon, and cornea is type I collagen. Type I is the prototype fibrillar collagen, and much of the information that has already been discussed in this chapter has been obtained from studies on this collagen type.

Type I molecules are about 300 nm long as deduced by electron microscopic analysis of SLS crystallites and rotary shadowed preparations. This is also the approximate length of the molecules of the other fibrillar collagens. As already discussed, each type I molecule is the heterotrimer $[\alpha 1(I)_2 \alpha 2(I)]$, being composed of two $\alpha 1(I)$ chains and one $\alpha 2(I)$ chain. When type I collagen is thermally denatured the $\alpha 1(I)$ and $\alpha 2(I)$ chains can be separated from one another either by ion-exchange chromatography on *O*-(carboxymethyl)-cellulose (CM-cellulose) columns or by SDS-PAGE. Both $\alpha 1(I)$ and $\alpha 2(I)$ have a net basic charge, but $\alpha 2(I)$ is more basic and thus elutes later from the column.

If type I molecules are obtained from animals or cultures that have been treated with β-aminopropionitrile to prevent cross-linking (Gross, 1974), after denaturation by far the major amounts of material correspond to single $\alpha 1(I)$ and $\alpha 2(I)$ chains. Stoichiometrically, these chains appear in a 2:1 ratio reflecting their distribution within each molecule. If, however, the molecules are extracted from sources in which cross-linking has been less well blocked or not blocked at all, there are increasing amounts of cross-linked dimers composed of two α chains (termed β chains), trimers composed of three α chains (termed γ chains), and even higher molecular weight aggregates. Dimers (β chains) can either be composed of two $\alpha 1(I)$ chains, in which case they are designated $\beta_{1,1}$ chains, or they can be composed of one $\alpha 1(I)$ chain and one $\alpha 2(I)$ chain, in which case they are designated $\beta_{1,2}$ chains. On SDS-PAGE the two types of β chains migrate much more slowly than α chains, and on CM-cellulose columns they elute in the positions designated in Fig. 1-6. On SDS-PAGE the various combinations of cross-linked α chains and higher-order aggregates migrate only slightly from the origin, or in some cases do not enter the gel at all.

The $\alpha 1(I)$ and $\alpha 2(I)$ chains represent separate gene products each slightly more than 1000 amino acids long, as determined by amino acid sequencing and cDNA sequencing. Glycine accounts for one-third of the total amino acids of both the $\alpha 1(I)$ and the $\alpha 2(I)$ chains, reflecting the presence of the repeating Gly-X-Y triplets. In addition, both chains have about half of their total proline content as hydroxyproline, and some of their lysine content as hydroxylysine. Characteristically, as for most fibrillar collagens, the two chains have no cysteine, and very small amounts of the aromatic amino acids tyrosine and tryptophan. They differ from one another, however, in their concentrations of a number of the other amino acids.

5.2.3. Type II Collagen

With the discovery that cartilage contains a collagen molecule quite different from type I, it became apparent that the fibrillar collagens constitute a class of closely related, yet genetically distinct molecules. The new molecule, termed type II collagen, was thought to be found only in embryonic and adult cartilage. Indeed, the synthesis of type II collagen was considered to be the sine qua non for chondrocyte differentiation, and it still is one of the better molecular criteria we have for differentiated cartilage. It is now known, however, that type II collagen is a major collagen type of the embryonic notochord (Linsenmayer *et al.*, 1986a; Fitch *et al.*, 1988b) and its adult derivative the nucleus pulposus (Eyre and Muir, 1976); the vitreous body of the eye (Smith *et al.*, 1976; Fitch *et al.*, 1988b); and of a number of epithelially derived, embryonic ma-

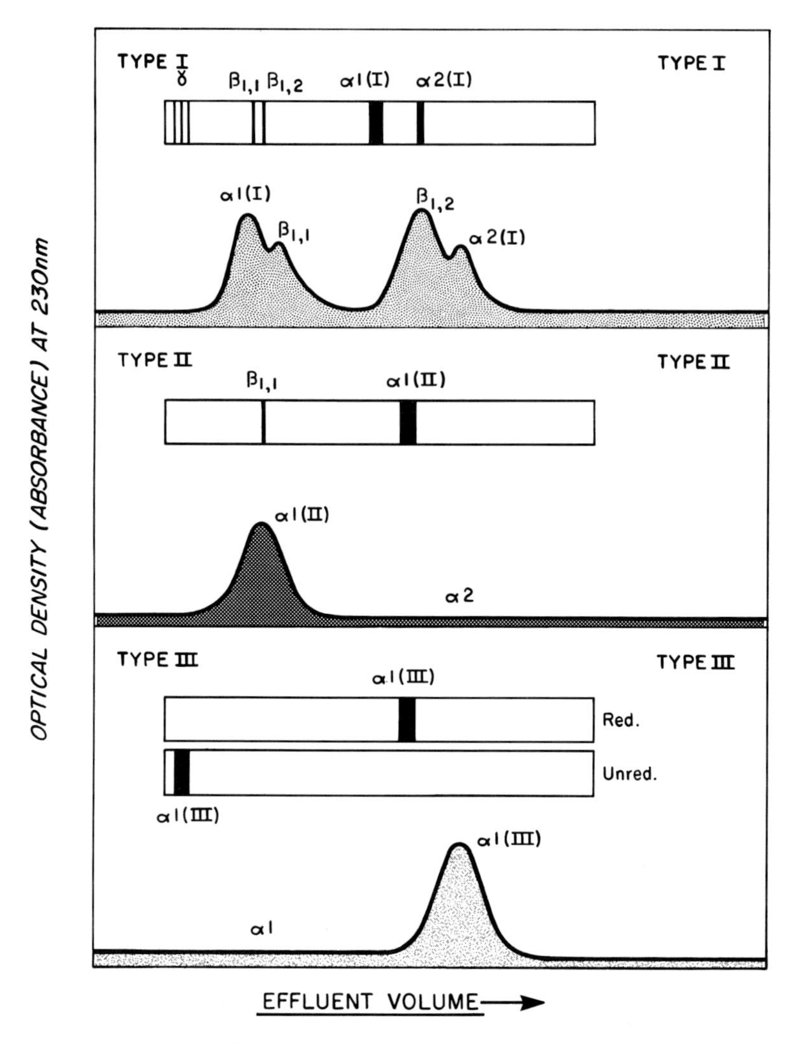

Figure 1-6. Diagram showing the positions of the composite chains of collagen types I, II, and III when separated by either CM-cellulose column chromatography or SDS-PAGE. The individual chains are labeled "α," the dimers "β," and the trimers "γ." Red., reduced; unred., unreduced.

trices (Fitch *et al.*, 1989; Kosher and Solursh, 1989), such as those of the primary corneal stroma of the early chick embryo.

Type II collagen can be solubilized by neutral salt extraction of tissues from lathyritic animals (i.e., ones that have received the cross-linking inhibitor β-aminopropionitrile), as is true for all of the fibrillar collagens and for some other collagen types. But studies of this collagen were aided when it was discovered that the intact collagenous domain of the molecule could be extracted by limited pepsin digestion of tissue (Miller, 1972), and that the molecule could then be cleanly separated from type I by a fractional salt precipitation technique (Trelstad *et al.*, 1970). The collagenous domains of numerous other collagens have subsequently been isolated using the combined procedures of pepsin solubilization and fractional salt separation.

Type II collagen molecules, $[\alpha1(II)]_3$, are homotrimers of a single type of α chain, termed α1(II). Thus, when the chains of type II collagen are analyzed by CM-cellulose chromatography or by SDS-PAGE (Fig. 1-6), only a single peak or band is seen, which is found in the same region as the α1(I) chain. Although the α1 chains from type I and II collagens have similar sizes and net charge characteristics, they are separate gene products, as are all the different α chains of collagens (Trelstad *et al.*, 1970). Also, the α1(II) chain of cartilage has undergone extensive posttranslational modification, with most of the lysines that are potentially hydroxylatable being converted to hydroxylysine, and many of these being glycosylated (Miller, 1972).

Within extracellular matrices, type II collagen molecules frequently occur in heterotypic fibrils with certain other fibrillar collagens, as demonstrated by immunoelectron microscopy. In the fibrils of the avian embryonic primary corneal stroma the molecules are co-assembled with those of collagen type I (Hendrix *et al.*, 1982; Linsenmayer *et al.*, 1990); and in cartilage they are complexed with those of collagen type XI (Mendler *et al.*, 1989). In addition, in both cartilage (Vaughan *et al.*, 1988) and vitreous (Wright and Mayne, 1988; Yada *et al.*, 1990), and possibly the primary corneal stroma (Fitch *et al.*, 1988b), these heterotypic fibrils are coated with the fibril-associated collagen, type IX (discussed later).

5.2.4. Type III Collagen

Type III collagen (Kuhn, 1987) is found in many stromal connective tissues (Keene *et al.*, 1987; Fleischmajer *et al.*, 1990), notable exceptions being bone and avian cornea in which it is either absent or present in extremely small amounts. The native molecule, $[\alpha1(III)]_3$, is a homotrimer composed of a single type of α chain termed α1(III). When type III collagen is analyzed by CM-cellulose chromatography, it elutes as a single peak near that of the α2 chain of type I collagen (Fig. 1-6). When analyzed by SDS-PAGE in the unreduced state, type III migrates as a trimeric, γ-chain-sized molecule; only under reducing conditions does it migrate as an α chain. The reason for this is that the chains of type III collagen, unlike those of other fibrillar collagens, are connected by intramolecular disulfide cross-links that occur within the triple helical portion of the molecule.

Type III collagen has several other unique features including high levels of 4-hydroxyproline, more than 333 glycines, and the presence of half-cystines, which participate in the intramolecular disulfide cross-links. The presence of more than 333 glycine residues in the α chain is due to some of the Gly-X-Y triplets having an additional glycine at the X or Y position. It has been suggested that these extra glycine residues may cause localized helix instability, resulting in increased susceptibility to proteolytic cleavage and more rapid turnover of matrices containing this collagen type. No definitive data, however, exist on the relative turnover rates of different types of collagens *in vivo*.

The supramolecular structural organization and tissue distribution of type III collagen are somewhat unclear. Early work with polyclonal antibodies suggested that type III was principally a component of the small argyrophilic collagen fibers that characterize reticular connective tissues (Gay and Miller, 1979). Recently, however, studies with monoclonal antibodies have suggested that type III is a component of the striated fibrils of certain stromal matrices such as dermis. There it occurs in heterotypic fibrils along with collagen type I (Keene *et al.*, 1987; Fleischmajer *et al.*, 1990). Also, both immunoelectron microscopic and metabolic studies have suggested that the amino propeptide of type III collagen is retained for some time after the molecule is incorporated into fibrils. Whether this can result in fibrils with different properties remains to be tested.

5.2.5. Type V Collagen

What is termed type V collagen is likely to constitute a class of distinct but related molecules (Fessler and Fessler, 1987). The type V molecule as originally isolated from a pepsin digest of human placenta contained two different α-chain-sized polypeptides termed αA and αB, now designated α2(V) and α1(V), respectively. These chains migrate on SDS gels with molecular weights approximately that of an α1(I) chain. The α2(V) chain migrates slightly slower than an α1(I) chain, and the α1(V) chain migrates somewhat slower yet. A third chain, the α3(V) chain, has also been reported to occur in some tissues, and yet another chain has been detected in radiolabeling studies of embryonic chick tendon.

With this number of α chains potentially available for molecular assembly, the type V collagen family of molecules is likely to have multiple members. The predominant molecular species of type V collagen found in most tissues is the heterotrimer $[\alpha1(V)]_2\alpha2(V)$. In some cases a portion of the α1(V) chains appears to be assembled in a triple-chained homopolymer $[\alpha1(V)]_3$; in one cell line this is the only molecule assembled. In tissues containing small amounts of α3(V) chain, it has been reported that molecules of composition α1(V)α2(V)α3(V) may occur.

The $[\alpha1(V)]_2\alpha2(V)$ form of the molecule is the only one whose tissue distribution and supramolecular organization have been extensively studied. Immunofluorescence and immunoelectron microscopic studies employing monoclonal antibodies directed against this molecule suggest that in most tissues it occurs in heterotypic fibrils along with type I collagen. In such fibrils, both types of collagens are arranged in the quarter-stagger array characteristic of a fibrillar collagen, but the precise relationships of one molecular type to the other remain to be elucidated. Interestingly, when type V molecules are assembled in this arrangement, in addition to being unavailable to antibody (Fig. 1-7 and discussed earlier) they are refractory to enzymatic cleavage by a type V-specific collagenolytic enzyme (Fitch *et al.*, 1988a; Chapter 8). Only if the type I collagen is removed (e.g., by a type I collagenase) is the type V collagen susceptible to cleavage. Thus, the turnover and degradation of at least some heterotypic fibrils is likely to be complex, involving several different enzymes.

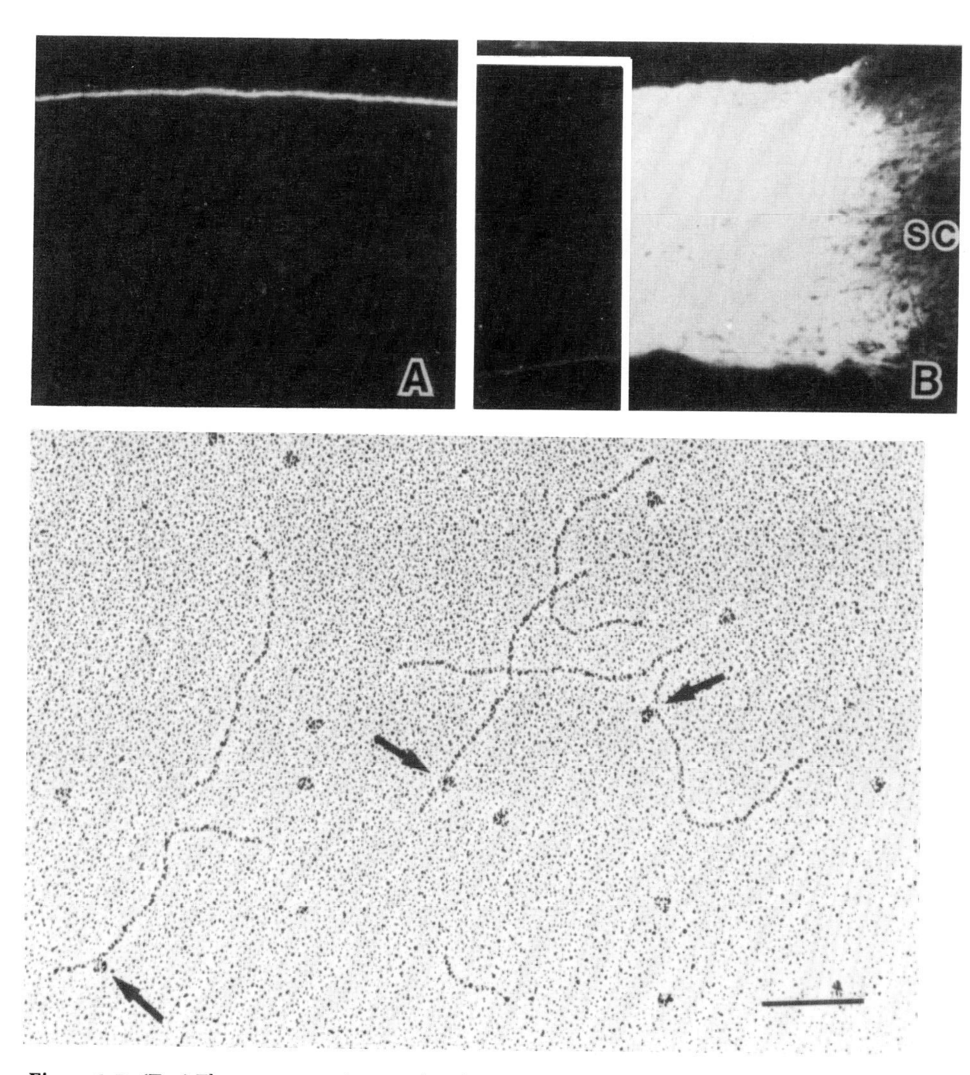

Figure 1-7. (Top) Fluorescence micrographs of sections of corneal stroma reacted with monoclonal antibodies against collagen type V. In (A) the section was directly reacted with the antibody; in (B) type I collagen was first removed by digestion with a vertebrate collagenase which degrades that collagen type but not type V. SC is the sclera and the insert (middle) is a control section. Similar results are obtained with agents which disrupt fibril structure (Linsenmayer *et al.*, 1985, 1989). ×100. (Modified after Fitch *et al.*, 1984.) (Bottom) Electron micrographs of a rotary-shadowed preparation of the anti-type V collagen monoclonal antibody (arrows) bound to type V collagen molecules. The epitope is located approximately 45 nm from one end. Bar = 100 nm. (Modified from Mayne *et al.*, 1984.)

The normal processing of type V procollagen may take considerably longer and be more complex than for that of other interstitial collagens. The final product(s) are somewhat larger than their counterparts in other collagens, suggesting that the type V molecules normally incorporated into tissues may be in a form(s) resembling "partially processed" procollagen molecules. With the degree of complexity displayed by type V collagen(s), the potential clearly exists for it to exhibit regulatory functions. That the molecule may be involved in regulating fibril diameter has been suggested from *in vitro* fibrillogenesis studies employing mixtures of collagen types V and I (Adachi and Hayashi, 1985; Birk *et al.*, 1990). The greater the percentage of type V collagen incorporated into a heterotypic fibril, the smaller is the diameter of fibril formed.

5.2.6. Type XI Collagen

Type XI collagen is thought to be the cartilage homologue of type V collagen, both in structure and in function (Eyre and Wu, 1987). This collagen is a heterotrimer (Morris and Bachinger, 1987), α1(XI)α2(XI)α3(XI), composed of three different α chains. The α1(XI) and α2(XI) chains are closely related to the α1(V) and α2(V) chains, respectively, but they represent different gene products. The α3(XI) chain, however, bears a striking resemblance to the α1(II) chain of type II collagen, and is thought to be derived from the same gene. The only difference appears to be that the α3(XI) chain is more extensively altered through posttranslational modifications. Type XI collagen, or at least its component chains, may not have absolute specificity for cartilage. Recently, the molecule has been detected in bone, placenta, and several tumor cell lines of noncartilage origin (Bernard *et al.*, 1988; Yoshioka and Ramirez, 1990). Bone may even contain a collagen fraction in which α1(XI), α1(V), and α2(V) chains are present in a 1:1:1 ratio, raising the possibility of hybrid molecules composed of chains from both collagen types. If so, collagen types XI and V may truly be members of one large collagen superfamily.

Recently, the correlation between collagen types XI and V has been taken to the level of fibril structure. Immunoelectron microscopic studies have demonstrated that in cartilage, type XI collagen is a component of heterotypic collagen fibrils (Mendler *et al.*, 1989), along with collagen types II and IX (discussed later). In a similar manner to the antibodies for type V collagen described above, those against type XI fail to react unless the fibrils have been disrupted by either physical or chemical means. Presumably, in this case masking of epitopes on the type XI collagen is effected by their association with the collagen types II or IX of the fibrils. Thus, there appears to be a common strategy for the heterotypic assembly of molecules, at least within some striated collagen fibrils.

5.3. The Fibril-Associated Collagens: Types IX and XII

It is now known that certain collagens are associated with the surface of striated collagen fibrils. These collagens, as they occur in matrix, usually have

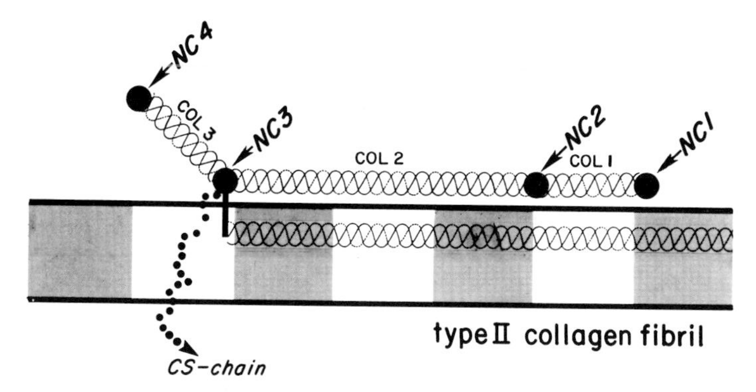

Figure 1-8. Diagram of a type IX collagen molecule (top) aligned along a type II collagen fibril. Shown are the collagenous domains. COL1–3; noncollagenous domains, NC1–4; and the chondroitin sulfate side chain (CS-chain). "Hole zones" are the clear areas along the fibril. A disulfide cross-link (vertical line) is shown between the NC3 domain of the type IX molecule and the amino telopeptide of a type II molecule within the fibril. (Modified after Nishimura *et al.*, 1989.)

multiple "functional" domains. It is thought that they provide another level at which fibrillar structure can be modified, and that they can effect interactions between fibrils, and between fibrils and other components of extracellular matrices. Until recently, it seemed that the fibril-associated collagens might all be members of a single family, of which collagen type IX is a prototype. But evidence now exists that other collagen types can also become fibril-associated (see Section 5.6.1).

5.3.1. Type IX Collagen

The best studied fibril-associated collagen is type IX (Van der Rest and Mayne, 1987; Reese and Mayne, 1981; Van der Rest *et al.*, 1985). In cartilage, where type IX occurs in abundance and has been extensively studied, the molecule is found along the surface of cartilage collagen (heterotypic type II/XI) fibrils (Vaughan *et al.*, 1988). The molecule, α(IX) α2(IX) α3(IX), is a heterotrimer composed of three different chains. It has three collagenous domains (COL1–3) and four noncollagenous ones (NC1–4)* (Fig. 1-8).

Structurally, type IX collagen seems eminently suited for simultaneous interaction with fibrils and other matrix components. Along the surface of cartilage collagen fibrils, the type IX collagen molecules are specifically arranged such that certain of their domains (NC1 through COL2) are in register with the D period of the fibril (Fig. 1-8). This allows covalent cross-linking to

*By convention, the collagenous and noncollagenous domains of type IX collagen, as well as those of other multidomain collagens, are designated by the term "COL" for a collagenous domain, and "NC" for a noncollagenous one. The order in which these are arranged within the molecule is designated by arabic numbers, starting at the COOH-terminal end. Thus, the carboxyl-most noncollagenous domain is designated NC1, the collagenous one is COL1, the next noncollagenous one is NC2, etc.

occur between a type IX molecule and a type II collagen molecule within the fibril (Van der Rest and Mayne, 1988). The NC3 domain functions as a hinge, allowing the molecule to flex such that the remaining domains (COL3 and NC4) extend outwards from the fibril surface. Potentially the extended domains may function as a bridge, linking fibrils to each other and/or to other components of the matrix. The NC3 domain also has a covalently bound chondroitin sulfate side chain (Van der Rest and Mayne, 1987; Yada *et al.*, 1990), which provides another modification potentially capable of facilitating interaction with other matrix components.

Type IX collagen was once considered a tissue-specific product of cartilage, but it is now known to be a component of the embryonic avian primary corneal stroma, the vitreous, and the notochordal sheath. Studies on the primary corneal stroma have revealed several other interesting characteristics of type IX. One is alternative splicing and promoter usage. Some of the type IX collage mRNA isolated from avian corneal epithelium (the source responsible for synthesis of this stroma) is smaller than that of cartilage (Svoboda *et al.*, 1988; Nishimura *et al.*, 1989), presumably resulting in a form of the molecule with a greatly truncated NC4 domain. Such a truncated molecule is likely to have different properties, or in the least, altered ones. Another study has reinforced the possibility that type IX collagen may stabilize matrix structure (Linsenmayer *et al.*, 1990). Just prior to the swelling of the primary corneal stroma, a naturally occurring event in avian corneal morphogenesis, immunohistochemically detectable type IX collagen disappears (Fitch *et al.*, 1988b). Thus, one function of the molecule might be to maintain the compactness of matrices which have a natural tendency to swell, such as the primary corneal stroma and cartilage.

Another potentially important characteristic of type IX collagen, and of certain other collagens, is their selective degradation. Type IX is not a substrate for conventional animal collagenases, but it is for stromelysin (Okada *et al.*, 1989), a matrix metalloprotease which does not cleave fibrillar collagens (Chapter 8). Thus, during embryonic development, collagens can be selectively removed.

5.3.2. Type XII Collagen

Type XII collagen (Gordon *et al.*, 1987, 1990) is a molecule with partial structural homology to collagen type IX. This, plus its occurrence in certain matrices rich in collagen type I (Sugrue *et al.*, 1989) have suggested that the molecule may be associated with certain type I collagen fibrils.

Much of the structure of type XII collagen has been deduced from a combination of cDNA sequencing, biochemical analyses, and rotary shadowing. The molecule most likely is a homotrimer with chains of 220 kDa. One of the collagenous domains (COL1) of type XII bears a considerable resemblance to the corresponding domain of collagen type IX, as do certain aspects of the exon–intron organization of its gene. While portions of the noncollagenous domains also have some similarities to those of collagen type IX, the other collagenous domain of type XII has no similarity to any in type IX.

A notable feature of the molecule is the presence of a large noncollagenous domain at its NH_2-terminal end. In rotary shadowed preparations, this domain consists of three relatively thick projections, each 60 nm long. The remaining collagenous and noncollagenous domains add an additional 75 nm to the overall length of the molecule. By weight the relative size of the globular domain is approximately six times that of the remainder of the molecule. This domain appears to have the potential for extensive interactions with other matrix components.

5.4. The Network-Forming Collagen: Type IV

Basement membranes are composed of a variety of different classes of matrix molecules, one of which is type IV collagen. This is the only collagen that has been definitively shown to be present in basement membranes. However, recent studies have suggested that type IV collagen may be a family of molecules, so the potential for extensive heterogeneity exists.

Type IV was the first nonfibrillar collagen to be extensively investigated, and the first in which approaches such as rotary shadowing electron microscopy and domain-specific monoclonal antibodies were employed in structural analyses (Timpl *et al.*, 1981; Linsenmayer *et al.*, 1984; Yurchenco and Schittny, 1990). Initially, some studies reported that limited proteolytic extraction of basement membrane-rich tissues released a collagen with chains clearly different from those of any known types (Kefalides, 1973). Other studies, however, reported isolating collagenous, or partially collagenous components, of various molecular weights.

Clearly, methods other than those of classic biochemistry would be required to determine the structure of this collagen and its organization within basement membranes. These include biosynthetic studies, electron microscopic observations on rotary-shadowed preparations (Timpl *et al.*, 1981), and in certain cases the use of domain-specific monoclonal antibodies (Mayne *et al.*, 1984). These approaches demonstrated that the variety of components observed in the proteolytic extracts was due in part to the supramolecular forms of type IV collagen which occur in tissues. These forms, as well as individual type IV molecules themselves, contain molecular domains including collagenase-sensitive triple-helical ones, pepsin-sensitive nonhelical ones, and ones that are resistant to both pepsin and collagenase.

It is thought that the most common molecular form of type IV collagen is a heteropolymeric molecule $[\alpha1(IV)]_2\ \alpha2(IV)$, but it has also been suggested that the homopolymeric forms $[\alpha1(IV)]_3$ and $[\alpha2(IV)]_3$ occur and even other chains exist. Biosynthetic studies showed that type IV collagen is secreted and assembled as a procollagenlike molecule in which each chain has an apparent molecular weight of 160,000–180,000. Pulse–chase experiments provided evidence that this intact procollagenlike molecule is incorporated directly into the basement membrane structure without undergoing detectable extracellular processing.

The intact procollagenlike molecule can be extracted in relatively large

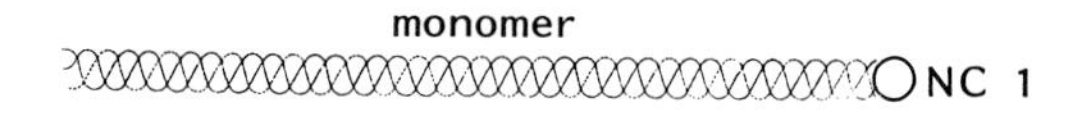

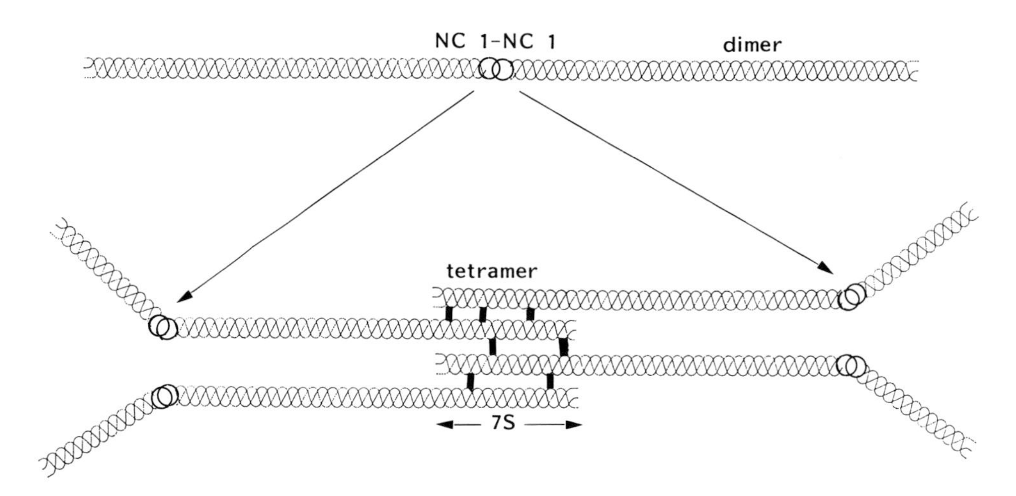

Figure 1-9. Diagram of a type IV collagen molecule and two of the types of intermolecular interactions that can occur. Dimers are formed by interactions at the COOH-terminal (NC1) domain. Tetramers are formed by interactions at the NH_2-terminal region, resulting in the highly disulfide-cross-linked (vertical lines) 7 S domain. The two types of interactions when combined (as shown by arrows) produce a networklike structure. (Modified after Timpl *et al.*, 1981.)

amounts from certain tumors and tumor cell lines. When molecules from these sources are viewed in the electron microscope as rotary-shadowed preparations, each molecule appears as a rod (~ 400 nm) with a knoblike nonhelical domain at the COOH-terminal end (see Fig. 1-9). In such preparations, two molecules can sometimes be seen to be associated by these nonhelical domains. Others, in groups of twos and fours, can be associated by their opposite (NH_2-terminal) ends which are also partially nonhelical. From these types of studies, a "network" model (Timpl *et al.*, 1981) for the supramolecular organization of type IV was derived (Fig. 1-9). The basic repeating unit consists of four individual molecules, overlapping and joined together at their NH_2-terminal ends. In this overlap region, termed "7 S," the molecules are highly disulfide cross-linked and are resistant to digestion both with bacterial collagenase and with pepsin. At their other end, two molecules become associated by their knoblike, nonhelical domains, termed NC1 (Fig. 1-9). The intervening, long rodlike portion of each molecule, extending out of the 7 S region as one of the four "arms," is largely in the form of a collagen triple helix that is interrupted, in places, by nonhelical sequences.

Recently, Yurchenco and co-workers (Yurchenco and Schittny, 1990) have obtained evidence suggesting that within basement membranes *in situ* lateral associations also occur between type IV collagen molecules, allowing the "network" to expand into the third dimension. Such associations could affect the thickness of a basement membrane and the pore size of its meshwork. Also, three additional type IV collagen chains are now known to exist (Hostikka *et al.*, 1990; Gunwar *et al.*, 1990), at least in kidney.

5.5. Beaded Filaments: Type VI

Type VI collagen (Timpl and Engel, 1987), originally termed intima collagen, has recently become the subject of considerable interest to cell biologists, due chiefly to its unique filamentous arrangement, its identification with the transformation-sensitive glycoprotein GP140, and its high degree of immunogenicity. It is found in most stromal connective tissues. In some, such as corneas, it is present in quantities approaching that of the fibrillar collagens (Zimmermann *et al.*, 1988).

Biochemical analyses of type VI collagen suggest that the molecule is composed of three different chains of 110–140 kDa. Each chain is composed of a central collagenous domain, with noncollagenous domains at each end. The combined mass of the noncollagenous domains is larger than that of the collagenous one. From rotary shadowing analyses, the currently accepted model for the type VI collagen molecule consists of a 105-nm rodlike central portion (the collagenous triple-helical domain) with a large globular domain at each end (Fig. 1-10). The supramolecular assembly of type VI collagen molecules within the beaded filament structure involves dimers, in which two type VI molecules are arranged antiparallel to one another, and tetramers in which two dimers laterally associate. The tetramers then associate linearly, giving rise to the beaded filaments found within most stromal matrices.

These type VI structures, as found within tissues *in situ* (Fig. 1-10), consist of alternating filamentous regions (65–75 nm long) and beaded regions (42–44 nm long), thus giving rise to the term *beaded filaments*. This periodicity is difficult to discern by routine electron microscopy, as are the beaded filaments themselves. However they can be readily visualized in tissues treated with ATP to produce SLS-type aggregates (Bruns, 1984), or by immunoelectron microscopy with monoclonal antibodies (Fig. 1-10).

Type VI collagen is highly immunogenic (Hessle and Engvall, 1984; Linsenmayer *et al.*, 1986b), especially when compared with other types of collagen molecules which generally exhibit weak antigenicity. While this is useful in generating antibodies against type VI, it is a serious concern when trying to generate specific antibodies against certain other collagen types. This problem is exacerbated by the observation that fragments of type VI collagen can copurify with other collagens such as type V (Linsenmayer *et al.*, 1986b). Even when type VI is present in amounts too small to be readily detectable, a large portion of the antibodies obtained may be against the contaminating fragments of type VI collagen. The use of monoclonal antibodies can circumvent this problem (Hessle and Engvall, 1984; Linsenmayer *et al.*, 1986b).

Studies have now established identity between type VI collagen and the

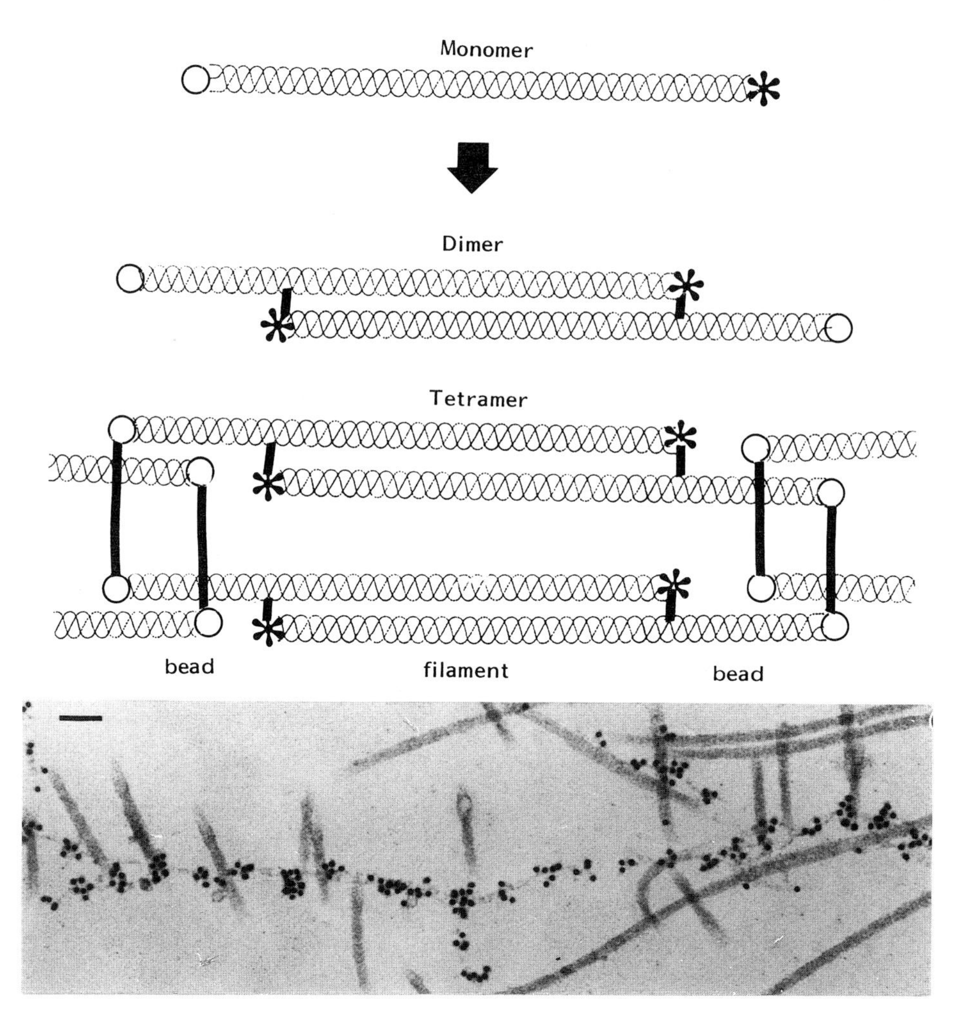

Figure 1-10. (Top) Diagram of the antiparallel association of type VI monomers into dimers, and the subsequent association of the dimers into the tetrameric unit which repeats within the beaded filament. The circles and asterisks designate the two ends of the molecule, and the vertical lines show regions thought to be involved in disulfide cross-linking. Domains giving rise to the beaded (bead) and filamentous (filament) regions are shown. (Modified after Timpl and Engel, 1987.) (Bottom) Electron micrograph of a beaded filament reacted with a monoclonal antibody against type VI collagen followed by a colloidal gold-labeled secondary antibody. The clusters of colloidal gold occur at the expected 100- to 110-nm intervals. The striated fibrils are unlabeled. Bar = 100 nm. (Modified from Timpl, 1976.)

cell matrix glycoprotein GP140. When present as a substratum, the molecule has been shown to promote cell attachment of human fibroblasts as well as adhering to itself and possibly other collagens. Those effects are possibly mediated by the numerous Arg-Gly-Asp units present in the helical domain, and the collagen binding domains in the globular regions (Koller *et al.*, 1989). The molecule is also transformation-sensitive, being undetectable in either the medium or cell layer of SV40-transformed fibroblasts.

5.6. The Short-Chain Collagens: Types VIII and X

As the name implies, the two collagens in the short-chain class are relatively small, due mainly to their single triple-helical domain which is only about half the size of that of a fibrillar collagen. We will first address type X collagen, since somewhat more is known about the structure and cell biology of this molecule, and since it can serve as a prototype for the short-chain family.

5.6.1. Type X Collagen

Type X collagen has a very restricted tissue distribution, being synthesized by hypertrophic chondrocytes during the process of endochondral bone formation (Schmid and Linsenmayer, 1987). In the fetal and adolescent skeleton, the molecule is a transient intermediate in the cartilage which will be replaced by bone (Fig. 1-4); in the adult it persists in the zone of calcified cartilage which separates the articular cartilage from the subchondral bone.

The molecule is a homotrimer, $[\alpha(X)]_3$. The $\alpha1(X)$ chain has an apparent molecular size of 59 kDa on SDS-PAGE, a value which subsequently has been verified by measurements of SLS crystallites (Fig. 1-11A) and rotary-shadowed preparations (Fig. 1-11B) of native molecules visualized in the electron microscope, and by sequencing of the gene. The molecule has three domains (Fig. 1-11, diagram). The smallest is a short nonhelical region at the NH_2 terminus; the largest is a single triple-helical region which, in rotary-shadowed preparations, is a 138-nm rod. Within this collagenous domain, the chains have several interruptions in the Gly-X-Y repeating structure. Some of these may be responsible for the molecule's high susceptibility to cleavage with vertebrate collagenase.

The third domain is a globular one at the COOH-terminal end. In rotary-shadowed preparations it has a ball-like appearance, and by biochemical analysis is composed of three identical peptides (one for each of the α chains) of 17–20 kDa. This region has a high concentration of tyrosine and other hydrophobic amino acids which may contribute to its high degree of stability. Although this domain has certain characteristics found in the procollagen forms of other collagens, it is unlikely that proteolytic processing of the molecule occurs naturally *in vivo*. Instead, it seems that the intact type X molecule becomes deposited within the extracellular matrix.

Within hypertrophic cartilage matrix *in situ*, type X collagen occurs in at least two different supramolecular forms. Recent immunoelectron microscopic studies suggest that some of the type X is found as pericellular mats of filamentous material which surround the hypertrophic chondrocytes (Schmid and Linsenmayer, 1990), and some is found along the surface of cartilage collagen fibrils (Fig. 1-11C) (Schmid and Linsenmayer, 1990; Poole and Pidoux, 1989).

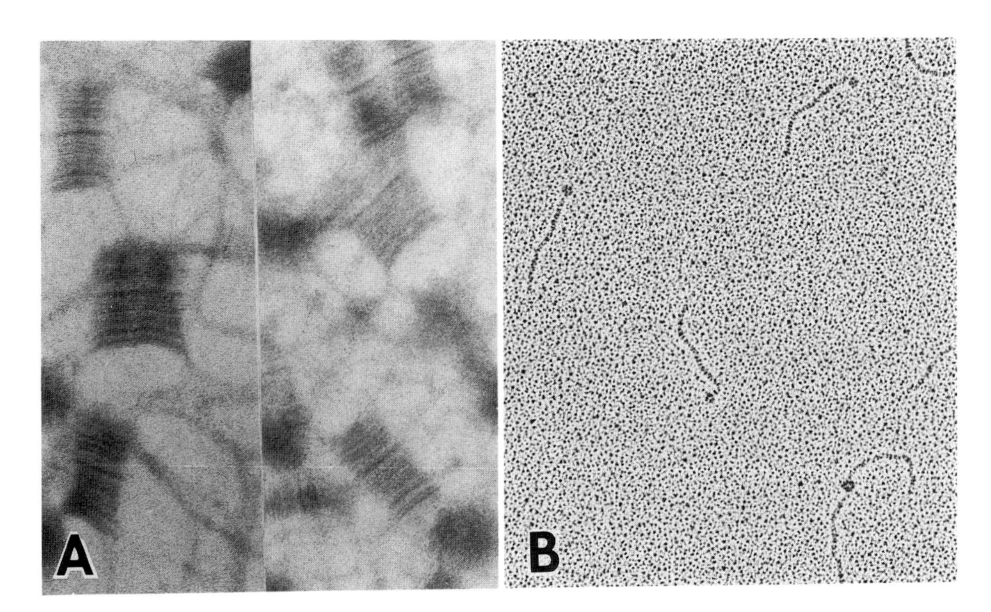
A
B

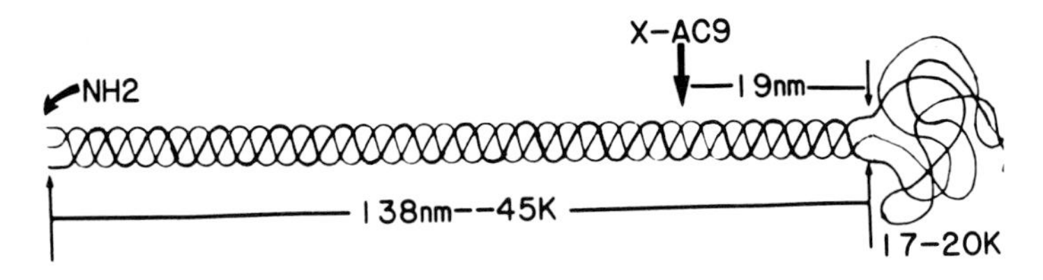
X-AC9
NH2
19nm
138nm--45K
17-20K

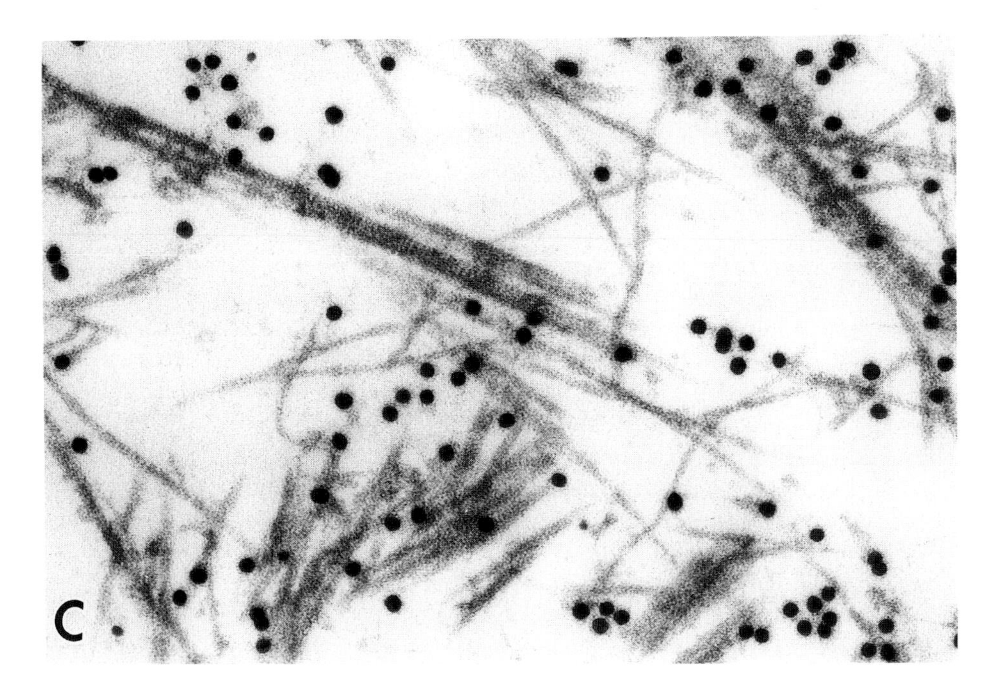
C

The latter form most likely arises by diffusion of new type X molecules away from the synthesizing chondrocyte and out into the matrix where they become associated secondarily with cartilage collagen fibrils assembled at earlier stages of cartilage formation. Possibly this association alters the calcification properties of the fibrils, or prepares them for subsequent removal—the fate of all components of hypertrophic cartilage matrix.

Consistent with this diffusion model, it has recently been demonstrated in a model system that type X molecules can rapidly and extensively move through cartilage matrix and bind to preexisting fibrils (Chen *et al.*, 1990). These observations raise the general proposition that during embryonic development, extracellular matrices can be assembled and modified at considerable distances from a synthesizing cell.

5.6.2. Type VIII Collagen

Type VIII collagen (Sage and Bornstein, 1987) was originally termed EC (endothelial cell) collagen for its synthesis by aortic and corneal endothelial cells. It probably should not, however, be considered a tissue-specific molecule since recent evidence has suggested that it has a considerably wider tissue distribution than was once thought (Sage and Iruela-Arispe, 1990).

The structure of the molecule, as deduced from biochemical studies, has been somewhat controversial, with at least two models being proposed (Sage and Bornstein, 1987; Benya and Padilla, 1986). Recent data from cDNA sequencing have suggested that the correct structure will be quite similar in both size and domain organization to that of collagen type X; the cDNAs for these two molecules are strikingly similar (Yamaguchi *et al.*, 1989). The only notable differences in the predicted structures of these two collagens are that the noncollagenous domain at the NH_2-terminal end of type VIII is considerably larger than that of type X, and a portion of the COOH-terminal globular domain shows some differences.

The supramolecular organization of type VIII has not yet been determined, but evidence suggests that it may be a component of the hexagonal model–intermodel network of Descemet's membrane, a subendothelial matrix located at the posterior cornea. It is unknown whether a supramolecular organization similar to that of Descemet's membrane exists within the matrices of the other tissues thought to contain type VIII collagen, such as those associated with

Figure 1-11. (A–C) Three different ways of visualizing a collagen molecule—here collagen type X. (A) SLS crystallites positively stained with phosphotungstate followed by uranyl acetate. (B) Rotary-shadowed preparation in which the collagenous, threadlike domain and the COOH-terminal globular domain are clearly visible. (C) Immunoelectron micrograph showing fibril-associated type X. The section is through the matrix deposited by hypertrophic chondrocytes *in vitro* and has been reacted with the anti-type X collagen-monoclonal antibody (X-AC9) complexed to colloidal gold. (A and B modified from Schmid *et al.*, 1984; C modified from Schmid and Linsenmayer, 1990.) (Middle) A diagram of a type X collagen molecule showing the triple-helical and globular domains, and their sizes. Also shown is the position of the epitope for monoclonal antibody X-AC9, as determined by rotary shadowing.

certain vascular epithelia. The suggestion has been raised, however, that the presence of type VIII in such a matrix may stabilize the endothelial phenotype.

5.7. The Long-Chain Collagen: Type VII

With respect to size, the converse of the short-chain collagens are the long-chain ones, a class of which type VII is the only known member (Burgeson *et al.*, 1990; Burgeson, 1987). This collagen is found solely within anchoring fibrils (Fig. 1-12), subepithelial structures which help stabilize the attachment of the basement membranes of stratified squamous epithelia to the underlying stroma.

Analyses by biochemical extraction and *in vitro* labeling studies suggest that the native type VII collagen molecule is a homotrimer, $[\alpha 1(\text{VII})]_3$, of three identical chains, each > 300 kDa. Rotary shadowing analyses suggest a mole-

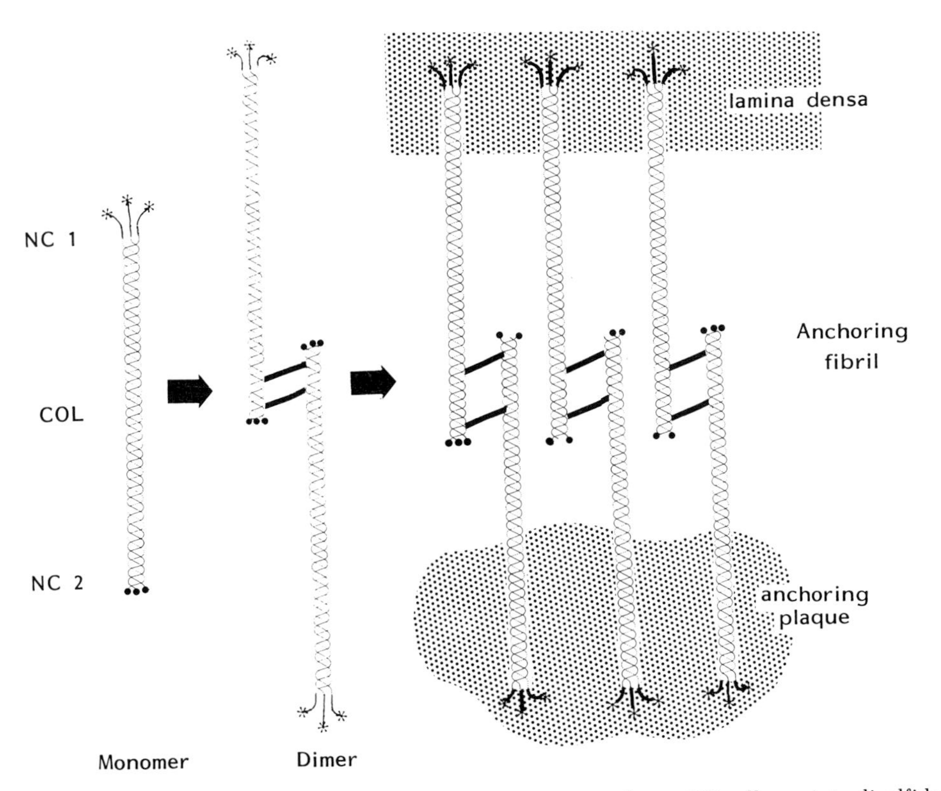

Figure 1-12. Diagram depicting the association of monomers of type VII collagen into disulfide cross-linked, antiparallel dimers, and the subsequent association of the dimers into the SLS crystallites of anchoring fibrils. The dotted areas depict the type IV collagen-containing lamina densa of the basement membrane and a anchoring plaque within the stroma. (Modified from Burgeson *et al.*, 1990.)

cule with a central 425-nm-long triple-helical domain and globular domains at each end. The COOH-terminal globular domain (Fig. 1-12, NC1) is itself a complex structure consisting of multiple subdomains. The NH_2-terminal globular domain (NC2) is smaller and consists of a single ball. Most likely, the molecule which occurs within extracellular matrix retains all of these domains intact; no evidence of proteolytic processing has been found.

In the supramolecular assembly of type VII, the repeating structural unit appears to be a dimer in which two type VII collagen molecules are arranged antiparallel to one another with an NH_2-terminal overlap of approximately 60 nm (Fig. 1-12). The region of overlap is stabilized by disulfide cross-links. Within tissues *in situ*, these dimers are arranged as SLS crystallites where they form the major, and perhaps the only component of anchoring fibrils, long, striated structures underlying some stratified squamous epithelia. Due to their molecular organization, both ends of an anchoring fibril contain a cluster of COOH-terminal domains of type VII collagen molecules available for interaction with other matrix components. Some anchoring filaments have the COOH-terminal domains at one end inserted into the lamina densa of the epithelial basement membrane while those at the other end are inserted into a subjacent tuft of lamina densa-like, type IV collagen-containing material within the stroma (anchoring plaques). In other anchoring filaments, both ends interact with and connect adjacent anchoring plaques (not shown). The interlocking network thus formed is thought to stabilize the adhesion of the epithelial basement membrane to the subjacent stromal matrix (see Burgeson, 1987).

6. Summary and Conclusions

In this chapter I have reviewed a number of aspects of collagen biochemistry and cell biology, including: some of the general properties of the triple-helical structure; the various supramolecular forms in which collagens can be assembled; the various methodologies used to analyze, identify, and characterize collagens; and the immunological properties of collagens and how these can be utilized to generate antibodies useful for studies of cell and developmental biology. I also surveyed some of the characteristics and properties of the different classes of collagens, which encompass most of the known types (I–XII).

I did not, however, address several of the newest collagens, such as: type XIII, which has been identified as an alternately spliced cDNA (Tikka *et al.*, 1988; Pihlajaniemi *et al.*, 1990; Pihlajaniemi and Tamminen, 1990); type XIV, which preliminary evidence suggests may be a member of the fibril-associated family; or several putative collagens with structures in which numerous short triple-helical domains are separated by noncollagenous ones (Marchant *et al.*, 1991) (and personal communications). Nor did I address any of the ever-increasing number of nonmatrix molecules that possess triple-helical domains, such as the complement component C1q, serum mannose binding protein, pulmonary surfactant apoprotein SP-A, the macrophage scavenger receptor,

Figure 1-13. Provided by a graduate student who shall go unnamed.

and even a viral coat protein. Clearly the Gly-X-Y triple-helix has a much wider distribution and more varied functions than previously thought. Possibly it should be considered in the same context as other structural forms of proteins, such as the alpha helix and the beta pleated sheet.

If so, one may ask what constitutes a "true" collagen. In historical context it would seem that at least two structural criteria must be met. The molecule must contain one or more domains in which the triple-helix is the predominant structure, and it must serve as a component of an extracellular matrix, for at least part of its existence. Even with these criteria, over the past several years the number of molecules which have been designated "true" collagens has burgeoned to the point where one graduate student in our laboratory, in his frustration at trying to learn the different collagens and their properties, started a campaign whose slogan is shown in Fig. 1-13. We obviously do not agree with him, since each of the new collagens that has been discovered has increased our understanding of the roles performed by extracellular matrices in cellular and developmental processes. We do, however, sympathize with this student and hope that this chapter makes his job easier. It, therefore, is dedicated to him and others like him.

ACKNOWLEDGMENTS. I thank Mary Currier for help in preparing the text of the manuscript, Lissy Linsenmayer for some of the figures, and Drs. John Fitch and Marion Gordon for their critical comments. The author's original research included here was supported by NIH grants EY05191 and HD23681.

References

Adachi, E., and Hayashi, T., 1985, In vitro formation of fine fibrils with a D-periodic banding pattern from type V collagen, *Collagen Relat. Res.* **5**:225–232.

Benya, P. D., and Padilla, S. R., 1986, Isolation and characterization of type VIII collagen synthesized by cultured rabbit corneal endothelial cells. A conventional structure replaces the interrupted-helix model, *J. Biol. Chem.* **261**:4160–4169.

Bernard, M., Yoshioka, H., Rodriguez, E., Van der Rest, M., Kimura, T., Ninomiya, Y., Olsen, B. R., and Ramirez, F., 1988, Cloning and sequencing of pro-alpha 1 (XI) collagen cDNA demonstrates that type XI belongs to the fibrillar class of collagens and reveals that the expression of the gene is not restricted to cartilagenous tissue, *J. Biol. Chem.* **263:**17159–17166.

Birk, D. E., Fitch, J. M., Babiarz, J. P., and Linsenmayer, T. F., 1988, Collagen type I and type V are present in the same fibril in the avian corneal stroma, *J. Cell Biol.* **106:**999–1008.

Birk, D. E., Fitch, J. M., Babiarz, J. P., Doane, K. J., and Linsenmayer, T. F., 1990, Collagen fibrillogenesis *in vitro:* Interaction of types I and V collagen regulates fibril diameter, *J. Cell Sci.* **95:** 649–657.

Bornstein, P., and Sage, H., 1980, Structurally distinct collagen types, *Annu. Rev. Biochem.* **49:**957–1003.

Bruns, R. R., 1984, Beaded filaments and long-spacing fibrils: Relation to type VI collagen, *J. Ultrastruct. Res.* **89:**136–145.

Burgeson, R. E., 1987, Type VII collagen, in: *Structure and Function of Collagen Types* (R. Mayne and R. E. Burgeson, eds.), pp. 145–172, Academic Press, New York.

Burgeson, R. E., Lunstrum, G. P., Rokosova, B., Rimberg, C. S., Rosenbaum, L. M., and Keene, D. R., 1990, The structure and function of type VII collagen, *Ann. N.Y. Acad. Sci.* **580:**32–43.

Chen, Q., Gibney, E., Fitch, J. M., Linsenmayer, C., Schmid, T. M., and Linsenmayer, T. F., 1990, Long-range movement and fibril association of type X collagen within embryonic cartilage matrix, *Proc. Natl. Acad. Sci. USA* **87:**8046–8050.

Eyre, D., and Muir, H., 1976, Types 1 and 2 collagens in intervertebral disc. Interchanging radial distributions in annuls fibrosus, *Biochem. J.* **157:**267–270.

Eyre, D., and Wu, J., 1987, Type XI collagen, in: *Structure and Function of Collagen Types* (R. Mayne and R. E. Burgeson, eds., pp. 261–282, Academic Press, New York.

Fessler, J. H., and Fessler, L. I., 1978, Biosynthesis of procollagen, *Annu. Rev. Biochem.* **47:**129–162.

Fessler, J. H., and Fessler, L. I., 1987, Type V collagen, in: *Structure and Function of Collagen Types* (R. Mayne and R. E. Burgeson, eds.), pp. 81–97, Academic Press, New York.

Fitch, J. M., Gross, J., Mayne, R., Johnson Wint, B., and Linsenmayer, T. F., 1984, Organization of collagen types I and V in the embryonic chicken cornea: Monoclonal antibody studies, *Proc. Natl. Acad. Sci. USA* **81:**2791–2795.

Fitch, J. M., Birk, D. E., Mentzer, A., Hasty, K. A., Mainardi, C., and Linsenmayer, T. F., 1988a, Corneal collagen fibrils: Dissection with specific collagenases and monoclonal antibodies, *Invest. Ophthalmol. Vis. Sci.* **29:**1125–1136.

Fitch, J. M., Mentzer, A., Mayne, R., and Linsenmayer, T. F., 1988b, Acquisition of type IX collagen by the developing avian primary corneal stroma and vitreous, *Dev. Biol.* **128:**396–405.

Fitch, J. M., Mentzer, A., Mayne, R., and Linsenmayer, T. F., 1989, Independent deposition of collagen types II and IX at epithelial–mesenchymal interfaces, *Development* **105:**85–95.

Fleischmajer, R., Perlish, J. S., Burgeson, R. E., Shaikh-Bahai, F., and Timpl, R., 1990, Type I and type III collagen interactions during fibrillogenesis, *Ann. N.Y. Acad. Sci.* **580:**161–175.

Gay, S., and Miller, E. J., 1979, *Collagen in the Physiology and Pathology of Connective Tissue,* Gustav Fischer Verlag, New York.

Gordon, M. K., Gerecke, D. R., and Olsen, B. R., 1987, Type XII collagen: Distinct extracellular matrix component discovered by cDNA cloning, *Proc. Natl. Acad. Sci. USA* **84:**6040–6044.

Gordon, M. K., Gerecke, D. R., Dublet, B., Van der Rest, M., Sugrue, S. P., and Olsen, B. R., 1990, The structure of type XII collagen, *Ann. N.Y. Acad. Sci.* **580:**8–16.

Grant, M. E., and Prockop, D. J., 1972, The biosynthesis of collagen, *N. Engl. J. Med.* **286:**194–300.

Gross, J., 1974, Collagen biology: Structure, degradation, and disease, *Harvey Lect.* **6B:**351–432.

Gunwar, S., Saus, J., Noelken, M. E., and Hudson, B. G., 1990, Glomerular basement membrane. Identification of a fourth chain, α4, of type IV collagen, *J. Biol. Chem.* **265:**5466–5469.

Hendrix, M. J., Hay, E. D. Von der Mark, K., and Linsenmayer, T. F., 1982, Immunohistochemical localization of collagen types I and II in the developing chick cornea and tibia by electron microscopy, *Invest. Ophthalmol. Vis. Sci.* **22:**359–375.

Hessle, H., and Engvall, E., 1984, Type VI collagen. Studies on its localization, structure, and biosynthetic form with monoclonal antibodies, *J. Biol. Chem.* **259:**3955–3961.

Hodge, A. J., and Petruska, J. A., 1963, Recent studies with the electron microscope on ordered aggregates of the tropocollagen molecule, in: *Aspects of Protein Structure* (G. N. Ramachandran, ed.), pp. 289–301, Academic Press, New York.

Hostikka, S. L., Eddy, R. L., Byers, M. G., Höyhtyä, M., Shows, T. B., and Tryggvason, K., 1990, Identification of a distinct type IV collagen α chain with restricted kidney distribution and assignment of its gene to the locus of X chromosome-linked Alport syndrome, *Proc. Natl. Acad. Sci. USA* **87:**1606–1610.

Keene, D. R., Sakai, L. Y., Bachinger, H. P., and Burgeson, R. E., 1987, Type III collagen can be present on banded collagen fibrils regardless of fibril diameter, *J. Cell Biol.* **105:**2393–2402.

Kefalides, N. A., 1973, Structure and biosynthesis of basement membranes, *Int. Rev. Connect. Tissue Res.* **6:**63–104.

Koller, E., Winterhalter, K. H., and Trueb, B., 1989, The globular domains of type VI collagen are related to the collagen-binding domains of cartilage matrix protein and von Willebrand factor, *EMBO J.* **8:**1073–1077.

Kosher, R. A., and Solursh, M., 1989, Widespread distribution of type II collagen during embryonic chick development, *Dev. Biol.* **131:**558–566.

Kuhn, K., 1987, The classical collagens: Types I, II, and III, in: *Structure and Function of Collagen Types* (R. Mayne and R. E. Burgeson, eds.), pp. 1–42, Academic Press, New York.

Linsenmayer, T. F., 1981, Collagen, in: *Cell Biology of Extracellular Matrix* (E. D. Hay, ed.), pp. 5–37, Plenum Press, New York.

Linsenmayer, T. F., and Hendrix, M. J. C., 1982, Production of monoclonal antibodies to collagens and their immunofluorescence localization in embryonic cornea and cartilage, in: *Immunochemistry of the Extracellular Matrix, Volume I* (H. Furthmayr, ed.), pp. 179–198, CRC Press, Boca Raton, Fla.

Linsenmayer, T. F., Fitch, J. M., and Mayne, R., 1984, Basement membrane structure and assembly: Inferences from immunological studies with monoclonal antibodies in: *The Role of Extracellular Matrix in Development* (R. L. Trelstad, ed.), pp. 146–172, Liss, New York.

Linsenmayer, T. F., Fitch, J. M., Gross, J., and Mayne, R., 1985, Are collagen fibrils in the developing avian cornea composed of two different collagen types? Evidence from monoclonal antibody studies, in: *Biochemistry, Chemistry and Pathology of Collagen* (R. Fleischmajer, B. R. Olsen, and K. Kuhn, eds.), pp. 232–245, New York Academy of Sciences, New York.

Linsenmayer, T. F., Gibney, E., and Schmid, T. M., 1986a, Segmental appearance of type X collagen in the developing avian notochord, *Dev. Biol.* **113:**467–473.

Linsenmayer, T. F., Mentzer, A., Irwin, M. H., Waldrep, N. K., and Mayne, R., 1986b, Avian type VI collagen. Monoclonal antibody production and immunohistochemical identification as a major connective tissue component of cornea and skeletal muscle, *Exp. Cell Res.* **165:**518–529.

Linsenmayer, T. F., Fitch, J. M., and Schmid, T. M., 1988, Multiple-reaction cycling: A method for enhancement of the immunochemical signal of monoclonal antibodies, *J. Histochem. Cytochem.* **36:**1075–1078.

Linsenmayer T. F., Fitch, J. M., Schmid, T. M., Birk, D. E., Bruns, R. R., and Mayne, R., 1989, Applications of anti-collagen monoclonal antibodies to studies of connective tissue structure and development, in: *Collagen: Biochemistry, Biotechnology and Molecular Biology,* Volume 4 (B. R. Olsen and M. E. Nimni, eds.), pp. 141–170, CRC Press, Boca Raton, Fla.

Linsenmayer, T. F., Fitch, J. M., and Birk, D. E., 1990, Heterotypic collagen fibrils and stabilizing collagens: Controlling elements in corneal morphogenesis, *Ann. N.Y. Acad. Sci.* **580:** 143–160.

Marchant, J. K., Linsenmayer, T. F., and Gordon, M. K., 1990, cDNA analysis predicts a novel cornea-specific collagen, *Proc. Natl. Acad. Sci. USA* **88:**1560–1564.

Mayne, R., and Burgeson, R. E. (eds.), 1987, *Structure and Function of Collagen Types,* Academic Press, New York.

Mayne, R., Wiedemann, H., Irwin, M. H., Sanderson, R. D., Fitch, J. M., Linsenmayer, T. F. and Kuhn, K., 1984, Monoclonal antibodies against chicken type IV and V collagens: Electron microscopic mapping of the epitopes after rotary shadowing, *J. Cell Biol.* **98:**1637–1644.

Mendler, M., Eich-Bender, S. G., Vaughan, L., Winterhalter, K. H., and Bruckner, P., 1989, Cartilage contains mixed fibrils of collagen types II, IX, and XI, *J. Cell Biol.* **108:**191–198.

Miller, E. J., 1972, Structural studies on cartilage collagen employing limited cleavage and solubilization with pepsin, *Biochemistry* **11:**4903–4909.

Morris, N. P., and Bachinger, H. P., 1987, Type XI collagen is a heterotrimer with the composition (1 alpha, 2 alpha, 3 alpha) retaining non-triple-helical domains, *J. Biol. Chem.* **262**:11345–11350.

Nishimura, I., Muragaki, Y., and Olsen, B. R., 1989, Tissue-specific forms of type IX collagen-proteoglycan arise from the use of two widely separated promoters, *J. Biol. Chem.* **264**:20033–20041.

Okada, Y., Konomi, H., Yada, T., Kimata, K., and Nagase, H., 1989, Degradation of type IX collagen by matrix metalloproteinase 3 (stromelysin) from human rheumatoid synovial cells, *FEBS Lett.* **244**:473–476.

Pihlajaniemi, T., and Tamminen, M., 1990, The α1 chain of type XIII collagen consists of three collagenous and four noncollagenous domains, and its primary transcript undergoes complex alternative splicing, *J. Biol. Chem.* **265**:16922–16928.

Pihlajaniemi, T., Tamminen, M., Sandberg, M., Hirvonen, H., and Vuorio, E., 1990, The α1 chain of type XIII collagen: Polypeptide structure, alternative splicing, and tissue distribution, *Ann. N.Y. Acad. Sci.* **580**:440–443.

Poole, A. R., and Pidoux, I., 1989, Immunoelectron microscopic studies of type X collagen in endochondral ossification, *J. Cell Biol.* **109**:2547–2554.

Prockop, D. J., Kivirikko, K. I., Tuderman, L., and Guzman, N. A., 1979, The biosynthesis of collagen and its disorders, *N. Engl. J. Med.* **301**:13–23.

Ramachandran, G. N., and Ramakrishnan, C. F., 1976, Molecular structure, in: *Biochemistry of Collagen* (G. N. Ramachandran and A. H. Reddi, eds., pp. 45–84, Plenum Press, New York.

Reese, C. A., and Mayne, R., 1981, Minor collagens of chicken hyaline cartilage, *Biochemistry* **20**:5443–5448.

Sage, H., and Bornstein, P., 1987, Type VIII collagen, in: *Structure and Function of Collagen Types* (R. Mayne and R. E. Burgeson, eds.), pp. 173–193, Academic Press, New York.

Sage, H., and Iruela-Arispe, M. -L., 1990, Type VIII collagen in murine development: Association with capillary formation *in vitro*, *Ann. N.Y. Acad. Sci.* **580**:17–31.

Schmid, T. M., and Linsenmayer, T. F., 1985, Developmental acquisition of type X collagen in the embryonic chick tibiotarsus, *Dev. Biol.* **107**:373–381.

Schmid, T. M., and Linsenmayer, T. F., 1987, Type X collagen, in: *Structure and Function of Collagen Types* (R. Mayne and R. E. Burgeson, eds.), pp. 223–259, Academic Press, New York.

Schmid, T. M., and Linsenmayer, T. F., 1990, Immunoelectron microscopy of type X collagen: Supramolecular forms within embryonic chick cartilage, *Dev. Biol.* **138**:53–62.

Schmid, T. M., Mayne, R., Bruns, R. R., and Linsenmayer, T. F., 1984, Molecular structure of short-chain (SC) cartilage collagen by electron microscopy, *J. Ultrastruct. Res.* **86**:186–191.

Siegel, R. C., 1979, Lysyl oxidase, *Int. Rev. Connect. Tissue Res.* **8**:73–118.

Smith, G. N., Jr., Linsenmayer, T. F., and Newsome, D. A., 1976, Synthesis of type II collagen in vitro by embryonic chick neural retina tissue, *Proc. Natl. Acad. Sci. USA* **73**:4420–4423.

Sugrue, S. P., Gordon, M. K., Seyer, J., Dublet, B., Van der Rest, M., and Olsen, B. R., 1989, Immunoidentification of type XII collagen in embryonic tissues, *J. Cell Biol.* **109**:939–945.

Summers, T. A., Irwin, M. H., Mayne, R., and Balian, G., 1988, Monoclonal antibodies to type X collagen. Biosynthetic studies using an antibody to the amino-terminal domain, *J. Biol. Chem.* **263**:581–587.

Svoboda, K. K., Nishimura, I., Sugrue, S. P., Ninomiya, Y., and Olsen, B. R., 1988, Embryonic chicken cornea and cartilage synthesize type IX collagen molecules with different amino-terminal domains, *Proc. Natl. Acad. Sci. USA* **85**:7496–7500.

Tanzer, M. L., 1973, Cross-linking of collagen, *Science* **180**:561–566.

Tanzer, M. L., 1976, Cross-linking, in: *Biochemistry of Collagen* (G. N. Ramachandran and A. H. Reddi, eds.), pp. 137–157, Plenum Press, New York.

Tikka, L., Pihlajaniemi, T., Henttu, P., Prockop, D. J., and Tryggvason, K., 1988, Gene structure for the alpha 1 chain of a human short-chain collagen (type XIII) with alternatively spliced transcripts and translation termination codon at the 5′ end of the last exon, *Proc. Natl. Acad. Sci. USA* **85**:7491–7495.

Timpl, R., 1976, Immunological studies on collagen, in: *Biochemistry of Collagen* (G. N. Ramachandran and A. H. Reddi, eds.), pp. 319–375, Plenum Press, New York.

Timpl, R., and Engel, E., 1987, Type VI collagen, in: *Structure and Function of Collagen Types* (R. Mayne and R. E. Burgeson, eds.), pp. 105–140, Academic Press, New York.

Timpl, R., Weidemann, H., VanDelden, V., Furthmayr, H., and Kuhn, K., 1981, A network model for the organization of type IV collagen molecules in basement membranes, *Eur. J. Biochem.* **120:**203–211.

Traub, W., and Piez, K. A., 1971, The chemistry and structure of collagen, *Adv. Protein Chem.* **25:**243–352.

Trelstad, R. L., Kang, A. H., Igarashi, S., and Gross, J., 1970, Isolation of two distinct collagens from chick cartilage, *Biochemistry* **9:**4993–4998.

Van der Rest, M., and Mayne, R., 1987, Type IX collagen, in: *Structure and Function of Collagen Types* (R. Mayne and R. E. Burgeson, eds.), pp. 195–219, Academic Press, New York.

Van der Rest, M. and Mayne, R., 1988, Type IX collagen proteoglycan from cartilage is covalently cross-linked to type II collagen, *J. Biol. Chem.* **263:**1615–1618.

Van der Rest, M., Mayne, R., Ninomiya, Y., Seidah, N. G., Chretien, M., and Olsen, B. R., 1985, The structure of type IX collagen, *J. Biol. Chem.* **260:**220–225.

Vaughan, L., Mendler, M., Huber, S., Bruckner, P., Winterhalter, K. H., Irwin, M. I., and Mayne, R., 1988, D-periodic distribution of collagen type IX along cartilage fibrils, *J. Cell Biol.* **106:**991–997.

Wright, D. W., and Mayne, R., 1988, Vitreous humor of chicken contains two fibrillar systems: An analysis of their structure, *J. Ultrastruct. Mol. Struct. Res.* **100:**224–234.

Yada, T., Suzuki, S., Kobayashi, K., Kobayashi, M., Hoshino, T., Horie, K., and Kimata, K., 1990, Occurrence in chick embryo vitreous humor of a type IX collagen proteoglycan with an extraordinarily large chondroitin sulfate chain and short α1 polypeptide, *J. Biol. Chem.* **265:**6992–6999.

Yamaguchi, N., Benya, P. D., Van der Rest, M., and Ninomiya, Y., 1989, The cloning and sequencing of α1(VIII) collagen cDNAs demonstrate that type VIII collagen is a short chain collagen and contains triple-helical and carboxyl-terminal non-triple-helical domains similar to those of type X collagen, *J. Biol. Chem.* **264:**16022–16029.

Yoshioka, H., and Ramirez, F., 1990, Pro-α1(XI) collagen. Structure of the amino-terminal propeptide and expression of the gene in tumor cell lines, *J. Biol. Chem.* **265:**6423–6426.

Yurchenco, P. D., and Schittny, J. C., 1990, Molecular architecture of basement membranes, *FASEB J.* **4:**1577–1590.

Zimmerman, D. R., Fischer, R. W., Winterhalter, K. H., Witmer, R., and Vaughan, L., 1988, Comparative studies of collagens in normal and keratoconus corneas, *Exp. Eye Res.* **46:**431–442.

Chapter 2

Proteoglycans

Structure and Function

THOMAS N. WIGHT, DICK K. HEINEGÅRD, and VINCENT C. HASCALL

1. Introduction

Proteoglycans (PGs) are marvelously complex macromolecules that each contain a core protein with one or more covalently bound glycosaminoglycan (GAG) chains. GAGs are linear polymers of repeating disaccharides that contain one hexosamine and either a carboxylate or a sulfate ester, or usually both. These simple definitions encompass an exceptionally large range of structures involving different core proteins (see Section 2.1), different classes of GAGs, and different numbers and lengths of individual GAG chains (see Section 2.2). This large structural diversity no doubt contributes to a wide variety of biological functions. PGs are found inside cells, on the cell surface, and in the extracellular matrix. The purpose of this chapter is to describe the structures and possible functions of representative examples of PGs found in each of these tissue compartments. In the last few years, investigators have shown creativity in naming core proteins of individual PGs under investigation. In this chapter, such names will be used only for those core proteins for which the entire primary sequence of amino acids is known. For further background information, see Evered and Whelan (1986) and Wight and Mecham (1987).

Abbreviations used in this chapter: CS, chondroitin sulfate; DS, dermatan sulfate; HS, heparan sulfate; KS, keratan sulfate; HA, hyaluronic acid (hyaluronan, hyaluronate); GAG, glycosaminoglycan; PG, proteoglycan; HSPG, heparan sulfate proteoglycan; CSPG, chondroitin sulfate proteoglycan; DSPG, dermatan sulfate proteoglycan; KSPG, keratan sulfate proteoglycan; bFGF, basic fibroblast growth factor; TGF-β, transforming growth factor-β; PDGF, platelet-derived growth factor.

THOMAS N. WIGHT • Department of Pathology, University of Washington, Seattle, Washington 98195. DICK K. HEINEGÅRD • Department of Physiological Chemistry, University of Lund, Lund, Sweden. VINCENT C. HASCALL • National Institute of Dental Research, National Institutes of Health, Bethesda, Maryland 20892.

Cell Biology of Extracellular Matrix, Second Edition, edited by Elizabeth D. Hay, Plenum Press, New York, 1991.

2. Proteoglycan Components

2.1. Core Proteins

Core proteins in mature PGs are often embedded in complex carbohydrate structures, both oligosaccharides and GAGs. For this reason, conventional protein sequencing has been difficult, and has provided primary structures for only two core proteins, those for the small interstitial PGs now referred to as decorin (Pearson *et al.*, 1983; Brennan *et al.*, 1984) and biglycan (Neame *et al.*, 1989). Also, key sequence information has been provided for some of the globular domains (Neame *et al.*, 1987) in the large aggregating PG from cartilage now referred to as aggrecan (Doege *et al.*, 1990).

The recent advances in molecular biology have now opened the treasure chest of core protein diversity and have provided deduced amino acid sequences for at least six core proteins. Figure 2-1 shows a schematic scale drawing of the core proteins from four distinctly different PGs, all of which have been cloned and sequenced. The smallest, ~ 20 kDa, is the core protein of a PG referred to as serglycin (Bourdon *et al.*, 1985) because it contains an embedded sequence with 24 consecutive repeats of serine-glycine, which determine the eventual location of the GAG side chains. Serglycin is found in storage granules of hematopoietic cells and mast cells (see Section 3.6) and is, therefore, an example of an intracellular PG. Decorin, named because it "decorates" collagen fibrils (Krusius and Ruoslahti, 1986), has a core protein of ~ 36 kDa. It is characterized by a motif of leucine-rich repeat sequences and interacts with fibrillar collagens in the organization of many connective tissues (see Section 3.3). Syndecan, derived from Greek *syndein* meaning to bind together (Saunders *et al.*, 1989), has a core protein of ~ 32 kDa. It contains a COOH-terminal cytoplasmic domain, a hydrophobic domain which intercalates in the plasma membrane and an extracellular domain which carries the GAG chains in the mature PG (see Section 3.5). Aggrecan, named because it forms aggregates (Doege *et al.*, 1990), has a very large core protein, ~ 210 kDa. It is divided into a number of domains, including three globular domains and intervening regions which carry oligosaccharides and GAG chains in the mature PG. This PG is a major structural macromolecule of cartilages (see Section 3.2). It is apparent that these four core proteins are not genetically related, although each may be a member of its own gene family. Because nature chose to assemble GAG chains on a large number of independent and extremely diverse core proteins, PGs cannot be classified into a single related supergene family.

2.2. Glycosaminoglycans

Figure 2-2 shows the structures of the backbone repeat disaccharides for the four classes of GAGs and illustrates some interesting features of their chemistry (for review see Rodén, 1980).

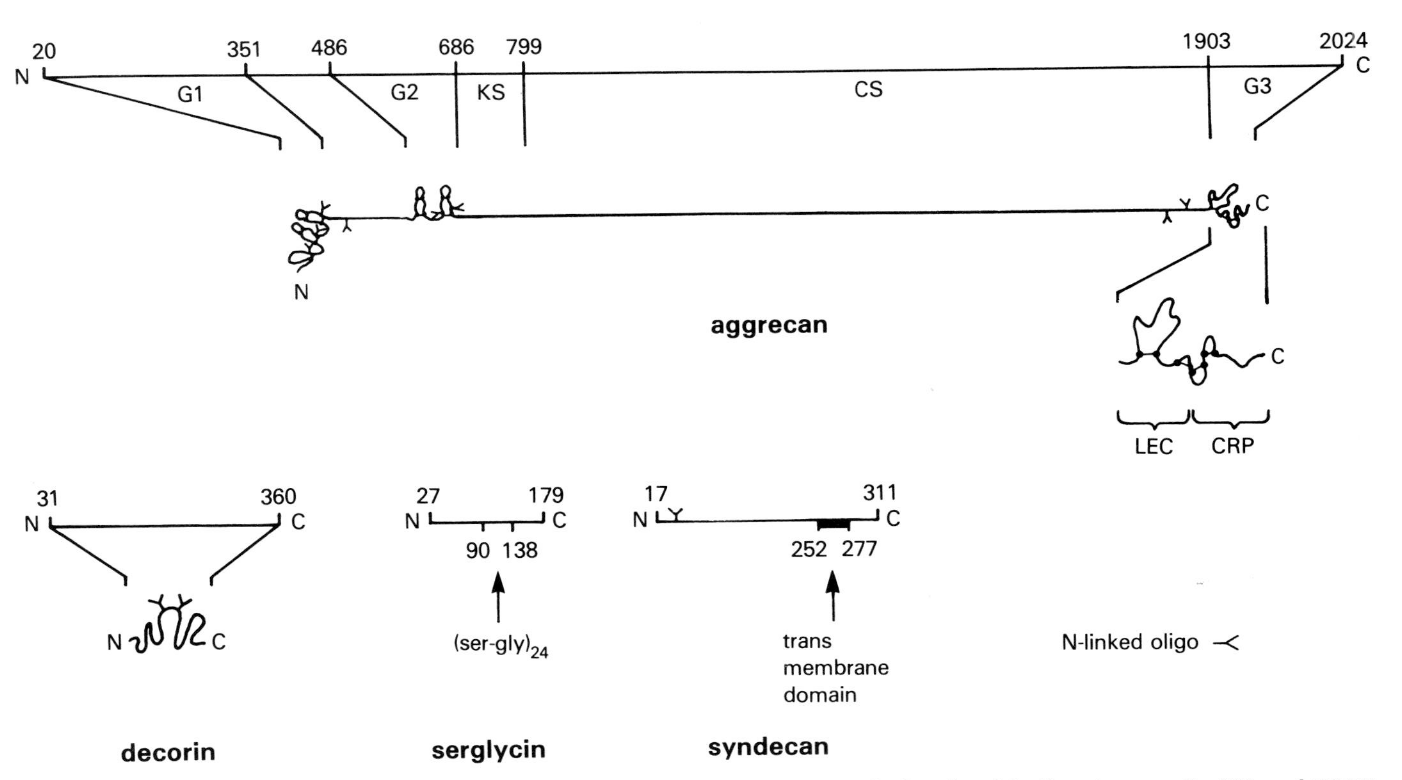

Figure 2-1. Schematic drawings of the core proteins for four distinctly different PGs. The lengths of the lines between the NH_2 and COOH termini are proportional to the number of amino acids in each protein. The straight lines indicate the extended lengths and the residue numbers separating different domains in the primary sequences. The known globular domains in aggrecan and decorin are folded in their core protein diagrams. The branched Y-structures indicate locations of asparagine-X-serine(threonine), the sequence required for adding *N*-linked oligosaccharides. The G3 domain of aggrecan is enlarged to indicate the lectinlike binding domain (LEC) and the domain with homology to complement regulatory protein (CRP). The CS and KS designations in aggrecan indicate domains enriched in CS chains and in KS chains and/or *O*-linked oligosaccharides respectively in the mature PG molecules (see Fig. 2-4).

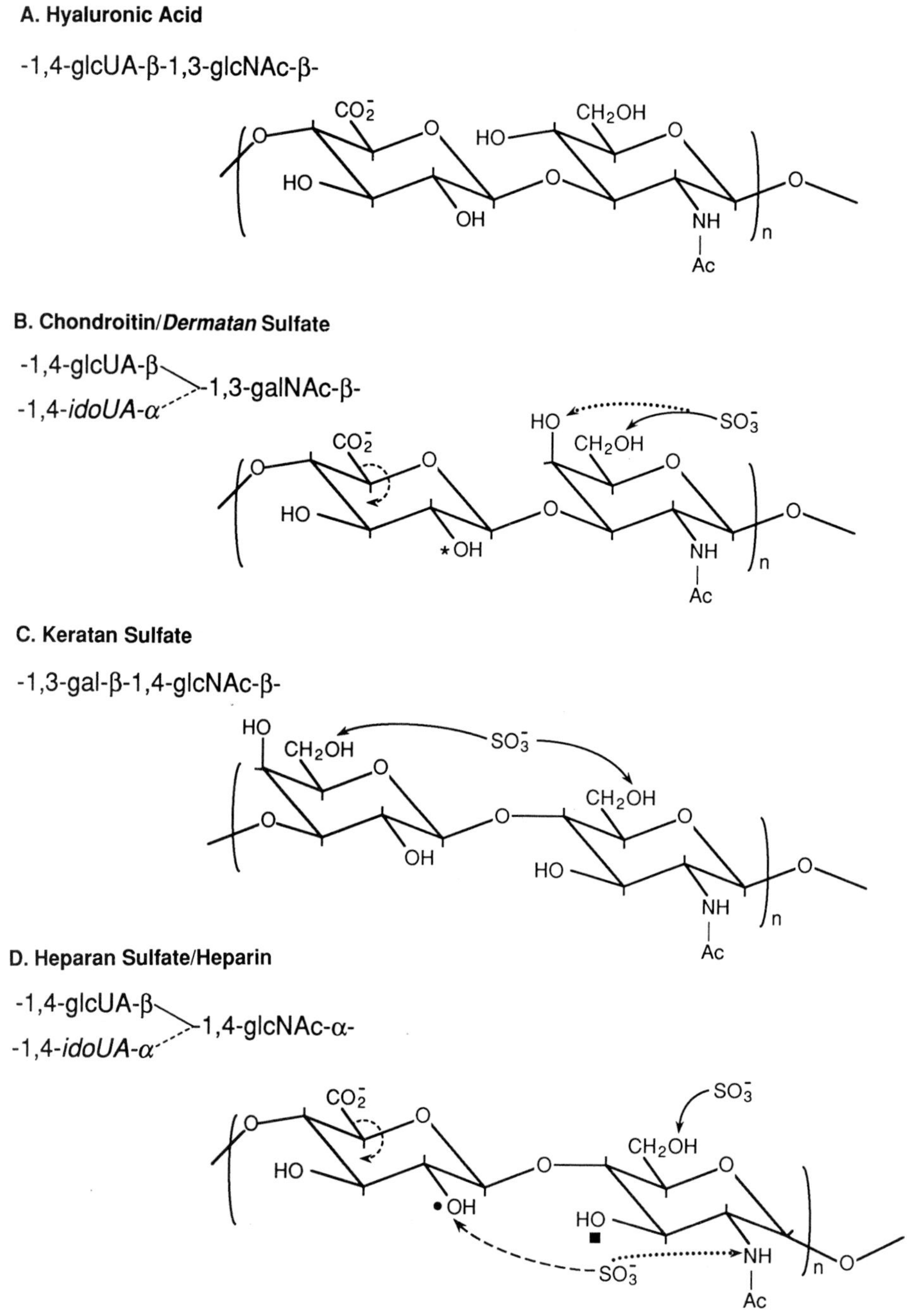

A. Hyaluronic Acid

-1,4-glcUA-β-1,3-glcNAc-β-

B. Chondroitin/*Dermatan* Sulfate

-1,4-glcUA-β- / -1,4-*idoUA-α*- → -1,3-galNAc-β-

C. Keratan Sulfate

-1,3-gal-β-1,4-glcNAc-β-

D. Heparan Sulfate/Heparin

-1,4-glcUA-β- / -1,4-*idoUA-α*- → -1,4-glcNAc-α-

Figure 2-2. Chemical structures of GAGs. The repeating disaccharide backbone structures of the four classes of GAGs are shown. See text for details.

2.2.1. Hyaluronic Acid (HA)

Hyaluronic acid (hyaluronan, hyaluronate) has the simplest GAG structure, consisting of an alternating polymer of *N*-acetylglucosamine and glucuronic acid linked by the specific bonds indicated in Fig. 2-2A. Unlike the other GAGs, HA is not synthesized covalently bound to a protein core and hence is classified simply as a GAG and not as a PG. A single molecule can have a molecular weight of up to 10 million (~ 25,000 repeat disaccharides) which would have an extended length of ~ 25 μm (~ 1 nm/disaccharide). In physiological solutions, HA is polyanionic from the carboxyl groups on the glucuronic acid residues and assumes a rather stiff, randomly kinked coil structure which occupies large solution domains. The individual molecules can entangle through specific associative interactions to form networks which may contribute to the properties of the matrices where they reside (see Section 3.1).

A number of molecules can bind to HA through specific, noncovalent interactions that extend over defined numbers of the disaccharide repeat units. These HA binding proteins include the link protein, which stabilizes the attachment of aggrecan to HA (see Section 3.2), and cell surface HA receptors (see Chapter 9).

2.2.2. Chondroitin Sulfate (CS)/Dermatan Sulfate (DS)

Chondroitin, the nonsulfated backbone of CS, has the same basic backbone structure as HA, but with *N*-acetylgalactosamine replacing the *N*-acetylglucosamine (Fig. 2-2B). Individual chains are seldom more than 100 kDa (~ 250 repeat disaccharides) although occasionally they may exceed 300 kDa (Wright and Mayne, 1988; Yada *et al.*, 1990). As the chains are elongated, specific enzymes add sulfate esters onto particular hydroxyl groups to form CS. The most common sites of sulfation are the 4 and 6 positions of the *N*-acetylgalactosamine residue (Fig. 2-2B, dotted and solid arrows, respectively), yielding chains referred to as chondroitin-4-sulfate and chondroitin-6-sulfate. More commonly, a given chain contains stretches of one or several disaccharides with sulfate in the 4 position followed by stretches having sulfate in the 6 position. To add to the complexity, not all disaccharides in a chain are sulfated, and in some cases disulfated disaccharides may be present.

Some cells have a specific enzyme which epimerizes some of the carboxyl groups on glucuronic acids in the elongated chain from an equatorial to an axial configuration (Fig. 2-2B, dashed arrow). This epimerization occurs at carbon 5, the one which determines the stereospecific structure (D or L family) of the uronic acid, and therefore converts the D-glucuronic acid to L-iduronic acid. By convention, this also changes the designation of the glycoside bond from β1,3 for the glucuronic acid to α1,3 for the iduronic acid. A CS chain in which such epimerization occurs to yield one or more iduronic acids is referred to as *dermatan sulfate*. Once formed, the L-iduronic acids can be sulfated at position 2 (Fig. 2-2B, asterisk).

2.2.3. Keratan Sulfate (KS)

KS, like HA and CS, consists of a backbone structure with alternating β1,3 and β1,4 bonds (Fig. 2-2C). However, the position of the hexuronic acid in the other GAGs is occupied by N-acetylglucosamine in KS, and that of the hexosamine by galactose. The disaccharide shown in Fig. 2-2C has the same structure as lactose, with N-acetylglucosamine replacing glucose, and is therefore referred to as lactosamine. For this reason, the nonsulfated backbone of KS, which is present on a number of glycoproteins, is often referred to as polylactosamine or lactosaminoglycan. In PGs, the lactosaminoglycan chains are normally sulfated on the 6 position of either or both sugars to form KS (Fig. 2-2C, arrows). Individual chains are seldom more than 40 kDa (~ 80 disaccharide repeats) and are commonly in the 10–20 kDa range.

2.2.4. Heparan Sulfate (HS)/Heparin

HS and heparin have the same backbone structure, one which is notably different from the other GAGs in its α1,4 linkage between the N-acetylglucosamine and the glucuronic acid (Fig. 2-2D). Individual chains can be 100 kDa or more but are usually below 50 kDa. A large number of changes occur in the backbone either during or immediately following chain elongation (see Chapter 5) (for review see Lindahl, 1989). First, a proportion of the acetyl groups are removed with concurrent addition of sulfates to form N-sulfated glucosamines (Fig. 2-2D, dotted arrow), thereby adding negative charges. This is a critical step because subsequent modifications appear to occur primarily in block regions along the backbone where N-sulfation has occurred. The degree of N-sulfation is one major characteristic that distinguishes HS from heparin. Less than 50% of the N-acetyl groups are normally converted to N-sulfates in HS whereas usually 70% or more are converted in heparin. Epimerization of glucuronic acid to iduronic acid with closely coupled 2-sulfation is prevalent (Fig. 2-2D, dashed arrows), as is 6-sulfation of the glucosamine (Fig. 2-2D, solid arrow). The 2-sulfation appears to prevent the back epimerization of iduronic acid to glucuronic acid which would otherwise be greatly favored at equilibrium. The final structure of heparin, then, is highly polyanionic, frequently carrying four potential anions per disaccharide: one carboxyl, one N-sulfate, and two O-sulfates.

Structural modifications of HS and heparin are often important in defining specific interactions of biological importance. Occasionally in some cell types, the 3 position of the glucosamine is sulfated (Fig. 2-2D, solid square). This substitution is required for tight binding of heparin to the heparin-binding site of antithrombin III, and therefore has an essential role in the anticoagulant activity of heparin (Lindahl *et al.*, 1980; Rosenberg, 1985; Edge and Spiro, 1990). Another infrequent substitution occurs when the 2 position of glucuronic acid is sulfated (Fig. 2-2D, solid circle). HS with this substitution is selectively concentrated in the nucleus of some cells which are arrested in their cell cycle, suggesting that this modified HS may play a role in inhibiting mitosis (Ishihara *et al.*, 1987; Fedarko *et al.*, 1989).

2.3. Linkage Structures/Oligosaccharides

The GAG chains are elongated on oligosaccharide linkage structures which attach them to the core protein (Fig. 2-3; for review see Rodén, 1980). The linkage region for CS/DS and for HS/heparin is identical, consisting of four sugars, glucuronosyl-β1,3-galactosyl-β1,3-galactosyl-β1,4-xylose, attached by a β-glycoside bond between the reducing terminal xylose and the hydroxyl of a serine residue in an appropriate sequence in the core protein, usually serine-glycine (Fig. 2-3A). Often the xylose is substituted at the 2 position with a phosphate ester in both CS (Oegema *et al.*, 1984) and HS (Fransson *et al.*, 1985). The backbone disaccharides of CS/DS or HS/heparin are then assembled on the glucuronosyl residue of the linkage structure (Fig. 2-3A, arrows).

The large CS/KSPG, aggrecan, from cartilage carries *O*-linked oligosaccharides (typical of mucins) attached to the core protein by α-glycoside bonds between reducing terminal *N*-acetylgalactosamines and hydroxyls of serine and threonine residues. The branched hexasaccharide indicated in Fig. 2-3B is the most common *O*-linked oligosaccharide found on PGs (Nilsson *et al.*, 1982). In cartilages where aggrecan contains KS chains, the sialic acid on the upper branch of the hexasaccharide is frequently replaced by a KS chain (Hopwood and Robinson, 1974; Dickenson *et al.*, 1990) (Fig. 2-3B, solid arrow); the KS chains usually contain a sialic acid at their nonreducing termini as indicated.

Many PGs contain *N*-linked oligosaccharides (typical of glycoproteins) attached to their core proteins by *N*-glycosylamine bonds to asparagine residues. A typical *N*-linked oligosaccharide, with a biantennary complex structure, is shown in Fig. 2-3B. In the small, interstitial KSPG of cornea, one (Stuhlsatz *et al.*, 1988) or possibly both (Nilsson *et al.*, 1983) of the sialic acids on *N*-linked oligosaccharides with this structure are replaced by KS chains (Fig. 2-3B, dashed arrows). Similarly, the small PG, fibromodulin, contains several *N*-glycosidically linked KS chains (Oldberg *et al.*, 1989) (see Section 3.3).

Once the linkage structures and chains are synthesized on the appropriate acceptor sites (see Chapter 5), the core proteins (Fig. 2-1) are converted to mature PGs as schematically indicated in Fig. 2-4. Serglycin contains clusters of ten or more heparin (or oversulfated CS) chains along its embedded serine-glycine repeat. Decorin contains only one CS or DS chain on the serine at position 4 from the NH_2 terminus. Syndecan contains three or four HS and one or two CS chains distributed on its extracellular domain. Aggrecan, with ~ 100 CS chains and 100 or more *O*-linked oligosaccharides plus KS chains, literally bristles with complex carbohydrate constituents. Each of these PGs is discussed in greater detail in Section 3.

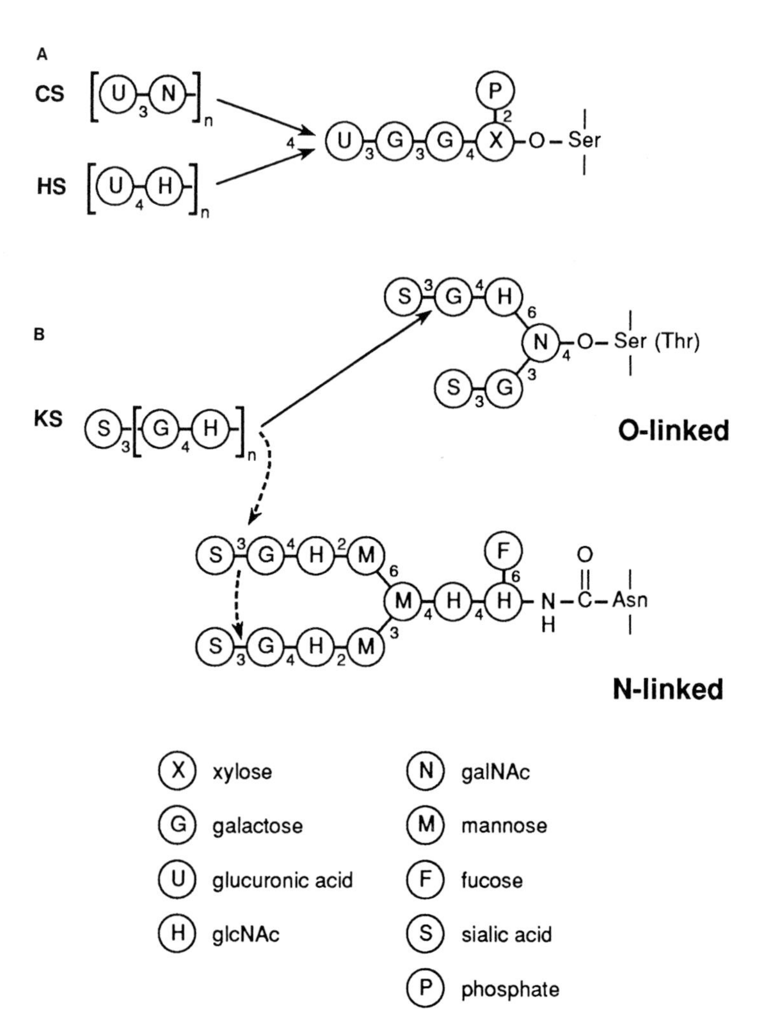

Figure 2-3. Structures of the linkage regions between GAG chains and the core protein. Panel A shows the linkage oligosaccharide common to all CS/DS and HS/heparin chains. Panel B shows the O-linked hexasaccharide found on many PGs, and which is elongated in aggrecan from many hyaline cartilages by replacing the terminal sialic acid with a KS chain (solid arrow). Panel B also shows the biantennary complex N-linked oligosaccharide found in many glycoproteins and which is elongated on the KSPG from cornea by replacing either or both of the terminal sialic acids with a KS chain (dashed arrows). The sugars are indicated in the key, and the numbers indicate the positions of the glycoside linkages.

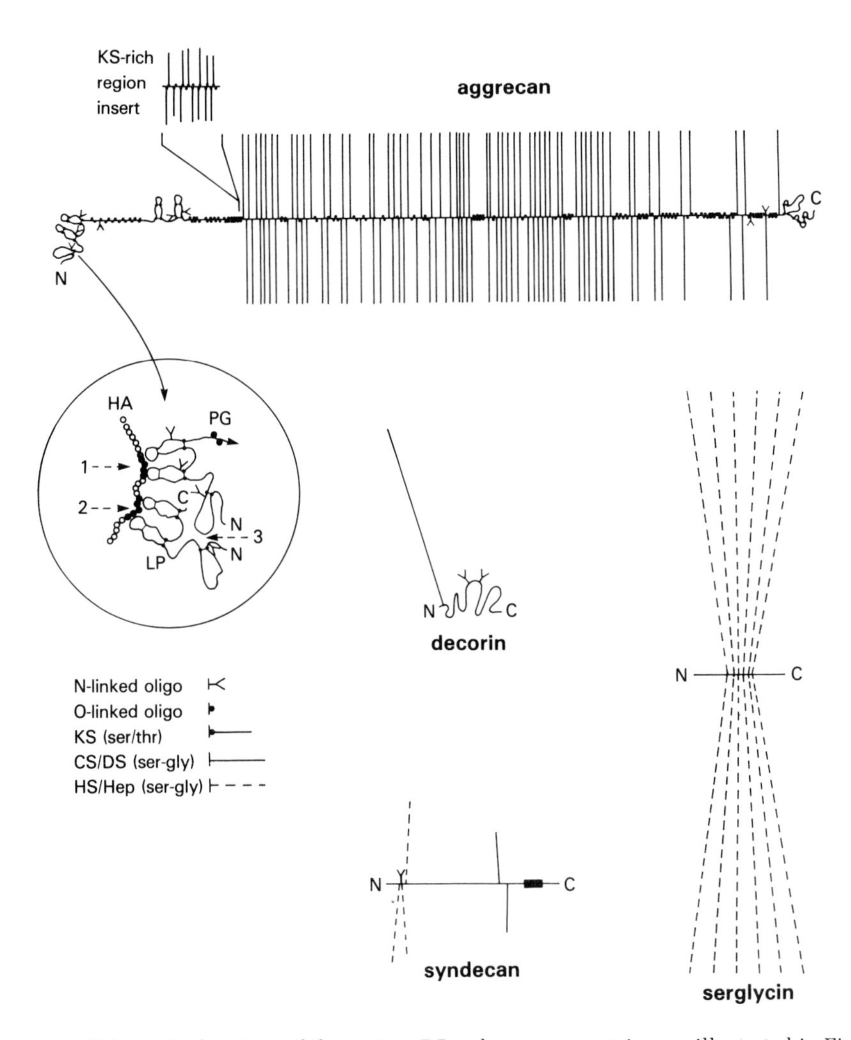

Figure 2-4. Schematic drawings of the mature PGs whose core proteins are illustrated in Fig. 2-1. The inset shows the G1 domain of aggrecan, highlighting its interactions with HA (arrow 1) and the link protein (LP) (arrow 3) in the formation of aggregates. The link protein interaction with HA is also indicated (arrow 2). Each circle indicates one disaccharide of HA of the structure shown in Fig. 2-2A, and the filled circles indicate those disaccharides in the specific HA-binding sites of the G1 domain and the link protein. The 108-amino-acid region that is absent in aggrecan from rats but present in aggrecan from cows is shown at the top (KS-rich region insert) carrying KS chains and *O*-linked oligosaccharides. The aggrecan from rats contains only *O*-linked oligosaccharides. For serglycin, only half of the serine-glycine sites in the midregion of the core are indicated by the side chains. Not all serine-glycine sites in the final aggrecan or serglycin PGs carry GAG chains. The relative lengths of the side chains represent typical averages for the respective PGs.

3. Distribution, Structure, and Function

3.1. Extracellular: Hyaluronic Acid

HA molecules have very high molecular weights. Their open, random coil structures occupy large solvent domains which give solutions of high viscosity. These macromolecules can also self-associate to form regions of ordered helical structure that can produce entangled networks with considerable elasticity. An example of a few entangled HA molecules is shown in Fig. 2-5E. In some tissues, particularly during early development and before tissue differentiation, HA with little or no associated PGs can constitute the major structural macromolecule in the extracellular matrix, where it can promote both cell proliferation and migration (see Chapter 9).

A fascinating example of the formation of a matrix composed primarily of HA occurs in the preovulatory follicle of most mammals when the oocyte resumes meiosis (Dekel and Phillips, 1979). At this stage, the oocyte is surrounded by a compact mass of cumulus cells in the cumulus cell–oocyte complex. In response to a gonadotropin surge and a factor produced by the oocyte (Salustri *et al.*, 1990), the cumulus cells are induced to synthesize large amounts of HA, and within 2–3 hr begin to deposit it in their surrounding matrix (McClean and Rowlands, 1942; Eppig, 1979). This leads to a rapid expansion of the complex, yielding a final concentration of HA in the matrix of ~ 0.5 mg/ml within 10–15 hr (Salustri *et al.*, 1989). The progression of this process can be visualized in the light microscope with a staining procedure specific for HA (Tarone and Underhill, 1988) (Fig. 2-5A–C). At higher magnification in the electron microscope, the matrix appears to be organized into an open lattice, rather like a three-dimensional net (Fig. 2-5D,F) (Yudin *et al.*, 1988). Interestingly, the extended length of an HA molecule of 1000 kDa, the minimum size synthesized by these cells, is considerably larger than the average lattice dimension (Fig. 2-5D). Thus, the organization of the large HA molecules, such as the ones shown at the same magnification in the insert (Fig. 1-5F), into the more ordered array in the matrix must require considerable folding. This probably involves specific interactions with other structural macromolecules present in the matrix at lower concentrations (Cherr *et al.*, 1990).

At ovulation, the expanded complex surrounding the oocyte is released from the follicle and picked up by the oviductal fimbria for the journey through the fallopian tube. For fertilization, the spermatozoon must penetrate this matrix (Fig. 2-5F), perhaps utilizing the hyaluronidase in its acrosomes to accomplish this task. The acrosome is located in front of the cell nucleus, and the cell body is propelled by a microtubule-driven flagellum as seen in the micrograph.

Overall, HA-based extracellular matrices such as described in this section have open structures that entrap large amounts of solvent. They are generally elastic, but easily deformed, as in the vitreous humor of the eye. They frequently provide pathways through which cells, such as neural crest cells and hematopoietic cells, move; and they define the space in which cells differentiate and form new matrices (see Chapter 9).

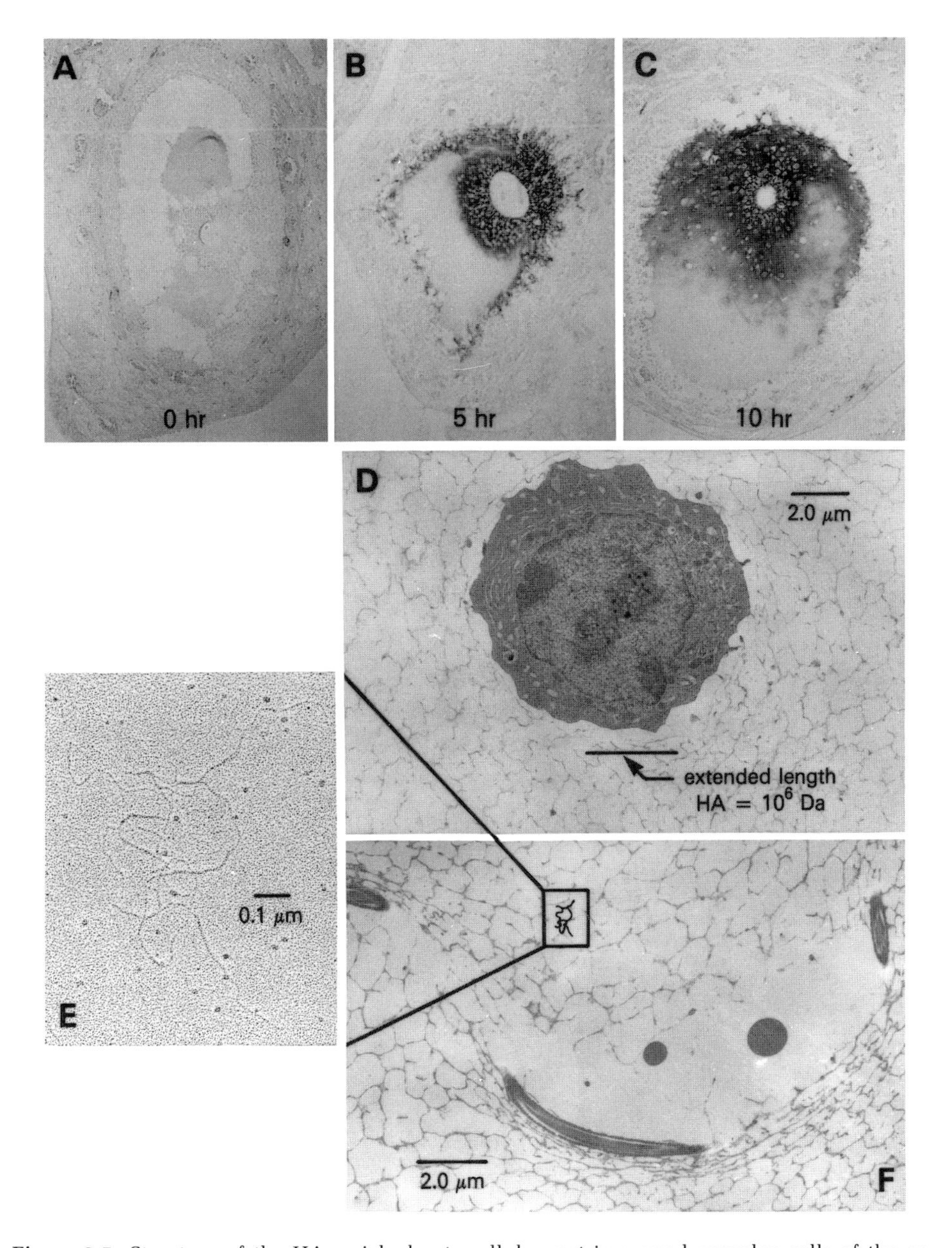

Figure 2-5. Structure of the HA-enriched extracellular matrix around cumulus cells of the expanded cumulus cell–oocyte complex. Panels A–C are light micrographs of sections of hamster ovaries stained with biotinylated HA-binding proteins (the G1 domain and link protein, Fig. 2-4). The animals were primed with gonadotropins 48 hr before an injection of luteinizing hormone to induce cumulus cell–oocyte complex expansion. The animals were sacrificed at 0 (A), 5 (B), and 10 (C) hr after the injection. The dark areas indicate the localization of HA and stain magenta in the original sections. The photographs were kindly provided by Dr. Charles Underhill. Panels D and F are electron micrographs of the expanded matrix around a cumulus cell (D) and the flagellum of a spermatozoon (F) penetrating the matrix. Panel E shows a rotary-shadowed picture of a few intertwined HA molecules and is reduplicated at the same magnification as the matrix in the inset of panel F. (Micrographs D and F were kindly provided by Dr. Gary Cherr from Yudin *et al.*, 1988.) Micrograph E was kindly provided by Dr. Richard Mayne and Dr. Randolph Brewton.

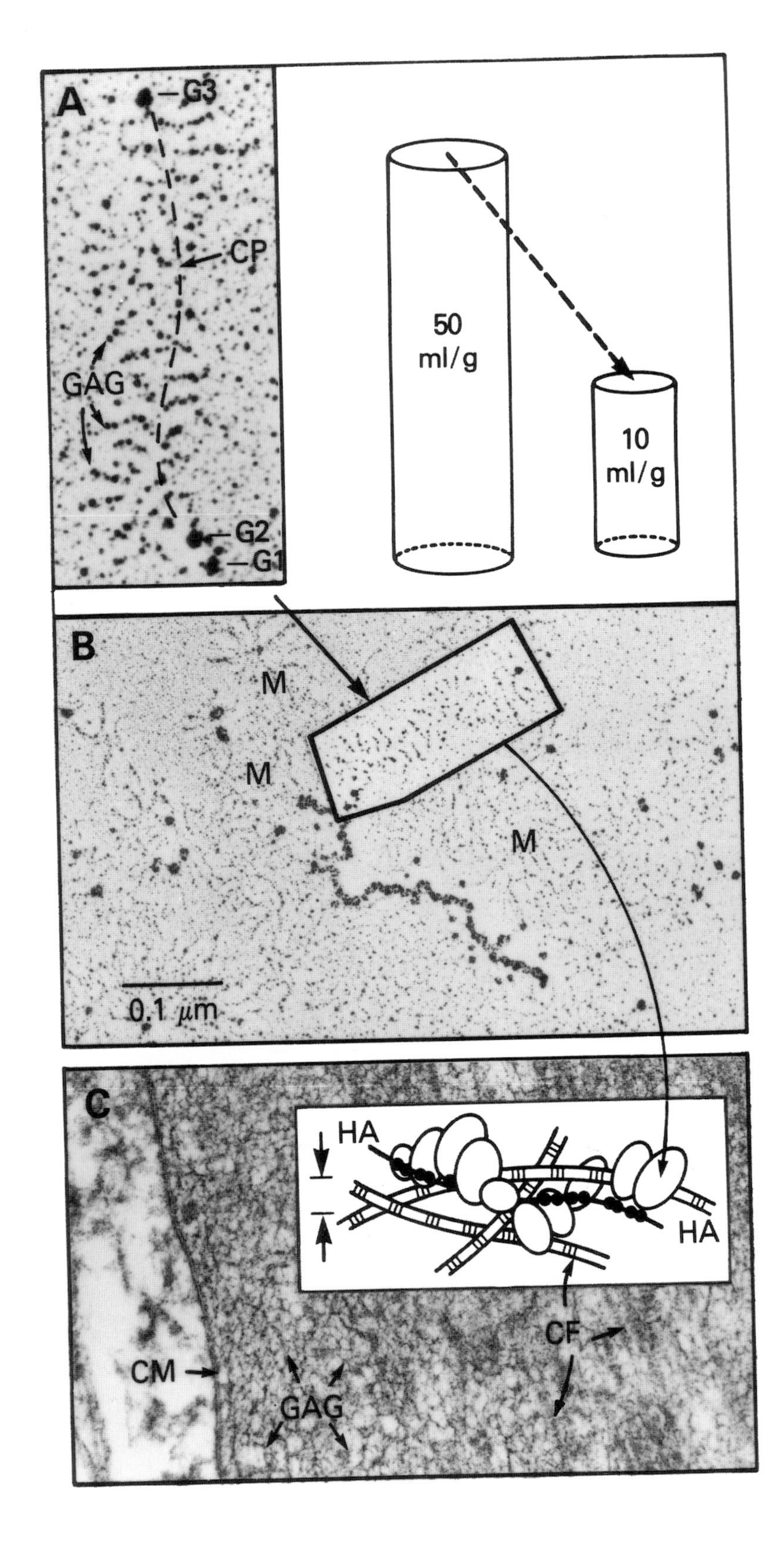
A
G3
CP
GAG
G2
G1
50
ml/g
10
ml/g
B
M
M
M
0.1 μm
C
HA
HA
CF
CM
GAG

3.2. Extracellular: Large Interstitial Proteoglycans

Aggrecan, the large CS/KSPG from cartilage, is an exceptionally complex macromolecule (for review see Hascall, 1988; Heinegård and Oldberg, 1989; Doege *et al.*, 1990). Its large core protein contains three globular domains, G1, G2, and G3 (Doege *et al.*, 1987) (Fig. 2-1). The intervening regions, particularly between G2 and G3, are heavily glycosylated in the mature PG (Fig. 2-4). The G1, G2, and G3 domains can be visualized as small spheres, and the GAG chains as feathery strands extending out from the core protein in the rotary-shadowed electron micrograph of a monomer (Fig. 2-6A) (Wiedemann *et al.*, 1984; Paulsson *et al.*, 1987b). The G1 domain is highly specialized. It contains one binding site in the tandemly repeated double loop of the COOH-terminal portion that interacts specifically with HA (Fig. 2-4 inset, dashed arrow 1) and another in the NH_2-terminal portion that interacts with link protein (dashed arrow 3) (Perin *et al.*, 1987). The link protein has a size of ~ 40 kDa and is highly homologous to the G1 domain of aggrecan (Neame *et al.*, 1986; Doege *et al.*, 1986; Deak *et al.*, 1986). It also contains a specific binding site for HA (Fig. 2-4 inset, dashed arrow 2). Each of the HA-binding sites extends over five repeat disaccharides (Fig. 2-4 inset, solid circles) of the type shown in Fig. 2-2A (Hardingham and Muir, 1973; Hascall and Heinegård, 1974). The ternary complex formed between the G1 domain, the link protein, and HA is very stable (Heinegård and Hascall, 1974) and provides the basis for anchoring many monomer PGs to individual strands of HA, thereby forming the aggregates found in the extracellular matrix. The backbone formed by this ternary complex along the HA can be seen as a beaded structure in the rotary-shadowed micrograph of an aggregate (Fig. 2-6B).

The G2 domain has a high degree of homology with the tandemly repeated double loop structure present in both G1 and the link protein. However, the G2 domain does not interact with HA and its function is presently unknown (Mörgelin *et al.*, 1988; Fosang and Hardingham, 1989). The NH_2-terminal portion of

Figure 2-6. Structure of the aggrecan aggregate-enriched extracellular matrix in hyaline cartilage. Panel A shows a rotary-shadowed electron micrograph of an aggrecan monomer (0.3 μm long) with the globular domains (G1, G2, G3) indicated. The location of the core protein (CP) between G2 and G3 is suggested by the dashed line and a region of extended GAG chains is noted. The micrograph is reduplicated in panel B to show the relative size of a monomer in the rotary-shadowed electron micrograph of an aggregate of aggrecan. Three partially resolved monomers (M) in the aggregate are noted. The cylinders in panel A represent the approximate volume the fully extended monomer would occupy in solution and the smaller the volume it would occupy in a typical cartilage matrix. Panel C shows an electron micrograph of a section of hyaline cartilage featuring the cell membrane (CM) boundary between a chrondrocyte (left) and the extracellular matrix (right). The PGs have been fixed and stained by procedures designed to retain their structures, and they appear as the interwoven filamentous material. A few collagen fibrils (CF) can be seen, but most are masked by the extensive GAG phase. The inset shows a schematic model illustrating the sizes of collagen fibrils (banded) and PGs (ellipsoids) aggregated to HA (solid line) through G1 domains and link proteins (solid spheres). The arrows in the inset indicate the approximate thickness of the thin section. Panels B and C are at the same magnification. (The micrographs in panels A and B were kindly provided by Dr. Matthias Mörgelin and are similar to those in Mörgelin *et al.*, 1988. The micrograph in panel C was kindly provided by Dr. Ernst Hunziker and is similar to those in Hunziker and Schenk, 1984, 1987.)

the G3 domain is homologous to the hepatic lectin and its COOH-terminal portion is homologous to complement regulatory protein (Fig. 2-1). The expressed G3 domain interacts specifically, but with low affinity, with several sugar ligands, preferring galactose and fucose (Halberg *et al.*, 1988). Thus, this region may interact with carbohydrate ligands on other matrix macromolecules and participate in the organization of the extracellular matrix.

In the major GAG attachment region between G2 and G3, the CS chains and the O-linked oligosaccharides and KS chains are closely spaced, averaging ~ 1 chain per 7 amino acid residues. The CS chains typically have average sizes of 20–30 kDa while the KS chains are generally smaller, 10–15 kDa. In aggrecan isolated from bovine cartilage, an extra polypeptide of 108 amino acid residues is inserted after residue 793 in the sequence of aggrecan from rat (Fig. 2-4). This region and adjacent regions on either side are composed of a predominant continuous hexapeptide motif, glutamate-glutamate/lysine-proline-phenylalanine-proline-serine, repeated 23 times (Antonsson *et al.*, 1989). This region corresponds to the protease-resistant KS-rich region, previously identified in bovine aggrecan, where KS chains are tightly clustered (Heinegård and Axelsson, 1977).

In dilute solution, the GAG chains extend maximally away from the core protein, like bristles on a brush, thereby creating a domain which entraps up to 50 ml of solvent per gram of dry weight (Fig. 2-6A, large cylinder). In cartilages the PGs can attain concentrations of 10% or more of the wet weight in the matrix, and therefore occupy volumes of only 10 ml/g or less (Fig. 2-6A, small cylinder). In this concentrated PG phase, the GAG chains are still segmentally highly mobile, as indicated by carbon-13 nuclear magnetic resonance studies of cartilage tissue (Torchia *et al.*, 1977), and they have a high anionic charge density, up to 0.2 M in fixed sulfate esters. Thus, the compressed PGs exert a swelling pressure on the inextensible collagen fibrillar network that forms the fabric of the tissue (see Chapter 1). This allows cartilages to withstand compressive load with minimal deformation, and hence to cushion and protect underlying bone during normal skeletal function.

Figure 2-6C is an electron micrograph of a thin section of cartilage prepared by freeze fixation and freeze substitution, which maintains PGs in their extended forms (for review see Hunziker and Schenk, 1987). The GAG chains on the PGs are evident as a fine, filamentous meshwork that extends uniformly from the cell surface throughout the extracellular space (Hunziker and Schenk, 1984). This meshwork surrounds and masks the collagen fibrils, which can occasionally be seen (Fig. 2-6C, CF). The micrograph of the section is at the same magnification as that for the isolated aggregate (Fig. 2-6B,C). Thus, each monomer in an aggregate would be compressed into the approximate volume depicted by the ellipsoids in the diagram overlaid on the matrix (Fig. 2-6C, inset). The thickness of the section, approximately equal to the distance between the two arrows in the inset, is less than the extended length of a monomer; thus, individual PGs cannot be distinguished in thin sections.

The cartilage matrix, like that in the expanded cumulus cell–oocyte complex (Fig. 2-5), contains an extensive HA network. Typically, the HA concentration in the cartilage matrix, ~ 1 mg/ml, would be as great or greater than in the cumulus cell–oocyte complex matrix (Fig. 2-5). Yet HA in cartilage only provides a scaffold for holding the much larger mass of the PGs, ~ 50 mg/ml, in the tissue.

Another HA-binding PG has been isolated from human fibroblasts, and the sequence of its core protein, ~ 260 kDa, deduced from overlapping cDNA clones (Zimmerman and Ruoslahti, 1989). This PG, referred to as versican, has homology to aggrecan in its NH_2-terminal G1 and COOH-terminal G3 domains, but lacks a G2 domain and differs greatly in the region which contains the CS/DS chains. The G3 region of versican contains a lectinlike sequence and a complement regulatory protein-like domain, as does aggrecan (Fig. 2-1). Additionally, unlike aggrecan, versican contains two epidermal growth factor-like repeats adjacent to the NH_2-terminal end of the lectinlike region. Sequence analysis shows that versican contains 15 serine-glycine and 16 glycine-serine sequences which are potential sites for CS/DS chains, 20 asparagine-X-serine (threonine) sequences for *N*-linked oligosaccharides, and more than 200 serines and threonines suitable for *O*-linked oligosaccharides.

Versican, or a structurally closely related PG, has been isolated from blood vessels (Salisbury and Wagner, 1981; Kapoor *et al.*, 1986; Mörgelin *et al.*, 1989) and from arterial smooth muscle cells (Wight and Hascall, 1983; Chang *et al.*, 1983). This PG has distinctly different immunoreactivity from the large aggregating PGs isolated from sclera, cartilage, and tendon (Heinegård *et al.*, 1985). Further, arterial smooth muscle cells contain transcripts for the core protein of versican (Schöenherr *et al.*, 1991). These large aggregating CSPGs can be visualized in electron micrographs of blood vessels fixed in the presence of ruthenium red, a cation which precipitates the polyanionic PGs and forms electron-dense granules (Fig. 2-7C). In the presence of the cation, the 20–25 GAG side chains in the extended structure of individual PGs, such as that shown in Fig. 2-7A, collapse to form individual granules (G. Hascall, 1980). The circles in Fig. 2-7C show the space that would be occupied by the extended PGs. The granules are interconnected by fine filamentous threads, possibly representing HA, which form an interlaced network that fills the space not occupied by elastic fibers and collagen fibrils. Figure 2-7B shows that these interlaced networks react selectively with antibodies against the large aggregating CSPG from blood vessels (Lark *et al.*, 1988). The function of these large CSPGs in blood vessels, like aggrecan in cartilage, appears to be to resist compression, in this case generated by pulsatile forces in the cardiovascular system.

3.3. Extracellular: Small Interstitial Proteoglycans

Most extracellular matrices contain low-molecular-weight PGs with one or a few GAG chains (Fig. 2-8). Two members of this group, decorin, 90–140 kDa, and biglycan, 150–240 kDa, contain CS or DS and have core proteins that are particularly rich in leucine and aspartic acid/asparagine (Heinegård *et al.*, 1985; Fisher *et al.*, 1987). After removing the CS or DS side chains with

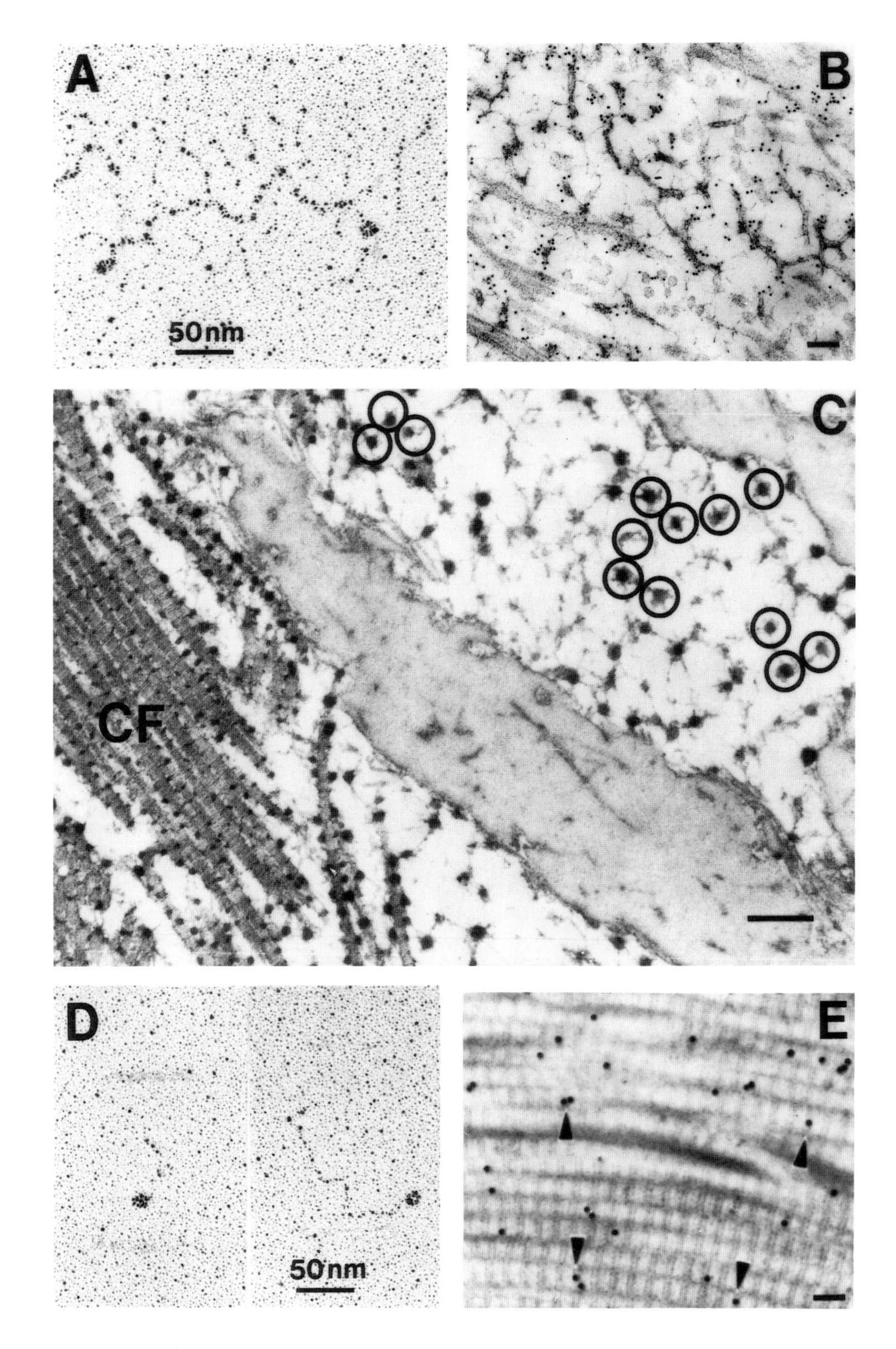
A
50nm
B
C
CF
D
50nm
E

chondroitinase, the core proteins from both PGs exhibit masses of ~ 45 kDa. Generally the CS/DS chains are longer than those on aggrecan, typically 30–40 kDa. N-linked oligosaccharides and in some cases O-linked oligosaccharides are present, although their structures remain to be determined. Fibromodulin, a third member of this family, ~ 60 kDa, with a very similar core protein lacks CS/DS chains, but contains KS chains attached through N-linked oligosaccharides, as for the corneal KSPG (see Section 2.2.3) (Oldberg *et al.*, 1989).

The amino acid sequences of these three core proteins have been determined by conventional and cDNA sequencing methods (Fisher *et al.*, 1989; for review see Heinegård and Oldberg, 1989). They contain highly conserved sequences with ten internal homologous repeats of ~ 25 amino acids featuring leucine-rich motifs (Fig. 2-8). Cysteine positions are conserved, with disulfide bridges between cysteines 1–4 and 5–6; cysteines 2 and 3 are separated by a single amino acid which makes it less likely that they are disulfide bonded (Neame *et al.*, 1989). Decorin contains one, while biglycan contains two serine-glycine sequences near their NH_2 termini in appropriate sequences for initiating CS/DS chains. Glycosylation appears rather complete as decorin contains one and biglycan usually contains two GAG chains (Mörgelin *et al.*, 1989; Neame *et al.*, 1989). In cartilage and bone, the chains are CS as they do not appear to contain iduronic acid, while in other tissues, such as sclera, skin, or the intima of aorta, they are DS as they contain up to 70% of the hexuronate as iduronic acid. Electron microscopy of decorin reveals a highly folded, globular protein structure with its GAG chain extending out from it (Fig. 2-7D).

Fibromodulin contains no serine-glycine sequence in its NH_2 terminus, but does have several sulfated tyrosines in this region (P. Antonsson, personal communication). KS chains are attached to the four N-linked oligosaccharide sites in the middle portion of the protein (Plaas *et al.*, 1990). Cornea contains a KSPG with a core protein that does not cross-react with antibodies to fibro-

Figure 2-7. Relationship of versican, or a closely related CSPG, and decorin to matrix structures in the intima of aortic tissue. Panel A shows the large CSPG extracted from bovine aorta as visualized in the electron microscope after glycerol spraying and rotary shadowing. Globular domains are seen at the opposite ends of the core protein which contains 15–20 side projections representing GAG chains (Morgelin *et al.*, 1989). Panel B shows an electron micrograph of a section of blood vessel immunostained with a monoclonal antibody against a versicanlike CSPG. The immunogold is localized primarily on short filamentous aggregates throughout the interstitial space (Lark *et al.*, 1988). Bar = 100 nm. Panel C shows an electron micrograph of a section of aortic tissue that was processed in the presence of ruthenium red. The dense, punctate precipitates represent collapsed PG molecules. The circles indicate the volume that a CSPG such as that pictured in panel A would occupy if the CS chains were fully extended. The distribution of these granules is very similar to the immunogold localization shown in panel B. The smaller ruthenium red-positive granules in the lower left are closely associated with collagen fibrils (CF). They represent the small DSPG decorin as indicated by immunolocalization in panel E (Mar and Wight, 1988). Bar = 50 nm. Panel D shows two decorin molecules as visualized in the electron microscope after glycerol spraying and rotary shadowing. The core protein appears as a globule and the side projection represents the single DS chain. Panel E shows an electron micrograph of a section of blood vessel immunostained with antibody against decorin. Numerous gold particles are localized to the periodic bands of collagen fibrils (arrowheads) (Mar and Wight, 1988). Bar = 100 nm. (The micrographs in panels A and D were kindly provided by Dr. Mats Morgelin.)

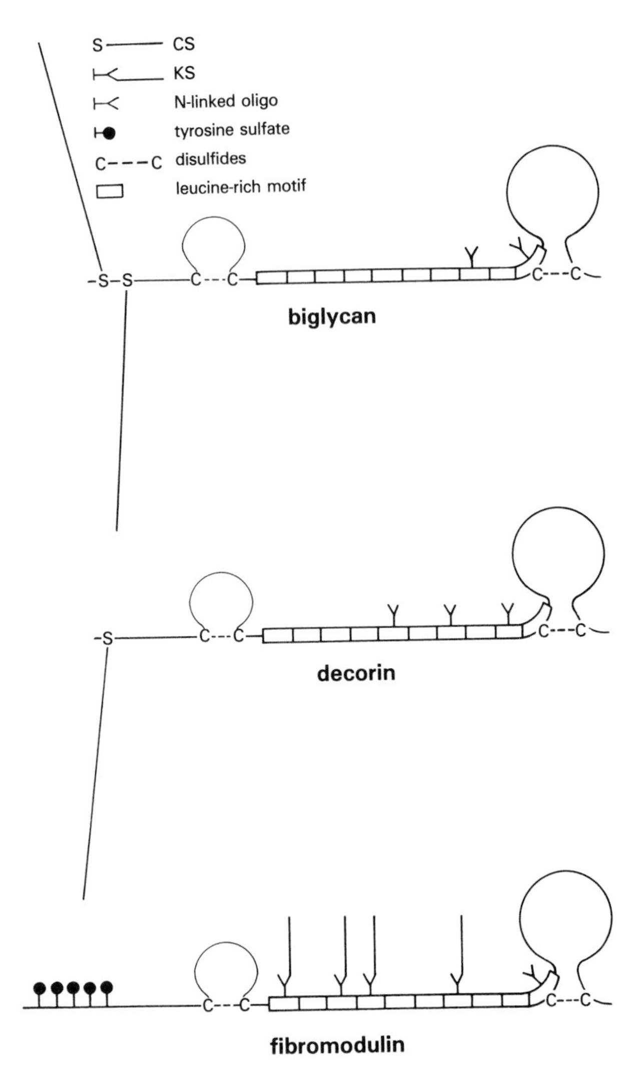

Figure 2-8. Schematic diagram comparing the homologous protein structures of biglycan, decorin, and fibromodulin. See text for details.

modulin, but which is similar in terms of size and high content of leucine and aspartic acid/asparagine (Axelsson and Heinegård, 1978). Thus, this core protein quite likely represents a fourth member of this family.

Decorin, fibromodulin, and the corneal KSPG appear to have functional roles in regulating collagen fibrillogenesis and organization in the matrix. Decorin and fibromodulin, but not biglycan, inhibit collagen fibrillogenesis *in vitro* (Vogel *et al.*, 1984; Hedbom and Heinegård, 1989). Further, decorin can be shown by histochemistry and immunohistochemistry (Fig. 2-7C,E) to be localized at specific sites along collagen fibrils *in vivo* (for review see Scott,

1988). The core protein, but not the GAG chain, is essential for binding to collagen (Vogel *et al.*, 1987; Hedbom and Heinegård, 1989). In cornea, selective degradation with chondroitinase or keratanase enzymes revealed that the KSPG preferentially localizes on the collagen surface near the *a* and *c* bands while decorin preferentially localizes near the *d* and *e* bands (Scott, 1988).

Immunohistochemical and *in situ* hybridization studies have shown that decorin and biglycan are usually localized and expressed in different regions of developing tissues (Bianco *et al.*, 1990). Decorin is always associated with fibrillar collagen in matrices from different tissues. In contrast, biglycan is usually associated with the cell surface and pericellular matrices in such cell types as skeletal myoblasts, endothelial cells, and differentiating keratinocytes. The distribution of decorin supports its probable function in regulating collagen fibril formation and organization, whereas that for biglycan indicates that this PG has quite different functions in tissue development (see Chapter 5, Section 4.3).

Decorin cDNA has been transfected into Chinese hamster ovary cells which normally synthesize little or no decorin (Yamaguchi and Ruoslahti, 1988). Positive transfectants expressing significant quantities of decorin exhibit flattened morphology and decreased growth rates. Although the mechanism by which decorin decreases cell growth is not understood, decorin can bind to transforming growth factor-β (TGF-β) and neutralize its growth-promoting activity (Yamaguchi *et al.*, 1990). Such studies indicate that these small PGs may have roles in key cellular events associated with both development and disease.

3.4. Extracellular: Basement Membrane Proteoglycans

Proteoglycans, both HSPGs and CSPGs, are integral components of the extracellular matrix of basement membranes, the specialized thin and sheetlike tissues found throughout the body closely associated with the cells that produce and maintain them. Basement membranes may be deposited in a polarized fashion at the basal side of epithelial or endothelial cells, or they may surround such cells as muscle cells, adipocytes, or nerve cells. The basement membrane surrounding cells is often referred to as a basal lamina.

The most widely studied basement membrane PG is a large HSPG isolated from a mouse basement membrane tumor, the Engelbreth–Holm–Swarm (EHS) sarcoma. This HSPG contains a large core protein, ~ 400 kDa, and 3–4 HS chains of ~ 60 kDa (for review see Hassell *et al.*, 1986; Paulsson *et al.*, 1986). Electron micrographs of rotary-shadowed molecules (Fig. 2-9) show a series of globular domains extending ~ 80 nm along the core protein with the HS chains near one end (Paulsson *et al.*, 1987a; Laurie *et al.*, 1988). Sequence analyses of cDNA clones (Noonan *et al.*, 1988) revealed that the core protein contains two domains with homology to the cysteine-rich regions in laminin chains and eight internal loop repeats, with two cysteines each, that show homology with neural-cell adhesion molecule (N-CAM), a member of the immunoglobulin

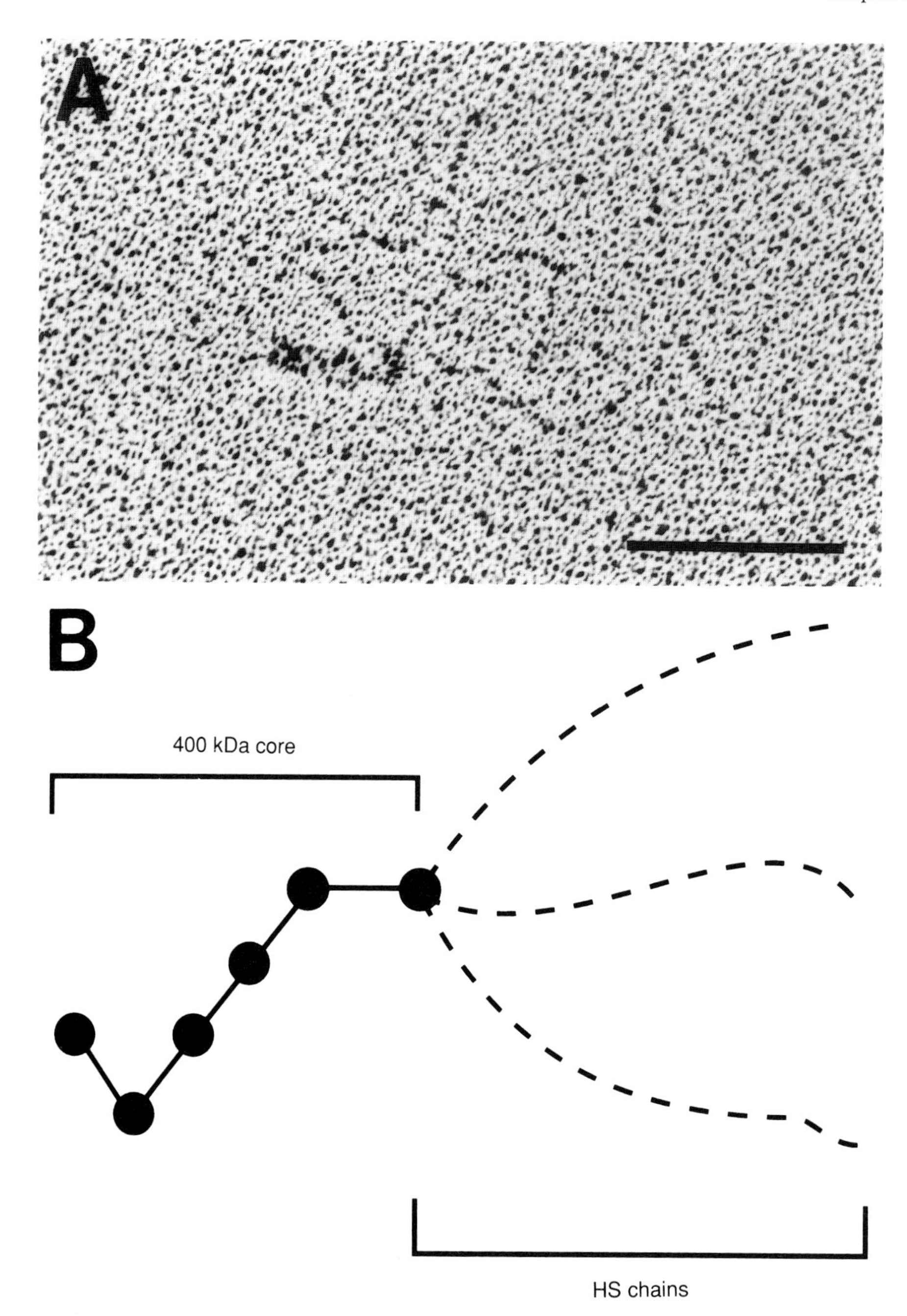

Figure 2-9. Structure of the EHS basement membrane PG. Panel A shows an electron micrograph of a glycerol-sprayed, low-angle rotary-shadowed replica of the low-density HSPG isolated from the mouse EHS basement membrane (Yurchenco and Schittny, 1990). Bar = 100 nm. Panel B shows a schematic model of the HSPG visualized in panel A. This model is based on chemical, physical, and electron microscopic data which best describe the PG as a tandem array of at least six globular domains with three GAG chains radiating from one pole of the protein core (Paulsson *et al.*, 1987a). (The micrograph was kindly provided by Dr. Peter Yurchenco.)

gene superfamily. This HSPG can interact with itself (Yurchenco *et al.*, 1987) and with other basement membrane macromolecules, including type IV collagen (see Chapter 1) and laminin (see Chapter 4). Antibodies against epitopes in this core protein localize mainly within the lamina densa in most basement membranes (Schittny *et al.*, 1988; Grant and LeBlond, 1988) whereas antibodies raised against a smaller HSPG isolated from glomerular basement membrane localize mainly in the lamina rarae interna and externa (Stow *et al.*, 1985a). This suggests that the core proteins of the basement membrane HSPGs are anchored in or at the surface of the lamina densa via specific interactions that contribute to basement membrane architecture. In addition, basement membrane HSPGs influence other properties such as permeability of glomerular basement membranes (see Chapter 11), anchoring acetylcholinesterase in the neuromuscular junction (Brandan *et al.*, 1985), binding protease inhibitors such as antithrombin III (Pejler *et al.*, 1987), and facilitating attachment of cells such as hepatocytes to their underlying basement membrane (Clement *et al.*, 1989).

HSPGs with similar numbers of HS chains but much smaller core proteins have been isolated from some tissues, including glomerular basement membranes and the mouse EHS tumor. Some of these smaller forms appear to have different protein structures (Paulsson *et al.*, 1987a; Kato *et al.*, 1987) perhaps generated from different genes or through alternate splicing of the same gene. Others may be derived from proteolytic processing of the larger HSPGs (Ledbetter *et al.*, 1985; Hassell *et al.*, 1986).

Some understanding of potential functions of these PGs has come from studies of their isolated GAG chains. Thus, the HS chains isolated from several basement membranes have more extensive *O*-sulfation than *N*-sulfation, ~ 3 to 1 (Pejler *et al.*, 1987; Gallagher and Lyon, 1989). Most of the HS from Reichert's membrane binds with high affinity to antithrombin III and contains substantial amounts of the infrequent 3-sulfated glucosamine as does the HS from glomerular basement membrane (Edge and Spiro, 1990) (see Section 2.2.4). These results suggest that antithrombin III binding capacity of HS in vascular basement membranes may contribute to the nonthrombogenic properties of the vascular system.

Basic fibroblast growth factor (bFGF), a cationic molecule with a pI of ~ 10, binds to HS chains in basement membrane HSPGs, and is released by treatment with an enzyme, heparinase, which degrades HS (Folkman *et al.*, 1988). Proteases such as plasmin release bFGF still complexed with HS peptides derived from the HSPGs. In the complex, bFGF retains its biological activities, such as stimulating cell chemotaxis and proliferation, and is protected from inactivation by proteases (for reviews see Rifkin and Moscatelli, 1989; Klagsbrun, 1990). Thus, HSPGs in basement membranes, on cell surfaces (Kiefer *et al.*, 1990), or in other extracellular matrices can sequester and/or concentrate particular growth factors important for regulating such cellular processes as proliferation, angiogenesis, and hemopoiesis (see Roberts *et al.*, 1988).

Heparin and HS chains can also interact directly with certain cell types and influence their growth properties. For example, HS from HSPGs synthesized by vascular endothelial cells can inhibit vascular smooth muscle cell proliferation (Castellot *et al.*, 1981; for review see Marcum *et al.*, 1987). Further, pulmonary endothelial cells synthesize HSPGs which exhibit growth-inhibiting activity in smooth muscle cells and which are recognized by antisera against the large HSPG from the mouse EHS tumor (Benitz *et al.*, 1990). It is possible, then, that HSPGs in endothelial basement membranes may regulate smooth muscle cell growth during development and vascular remodeling, and/or cell proliferation, cell migration, and matrix production in response to vascular damage as in atherosclerosis (for review, see Wight, 1989).

Many basement membranes also contain PGs with CS chains. For example, a large CSPG, greater than 1000 kDa, has been identified in the electrocyte and Schwann cell basement membranes of the elasmobranch electric fish (Carlson and Wight, 1987). Electron microscopy of this PG revealed a core protein ~ 350 nm long with ~ 20 CS chains of 50 kDa average size (Fig. 2-10). This PG is quite large relative to the dimensions of the synaptic cleft and its basal lamina. This places significant constraints on how it is organized into the structure of the basement membrane (Fig. 2-10). Another CSPG with a core protein of ~ 150 kDa and ~ 20 CS chains of ~ 17 kDa average size has been isolated from Reichert's membrane (McCarthy *et al.*, 1989). Immunolocalization of this CSPG with antiserum that does not recognize HSPGs, stains basement membranes of nerve fibers, blood vessels, dermal–epidermal junctions, striated muscle, kidney tubules, and glomerular mesangium, but not, interestingly, glomerular basement membrane.

3.5. Cell Surface Proteoglycans

A variety of PGs, most commonly HSPGs, are associated with the plasma membrane of most cells by at least three types of interactions: (1) transmembrane intercalation of a hydrophobic peptide region of the core protein; (2) covalent linkage of the core protein with phosphatidylinositol (Ishihara *et al.*, 1987; Carey and Evans, 1989; Yanagishita and McQuillan, 1989); and (3) non-covalent association with plasma membrane binding proteins (Kjellen *et al.*, 1980; Biwas, 1988). Generally, these PGs have distinctly different antigenic properties compared with PGs isolated from basement membranes (Stow *et al.*, 1985a,b). A cell surface HSPG was initially isolated from liver cells (Oldberg *et al.*, 1979). This PG had a core protein of ~ 40 kDa with ~ 4 short HS chains of ~ 14 kDa attached, and it appeared to be intercalated in plasma membrane.

The most widely studied intercalated HSPG is syndecan, which is located primarily on the surfaces of epithelial cells (for review see Rapraeger *et al.*, 1987; Bernfield and Sanderson, 1990). The primary structure of the core protein has been deduced from its full-length cDNA (Saunders *et al.*, 1989; Mali *et al.*, 1990) and contains: (1) a 34-amino-acid cytoplasmic domain at the COOH terminus; (2) a 25-amino-acid hydrophobic transmembrane domain; and (3) a 235-amino-acid extracellular domain (Fig. 2-4). The extracellular domain contains five serine-glycine sites substituted with up to three HS and two CS/DS chains. It also has a protease cleavage site, arginine–lysine, adjacent to the

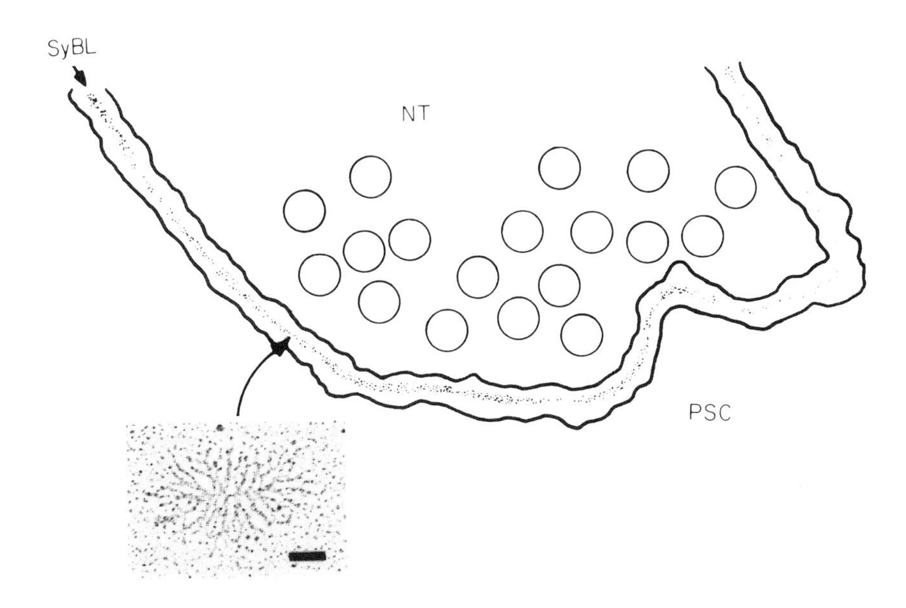

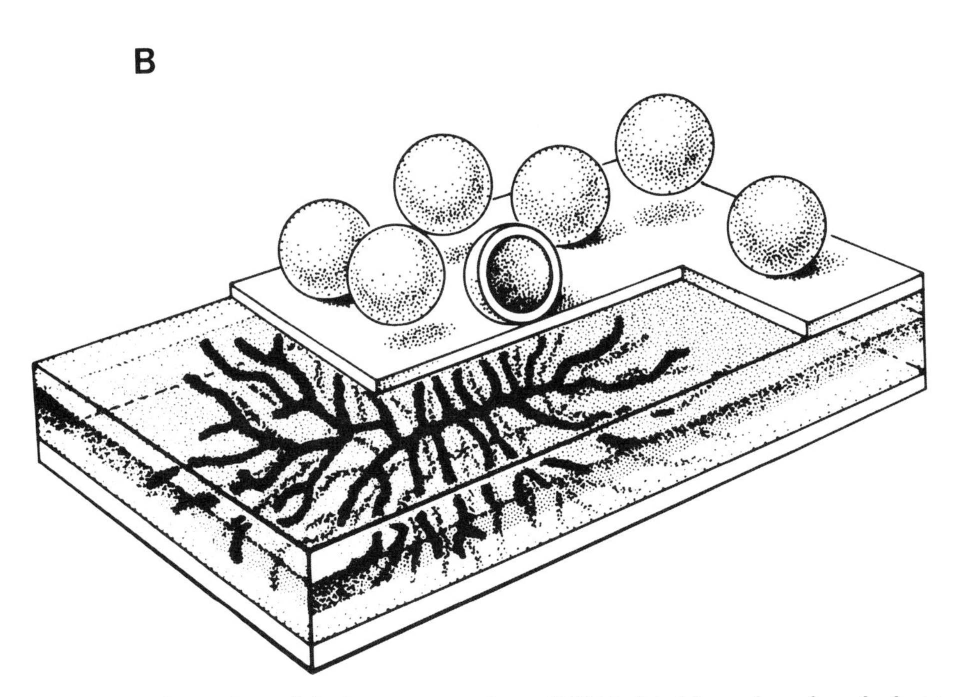

Figure 2-10. Comparison of the basement membrane CSPG isolated from elasmobranch electric organ and its localization in the synaptic cleft. An electron micrograph of the basement membrane CSPG is shown in panel A beneath a schematic diagram of a cross section of the electric organ synapse, which shows the synaptic basal lamina (SyBL) separating the nerve terminal (NT) with its synaptic vesicles from the postsynaptic cell (PSC). The schematic enlargement (B) shows a hypothetical placement of the PG in the synaptic cleft. (The figure is modified from Carlson and Wight, 1987, and was kindly provided by Dr. Steve Carlson.)

transmembrane domain which, when cleaved, releases the extracellular domain from the cell surface, a "shedding" process frequently observed for this class of PG.

The HS chains on syndecan interact with extracellular matrix proteins such as collagens type I, III, and V, thrombospondin, and fibronectin; and the cytoplasmic domain appears to be involved either directly or indirectly in interactions with intracellular actin cables (for review see Woods and Couchman, 1988; Rapraeger *et al.*, 1987; Bernfield and Sanderson, 1990). Thus, syndecan may act as a bridge between the cytoskeleton and the extracellular matrix, and thus have a role in maintaining cell shape. For example, simple epithelial cells grow in culture as islands of closely adherent polygonal cells. When transfected with antisense cDNA to syndecan, transfectants expressing low levels of syndecan grow as fusiform, overlapping cells with fibroblastic morphology (Bernfield and Sanderson, 1990). Different cells produce syndecan with different sizes, numbers, and proportions of HS and CS/DS chains. For example, syndecan from simple epithelia has a size of ~ 160 kDa and localizes to basolateral cell surfaces in contact with basement membranes, whereas syndecan from stratified epithelia is ~ 90 kDa and distributes over the entire cell surface (Bernfield and Sanderson, 1990).

Syndecan appears to be developmentally regulated and is transiently expressed during organogenesis, especially at epithelial–mesenchymal interfaces and in the associated condensed mesenchyme such as in the embryonic tooth bud (Thesleff *et al.*, 1988) and the embryonic kidney (Vainio *et al.*, 1989). Syndecan is also present on immature B and pre-B lymphocytes, absent on circulating and peripheral B lymphocytes, and reappears when these cells differentiate into plasma cells within the extracellular matrix (Bernfield and Sanderson, 1990). This suggests that syndecan expression correlates with the ability of the expressing cells to associate with the extracellular matrix.

Several other intercalated cell surface PGs have been identified. Heparitinase digestion of cell surface HSPGs isolated from lung fibroblasts reveals core glycoproteins with apparent sizes of 125, 90, 64, 48, and 35 kDa, and specific antisera against the core proteins clearly distinguish at least three antigenically different types (Lories *et al.*, 1987; Marynen *et al.*, 1989). A partial amino acid sequence was deduced from cDNA clones for the HSPG with the ~ 48 kDa core. The cytoplasmic and transmembrane domains show a high degree of homology with those domains in syndecan, but the extracellular domain differs greatly. Thus, these two cell surface HSPGs may be representatives of a family with common cytoplasmic and transmenbrane domains that direct them to specific cell membrane compartments, but which have distinct extracellular domains with potentially different functions. Molecular cloning of the 64 kDa form revealed a 558 amino acid polypeptide with six serine–glycine sequences, four of which conform to the proposed consensus sequence for GAG attachment (David *et al.*, 1990). One of these occurs near the N terminus while the other three occur as a triplet near the C terminus of the protein. Unlike the 48 kDa form, the 64 kDa species does not contain a hydrophobic transmembrane domain and appears to be linked to the plasma membrane by a phosphatidylinositol anchor.

Some cells, such as nerve cells (Zaremba *et al.*, 1989) and melanoma cells (for review see Harper and Reisfeld, 1987), have CSPGs on their cell surfaces. The CSPG isolated from the melanoma cells is large, ~ 800 kDa, has a core glycoprotein of ~ 250 kDa after chondroitinase digestion, and contains 3–4 CS chains of ~ 60 kDa average size. These PGs are localized to microspikes which form attachment sites between the cell surface and the substratum (Garrigues *et al.*, 1986), and they influence the ability of these cells to adhere and proliferate. Antibodies to this PG immunostain tumor cells in human malignant melanoma (Hellström *et al.*, 1983).

3.6. Intracellular: Storage Granule Proteoglycans

The connective tissue mast cell (Fig. 2-11) is an example of one of the different types of hematopoietic cells that participate in immune and inflammatory responses (for review see Stevens, 1987). Most of these cells have storage granules which contain a distinct PG, serglycin. The primary structure of the core protein synthesized by a rat basophilic leukemic cell line (RBL-1) has been deduced from cDNA clones (Avraham *et al.*, 1988) and is equivalent to that synthesized by a rat yolk sac tumor cell line (L2) (Bourdon *et al.*, 1985). After removal of the signal peptide of 26 amino acids, the protein contains 153 amino acids with 24 serine-glycine repeats located between residues 89 and 137 (Fig. 2-1). These are attachment sites for the ~ 15 GAG chains on the mature PG (Fig. 2-4). The GAG chains are either heparin or an oversulfated form of CS (see below). The core protein sequence of the human analogue has been determined from cDNA clones prepared from a human promyelocytic leukemic cell line (HL-60) (Avraham *et al.*, 1988). The NH_2-terminal portion preceding the serine-glycine repeat region was highly homologous with rat serglycin while the COOH-terminal portion showed very little homology. The serine-glycine repeat region consisted of only nine repeats with one of the serines replaced by a phenylalanine. Biosynthetic studies with the HL-60 cells showed that the average PG molecule contained 7–8 CS chains of ~ 10 kDa average size and no O-linked oligosaccharides, indicating that essentially all of the available serine residues in the serine-glycine repeat are substituted with CS chains (Lohmander *et al.*, 1990).

In many cases, such as mucosal mast cells, basophils, natural killer cells, and platelets, the serglycin core protein contains CS chains, frequently with an average of more than one sulfate ester per disaccharide. For example, in mouse mucosal mast cells, the CS contains a high proportion of disaccharides with sulfate on both the 4 and 6 positions of the GalNAc residue (Fig. 2-2B), a form referred to as chondroitin sulfate E. The connective tissue mast cell synthesizes heparin chains on the core protein, clustering ten or more heparin chains of ~ 100 kDa in the repeating serine-glycine region (Fig. 2-4). The high number of chains per length of protein and the exceptionally high number of anionic groups along each chain yield a PG with a higher concentration of anionic groups than any other known biopolymer. The mast cell is the ultimate source of the heparin used widely in clinical applications.

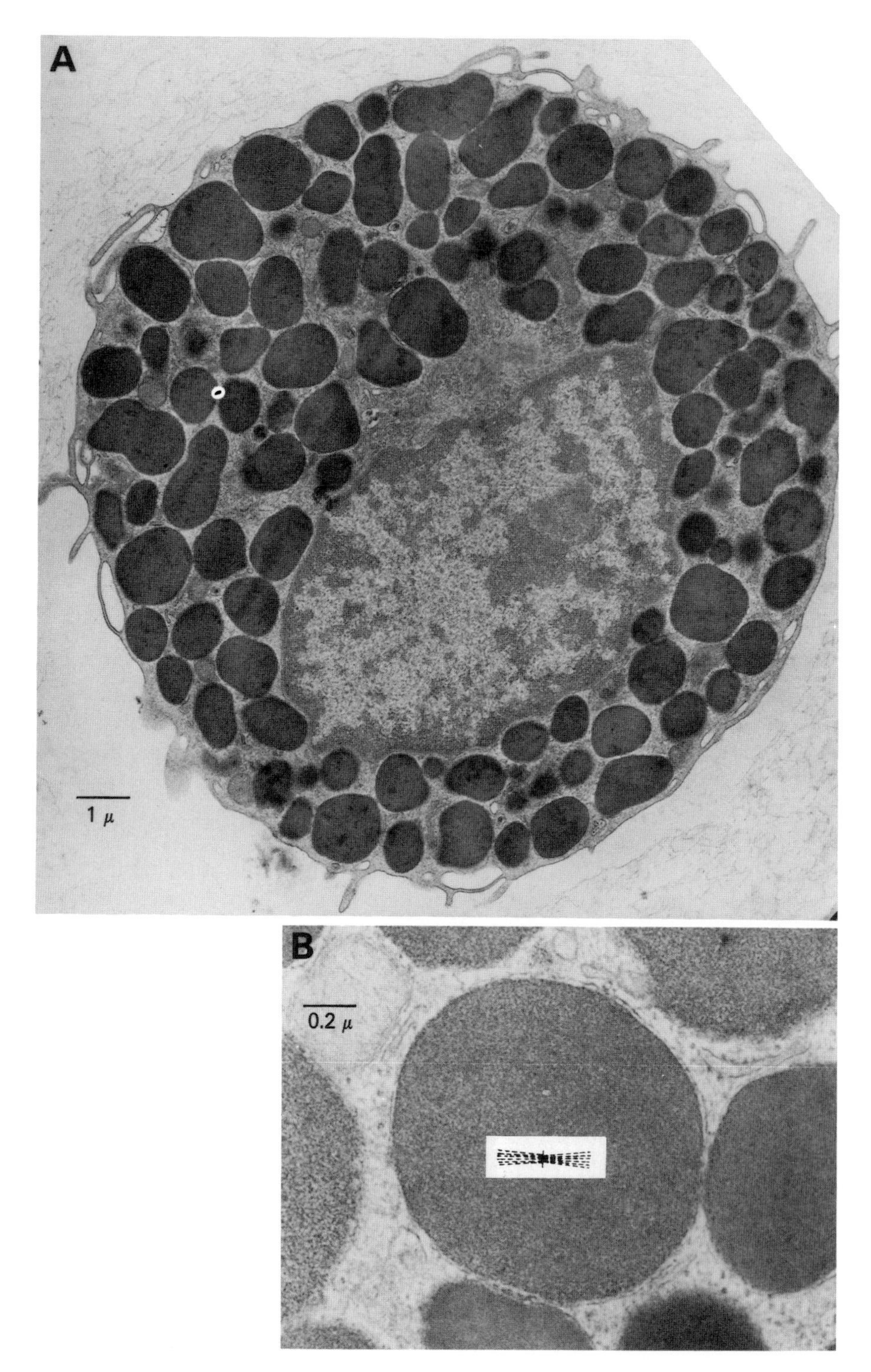

Figure 2-11. Comparison of the heparin form of serglycin with the storage granules from a connective tissue mast cell. The micrographs show a thin section of a connective tissue mast cell with numerous darkly stained storage granules in the intracellular space and an enlargement of a granule. The schematic drawing in the inset on the storage granule shows a heparin serglycin PG (see Fig. 2-4) at the same scale. (The micrographs were kindly provided by Dr. Mark Sweiter.)

A major function of the high concentration of the heparin PG in the storage granule is to interact with the cationic proteases, carboxypeptidases, histamines, and other molecules which the cell stores for release when participating in a host defense reaction. The heparin chains provide an effective cation exchange matrix which organizes and concentrates the bioactive, basic molecules in the storage granule (Fig. 2-11). Further, this ionic interaction appears important to prevent autolysis in the storage granule and may provide a slow release mechanism once the storage granules are discharged and begin to dissolve, thereby keeping the bioactive molecules near the site of action.

A number of immortalized mast cell lines have been developed which are intermediate between the mucosal and connective tissue mast cells in that they express mixtures of the CSPG and heparin PG forms. Cell lines with the highest carboxypeptidase A content also contained the highest heparin content (Reynolds *et al.*, 1988). This suggests that the spectrum of proteins in the storage granule may be closely correlated with the type of GAG synthesized on the serglycin core protein. The mechanism in the Golgi complex that directs the same core protein to a site of assembly of CS in one form of the mast cell but to a site of assembly of heparin in another is unknown (see Chapter 5).

4. Concluding Remarks

This chapter has described several distinct classes of proteoglycans which differ widely in their structures and functions. In addition to these, other proteins may in certain circumstances have a proportion of their population substituted with a GAG chain(s). These molecules have been referred to as "part-time" PGs (Fransson, 1987) and include: collagen type IX (McCormick *et al.*, 1987), the invariant chains of immunoglobulins (Miller *et al.*, 1988), the high-molecular-weight receptor for TGF-β (Cheifetz *et al.*, 1988), thrombomodulin (Bourin *et al.*, 1986), and a lymphocyte homing receptor (Jalkanen *et al.*, 1988). PGs have prominent roles in many aspects of cell biology, tissue organization, and tissue properties which are now becoming apparent. Future research analyzing their structures, expression, metabolism (see Chapter 5), and location in the cell and tissue will continue to define their amazing repertoire of functions.

ACKNOWLEDGMENTS. The original research reported in this chapter was supported by NHLBI-18645 and the National Institute of Dental Research, National Institutes of Health, Bethesda, Maryland.

References

Antonsson, P., Heinegård, D., and Oldberg, Å., 1989, The keratan sulfate enriched region of bovine cartilage proteoglycan consists of a consecutively repeated hexapeptide motif, *J. Biol. Chem.* **264:**16170–16173.

Avraham, S., Stevens, R. L., Gartner, M. C. Austen, K. F., Lalley, P. A., and Weis, J. H., 1988, Isolation of a cDNA that encodes the peptide core of the secretory granule proteoglycan of rat basophilic leukemia-1 cells and assessment of its homology to the human analogue, *J. Biol. Chem.* **263**:7292–7296.

Axelsson, I., and Heinegård, D., 1978, Characterization of the keratan sulphate proteoglycans from bovine corneal stroma, *Biochem. J.* **169**:517–530.

Benitz, W. E., Kelly, R. T., Anderson, C. M., Lorant, D. E., and Bernfield, M., 1990, Endothelial heparan sulfate proteoglycan. I. Inhibitory effects on smooth muscle cell proliferation, *Amer. J. Respir. Cell Mol. Biol.* **2**:13–24.

Bernfield, M., and Sanderson, R. D., 1990, Syndecan, a developmentally regulated cell surface proteoglycan that binds extracellular matrix and growth factors, *Philos. Trans. R. Soc. London Ser. B* **327**:171–186.

Bianco, P., Fisher, L. W., Young, M. F., Termine, J. D., and Gehron Robey, P., 1990, Expression and localization of the two small proteoglycans biglycan and decorin in developing human skeletal and non-skeletal tissues, *J. Histochem. Cytochem.* **38**:1549–1563.

Biswas, C., 1988, Heparin and heparan binding sites on B-16 melanoma cells, *J. Cell. Physiol.* **136**:147–153.

Bourdon, M. A., Oldberg, Å., Pierschbacher, M., and Ruoslahti, E., 1985, Molecular cloning and sequence analysis of a chondroitin sulfate proteoglycan cDNA, *Proc. Natl. Acad. Sci. USA* **82**:1321–1325.

Bourin, M. C., Boffa, M. C., Björk, I., and Lindahl, U., 1986, Functional domains of rabbit thrombomodulin, *Proc. Natl. Acad. Sci. USA* **88**:5924–5928.

Brandan, E., Maldonado, M., Garrido, J., and Inestrosa, N. C., 1985, Anchorage of collagen-tailed acetylcholinesterase to the extracellular matrix is mediated by heparan sulfate proteoglycans, *J. Cell. Biol.* **101**:985–992.

Brennan, M. J., Oldberg, Å., Pierschbacher, M. D., and Ruoslahti, E., 1984, Chondroitin/dermatan sulfate proteoglycan in human fetal membranes, *J. Biol. Chem.* **259**:13743–13750.

Carey, D. J., and Evans, D. M., 1989, Membrane anchoring of heparan sulfate proteoglycans by phosphatidylinositol and kinetics of synthesis of peripheral and detergent-solubilized proteoglycans in Schwann cells, *J. Cell Biol.* **108**:1897.

Carlson, S. S., and Wight, T. N., 1987, Nerve terminal anchorage protein 1 (TAP-1) is a chondroitin sulfate proteoglycan: Biochemical and electron microscopic characterization, *J. Cell Biol.* **105**:3075–3086.

Castellot, J. J., Addonizio, M. L., Rosenberg, R., and Karnovsky, M. J., 1981, Cultured endothelial cells produce a heparin-like inhibitor of smooth muscle cell growth, *J. Cell Biol.* **90**:372–379.

Chang, Y., Yanagishita, M., Hascall, V. C., and Wight, T. N., 1983, Proteoglycans synthesized by smooth muscle cells derived from monkey (Macaca nemistrina) aorta, *J. Biol. Chem.* **258**:5679–5688.

Cheifetz, S., Andres, J. L., and Massagué, J., 1988, The transforming growth factor-beta receptor type III is a membrane proteoglycan. Domain structure of the receptor, *J. Biol. Chem.* **263**:16984–16991.

Cherr, G. N., Yudin, A. I., and Katz, D. F., 1990, Organization of the hamster cumulus extracellular matrix: A hyaluronate-glycoprotein gel which modulates sperm access to the oocyte, *Dev. Growth Differ.* **32**:353–365.

Clement, B., Segui-Real, B., Hassell, J., Martin, G., and Yamada, Y., 1989, Identification of a cell surface-binding protein for the core protein of the basement membrane proteoglycan, *J. Biol. Chem.* **262**:12467–12471.

David, G., Lories, V., Decock, B., Marynen, P., Cassiman, J.-J., and Van der Berghe, H., 1990, Molecular cloning of a phosphatidylinositol-anchored membrane heparan sulfate proteoglycan from human lung fibroblasts, *J. Cell. Biol.* **111**:3165–3176.

Deak, F., Kiss, I., Sparks, K. J., Argraves, W. S., Hampikian, G., and Goetinck, P. F., 1986, Complete amino acid sequence of chicken cartilage link protein deduced from cDNA clones, *Proc. Natl. Acad. Sci. USA* **83**:3766–3770.

Dekel, N., and Phillips, D. M., 1979, Maturation of rat cumulus oophorus: A scanning electron microscopy study, *Biol. Reprod.* **21**:9–18.

Dickenson, J. M., Hucherby, T. N. and Nieduszynski, I. A., 1990, Two linkage-region fragments isolated from skeletal keratan sulphate contain a sulphated N-acetylglucosamine residue, *Biochem. J.* **269**:55–59.

Doege, K., Hassell, J. R., Caterson, B., and Yamada, Y., 1986, Link protein cDNA sequence reveals a tandemly repeated protein structure, *Proc. Natl. Acad. Sci. USA* **83**:3761–3765.
Doege, K., Sasaki, M., Horigan, E., Hassell, J. R., and Yamada, Y., 1987, Complete primary structure of the rat cartilage proteoglycan core protein deduced from cDNA clones, *J. Biol. Chem.* **262**:17757–17767.
Doege, K., Rhodes, C., Sasaki, M., Hassell, J. R., and Yamada, Y., 1990, Molecular biology of cartilage proteoglycan (aggrecan) and link protein, in: *Extracellular Matrix Genes* (L. Sandell and C. Boyd, eds.), pp. 137–155, Academic Press, New York.
Edge, A. B., and Spiro, R. G., 1990, Characterization of novel sequences containing 3-0-sulfated glucosamine in glomerular basement membrane heparan sulfate and localization of sulfated disaccharides to a peripheral domain, *J. Biol. Chem.* **265**:15874–15881.
Eppig, J. J., 1979, FSH stimulates hyaluronic acid synthesis by oocyte–cumulus cell complexes from mouse preovulatory follicles, *Nature* **281**:483–484.
Evered, D., and Whelan, J. (eds.), 1986, *Functions of Proteoglycans, CIBA Found. Symp.* **124**:Wiley, New York.
Fedarko, N. S., Ishihara, M., and Conrad, H. E., 1989, Control of cell division in hepatoma cells by exogenous heparan sulfate proteoglycan, *J. Cell. Physiol.* **139**:287–294.
Fisher, L. W., Hawkins, G. R., Tuross, N., and Termine, J. D., 1987, Purification and partial characterization of small proteoglycans I and II, bone sialoproteins I and II, and osteonectin from the mineral compartment of developing human bone, *J. Biol. Chem.* **262**:9702–9708.
Fisher, L. W., Termine, J. D., and Young, M. F., 1989, Deduced protein sequence of bone small proteoglycan I (biglycan) shows homology with proteoglycan II (decorin) and several nonconnective tissue proteins in a variety of species, *J. Biol. Chem.* **264**:4571–4576.
Folkman, J., Klagsbrun, M., Sasse, J., Wadzinski, M., Ingber, D., and Vlodavsky, I., 1988, A heparin-binding angiogenic protein-basic fibroblast growth factor is stored within basement membrane, *Am. J. Pathol.* **130**:393–400.
Fosang, A. J., and Hardingham, T. E., 1989, Isolation of the N-terminal globular protein domains from cartilage proteoglycans, identification of G2 domain and its lack of interaction with hyaluronic acid and link protein, *Biochem. J.* **261**:801–809.
Fransson, L. Å., 1987, Structure and function of cell-associated proteoglycans, *Trends Biochem. Sci.* **12**:406–411.
Fransson, L. Å., Silverberg, L., and Carlstedt, I., 1985, Structure of the heparan sulfate-protein linkage region. Demonstration of the sequence galactosyl–galactosyl–xylose-2-phosphate, *J. Biol. Chem.* **260**:14722–14726.
Gallagher, J. T., and Lyon, M., 1989, Molecular organization and functions of heparan sulfate, in: *Heparin* (D. Lane and U. Lindahl, eds.), pp. 135–158, Arnold, London.
Garrigues, H. J., Lark, M. W., Lara, S., Hellström, I., Hellström, K. E., and Wight, T. N., 1986, The melanoma proteoglycan: Restricted expression on microspikes, a specific microdomain of the cell surface, *J. Cell Biol.* **103**:1699–1710.
Grant, D. S., and LeBlond, C. P., 1988, Immunogold quantitation of laminin, type IV collagen and heparan sulfate proteoglycan in a variety of basement membranes, *J. Histochem. Cytochem.* **36**:271–283.
Halberg, D. F., Proulx, G., Doege, K., Yamada, Y., and Drickamer, K. A., 1988, A segment of the cartilage proteoglycan core protein has lectin-like activity, *J. Biol. Chem.* **263**:9486–9490.
Hardingham, T. E., and Muir, H., 1973, Binding of oligosaccharides of hyaluronic acid to proteoglycans, *Biochem. J.* **135**:905–908.
Harper, J. R., and Reisfeld, R. A., 1987, Cell associated proteoglycans in human malignant melanoma, in: Biology of Proteoglycans (T. Wight and R. Mecham, eds.), pp. 345–366, Academic Press, New York.
Hascall, G. K., 1980, Cartilage proteoglycans: Comparison of sectioned and spread whole molecules, *J. Ultrastruct. Res.* **70**:369–375.
Hascall, V. C., 1988, Proteoglycans: The chondroitin sulfate/keratan sulfate proteoglycan of cartilage, *ISI Atlas of Science: Biochemistry* **1**:189–198.
Hascall, V. C., and Heinegård, D., 1974, Aggregation of cartilage proteoglycans: Oligosaccharide competitors of the proteoglycan–hyaluronic acid interaction, *J. Biol. Chem.* **249**:4242–4249.
Hassell, J. M., Noonan, D. M., Ledbetter, S. R., and Laurie, G. W., 1986, Biosynthesis and structure of

the basement membrane proteoglycan containing heparan sulphate side-chains, in: *Functions of the Proteoglycans* (D. Evered and J. Whelan, eds.), pp. 204–214, Wiley, New York.
Hedbom, E., and Heinegård, D., 1989, Interaction of a 59 kDa connective tissue matrix protein with collagen I and collagen II, *J. Biol. Chem.* **264:**6898–6905.
Heinegård, D., and Axelsson, I., 1977, Distribution of keratan sulfate in cartilage proteoglycans, *J. Biol. Chem.* **252:**1971–1979.
Heinegård, D., and Hascall, V. C., 1974, Aggregation of cartilage proteoglycans: Characteristics of the proteins isolated from trypsin digests of aggregates, *J. Biol. Chem.* **249:**4250–4256.
Heinegård, D., and Oldberg, Å., 1989, Structure and biology of cartilage and bone noncollagenous macromolecules, *FASEB J.* **3:**2042–2051.
Heinegård, D., Björne-Persson, A., Cöster, L., Franzén, A., Gardell, S., Malmström, A., Paulsson, M., Sandfalk, R., and Vogel, K., 1985, The core protein of large and small interstitial proteoglycans from various connective tissues form distinct subgroups, *Biochem. J.* **230:**181–194.
Hopwood, J. J., and Robinson, H. C., 1974, The alkali-labile linkage between keratan sulphate and protein, *Biochem. J.* **141:**57–69.
Hunziker, E. B., and Schenk, R. K., 1984, Cartilage ultrastructure after high pressure freezing, freeze substitution, and low temperature embedding. II. Intercellular matrix ultrastructure—Preservation of proteoglycans in their native state, *J. Cell Biol.* **98:**277–282.
Hunziker, E. B., and Schenk, R. K., 1987, Structural organization of proteoglycans in cartilage, in: *Biology of Proteoglycans* (T. Wight and R. Mecham, eds.), pp. 155–185, Academic Press, New York.
Ishihara, M., Fedarko, N. S., and Conrad, H. E., 1987, Involvement of phosphatidylinositol and insulin in the co-ordinate regulation of proteoheparan sulfate metabolism and hepatocyte growth, *J. Biol. Chem.* **262:**4708–4716.
Jalkanen, S., Jalkanen, M., Bargatze, R., Tammi, M., and Butcher, E. C., 1988, Biochemical properties of glycoproteins involved in lymphocyte recognition of high endothelial venules in man, *J. Immunol.* **141:**1615–1623.
Kapoor, R., Phelps, C. F., and Wight, T. N., 1986, Physical properties of chondroitin sulfate/dermatan sulfate proteoglycans from bovine aorta, *Biochem. J.* **240:**575–583.
Kato, M., Koike, Y., Ito, Y., Suzuki, S., and Kimata, K., 1987, Multiple forms of heparan sulfate proteoglycans in the Engelbreth–Holm–Swarm mouse tumor: The occurrence of high density forms bearing both heparan sulfate and chondroitin sulfate side chains, *J. Biol. Chem.* **262:**7180–7188.
Kiefer, M. C., Stephans, J. C., Crawford, K., Okino, K., and Barr, P. J., 1990, Ligand-affinity cloning and structure of a cell surface heparan sulfate proteoglycan that binds basic fibroblast growth factor, *Proc. Natl. Acad. Sci. USA* **87:**6985–6989.
Kjellén, L., Oldberg, Å., and Höök, M., 1980, Cell surface heparan sulfate: Mechanisms of proteoglycan–cell association, *J. Biol. Chem.* **255:**10407–10413.
Klagsbrun, M., 1990, The affinity of fibroblast growth factors (FGFs) for heparin; FGF–heparan sulfate interactions in cells and extracellular matrix, *Curr. Opin. Cell Biol.* **2:**857–863.
Krusius, T., and Ruoslahti, E., 1986, Primary structure of an extracellular matrix proteoglycan core protein deduced from cloned cDNA, *Proc. Natl. Acad. Sci. USA* **83:**7683–7687.
Lark, M. W., Yeo, T. K., Mar, H., Lara, S., Hellström, I., Hellström, K. E., and Wight, T. N., 1988, Arterial chondroitin sulfate proteoglycan: Localization with a monoclonal antibody, *J. Histochem. Cytochem.* **36:**1211–1221.
Laurie, G. W., Inoue, S., Bing, J. T., and Hassell, J. R., 1988, Visualization of the large heparan sulfate proteoglycan from basement membrane, *Am. J. Anat.* **181:**320–326.
Ledbetter, S., Tyree, B., Hassell, J. R., and Horigan, E. A., 1985, Identification of the precursor protein to basement membrane heparan sulfate proteoglycans, *J. Biol. Chem.* **260:**8106–8113.
Lindahl, U., 1989, Biosynthesis of heparin and related structures, in: *Heparin: Chemical and Biological Properties, Clinical Applications* (D. Lane and U. Lindahl, eds.), pp. 159–189, Arnold, London.
Lindahl, U., Bäckström, G., Thunberg, L., and Leder, I. G., 1980, Evidence for a 3-0-sulfated D-glucosamine residue in the antithrombin-binding sequence of heparin, *Proc. Natl. Acad. Sci. USA* **77:**6551–6555.

Lohmander, L. S., Arnljots, K., and Yanagishita, M., 1990, Structure and synthesis of intracellular proteoglycan in HL-60 human leukemic promyelocytes, *J. Biol. Chem.* **265:**5802–5808.

Lories, V., DeBoeck, H. David., Jacques Casiman, J., and Van der Berghe, H., 1987, Heparan sulfate proteoglycans of human lung fibroblasts. Structural heterogeneity of the core proteins of the hydrophobic cell associated forms, *J. Biol. Chem.* **262:**854–859.

Mali, M., Jaakkola, P., Arvilommi, A.-M., and Jalkanen, M., 1990, Sequence of human syndecan indicates a novel gene family of integral membrane proteoglycans, *J. Biol. Chem.* **265:**6884–6889.

Mar, H., and Wight, T. N., 1988, Colloidal gold immunostaining on deplasticized ultrathin sections, *J. Histochem. Cytochem.* **36:**1387–1395.

Marcum, J. A., Reilly, C. F., and Rosenberg, R. D., 1987, Heparan sulfate species and blood vessel wall function, in: *Biology of Proteoglycans* (T. Wight and R. Mecham, eds.) pp. 301–343, Academic Press, New York.

Marynen, P., Zhang, J., Cassiman, J.-J., Van der Berghe, H., and David, G., 1989, Partial primary structure of the 48- and 90-kilodalton core proteins of cell surface-associated heparan sulfate proteoglycans of lung fibroblasts. Prediction of an integral membrane domain and evidence for multiple distinct core proteins at the cell surface of human lung fibroblasts, *J. Biol. Chem.* **262:**7017–7024.

McCarthy, K. J., Accavitti, M. A., and Couchman, J. R., 1989, Immunological characterization of a basement membrane-specific chondroitin sulfate proteoglycan, *J. Cell Biol.* **109:**3187–3198.

McClean, D., and Rowlands, I. W., 1942, Role of hyaluronidase in fertilization, *Nature* **150:**627–628.

McCormick, D., Van der Rest, M., Goodship, J., Lozano, G., Ninomiya, Y., and Olsen, B. R., 1987, Structure of the glycosaminoglycan domain in the type IX collagen-proteoglycan, *Proc. Natl. Acad. Sci. USA* **84:**4044–4048.

Miller, J., Hatch, J. A., Simonis, S., and Cullen, S., 1988, Identification of the glycosaminoglycan-attachment site of mouse invariant-chain proteoglycan core protein by site-directed mutagenesis, *Proc. Natl. Acad. Sci. USA* **85:**1359–1363.

Mörgelin, M., Paulsson, M., Hardingham, T., Heinegård, D., and Engel, J., 1988, Cartilage proteoglycans: Assembly with hyaluronate and link proteins as studied by electron microscopy, *Biochem. J.* **253:**175–185.

Mörgelin, M., Paulsson, M., Malmström, A., and Heinegård, D., 1989, Shared and distinct structural features of interstitial proteoglycans from different bovine tissues revealed by electron microscopy, *J. Biol. Chem.* **264:**12080–12090.

Neame, P. J., Christner, J. E., and Baker, J. R., 1986, The primary structure of link protein from rat chondrosarcoma proteoglycan aggregate, *J. Biol. Chem.* **261:**3519–3535.

Neame, P. J., Christner, J. E., and Baker, J. R., 1987, Cartilage proteoglycan aggregates: The link protein and proteoglycan amino-terminal globular domains have similar structures, *J. Biol. Chem.* **262:**17768–17778.

Neame, P. J., Choi, H. U., and Rosenberg, L. C., 1989, The primary structure of the core protein of the small, leucine-rich proteoglycan (PG I) from bovine articular cartilage, *J. Biol. Chem.* **264:**8653–8660.

Nilsson, B., De Luca, S., Lohmander, L. S., and Hascall, V. C., 1982, Structures of N-linked and O-linked oligosaccharides on proteoglycan monomer isolated from the Swarm rat chondrosarcoma, *J. Biol. Chem.* **257:**10920–10927.

Nilsson, B., Nakazawa, K., Hassell, J. R., Newsome, D. A., and Hascall, V. C., 1983, Structure of oligosaccharides and the linkage region between keratan sulfate and the core protein on proteoglycans from monkey cornea, *J. Biol. Chem.* **258:**6056–6063.

Noonan, D. M., Horigan, E. A., Ledbetter, S., Vogeli, G., Sasaki, M., Yamada, Y., and Hassell, J. R., 1988, Identification of cDNA clones encoding different domains of the basement membrane heparan sulfate proteoglycans, *J. Biol. Chem.* **263:**16379–16387.

Oegema, T. R., Kraft, E. L., Jourdian, G. W., and Van Valen, T. R., 1984, Phosphorylation of chondroitin sulfate in proteoglycans from the Swarm rat chondrosarcoma, *J. Biol. Chem.* **259:**720–726.

Oldberg, Å., Kjellén, L., and Höök, M., 1979, Cell-surface heparan sulfate: Isolation and characterization of a proteoglycan from rat liver membranes, *J. Biol. Chem.* **254:**8505–8510.

Oldberg, Å., Antonsson, P., Lindblom, K., and Heinegård, D., 1989, A collagen-binding 59 kDa protein (fibromodulin) is structurally related to the small interstitial proteoglycans PG-S1 and PG-S2 (decorin), *EMBO J.* **8**:2601–2604.

Paulsson, M., Fujiwara, S., Dziadek, M., Timpl, R., Pejler, G., Bäckström, G., Lindahl, U., and Engel, J., 1986, Structure and function of basement membrane proteoglycans, in: *Functions of the Proteoglycans* (D. Evered and J. Whelan, eds.), pp. 189–200, Wiley, New York.

Paulsson, M., Yurchenco, P. D., Ruben, G. C., Engel, J., and Timpl, R., 1987a, Structure of low density heparan sulfate proteoglycan isolated from a mouse tumor basement membrane, *Mol. Biol.* **197**:297–313.

Paulsson, M., Mörgelin, M., Wiedemann, H., Beardmore-Gray, M., Dunham, D., Hardingham, T., Heinegård, D., Timpl, R., and Engel, J., 1987b, Extended and globular protein domains in cartilage proteoglycans, *Biochem. J.* **245**:763–772.

Pearson, C. H., Winterbottom, N., Fackre, D. S., Scott, P. G., and Carpenter, M. R., 1983, The NH_2-terminal amino acid sequence of bovine skin proteodermatan sulfate, *J. Biol. Chem.* **258**:15101–15104.

Pejler, G., Backström, G., Lindahl, U., Paulsson, M., Dziadek, M., Fujiwara, S., and Timpl, R., 1987, Structure and affinity for antithrombin of heparan sulfate chains derived from basement membrane proteoglycans, *J. Biol. Chem.* **262**:5038–5043.

Pèrin, J.-P., Bonnet, F., Thurieau, C., and Jollès, P., 1987, Link protein interactions with hyaluronate and proteoglycans: Characterization of two distinct domains in bovine cartilage link proteins, *J. Biol. Chem.* **262**:13269–13272.

Plaas, A. H. K., Neame, P. J., Nivens, C. M., and Reiss, L., 1990, Identification of the keratan sulfate attachment sites on bovine fibromodulin, *J. Biol. Chem.* **265**:20634–20640.

Rapraeger, A., Jalkanen, M., and Bernfield, M., 1987, Integral membrane proteoglycans as matrix receptors: Role in cytoskeleton and matrix assembly at the epithelial cell surface, in: *Biology of Proteoglycans* (T. Wight and R. Mecham, eds.), pp. 129–154, Academic Press, New York.

Reynolds, D. S., Serafin, W. E., Faller, D. V., Wall, D. A., Abbas, A. K., Dvorak, A. M. Austen, K. F., and Stevens, R. L., 1988, Immortalization of murine connective tissue-type mast cells at multiple stages of their differentiation by coculture of splenocytes with fibroblasts that produce Kirsten sarcoma virus, *J. Biol. Chem.* **263**:12783–12791.

Rifkin, D. B., and Moscatelli, D., 1989, Recent developments in the cell biology of basic fibroblast growth factor, *J. Cell Biol.* **109**:4060–4066.

Roberts, R., Gallagher, J., Spooncer, E., Allen, T. D., Bloomfield, F., and Dexter, T. M., 1988, Heparan sulphate bound growth factors: A mechanism for stromal cell mediated haemopoiesis, *Nature* **332**:376–378.

Rodén, L., 1980, Structure and metabolism of connective tissue proteoglycans, in: *The Biochemistry of Glycoproteins and Proteoglycans* (W. Lennarz, ed.), pp. 267–371, Plenum Press, New York.

Rosenberg, R. D., 1985, Role of heparin and heparinlike molecules in thrombosis and atherosclerosis, *Fed. Proc.* **44**:404–409.

Salisbury, B., and Wagner, W., 1981, Isolation and preliminary characterization of proteoglycans dissociatively extracted from human aorta, *J. Biol. Chem.* **256**:8050–8057.

Salustri, A., Yanagishita, M., and Hascall, V. C., 1989, Synthesis and accumulation of hyaluronic acid and proteoglycans in the mouse cumulus cell–oocyte complex during FSH-induced mucification, *J. Biol. Chem.* **264**:13840–13847.

Salustri, A., Yanagishita, M., and Hascall, V. C., 1990, Mouse oocytes regulate hyaluronic acid synthesis and mucification by FSH-stimulated cumulus cells, *Dev. Biol.* **138**:26–32.

Saunders, S., Jalkanen, M., O'Farrell, S., and Bernfield, M., 1989, Molecular cloning of syndecan, a membrane proteoglycan, *J. Cell Biol.* **108**:1547–1556.

Schittny, J. C., Timpl, R., and Engel, J., 1988, High resolution immunoelectron microscopic localization of functional domains of laminin, nidogen and heparan sulfate proteoglycan in epithelial basement membrane of mouse cornea reveals different topographical orientations, *J. Cell Biol.* **107**:1599–1610.

Schönherr, E., Järveläinen, H., Sandell, L. J., and Wight, T. N., 1991, Effects of PDGF and TGF-β on the synthesis of a large versican-like chondroitin sulfate proteoglycan by arterial smooth muscle cells, *J. Biol. Chem.*, in press.

Scott, J. E., 1988, Proteoglycan–fibrillar collagen interactions, *Biochem. J.* **252**:313–323.

Stevens, R. L., 1987, Intracellular proteoglycans in cells of the immune system, in: *Biology of Proteoglycans* (T. Wight and R. Mecham, eds.), pp. 367–387, Academic Press, New York.

Stevens, R. L., Nicodemus, C. F., and Avraham, S., 1990, The gene that encodes the peptide core of secretory granule PGs of haematopoietic cells, *Br. Biochem. Soc. Trans.* **18**:810–812.

Stow, J. L., Kjellén, L., Unger, E., Höök, M., and Farquhar, M. G., 1985a. Heparan sulfate proteoglycans are concentrated on the sinusoidal plasmalemmal domain and in intracellular organelles of hepatocytes, *J. Cell Biol.* **100:** 975–980.

Stow, J. L., Sawada, H., and Farquhar, M. G., 1985b, Basement membrane heparan sulfate proteoglycans are concentrated in the lamine rarae and in podocytes of the rat renal glomerulus, *Proc. Natl. Acad. Sci. USA* **82**:3296–3300.

Stuhlsatz, H. W., Keller, R., Becker, G., Oeben, M., Lennarts, L., Fisher, D. C., and Greiling, H., 1988, Structure of keratan sulphate proteoglycans: Core proteins, linkage regions, carbohydrate chains, in: *Keratan Sulphate: Chemistry, Biology, Chemical Pathology* (H. Greiling and J. Scott, eds.), pp. 1–11, The Biochemical Society, London.

Tarone, G., and Underhill, C. B., 1988, Distribution of hyaluronate and hyaluronate receptors in the adult lung, *J. Cell Sci.* **89**:145–156.

Thesleff, I., Jalkanen, M., Vainio, S., and Bernfield, M., 1988, Cell surface proteoglycan expression correlates with epithelial–mesenchymal interaction during tooth morphogenesis, *Dev. Biol.* **129**:565–572.

Torchia, D. A., Hasson, M. A., and Hascall, V. C., 1977, Investigation of molecular motion of proteoglycans in cartilage by 13-C magnetic resonance, *J. Biol. Chem.* **251**:3617–3625.

Vainio, S., Lehtonen, E., Jalkanen, M., Bernfield, M., and Saxén, L., 1989, Epithelial–mesenchymal interactions regulate the stage specific expression of a cell surface proteoglycan, syndecan, in developing kidney, *Dev. Biol.* **134**:382–391.

Vogel, K. G., Paulsson, M., and Heinegård, D., 1984, Specific inhibition of type I and type II collagen fibrillogenesis by the low molecular mass proteoglycan of tendon, *Biochem. J.* **223**:587–597.

Vogel, K. G., Koob, T. J., and Fisher, L. W., 1987, Characterization and interactions of a fragment of the core protein of the small proteoglycan (PGII) from bovine tendon, *Biochem. Biophys. Res. Commun.* **148**:658–663.

Wiedemann, H., Paulsson, M., Timpl, R., Engel, J., and Heinegård, D., 1984, Domain structure of cartilage proteoglycans revealed by rotary shadowing of intact and fragmented molecules, *Biochem. J.* **224**:331–333.

Wight, T. N., 1989, Cell biology of arterial proteoglycans, *Arteriosclerosis* **9**:1–20.

Wight, T., and Mecham, R. P. (eds.), 1987, *Biology of Proteoglycans*, Academic Press, New York.

Wight, T. N., and Hascall, V. C., 1983, Proteoglycans in primate arteries. III. Characterization of the proteoglycans synthesized by arterial smooth muscle cells in culture, *J. Cell Biol.* **96**:167–176.

Woods, A., and Couchman, J. R., 1988, Focal adhesions and cell–matrix interactions, *Collagen Relat. Res.* **8**:155–182.

Wright, D. W., and Mayne, R., 1988, Vitreous humor of chicken contains two fibrillar systems: An analysis of their structure, *J. Ultrastruct. Mol. Struct. Res.* **100**:224–234.

Yada, T., Suzuki, S., Kobayashi, K., Kobayashi, M., Hoshino, T., Horie, K., and Kimata, K., 1990, Occurrence in chick embryo vitreous humor of a type IX collagen proteoglycan with an extraordinarily large chondroitin sulfate chain and short α1 polypeptide, *J. Biol. Chem.* **265**:6992–6999.

Yamaguchi, Y., and Ruoslahti, E., 1988, Expression of human proteoglycan in Chinese hamster ovary cells inhibits cell proliferation, *Nature* **336**:244–246.

Yamaguchi, Y., Mann, D. M., and Ruoslahti, E., 1990, Negative regulation of transforming growth factor-β by the proteoglycan decorin, *Nature* 346:281–284.

Yanagishita, M., and McQuillan, D. J., 1989, Two forms of plasma membrane-intercalated heparan sulfate proteoglycan in rat ovarian granulosa cells: Labeling of proteoglycans with a photoactivatable hydrophobic probe and effect of the membrane anchor-specific phospholipase C *J. Biol. Chem.* **264**:17551–17558.

Yudin, A. I., Cherr, G. N., and Katz, D. F., 1988, Structure of the cumulus matrix and zona pellucida in the golden hamster: A new view of sperm interaction with oocyte-associated extracellular matrix, *Cell Tissue Res.* **251**:555–564.

Yurchenco, P. D., and Schittny, J. C., 1990, Molecular architecture of basement membranes, *FASEB J.* **4**:1577–1590.

Yurchenco, P. D., Cheng, Y. S., and Reuben, G. C., 1987, Self-assembly of a high molecular weight basement membrane heparan sulfate proteoglycan into dimers and oligomers, *J. Biol. Chem.* **262**:17668–17676.

Zaremba, S., Guimaraes, A., Kalb, R. G., and Ho, S., 1989, Characterization of an activity dependent neuronal surface proteoglycan identified with monoclonal antibody Cat-301, *Neuron* **2**:1207–1219.

Zimmerman, D. R., and Ruoslahti, E., 1989, Multiple domains of the large fibroblast proteoglycan, versican, *EMBO J.* **8**:2975–2981.

Chapter 3

The Elastic Fiber

ROBERT P. MECHAM and JOHN E. HEUSER

1. Introduction

With the development of higher organisms came the need for tissue flexibility and the necessity to withstand stretch. Several unrelated proteins have evolved to satisfy this requirement, including resilin in arthropods, abductin in molluscs, and elastin in the vertebrates (Sage and Gray, 1979). Phylogenetic studies have shown that elastin occurs only in vertebrates and arose first in cartilaginous fish where its appearance coincides with the achievement of a fully closed circulatory system. The occurrence of elastin in both primitive and modern sharks, and in chondrostean fishes suggests that the protein appeared in an early Devonian ancestor of all present-day fish, at some point after the divergence of the cyclostome and gnathostome lines (Sage, 1983).

In discussion of elastic tissues it is important to distinguish between "elastin" and "elastic fibers." The elastic fiber is the complex structure found in the extracellular matrix that contains elastin, microfibrillar proteins, lysyl oxidase, and, perhaps, proteoglycans. Elastin is the predominant protein of mature elastic fibers and endows the fiber with the characteristic property of elastic recoil. It is a chemically inert, extremely hydrophobic protein whose resistance to dissolution is such that, prior to the discovery of the soluble precursor molecule, it was possible to define elastin only in operational and qualitative terms as the insoluble protein residue remaining after all other tissue components had been solubilized (Partridge, 1962). In addition to its unusual physical properties, insoluble elastin is stable under normal physiological conditions, perhaps lasting for the life of the organism (Lefevre and Rucker, 1980; Shapiro *et al.*, 1991).

Biosynthesis of elastic fibers presents complex problems for the elastin-producing cell. Elastin must be synthesized in soluble form, post-transla-

ROBERT P. MECHAM • Department of Cell Biology and Physiology, Washington University School of Medicine, and Department of Medicine, Respiratory and Critical Care Division, Jewish Hospital at Washington University Medical Center, St. Louis, Missouri 63110. JOHN E. HEUSER • Department of Cell Biology and Physiology, Washington University School of Medicine, St. Louis, Missouri 63110.

Cell Biology of Extracellular Matrix, Second Edition, edited by Elizabeth D. Hay, Plenum Press, New York, 1991.

tionally modified, packaged, and transported to unique sites on the cell surface for cross-linking and fiber assembly. During development, the cell must adjust the temporal, qualitative, and quantitative pattern of its synthetic activities to coordinate the amount, type, and spatial organization of elastic fibers and other matrix components, so as to maintain the phenotypic composition typical of a particular connective tissue. The cell must also remain responsive to physiological and pathological stresses that necessitate modification of the matrix. It is now evident that this elastogenic pathway is an integrated system regulated at many points by the cell, which can respond to common biological signals (such as hormones or macromolecules from the extracellular matrix) thus permitting timing and control of elastogenesis by external factors. The purpose of this chapter is to focus on the "cell biology" of elastin and the interesting strategies used by the cell to organize and assemble this complex, insoluble extracellular structure. More detailed information on the chemical and physical properties of elastin can be found in reviews by Partridge (1962), Franzblau (Franzblau, 1971; Franzblau and Faris, 1981), and Gosline and Rosenbloom (1984). Regulation of elastin synthesis and the relationship between elastin and disease have been reviewed by Davidson (Davidson and Giro, 1986) and Sandberg (Sandberg *et al.*, 1981), respectively. A discussion of elastin gene structure and alternative splicing can be found in reviews by Rosenbloom *et al.* (Indik *et al.*, 1990) and Parks and Deak (1990).

2. Elastic Fibers

2.1. Tissue Organization

Histochemical stains that selectively bind elastic fibers have enabled the visualization of elastic structures in many different tissues. It is clear from these studies that elastin is distributed as interconnected fibers in three morphologically distinct forms. In elastic ligaments, lung, and skin, the fibers are small, ropelike, and variable in length. In major arteries, such as the aorta, elastic fibers form concentric sheets or lamellae while in elastic cartilage a three-dimensional honeycomb arrangement of very large anastomosing fibers is apparent. These differing and complex structures are thought to arise as a consequence of the strength and direction of forces put upon the tissue. Figure 3-1 is a scanning electron micrograph of thoracic aorta showing layers of elastin in the medial layer of the vessel after other connective tissue components have been removed by sodium hydroxide. The elastic lamellae are interconnected by radially oriented interlaminar elastic fibers that transfer stress throughout the vessel wall.

2.2. Elastic Fiber Composition

Ultrastructural examination of elastic fibers has shown that they comprise two morphologically distinct components: a core of polymeric insoluble elas-

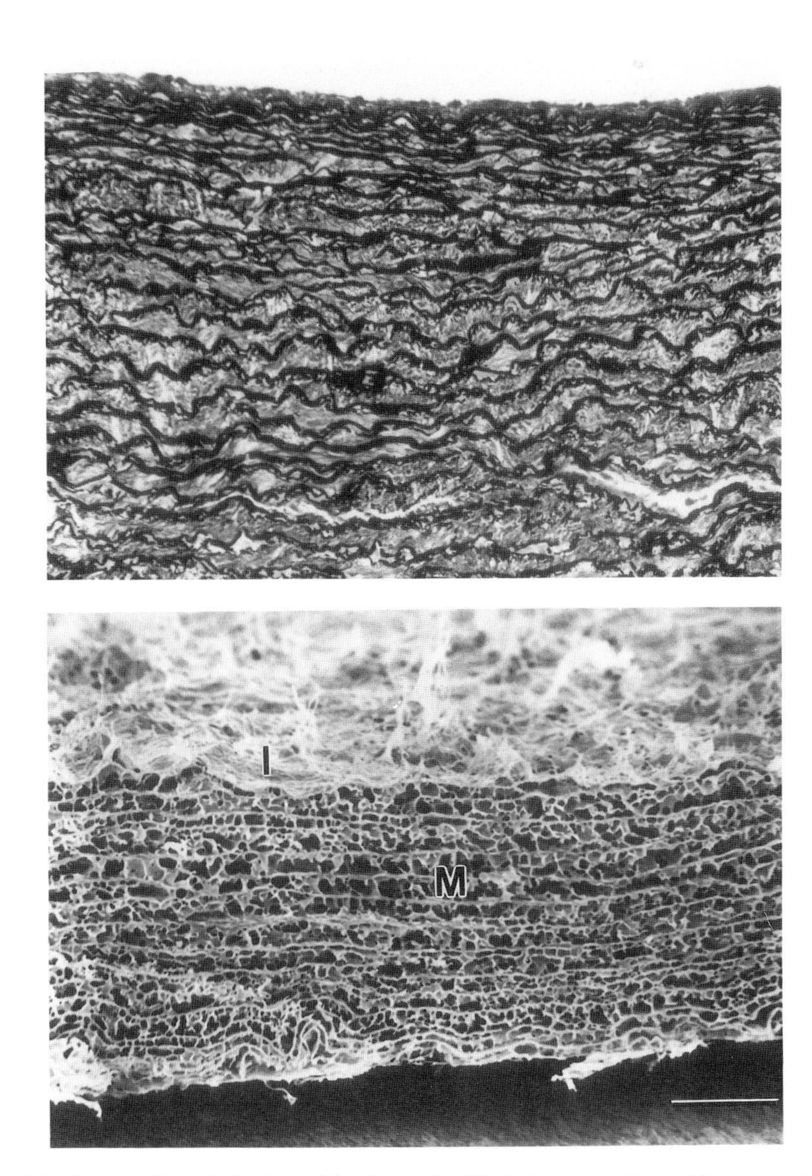

Figure 3-1. Organization of elastin in blood vessels. (Top) A cross section of bovine pulmonary artery stained with Verhoeff–Van Gieson elastic stain showing successive layers, or lamellae, of elastin organized in concentric sheets around the blood vessel. The lumen of the blood vessel is oriented toward the top of the figure. ×80. (Bottom) The three-dimensional organization of the elastic lamellae can be visualized in adult rat aorta by scanning electron microscopy following removal of nonelastin matrix by sodium hydroxide digestion (Crissman, 1987). The innermost elastic lamellae (I) and other lamellae of the tunica media (M) are interconnected by fine elastic fibers and incomplete septa. Bar = 50 μm. (Figures courtesy of Dr. Edmond Crouch and Dr. Robert Crissman.)

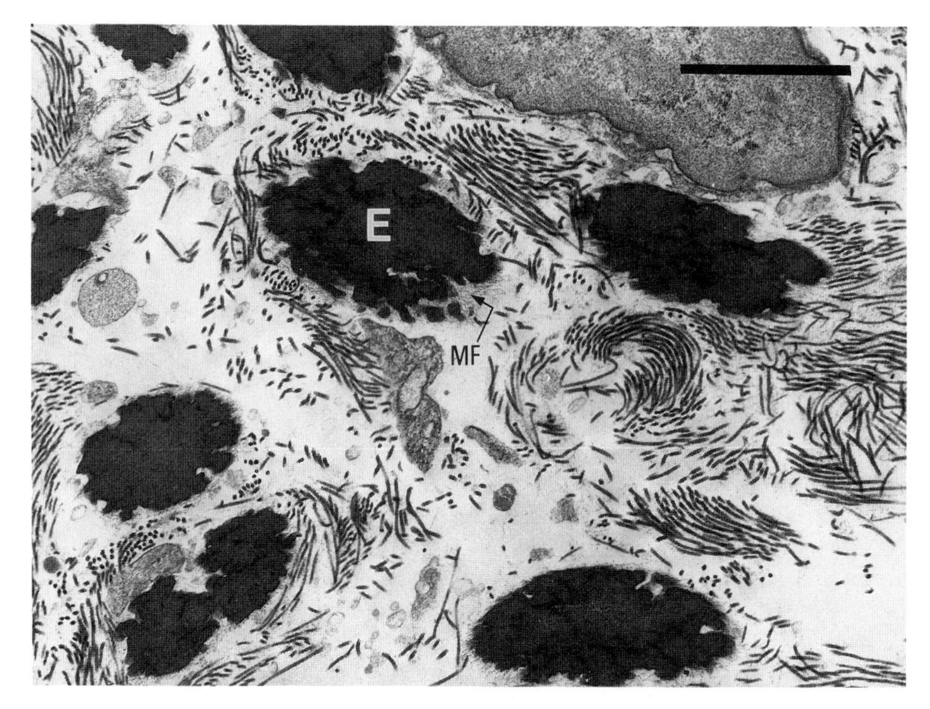

Figure 3-2. Transverse section of developing elastic fibers in the ligamentum nuchae of a 230-day gestation fetal calf. The elastin component (E) forms a central core with the microfibrils (MF) at the periphery of the growing fiber. The section was treated with tannic acid to enhance the staining of elastin. Bar = 0.3 μm. (Figure courtesy of Dr. Giuliano Quintarelli.)

tin, which appears amorphous and stains lightly with cationic heavy-metal stains, and a peripheral mantle of tubular-appearing microfibrils which have a beadlike periodicity in longitudinal sections and display a strong affinity for such stains (Fig. 3-2).

2.2.1. The Microfibrillar Component

Elastic fibers first appear in fetal development as aggregates of 10- to 12-nm microfibrils arranged in parallel array, often occupying infoldings of the cell membrane. Definitive elastin is then deposited as small clumps of amorphous material within these bundles of microfibrils, which subsequently coalesce to form true elastic fibers. The relative proportion of microfibrils to elastin declines with increasing age with adult elastic fibers having only a sparse peripheral mantle of microfibrillar material. Despite advances in understanding the structural complexity of microfibrils, the mechanism(s) by which they participate in fibrillogenesis remain speculative. The observation that microfibrillar aggregates take the form and orientation of presumptive elastic fibers led Ross *et al.* (1977) to suggest that they direct the morphogenesis of elastic fibers by acting as a "scaffold" on which elastin is deposited. It is likely that microfibrils serve to align tropoelastin molecules, the soluble secreted form of elastin, in precise register, so that cross-linking regions are juxtaposed prior to oxidation by lysyl oxidase.

Isolation and characterization of microfibrillar components has proven a difficult problem. Recently, several molecules have been isolated and shown by immunoelectron microscopy techniques to be associated with microfibrils. One that has been characterized is a 31-kDa glycoprotein termed microfibril-associated glycoprotein (MAGP) which has recently been cloned and sequenced (Gibson *et al.*, 1991). The amino acid sequence indicates that MAGP contains two structurally distinct regions. The amino-terminal half of the protein is rich in glutamine, proline, and acidic amino acids whereas the carboxy-terminal half contains all 13 of the cysteine residues and most of the basic amino acids. The odd number of cysteine residues is consistent with the finding that MAGP forms intermolecular disulfide bonds. MAGP contains no Arg-Gly-Asp (RGD) sequences of Asn-X-Ser/Thr sites. Thus, the carbohydrate components of MAGP are most likely *O*-linked through serine or threonine.

Another component of microfibrils is fibrillin, a 350,000 dalton glycoprotein that is a major constituent of both elastin-containing and elastin-free microfibrils (Sakai *et al.*, 1986). Characterization of fibrillin cDNAs reveals a structure consisting of large regions of EGF-like repeats and one RGD sequence. There is also homology to the Drosophila *Notch* gene, the *lin-12* gene of *C. elegans*, and transforming growth factor beta-binding protein (Maslen *et al.*, 1991). Sequence analysis of genomic clones has identified multiple fibrillin genes (Lee *et al.*, 1991), suggesting genetic heterogeneity typical of a gene family. Hollister and Sakai and their co-workers have recently presented evidence that the fibrillin gene localized on chromosome 15 is linked to Marfan's syndrome [reviewed by McKusick (1991)]. A second fibrillin gene on chromosome 5 is linked to congenital contractural arachnodactyly (Lee *et al.*, 1991), a disease that shares some of the features of Marfan syndrome.

Several other proteins have been shown by antibody studies to be associated with microfibrils, including (1) a 34-kDa glycoprotein with amine oxidase activity (Serafini-Fracassini *et al.*, 1981), (2) lysyl oxidase (Kagan *et al.*, 1986), and (3) several glycoproteins extracted from developing ligamentum nuchae (Gibson *et al.*, 1989). Although the precise relationship between these molecules and microfibrillar structure and function is yet to be resolved, it appears that microfibrils are composed of a number of subcomponents of widely differing sizes.

While developing elastic fibers always contain microfibrils, microfibrils have been identified in tissues that do not contain elastin. In periodontal ligament, ocular zonule, and mesangium of renal glomerulus, for example, microfibrils have been described that contain no visible or immunoreactive elastin (Kumaratilake *et al.*, 1989). In other tissues, limited amounts of immune-reacting elastin in apposition to the microfibrils have been demonstrated. In the dermis, the dermal microfibrillary bundles which connect the deep dermal elastic plexus with the dermo-epidermal junction region are seen to consist of microfibrils, indistinguishable from those associated with elastic tissue. In the superficial distribution, they contain no associated immunoreactive elastin, but as they traverse the dermis they acquire an increasing amorphous elastic component. Tissue localization with antibodies to MAGP and fibrillin shows that all of these microfibrillar structures express antigenic epitopes for both proteins, whether or not they contain immunoreactive elastin.

2.2.2. Elastin

Elastin is generally isolated from elastic tissues by removing all other connective tissue components by denaturation or degradation. The most successful purification procedures rely upon elastin's extreme inertness to protein solvents and its resistance to hydrolysis during treatment with dilute acids or alkali. Lansing *et al.* (1951) described a purification of elastin from aortic tissue that included 0.1 N NaOH at 95°C for 45 min. Partridge *et al.* (1955) utilized an exhaustive autoclave procedure for purifying elastin from ligamentum nuchae that included successive 1-hr autoclave periods until no further protein appeared in the supernatant. To avoid the extensive peptide bond cleavage that occurs with hot alkali and repeated autoclaving, several alternative purification methods have been proposed that include treatment with a combination of proteases, chaotropic and reducing agents, and cyanogen bromide. These milder extraction methods are successful to varying degrees, depending on the starting tissue. Generally speaking, however, hot alkali and autoclave techniques give the cleanest elastin.

Table I. Amino Acid Composition of Bovine Tropoelastin Predicted from the Full-Length Bovine Tropoelastin cDNA (Yeh *et al.*, 1987)[a]

Amino acid		Residues/molecule
Cys		2
Asx		3
Thr		8
Ser		7
Glx		10
Pro	(Pro + Hyp)	86
Gly		230
Ala		152
Val		91
Met		0
Ile		18
Leu		43
Tyr		7
Phe		21
Hyl		0
His		0
Lys		38
Arg		5

[a]Values do not include the signal peptide.

Elastin has an unusual and highly characteristic amino acid composition (Table I). It has a low content of acidic and basic amino acids and is correspondingly rich in hydrophobic amino acids, particularly valine (~15%). One-third of the residues are glycine and one-ninth are proline. Elastin contains no hydroxylysine but small amounts of hydroxyproline are present. In addition, elastin contains several lysine derivatives that serve as covalent cross-linkages between protein monomers. These cross-links, as pointed out below, are important in imparting rubberlike properties to elastin. Their structure is shown in Fig. 3-3. The initial step in their formation is the deamination of the ε-amino group of lysine side chains by the copper-requiring enzyme lysyl oxidase. The reactive aldehyde that is formed (α-amino adipic δ-semialdehyde, or allysine) can then condense with a second aldehyde residue to form allysine aldol or with an ε-amino group on lysine to form dehydrolysinonorleucine. Allysine aldol and dehydrolysinonorleucine can then condense to form the pyridinium cross-links called desmosine and isodesmosine, both of which in vertebrates are unique to elastin and can be used as distinctive markers for the protein. Although desmosine and isodesmosine are tetrafunctional, current evidence suggests that they join only two chains, with lysine residues from each chain contributing to their formation.

Oxidation–reduction reactions are important for cross-link biosynthesis and maturation. Condensation of desmosine occurs at the oxidation level of dihydrodesmosine yet there are naturally present in elastin ring compounds at

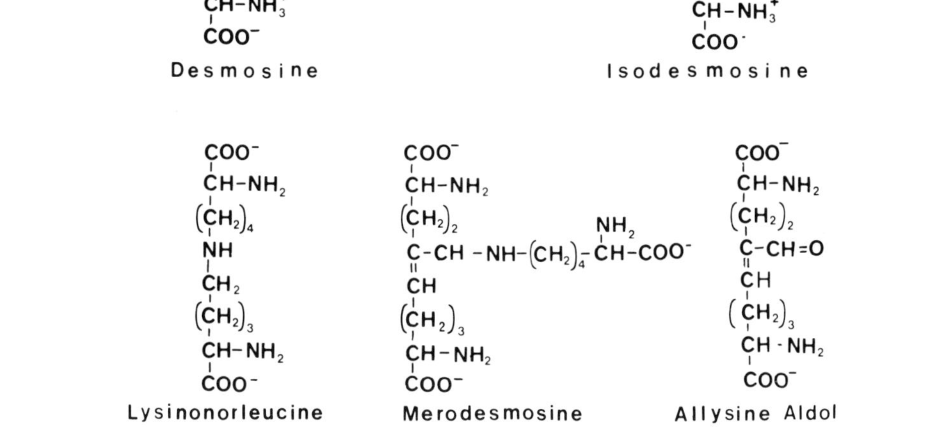

Desmosine

Isodesmosine

Lysinonorleucine

Merodesmosine

Allysine Aldol

Figure 3-3. Cross-links in elastin formed by the oxidation of lysyl epsilon-amino groups by the enzyme lysyl oxidase.

three oxidation levels containing one, two, or three double bonds and additional isomers with one of the double bonds outside the ring. Also found in elastin are lysinonorleucine and merodesmosine which are the reduced forms of the original condensation products dehydrolysinonorleucine and dehydromerodesmosine. The biochemistry and importance of cross-link reduction is obscure and nothing is known about the reducing agent *in vivo*.

All but 5–8 of the 38 lysine residues of bovine tropoelastin are modified during elastin maturation but not all of the modified lysyl residues have been accounted for as cross-links. In bovine ligamentum nuchae elastin, for example, there is estimated to be 2 desmosines, 1 isodesmosine, 3 lysinonorleucines, and 4 aldol condensation products. This translates into, at most, 26 lysine equivalents per monomer, or 70% of the available modified lysines. The difference may be that insoluble elastin contains as yet unidentified cross-links or that peptides from the tropoelastin precursor are cleaved from the molecule before its incorporation into the fiber.

3. Why Is Elastin Elastic?

Elastin is characterized by a high degree of reversible distensibility, including the ability to deform to large extensions with small forces. The precise physicochemical properties that account for elastin's rubberlike characteristics, however, have not been fully characterized. Biophysical studies of purified elastic fibers have suggested that elastin behaves as a classical "rubber" (Hoeve, 1974) modeled after a cross-linked network of randomly oriented chains that lack both short- and long-range order. According to the rubber theory, an unstretched rubber network is in a state of maximum disorder and, hence, maximum entropy. Stretching the rubber induces order in the direction of the extension, thereby decreasing entropy. When the external force is removed, elastic recoil will occur because the partially ordered chains return spontaneously to their initial, random state in order to return to a level of maximum entropy. Elastin exhibits the physical features characteristic of a polymeric rubber, including interchain cross-linkages and chains that behave as a kinetically free, random chain network (Fleming *et al.*, 1980).

While the dominant pattern of elastin is that of a random structure, the elastin system is more complicated than a simple three-dimensional array of random chains. Several physiochemical studies suggest that there are regions of the molecule that exhibit areas of local order. For example, the alanine-rich sequences that define cross-linking domains have been shown to be in an extended helix conformation and Urry and co-workers have shown extensive secondary structure (e.g., beta-turns giving rise to beta-spirals) associated with several repeating hydrophobic sequences (Urry, 1983). These microdomains of secondary order could serve to restrict movement around chemical bonds, and therefore the rotational freedom assumed for an ideal rubber may not always hold for all areas of elastin.

Elastin is an extremely hydrophobic protein with approximately 70% of its amino acid side chains being nonpolar. This unusual property has led to models of elasticity that take into account the possibility that nonpolar regions of the molecule are exposed to water when elastin is stretched. Recoil then occurs when the nonpolar groups reaggregate and expel the absorbed water after the distending force is removed (Gosline, 1978). Based upon studies which examined the ability of sugars and polyglycols to penetrate an elastin gel, Partridge (1969) proposed that elastin is a network of 5-nm protein globules having hydrophobic interiors surrounded by free solvent spaces of about 3 nm. Weis-Fogh and Andersen (1970) expanded on Partridge's globular model proposing a two-phase "liquid drop" model in which deformation of the subunits exposes the hydrophobic interiors of the globules, creating additional interfacial surfaces. Gray *et al.* (1973) have proposed an alternative model which, like the liquid drop model, emphasizes hydrocarbon–water interactions, but views the monomeric units as fibrillar rather than globular. In this model, each fibrillar monomer is made up of alternating segments of cross-link regions and "oiled coil." The oiled coil region is a broad coil in which glycines occupy the exterior positions of beta-turns exposed to solvent, while proline, valine, and other hydrophobic residues are buried. The principal site of hydration is the peptide chain rather than amino acid side chains. Stretching of elastin would expose the hydrocarbon core of the polar aqueous surroundings resulting in a high negative entropy change. Urry *et al.* (1974) have proposed a fibrillar model which, like the oiled coil, is a sequential arrangement of beta-spirals and alpha-helix. It differs from the oiled coil in that the hydrophobic side chains constitute hydrophobic sides of the spiral. The individual chains can associate by cross-linking and by hydrophobic interactions which may also serve to align the chains before cross-linking takes place.

The liquid drop, oiled coil, and fibrillar models all rely on water–protein interactions as an integral aspect of the thermodynamics involved in elastin's elasticity. The relative proportion of energy stored in the reversible exposure of hydrophobic groups to water versus entropy effects associated with ordering of random chains upon stretching, however, is still an open question. Whether elastin fibers exist in a random configuration or whether they can best be described by one of the models relying on protein–water interactions remains unclear. The actual mechanism is probably a dynamic blend of these two extremes.

4. Tropoelastin Structure

A significant advance in our knowledge of elastin structure was the identification and characterization of cDNAs which encoded the elastin molecule. From the analysis of tryptic peptides derived primarily from porcine and chick tropoelastin, the soluble secreted form of elastin, it was known that elastin contained segments enriched in hydrophobic amino acids as well as alanine-rich regions which incorporated several lysine residues that serve as cross-links. From these data, Gray *et al.* (1973) predicted that the molecule was made up of alternating hydrophobic and cross-linking domains. As with many other proteins, recombinant DNA techniques presented an easier solution to the problem of primary structure. In the past few years, the amino acid sequences of human (Indik *et al.*, 1987b), bovine (Raju and Anwar, 1987; Yeh *et al.*, 1987), chick (Bressan *et al.*, 1987; Tokimitsu *et al.*, 1987), and rat (Deak *et al.*, 1988) tropoelastins have been determined. The deduced maximum size of bovine tropoelastin (760 residues) is consistent with previous estimates of molecular weight, corresponding to a coding region of approximately 2.5 kb in an overall elastin mRNA size of 3.5 kb. The 3′ end of the elastin message has been found to contain a large (1.0–1.2 kb) untranslated region, which appears to be highly conserved in all species (Rosenbloom, 1987). This region is completely contained within a single exon in both the bovine and human genes (Indik *et al.*, 1987b). It is interesting to note that a 258-bp segment in the 3′-untranslated region of lysyl oxidase is 93% identical with the 3′-untranslated region of elastin (Trackman *et al.*, 1990).

The cDNA confirmed the prediction that the primary structure of tropoelastin consists of alternating hydrophobic and cross-linking domains (Fig. 3-4). In the cross-linking regions, the lysine residues usually occur in pairs in cluster of alanine residues, although in two regions near the center of the molecule three lysine residues are found. In addition, potential cross-linking sequences in the first 200 residues frequently contain prolyl or other residues between the lysines instead of the usual alanines (Indik *et al.*, 1990). Overall, there is extensive homology between mammalian tropoelastins with most substitutions being conservative in nature. There are, however, some interesting differences. For example, the number of amino acids that may be found in a particular hydrophobic region may differ between species, suggesting that hydrophobic domains tolerate changes in either the number or sequence of amino acids. Chicken tropoelastin shows extensive homology with mammalian sequences for the first 300 residues and for the last 57 residues. Major differences exist in the central portion, however, with sequences that suggest duplication and deletion events (Bressan *et al.*, 1987; Indik *et al.*, 1990; Tokimitsu *et al.*, 1987).

In contrast to the ability of tropoelastin to tolerate sequence variability in the hydrophobic domains, there appears to be absolute conservation, in elastins from all species, in the number of alanine residues between lysines in the alanine-rich cross-linking domains. In these sequences, two lysines are always separated by two or three alanines, never only one or more than three. The reason for this is obvious when the limitations of secondary structure and requirements for cross-linking are considered (Fig. 3-5). The conformation of the alanine-rich cross-linking domains is essentially α-helical, forming a rodlike structure with side chains extended outward in a helical array. In an α-helix, each residue is related to the next one by a translation of 1.5 Å along the helix axis and a rotation of 100°, which gives 3.6 amino acid residues per turn of helix. Thus, the side chains of amino acids separated by two or three residues in the linear sequence are spatially close to one another on the same side of the helix whereas side chains of amino acids separated by one or four residues are situated on opposite sides of the helix and are unlikely to make contact. Gray (1977) has proposed that the first step in desmosine formation is

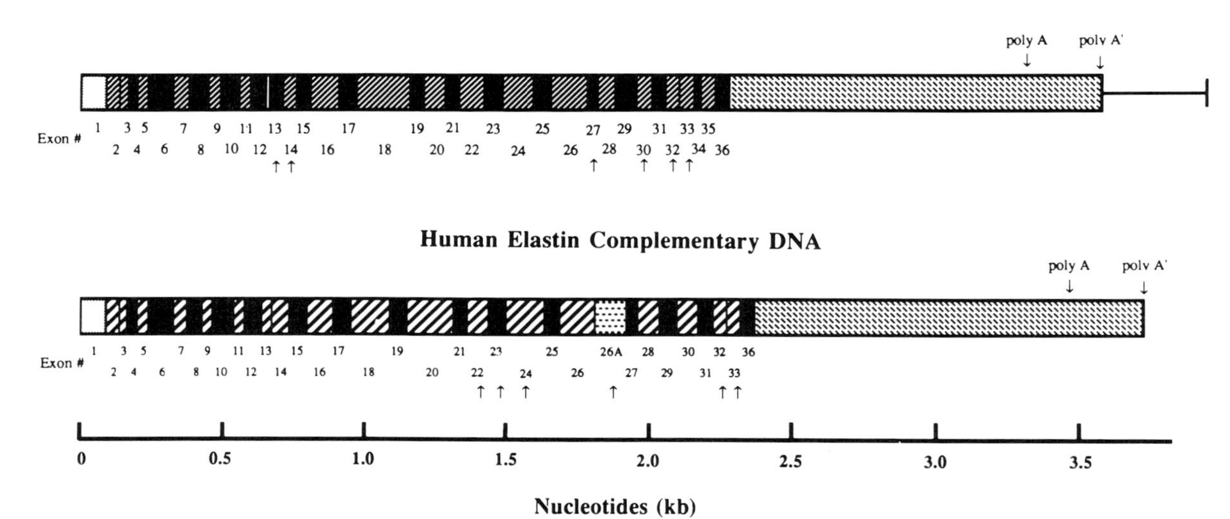

Figure 3-4. Schematic diagram of cDNAs for human and bovine tropoelastin. The cDNAs are divided into exons which are numbered. ▨, hydrophobic sequences; ■, potential cross-linking sequences; □, signal sequence; ▤, 3′-untranslated region. Arrows mark the exons subject to alternative splicing. (From Indik *et al.*, 1990.)

Present in Elastin

1 2 3 4 5

...Lys-Ala-Ala-Lys...

...Lys-Ala-Ala-Ala-Lys...

Never Found in Elastin

1 2 3 4 5 6

...Lys-Ala-Lys...

...Lys-Ala-Ala-Ala-Ala-Lys...

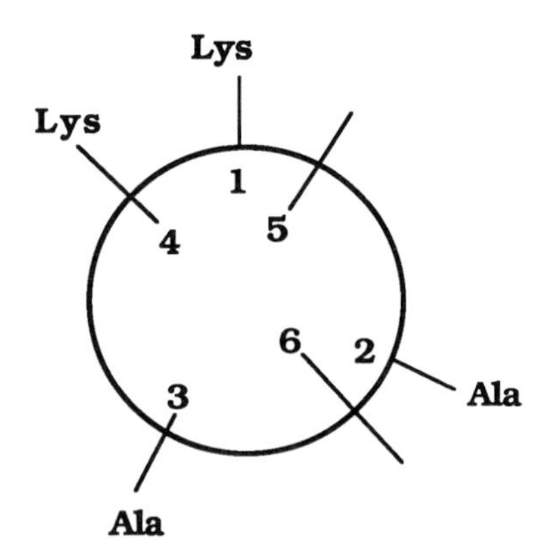

Figure 3-5. Cross-sectional view of the cross-linking region of elastin drawn as an α-helix. The polypeptide main chain forms the inner part of the helix with side chains extending outward in a helical array. When the sequences found in elastin are substituted into the side chain positions, beginning with position 1, it can be seen that lysine side chains in the 4 or 5 position are close to the side chain of lysine 1, able to form the "within chain" cross-link that is the precursor to the tetrafunctional desmosine cross-link. Lysines in positions 3 and 6 are on the opposite side of the helix from lysine 1 and, hence, cannot interact.

the formation of bifunctional "within chain" cross-links which then condense "between chains" to form the tetrafunctional desmosines. If the elastin sequences are considered in this context, the lysine side chains in the sequences . . . K-A-A-K . . . and K-A-A-A-K . . . are positioned on the same side of the helix where they can interact one with another following oxidation by lysyl oxidase. Separation by one or four alanines (. . . K-A-K . . . or . . . K-A-A-A-A-K . . .) would position the side chains on opposite sides of the helix too far apart for subsequent condensation reactions.

Another particularly interesting finding from cloning studies was the identification of a highly basic, cysteine-containing sequence at the carboxy-terminus of the tropoelastin molecule (Indik *et al.*, 1987b; Yoon *et al.*, 1985). This 58-residue sequence, which had not been observed in earlier sequencing studies of porcine tropoelastin peptides, appears to be highly conserved in the spliced elastin RNA molecule and is atypical of the amino acid composition of the rest of the protein. The function of this region is unknown, but it has been speculated that it may play a role in assembly of the elastic fiber (Wrenn *et al.*, 1987), perhaps by acting as a mediator to complex tropoelastin with the microfibrillar framework.

5. Tropoelastin Biosynthesis and Secretion

Tropoelastin is synthesized on the rough endoplasmic reticulum and appears to undergo little intracellular posttranslational modification. Although some hydroxylation of proline residues occurs, the functional significance of the presence of hydroxyproline in elastin is uncertain. In contrast to collagen, posttranslational hydroxylation of proline residues appears to be unimportant in elastin fibrillogenesis (Kao *et al.*, 1982; Rosenbloom and Cywinsky, 1976; Uitto *et al.*, 1976). It has been suggested that tropoelastin may be "accidentally" hydroxylated by collagen prolyl hydroxylase as a consequence of the concomitant synthesis of procollagen by the cell (Rosenbloom, 1982). There is evidence, however, that overhydroxylation can compromise the ability of tropoelastin to form mature fibers (Barone *et al.*, 1985).

As will be discussed in more detail below, elastin fiber assembly occurs at distinct sites on the cell membrane, implying that tropoelastin is not secreted from all surfaces of the cell but is specifically targeted to specific areas. The processes involved in intracellular segregation and secretion of proteins are complex and poorly characterized. Specific sugar residues have been implicated in the segregation of lysosomal proteins and specific amino acid sequences have been shown to direct the retention of proteins in the endoplasmic reticulum. Lacking glycosylation, tropoelastin is unable to utilize sugar residues as an intracellular trafficking signal. Nor does there appear to be intrinsic transport information contained within the tropoelastin amino acid sequence even though the carboxy-terminal domain has structural similarities to "stop translocation" motifs that have been implicated as trafficking signals for several membrane-associated proteins (Grosso and Mecham, 1988). There is increasing evidence, however, that tropoelastin is carried through the secretory pathway by its own receptor which acts as a transporter and matrix assembly protein.

6. Elastic Fiber Assembly

In the extracellular space, tropoelastin appears to be directly and rapidly accreted to the surface of elastic fibers in a monomeric form; there is no direct evidence that the molecule undergoes cleavage prior to cross-linking into the developing fiber. Under normal conditions, the cross-linking reaction is so efficient that no pool of soluble elastin can be detected in the extracellular matrix. Cross-linking of elastin is initiated by the action of lysyl oxidase which catalyzes the oxidative deamination of lysine to allysine. This appears to be the only enzymatic step involved in elastin cross-linking. Subsequent formation of elastin cross-links (including the tetrafunctional amino acid isomers desmosine and isodesmosine) probably occurs as a series of spontaneous condensation reactions (Paz *et al.*, 1982).

Elastin fibrillogenesis takes place close to the cell membrane, generally situated in infoldings of the cell surface (Serafini-Fracassini, 1984). As discussed above, microfibrils are the first visible components of elastic fibers and are found grouped in small bundles near the plasma membrane. As the fiber develops, elastin appears as an amorphous material in discrete loci within each microfibrillar bundle. These amorphous areas gradually coalesce and generate the central core of elastin. The majority of microfibrils are progressively displaced to the outer aspect of the fiber, a position they retain in the mature tissue.

Understanding the supramolecular organization of mature elastic fibers has been a difficult problem. When stained with uranyl acetate and lead citrate, elastic fiber microfibrils become electron-dense whereas the elastin component remains completely electron-transparent and has an amorphous appearance. Suggestions of a fibrillar substructure were obtained by negative staining of chemically purified elastic fibers, α-elastin, and coacervated tropoelastin where filaments 5 nm in diameter were observed. Optical diffraction as well as low-angle X-ray scattering found evidence for similarly sized filaments. Filaments of somewhat larger diameter (6–10 nm) were observed with freeze-fracture images of ligamentum nuchae (Morocutti *et al.*, 1988; Pasquali-Ronchetti and Fornieri, 1984).

A more detailed picture of elastin structure has been recently obtained using quick-freeze, deep-etch electron microscopy. This technique obviates the need for chemical fixation, dehydration, or staining, and hence minimizes the possibility of structural alterations induced by these treatments. The method also provides three-dimensional surface views of macromolecular structure that enable a more accurate reconstruction of matrix organization (Heuser, 1981; Mecham and Heuser, 1990). Figure 3-6 shows a freeze-etched view of fetal bovine ear cartilage in which a portion of a chondrocyte and its surrounding extracellular matrix are displayed. The plane of fracture passed through the cell, revealing a densely packed cytoplasm containing numerous microtubules and microfilaments. Also apparent are the plasma membrane, the nuclear envelope, and other intracellular membranes. When the cell–matrix interface is examined at higher magnification (Fig. 3-7), the "structural and functional continuum" formed between the cell interior, the cell membrane, and the molecules of the matrix referred to in the Introduction to the first edition of this book (Hay, 1981) is clear. The plasma membrane is embedded in a meshwork on both sides. On the cytoplasmic side, many fine fibrils radiate orthogonally from the surface of the membrane and appear to attach to 25-nm microtubules and 9-nm actin filaments deeper in the cytoplasm. On the extracellular side, matrix fibrils attach to the exterior face of the plasma membrane.

Closer inspection of the proteoglycan-rich elastic ear cartilage matrix in regions away from the cell membrane revealed a tightly woven network of fibrils with extensive lateral associations that formed a trabeculum or honeycomblike lattice (Fig. 3-8). The matrix fibrils appear to branch and anastomose at irregular intervals and often have a granular appearance along their surface. Within the matrix was also found an extensive array of type II collagen filaments of approximately 20-nm diameter that traverse large distances without branching and have few lateral associations with similar fibers. In contrast to the uniform diameter exhibited by the collagen fibers, elastin fibers appeared to have large, irregular shapes and to be of variable diameter and length.

What is immediately evident from the micrograph in Fig. 3-8 is the clear structural definition of the elastin component (especially when compared with the image of elastin in Fig. 3-2 obtained by transmission electron microscopy). The major structural feature is a densely packed, randomly arranged network of fine filaments. The filamentous substructure of elastin is again evident in Fig. 3-9, which shows an elastin fiber close to the plasma membrane of a

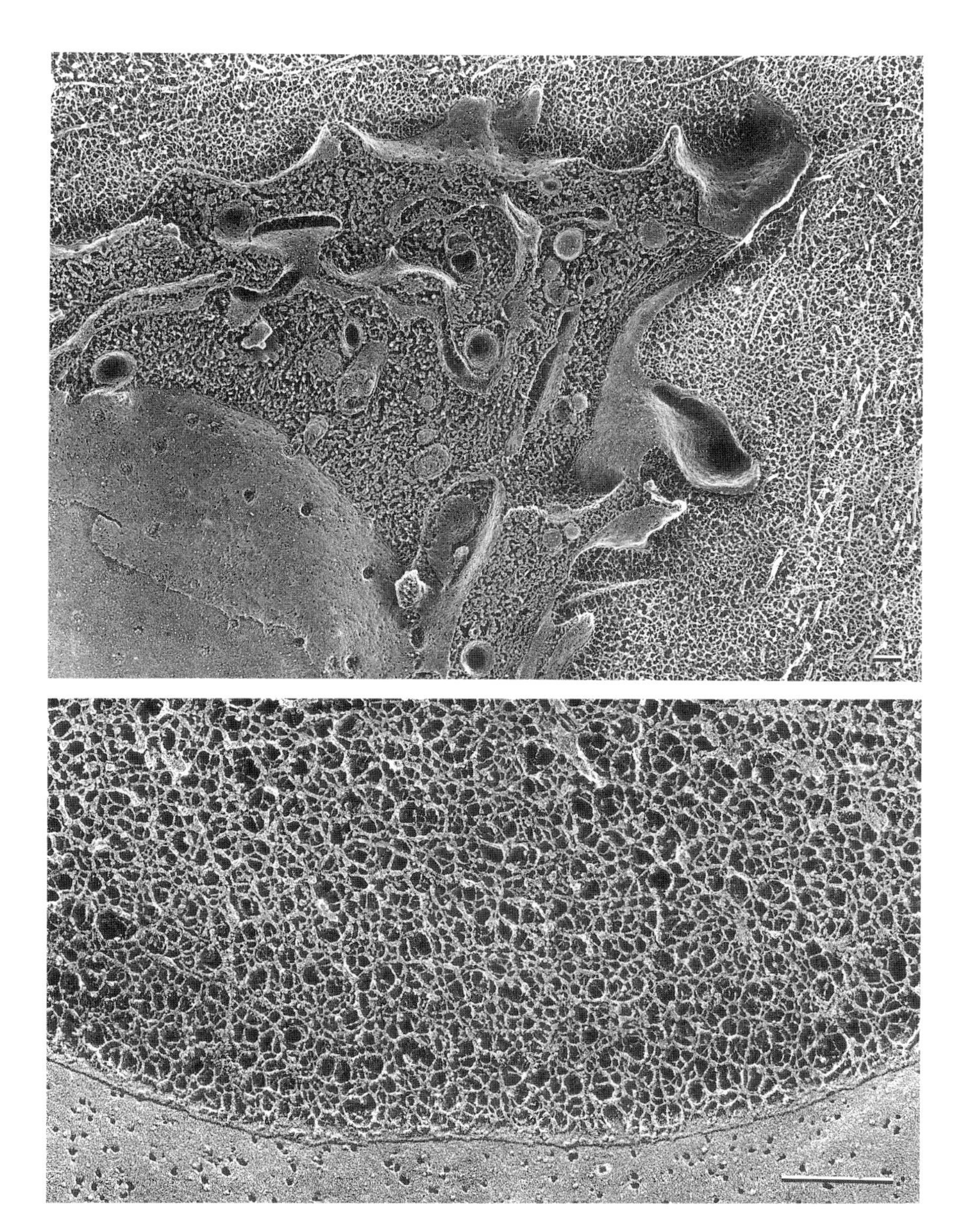

Figure 3-6. (Upper) Freeze-etch view of 150-day gestation fetal bovine ear cartilage quick-frozen without fixation and freeze-fractured before etching. Shown is a chondroblast with surrounding extracellular matrix. The plasma membrane of the cell is highly convoluted with numerous projections and infoldings. Nuclear pores are evident on the nuclear membrane which is continuous with the cytoplasmic membrane system. The extracellular matrix consists of a finely woven meshwork that adjoins the cell membrane and is continuous throughout the extracellular space. (Lower) At higher magnification, the woven network of the cartilaginous matrix immediately adjacent to the cell surface can be seen. The number of interactions between the cell and extracellular matrix can also be appreciated. Bars = 0.2 μm. (From Mecham and Heuser, 1990, with permission.)

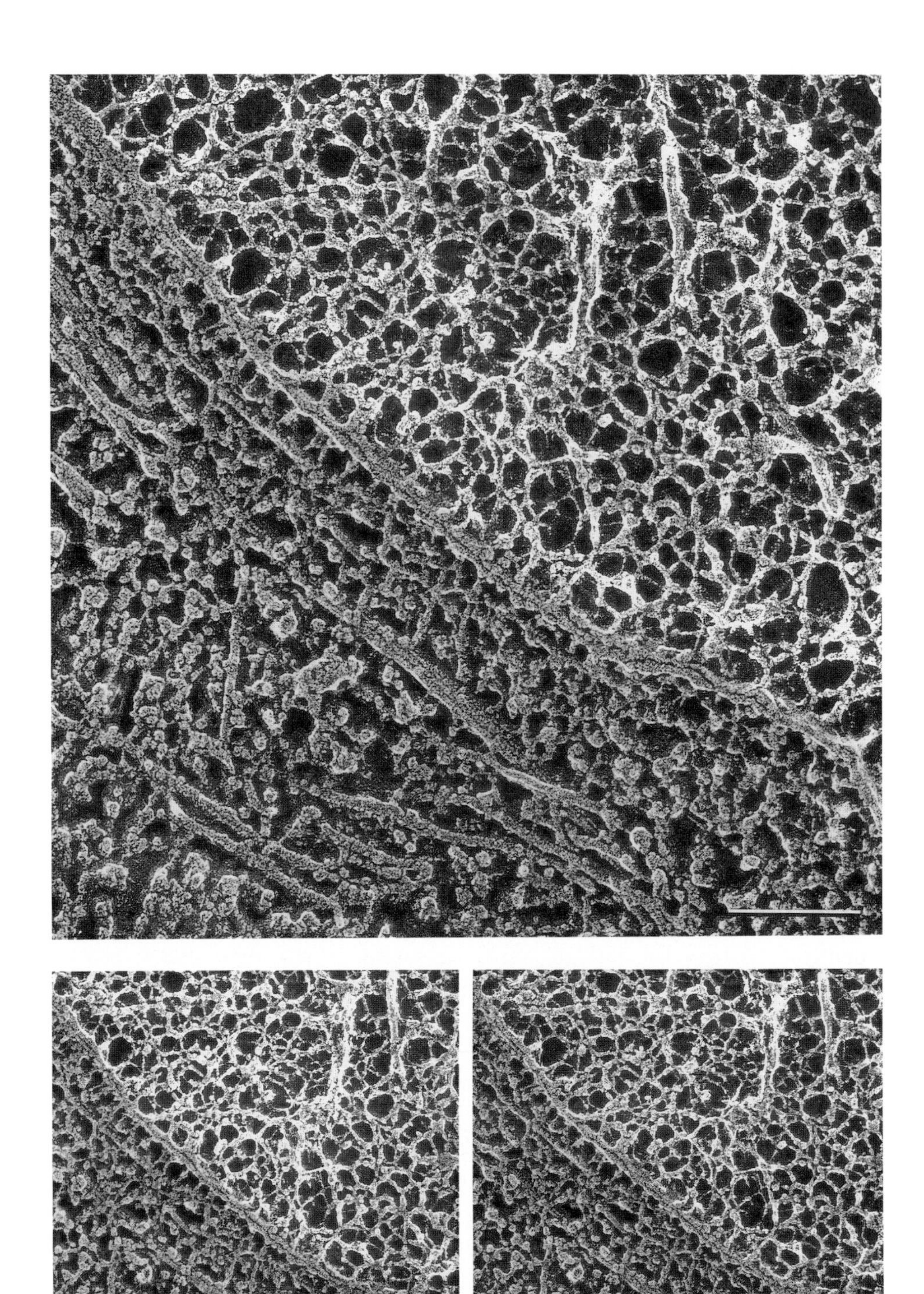

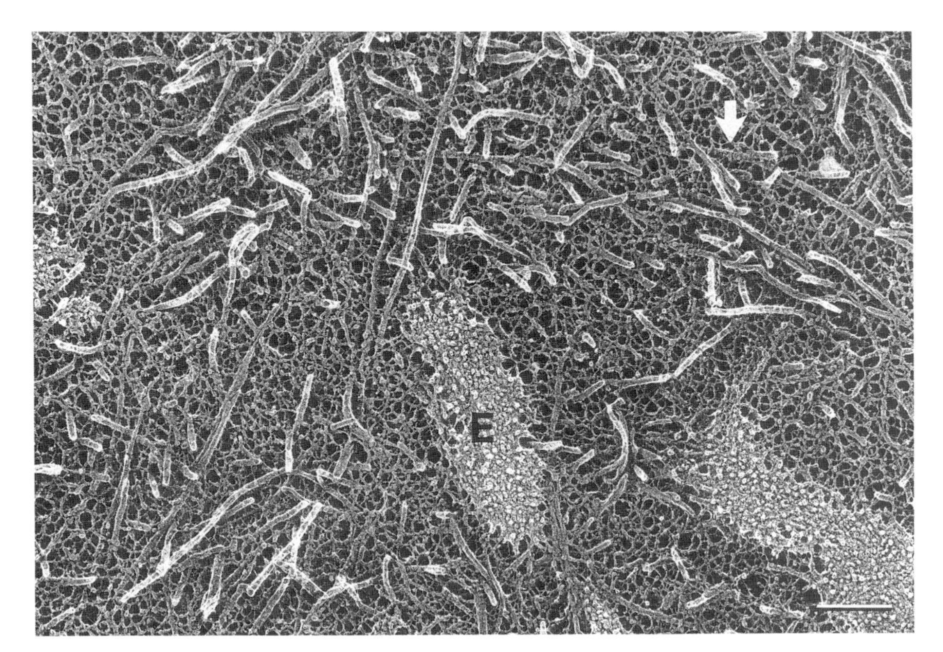

Figure 3-8. Collagen and elastin (E) fibers in the extracellular matrix of ear cartilage. The arrow (upper right) indicates lateral attachments of cartilage matrix to collagen filaments that dip below the plane of fracture. Bar = 0.2 μm. (From Mecham and Heuser, 1990, with permission.)

chondroblast. When viewed in stereo, fine filaments having a mean diameter of 7 nm (9 nm minus 2 nm for shadowing) an be seen protruding from the sectioned fiber.

The disordered arrangement of microfilaments comprising the elastin component of an unstretched elastic fiber teased out of adult bovine ligamentum nuchae is evident in Fig. 3-10. In these sections, the plane of fracture has exposed filaments with an approximate diameter of 7 nm within the core of the elastin fiber that are so tightly packed that little or no etching occurs during sample preparation. At the bottom of Fig. 3-10 is an image showing a linear orientation to the filamentous substructure where the sectioning knife cut into the mature fiber, producing a stretching effect. The images in Fig. 3-10 suggest that in relaxed elastic fibers, the filaments are randomly arranged and that by stretching they are oriented in the direction of applied force.

The origin of the filamentous substructure in elastin became apparent when purified tropoelastin monomers were visualized using freeze-etching and

Figure 3-7. High magnification of the cell–matrix interface showing attachment of extracellular matrix (upper right) to the external face of the chondroblast plasma membrane (running from the upper left to lower right corners). Numerous filaments are evident within cytoplasm of the cell (lower left), including microfilaments and larger microtubules. The spatial arrangements of the cell–matrix junctions can best be appreciated when viewed in stereo (bottom panels). Bars = 0.2 μm. (From Mecham and Heuser, 1990, with permission.)

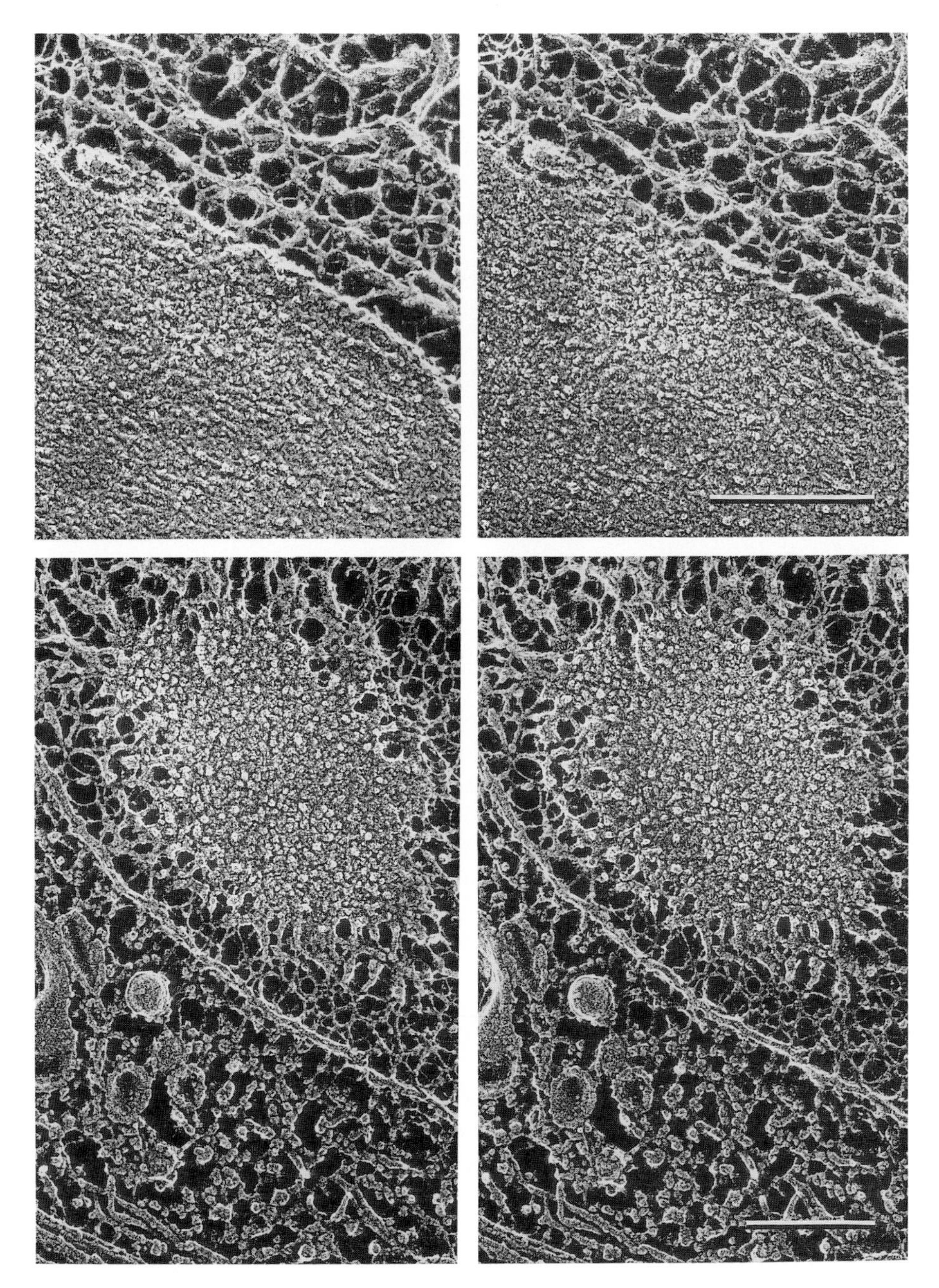

Figure 3-9. (Upper) Freeze-etch view of elastic fiber in mature ligamentum nuchae showing a densely packed, fibrous substructure. (Lower) Fetal bovine auricular cartilage showing a developing elastic fiber next to the surface of a chondroblast. Bars = 0.2 μm.

rotary shadowing. Tropoelastin is a globular protein (Fig. 3-10) having a diameter of 5–7 nm, the same diameter as filaments in the mature fiber. This finding suggest that tropoelastin molecules join one to another, in three dimensions, to form the network visible in the freeze-etch images. To explain these images, we propose the analogy of filling a cylinder with rubber balls and covalently join the balls with each other wherever they touch (presumably on several different surfaces). When the cylinder is removed, one would visualize a uniform network of joined spheres of similar diameter in three dimensions (Fig. 3-11). If the joined balls could then be sectioned and the split surfaces observed, because of the spherical nature of the balls one would expect to see the same image for every plane of section. If the rubberlike properties of the balls result in tearing during sectioning instead of a clean planar break, then filamentous structures (produced by shadowing ragged ends) would be evident along the surface. This, we feel explains the origin of the filaments within the elastin core observed in the freeze-etch images.

The process of elastin assembly along microfibrils is illustrated in Fig. 3-12. At early gestational ages, microfibrils with a mean diameter of 10–12 nm are evident in the extracellular matrix close to the cell surface. With increasing fetal age, the microfibrils first become coated or encrusted with globular structures of about 7-nm diameter and then become imbedded in the dense, growing network that is characteristic of mature, insoluble elastin. The microfibrils seem to be excluded to the periphery of the growing insoluble elastin component as the tissue matures.

7. Elastin Receptors

The first evidence for elastin receptors on mammalian cells came from studies documenting a chemotactic response by monocytes and fibroblasts to tropoelastin and peptide fragments of insoluble elastin (Senior *et al.*, 1980). Subsequently, three cell surface elastin-binding proteins have been described: a protein of 120 kDa (elastonectin) whose expression on the surface of smooth muscle cells is inducible by elastin peptides (Hornebeck *et al.*, 1986), a 59-kDa protein found on tumor cells that is coupled to protein kinase C (Blood *et al.*, 1988; Blood and Zetter, 1989), and a 67-kDa protein found on most cell types that bind elastin (Hinek *et al.*, 1988; Mecham *et al.*, 1989a).

Elastin contains no Arg-Gly-Asp (RGD) sequences and there is no evidence that elastin or tropoelastin interacts with integrins. Instead, both the 67- and 59-kDa elastin-binding proteins recognize a hydrophobic hexapeptide, VGVAPG, which repeats several times in the human and bovine molecule. Recent studies suggest that receptor recognition is not sequence-specific but involves a conformational determinant that tolerates several different amino acids as long as the overall hydrophobic nature of the peptide is maintained. The binding properties of the 67-kDa protein are somewhat promiscuous in that laminin (Mecham *et al.*, 1989b) and type IV collagen (Senior *et al.*, 1989) compete with elastin for receptor binding.

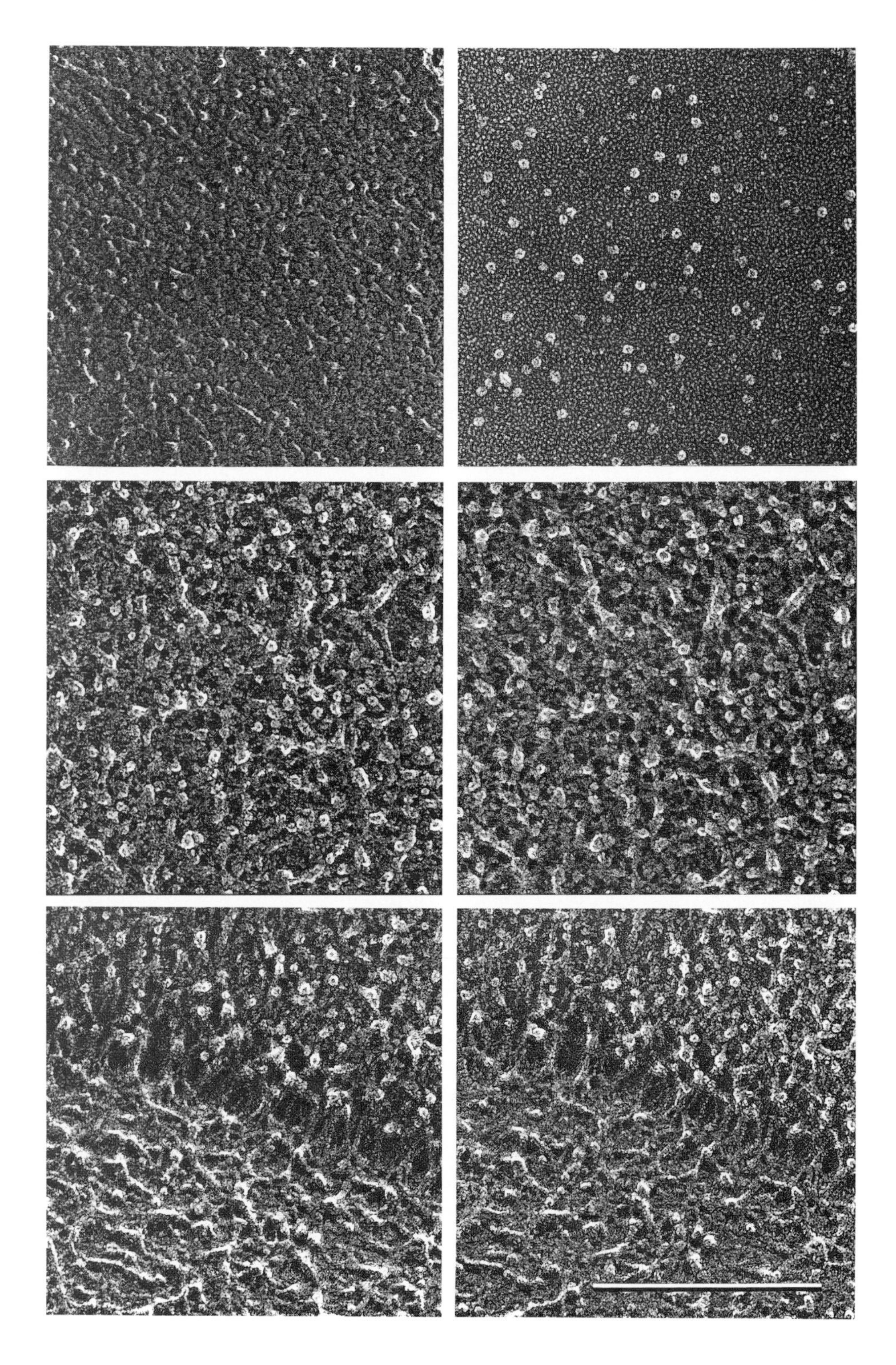

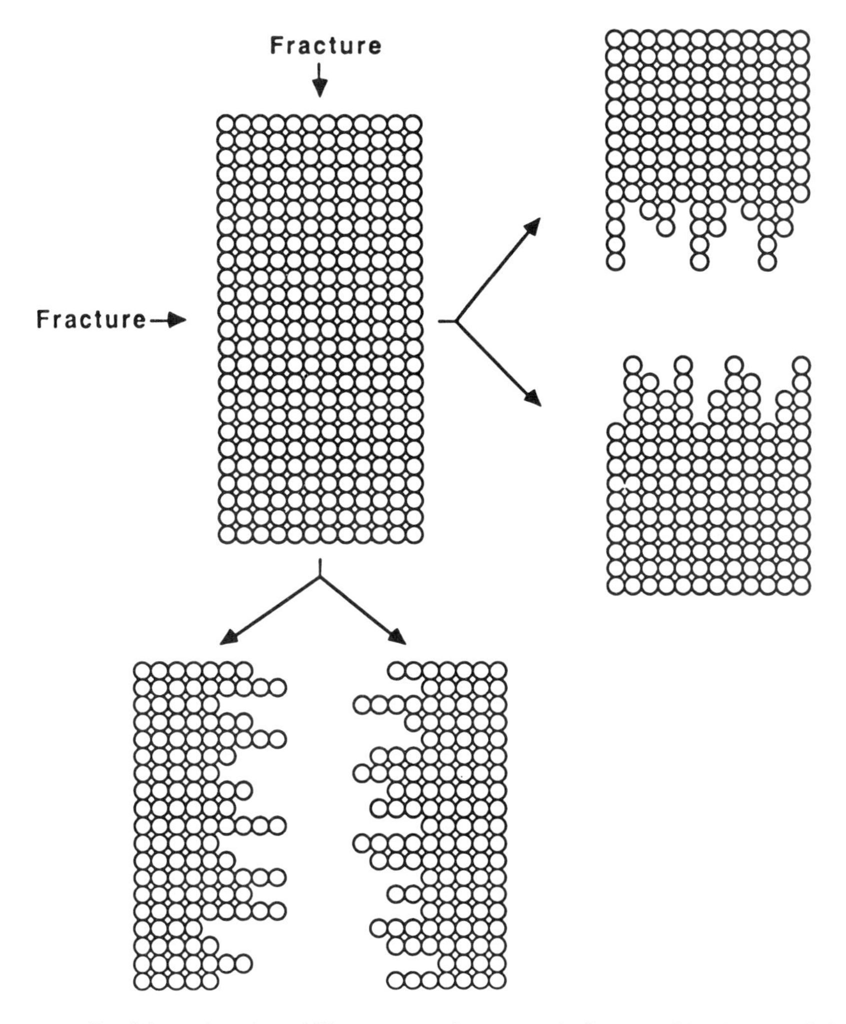

Figure 3-11. Possible explanation of filamentous substructure in freeze-etching images of elastin. Globular tropoelastin molecules are stacked in three dimensions and cross-linked one to another. During sectioning, the rubberlike properties of tropoelastin result in tearing instead of a clean planar break. When the shadowed section is then visualized, the ragged ends show up as filaments with a diameter equal to that of individual tropoelastin molecules. Because of the spherical nature of tropoelastin, one would expect to see the same image for every plane of section.

Figure 3-10. (Top, left) Elastic fiber in bovine ligamentum nuchae showing a filamentous substructure that is so tightly packed that little or no etching occurs. (Top right) Purified tropoelastin from bovine ligamentum nuchae viewed by freeze-etching and rotary shadowing. The diameter of individual tropoelastin molecules is 5–7 nm. (Center) Stereo pair showing the disordered arrangement of elastin microfilaments in an unstretched elastic fiber teased out of adult bovine ligamentum nuchae. The filaments have a diameter of approximately 7 nm. (Bottom) Mature elastic fiber showing a linear orientation of the filamentous elastin component at the bottom of a groove where the sectioning knife cut into the fiber, producing a stretching effect. Bar= 0.2 μm.

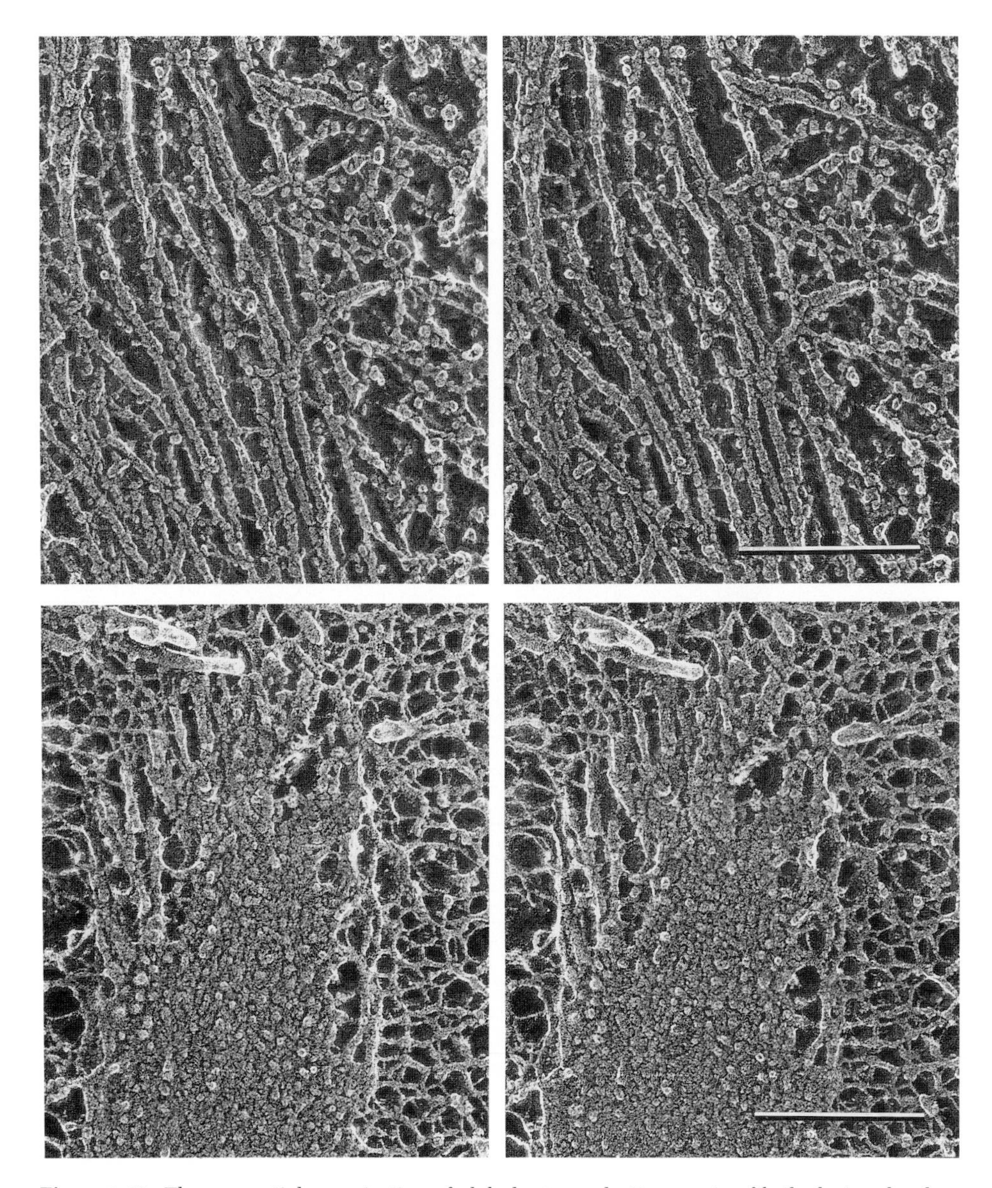

Figure 3-12. The sequential organization of globular tropoelastin on microfibrils during development. Freeze-etched images are from the ligamentum nuchae of 120- and 200-day gestation bovine fetuses. Bars = 0.2 μm.

Recent studies have suggested that several pathogenic bacteria specifically bind to elastin. Interestingly, a prominent bacterial binding domain has been mapped to a 30-kDa region at the amino end of the elastin molecule (Park *et al.*, 1991), obviously different from the mammalian receptor binding region which is located about one-third the distance in from the carboxy-terminus. Agonists known to perturb elastin interaction with the mammalian receptor do not influence bacterial binding. Many bacteria, including *Staphylococcus aureus*, are known to infect and colonize elastin-rich organs like lung, skin, and blood vessels. Because invasive bacteria have a high probability of interacting with the elastin matrix of the vascular wall, the existence of a binding protein for elastin on the bacterial surface could facilitate bacterial invasion in addition to adherence during colonization.

The 67-kDa Elastin-Binding Protein: Receptor or Matrix Assembly Protein?

The 67-kDa elastin-binding protein has properties of a peripheral and not an integral membrane protein. Biochemical studies have shown that the receptor is a galactoside-binding protein that has immunological similarity with a family of low-molecular-weight galactoside lectins (Barondes, 1988; Hinek *et al.*, 1988). A striking feature of this receptor is that its affinity for elastin is greatly influenced by its lectin domain (Fig. 3-13). In the absence of carbohydrate, the receptor binds to elastin with high affinity whereas in the presence of galactoside sugar, little, if any, binding occurs (Hinek *et al.*, 1988). Using video-enhanced microscopy to study the binding of elastin-coated beads to the surface of cells, it was possible to show that binding of carbohydrate lowers receptor affinity for elastin by increasing the dissociation constant at the protein binding site (Mecham *et al.*, 1991).

With multiple binding domains, the 67-kDa protein is ideally suited to direct the specific association between tropoelastin and highly glycosylated microfibrillar components. Immunolocalization studies with elastin-producing cells document that tropoelastin and the 67-kDa protein colocalize intracellularly and suggest that tropoelastin is bound to the 67-kDa protein in the secretory pathway. These findings raise the interesting possibility that the 67-kDa protein provides trafficking signals that direct the proper movement of the receptor–tropoelastin complex to sites of elastin fiber formation on the cell surface. At the plasma membrane, tropoelastin remains bound to the 67-kDa protein until an interaction with a microfibril-associated galactoside sugar induces the transfer of tropoelastin to an acceptor site on the microfibril (Fig. 3-13). Supporting a matrix-assembly function for the 67-kDa protein are studies showing inhibition of elastin fiber assembly by the addition of lactose or galactose sugars to the culture medium of elastin-producing cells (Hinek *et al.*, 1988). As explained by the model in Fig. 3-13, lactose competes with glycoconjugates on microfibrils for the sugar binding site on the receptor and causes premature release of tropoelastin. Biochemical studies showed that in the presence of lactose the majority of newly synthesized tropoelastin was released directly into the medium with only low levels remaining associated with the cell layer. Electron microscopy confirmed the inhibitory effects of lactose on fiber assembly and showed that the small amounts of elastin associated with the cell layer accumulated in the matrix as small, irregularly distributed

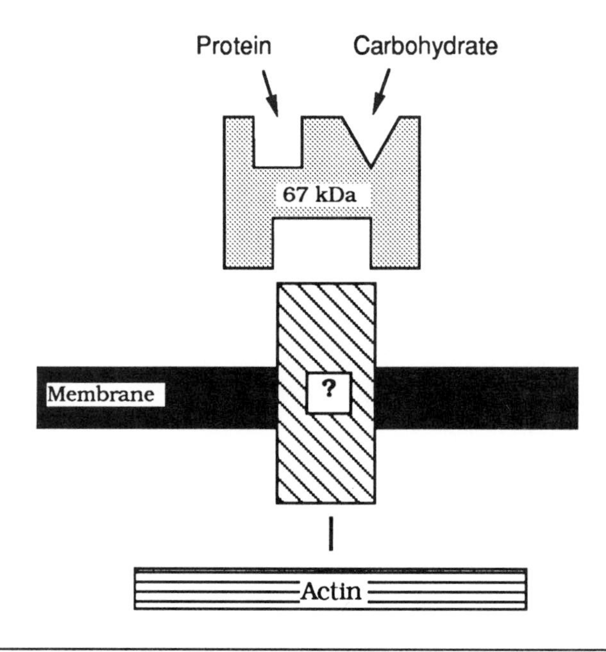

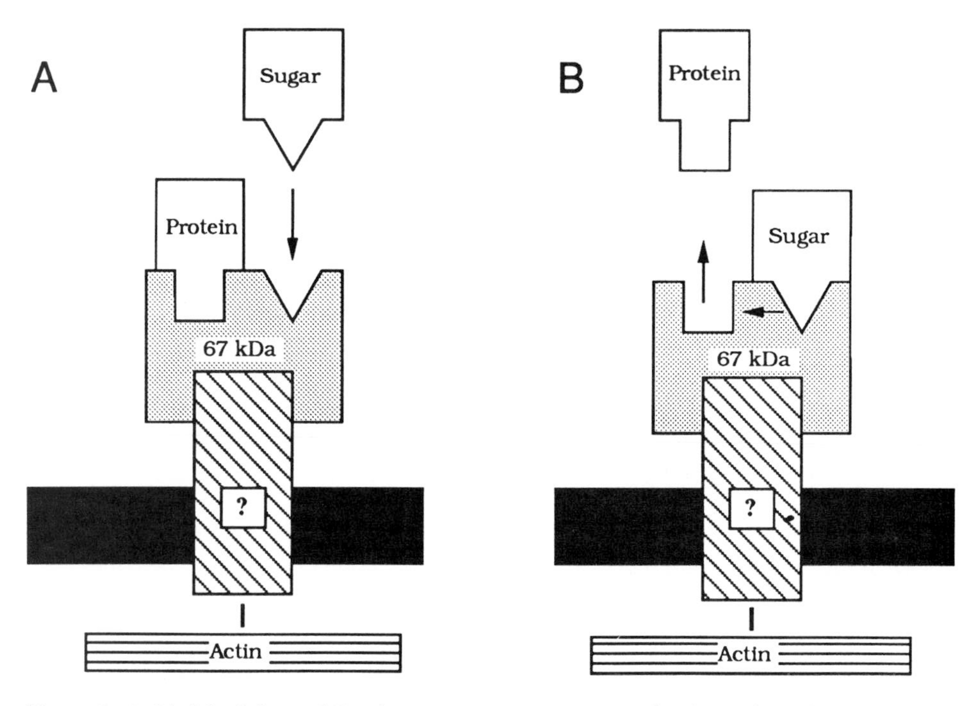

Figure 3-13. Model of the 67-kDa elastin receptor. (Upper panel) The 67-kDa elastin receptor contains a protein binding site that recognizes hydrophobic sequences in elastin, a carbohydrate binding site that interacts with galactoside sugars, and a membrane binding site that mediates attachment to the plasma membrane. (Lower panel) Tropoelastin binds with high affinity to the receptor at the protein binding site. Binding of carbohydrate at the sugar binding site, however, lowers the affinity at the protein binding site resulting in the release of bound protein.

globules that did not coalesce to form the large amorphous structures typical of mature fibers. The inhibitory effect was specific for galactose-containing sugars since fiber formation was not altered by fucose, mannose, or glucose.

By facilitating the proper association of tropoelastin and microfibrils, the 67-kDa protein fits the definition of a "molecular chaperone" which is a protein whose function is to ensure the folding and assembly of other proteins into oligomeric structure (Ellis, 1987). In addition to providing transport signals and facilitating fiber assembly, another function of the receptor may be to prevent intracellular coacervation of tropoelastin. Because of its extreme hydrophobicity, tropoelastin in solution tends to undergo a phase separation (coacervation) under physiological conditions. While this may be an important aspect of fiber assembly in the extracellular space, aggregation within the secretory pathway could be detrimental to the cell. As a transporter protein, the receptor would prevent coacervation and ensure that aggregation does not occur until tropoelastin is at the correct site on the cell surface.

The distinction between a "receptor" and "binding protein" is sometimes difficult, but implicit in the definition of a receptor is a functional involvement in signal transduction. There is, in fact, accumulating evidence that binding of elastin to the cell surface initiates a biological response within the cell. For example, extracellular elastin peptides have been reported to induce phosphatidylinositol breakdown with mobilization of intracellular calcium, to alter protein kinase C activity, and to provide chemotactic signals to several cell types (Blood and Zetter, 1989; Jacob *et al.*, 1987; Senior *et al.*, 1984; Varga *et al.*, 1990). Whether all of the elastin-binding proteins are capable of generating intracellular signals is unknown, as is how the signals influence cellular metabolism.

8. The Elastin Gene

Delineating the organization and control of the elastin gene is providing important new insights into regulation of the elastin phenotype. Most evidence currently favors the presence of a single elastin gene (Davidson and Giro, 1986; Indik *et al.*, 1987a; Olliver *et al.*, 1987) with a remarkably high intron-to-coding-exon ratio of about 20:1. This degree of dilution is unusual even in comparison to other extracellular matrix proteins such as the fibrillar collagens which have a ratio of 8:1. Maintenance of such a degree of dilution throughout the elastin gene would suggest an overall size greater than 45 kb, coding for an mRNA of only 3.5 kb (Fazio *et al.*, 1990; Indik *et al.*, 1990; Kähäri *et al.*, 1990).

The coding sequences for cross-linking and hydrophobic domains of the protein are encoded in the gene by separate exons, so that the domain structure of the protein is a reflection of the exon organization. The gene is characterized by small exons (27–69 bp) interspersed between long intron sequences. Exons are dissimilar in size with glycine being found consistently at the exon–intron junctions. During splicing, the second and third nucleotides of a codon at the 5′ border of an exon are included while the first nucleotide of the new codon is contributed by the 3′ border of the previous exon. This motif permits extensive alternative splicing of the primary transcript in a cassettelike fashion while maintaining the correct reading frame.

The promoter region of the elastin gene contains several segments that may have important functional roles. There is no canonical TATA box and the promoter is G + C rich (66%) with a high frequency of CpG dinucleotides. Three major transcription initiation sites have been identified, in agreement with the absence of a TATA box. Also included in the promoter region are multiple SP1 and AP2 binding sites, glucocorticoid-responsive elements, and TPA- and cyclic AMP (CRE)-responsive elements. Both up- and down-regulatory elements have been identified within the 5′ flanking region by functional analysis using promoter/chloramphenicol acetyl transferase gene chimeric constructs (Fazio *et al.*, 1990; Kähäri *et al.*, 1990).

Although the factors controlling transcription of the elastin gene are unknown, it is evident that a great deal of posttranscriptional splicing is necessary to produce functional mRNA. Recent evidence suggests that considerable alternative splicing results in the deletion of either complete exons or portions thereof (Indik *et al.*, 1987a; Raju and Anwar, 1987; Yeh *et al.*, 1989) in a manner analogous to that seen in the biosynthetic pathways of many other proteins, including the matrix protein fibronectin. Translation of such elastin mRNA molecules would result in significant variation in amino acid sequence. In fact, multiple tropoelastin isoforms have been identified in several animal tissues (Burnett and Rosenbloom, 1979; Foster *et al.*, 1980; Rich and Foster, 1987). In bovine tissues, at least three distinct protein isoforms have been detected, the proportions of which appear to vary in a congruent manner according to the stage of development (Parks *et al.*, 1988; Wrenn *et al.*, 1987). These observations suggest that microheterogeneity of tropoelastin molecules may be genetically "programmed" by the cell during the laying down of elastic connective tissues as a necessary part of elastic fiber assembly (Parks and Deak, 1990).

9. Degradation and Turnover of Elastin

Physiological turnover of insoluble elastin appears to be very slow, with a half-life approaching that of the animal (Lefevre and Rucker, 1980; Shapiro *et al.*, 1991). Thus, under normal conditions, very little remodeling of elastic fibers occur in adult life. It has been shown that some elastin-synthesizing cells, such as human skin fibroblasts and rat aortic smooth muscle cells, possess small amounts of elastase activity. While this activity is very low in comparison to that found in inflammatory cells, it is of potential importance in allowing cell migration in the developing extracellular matrix and in modulating morphogenesis of newly synthesized elastic fibers.

Although tropoelastin has been shown to be sensitive to proteolytic degradation before cross-linking into insoluble elastin, there is no evidence that significant intracellular degradation occurs prior to secretion. Studies in cultured cells suggest that only about 1% of newly synthesized elastin is degraded per hour, in contrast to the situation observed in collagen synthesis where a larger percentage of newly synthesized interstitial collagen undergoes intracellular catabolism.

Increased elastin destruction does take place in certain pathological conditions, either as a result of the release of powerful elastinolytic serine proteases (elastases) from inflammatory cells or bacteria, or because of a genetic deficiency in the naturally occurring elastase inhibitor α1-antitrypsin. Although inflammatory cell-derived elastases are unlikely to play a significant role in normal regulation of the elastic matrix, it is clear that they contribute substantially to the pathogenesis of some human diseases, including pulmonary emphysema, pollutional lung disease, atherosclerosis, and rheumatoid arthritis. Neutrophil and macrophage elastases are of particular importance in this regard, and have been the subject of much study. Of significant interest is the recent finding that human 92-kDa and 71-kDa type IV collagenases can degrade elastin (Senior *et al.*, 1991). These results extend the spectrum of human proteinases with elastolytic activity to metalloproteinases and suggest a role for these enzymes in the pathogenesis of diseases involving damage to elastic fibers, such as pulmonary emphysema. Many mesenchymal cell types, including fibroblasts and smooth muscle cells, produce 72-kDa type IV collagenase as a major gene product (Wilhelm *et al.*, 1989). The 92-kDa type IV collagenase is an abundant secretory product of human alveolar macrophages (Hibbs *et al.*, 1987; Welgus *et al.*, 1990) and some metastatic tumor cell lines (Khokha and Denhardt, 1989; Liotta *et al.*, 1988). A full discussion of elastases and elastin turnover can be found in a review by Bieth (Bieth, 1986).

10. Concluding Remarks

Our understanding of elastin structure and elastic fiber assembly is expanding rapidly due in large part to advances in cell and molecular biology. The recent delineation of the elastin gene, and the concomitant development of appropriate probes for gene expression, have provided precise and powerful tools with which to dissect the control of elastogenesis at a molecular level. In addition, steady progress has been made in developing *in vitro* models to study the interaction of elastin-producing cells with environmental influences. Such models have offered insights into the mechanisms by which the cell interacts with, and is influenced by, the extracellular matrix. The future thus holds considerable promise for rapid advances in our understanding of normal elastin biosynthesis and its pathological alterations. Many important and interesting questions remain, however, such as determining the structure and composition of microfibrils and resolving the pathway of tropoelastin secretion and fiber assembly.

Acknowledgment. The original research described in this review was supported in part by NIH research grants HL-26499 and HL-41926 to R.P.M. and GM-29647 to J.E.H.

References

Barondes, S. H., 1988, Bifunctional properties of lectins: Lectins redefined, *Trends Biochem. Sci.* **13:**480–482.

Barone, L. M., Faris, B., Chipman, S. D., Toselli, P., Oakes, B. W., and Franzblau, C., 1985, Alteration of the extracellular matrix of smooth muscle cells by ascorbate treatment, *Biochim. Biophys. Acta* **840:**245–254.

Bieth, J. G., 1986, Elastases: Catalytic and biological properties, in: *Regulation of Matrix Accumulation* (R. P. Mecham, ed.), pp. 217–320, Academic Press, New York.

Blood, C. H., and Zetter, B. R., 1989, Membrane-bound protein kinase C modulates receptor affinity and chemotactic responsiveness of Lewis lung carcinoma sublines to an elastin-derived peptide, *J. Biol. Chem.* **264:**10614–10620.

Blood, C. H., Sasse, J., Brodt, P., and Zetter, B. R., 1988, Identification of a tumor cell receptor for VGVAPG, an elastin-derived chemotactic peptide, *J. Cell Biol.* **107:**1987–1993.

Bressan, G. M., Argos, P., and Stanley, K. K., 1987, Repeating structures of chick tropoelastin revealed by complementary DNA cloning, *Biochemistry* **26:**1497–1503.

Burnett, W., and Rosenbloom, J., 1979, Isolation and translation of elastin mRNA from chick aorta, *Biochem. Biophys. Res. Commun.* **86:**478–484.

Crissman, R. S., 1987, Comparison of two digestive techniques for preparation of vascular elastic networks for SEM observation, *J. Electron Microsc. Tech.* **6:**335–348.

Davidson, J. M., and Giro, M. G., 1986, Control of elastin synthesis: Molecular and cellular aspects, in: *Regulation of Matrix Accumulation* (R. P. Mecham, ed.), pp. 177–216, Academic Press, New York.

Deak, S. B., Pierce, R. A., Belsky, S. A., Riley, D. J., and Boyd, C. D., 1988, Rat tropoelastin is synthesized from a 3.5 kilobase mRNA, *J. Biol. Chem.* **263:**13504–13507.

Ellis J., 1987, Proteins as molecular chaperones, *Nature* **328:**378–379.

Fazio, M. J., Kähäri, V.-M., Bashir, M. M., Saitta, B., Rosenbloom, J., and Uitto, J., 1990, Regulation of elastin gene expression: Evidence for functional promoter activity in the 5′-flanking region of the human gene, *J. Invest. Dermatol.* **94:**191–196.

Fleming, W. W., Sullivan, C. E., and Torchia, D. A., 1980, Characterization of molecular motions in ^{13}C-labelled elastin ^{13}C-^{1}H magnetic double resonance, *Biopolymers* **19:**597–618.

Foster, J. A., Rich, C. B., Fletcher, S., Karr, S. R., and Przybyla, A., 1980, Translation of chick aortic mRNA. Comparison to elastin synthesis in chick aorta organ culture, *Biochemistry* **19:**857–864.

Franzblau, C., 1971, Elastin, *Compr. Biochem.* **26c:**659–712.

Franzblau, C., and Faris, B., 1981, Elastin, in: *Cell Biology of Extracellular Matrix* (E. D. Hay, ed.), pp. 65–94, Plenum Press, New York.

Gibson, M. A., Kumaratilake, J. S., and Cleary, E. G., 1989, The protein components of the 12-nanometer microfibrils of elastic and non-elastic tissues, *J. Biol. Chem.* **264:**4590–4598.

Gibson, M. A., Sandberg, L. B., Grosso, L. E., and Cleary, E. G., 1990, Complementary DNA cloning establishes microfibril-associated glycoprotein (MAGP) to be a discrete component of the elastin-associated microfibrils, *J. Biol. Chem.* **266:**7596–7601.

Gosline, J. M., 1978, Hydrophobic interaction and a model for the elasticity of elastin, *Bioplymers* **17:**677–695.

Gosline, J. M., and Rosenbloom, J., 1984, Elastin, in: *Extracellular Matrix Biochemistry* (K. A. Piez and A. H. Reddi, eds.), pp. 191–228, Elsevier, Amsterdam.

Gray, W. R., 1977, Some kinetic aspects of crosslink biosynthesis, *Adv. Exp. Med. Biol.* **79:**285–290.

Gray, W. R., Sandberg, L. B., and Foster, J. A., 1973, Molecular model for elastin structure and function, *Nature* **246:**461–466.

Grosso, L., and Mecham, R. P., 1988, In vitro processing of tropoelastin: Investigation of a possible transport function associated with the carboxy-terminal domain, *Biochem. Biophys. Res. Commun.* **151:**822–826.

Hay, E. D., 1981, Introductory remarks, in: *Cell Biology of Extracellular Matrix* (E. D. Hay, ed.), pp. 1–4, Plenum Press, New York.

Heuser, J. E., 1981, Preparing biological samples for stereomicroscopy by the quick-freeze, deep-etch, rotary-replication technique, *Methods Cell Biol.* **22:**97–122.

Hibbs, M. S., Hoidal, J. R., and Kang, A. H., 1987, Expression of a metalloproteinase that degrades native type V collagen and denatured collagens by cultured human alveolar macrophages, *J. Clin. Invest.* **80:**1644–1650.

Hinek, A., Wrenn, D. S., Mecham, R. P., and Barondes, S. H., 1988, The elastin receptor is a galactoside binding protein, *Science* **239:**1539–1541.

Hoeve, C. A. J., 1974, The elastic properties of elastin, *Biopolymers* **13:**677–686.

Hornebeck, W., Tixier, J. M., and Robert, L., 1986, Inducible adhesion of mesenchymal cells to elastic fibers: Elastonectin, *Proc. Natl. Acad. Sci. USA* **83:**5517–5520.

Indik, Z., Yeh, H., Ornstein-Goldstein, N., Sheppard, P., Anderson, N., Rosenbloom, J. C., Peltonen, L., and Rosenbloom, J., 1987a, Alternative splicing of human elastin messenger-RNA indicated by sequence analysis of cloned genomic and complementary DNA, *Proc. Natl. Acad. Sci. USA* **84:**5680–5684.

Indik, Z., Yoon, K., Morrow, S. D., Cicila, B., Rosenbloom, J. C., Rosenbloom, J., and Ornstein-Goldstein, N., 1987b, Structure of the 3′ region of the human elastin gene: Great abundance of Alu repetitive sequences and few coding sequences, *Connect. Tissue Res.* **16:**197–211.

Indik, Z., Yeh, H., Ornstein-Goldstein, N., and Rosenbloom, J., 1990, Structure of the elastin gene and alternative splicing of elastin mRNA, in: *Genes for Extracellular Matrix Proteins* (L. Sandell and C. Boyd, eds.), pp. 221–250, Academic Press, New York.

Jacob, M. P., Fulop, T., Jr., Foris, G., and Robert, L., 1987, Effect of elastin peptides on ion fluxes in mononuclear cells, fibroblasts, and smooth muscle cells, *Proc. Natl. Acad. Sci. USA* **84:**995–999.

Kagan, H. M., Vaccaro, C. A., Bronson, R. E., Tang, S. S., and Brody, J. S., 1986, Ultrastructural immunolocalization of lysyl oxidase in vascular connective tissue, *J. Cell Biol.* **103:**1121–1128.

Kähäri, V.-M., Fazio, M. J., Chen, Y. Q., Bushir, M. M., Rosenbloom, J., and Uitto, U., 1990, Deletion analysis of the 5′-flanking region of the human elastin gene. Delineation of functional promoter and regulatory cis-elements, *J. Biol. Chem.* **265:**9485–9490.

Kao, W. W. Y., Bressan, G. W., and Prockop, D. J., 1982, Kinetics of the incorporation of tropoelastin into elastic fibers in embryonic chick aorta, *Connect. Tissue Res.* **10:**263–274.

Khokha, R., and Denhardt, D. T., 1989, Matrix metalloproteinases and tissue inhibitor of metalloproteinases: A review of their role in tumorigenesis and tissue invasion, *Invasion Metastasis* **9:**391–405.

Kumaratilake, J. S., Gibson, M. A., Fanning, J. C., and Cleary, E. G., 1989, The tissue distribution of microfibrils reacting with a monospecific antibody to MAGP, the major glycoprotein antigen of elastin-associated microfibrils, *Eur. J. Cell Biol.* **50:**117–127.

Lansing, A. I., Roberts, E., Ramasarma, G. B., Rosenthal, T. B., and Alex, M., 1951, Changes with age in amino acid composition of arterial elastin, *Proc. Soc. Exp. Biol. Med.* **76:**714–717.

Lee, B., Goodfrey, M., Vitale, E., Hori, H., Mattei, M.-G., Sarfarazi, M., Tsipouras, P., Ramirez, F., and Hollister, D. W., 1991, Linkage of Marfan syndrome and a phenotypically related disorder to two different fibrillin genes, *Nature (Lond.)* **352:**330–334.

Lefevre, M., and Rucker, R. B., 1980, Aorta elastin turnover in normal and hypercholesterolemic Japanese quail, *Biochim. Biophys. Acta* **630:**519–529.

Liotta, L. A., Wewer, U., Rao, N. C., Schiffmann, E., Stracke, M., Guirguis, R., Thorgeirsson, U., Muschel, R., and Sobel, M., 1988, Biochemical mechanisms of tumor invasion and metastases, *Adv. Exp. Med. Biol.* **233:**161–169.

Maslen, C. L., Corson, G. M., Maddox, B. K., Glanville, R. W., and Sakai, L. Y., 1991, Partial sequence of a candidate gene for the Marfan syndrome, *Nature (Lond.)* **352:**334–337.

McKusick, V. A., 1991, The defect in Marfan syndrome, *Nature (Lond.)* **352:**279–281.

Mecham, R. P., and Heuser, J., 1990, Three-dimensional organization of extracellular matrix in elastic cartilage as viewed by quick freeze, deep etch electron microscopy, *Connect. Tissue Res.* **24:**83–93.

Mecham, R. P., Hinek, A., Entwistle, R., Wrenn, D. S., Griffin, G. L., and Senior, R. M., 1989a, Elastin binds to a multifunctional 67 kD peripheral membrane protein, *Biochemistry* **28:**3716–3722.

Mecham, R. P., Hinek, A., Griffin, G. L., Senior, R. M., and Liotta, L., 1989b, The elastin receptor shows structural and functional similarities to the 67-kDa tumor cell laminin receptor, *J. Biol. Chem.* **264:**16652–16657.

Mecham, R. P., Whitehouse, L., Hay, M., Hinek, A., and Sheetz, M., 1991, Ligand affinity of the 67 kDa elastin/laminin binding protein is modulated by the protein's lectin domain: Visualization of elastin/laminin receptor complexes with gold-tagged ligands, *J. Cell Biol.* **113**:187–194.

Morocutti, M., Raspanti, M., Govoni, P., Kadar, A., and Ruggeri, A., 1988, Ultrastructural aspects of freeze-fractured and etched elastin, *Connect. Tissue Res.* **18**:55–64.

Olliver, L., Luvalle, P. A., Davidson, J. A., Rosenbloom, J., Matthew, C. G., Bester, A. J., and Boyd, C. D., 1987, Copy number of the sheep elastin gene, *Collagen Relat. Res.* **7**:77–89.

Park, P. W., Roberts, D. D., Grosso, L. E., Parks, W. C., Rosenbloom, J., Abrams, W. R., and Mecham, R. P., 1991, Binding of tropoelastin to *Staphyloccus aureus*, *J. Biol. Chem.*, in press.

Parks, W. C., and Deak, S. B., 1990, Tropoelastin heterogeneity: Implications for protein function and disease, *Am. J. Respir. Cell Mol. Biol.* **2**:399–406.

Parks, W. C., Secrist, H., Wu, L. C., and Mecham, R. P., 1988, Developmental regulation of tropoelastin isoforms, *J. Biol. Chem.* **263**:4416–4423.

Partridge, S. M., 1969, Isolation and characterization of elastin, in: *Chemistry and Molecular Biology of the Intracellular Matrix, Volume 1*, (E. A. Balazs, ed.), pp. 593–616, Academic Press, New York.

Partridge, S. M., 1962, Elastin, *Adv. Prot. Chem.* **17**:227–302.

Partridge, S. M., Davis, H. F., and Adair, G. S., 1955, The chemistry of connective tissues. 2. Soluble proteins derived from partial hydrolysis of elastin, *Biochem. J.* **61**:11–21.

Pasquali-Ronchetti, I., and Fornieri, C., 1984. The ultrastructural organization of the elastin fibre, in: *Ultrastructure of the Connective Tissue Matrix* (A. Ruggeri and P. M. Motta, eds.), pp. 126–139, Nijhoff, The Hague.

Paz, M. A., Keith, D. A., and Gallop, P. M., 1982, Elastin isolation and cross-linking, *Methods Enzymol.* **82**:571–587.

Raju, K., and Anwar, R. A., 1987, Primary structures of bovine elastin a, b, and c deduced from the sequences of cDNA clones, *J. Biol. Chem.* **262**:5755–5762.

Rich, C. B., and Foster, J. A., 1987, Evidence for the existence of three chick lung tropoelastins, *Biochem. Biophys. Res. Commun.* **146**:1291–1295.

Rosenbloom, J., 1982, Elastin: Biosynthesis, structure, degradation and role in disease processes, *Connect. Tissue Res.* **10**:73–91.

Rosenbloom, J., 1987, Molecular cloning and gene structure of elastins, *Methods. Enzymol.* **144(D)**:259–288.

Rosenbloom, J., and Cywinsky, A., 1976, Inhibition of proline hydroxylase does not inhibit secretion of tropoelastin by chick aorta cells, *FEBS Lett.* **65**:246–250.

Ross, R., Fialkow, P. J., and Altman, K., 1977, The morphogenesis of elastic fibers *Adv. Exp. Med. Biol.* **79**:7–17.

Sage, H., 1983, The evolution of elastin: Correlation of functional properties with protein structure and phylogenetic distribution, *Comp. Biochem. Physiol.* **74B**:373–380.

Sage, H., and Gray, W. R., 1979, Studies on the evolution of elastin. I. Phylogenetic distribution, *Comp. Biochem. Physiol.* **64B**:313–327.

Sakai, L. Y., Keene, D. R., and Engvall, E., 1986, Fibrillin, a new 350-kD glycoprotein, is a component of extracellular microfibrils, *J. Cell Biol.* **103**:2499–2509.

Sandberg, L. B., Soskel, N. T., and Wolt, T. B., 1981, Elastin structure, biosynthesis and relation to disease states, *N. Engl. J. Med.* **304**:566–579.

Senior, R. M., Griffin, G. L., and Mecham, R. P., 1980, Chemotactic activity of elastin-derived peptides, *J. Clin. Invest.* **66**:859–862.

Senior, R. M., Griffin, G. L., Mecham, R. P., Wrenn, D. S., Prasad, K. U., and Urry, D. W., 1984, Val-Gly-Val-Ala-Pro-Gly, a repeating peptide in elastin, is chemotactic for fibroblasts and monocytes, *J. Cell Biol.* **99**:870–874.

Senior, R. M., Hinek, A., Griffin, G. L., Pipoly, D. J., Crouch, E. C., and Mecham, R. P., 1989, Neutrophils show chemotaxis to type IV collagen and its 7S domain and contain a 67 kD type IV collagen-binding protein with lectin properties, *Am. J. Respir. Cell Mol. Biol.* **1**:479–487.

Senior, R. M., Griffin, G. L., Fliszar, C. J., Shapiro, S. D., Goldberg, G. I., and Welgus, H. G., 1991, Human 92-kilodalton and 72-kilodalton type IV collagenases are elastases, *J. Biol. Chem.* **266**:7870–7875.

Serafini-Fracassini, A., 1984, Elastogenesis in embryonic and post-natal development, in: *Ultra-*

structure of the Connective Tissue Matrix (A. Ruggeri and P. M. Motta, eds.), pp. 140–150, Nijhoff, The Hague.

Serafini-Fracassini, A., Ventrella, G., Field, M. J., Hinnie, J., Onyezili, N. I., and Griffiths, R., 1981, Characterization of a structural glycoprotein from bovine ligamentum nuchae exhibiting dual amine oxidase activity, *Biochemistry* **2**:5424–5429.

Shapiro, S. D., Endicott, S. K., Province, M. A., Pierce, J. A., and Campbell, E. J., 1991, marked longevity of human lung parenchymal elastic fibers deduced from prevalence of D-aspartate and nuclear weapons-related radiocarbon, *J. Clin. Invest.* **87**:1828–1834.

Tokimitsu, I., Tajima, S., Nishikawa, T., Tajima, M., and Fukasawa, T., 1987, Sequence analysis of elastin cDNA from chick aorta and tissue-specific transcription of the elastin gene in developing chick embryo, *Arch. Biochem. Biophys.* **256**:455–461.

Trackman, P. C., Pratt, A. M., Wolanski, A., Tang, S. S., Offner, G. D., Troxler, R. F., and Kagan, H. M., 1990, Cloning of rat aorta lysyl oxidase cDNA: Complete codons and predicted amino acid sequence, *Biochemistry* **29**:4863–4870.

Uitto, J., Hoffman, H. P., and Prockop, D. J., 1976, Synthesis of elastin and procollagen by cells from embryonic aorta. Differences in the role of hydroxyproline and the effects of proline analogs on the secretion of the two proteins *Arch. Biochem. Biophys.* **173**:187–200.

Urry, D. W., 1983, What is elastin; what is not, *Ultrastruct. Pathol.* **4**:227–251.

Urry, D. W., Long, M. M., Cox, B. A., Oshnishi, T., Mitchell, L. W., and Jacobs, M., 1974, The synthetic polypentapeptide of elastin coacervates and forms filamentous aggregates, *Biochim. Biophys. Acta* **371**:597–602.

Varga, Z., Jacob, J.-P., Csongor, J., Robert, L., Leövey, A., and Fülöp, T. J., 1990, Altered phosphatidylinositol breakdown after k-elastin stimulation in PMNLs of elderly, *Mech. Ageing Dev.* **52**:61–70.

Weis-Fogh, T., and Andersen, S. O., 1970, New molecular model for the long-range elasticity of elastin, *Nature* **227**:718–721.

Welgus, H. G., Campbell, E. J., Cury, J. D., Eisen, A. Z., Senior, R. M., Wilheim, S. M., and Goldberg, G. I., 1990, Neutral metalloproteinases produced by human mononuclear phagocytes. Enzyme profile, regulation, and expression during cellular development, *J. Clin. Invest.* **86**:1496–1502.

Wilhelm, S. M., Collier, I. E., Marmer, B. L., Eisen, A. Z., Grant, G. A., and Goldberg, G. I., 1989, SV40-transformed human lung fibroblasts secrete a 92 kDa type IV collagenase which is identical to that secreted by normal human macrophages, *J. Biol. Chem.* **264**:17213–17221.

Wrenn, D. S., Parks, W. C., Whitehouse, L. A., Crouch, E. C., Kucich, U., Rosenbloom, J., and Mecham, R. P., 1987, Identification of multiple tropoelastins secreted by bovine cells, *J. Biol. Chem.* **262**:2244–2249.

Yeh, H., Ornstein-Goldstein, N., Indik, Z., Sheppard, P., Anderson, N., Rosenbloom, J. C., Cicila, G., Yoon, K., and Rosenbloom, J., 1987, Sequence variation of bovine elastin mRNA due to alternative splicing, *Collagen Relat. Res.* **7**:235–247.

Yeh, H., Anderson, N., Ornstein-Goldstein, N., Bashir, M. M., Rosenbloom, J. C., Abrams, W., Indik, Z., Yoon, K., Parks, W., Mecham, R., and Rosenbloom, J., 1989, Structure of the bovine elastin gene and S1 nuclease analysis of alternative splicing of elastin mRNA in the bovine nuchal ligament, *Biochemistry* **28**:2365–2370.

Yoon, K., Davidson, J. M., Boyd, C., May, M., LuValle, P., Ornstein-Goldstein, N., Smith, J., Indik, Z., Ross, A., Golub, E., and Rosenbloom, J., 1985, Analysis of the 3′ region of the sheep elastin gene, *Arch. Biochem. Biophys.* **241**:684–691.

Chapter 4

Fibronectin and Other Cell Interactive Glycoproteins

KENNETH M. YAMADA

1. Introduction

Fibronectin, laminin, and other recently characterized noncollagenous extracellular proteins play important roles in many cell surface interactions. For example, fibronectin helps to mediate cell adhesion, embryonic cell migration, and wound healing, while laminin can promote processes as diverse as axonal outgrowth, maintenance of the polarized, differentiated phenotype of epithelial cells on a basement membrane, and metastasis. Each of these glycoproteins can participate in a variety of functions by using different specialized domains or peptide recognition sequences for binding to specific cell surface receptors, or to collagens, proteoglycans, or other extracellular molecules. In the decade since the first edition of this book, there has been remarkable progress in our understanding of the structures, domain organization, and biological roles of these multifunctional cell interactive proteins.

As will be reviewed in detail for each cell interactive protein, several major themes appear repeatedly in recent studies of these proteins:

- These molecules generally contain distinct, functionally active polypeptide domains specialized for binding to specific cell surface receptors, as well as to other extracellular molecules, e.g., collagen-binding domains and heparin/heparan sulfate-binding domains (Fig. 4-1).
- Most contain specific, short peptide sequences that are bound by cell surface receptors. Such surprisingly short amino acid sequences are now thought to serve often as adhesive recognition signals for receptors for these proteins, and some of them can even provide cell-type specificity.
- Many can form oligomers or polymers, either by formation of interchain disulfide bonds at specific residues or by noncovalent self–self association.

KENNETH M. YAMADA • Laboratory of Developmental Biology, National Institute of Dental Research, National Institutes of Health, Bethesda, Maryland 20892.

Cell Biology of Extracellular Matrix, Second Edition, edited by Elizabeth D. Hay, Plenum Press, New York, 1991.

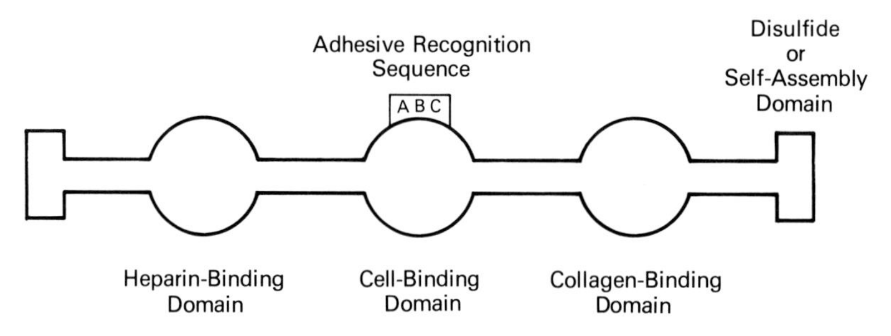

Figure 4-1. General structural model of cell interactive glycoproteins. All proteins of this type contain a region specialized for binding to the cell surface (frequently termed the cell-binding domain). Most of these domains contain a short, specific amino acid sequence recognized by a cell surface receptor (Adhesive Recognition Sequence; letters ABC represent specific amino acid residues). Most interaction proteins also contain a domain that binds to heparin and to related glycosaminoglycans, and some also contain a collagen-binding domain. A number of the larger glycoproteins of this class also have sites for formation of polymers, either by intersubunit disulfide bond formation at specific cysteine residues or by means of self-assembly domains in the protein that mediate noncovalent polymerization.

- Specific transmembrane receptors can account for many of the interactions of extracellular matrix (ECM) molecules with cells.
- Repetitive structural units such as fibronectin- or EGF-like repeats often comprise major parts of these proteins; local domains of these repeats can become specialized for different functions.
- Families of closely related proteins can be generated by alternative splicing or gene duplication, e.g., to form variant fibronectins, laminins, and tenascins.
- These ECM proteins and their receptors are regulated in location, type, and quantity during embryonic development and tissue remodeling.

These properties of cell interactive proteins and their receptors provide a considerable amount of complexity and sophistication to the repertoire of potential interactions between cells and the ECM, as well as between various matrix molecules. Although most studies have tended to focus on only one of these proteins at a time, it is important to recall that a cell often interacts with a complex mixture of ECM proteins, each of which may contribute to regulating cell behavior. In the following review, more recent references are often cited in preference to older references, in order to permit access to the additional references listed in each paper; in addition, newer findings are emphasized more than older findings, since the latter can be found in numerous previous reviews and in the first edition of this book.

2. Fibronectin

2.1. Introduction

Fibronectin is a well-characterized, multifunctional cell interactive protein present in a variety of extracellular matrices (reviewed recently in several

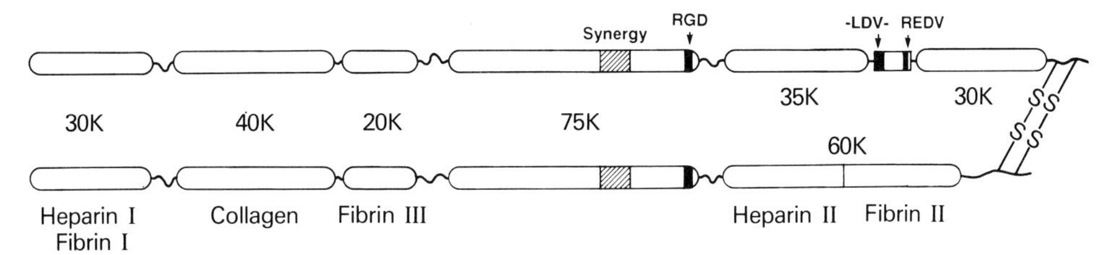

Figure 4-2. Functional domains of fibronectin. This large glycoprotein is depicted as a dimer of similar but nonidentical subunits linked by a pair of carboxyl-terminal disulfide bonds. The numbers indicate the approximate molecular weights of each domain. The major ligand(s) bound by each domain are listed along the bottom; there are two heparin-binding domains and three fibrin-binding domains (although the fibrin III domain is normally cryptic and is exposed only by proteolysis). Cell-binding regions or sequences are indicated along the top; the 75K central cell-binding domain contains the Arg-Gly-Asp (RGD) sequence and a "synergy" region that interacts synergistically with the RGD sequence to mediate adhesion; the IIICS region contains the minimal active sequence Leu-Asp-Val (LDV), as well as Arg-Glu-Asp-Val (REDV).

books: Mosher, 1989; Carsons, 1989; Hynes, 1990). Although it is encoded by only one gene, fibronectin exists in a number of variant forms that differ in sequence at three general regions of alternative splicing of its precursor mRNA; for example, there can be 20 different forms of human fibronectin subunits. Although fibronectin molecules can display a number of biological activities (reviewed in the first edition of this book and the texts listed above), they have particularly important roles in both cell adhesion and cell migration. All fibronectin molecules appear to contain the same basic functional domains shown in Fig. 4-2, but some of the alternative splicing is directed at cell adhesion sequences, which provides a posttranscriptional mechanism for potentially regulating fibronectin cell-type specificity (see Section 2.3.1b).

Fibronectin is found in blood as a soluble plasma glycoprotein at the substantial concentration of 0.3 g/liter; the form found in plasma is termed "plasma fibronectin." "Cellular" fibronectins are produced by a wide variety of cell types, which secrete them and often organize them into extensive extracellular matrices (Fig. 4-3). "Cellular" fibronectins vary in composition, but they characteristically contain considerably higher proportions of alternatively spliced sequences than "plasma" fibronectin (see Section 2.2).

A rather daunting number of descriptive studies have localized fibronectin in many tissues during many morphogenetic and pathogenic events. Summaries of such studies can be found in texts including the first edition of this book and texts alluded to above. Fibronectin (along with certain other extracellular proteins) is particularly prominent in migratory pathways for embryonic cells, such as during gastrulation, neural crest cell migration, and precardiac mesoderm cell migration. Fibronectin is also characteristically present in loose connective tissue and in embryonic (but often not in adult) basement membranes. Large quantities of fibronectin are also present in clots in association with fibrin, where it can serve as a substrate for migratory cells during wound healing; it is also increased during fibrotic processes.

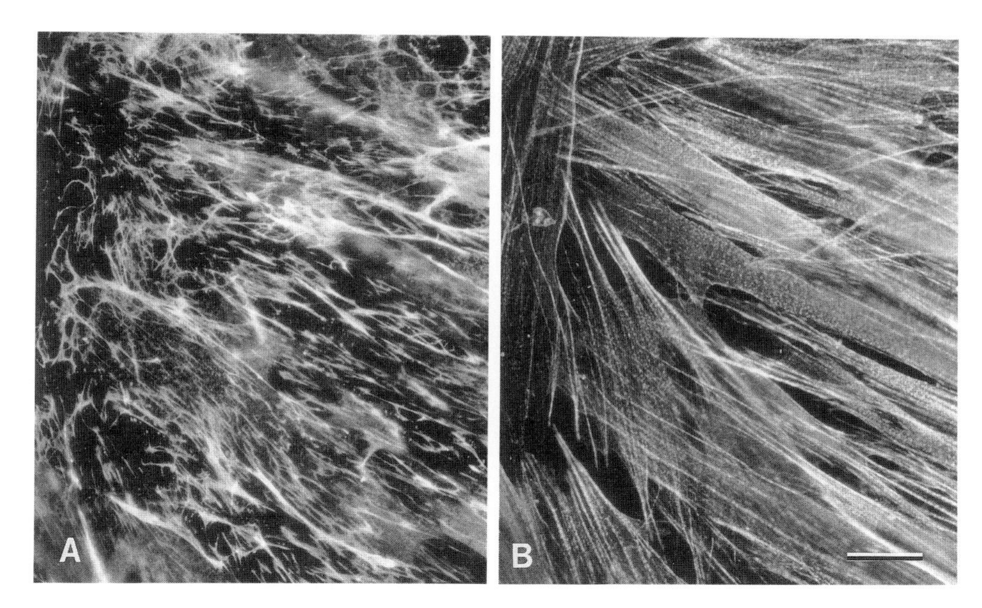

Figure 4-3. Comparison of the organization of fibronectin and intracellular microfilament bundles. Panel A shows fibronectin organized as a fibrillar matrix surrounding human embryonic lung cells *in vitro*. Panel B shows for comparison the localization of actin-containing microfilament bundles, which play important cytoskeletal roles. Fibronectin is visualized by immunofluorescence using rabbit antifibronectin antibodies labeled with fluorescein isothiocyanate, while F-actin in microfilament bundles is visualized using rhodamine-labeled phalloidin. Bar = 20 μm.

2.2. Gene and Protein Structure of Fibronectin

2.2.1. The Fibronectin Gene

The fibronectin gene is relatively large, with an estimated length of about 50 kb. Its coding sequence is subdivided into about 50 exons of similar size, several of which undergo alternative splicing (Hirano *et al.*, 1983; Hynes, 1990). Fibronectin is comprised of three general types of homologous repeating units or modules, termed types I, II, and III (Fig. 4-4). There are 12 type I repeats, 2 type II repeats, and 15–17 type III repeats. In the gene, each exon encodes one repeating module of the type I or type II homology unit, but the larger type III repeats generally require the contribution of two exons each; two type III repeats exist as a single exon, and both can undergo alternative splicing (Patel *et al.*, 1987). This molecule provides an interesting example of how a large, multifunctional structural protein can be derived from presumptive gene duplications of a few basic modular units.

These modular units seem to be used as basic building blocks for forming domains that develop distinct functions independent of their original derivation. For example, type I modules are used in domains that bind to heparin or fibrin, to fibrin alone, or even to collagen (Mosher, 1989). Type II modules are

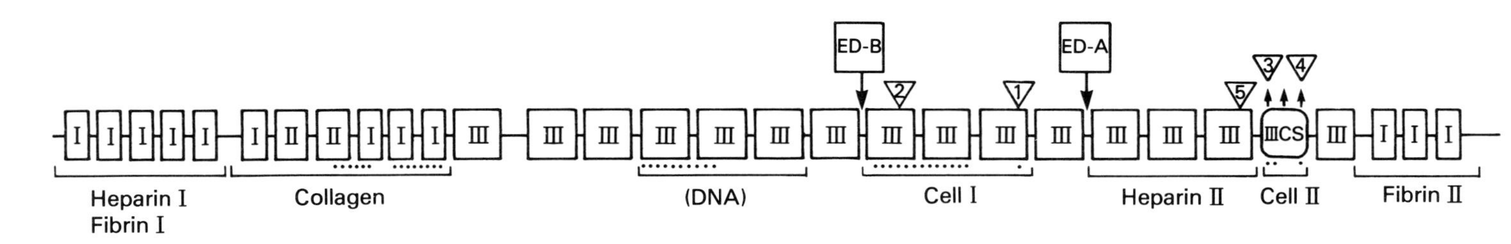

Figure 4-4. Detailed structural map of fibronectin showing repeating homologous units. Fibronectin is composed of multiple repeats of three types of structural module termed types I, II, or III. Large arrows (down) indicate sites where alternative splicing of mRNA results in the insertion or omission of type III modules named ED-A and ED-B, or of portions of the IIICS region (small upward arrows indicate three potential sites of alternative splicing). The labels along the bottom of the figure indicate functional binding domains, and the dotted regions indicate the minimal active size of binding regions identified to date. The numbers inside the triangles indicate the following cell interaction sites: 1 = Arg-Gly-Asp-(Ser); 2 = synergy region; 3 = CS1 sequence containing Leu-Asp-Val; 4 = Arg-Glu-Asp-Val sequence; 5 = two active cell binding sequences in the heparin-binding domain.

found only in the collagen-binding domain, but they may not be essential for binding of collagen (Ingham *et al.*, 1989). Type III repeats are used in domains that bind to cells or to heparin, but not to collagen (Mosher, 1989; Hynes, 1990).

The three basic protein structural modules of fibronectin are not unique to this molecule (reviewed in Mosher, 1989). For example, repeats homologous to type I modules are also found in tissue plasminogen activator. The structure of this type of module has been determined recently by NMR approaches (Baron *et al.*, 1990). Type II modules of fibronectin are homologous to the Kringle domains of blood clotting and fibrinolytic proteins. Finally, fibronectin type III modules are found in tenascin and in cell–cell adhesion molecules such as L1 (also known as "NILE" or "Ng-CAM"; Moos *et al.*, 1988). The promoter region of the fibronectin gene has been characterized, and it contains a cyclic AMP response element sequence responsible for fibronectin induction via nuclear factors (Dean *et al.*, 1990). A number of growth factors and cytokines can regulate fibronectin synthesis, including transforming growth factor beta (reviewed in Yamada, 1989). The mechanisms by which these factors affect gene expression, e.g., the identity of the transcription factors needed for regulation of fibronectin expression, remain to be determined.

2.2.2. Alternative Splicing

The patterns of alternative splicing of fibronectin are particularly interesting, since they can produce the insertion or deletion of cell-type-specific adhesion sites. The IIICS (also termed V) region of fibronectin shown in Fig. 4-4 can undergo complex patterns of alternative splicing that can result in human fibronectin molecules with five different sequence patterns in this region (reviewed in Mosher, 1989; Hynes, 1990). Two of these spliced sites, termed CS1 and CS5, contain distinct adhesive recognition sequences (discussed in Section 2.3.1b). The sequence of this complex spliced region differs from the repeating structural units that comprise most of fibronectin. In addition, however, there are two other sites of alternative splicing in fibronectin that result from insertion or deletion of entire type III repeating units (Fig. 4-4). The presence of these spliced sites, termed ED-A (ED-1) and ED-B (ED-2), can correlate temporally and spatially with periods of cell migration and morphogenesis, but their functions are not yet clear. They may also promote fibril formation by fibronectins (Guan *et al.*, 1990).

Differential splicing within all three of the general regions of alternative splicing in human fibronectin can produce a total of 20 different variants of each fibronectin chain of approximately 250,000 Da. Since fibronectin is a dimer, or even a multimer when associated with cell surfaces, there is a considerable complexity of fibronectin variant mixtures. Little is known about the regulation of this alternative splicing, although spliced sequences tend to be retained in "cellular" fibronectins, especially by certain embryonic and malignant cells (Vartio *et al.*, 1987; Oyama *et al.*, 1989). Plasma fibronectin, considerable proportions of which are produced by the liver, generally lacks ED-A and

ED-B sequences, and can contain 50% or less of the IIICS type of spliced region (Hershberger and Culp, 1990).

2.2.3. Posttranslational Modifications

Fibronectin can be glycosylated, phosphorylated, and sulfated (reviewed in Mosher, 1989). Glycosylation of fibronectin is primarily of the complex N-linked oligosaccharide type. Although nonglycosylated fibronectin still displays much of the activity of the native protein, it is much more sensitive to proteases; carbohydrates therefore appear to stabilize the protein against proteolysis (Olden *et al.*, 1979). Moreover, the presence of carbohydrate moieties can modulate binding to collagen and slightly alter interactions with cells; for example, higher amounts of glycosylation cause a decreased avidity for collagen (Zhu and Laine, 1985).

2.2.4. Overall Shape

Soluble plasma fibronectin exists as a disk-shaped molecule approximately 30 nm in diameter and 2 nm thick (Sjöberg *et al.*, 1989; Benecky *et al.*, 1990). Increased ionic strength opens or unfolds the molecule to a more linear structure. Binding of fibronectin to an artificial substrate and possibly binding to ligands may also cause such opening from a more compact shape to a linear shape that could theoretically be more accessible to other ligands (e.g., see Sjöberg *et al.*, 1989; Wolff and Lai, 1989).

Both "cellular" and plasma fibronectins can become incorporated into fibronectin fibrils, but they tend to remain segregated, suggesting that alternative splicing may help regulate such supramolecular organization (Pesciotta-Peters *et al.*, 1990). When incorporated into fibrils, fibronectin becomes cross-linked covalently by disulfide bonds and by transglutaminase cross-linking (reviewed in Mosher, 1989). Such cross-linking would stabilize the fibrillar form.

2.3. Organization of Functional Domains

Each fibronectin polypeptide chain consists of a series of structural and functional domains (Fig. 4-2). These domains have been defined by proteolytic fragmentation studies (e.g., see the first edition of this book) and more recently by recombinant DNA analyses.

2.3.1. Cell Interaction Sequences

Fibronectin contains at least six peptide sites capable of mediating cell adhesion. They can be classified into three general regions, the central cell-binding domain, the alternatively spliced IIICS region, and the heparin-binding domain. In most cases, the sequences involved in recognition by cell surface receptors are surprisingly short, suggesting the importance of linear amino acid sequences either alone or in combination to mediate processes such as cell adhesion. Ongoing studies, however, also seem to indicate the importance of another, less well-defined polypeptide information in fibronectin that is necessary for function of the short, linear sequences; this requirement for a second part of the protein could be due to the need for another recognition sequences(s) that binds to the receptor, or to a requirement for a protein framework that can present a specific conformation of the recognition sequence to the receptor.

2.3.1a. Central Cell-Binding Domain. Most cells can adhere to fibronectin, at least in part by binding to the centrally located "cell-binding" domain (Figs. 4-2, 4-4). Some cells, such as fibroblasts, can bind to isolated 37- to 120-kDa proteolytic fragments containing this domain as well as they can to intact fibronectin (equal molar activities; e.g., see Akiyama *et al.*, 1985). A crucial sequence in this domain is Arg-Gly-Asp-Ser (abbreviated RGDS using the single-letter amino acid code; Pierschbacher and Ruoslahti, 1984, 1987; Yamada and Kennedy, 1984). The first three amino acids are particularly important, and they define the now-classical "RGD" recognition motif (reviewed in Chapter 10). Deletion of this sequence, or even mutation of the aspartate to a glutamate residue (which retains the same charge) in an otherwise intact domain, leads to the loss of nearly all adhesive activity (Obara *et al.*, 1988).

Adhesive activity is not completely abolished, however, suggesting the existence of a second interaction site in this domain. Conversely, the estimated affinity of interaction of RGD-containing fibronectin peptides with cell surface receptors is much lower than that of larger fragments (e.g., see Akiyama and Yamada, 1985). Mutagenesis experiments in fact define a second key region of this domain located about 14,000–28,000 Da toward the amino terminus of the protein, and monoclonal antibody inhibition studies appear to confirm that this region contains a second binding region that functions in synergy with the RGD sequence in cell adhesion, migration, and fibronectin matrix assembly (Obara *et al.*, 1988; Nagai *et al.*, 1991).

It has been proposed that this synergistic system can account for specificity and for full affinity of fibronectin toward the $\alpha_5\beta_1$ integrin receptor; however, the potential importance of the three-dimensional conformation of the RGD sequence by itself as the sole determinant of specificity for fibronectin receptors remains uncertain (in contrast to the much stronger evidence for such constraints for certain other integrin receptors as noted below). These biochemical distinctions are important for both evolutionary and functional considerations, since classical biological questions such as the basis of adhesive specificity could be explained either by the use of unique combinations of such sequences, or by more lock-and-key mechanisms requiring unique conformations of these sequences (reviewed in Yamada, 1989).

RGD-containing peptides specifically inhibit a variety of migratory processes (reviewed in Mosher, 1989, and Chapters 10 and 12).

2.3.1b. Alternatively Spliced Adhesion Sites. The alternatively spliced sequences of the IIICS region are unique in that they encode cell-type-specific adhesion sequences. Although cells such as fibroblasts cannot adhere to this region of fibronectin, cells derived from the embryonic neural crest and certain lymphocytes adhere readily. For example, embryonic neural crest cells, neurons from sympathetic and sensory ganglia, and melanoma cells adhere, as do activated T cells and certain B lymphocytes. All of these cells adhere to one or both of the two binding sequences in this region by means of the $\alpha_4\beta_1$ integrin receptor (see Mould *et al.*, 1991, and references therein). IIICS sequences are used by neural crest cells for migration in conjunction with several other sites elsewhere in the protein (Dufour *et al.*, 1988a).

The stronger of the two peptide adhesion sequences in this region is unusual in its potency: even a 25-residue synthetic peptide corresponding to this site, termed CS1, still retains 40% of the total molar activity of fibronectin itself (Humphries *et al.*, 1987); this represents a surprisingly high retention of function by a synthetic peptide. By contrast, RGD-containing peptides appear to be 100–200 times less active for recognition by the $\alpha_5\beta_1$ fibronectin receptor on a molar basis (Akiyama and Yamada, 1985). The minimal essential peptide sequence required for function of this CS1 region has recently been determined to be the tripeptide Leu-Asp-Val (Komoriya *et al.*, 1991). Other information elsewhere in the CS1 peptide region appears to be required for full activity of this region, perhaps due to its effects on conformation of the Leu-Asp-Val sequence.

An independent sequence that is also alternatively spliced consists of the sequence Arg-Glu-Asp-Val (Humphries *et al.*, 1986). This sequence is two orders of magnitude less active than the CS1 sequence, but it appears to display identical cell-type specificity and also interacts with the $\alpha_4\beta_1$ integrin receptor (Mould *et al.*, 1991). An interesting speculation concerning the nature of adhesive recognition sequences comes from inspection of these sequences compared with the canonical Arg-Gly-Asp sequence, all of which appear to contain a critical Asp residue: it has been proposed that this residue participates in a general binding mechanism based on cation binding to a functional site contributed by both this Asp and other sequences located in the integrin receptor (e.g., see Loftus *et al.*, 1990). Experiments in which this sequence is mutated in either proteins or synthetic peptides all tend to support this hypothesis to date, but whether it is the underlying mechanism for this class of adhesive interactions remains to be determined.

2.3.1c. Heparin-Binding Sites. Although binding of fibronectin to cell surface heparan sulfate (see Chapter 9) can contribute to fibronectin-mediated adhesion, a more direct interaction between cell surface receptors and this type of domain may occur. Two peptide sequences from the high-affinity heparin-binding domain that can bind heparin directly can also mediate cell attachment (Fig. 4-4, site 5), and the receptor used for this interaction appears to be the $\alpha_4\beta_1$ integrin (McCarthy *et al.*, 1988, 1990). The relative strengths, cell-type specificities, and functional interrelationships of these sites to the central cell-binding domain sites and to the alternatively spliced sites remain to be determined.

2.3.1d. Distribution and Transmembrane Signaling of Receptors That Bind Adhesive Recognition Sequences. Although these fibronectin adhesive recognition sequences are bound by specific integrin receptors, the strength of their binding and the consequences of their interactions with the plasma membrane appear to be affected by the spatial organization of the receptors. For example, rapidly migratory embryonic cells and many malignant cells tend to be unable to retain and organize fibronectin into extracellular fibrils, and have diffuse patterns of receptors. In contrast, more slowly moving fibroblastic cells can elaborate complex networks of fibronectin in association with acquisition of strong adhesion and specialized adhesion sites, and clusters of receptors are arrayed parallel to intracellular cytoskeletal elements such as talin, α-actinin, and microfilament bundles (reviewed in Dufour *et al.*, 1988b; Yamada, 1989). The existence of two alternatively spliced regions known to interact with $\alpha_4\beta_1$ integrins provides a potential means for regulating the strength of these interactions by their extent of expression, which might be important for processes such as rates of cell migration or neurite extension (e.g., see Humphries *et al.*, 1988).

If provided in high enough amounts, either the RGD sequence or the CS1 peptide sequence immobilized alone on a substrate can produce cell spreading and promote microfilament bundle organization (Singer *et al.*, 1987; K. Yamada, unpublished results). Besides transmembrane effects on cytoskeletal organization that can be attributed to transmission of physical information (e.g., clustering of a transmembrane receptor), specific binding sequences in fibronectin might also produce direct signal transduction. For example, the binding of synthetic peptides containing the RGD sequence or antibody to the $\alpha_5\beta_1$ receptor results in enhanced protease expression (see Chapter 8), although fibronectin itself is puzzlingly not active. A possibly similar disparity between effects of the intact molecule and this peptide sequence, however, is seen in an embryonic system, where RGD peptides but not fibronectin can induce cell–cell aggregation and compaction of segmental plate cells, possibly through the induction of cell surface expression of the N-cadherin molecule (Lash *et al.*, 1987; J. Lash, personal communication). It will be of interest to learn which receptors are involved in these signal transduction processes.

Even cell proliferation can be regulated by either the RGD or CS1 domains. Binding of cells either to the central cell-binding domain of fibronectin containing RGD or to the LDV-containing CS1 peptide can help induce T lymphocyte proliferation. In the intact fibronectin molecule, both sites can contribute to mitogenesis (e.g., see Nojima *et al.*, 1990, and references therein). These and other findings suggest that the intracellular sequelae of binding to different fibronectin sequences can be similar; whether there are subtle sequence-dependent differences, or whether they simply provide additive regulatory effects, still needs to be established.

2.3.2. Collagen-Binding Domain

The first functional domain to be identified on fibronectin was the "collagen-binding" domain. The domain seems to account for all of fibronectin's capacity to bind to collagen/gelatin. Although this domain initially appeared to interact with native, nondenatured collagen, model studies indicate that the interaction may be truly significant only at subphysiological ionic strength (Ingham *et al.*, 1985). It is clear that this domain binds far more effectively to denatured collagen (e.g., gelatin), and thus its interactions with collagens in general may be due to its binding to unfolded regions of the collagen triple helix. Interestingly, the most preferred binding site on type I collagen is in the same region cleaved by bacterial collagenase, a region that is slightly less helical and thus might be susceptible to binding of a fibronectin domain that recognizes only denatured collagen chains (see first edition of this book).

The evolutionary conservation of the collagen-binding function in fibronectins from a wide variety of species suggests its biological importance. The marked preference of this domain for binding to unfolded collagen, however, is puzzling when considering its potential biological function *in vivo*. For example, although fibronectin can readily mediate cell adhesion to collagen *in vitro*, it is not known whether this function is important *in vivo*. It is conceivable that this domain relates more to binding and clearance of denatured collagenous materials from blood, or to forming noncovalent cross-links with loose meshworks of collagen in migratory pathways, than to mediate any direct adhesion of cells to collagen. In this sense, it is possible fibronectin may not function *in vivo* as the intermediary in cell adhesion to native collagen. Moreover, the recent identification of several distinct integrin receptors that can mediate directly the adhesion and migration of cells on collagen supports this notion.

The polypeptide regions of fibronectin capable of mediating its binding to collagen have been variously identified as including a type II repeat (Owens and Baralle, 1986) or only type I repeats (Ingham *et al.*, 1989). The results obtained to date seem to indicate that several parts of the collagen-binding region contribute to full avidity of binding, although the relative contributions to binding of combining independent binding sites versus producing a better conformation of the binding region remain to be determined.

2.3.3. Heparin-Binding Domains

Fibronectin contains two heparin-binding domains (Figs. 4-2, 4-4), which are thought to interact most often with heparan sulfate proteoglycans. The binding of heparin by intact fibronectin is of relatively high affinity, with two classes of affinities with $K_d = 10^{-7}$ to 10^{-9} M (reviewed in the first edition of this book; Mosher, 1989; Hynes, 1990). The specific heparin-binding domains are located toward opposite ends of the protein, and they differ in both affinities and sensitivity to calcium ion. The strongest heparin-binding site is in the carboxyl-terminal third of the protein (Fig. 4-2). A weaker binding domain is located in the amino-terminal end of the protein. This latter domain also binds to fibrin and can modulate cell spreading. Heparin/heparan binding by this amino-terminal domain can be regulated by the extracellular concentration of calcium. Inhibition of binding occurs when levels of Ca^{2+} rise above average physiological concentrations in blood; thus, tissue sites with high local divalent cation concentrations could have decreased binding by this domain (Hayashi and Yamada, 1982).

As noted in Section 2.3.1c, heparin-binding regions of fibronectin can also modulate cell adhesion *in vitro*. The importance of these heparin-binding domains for cell interactions has been demonstrated using fibronectin fragments. Certain cells could not organize microfilament bundles efficiently unless a heparin-binding domain was present, supporting the hypothesis that cell surface heparan sulfate is important for part of fibronectin's effects on cell behavior (Woods *et al.*, 1986; Izzard *et al.*, 1986). It is of interest from an evolutionary standpoint that heparin/heparan binding activity in fibronectin arose in two structurally diverse domains. One domain consists of type III repeats lacking any disulfide bonds, whereas the other domain is based on the type I repeat and contains five disulfide bonds, forming a compact domain.

2.3.4. Fibrin-Binding Domains

Fibronectin contains at least two fibrin-binding domains; a third is detectable after proteolysis of the protein (Figs. 4-2, 4-4; reviewed in the first edition of this book; Mosher, 1989; Hynes, 1990). The major binding domains bind to fibrin or fibrinogen, but binding tends to be relatively weak. Nevertheless, these interactions are probably important for fibronectin's role in mediating migration of fibroblastic cells in fibrin clots, especially after being cross-linked in place by activated factor XIII transglutaminase.

2.4. Matrix-Assembly Regions

At least two distinct regions in fibronectin are involved in cellular organization of secreted fibronectin molecules into extracellular fibrils. This matrix-assembly process appears to be an active event, involving as yet poorly defined contributions from living cells rather than occurring by a simple self-assembly process (McDonald, 1988; Mosher, 1989). Fibronectin matrix assembly involves the binding of fibronectin by a receptorlike system, followed by its conversion to a detergent-insoluble form during fibrillogenesis. Not surprisingly, the central cell-binding domain of fibronectin seems to be involved in matrix assembly: the process of *in vitro* fibronectin matrix assembly is inhibited by antibodies against the cell-binding domain, by antibodies against the $\alpha_5\beta_1$ fibronectin receptor, or by competition with excess free proteolytic fragments containing this region (McDonald *et al.*, 1987; Fogerty *et al.*, 1990). Besides preventing formation of a fibronectin-based matrix, inhibiting this matrix assembly process can block events such as embryonic cell migration during gastrulation (Darribére *et al.*, 1990).

In addition to the cell-binding domain, the amino-terminal end of the protein is surprisingly also implicated in the matrix-assembly process. Fragments of 70 kDa or, somewhat less effectively, the amino-terminal domain of 25 kDa, can be bound to cells in the first step of matrix assembly (McKeown-Longo and Mosher, 1985; McDonald *et al.*, 1987). An excess of the 70-kDa domain can competitively inhibit incorporation of fibronectin into fibrils. This fragment, however, cannot complete assembly by itself, remaining detergent soluble even while intact fibronectin becomes assembled. Since fibrillogenesis involves self–self association of one fibronectin molecule with another, the report of a fibronectin self-binding region may be relevant (Ehrismann *et al.*, 1982). Interestingly, plasma and "cellular" (i.e., containing alternatively spliced ED-A and ED-B sites) forms of fibronectin tend not to co-assemble in fibrils, but instead segregate into short alternating stripes of assembly on fibrils (Pesciotta-Peters *et al.*, 1990).

The molecular nature of this fibronectin matrix assembly process remains undefined. While the $\alpha_5\beta_1$ integrin is implicated in part of the process, some other β_1 integrin seems to be required as part of the receptor mechanism (Fogerty *et al.*, 1990). Moreover, gangliosides appear to considerably augment fibronectin matrix assembly (Spiegel *et al.*, 1986), and gangliosides bind effectively ($K_D = 10^{-8}$ M) to the same 25- to 31-kDa amino-terminal domain implicated in matrix assembly (Thompson *et al.*, 1986). Although binding interactions between fibronectin domains have been reported, as well as occasional success in inducing *in vitro* fibril formation, current evidence suggests that the process requires major cellular inputs. This cell-catalyzed or -mediated assembly mechanism contrasts markedly with collagen fibrillogenesis, which proceeds dramatically in the absence of cells *in vitro*. Nevertheless, fibronectin fibrils are major components of the ECM produced by cells in tissue culture, as well as *in vivo* pathways of embryonic cell migration, e.g., for gastrulation and neural crest cell migration. The requirement for a cellular matrix-assembly system permits more precise regulation of the sites of fibronectin fibril formation than if the fibrils were self-assembling, and probably accounts for the fact that even though high levels of fibronectin circulate in blood (0.3 g/liter), it remains freely soluble until incorporated into blood clots or tissue matrices.

2.5. Other Molecular and Functional Associations of Fibronectin

Fibronectin can sometimes bind to other molecules, which then co-purify with it. These fibronectin complexes can then display properties dependent on both components. Partially purified preparations of fibronectin produced by cultured cells contain the molecule tenascin, which accounts for hemagglutinating activity of such "cell surface protein" preparations. Further routine purification by monoclonal antibody affinity chromatography or molecular sieve chromatography at elevated pH can separate the fibronectin dimer from the tenascin hexabrachion (reviewed in Chiquet-Ehrismann, 1990).

Similarly, a complexed mixture of fibronectin and proteoglycan M can bind hyaluronic acid. This same proteoglycan can inhibit cell adhesion to fibronectin by an as yet undefined interaction with the cell surface using its chondroitin sulfate side chains (Yamagata *et al.*, 1986).

Growth factor activity is sometimes reported in purified fibronectin preparations, although some studies report that activity must be released by proteolysis. In addition, however, a factor such as TGF-β can bind tightly to fibronectin; this factor can be recovered by acid extraction (Fava and McClure,

1987). This type of binding of a factor by an extracellular molecule may be a general mechanism of providing an immobilized store of growth factors or cytokines that can be released when needed.

Fibronectin can also bind molecules such as complement proteins, bacteria, DNA, and denatured actin. The biological roles of these interactions remain to be established conclusively, although one function may be as a scavenger or opsonic molecule that promotes the clearing of such materials from blood by the reticuloendothelial system (reviewed in Carsons, 1989).

3. Laminin

3.1. Introduction

Laminin is a major component of basement membranes (reviews include Martin *et al.*, 1988; Timpl, 1989; Beck *et al.*, 1990; Yurchenco *et al.*, 1990; Kleinman and Weeks, 1991). It has been implicated in a host of biological processes, including cell adhesion, migration, and differentiation. Laminin can contain several different subunits linked together by disulfide bonds to form a final structure that can approach 1 million daltons in size. The best-studied type of laminin molecule is shown in Fig. 4-5, which consists of a 400-kDa A chain, a B1 chain of 210 kDa, and a B2 chain of 200 kDa. This particular form of laminin is ubiquitous in basement membranes. It is a cross-shaped molecule with a large, multilobulated globular domain at the base of the cross, and smaller globular domains on each arm.

Unlike molecules such as fibronectin that undergo alternative splicing to form multiple variants, this and other forms of laminin appear to be assembled from closely related genes encoding isoforms of certain chains (see Kleinman and Weeks, 1991, for a recent proposed classification scheme). At least two distinct A chains can be found in laminins, including the classical A chain shown in Fig. 4-5 and a distinct variant A chain termed merosin. Merosin is particularly enriched in the brain and in the peripheral nervous system (Ehrig *et al.*, 1990). Moreover, at least one B chain variant is known, which is somewhat confusingly termed "s-laminin." This B chain variant is homologous to the B1 chain of laminin ($>$ 50% amino acid identity), and is enriched at synaptic sites (Hunter *et al.*, 1989). Since different forms of laminin are found at different stages in development and in different tissues, distinct functions for these different types of laminin appear likely. Determining how laminin function is regulated by the nature of its subunits will be an important area for future research.

3.2. Cell Adhesion Sites in Laminin and Their Receptors

The mechanisms by which cells interact with laminin appear to be complex, and at times confusing. The simplest interpretation at present is that laminin can interact with cells by a host of mechanisms, including a variety of adhesive recognition sequences and receptors.

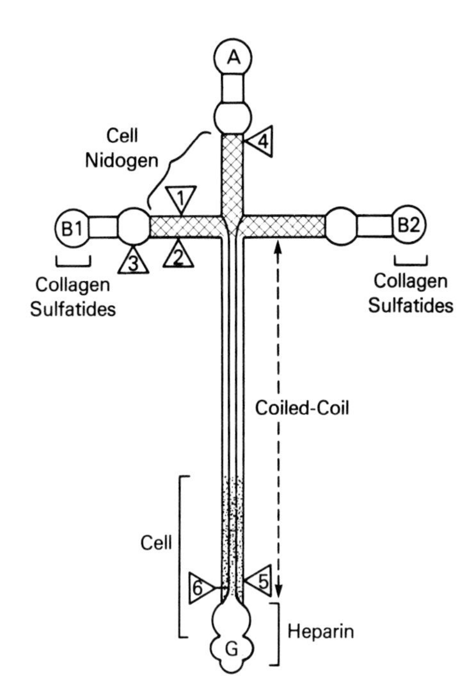

Figure 4-5. Organization of the "classical" isoform of laminin. The cross-shaped laminin molecule generally consists of three subunits linked together by disulfide bonds. The A subunit is the largest, extending the length of the molecule and terminating in a lobulated globule (G) that binds to heparin and type IV collagen. The smaller B1 and B2 chains are associated by disulfide bonds and a coiled-coil region of polypeptide structure that includes the A chain; all three combine to form the long arm. Putative functional binding regions are labeled. The numbers in triangles represent proposed short peptide recognition sequences that are apparently bound by specific cell-surface laminin receptors: 1 = Tyr-Ile-Gly-Ser-Arg (YIGSR in the B1 chain, requiring a carboxyl-terminal amino acid or an amide for function); 2 = Pro-Asp-Ser-Gly-Arg (PDSGR in the B1 chain); 3 = Arg-Tyr-Val-Val-Leu-Pro-Arg (RYVVLPR in F-9, a heparin-binding peptide in the B1 chain); 4 = Arg-Gly-Asp-Asn (an RGD peptide in the A chain); 5 = Arg-Asn-Ile-Ala-Glu-Ile-Ile-Lys-Asp-Ile (peptide P20 in the B2 chain); 6 = Ile-Lys-Val-Ala-Val (IKVAV in the A chain). A Leu-Arg-Glu (LRE) site is found in the B1 chain isoform (homologue) termed s-laminin. Although the short arm of the A chain is depicted as containing two terminal globules as seen by electron microscopy, sequence analysis suggests that it contains three such polypeptide domains. See the text for details and references.

3.2.1. "Neurite Interaction" Site

A major site for cell interactions is present near the junction of the large globular domain with the rigid coiled-coil domain of the long arm of laminin. Although originally associated with the adhesion and migration of neuronal cells on laminin, this region can also be used for cell adhesion by a number of other cell types, including muscle, carcinoma, and sarcoma cells (Dillner *et al.*, 1988). Work from different laboratories has so far implicated the peptide sequence IKVAV on the A chain (Tashiro *et al.*, 1989; Fig. 4-5, site 6), RNIAEIIKDA on the carboxyl-terminus of the B2 chain (Liesi *et al.*, 1989; Fig. 4-5, site 5), or a native complex of A, B1, and B2 chains (Deutzmann *et al.*, 1990) in function of this region for different neuronal cell types. In addition, the "s-laminin" homologue of the B1 chain can mediate adhesion of ciliary ganglion neurons, but not other neurons, using a Leu-Arg-Glu (LRE) site (Hunter *et al.*, 1989).

The "neurite" region of laminin is recognized by the $\alpha_6\beta_1$ integrin, and inhibition of this interaction with specific monoclonal antibodies blocks the adhesion of certain cells to laminin (Hall *et al.*, 1990). On the other hand, other evidence suggests that the $\alpha_3\beta_1$ integrin also binds in this area (Gehlsen *et al.*, 1989), and the IKVAV sequence is bound by nonintegrin laminin-binding proteins (e.g., see Kleinman and Weeks, 1991). Which peptide sequences bind to which receptor(s), and which of the discrepancies are due to cell type and which are due to assay methods remains to be determined.

3.2.2. Central Cell-Binding Site

A second major region of cell interaction with laminin is located near the center of the laminin cross. Proteolytic fragments containing this region can mediate adhesion of a variety of cell types including tumor cells, but are often generally less active for neuronal cells. Interestingly, the function of this putative major adhesion region has been claimed to be relatively cryptic in intact laminin (Aumailley *et al.*, 1990). The role of proteolysis in activating cell adhesion to laminin is therefore of potential importance and needs further evaluation. Several active peptide sequences in this region of laminin include YIGSR and PDSGR located adjacent to one another on the B1 chain (Graf *et al.*, 1987; Kleinman *et al.*, 1989; Fig. 4-5, sites 1 and 2), and RYVVLPR further distal on the B1 chain (Skubitz *et al.*, 1990; Fig. 4-5, site 3). In addition, there is an RGD(N) sequence in the A chain above the center of the cross (Sasaki *et al.*, 1988; Fig. 4-5, site 4). Distal to the RYVVLPR sequence on the B1 chain is an elastinlike sequence LGTIPG, which is also implicated in cell interactions with laminin (Mecham *et al.*, 1989).

Synthetic peptides from this central region of laminin can inhibit certain key biological steps. For example, the YIGSR peptide can inhibit migration of neural crest cells and of cardiac mesenchyme cells (see Chapter 12), as well as experimental metastasis (see Kleinman *et al.*, 1989). However, other studies do not see effects of YIGSR in certain adhesive or migratory processes, suggesting a complex variety of cell interaction mechanisms depending on the cell type and culture conditions. One cell type can respond differently to different sites: endothelial cell formation of tubes *in vitro* is induced by laminin in a process apparently dependent on attachment via the RGD sequence in the A chain and tube formation driven by the YIGSR site in the B1 chain (Grant *et al.*, 1989).

The complexity of laminin functions is further underscored by recent findings that one sequence may have different effects depending on the cell type and on the actual sequence used. The IKVAV-containing sequence from the neurite extension region of laminin can promote neurite outgrowth, but it appears to require either an artificial amino-terminal cysteine residue or association with other chains for maximal activity (Deutzmann *et al.*, 1990). The same peptide can serve to promote experimental metastasis of melanoma cells, accompanied by increased production of type IV collagenase (Kanemoto *et al.*, 1990).

3.3. Heparin-Binding Domains

The dominant heparin-binding domain of laminin is the large, multilobulated, globular domain at the carboxyl-terminal end of the A chain (Fig. 4-5; Martin *et al.*, 1988; Timpl, 1989). Heparin binds to laminin with moderate affinity ($K_D = 10^{-7}$ M), and promotes laminin aggregation into a noncovalently associated polymer (Yurchenco *et al.*, 1990). High concentrations of heparin, in contrast, inhibit laminin self-assembly (Kouzi-Koliakos *et al.*, 1989). Heparin-binding sites other than the large globule on the A chain have been identified on laminin. These additional sites correspond to the globular domains on the short arms; some of these can be mimicked by synthetic peptides. One such hydrophilic, basic peptide can both bind heparin and mediate cell attachment if attached to substrates, as well as inhibiting experimental metastasis when incubated in solution with certain tumor cells (Skubitz *et al.*, 1990 and references therein). The heparin/heparan binding activity of laminin appears to be particularly important for its interaction with the large heparan sulfate proteoglycan that is present in basement membranes in large quantities. Complexes of laminin, heparan sulfate proteoglycan, and other molecules provide structural organization to basement membranes (e.g., see Chapter 11).

3.4. Interactions with Collagen and Entactin

Laminin can interact with type IV collagen, and the domains involved in this binding are currently thought to be located near the ends of the short arms and in the large, carboxyl-terminal globular domain at the end of the long arm. These interactions remain to be characterized in more molecular detail. Laminin forms a complex with entactin (also known as nidogen) that is difficult to disrupt during isolation. As a consequence, most studies reported in the literature concerning laminin actually concern the laminin–entactin (or nidogen) complex.

3.5. Growth Factor Domain

Since the laminin molecule contains a number of EGF-like repeats, and since laminin can promote growth of certain cells, the question has arisen whether some of these repeats may have biological activity. Cell proliferation can be stimulated in cells possessing EGF receptors by a fragment of laminin that includes EGF-like repeats, at concentrations comparable to those of EGF itself (Panayotou *et al.*, 1989). It is therefore possible that proteolytic degradation of laminin *in vivo* can lead to liberation of growth factor activity that stimulates growth in regions of tissue damage. Since at least one type of adhesive activity present in the junction of the arms can also be exposed by proteases, it is conceivable that both cell adhesive and proliferative activities could be unleashed at local regions of tissue damage or inflammation to result in tissue remodeling involving only those cells sensitive to these signals.

4. Vitronectin

4.1. Introduction

Vitronectin (S-protein) is a multifunctional glycoprotein found in plasma and in extracellular matrices. First identified as a cell attachment factor with high avidity for glass surface (vitro = glass), vitronectin has been investigated independently under the names "serum spreading factor," "S-protein," and "epibolin." Besides mediating cell adhesion, vitronectin can also protect cells from cytolytic destruction by released activated complement complexes and protect thrombin from inactivation by antithrombin III (reviewed by Tomasini and Mosher, 1991).

4.2. Structure and Location

The vitronectin molecule is a 75-kDa monomer that can be proteolytically clipped to a disulfide-bonded 65- + 10-kDa form; plasma contains both forms in varying ratios in different individuals, at a total concentration of 0.2–0.4 g/liter (Kubota *et al.*, 1988, and references therein). The form of vitronectin in tissues remains to be determined. The vitronectin gene is relatively small, extending only 3 kb in length and containing eight exons (Suzuki *et al.*, 1985; Jenne and Stanley, 1985, 1987). Exon 2 encodes the sequence for somatomedin B. Proteolysis of vitronectin apparently releases this polypeptide, whose function remains unknown. Both highly purified somatomedin B and vitronectin lack growth factor activity (Barnes *et al.*, 1984).

Vitronectin is also present in fibrillar form in the extracellular matrices of a variety of tissues, where it sometimes colocalizes with fibronectin. In adult skin, it is located at the periphery of dermal elastic fibers; however, it does not necessarily colocalize with fibrillin in microfibrils. Surprisingly, vitronectin is absent from the skin of children (Dahlbäck *et al.*, 1989).

4.3. Cell Attachment Domain

The cell attachment activity of vitronectin is encoded by exon 3, and is based on the sequence Arg-Gly-Asp (Suzuki *et al.*, 1985; Jenne and Stanley, 1985, 1987). This attachment activity is recognized by a wide variety of cell types. In fact, most of the cell adhesive activity of serum used for tissue culture at 5–10% can be attributed to vitronectin (fibronectin becomes more important only at low concentrations of serum; Knox, 1984; Grinnell and Phan, 1983; Bale *et al.*, 1989).

Attachment to vitronectin occurs via any of several vitronectin receptors in the integrin family, including $\alpha_v\beta_3$, $\alpha_{IIb}\beta_3$, and $\alpha_v\beta_5$. It is important to note that these receptors also bind to fibronectin and in some cases to other proteins; for example, $\alpha_{IIb}\beta_3$ is the major platelet aggregation receptor that binds to the glycoproteins fibronectin, vitronectin, von Willebrand factor, and fibrinogen, all of which contain the canonical Arg-Gly-Asp (RGD) recognition sequence. The binding site on the first two integrin receptors has been identified by cross-linking experiments using synthetic peptide ligands; current data suggest that both α and β chains contribute to binding of this peptide sequence (D'Souza *et al.*, 1990). These receptors appear to have variations of a binding specificity that is strongly focused on the RGD sequence itself. Unlike the $\alpha_5\beta_1$ receptor, these receptors can bind readily to immobilized synthetic peptides containing this sequence, and there is good evidence that the local sequence and conformation of such peptides contribute substantially to binding; these results indicate the importance of RGD conformation in recognition by these particular receptors (Pierschbacher and Ruoslahti, 1987). Nevertheless, even the classically RGD-recognizing $\alpha_{IIb}\beta_3$ platelet receptor appears to require additional sequence information for binding to fibronectin (Bowditch *et al.*, 1990); whether other sequences in vitronectin and its other ligands also contribute to binding by this receptor remains to be determined.

In terms of cell surface localization, vitronectin–vitronectin receptor complexes are prominently located at "focal contact" (focal adhesion) regions of cells in association with intracellular cytoskeletal elements such as vinculin, talin, paxillin, and the ends of actin microfilament bundles (Burridge *et al.*, 1988; Dejana *et al.*, 1988; Singer *et al.*, 1988). This *in vivo* relevance of this striking type of transmembrane complex observed *in vitro* will be interesting to unravel.

4.4. Heparin-Binding and Other Binding Activities

Vitronectin contains a cryptic heparin-binding domain that is exposed (along with certain antibody epitopes) after adsorption onto a surface or after denaturation by agents such as urea or guanidine (reviewed in Tomasini and Mosher, 1991). This property has been used as the basis for a simple, efficient purification protocol for this glycoprotein (Yatohgo *et al.*, 1988). In the absence of denaturing treatments, only small amounts of the heparin-binding form of vitronectin are found in plasma and in serum (2 and 7%, respectively) in the form of large aggregates, which have similar cell adhesion activities as monomeric vitronectin (Izumi *et al.*, 1988). Activated vitronectin binds to heparin with high affinity (10^{-8} M), and can neutralize heparin anticoagulant activity (Preissner and Müller-Berghaus, 1987). The expression of heparin-binding activity after activation by denaturants is accompanied by the exposure of new epitopes. The current interpretation of these findings is that the vitronectin molecule is normally folded to conceal this region (Fig. 4-6), and unfolding

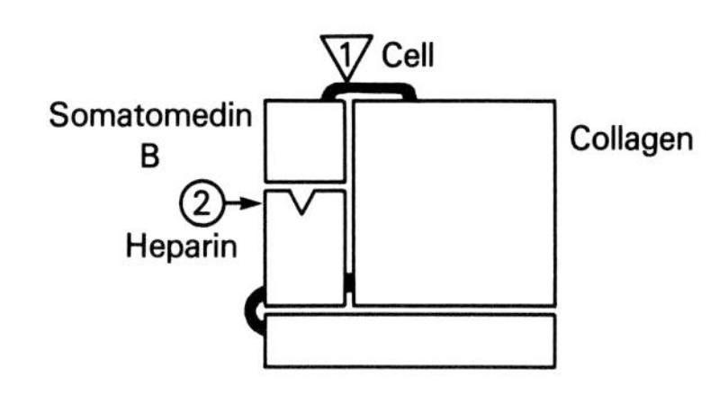

Figure 4-6. Schematic model of the major native form of vitronectin. The glycoprotein vitronectin usually exists as a relatively globular, monomeric protein containing an amino-terminal region that can be proteolytically cleaved to release the plasma protein somatomedin B. A cell-binding sequence (numeral 1) consists of the amino acid sequence Arg-Gly-Asp-(Val). A putative collagen-binding domain can be cleaved proteolytically from an adjacent part of the protein at protease-sensitive connecting strands indicated by the thicker lines. The heparin-binding domain (numeral 2) is usually cryptic (V-shaped groove), and it generally becomes exposed or available only after denaturation or unfolding of the molecule.

exposes this cryptic site. Additionally, urea-treated vitronectin binds to native collagen types I–VI, although this binding is minimal at physiological salt concentrations (see Izumi *et al.*, 1988). The biological relevance of this interaction therefore remains to be determined.

Vitronectin also binds plasminogen activator inhibitor (PAI-1) and stabilizes its activity (reviewed in Tomasini and Mosher, 1991). These functions may help regulate the localization and activity of this important inhibitor of plasminogen activator, a protease implicated in a host of tissue remodeling events including implantation, cell migration, and tumor cell invasion (see Chapter 8).

5. Thrombospondin

5.1. Introduction

Thrombospondin is a large, multifunctional glycoprotein that is released when platelets are activated, but it is also secreted continuously by a variety of other cell types during growth (reviewed by Lawler, 1986; Frazier, 1987; Asch and Nachman, 1989; Mosher, 1990). It interacts with cells and binds to several ECM molecules. Major activities of this protein include mediating or inhibiting cell adhesion, and regulation of growth of certain cells. A causal role for thrombospondin in cell growth has been suggested from experiments with smooth muscle cells, in which antibodies against thrombospondin inhibit cell proliferation by arresting cells in the G1 phase of the cell cycle. Heparin-mediated inhibition of the interaction of thrombospondin with these cells also inhibits proliferation (Majack *et al.*, 1988).

5.2. Structure and Location

Thrombospondin is a trimeric glycoprotein comprised of three identical 140-kDa subunits joined together by a local region of interchain disulfide bonding (Fig. 4-7). The molecule contains an amino-terminal heparin-binding domain that binds avidly to heparin (K_D = 80 nM). The next structural feature on each chain is the site of interchain disulfide bonding, which links each subunit into a trimer. The center of the molecule consists of a relatively linear domain, and the protein terminates in a large carboxyl-terminal domain that binds Ca^{2+}. Thrombospondin is the most abundant protein of platelet alpha granules, but is also a component of a variety of extracellular matrices, located in embryonic basement membranes, around epithelial cells, and associated with peripheral nerves, myoblasts, and chondroblasts. After differentiation, levels of this protein decrease (O'Shea and Dixit, 1988).

Thrombospondin synthesis can be stimulated by transforming growth factor beta (TGF-β), platelet-derived growth factor (PDGF), heparin, and heat shock (Penttinen *et al.*, 1988; Lyons-Giordano *et al.*, 1989; Ketis *et al.*, 1988). Analysis

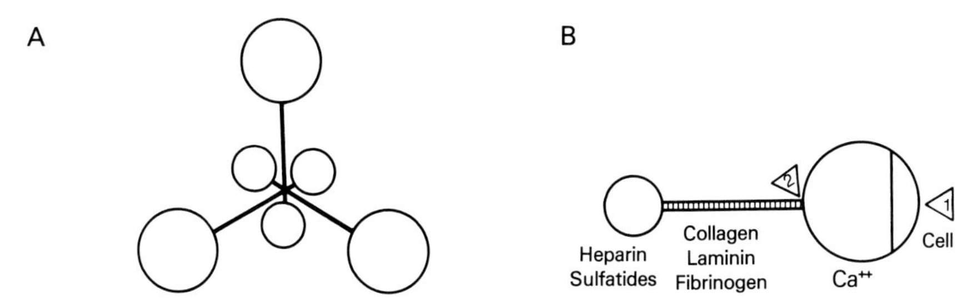

Figure 4-7. Schematic models of the overall morphology and structure of thrombospondin. A shows the general shape of the glycoprotein thrombospondin, which is a trimer of identical subunits linked together by disulfide bonds located near the smaller globular domain. The globular domain is attached to a rodlike domain consisting of repeats (panel B, vertical hatching); the other end of the rod connects to a larger globular domain. There are at least three cell-binding sites: the first is on the large globule, which also binds Ca^{2+}, and the second contains the sequence Arg-Gly-Asp-; the third is the heparin-binding domain (which can be the strongest adhesion site in certain cells). Other regions for binding specific ligands are indicated along the bottom of the figure.

of the promoter region of the thrombospondin gene has been initiated to elucidate this complexity of regulation (Donoviel *et al.*, 1988; Laherty *et al.*, 1989).

5.3. Cell-Binding Sites

Thrombospondin can bind to cell surface receptors, mediating or modulating cell adhesion. Depending on the cell type, this interaction can lead to cell spreading or aggregation, or even to negative effects on strength of adhesion, e.g., to an inhibition of cell adhesion to fibronectin. For example, thrombospondin can mediate or modulate platelet aggregation (reviewed in Asch and Nachman, 1989; Mosher, 1990). Moreover, substrates coated with thrombospondin mediate attachment and spreading of keratinocytes (Varani *et al.*, 1988). In contrast, however, thrombospondin inhibits the adhesion of endothelial cells to serum-coated substrates or to fibronectin (Lahav, 1988). In cultured fibroblasts, thrombospondin can partially inhibit cell adhesion to fibronectin, producing a decrease in focal contacts (Murphy-Ullrich and Höök, 1989). This complexity of effects of thrombospondin may be due at least in part to the existence of at least two distinct cell interaction sites on this protein.

Strong cell adhesive interactions with thrombospondin are mediated by the amino-terminal heparin-binding domain, which can bind to cell surface sulfated lipids (sulfatides) or to membrane-inserted heparan sulfate proteoglycan (Fig. 4-7B; Roberts and Ginsburg, 1988; Kaesberg *et al.*, 1989). This heparin-binding domain can also mediate the incorporation of thrombospondin into ECM (e.g., see Prochownik *et al.*, 1989). Addition of exogenous heparin or antibodies to this domain can often block cell adhesion.

Nevertheless, other studies using different adhesion assays or cells suggest that other regions of thrombospondin can also mediate cell adhesion. For example, keratinocytes and hematopoietic progenitor cells do not adhere to the 25-

kDa amino-terminal heparin-binding domain, but instead appear to use carboxyl-terminal sequence(s) in the large residual fragment (Varani *et al.*, 1988; Long and Dixit, 1990). The best-characterized alternative adhesion site for certain cells is the Arg-Gly-Asp sequence present in a predicted calcium-sensitive loop region of the protein near the carboxyl-terminal end of the connecting rod (Fig. 4-7B, site 2). In some assays, attachment of several cell types to thrombospondin could be inhibited by a synthetic peptide containing this adhesive recognition sequence (Lawler *et al.*, 1988). This site probably accounts for the binding of thrombospondin by the integrin receptors $\alpha_v\beta_3$ and $\alpha_{IIb}\beta_3$ (Lawler and Hynes, 1989). Consistent with the complexity of thrombospondin function, however, another proposed receptor is platelet glycoprotein IV (Silverstein *et al.*, 1989).

5.4. Heparin-, Calcium-, and Fibronectin-Binding Domains

Thrombospondin is secreted by cells and eventually becomes incorporated into a fibrillar matrix surrounding cells. Depending on the cell type and time in culture, it can be organized into 100- to 300-nm spherical granules together with heparan sulfate proteoglycan (Vischer *et al.*, 1988) in patterns distinct from fibronectin, or can become organized into fibrils that colocalize with fibronectin. The latter organization appears to require a preexisting matrix as a scaffolding for assembly, and involves one or more heparin-binding domains. Such domains also bind to fibronectin (Dardik and Lahav, 1989). The major heparin-binding region is a globular domain at the amino-terminus of thrombospondin that binds to heparin with relatively high affinity ($K_D = 7–8 \times 10^{-8}$ M). In mutagenesis studies, heparin-binding and incorporation of this domain into the ECM can occur as long as it retains a critical intrachain disulfide bridge region immediately adjacent to the heparin-binding domain (Prochownik *et al.*, 1989). Thrombospondin from endothelial cells binds to fibronectin with modest affinity ($K_D = 0.7 \times 10^{-7}$ M) using two distinct domains, a 70-kDa core fragment similar to that in platelet thrombospondin ($K_D = 3 \times 10^{-7}$ M) and another domain of 27 kDa that is reportedly unique to endothelial cell thrombospondin ($K_D = 9 \times 10^{-7}$ M). Heparin competitively inhibits binding of intact thrombospondin and of the 27-kDa fragment of fibronectin (Dardik and Lahav, 1989).

6. Tenascin

6.1. Introduction

Tenascin (also known as cytotactin, J1, or hexabrachion) exists as a striking, six-armed, star-shaped extracellular complex of about 1.9 million daltons consisting of similar subunits linked by interchain disulfide bonds (Fig. 4-8; recent reviews include Erickson and Bourdon, 1989; Chiquet-Ehrismann, 1990;

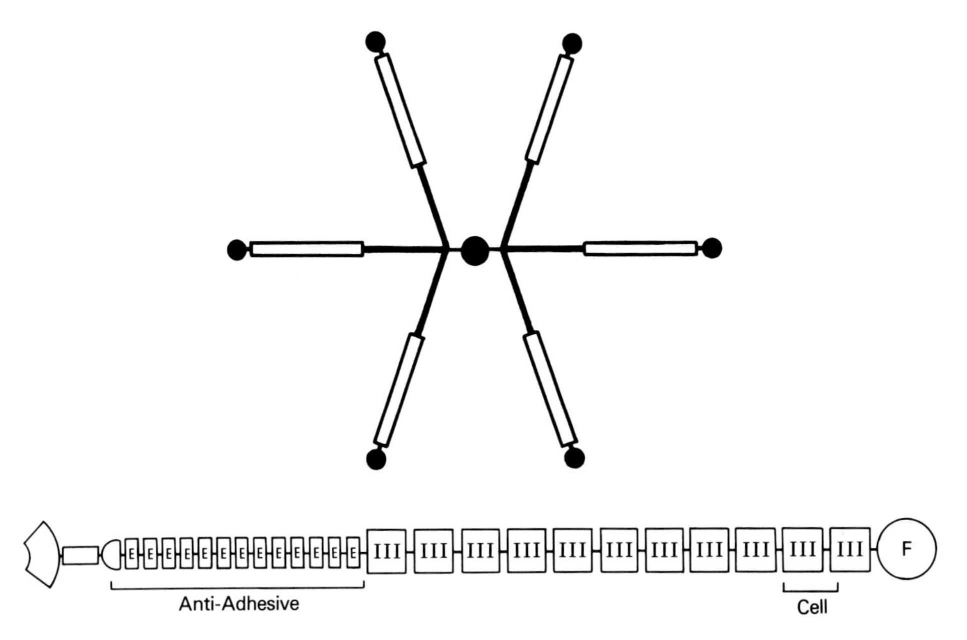

Figure 4-8. Schematic models of the overall morphology and structure of tenascin (also termed cytotactin or J1). This unusually large glycoprotein is a disulfide-linked complex comprised of six similar subunits joined at their amino-termini to form a central knot (top). The structure of a single tenascin subunit shown below includes EGF repeats (E) associated with a polypeptide region that mediates antiadhesive function. Next are at least 11 fibronectin type III repeats including a region of relatively weak cell adhesive activity based on an Arg-Gly-Asp- sequence. Some of these type III units can be deleted in tenascin isoforms due to alternative slicing of precursor tenascin mRNA. The carboxyl-terminus consists of a fibrinogen-like knob located at the distal end of each arm of the intact molecule.

Hoffman *et al.*, 1990). Tenascin is an unusual cell interaction protein, capable of mediating both adhesive and repulsive (antiadhesive) interactions, as well as binding to certain proteoglycans and fibronectin. Interest in this protein is further increased by the finding that its expression is particularly closely associated with morphogenetic events, including embryonic cell migration, wound healing, and tumorigenesis. As described in more detail below, its apparent complexities of function may permit tenascin to play important roles in helping to regulate morphogenetic movements.

6.2. Structure and Location

Native tenascin molecules are predominantly six-subunit oligomers, in which interchain disulfide bonds near the amino-terminus link subunits of about 190 to 230 kDa into a large, six-armed complex or hexabrachion (Fig. 4-8A; Erickson and Bourdon, 1989; Hoffman *et al.*, 1990). Tenascin is characterized by structural units homologous to those of other proteins: it includes 13 EGF-like repeats, 8–11 fibronectin type III repeats, and a globular carboxyl-terminus homologous to fibrinogen that includes a calcium-binding region (Fig. 4-8B). Tenascin undergoes alternative splicing of its precursor mRNA to produce at least three variant molecules, due to removal of sequences encoding three type III fibronectin repeats (Jones *et al.*, 1989; Spring *et al.*, 1989; Chiquet-Ehrismann, 1990). Although its three-dimensional structure remains to be determined, a predominance of beta structure is suggested by its far-UV circular dichroism spectrum (Taylor *et al.*, 1989).

Tenascin is frequently found at sites of tissue remodeling. In embryonic development, its pattern of localization may correlate with pathways of migration (see Chapter 12). *In vitro* analyses of tenascin effects on neural crest cell migration reveal that crest cells may migrate more rapidly on tenascin than on fibronectin. Interestingly, crest cells on tenascin appear rounded and lack lamellipodia, whereas they are more flattened (and presumably more adhesive) on fibronectin or laminin. When added to medium, however, tenascin can be an inhibitor of cell migration (Chapter 12). These differing effects suggest that tenascin functions to decrease cell–substrate interactions and to weaken tractional forces on migrating cells.

In adults, tenascin reappears at the edges of healing wounds, particularly beneath migrating, proliferating epidermal cells at the dermal–epidermal junction; its distribution differs from that of fibronectin, which is also thought to be involved in migration during wound healing (Mackie *et al.*, 1988). Levels of tenascin are also increased during nerve regeneration and in association with various tumors (see Erickson and Bourdon, 1989; Hoffman *et al.*, 1990; Chiquet-Ehrismann, 1990). Tenascin synthesis is stimulated by serum and by TGF-β (Pearson *et al.*, 1988), suggesting that it may be regulated in part by growth factors associated with wounds, morphogenetic events, or tumors.

6.3. Cell Interactions

The interaction of cells with tenascin is quite weak, and unlike adhesion to fibronectin, does not strengthen over time (Lotz *et al.*, 1989). A major cell-binding site of tenascin involved in such adhesion has been localized by antibody inhibition studies and proteolytic fragments, and is present in the two fibronectin type III repeating units toward the carboxyl-terminus of the protein in the tenth and eleventh fibronectin type III repeats (Fig. 4-8; Friedlander *et al.*, 1988; Spring *et al.*, 1989). A monoclonal antibody against this region can block cell adhesion to tenascin, supporting the importance of this site in tenascin-mediated cell attachment. Besides this site, there is an Arg-Gly-Asp (RGD) sequence in the third fibronectin type III repeat that may also function in some cells by interaction with an integrin receptor (e.g., see Bourdon and Ruoslahti, 1989).

Tenascin can, however, also exert antiadhesive activity. Mixing tenascin with fibronectin when substrates are prepared for adhesion assays results in an inhibition of fibronectin function; tenascin can also inhibit adhesion to laminin and to the GRGDS peptide (Chiquet-Ehrismann *et al.*, 1988; Lotz *et al.*, 1989). The mechanism of this inhibition is not yet clear. A distinct antiadhesive region involved in this function can be mapped on the tenascin molecule to the region of EGF-like repeats adjacent to the amino-terminal globular domain (Chiquet-Ehrismann *et al.*, 1988; Spring *et al.*, 1989). On the other hand, another study suggests that the effect is due to simple steric inhibition of cell access to adhesion proteins, since tenascin can also block access of antibodies to the substrate (Lightner and Erickson, 1990).

Whatever its mechanism, inhibition of cell interactions by tenascin can also be demonstrated *in vivo*. Exogenous tenascin injected into developing amphibian gastrulas causes an arrest of gastrulation, an effect attributable to its inhibitory effect on mesodermal cell migration (Riou *et al.*, 1990). Further studies of the mechanism of such tenascin-induced inhibition will be important in light of the general conceptual importance of possible repulsive activities in cell interaction molecules during embryonic development (Keynes and Cook, 1990).

6.4. Interactions with Other Extracellular Molecules

Tenascin binds to a particular chondroitin sulfate proteoglycan, and this interaction is dependent on divalent cations. The localization of this interacting pair of molecules in embryos is similar, but not always coincident; in nervous tissue, tenascin is synthesized by glia, whereas its binding proteoglycan is specifically synthesized by neurons (Hoffman *et al.*, 1988). The precise biological importance of the pairing of these two extracellular molecules remains to be explored further.

Tenascin also binds weakly to fibronectin, and in fact contaminates fibronectin preparations that are not subjected to gel filtration. The interaction is readily reversible, but it appears to be specific in that tenascin was not found to bind to laminin or to collagens (Lightner and Erickson, 1990). This interaction may be functionally important, since it is thought to inhibit cell interaction with fibronectin (Section 6.3).

7. Entactin (Nidogen)

7.1. Introduction

Entactin (also known as nidogen) is a 150-kDa sulfated glycoprotein found ubiquitously in basement membranes (see Chapter 11). Entactin binds tightly to laminin, and in fact nearly all laminin preparations contain the glycoprotein bound noncovalently in a highly stable 1:1 complex (the laminin–entactin complex).

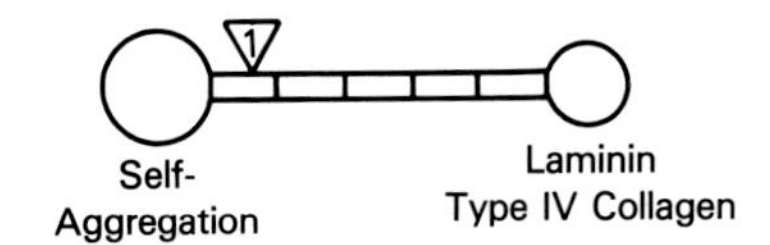

Figure 4-9. Schematic model of entactin. This dumbbell-shaped glycoprotein consists of two globular domains with distinct functions (listed underneath each) connected by five repeats. The numbered triangle indicates an Arg-Gly-Asp-(Gly) adhesive recognition site. Entactin molecules can form clusters of variable sizes by self-aggregating using their amino-terminal globular domains. The laminin- and type IV collagen-binding domains appear to promote association of these larger structural molecules in basement membranes.

7.2. Structure

Entactin is a dumbbell-shaped molecule (Fig. 4-9), with a large amino-terminal globular domain of 70 kDa (641 amino acid residues) separated by a cysteine-rich connecting domain of 28 kDa and a carboxyl-terminal globular domain of 36 kDa (328 residues). The rodlike connecting domain consists of five or six EGF-type cysteine-rich repeat units and one copy of a cysteine-rich motif found in thyroglobulin (Durkin *et al.*, 1988; Mann *et al.*, 1989).

7.3. Functional Sites

Entactin can mediate cell adhesion by a variety of cell types (Chakravarti *et al.*, 1990). The functional sequence appears to be the Arg-Gly-Asp sequence present in one of the EGF-like repeats toward the carboxyl-terminal end of the rodlike domain.

Entactin binds to laminin to form a stable noncovalent complex; the interaction is of relatively high affinity compared with other interactions of ECM molecules (K_D = 10 nM). The laminin-binding activity of this protein is present in the carboxyl-terminal (smaller) globular domain. Entactin appears to bind to the short arm(s) of laminin (Mann *et al.*, 1988).

Purified entactin and the laminin–entactin complex both bind readily to collagen type IV. The collagen IV binding site on entactin is present in the carboxyl-terminal globular domain. Purified entactin binds to a site approximately 80 nm from the carboxyl-terminus of the collagen IV triple helix, although a second site may also exist; the entactin–laminin complex has a less localized binding pattern, perhaps due to additional binding sites on laminin itself (Aumailley *et al.*, 1989). The association of the major basement membrane proteins laminin and type IV collagen may therefore result from a combination of weaker direct interactions and stronger bridging between them by entactin.

8. Other Cell Interactive Molecules

A rapidly increasing number of extracellular glycoproteins that are involved in cell surface interactions are being identified each year. They include a number of putative adhesive molecules that function by mechanisms that remain to be elucidated, as well as others whose function can be attributed to the presence of an Arg-Gly-Asp (RGD) sequence. Potentially quite interesting molecules with unknown receptors or mechanisms of action include epinectin, which can mediate epithelial cell adhesion and bind heparin (Enenstein and Furcht, 1988), chondronectin (reviewed in the first edition of this book), SPARC (secreted protein acidic and rich in cysteine; Lane and Sage, 1990), and fibrillin (Sakai *et al.*, 1986.).

Another, major class of molecules contains the RGD sequence. It is important to note that the mere presence of this sequence in a protein does not mean that the protein can function as an adhesion protein: immunoglobulin G, β-galactosidase, and hundreds of other proteins contain this sequence, yet do not display adhesive activity. In fact, the frequency of this sequence in data banks appears to be close to that expected if its distribution were random. Nevertheless, a rapidly growing number of interesting proteins contain this tripeptide sequence and display adhesive activity. Activity shown by *in vitro* attachment assays, of course, does not necessarily guarantee functions *in vivo*.

Perhaps the best-characterized and physiologically important group of RGD-dependent molecules are those involved in platelet aggregation. This group includes fibrinogen, fibronectin, von Willebrand factor, vitronectin, and possibly thrombospondin (e.g., see review by Ginsberg *et al.*, 1988). Interestingly, viper venoms can contain a number of RGD-containing, cysteine-rich peptides with specific conformations that render the peptides unusually active as inhibitors of $\alpha_{IIb}\beta_3$; they have been termed "disintegrins," since they block platelet integrins and disrupt platelet adhesion (Musial *et al.*, 1990).

Another set of RGD-containing molecules is found in bone, including osteopontin and bone sialoprotein (reviewed in Heinegård and Oldberg, 1989). These molecules are hypothesized to mediate osteoblast and/or osteoclast adhesive interactions, probably by the use of a vitronectin receptor (reviewed in Heinegård and Oldberg, 1989).

Yet another type of RGD-containing molecule is found on microorganisms that use this sequence for attaching to target cells as a prelude to infection. For example, an RGD tripeptide embedded in the sequence of filamentous hemagglutinin is involved in the adherence of Bordetella pertussis to ciliated eukaryotic cells using the $\alpha_M\beta_2$ (Mac-1) integrin receptor (Relman *et al.*, 1990). Similarly, foot-and-mouth disease virus has been shown to use its RGD sequence for attachment to fibroblasts (Fox *et al.*, 1989). In these cases, pathogenic microorganisms take advantage of eukaryotic cell surface receptors for ECM molecules in order to attach to target cells, a key first step in infection.

9. Concluding Remarks

We can expect that the explosive growth in our understanding of cell interactive molecules such as fibronectin and laminin that has occurred during the decade since the first edition of this book will continue for many more years. The way in which cell surface receptors use simple peptide recognition sequences such as RGD, LDV, and others for cell attachment and subsequent biological events provides a simple explanation for how cells interact with ECM molecules. This basic paradigm for cell adhesive interactions will need further development, since it provides only a partial explanation: the conformation of these sequences and the possible involvement of more than one peptide recognition sequence in binding to receptors both need evaluation as explanations for the observation that synthetic peptides often display lower affinity and specificity than the intact protein. Moreover, many adhesive proteins contain the RGD sequence, yet receptors can often readily distinguish between these proteins.

Unique conformations or specific synergy sequences may explain how the fibronectin receptor binds so well to fibronectin, yet not to many other RGD-containing molecules, and the use of such mechanisms by other ECM proteins needs evaluation. Adhesive specificity in cell interactions with ECM molecules may depend not only on the composition and conformation of the minimal sequence binding to the active site of the receptor, but also on additional interaction sites elsewhere in the protein that provide specificity. It is important to emphasize that protein–protein interactions usually involve more than a simple linear sequence. In this regard, determination of the three-dimensional structure of crucial regions of cell interactive proteins and their receptors by X-ray crystallography or three-dimensional NMR will be important for future understanding of their functions.

The recognition of adhesion and cell interactive proteins by rapidly growing families of receptors is now well established. More members of each family, and probably even new families of receptors, are to be expected. A sobering finding seen more and more often recently is that there is a high degree of redundancy or overlap in receptor specificities. There are at least five fibronectin-binding receptors and a similar number of distinct laminin-binding receptors; more than one can be expressed at any time on the same cell. Such redundancy may be important for subtle regulation of cell interactions with each adhesive protein, and to permit cell interactions even in the presence of mutations or inhibitor of a single receptor, but it complicates our understanding of the function of each receptor.

Studies on various types of regulation of these molecules and their receptors will be a major area of future research, since many cell biological processes and disease states will probably show alterations in ECM expression and differential splicing, as well as alterations in the type and posttranslational modifications of receptors. Two basic questions that will need to be answered concern whether these regulated alterations are (1) necessary or merely associated, and (2) permissive or instructive, for biological processes. That is, does regulation of these molecules directly regulate morphogenetic or pathological processes, or is it usually just associated or contributory? The possibility that abnormal levels of some of these proteins or of their receptors can serve as markers of specific diseases, and thus as diagnostic tools, remains to be established.

Elucidating the evolution of these ECM proteins and their receptors should be of great interest. Were short peptide sequences inserted into cell interactive proteins to provide recognition sites? For example, alignment of the type III repeats of fibronectin can be interpreted as showing that the RGD sequence is an extra addition (Kornblihtt *et al.*, 1985). Alternatively, since tripeptide sequences occur relatively frequently by chance, existing sequences that initially had no function might have evolved into binding sites by changes in exposure of these sites and in their conformation, and/or by the addition of functionally synergistic sequences.

The identification of key sequences in cell interactive proteins has provided tools for studying their functions *in vivo* (e.g., Boucaut *et al.*, 1984; discussed in Chapter 12). Since short peptides tend to have lower specificity and affinity than native molecules, more sophisticated experimental probes need to be developed. Another uncertainty concerns whether each of the peptide recognition sequences identified to date is actually functional in the intact molecule, or whether it requires proteolysis for its expression, or might even be an *in vitro* artifact. Using short peptides *in vivo* may elicit unrelated side effects, and the relationship of peptide effects to physiological effects of the original cell interaction protein needs confirmation by other methods, e.g., by immunological inhibition studies or by correlations between the activities of a series of inhibitors *in vitro* and their activities *in vivo*.

As the field progresses, more precise inhibitors of function will appear, and they should prove valuable in further analyses of the cell biology of embryonic development, wound healing, and pathogenesis of diseases such as fibrotic disorders. Application of these principles to clinical therapy might even be possible if highly specific, stable inhibitors or receptor-specific synthetic adhesive substrates can be developed.

ACKNOWLEDGMENTS. The original research reported in this chapter was performed at and was supported by the National Institute of Dental Research and the National Cancer Institute, National Institutes of Health, Bethesda, Maryland. I thank Susan Yamada for valuable assistance, and Hynda Kleinman and Yoshi Yamada for helpful comments on the manuscript.

References

Akiyama, S. K., and Yamada, K. M., 1985, Synthetic peptides competitively inhibit both direct binding to fibroblasts and functional biological assays for the purified cell-binding domain of fibronectin, *J. Biol. Chem.* **260**:10402–10405.

Akiyama, S. K., Hasegawa, E., Hasegawa, T., and Yamada, K. M., 1985, The interactions of fibronectin fragments with fibroblastic cells, *J. Biol. Chem.* **260**:13256–13260.

Asch, A. S., and Nachman, R. L., 1989, Thrombospondin: Phenomenology to function, *Prog. Hemost. Thromb.* **9**:157–176.

Aumailley, M., Wiedemann, H., Mann, K., and Timpl, R., 1989, Binding of nidogen and the laminin–nidogen complex to basement membrane collagen type IV, *Eur. J. Biochem.* **184**:241–248.

Aumailley, M., Gerl, M., Sonnenberg, A., Deutzmann, R., and Timpl, R., 1990, Identification of the Arg-Gly-Asp sequence in laminin A chain as a latent cell-binding site being exposed in fragment P1, *FEBS Lett.* **262**:82–86.

Bale, M. D., Wohlfahrt, L. A., Mosher, D. F., Tomasini, B., and Sutton, R. C., 1989, Identification of vitronectin as a major plasma protein absorbed on polymer surfaces of different copolymer composition, *Blood* **74:**2698–2706.

Barnes, D. W., Foley, T. P., Shaffer, M. C., and Silnutzer, J. E., 1984, Human serum spreading factor: Relationship to somatomedin B, *J. Clin. Endocrinol. Metab.* **59:**1019–1021.

Baron, M., Norman, D., Willis, A., and Campbell, I. D., 1990, Structure of the fibronectin type 1 module, *Nature* **345:**642–646.

Beck, K., Hunter, I., and Engel, J., 1990, Structure and function of laminin: Anatomy of a multidomain glycoprotein, *FASEB J.* **4:**148–160.

Benecky, M. J., Kolvenbach, C. G., Wine, R. W., DiOrio, J. P., and Mosesson, M. W., 1990, Human plasma fibronectin structure probed by steady-state fluorescence polarization: Evidence for a rigid oblate structure, *Biochemistry* **29:**3082–3091.

Boucaut, J.-C., Darribére, T., Poole, T. J., Aoyama, H., Yamada, K. M., and Thiery, J. P., 1984, Biologically active synthetic peptides as probes of embyonic development: A competitive peptide inhibitor of fibronectin function inhibits gastrulation in amphibian embryos and neural crest cell migration in avian embryos, *J. Cell Biol.* **99:**1822–1830.

Bourdon, M. A., and Ruoslahti, E., 1989, Tenascin mediates cell attachment through an RGD-dependent receptor, *J. Cell Biol.* **108:**1149–1155.

Burridge, K., Fath, K., Kelly, T., Nuckolls, G., and Turner, C., 1988, Focal adhesions: Transmembrane junctions between the extracellular matrix and the cytoskeleton, *Annu. Rev. Cell Biol.* **4:**487–525.

Carsons, S. E. (ed.), 1989, *Fibronectin in Health and Disease,* CRC Press, Boca Raton.

Chakravarti, S., Tam, M. F., and Chung, A. E., 1990, The basement membrane glycoprotein entactin promotes cell attachment and binds calcium ions, *J. Biol. Chem.* **265:**10597–10603.

Chiquet-Ehrismann, R., 1990, What distinguished tenascin from fibronectin? *FASEB J.* **4:**2598–2604.

Chiquet-Ehrismann, R., Kalla, P., Pearson, C. A., Beck, K., and Chiquet, M., 1988, Tenascin interferes with fibronectin action, *Cell* **53:**383–390.

Dahlbäck, K., Löfberg, H., Alumets, J., and Dahlbäck, B., 1989, Immunohistochemical demonstration of age-related deposition of vitronectin (S-protein of complement) and terminal complement complex on dermal elastic fibers, *J. Invest. Dermatol.* **92:**727–733.

Dardik, R., and Lahav, J., 1989, Multiple domains are involved in the interaction of endothelial cell thrombospondin with fibronectin, *Eur. J. Biochem.* **185:**581–588.

Darribére, T., Guida, K., Larjava, H., Johnson, K. E., Yamada, K. M., Thiery, J. P., and Boucaut, J.-C., 1990, In vivo analyses of integrin beta 1 subunit function in fibronectin matrix assembly, *J. Cell Biol.* **110:**1813–1823.

Dean, D. C., McQuillan, J. J., and Weintraub, S., 1990, Serum stimulation of fibronectin gene expression appears to result from rapid serum-induced binding of nuclear proteins to a cAMP response element, *J. Biol. Chem.* **265:**3522–3527.

Dejana, E., Colella, S., Conforti, G., Abbadini, M., Gaboli, M., and Marchisio, P. C., 1988, Fibronectin and vitronectin regulate the organization of their respective Arg-Gly-Asp adhesion receptors in cultured human endothelial cells, *J. Cell Biol.* **107:**1215–1223.

Deutzmann, R., Aumailley, M., Wiedemann, H., Pysny, W., Timpl, R., and Edgar, D., 1990, Cell adhesion, spreading and neurite stimulation by laminin fragment E8 depends on maintenance of secondary and tertiary structure in its rod and globular domain, *Eur. J. Biochem.* **191:**513–522.

Dillner, L., Dickerson, K., Manthorpe, M., Ruoslahti, E., and Engvall, E., 1988, The neurite-promoting domain of human laminin promotes attachment and induces characteristic morphology in non-neuronal cells, *Exp. Cell Res.* **177:**186–198.

Donoviel, D. B., Framson, P., Eldridge, C. F., Cooke, M., Kobayashi, S., and Bornstein, P., 1988, Structural analysis and expression of the human thrombospondin gene promoter, *J. Biol. Chem.* **263:**18590–18593.

D'Souza, S. E., Ginsberg, M. H., Burke, T. A., and Plow, E. F., 1990, The ligand binding site of the platelet integrin receptor GPIIb-IIIa is proximal to the second calcium binding domain of its alpha subunit, *J. Biol. Chem.* **265:**3440–3446.

Dufour, S., Duband, J.-L., Humphries, M. J., Obara, M., Yamada, K. M., and Thiery, J. P., 1988a, Attachment, spreading and locomotion of avian neural crest cells are mediated by multiple adhesion sites on fibronectin molecules, *EMBO J.* **7**:2661–2671.

Dufour, S., Duband, J.-L., Kornblihtt, A. R., and Thiery, J. P., 1988b, The role of fibronectins in embryonic cell migrations, *Trends Genet.* **4**:198–203.

Durkin, M. E., Chakravarti, S., Bartos, B. B., Liu, S. H., Friedman, R. L., and Chung, A. E., 1988, Amino acid sequence and domain structure of entactin. Homology with epidermal growth factor precursor and low density lipoprotein receptor, *J. Cell Biol.* **107**:2749–2756.

Ehrig, K., Leivo, I., Argraves, W. S., Ruoslahti, E., and Engvall, E., 1990, Merosin, a tissue-specific basement membrane protein, is a laminin-like protein, *Proc. Natl. Acad. Sci. USA* **87**:3264–3268.

Ehrismann, R., Roth, D. E., Eppenberger, H. M., and Turner, D. C., 1982, Arrangement of attachment-promoting, self-association, and heparin-binding sites in horse serum fibronectin, *J. Biol. Chem.* **257**:7381–7387.

Enenstein, J., and Furcht, L. T., 1988, Epithelial and neural localization and heparin binding of the cell-substratum adhesion molecule, epinectin, *J. Invest. Dermatol.* **91**:34–38.

Erickson, H. P., and Bourdon, M. A., 1989, Tenascin: An extracellular matrix protein prominent in specialized embryonic tissues and tumors, *Annu. Rev. Cell Biol.* **5**:71–92.

Fava, R. A., and McClure, D. B., 1987, Fibronectin-associated transforming growth factor, *J. Cell. Physiol.* **131**:184–189.

Fogerty, F. J., Akiyama, S. K., Yamada, K. M., and Mosher, D. F., 1990, Inhibition of binding of fibronectin to matrix assembly sites by anti-integrin ($\alpha_5\beta_1$) antibodies, *J. Cell Biol.* **111**:699–708.

Fox, G., Parry, N. R., Barnett, P. V., McGinn, B., Rowlands, D. J., and Brown, F., 1989, The cell attachment site on foot-and-mouth disease virus includes the amino acid sequence RGD (arginine-glycine-aspartic acid), *J. Gen. Virol.* **70**:625–637.

Frazier, W. A., 1987, Thrombospondin: A modular adhesive glycoprotein of platelets and nucleated cells, *J. Cell Biol.* **105**:625–632.

Friedlander, D. R., Hoffman, S., and Edelman, G. M., 1988, Functional mapping of cytotactin: Proteolytic fragments active in cell–substrate adhesion, *J. Cell Biol.* **107**:2329–2340.

Gehlsen, K. R., Dickerson, K., Argraves, W. S., Engvall, E., and Ruoslahti, E., 1989, Subunit structure of a laminin-binding integrin and localization of its binding site on laminin, *J. Biol. Chem.* **264**:19034–19038.

Ginsberg, M. H., Loftus, J. C., and Plow, E. F., 1988, Cytoadhesins, integrins, and platelets, *Thromb. Haemost.* **59**:1–6.

Graf., J., Iwamoto, Y., Sasaki, M., Martin, G. R., Kleinman, H. K., Robey, F. A., and Yamada, Y., 1987, Identification of an amino acid sequence in laminin mediating cell attachment, chemotaxis, and receptor binding, *Cell* **48**:989–996.

Grant, D. S., Tashiro, K., Segui-Real, B., Yamada, Y., Martin, G. R., and Kleinman, H. K., 1989, Two different laminin domains mediate the differentiation of human endothelial cells into capillary-like structures in vitro, *Cell* **58**:933–943.

Grinnell, F., and Phan, T. V., 1983, Deposition of fibronectin on material surfaces exposed to plasma: Quantitative and biological studies, *J. Cell. Physiol.* **116**:289–296.

Guan, J. L., Trevithick, J. E., and Hynes, R. O., 1990, Retroviral expression of alternatively spliced forms of rat fibronectin, *J. Cell Biol.* **110**:833–847.

Hall, D. E., Reichardt, L. F., Crowley, E., Holley, B., Moezzi, H., Sonnenberg, A., and Damsky, C. H., 1990, The α_1/β_1 and α_6/β_1 integrin heterodimers mediate cell attachment to distinct sites on laminin, *J. Cell Biol.* **110**:2175–2184.

Hayashi, M., and Yamada, K. M., 1982, Divalent cation modulation of fibronectin binding to heparin and to DNA, *J. Biol. Chem.* **257**:5263–5267.

Heinegård, D., and Oldberg, A., 1989, Structure and biology of cartilage and bone matrix noncollagenous macromolecules, *FASEB J.* **3**:2042–2051.

Hershberger, R. P., and Culp, L. A., 1990, Cell-type-specific expression of alternatively spliced human fibronectin IIICS mRNAs, *Mol. Cell. Biol.*, **10**:662–671.

Hirano, H., Yamada, Y., Sullivan, M., de Crombrugghe, B., Pastan, I., and Yamada, K. M., 1983, Isolation of genomic DNA clones spanning the entire fibronectin gene, *Proc. Natl. Acad. Sci. USA* **80:**46–50.

Hoffman, S., Crossin, K. L., and Edelman, G. M., 1988, Molecular forms, binding functions, and developmental expression patterns of cytotactin and cytotactin-binding proteoglycan, an interactive pair of extracellular matrix molecules, *J. Cell Biol.* **106:**519–532.

Hoffman, S., Crossin, K. L., Jones, F. S., Friedlander, D. R., and Edelman, G. M., 1990, Cytotactin and cytotactin-binding proteoglycan. An interactive pair of extracellular matrix proteins, *Ann. N.Y. Acad. Sci.* **580:**288–301.

Humphries, M. J., Akiyama, S. K., Komoriya, A., Olden, K., and Yamada, K. M., 1986, Identification of an alternatively spliced site in human plasma fibronectin that mediates cell type-specific adhesion, *J. Cell Biol.* **103:**2637–2647.

Humphries, M. J., Komoriya, A., Akiyama, S. K., Olden, K., and Yamada, K. M., 1987, Identification of two distinct regions of the type III connecting segment of human plasma fibronectin that promote cell type-specific adhesion, *J. Biol. Chem.* **262:**6886–6892.

Humphries, M. J., Akiyama, S. K., Komoriya, A., Olden, K., and Yamada, K. M., 1988, Neurite extension of chicken peripheral nervous system neurons on fibronectin: Relative importance of specific adhesion sites in the central cell-binding domain and the alternatively spliced type III connecting segment, *J. Cell Biol.* **106:**1289–1297.

Hunter, D. D., Porter, B. E., Bulock, J. W., Adams, S. P., Merlie, J. P., and Sanes, J. R., 1989, Primary sequence of a motor neuron-selective adhesive site in the synaptic basal lamina protein s-laminin, *Cell* **59:**905–913.

Hynes, R. O., 1990, *Fibronectins*, Springer-Verlag, Berlin.

Ingham, K. C., Landwehr, R., and Engel, J., 1985, Interaction of fibronectin with C1q and collagen. Effects of ionic strength and denaturation of the collagenous component, *Eur. J. Biochem,* **148:**219–224.

Ingham, K. C., Brew, S. A., and Migliorini, M. M., 1989, Further localization of the gelatin-binding determinants within fibronectin. Active fragments devoid of type II homologous repeat modules, *J. Biol. Chem.* **264:**16977–16980.

Izumi, M., Shimo-Oka, T., Morishita, N., Ii, I., and Hayashi, M., 1988, Identification of the collagen-binding domain of vitronectin using monoclonal antibodies, *Cell Struct. Funct.* **13:**217–225.

Izzard, C. S., Radinsky, R., and Culp, L. A., 1986, Substratum contacts and cytoskeletal reorganization of BALB/c 3T3 cells on a cell-binding fragment and heparin-binding fragments of plasma fibronectin, *Exp. Cell Res.* **165:**320–336.

Jenne, D., and Stanley, K. K., 1985, Molecular cloning of S-protein, a link between complement, coagulation and cell–substrate adhesion, *EMBO J.* **4:**3153–3157.

Jenne, D., and Stanley, K. K., 1987, Nucleotide sequence and organization of the human S-protein gene: Repeating peptide motifs in the "pexin" family and a model for their evolution, *Biochemistry* **26:**6735–6742.

Jones, F. S., Hoffman, S., Cunningham, B. A., and Edelman, G. M., 1989, A detailed structural model of cytotactin: Protein homologies, alternative RNA splicing, and binding regions, *Proc. Natl. Acad. Sci. USA* **86:**1905–1909.

Kaesberg, P. R., Ershler, W. B., Esko, J. D., and Mosher, D. F., 1989, Chinese hamster ovary cell adhesion to human platelet thrombospondin is dependent on cell surface heparan sulfate proteoglycan, *J. Clin. Invest.* **83:**994–1001.

Kanemoto, T., Reich, R., Royce, L., Greatorex, D., Adler, S. H., Shiraishi, N., Martin, G. R., Yamada, Y., and Kleinman, H. K., 1990, Identification of an amino acid sequence from the laminin A chain that stimulates metastasis and collagenase IV production, *Proc. Natl. Acad. Sci. USA* **87:**2279–2283.

Ketis, N. V., Lawler, J., Hoover, R. L., and Karnovsky, M. J., 1988, Effects of heat shock on the expression of thrombospondin by endothelial cells in culture, *J. Cell Biol.* **106:**893–904.

Keynes, R., and Cook, G., 1990, Cell–cell repulsion: Clues from the growth cone? *Cell* **62:** 609–610.

Kleinman, H. K., and Weeks, B. S., 1991, The neural cell response to laminin: Active sites, receptors, and intracellular signals, *Comments Develop. Neurobiol.* in press.

Kleinman, H. K., Graf, J., Iwamoto, Y., Sasaki, M., Schasteen, C. S., Yamada, Y., Martin, G. R., and

Robey, F. A., 1989, Identification of a second active site in laminin for promotion of cell adhesion and migration and inhibition of in vivo melanoma lung colonization, *Arch. Biochem. Biophys.* **272:**39–45.

Knox, P., 1984, Kinetics of cell spreading in the presence of different concentrations of serum or fibronectin-depleted serum, *J. Cell Sci.* **71:**51–59.

Komoriya, A., Green, L. J., Mervic, M., Yamada, S. S., Yamada, K. M., and Humphries, M. J., 1991, The minimal essential sequence for a major cell type-specific adhesion site (CS1) within the alternatively spliced IIICS domain of fibronectin is Leu-Asp-Val, *J. Biol. Chem.* **266:** 15075–15079.

Kornblihtt, A. R., Umezawa, K., Vibe-Pedersen, K., and Baralle, F. E., 1985, Primary structure of human fibronectin: Differential splicing may generate at least 10 polypeptides from a single gene, *EMBO J.* **4:**1755–1759.

Kouzi-Koliakos, K., Koliakos, G. G., Tsilibary, E. C., Furcht, L. T., and Charonis, A. S., 1989, Mapping of three major heparin-binding sites on laminin and identification of a novel heparin-binding site on the B1 chain, *J. Biol. Chem.* **264:**17971–17978.

Kubota, K., Katayama, S., Matsuda, M., and Hayashi, M., 1988, Three types of vitronectin in human blood, *Cell Struct. Funct.* **13:**123–128.

Lahav, J., 1988, Thrombospondin inhibits adhesion of endothelial cells, *Exp. Cell Res.* **177:** 199–204.

Laherty, C. D., Gierman, T. M., and Dixit, V. M., 1989, Characterization of the promoter region of the human thrombospondin gene. DNA sequences within the first intron increase transcription, *J. Biol. Chem.* **264:**11222–11227.

Lane, T. F., and Sage, E. H., 1990, Functional mapping of SPARC: Peptides from two distinct Ca^{++}-binding sites modulate cell shape, *J. Cell Biol.* **111:**3065–3076.

Lash, J. W., Linask, K. K., and Yamada, K. M., 1987, Synthetic peptides that mimic the adhesive recognition signal of fibronectin: Differential effects on cell–cell and cell–substratum adhesion in embryonic chick cells, *Dev. Biol.* **123:**411–420.

Lawler, J., 1986, The structural and functional properties of thrombospondin, *Blood* **67:**1197–1209.

Lawler, J., and Hynes, R. O., 1989, An integrin receptor on normal and thrombasthenic platelets that binds thrombospondin, *Blood* **74:**2022–2027.

Lawler, J., Weinstein, R., and Hynes, R. O., 1988, Cell attachment to thrombospondin: The role of ARG-GLY-ASP, calcium, and integrin receptors, *J. Cell Biol.* **107:**2351–2361.

Liesi, P., Narvanen, A., Soos, J., Sariola, H., and Snounou, G., 1989, Identification of a neurite outgrowth-promoting domain of laminin using synthetic peptides, *FEBS Lett.* **244:**141–148.

Lightner, V. A., and Erickson, H. P., 1990, Binding of hexabrachion (tenascin) to the extracellular matrix and substratum and its effect on cell adhesion, *J. Cell Sci.* **95:**263–277.

Loftus, J. C., O'Toole, T. E., Plow, E. F., Glass, A., Frelinger, A. L., and Ginsberg, M. H., 1990, A β_3 integrin mutation abolishes ligand binding and alters divalent cation dependent conformation, *Science* **249:**915–918.

Long, M. W., and Dixit, V. M., 1990, Thrombospondin functions as a cytoadhesion molecule for human hematopoietic progenitor cells, *Blood* **75:**2311–2318.

Lotz, M. M., Burdsal, C. A., Erickson, H. P., and McClay, D. R., 1989, Cell adhesion to fibronectin and tenascin: Quantitative measurements of initial binding and subsequent strengthening response, *J. Cell Biol.* **109:**1795–1805.

Lyons-Giordano, B., Brinker, J. M., and Kefalides, N. A., 1989, Heparin increases mRNA levels of thrombospondin but not fibronectin in human vascular smooth muscle cells, *Biochem. Biophys. Res. Commun.* **162:**1100–1104.

Mackie, E. J., Halfter, W., and Liverani, D., 1988, Induction of tenascin in healing wounds, *J. Cell Biol.* **107:**2757–2767.

Majack, R. A., Goodman, L. V., and Dixit, V. M., 1988, Cell surface thrombospondin is functionally essential for vascular smooth muscle cell proliferation, *J. Cell Biol.* **106:**415–422.

Mann, K., Deutzmann, R., Aumailley, M., Timpl, R., Raimondi, L., Yamada, Y., Pan, T. C., Conway, D., and Chu, M. L., 1989, Amino acid sequence of mouse nidogen, a multidomain basement membrane protein with binding activity for laminin, collagen IV and cells, *EMBO J.***8:**65–72.

Martin, G. R., Timpl, R., and Kühn, K., 1988, Basement membrane proteins: Molecular structure and function, *Adv. Prot. Chem.* **39:**1–50.

McCarthy, J. B., Chelberg, M. K., Mickelson, D. J., and Furcht, L. T., 1988, Localization and chemical synthesis of fibronectin peptides with melanoma adhesion and heparin binding activities, *Biochemistry* **27**:1380–1388.

McCarthy, J. B., Skubitz, A. P., Qi, Z., Yi, X. Y., Mickelson, D. J., Klein, D. J., and Furcht, L. T., 1990, RGD-independent cell adhesion to the carboxy-terminal heparin-binding fragment of fibronectin involves heparin-dependent and -independent activities, *J. Cell Biol.* **110**:777–787.

McDonald, J. A., 1988, Extracellular matrix assembly, *Annu. Rev. Cell Biol.* **4**:183–207.

McDonald, J. A., Quade, B. J., Broekelmann, T. J., LaChance, R., Forsman, K., Hasegawa, E., and Akiyama, S., 1987, Fibronectin's cell-adhesive domain and an amino-terminal matrix assembly domain participate in its assembly into fibroblast pericellular matrix, *J. Biol. Chem.* **262**:2957–2967.

McKeown-Longo, P. J., and Mosher, D. F., 1985, Interaction of the 70,000-mol-wt amino fragment of fibronectin with the matrix-assembly receptor of fibroblasts, *J. Cell Biol.* **100**:364–374.

Mecham, R. P., Hinek, A., Griffin, G. L., Senior, R. M., and Liotta, L. A., 1989, The elastin receptor shows structural and functional similarities to the 67-kDa tumor cell laminin receptor, *J. Biol. Chem.* **264**:16652–16657.

Moos, M., Tacke, R., Scherer, H., Teplow, D., Früh, K., and Schachner, M., 1988, Neural adhesion molecule L1 as a member of the immunoglobulin superfamily with binding domains similar to fibronectin, *Nature* **334**:701–703.

Mosher, D. F. (ed.,), 1989, *Fibronectin,* Academic Press, New York.

Mosher, D. F., 1990, Physiology of thrombospondin, *Annu. Rev. Med.* **41**:85–97.

Mould, A. P., Komoriya, A., Yamada, K. M., and Humphries, M. J., 1991, Affinity chromatographic isolation of the melanoma adhesion receptor for the IIICS region of fibronectin and its identification as the integrin $\alpha_4\beta_1$, *J. Biol. Chem.* **265**:4020–4024.

Murphy-Ullrich, J. E., and Höök, M., 1989, Thrombospondin modulates focal adhesions in endothelial cells, *J. Cell Biol.* **109**:1309–1319.

Musial, J., Niewiarowski, S., Rucinski, B., Stewart, G. J., Cook, J. J., Williams, J. A., and Edmunds, L. H., 1990, Inhibition of platelet adhesion to surfaces of extracorporeal circuits by disintegrins. RGD-containing peptides from viper venoms, *Circulation* **82**:261–273.

Nagai, T., Yamakama, N., Aota, S., Yamada, S. S., Akiyama, S. K., Olden, K., and Yamada, K. M., 1991, Monoclonal antibody characterization of two distant sites required for function of the central cell-binding domain of fibronectin in cell adhesion, cell migration, and matrix assembly, *J. Cell Biol.* in press.

Nojima, Y., Humphries, M. J., Mould, A. P., Komoriya, A., Yamada, K. M., Schlossman, S. F., and Morimoto, C., 1990, VLA-4 mediates CD3-dependent CD4 + T cell activation via the CS1 alternatively spliced domain of fibronectin, *J. Exp. Med.* **172**:1185–1192.

Obara, M., Kang, M. S., and Yamada, K. M., 1988, Site-directed mutagenesis of the cell-binding domain of human fibronectin: Separable, synergistic sites mediate adhesive function, *Cell* **53**:649–657.

Olden, K., Pratt, R. M., and Yamada, K. M., 1979, Role of carbohydrate in biological function of the adhesive glycoprotein fibronectin, *Proc. Natl. Acad. Sci. USA* **76**:3343–3347.

O'Shea, K. S., and Dixit, V. M., 1988, Unique distribution of the extracellular matrix component thrombospondin in the developing mouse embryo, *J. Cell Biol.* **107**:2737–2748.

Owens, R. J., and Baralle, F. E., 1986, Mapping the collagen-binding site of human fibronectin by expression in Escherichia coli, *EMBO J.* **5**:2825–2830.

Oyama, F., Murata, Y., Suganuma, N., Kimura, T., Titani, K., and Sekiguchi, K., 1989, Patterns of alternative splicing of fibronectin pre-mRNA in human adult and fetal tissues, *Biochemistry* **28**:1428–1434.

Panayotou, G., End, P., Aumailley, M., Timpl, R., and Engel, J., 1989, Domains of laminin with growth-factor activity, *Cell* **56**:93–101.

Patel, R. S., Odermatt, E., Schwarzbauer, J. E., and Hynes, R. O., 1987, Organization of the fibronectin gene provides evidence for exon shuffling during evolution, *EMBO J.* **6**:2565–2572.

Pearson, C. A., Pearson, D., Shibahara, S., Hofsteenge, J., and Chiquet-Ehrismann, R., 1988, Tenascin: cDNA cloning and induction by TGF-beta, *EMBO J.* **7**:2977–2982.

Penttinen, R. P., Kobayashi, S., and Bornstein, P., 1988, Transforming growth factor beta increases mRNA for matrix proteins both in the presence and in the absence of changes in mRNA stability, *Proc. Natl. Acad. Sci. USA* **85:**1105–1108.

Pesciotta-Peters, D. M., Portz, L. M., Fullenwider, J., and Mosher, D. F., 1990, Co-assembly of plasma and cellular fibronectins into fibrils in human fibroblast cultures, *J. Cell Biol.* **111:**249–256.

Pierschbacher, M. D., and Ruoslahti, E., 1984, Cell attachment activity of fibronectin can be duplicated by small synthetic fragments of the molecule, *Nature* **309:**30–33.

Pierschbacher, M. D., and Ruoslahti, E., 1987, Influence of stereochemistry of the sequence agr-gly-asp-xaa on binding specificity in cell adhesion, *J. Biol. Chem.* **262:**17294–17298.

Preissner, K. T., and Müller-Berghaus, G., 1987, Neutralization and binding of heparin by S protein/vitronectin in the inhibition of factor Xa by antithrombin III. Involvement of an inducible heparin-binding domain of S protein/vitronectin, *J. Biol. Chem.* **262:**12247–12253.

Prochownik, E. V., O'Rourke, K., and Dixit, V. M., 1989, Expression and analysis of COOH-terminal deletions of the human thrombospondin molecule, *J. Cell Biol.* **109:**843–852.

Relman, D., Tuomanen, E., Falkow, S., Golenbock, D. T., Saukkonen, K., and Wright, S. D., 1990, Recognition of a bacterial adhesion by an integrin: Macrophage CR3 (alpha M beta 2, CD11b/CD18) binds filamentous hemagglutinin of Bordetella pertussis, *Cell* **61:**1375–1382.

Riou, J. F., Shi, D. L., Chiquet, M., and Boucaut, J.-C., 1990, Exogenous tenascin inhibits mesodermal cell migration during amphibian gastrulation, *Dev. Biol.* **137:**305–317.

Roberts, D. D., and Ginsburg, V., 1988, Sulfated glycolipids and cell adhesion, *Arch. Biochem. Biophys.* **267:**405–415.

Sakai, L. Y., Keene, D. R., and Engvall, E., 1986, Fibrillin, a new 350-kD glycoprotein is a component of extracellular microfibrils, *J. Cell Biol.* **103:**2499–2509.

Sasaki, M., Kleinman, H. K., Huber, H., Deutzmann, R., and Yamada, Y., 1988, Laminin, a multidomain protein. The A chain has a unique globular domain and homology with the basement membrane proteoglycan and the laminin B chains, *J. Biol. Chem.* **263:**16536–16544.

Silverstein, R. L., Asch, A. S., and Nachman, R. L., 1989, Glycoprotein IV mediates thrombospondin-dependent platelet–monocyte and platelet–U937 cell adhesion, *J. Clin. Invest.* **84:**546–552.

Singer, I. I., Kawka, D. W., Scott, S., Mumford, R. A., and Lark, M. W., 1987, The fibronectin cell attachment sequence Arg-Gly-Asp-Ser promotes focal contact formation during early fibroblast attachment and spreading, *J. Cell Biol.* **104:**573–584.

Singer, I. I., Scott, S., Kawka, D. W., Kazazis, D. M., Gailit, J., and Ruoslahti, E., 1988, Cell surface distribution of fibronectin and vitronectin receptors depends on substrate composition and extracellular matrix accumulation, *J. Cell Biol.* **106:**2171–2182.

Sjöberg, B., Eriksson, M., Osterlund, E., Pap, S., and Osterlund, K., 1989, Solution structure of human plasma fibronectin as a function of NaCl concentration determined by small-angle X-ray scattering, *Eur. Biophys. J.* **17:**5–11.

Skubitz, A. P. N., McCarthy, J. B., Zhao, Q., Yi, X., and Furcht, L. T., 1990, Definition of a sequence, RYVVLPR, within laminin peptide F-9 that mediates metastatic fibrosarcoma cell adhesion and spreading, *Cancer Res.* **50:**7612–7622.

Spiegel, S., Yamada, K. M., Hom, B. E., Moss, J., and Fishman, P. H., 1986, Fibrillar organization of fibronectin is expressed coordinately with cell surface gangliosides in a variant murine fibroblast, *J. Cell Biol.* **102:**1898–1906.

Spring, J., Beck, K., and Chiquet-Ehrismann, R., 1989, Two contrary functions of tenascin: Dissection of the active sites by recombinant tenascin fragments, *Cell* **59:**325–334.

Suzuki, S., Oldberg, A., Hayman, E. G., Pierschbacher, M. D., and Ruoslahti, E., 1985, Complete amino acid sequence of human vitronectin deduced from cDNA. Similarity of cell attachment sites in vitronectin and fibronectin, *EMBO J.* **4:**2519–2524.

Tashiro, K., Sephel, G. C., Weeks, B., Sasaki, M., Martin, G. R., Kleinman, H. K., and Yamada, Y., 1989, A synthetic peptide containing the IKVAV sequence from the A chain of laminin mediates cell attachment, migration, and neurite outgrowth, *J. Biol. Chem.* **264:**16174–16182.

Taylor, H. C., Lightner, V. A., Beyer, W. F., McCaslin, D., Briscoe, G., and Erickson, H. P., 1989, Biochemical and structural studies of tenascin/hexabrachion proteins, *J. Cell. Biochem.* **41:**71–90.

Thompson, L. K., Horowitz, P. M., Bentley, K. L., Thomas, D. D., Alderete, J. F., and Klebe, R. J.,

1986, Localization of the ganglioside-binding site of fibronectin, *J. Biol. Chem.* **261**:5209–5214.
Timpl, R., 1989, Structure and biological activity of basement membrane proteins, *Eur. J. Biochem.* **180**:487–502.
Tomasini, B. R., and Mosher, D. F., 1991, Vitronectin, *Prog. Hemost. Thromb.* **10**:269–305.
Varani, J., Nickoloff, B. J., Riser, B. L., Mitra, R. S., O'Rourke, K., and Dixit, V. M., 1988, Thrombospondin-induced adhesion of human keratinocytes, *J. Clin. Invest.* **81**:1537–1544.
Vartio, T., Laitinen, L., Närvänen, O., Cutolo, M., Thornell, L. E., Zardi, L., and Virtanen, I., 1987, Differential expression of the ED sequence-containing form of cellular fibronectin in embryonic and adult human tissues, *J. Cell Sci.* **88**:419–430.
Vischer P., Völker, W., Schmidt, A., and Sinclair, N., 1988, Association of thrombospondin of endothelial cells with other matrix proteins and cell attachment sites and migration tracks, *Eur. J. Cell Biol.* **47**:36–46.
Wolff, C., and Lai, C. S., 1989, Fluorescence energy transfer detects changes in fibronectin structure upon surface binding, *Arch. Biochem. Biophys.* **268**:536–545.
Woods, A., Couchman, J. R., Johansson, S., and Höök, M., 1986, Adhesion and cytoskeletal organisation of fibroblasts in response to fibronectin fragments, *EMBO J.* **5**:665–670.
Yamada, K. M., 1989, Fibronectin structure, functions and receptors, *Curr. Opinion Cell Biol.* **1**:956–963.
Yamada, K. M., and Kennedy, D. W., 1984, Dualistic nature of adhesive protein function: Fibronectin and its biologically active peptide fragments can autoinhibit fibronectin function, *J. Cell Biol.* **99**:29–36.
Yamagata, M., Yamada, K. M., Yoneda, M., Suzuki, S., and Kimata, K., 1986, Chondroitin sulfate proteoglycan (PG-M-like proteoglycan) is involved in the binding of hyaluronic acid to cellular fibronectin, *J. Biol. Chem.* **261**:13526–13535.
Yatohgo, T., Izumi, M., Kashiwagi, H., and Hayashi, M., 1988, Novel purification of vitronectin from human plasma by heparin affinity chromatography, *Cell Struct. Funct.* **13**:281–292.
Yurchenco, P. D., Cheng, Y. S., and Schittny, J. C., 1990, Heparin modulation of laminin polymerization, *J. Biol. Chem.* **265**:3981–3991.
Zhu, B. C., and Laine, R. A., 1985, Polylactosamine glycosylation on human fetal placental fibronectin weakens the binding affinity of fibronectin to gelatin, *J. Biol. Chem.* **260**:4041–4045.

Part II

How Do Cells Produce the Matrix?

Chapter 5

Proteoglycans

Metabolism and Pathology

VINCENT C. HASCALL, DICK K. HEINEGÅRD, and THOMAS N. WIGHT

1. Introduction

Proteoglycan metabolism in most cells is a highly regulated, dynamic process that contributes directly to cell and tissue functions. In many, perhaps most cases, PG metabolism is in steady state, i.e., biosynthesis of new molecules balances the catabolism of older molecules such that a constant concentration of PGs is maintained in a particular compartment over time. Cells in mature connective tissues devote a large proportion of their metabolic energy to PG synthesis. Chondrocytes, for example, can devote 5% or more of their total protein synthesis to making the core protein of aggrecan; and the subsequent assembly of the complex carbohydrate structures on the core protein requires dozens of enzymes and the formation of an average of more than 25,000 covalent bonds. Catabolism of PGs is also very active. The half-life of PGs in extracellular matrices can range from a few days to a few weeks, and cell surface PGs generally have half-lives of only a few hours. Changes in the metabolism and structure of PGs can have profound effects on the pathobiology of many diseases. This chapter, then, summarizes current concepts and problems relating to biosynthesis, catabolism, and pathology of PGs. The structures and functions of the major PGs described below are presented in detail in Chapter 2.

Abbreviations used in this chapter: RER, rough endoplasmic reticulum; GAG, glycosaminoglycan; HA, hyaluronic acid (hyaluronate, hyaluronan); PG, proteoglycan; PAPS, phosphoadenosine-phosphosulfate; GlcNAc, *N*-acetylglucosamine; GalNAc, *N*-acetylgalactosamine; CS, chondroitin sulfate; DS, dermatan sulfate; HS, heparan sulfate; KS, keratan sulfate; CSPG, chondroitin sulfate proteoglycan; HSPG, heparan sulfate proteoglycan; KSPG, keratan sulfate proteoglycan; PDGF, platelet-derived growth factor; TGF-β, transforming growth factor β; UDP, uridine diphosphate.

VINCENT C. HASCALL • National Institute of Dental Research, National Institutes of Health, Bethesda, Maryland 20892. DICK K. HEINEGÅRD • Department of Physiological Chemistry, University of Lund, Lund, Sweden. THOMAS N. WIGHT • Department of Pathology, University of Washington, Seattle, Washington 98195.

Cell Biology of Extracellular Matrix, Second Edition, edited by Elizabeth D. Hay, Plenum Press, New York, 1991.

2. Biosynthesis

2.1. Core Proteins

The core proteins of PGs, like other proteins, are synthesized in the RER (Fig. 5-1). All those whose primary sequence is known have N-terminal, hydrophobic signal sequences which are removed cotranslationally. Some, such as the core proteins of aggrecan and decorin, are released into the lumen while others, such as that for the cell surface PG, syndecan, remain embedded in membrane. All, except serglycin, have one or more asparagine-X-serine(threonine) sites, the consensus sequence required to initiate synthesis of N-linked oligosaccharides. Such core proteins are classified as glycoproteins.

2.2. Oligosaccharides

2.2.1. N-Asparagine Linked

N-Asparagine-linked oligosaccharide synthesis utilizes a lipid-linked oligosaccharide precursor. A high-mannose oligosaccharide, with nine mannoses and three glucoses linked to two GlcNAc residues (Fig. 5-2, structure 1), is assembled on dolichol pyrophosphate (for review see Kornfeld and Kornfeld, 1985; Schachter, 1986). The dolichol moiety is embedded in the membrane of the RER with the oligosaccharide projecting into the lumen. A specific enzyme transfers the entire oligosaccharide onto the carboxyamido side chain of an asparagine in a nascent core protein provided the recognition amino acid sequence, asparagine-X-serine(threonine), is accessible. An N-glycosylamine bond is formed (Fig. 5-2, structure 2) with the release of the dolichol pyrophosphate. Once polypeptide synthesis is complete, the resulting glycoprotein moves through the RER to the location where it will be packaged into vesicles for transport into the *cis* Golgi compartment. During this movement through the RER, specific glycosidases remove the three glucose residues and one of the mannose residues to yield a $Man_8GlcNAc_2$ oligosaccharide (Fig. 5-2, structure 3).

The glycoprotein is now shuttled by vesicular transport to the *cis* Golgi where the oligosaccharide can be processed further to the $Man_5GlcNAc_2$ oligosaccharide (Fig. 5-2, structure 4) by enzymatic removal of three additional mannose residues. After vesicular shuttling to the lumen of the medial Golgi, the processes for converting the high-mannose structures to the complex forms begin. First, a GlcNAc is transferred from a high-energy nucleotide sugar, UDP-GlcNAc (see Section 2.3.1), to the mannose on the unbranched arm by a specific glycosyltransferase (Fig. 5-2, structure 5). This is an obligatory step before a specific mannosidase can remove the two outer mannoses on the other branch. Once they are removed, a GlcNAc can be transferred to the uncovered mannose and fucose can be added to the innermost GlcNAc of the $GlcNAc_2$ linkage region (Fig. 5-2, structure 6). In all cases, specific glycosyltransferases utilize

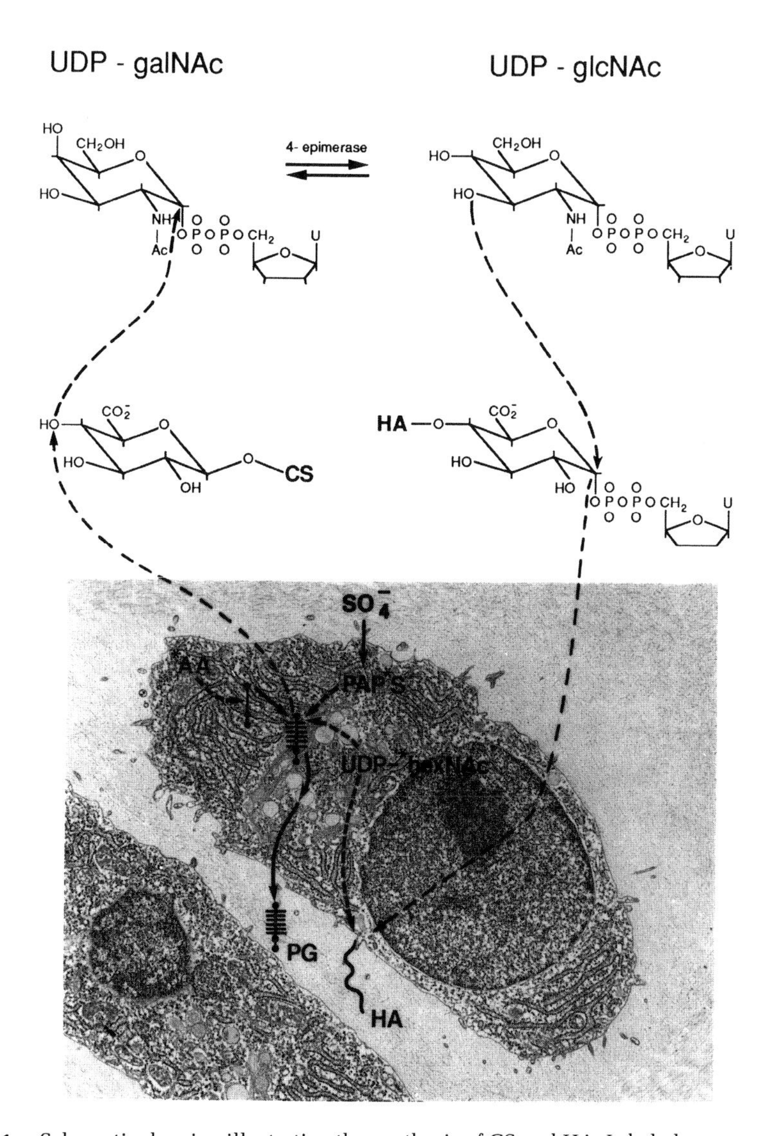

Figure 5-1. Schematic drawing illustrating the synthesis of CS and HA. Labeled sugar precursors such as [^{3}H]glucosamine are converted to nucleotide-sugar precursors in the cell cytosol. The UDP-N-acetylhexosamines are interconverted between the GlcNAc and GalNAc forms by a 4-epimerase enzyme. These are used to synthesize CS in the Golgi and HA at the cell surface by the mechanisms discussed in the text. Amino acid precursors (AA) label core proteins, such as that schematically indicated for aggrecan, in the RER. The core protein is translocated to the Golgi for adding the GAG chains. Sulfate in the extracellular pool equilibrates with PAPS in the cytosol, and the PAPS donates sulfate residues to the growing GAG chains.

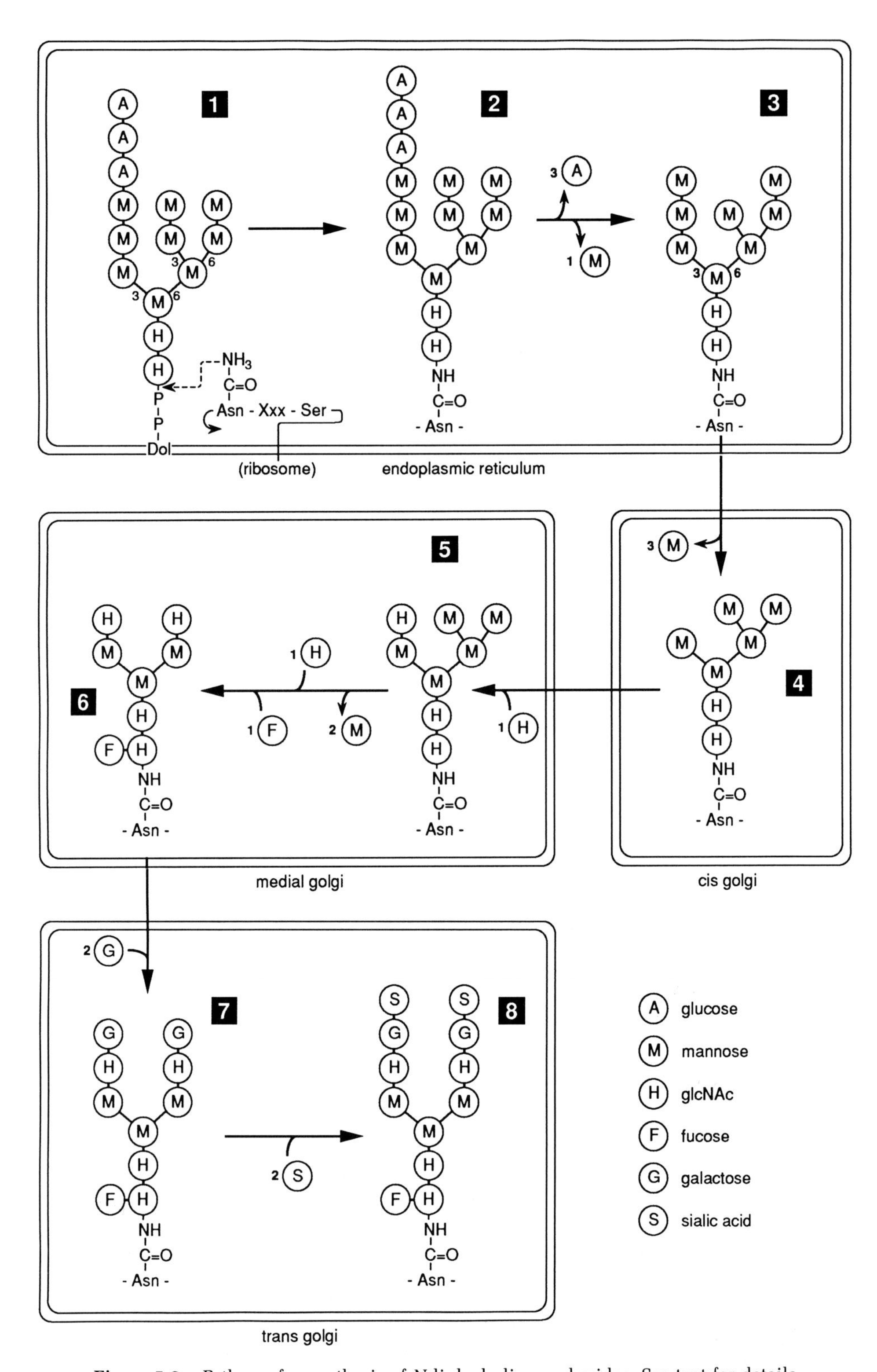

Figure 5-2. Pathway for synthesis of *N*-linked oligosaccharides. See text for details.

appropriate high-energy nucleotide-sugars. After vesicular shuttling to the *trans* Golgi, the building process continues as each branch is elongated by adding first a galactose (Fig. 5-2, structure 7) and then a sialic acid to give the final product, in this case the simple biantennary complex oligosaccharide (Fig. 5-2, structure 8) found on many glycoproteins.

Key intermediates in the medial Golgi can be modified in various ways to give a large variety of complex forms. For example, after removing the two outer mannoses from structure 5, additional GlcNAc residues can be added to available hydroxyl residues on each of the three remaining mannoses to give tri, tetra, penta, or even higher antennary structures. In other cases, the two outer mannoses in structure 5 may not be removed, which prevents synthesis of complex structures on this branch and yields oligosaccharides with only one arm being processed to a complex form. These are referred to as hybrid oligosaccharides. If the 3-branch mannose on structure 3 (Fig. 5-2) is not uncovered in the *cis* Golgi, then the oligosaccharide cannot be processed further, and the final glycoprotein will contain high-mannose oligosaccharide structures.

While most PGs contain *N*-linked oligosaccharides, the structures are known in only two cases. Aggrecan isolated from a rat chondrosarcoma contains predominantly triantennary complex *N*-linked oligosaccharides (Nilsson *et al.*, 1982), whereas the KSPG from monkey cornea contains two high-mannose *N*-linked oligosaccharides, a $Man_6GlcNAc_2$ form in which the 3-branch is not uncovered (Fig. 5-2, structure 4 with an extra mannose on the left branch), and an unusual $Man_4GlcNAc_2$ in which the mannose on the 3-branch remains while the two on the 6-branch have been removed (Nilsson *et al.*, 1983).

The asialo form of the biantennary complex oligosaccharide (Fig. 5-2, structure 7) provides the linkage region for adding lactosaminoglycan chains, $(Gal\text{-}GlcNAc\text{-})_n$, in the synthesis of KS on fibromodulin and the corneal KSPG (see Fig. 2-3B, Chapter 2). As this can only occur after the initial galactoses are added in the *trans* Golgi compartment (Fig. 5-2, structure 7), KS synthesis should occur in this compartment or in the closely associated *trans* Golgi network.

2.2.2. *O*-Serine/Threonine Linked

The *O*-serine/threonine-linked oligosaccharides are also synthesized in a late Golgi compartment (Hanover *et al.*, 1982; Thonar *et al.*, 1983; Lohmander *et al.*, 1986). These oligosaccharides are abundant in mucins and are found on many PGs. They are initiated by a glycosyltransferase that transfers GalNAc from UDP-GalNAc onto the hydroxyl of either serine or threonine in the core protein to form an α-glycoside bond. Specific sugars are then added sequentially to appropriate hydroxyl groups on nonreducing terminal sugars to form such structures as the hexasaccharide (Fig. 5-3), which is the largest and most prominent *O*-linked oligosaccharide on aggrecan isolated from a rat chondrosarcoma (Nilsson *et al.*, 1982). When either [^{3}H]glucose, a precursor for all the sugars in the hexasaccharide and in CS chains, or [^{3}H]glucosamine was

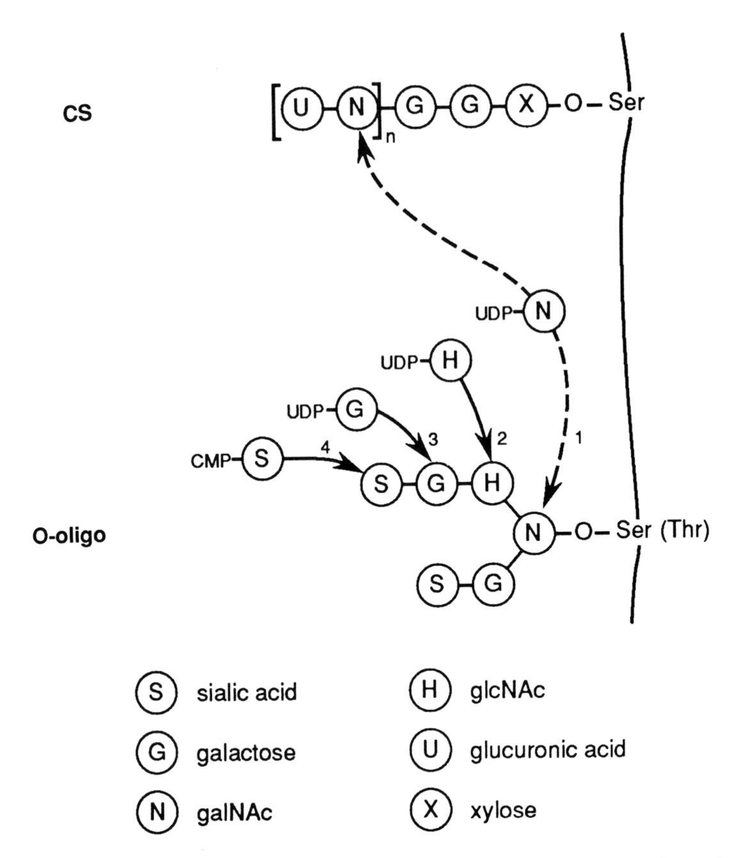

Figure 5-3. Pathway for synthesis of *O*-linked oligosaccharides (bottom). The indicated sugar residues on the upper branch of the oligosaccharide are transferred in the order indicated from 1 to 4 onto the growing hexasaccharide from their respective nucleotide-sugars. The sugars on the lower branch would also be added sequentially, first the galactose and then the sialic acid, but their order with respect to the synthesis of the upper branch is not known. The dashed arrows indicate the addition of GalNAc to both the hydroxyl of the serine on the *O*-linked oligosaccharide (bottom) and into the backbone of the CS chain (top). Kinetic analyses indicate that these transfers occur at essentially the same time during maturation of the PG structure.

used to study biosynthesis of the complex carbohydrate constituents on aggrecan, the kinetics of entry of label into both the linkage GalNAc and the GlcNAc of the hexasaccharide and into GalNAc residues in CS chains were identical (Fig. 5-3, dashed arrows) (Thonar *et al.*, 1983; Lohmander *et al.*, 1986). Thus, within experimental error, initiation and completion of the *O*-linked oligosaccharides occurs closely linked in time with CS chain elongation, and hence most likely occurs in the same *trans* Golgi compartment (see Section 2.3.1). In hyaline cartilages where KS chains are added to aggrecan, the lactosaminoglycan chains are elongated on the asialo form of the upper branch of the hexasaccharide (see Chapter 2, Fig. 2-3B); and hence, KS synthesis must also occur in the *trans* Golgi or in the closely associated *trans* Golgi network.

2.3. Glycosaminoglycans

2.3.1. Sulfated GAGs

The CS/DS and heparin/HS chains are added to the core protein in the Golgi. With the possible exception of the xylosyltransferase, which adds xyloses to hydroxyls on appropriate serines (Fig. 5-4, right) to initiate these GAG chains, all the enzymes required to synthesize a particular GAG appear to be organized into multienzyme complexes in the *trans* Golgi. For CS in aggrecan molecules, at least six specific glycosyltransferases and two sulfotransferases are normally required (for review see Rodén, 1980) (Fig. 5-4). Analyses of the entry of label into xylose in CS chains using [^{3}H]glucose as a precursor showed that the xylose may be added to the aggrecan core protein up to ~ 10 min before the rest of the chain (Lohmander *et al.*, 1986). Thus, the xylosyltransferase (Fig. 5-4, E_X) may exert its activity in an early Golgi, or possibly a late RER compart-

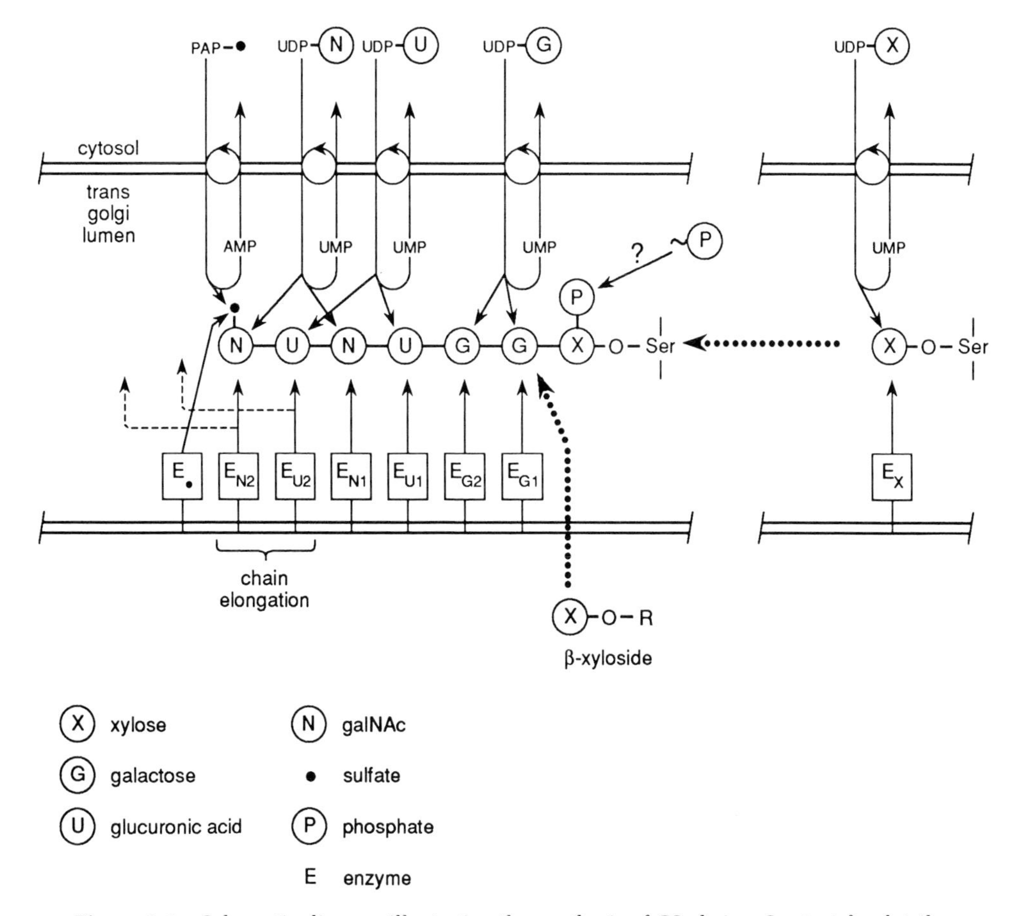

Figure 5-4. Schematic diagram illustrating the synthesis of CS chains. See text for details.

ment. This enzyme adds xylose from UDP-xylose preferentially onto serine residues in glutamic acid/aspartic acid-X-serine-glycine sequences (for review see Bourdon, 1990) or in serine-glycine repeats after such a sequence, as in serglycin (Bourdon *et al.*, 1985). However, the inverted sequence glycine-serine is also used to initiate CS chains in some cases, such as in the hinge region of the α_2(IX) chain of type IX collagen. The xylosyltransferase is not 100% efficient; 10–15% of serines in suitable sequences in aggrecan do not appear to carry CS chains (for review see V. Hascall, 1988), and in many tissues a small proportion (less than 5%) of biglycan molecules carry a GAG chain at only one of the two available sites (Neame *et al.*, 1989; Mörgelin *et al.*, 1989) (see Fig. 2-8, Chapter 2).

After xylose addition, each of the enzymes required to complete the linkage region [namely, galactosyltransferase 1 (E_{G1}), galactosyltransferase 2 (E_{G2}), and glucuronic acid transferase 1 (E_{U1})] and then to elongate the CS backbone [namely, GalNAc transferase 1 (E_{N1}) and the distinct glucuronic acid transferase 2 (E_{U2})], adds an appropriate sugar from a UDP-sugar donor (Fig. 5-4, left). Some evidence suggests that a second GalNAc transferase (E_{N2}) is actually required for chain elongation (Rohrmann, *et al.*, 1985), and this is the model shown (Fig. 5-4). The sugars are transferred sequentially to the nonreducing end of the growing chain to form glycoside bonds of the right anomeric configurations (α or β), and in the correct linkages (as defined in Chapter 2, Fig. 2-2B) with release of UDP. Figure 5-1 illustrates an example of this mechanism in which GalNAc transferase 2 adds GalNAc to glucuronic acid on the nonreducing end of a growing CS chain to form a β-1,3 glycoside bond, a reaction schematically indicated by the dashed arrow from E_{N2} in Fig. 5-4.

The UDP-sugars are synthesized in the cytosol of the cell and transported by an antiport mechanism into the Golgi lumen (Fig. 5-4); for example, the UDP-xylose transporter moves one UDP-xylose into the lumen while moving one UMP out (Nuwayhid *et al.*, 1986). This mechanism efficiently recycles the UMP and supplies sufficient amounts of the activated sugars in the Golgi to meet the biosynthetic demands (for review see Hirschberg and Snider, 1987). Phosphoadenosinephosphosulfate (PAPS) is the metabolically activated form of sulfate required by sulfotransferases to add sulfate esters to the elongating chains; it is also synthesized in the cytosol and transported into the Golgi by an antiport mechanism which exchanges 1 PAPS for 1 AMP (Fig. 5-4). Two sulfate transferases, one to form 4-*O*-sulfate and the other 6-*O*-sulfate on the GalNAc, are the predominant ones for synthesis of CS on most CSPGs.

For DS and heparin/HS, additional modifications in the structure occur at the same time as, or immediately after chain elongation (see structures in Fig. 2-2D, and for review see Lindahl, 1989). These include 5-epimerases for converting some of the D-glucuronic acids to L-iduronic acids and 2-*O*-sulfation of some of the iduronic acids in both classes of GAGs, and *N*-deacetylation coupled with *N*-sulfation for heparin/HS. Further, specific enzymatic 3-*O*-sulfation on GlcNAc and 2-*O*-sulfation on glucuronic acid in heparin/HS, while rare, appear to have important biological consequences, namely potential anticoagulant and antimitotic activities, respectively (see Chapter 2, Section 2.2.4).

Once GAG chain elongation begins, the completion of the PG structure is rapid, e.g., within a minute for aggrecan. The mature PGs are then packaged into vesicles and translocated from the *trans* Golgi to their final sites of function, such as storage in secretory granules, insertion in plasma membrane, or constitutive secretion into the extracellular matrix. Many cells synthesize more than one type of PG and ones with different types of GAG chains. Thus, a core protein must follow a pathway through the Golgi such that it receives its full complement of the correct type of GAG chains. How this traffic pattern is regulated is not known, but compartmentalization in the Golgi is one likely possibility.

Figure 5-5 shows morphological components in chondrocytes involved in packaging and secreting completed PGs and other extracellular matrix molecules such as collagen (G. Hascall, 1980). The tissue was stained with ruthenium red, a cationic dye that precipitates PGs to form the dark, punctate granules in the sections. Filamentous material (Fig. 5-5b,d,e), probably representing collagen molecules, is present in some exocytic vesicles along with dense granules, such as those which are spaced at ~ 70-nm intervals in Fig. 5-5e. This suggests that some PGs and collagen may be packaged in the same secretory vesicles. In this case, the distribution of dense granules is similar to that frequently observed on collagen fibrils in the extracellular matrix (Fig. 2-7C) and suggests that these PGs may be decorin molecules. However, it is possible that PGs, such as aggrecan in chondrocytes, may not be processed in the same Golgi vesicles or RER as is collagen (Vertel *et al.*, 1989).

Several experimental conditions have been used to modify PG synthesis. Tunicamycin inhibits the addition of GlcNAc-1-phosphate from UDP-GlcNAc to dolicholphosphate (Duskin and Mahoney, 1982), the first step in the formation of the dolicholpyrophosphoryl-oligosaccharide required for *N*-asparagine-linked oligosaccharide synthesis (see Fig. 5-2, structure 1). This prevents addition of KS chains to the corneal KSPG since *N*-linked oligosaccharides provide the linkage structure for this GAG (Hart and Lennarz, 1978). In chondrocytes, tunicamycin prevents the addition of *N*-linked oligosaccharides onto aggrecan and link protein without altering the rate of PG synthesis, secretion, or aggregation during the first several hours (Lohmander *et al.*, 1983), whereas in other cells, such as embryonic fibroblasts (Pratt *et al.*, 1979) or ovarian granulosa cells (Yanagishita, 1986), PG synthesis decreases rapidly, perhaps because translocation of the core glycoproteins from the RER to the Golgi is inhibited.

Chlorate at appropriate concentrations in growth medium inhibits the formation of PAPS without altering protein synthesis or GAG chain elongation (Humphries and Silbert, 1988; for review see Huttner and Baeuerle, 1988). This reagent, then, causes undersulfation of PGs and can be used to define possible biological effects of the sulfate groups on different PGs.

Certain β-xylosides effectively enter the Golgi compartment where CS/DS chains are synthesized and act as exogenous acceptors for the addition of the first linkage region galactose by galactose transferase I (Okayama *et al.*, 1973; Schwartz *et al.*, 1974; Robinson *et al.*, 1975). This diverts chain elongation from the xylosylated core protein onto the exogenous xyloside (Fig. 5-4, dotted

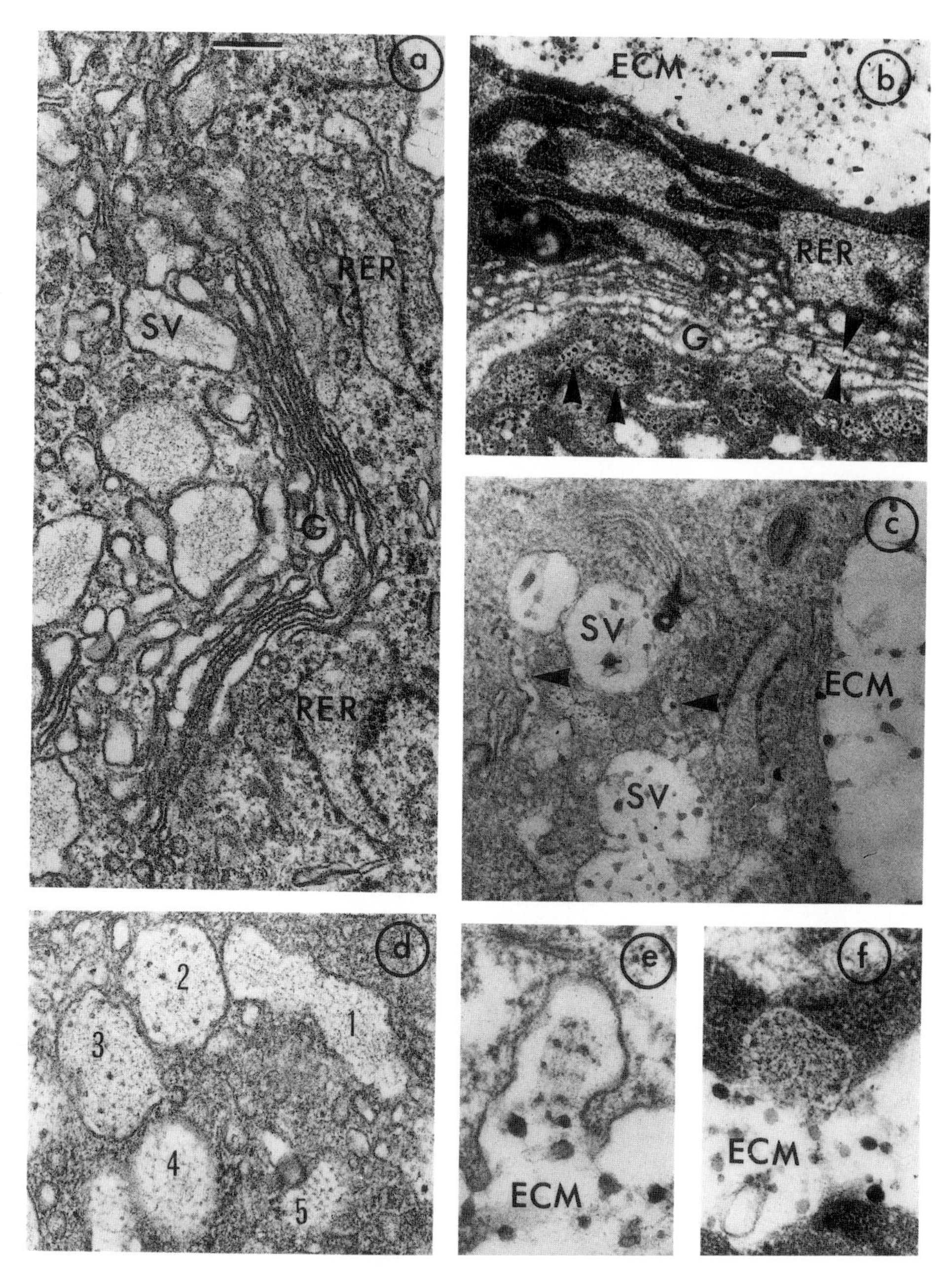

Figure 5-5. Morphological components involved in synthesis and secretion of PGs. (a) The Golgi area of a chondrocyte shows a stack of Golgi saccules (G), some of which are flat and others dilated. Cisternae of RER lie on the *cis* side of the stack, and secretory vacuoles (SV) on the *trans* side. (b) Some secretory vacuoles contain dense granules (arrowheads) which resemble PG matrix granules in the ECM. (c) These granules also occur in flattened Golgi saccules (arrowheads). (d) Contents of secretory vacuoles range from fine filaments in longitudinal array (vacuole 1) through tangles of filaments and granules (vacuoles 2–4) to occasional images of granules aligned on fine filaments

lines). The reagent also stimulates net CS synthesis in most cells, thereby showing that the capacity of cells for CS synthesis generally exceeds that needed to synthesize chains on endogenous core proteins. Chondrocytes chronically exposed to β-xyloside for 8 days synthesized and deposited the same number of PGs as untreated chondrocytes into their extracellular matrix. However, the PGs had fewer (~ 30% of control) and shorter (~ 75% of control) CS chains (Lohmander *et al.*, 1979). This greatly reduced the effective size of each PG domain and led to a dramatic condensation of the extracellular matrix.

In another example of the use of xylosides, connective tissue mast cells, which make only the heparin form of serglycin (see Chapter 2, Section 3.6), synthesize large amounts of chondroitin sulfate E on exogenously added xylosides, indicating that these cells contain a full complement of the enzymes required to make this latter GAG (Stevens *et al.*, 1983). In contrast, the mucosal mast cell, which is probably the precursor for the connective tissue mast cell, synthesizes only chondroitin sulfate E on serglycin. These results emphasize the problem: how do these cells regulate serglycin core protein traffic in the *trans* Golgi to obtain the appropriate complement of either heparin or chondroitin sulfate E chains?

While most β-xylosides stimulate synthesis of CS chains, they rarely stimulate synthesis of HS chains, even though HS/heparin chains are linked to core proteins by the same linkage region structure (Fig. 2-2A). However, an estradiol-β-xyloside has been developed which stimulates HS synthesis as well as CS synthesis (Lugemwa and Esko, 1991), and this may provide a convenient probe for altering HSPG production and studying biological responses in cells which synthesize primarily HSPGs.

2.3.2. Hyaluronic Acid

HA synthesis is not localized in the Golgi; rather it appears to occur in a compartment associated with the cell surface. Thus, HA synthetase activity in microsomal preparations colocalizes with cell surface enzyme markers (Philipson and Schwartz, 1984; Ng and Schwartz, 1989). Available evidence favors a mechanism in which the HA is elongated at its reducing end by adding UDP-N-acetylglucosamine (or UDP-glucuronic acid) with displacement of UDP from the UDP-glucuronosyl (or UDP-N-acetylglucosaminyl) residue which occupies the reducing end (Fig. 5-1) (for review see Prehm, 1989). The elongating HA molecule appears to be extruded directly into the extracellular space. Such a mechanism has several attractive features. First, it circumvents the problem of packaging the extremely large HA molecules, up to 10 million daltons, in an intracellular compartment (see Fig. 2-5); second, it is consistent with the in-

(vacuole 5). (e) Granules are spaced ~70 nm apart on a bundle of filaments ~300 nm long in this exocytic vacuole, suggesting PGs associated with collagen fibrils. (f) More commonly, exocytic vesicles contain a tangle of filaments and granules. The bars in panels a and b are 200 nm, and all panels except b are at the same magnification. The tissue was fixed in ruthenium red to enhance staining of PGs. (Reprinted from Hascall and Hascall, 1981.)

ability of most studies to provide convincing evidence for a covalently bound core protein (Mason *et al.*, 1982); and third, it provides a reasonable mechanism for segregating HA from aggrecan and link protein such that PG aggregation occurs in the extracellular compartment in cartilages (Kimura *et al.*, 1980) possibly involving HA closely associated with the cell surface (Sommarin and Heinegård, 1983).

Studies of HA synthesis are hampered because convenient labeling precursors, such as [^{3}H]glucosamine, are diluted by intracellular carbohydrate pools. Thus, the specific activity of UDP-GlcNAc after equilibration is much less than that of the labeling precursor in the culture medium, often by several hundredfold; and it is influenced variably by the metabolic status of the cells (Yanagishita *et al.*, 1989). Conversely, metabolic sources of sulfate, cysteine and methionine, contribute negligible amounts to the intracellular sulfate pool, with one notable exception, namely a mutant Chinese hamster ovary (CHO) cell line which is defective in sulfate transport into the cell (Esko *et al.*, 1986). Thus, the specific activity of sulfate in PAPS rapidly equilibrates with that of the sulfate in the culture medium. This permits a double-label protocol, [^{35}S]sulfate plus [^{3}H]glucosamine, to be used to estimate HA synthesis. The hexosamine precursor is diluted by intracellular sources of unlabeled hexosamine and equilibrates with the UDP-GlcNAc and UDP-GalNAc pools in the cytoplasm through the reversible 4-epimerase enzyme reaction depicted in Fig. 5-1. Therefore, the specific activity of the GlcNAc and the GalNAc incorporated into HA and CS/DS respectively will be the same. After labeling, the incorporation of ^{35}S activity into CS/DS provides a measure of the mass of this GAG synthesized, and the ratio of ^{3}H activity in HA to that in CS/DS provides an estimate of the mass of HA synthesized (Yanagishita *et al.*, 1989).

3. Catabolism

3.1. Extracellular Matrix Proteoglycans

Catabolism of PGs in most, if not all, connective tissues is dynamic and actively regulated by the resident cells. If the tissue is in steady state, then synthesis of PGs in a given compartment will equal catabolism. Typically, PGs in the metabolically active pool have half lives in the tissue of a few days to several weeks. In hyaline cartilage explants from young animals, aggrecan has a half-life of 3–4 weeks (for review see V. Hascall, 1988), and in glomerular basement membranes of rats, the HSPG has a half-life of 2–4 days *in vivo* (Beaven *et al.*, 1989). To balance catabolism and maintain constant concentrations of PGs in the matrix, some cells devote a large proportion of their synthetic activity to synthesis of PGs. Aggrecan can account for 5–10% of the total protein synthesis by articular cartilage chondrocytes. This rate produces about 20 pg of PG per cell per day and is sufficient to replace up to 3% of the total in the tissue per day (Hascall *et al.*, 1983, 1990). Similar rates of synthesis and catabolism must occur for PGs in corneal stroma (Midura and Hascall, 1989).

Little is known about the mechanisms for catabolism of PGs in extracellular matrices. Most evidence favors the involvement of specific proteases, such as the neutral protease stromelysin produced by many connective tissue cells (Chin *et al.*, 1985), which would degrade the core protein and release GAG-peptide fragments. How the cell regulates the activity of such enzymes to prevent widespread damage to other proteins in the matrix structure is not understood. However, most connective tissue cells also produce protease inhibitors, such as tissue-inhibitors of metalloproteases (TIMP), and a balance between proteases and inhibitors may be a critical factor.

Work with explant cultures of cartilage has provided some insight into how this degradation process may occur in cartilage. Metabolically labeled CS and HA in intact PG aggregates are catabolized from the tissue at the same rate and with nearly first-order kinetics (Morales and Hascall, 1988). Almost 95% of the labeled CS is released from the tissue as macromolecular GAG-peptide fragments, while the rest is internalized and degraded completely in lysosomes. In contrast, almost all of the HA in aggregates appears to be internalized and degraded. These results suggest that proteases at or near the cell surface selectively cleave bonds in the core protein to release fragments with intact GAG chains that can diffuse out of the tissue while the HA in the backbone of the aggregate is being internalized for transport to the lysosome.

Consistent with this model is the observation that degradation products from aggrecan molecules are continuously released from the cartilage into synovial fluid (Saxne *et al.*, 1985). A proportion of these fragments (~20%) eventually traverse the lymphatics to reach the circulation. The concentration of PG fragments in the synovial fluid, as monitored by antibodies which recognize protein epitopes, has been shown to be influenced by the metabolic status of PGs in cartilaginous tissues. For example, in patients with active inflammatory or degenerative joint disease, synovial fluid levels of PG fragments are elevated (Saxne *et al.*, 1985; Lohmander *et al.*, 1989). Most of the PG fragments that reach the general circulation are rapidly cleared by liver endothelial cells, which carry receptors with high affinity for PGs and HA, and they are completely degraded in lysosomes (for review see Engström-Laurent, 1989) (see Section 4.5). However, the serum level of PG fragments which contain KS is high enough to be measured accurately with antibodies that specifically recognize epitopes on KS. This assay has been used to show that patients with osteoarthritis, as a group, have a higher average KS concentration in their serum than does a normal population (Thonar *et al.*, 1985). This suggests that chondrocytes in osteoarthritic cartilages have higher PG metabolic rates.

As connective tissues age, the cells can lose their ability to regulate the metabolic processes. In cartilage, for example, a large proportion of the PGs may no longer be in the metabolically active pool. Hence, degradation over time caused by errant proteases, oxygen radicals, or wear and tear can damage these PGs. In the absence of active replacement of damaged molecules, tissue function will deteriorate. With age, the ability of the cells to synthesize the same amounts of PGs with their full complement of GAG chains of normal length diminishes, making the replacement molecules in the metabolically active compartment less functional. The net result of these processes can be a thinning of articular cartilage and loss of resiliency with age.

Connective tissue cells can alter their metabolism of proteoglycans in response to changes in biomechanical stimuli and remodel their extracellular matrix. Tendon tissue provides an impressive example (see Vogel and Koob, 1989, for review). Tendons, such as the bovine flexor tendon, which "wrap around" joints, often have regions that experience compression from weight bearing in addition to tension, and other regions that experience only tension. The extracellular matrix in regions of compression are fibrocartilaginous. They contain a high proportion of large CSPGs, which can bind to hyaluronic acid, and smaller amounts of the small interstitial DSPGs. Regions that experience only tension, however, contain almost exclusively the small DSPGs. In a rabbit model, tendons were surgically displaced such that areas formerly under compressive loading subsequently experienced only tension, and areas formerly only in tension subsequently experienced compressive loading (Gillard *et al.*, 1979). During the first two weeks after surgery, the CSPG and HA content of the tissue experiencing compressive loading for the first time increased dramatically while the DSPG remained constant. Conversely, the tissue no longer experiencing compressive loading rapidly lost the CSPGs within the same time-frame.

3.2. Cell Surface Proteoglycans

Catabolism of cell surface PGs is in general much faster than for extracellular matrix PGs; typical half-life values are 5–20 hr (for review see Yanagishita and Hascall, 1987). Cell surface PGs, such as syndecan, which bridge between extracellular matrix elements and intracellular cytoskeleton elements probably have longer half-lives that approach those typical for extracellular matrix PGs. Others, which are localized to certain membranous compartments, will have half-lives characteristic of the flow of membrane from the Golgi through that compartment to lysosomes for degradation. For example, a large proportion (50–60%) of the HSPG synthesized by a parathyroid cell line enter an endosome compartment ~ 20 min after maturation in the Golgi and remain in this compartment for 2–4 hr before entering the lysosome for rapid degradation (Takeuchi *et al.*, 1990). If the cells are exposed to reduced calcium levels, these HSPGs rapidly recycle between the endosomes and the cell surface with kinetics typical for receptor-mediated uptake processes, ~ 10 min per cycle. If the calcium concentration is raised to physiological levels, the HSPG are sequestered in an intracellular compartment and do not recycle. In contrast, the transferrin receptors on these cells cycle between the cell surface and an endosome compartment independent of the calcium concentration (Y. Takeuchi, personal communication). The results suggest that these HSPGs are located in a membrane compartment that initiates the biological responses of these parathyroid cells to changes in environmental calcium levels.

In many cells, once the membrane-bound PGs enter the degradation pathway, they are rapidly (half-life ~ 20 min) translocated into the lysosome compartment where both the protein and GAG constituents are completely depolymerized. In some cells, however, there are alternate intracellular pathways for catabolism of PGs. An example of this has been studied in ovarian granulosa cells (Fig. 5-6) (Yanagishita and Hascall, 1987; M. Yanagishita, personal communication). In these cells, ~ 30% of the HSPGs are anchored in the plasma membrane by phosphatidylinositol (Yanagishita and McQuillan, 1989) and this subpopulation has a half-life on the cell surface of ~ 4 hr before being internalized and rapidly delivered, within ~ 20 min, to the lysosome for total degradation. The remaining ~ 70% of the HSPGs are intercalated in the plasma membrane, and they also have a half-life on the surface of ~ 4 hr. However, ~ 30% of this subpopulation is "shed" into the medium after a proteolytic cleavage of the core protein, while the remainder is internalized to enter a complex degradation pathway. Within 1 hr after leaving the surface, the core protein is degraded by proteases in a step that can be slowed by leupeptin. At nearly the same time, the intact HS chains, ~ 30 kDa, are cleaved by an endoglycosidase to fragments of ~ 10 kDa. About 2 hr later, these fragments are cleaved by a second endoglycosidase to smaller fragments, ~ 5 kDa, which are subsequently translocated to lysosomes about 1 hr later for complete degradation. This pathway may pass through the nucleus which is known to concentrate selectively HS fragments with 2-sulfated glucuronic acid residues in some nondividing cells (Ishihara *et al.*, 1987).

4. Pathology

4.1. Hyaluronic Acid

Increased levels of serum HA are found in patients with inflammatory rheumatic diseases, cirrhotic liver diseases, or malignant mesotheliomas (for review see Engström-Laurent, 1989). With the exception of liver cirrhosis, where the elimination of HA from the serum is reduced, such elevations usually reflect increased synthesis of HA within a given tissue. For example, HA synthesis by synovial cells in an arthritic joint is stimulated by inflammatory mediators such as interleukin 1 and prostaglandin E_2, and by growth factors such as PDGF and EGF (for review see Laurent and Fraser, 1986). Similarly, inflammatory mediators released during lung injury probably stimulate HA synthesis by lung fibroblasts, leading to HA accumulation in such diseases as scleroderma, asthma, emphysema, idiopathic pulmonary fibrosis, and adult respiratory syndrome (Juul *et al.*, 1991). Because HA occupies large volumes of interstitial fluid, HA accumulation probably contributes significantly to interstitial edema, which is characteristic of most of these diseases. In these examples, increased HA deposition is the result of elevated synthesis by normal connective tissue cells responding to the mediators. No disease has yet been identified in which there is a primary defect in HA metabolism.

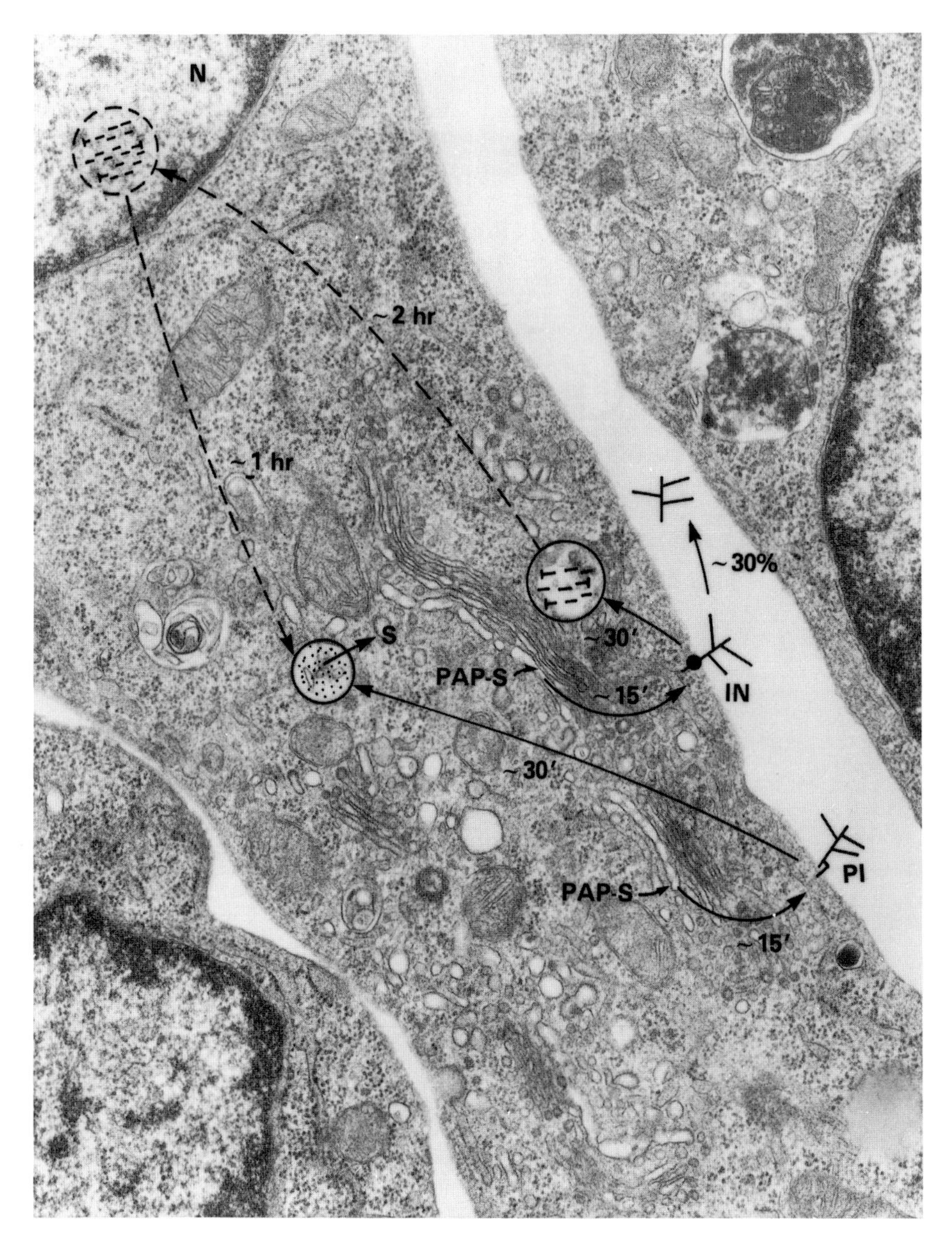

Figure 5-6. Schematic diagram illustrating the degradation pathways for cell surface HSPGs in ovarian granulosa cells. The HSPGs are anchored to the plasma membrane by phosphatidylinositol (PI) or by intercalation (IN). Both forms are completed in the Golgi, as indicated by the sulfation from PAPS, and appear on the cell surface ~20 min later. Their half lives on the surface are the same, ~4 hr. All of the PI-anchored HSPGs are degraded by internalization and translocation directly to lysosomes. The intercalated HSPGs exhibit more complex degradation pathways involving "shedding" of ~30% into the extracellular compartment and internalization of the remainder for stepwise degradation processes discussed in the text. The dashed lines indicate a possible pathway through the nucleus. (The micrograph was kindly provided by Dr. Jenifer Stow.)

4.2. Extracellular: Large Interstitial Proteoglycans

Cartilage growth plates provide the template for long bone formation, and thus define to a large extent the final shape of the skeleton. Two lethal genetic anomalies, nanomelia in chicken (Pennypacker and Goetinck, 1976) and cartilage matrix deficiency in mouse (Kimata *et al.*, 1981), have been found in which hyaline cartilages contain less than 5% of their normal content of PGs, but normal amounts of type II collagen. In the absence of the PGs, the volume of the cartilage matrix is greatly reduced. The consequences of this genetic deficiency for the mouse mutant are shown in Fig. 5-7 which compares homozygous mutant embryos with heterozygous littermates. Cartilages in the homozygote stain weakly with Alcian blue, a dye that interacts with the polyanionic PGs, in contrast with the intensely stained cartilages in the heterozygote. The bones formed on the cartilage growth plates, as shown by the alizarin red stain, are foreshortened and malformed in the homozygote.

In the brachymorphic mouse, another genetic anomaly, biosynthesis of PAPS is impaired (Orkin *et al.*, 1976; Sugahara and Schwartz, 1979). In growing cartilages, where PG synthesis is high, the concentration of PAPS becomes limiting, and the GAG chains on the PGs are ~ 15% undersulfated. Such PGs occupy smaller volumes under the same compressive load as their normal counterparts because they have lower anionic charge density within their solvent domains. This accounts in large part for the 15–20% reduction in the width of the growth plates, and thus for the shortened limbs characteristic of these mice (Orkin *et al.*, 1977).

There are several models showing that an altered mechanical load on connective tissues will induce the cells to change their surrounding matrix. For example, the stability of the knee joint can be altered by osteotomy, by removal of the meniscus, or by transsection of the anterior cruciate ligament (for review see Inerot *et al.*, 1991). In these cases, joint laxity increases, and the normal apposition and pressure relationships of the cartilages change without impairing the ability of the animal to walk or run. This invariably leads to osteoarthritic lesions in defined locations on the cartilages. In these locations, both PG synthesis and catabolism increase significantly before a lesion becomes apparent, indicating that altered metabolism of PGs may be a major factor in the progression of the osteoarthritic process. The structure of the CS chains on the PGs also changes. A monoclonal antibody which recognizes nonreducing terminal glucuronosyl-*N*-acetylgalactosamine-6-sulfate reacts much more strongly with PGs from osteoarthritic cartilages than with PGs from the same sites in normal cartilages (Caterson *et al.*, 1991).

Large CSPGs (> 1000 kDa) that can aggregate with HA are present in the interstitial matrix of blood vessels (see Section 2.3.2), where they accumulate in early lesions of atherosclerosis (Fig. 5-8) (for review see Wight, 1989). This increase may predispose the blood vessel to lipid deposition, a major complication of this disease, by virtue of the ability of PGs to interact with lipoproteins (for review see Camejo, 1982). Although it is not clear which PGs accumulate in this tissue, growth factors, such as TGF-β and PDGF, that are involved in ar-

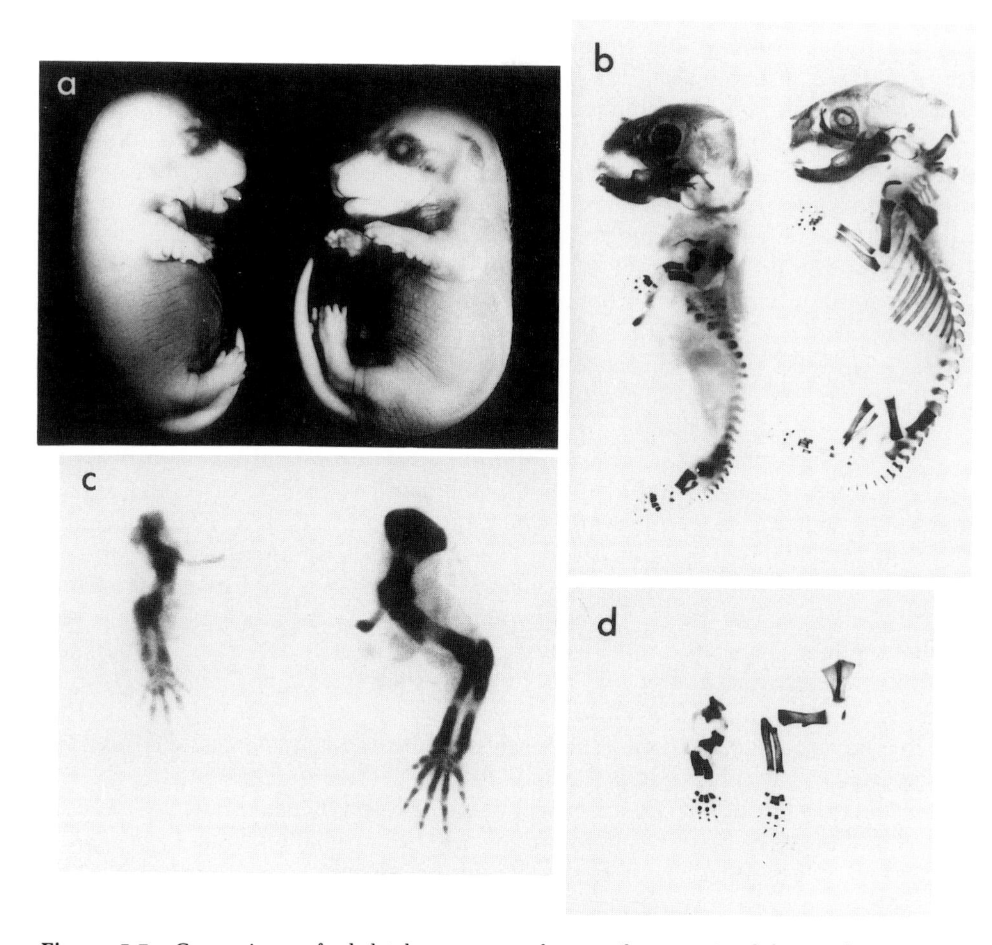

Figure 5-7. Comparison of skeletal structures for cartilage-matrix deficient homozygous (*cmd/cmd*) and heterozygous (*cmd/+*) mice. The homozygous samples are on the left and the phenotypically normal heterozygous samples are on the right. (a) Two littermates on the 18th day of gestation are shown, with the homozygote exhibiting stunted trunk and limbs, protruding tongue and abdomen. The fetuses were fixed in Bouin's solution before photography. (b) The fetuses were fixed in ethanol, cleared, and the bones stained with alizarin red S dye. The homozygote shows shorter ribs that are more flared and thicker; well-developed skull bones which are reduced along the anteroposterior axis; and severe stunting of the girdle and the long bones of both the upper and lower limbs. (c) Upper limbs from the 16th day of gestation were treated with Alcian blue to stain cartilages. The homozygote shows much fainter staining. The extreme reduction in linear growth is already apparent, yet the primary centers of ossification in the diaphysis have appeared, as is normal at this stage of development. (d) Upper limbs in this case have been stained with alizarin red S dye to show the effect on the bones in more detail. (The photographs were kindly provided by Dr. David Kochar, and the figure is from Hascall and Hascall, 1981.)

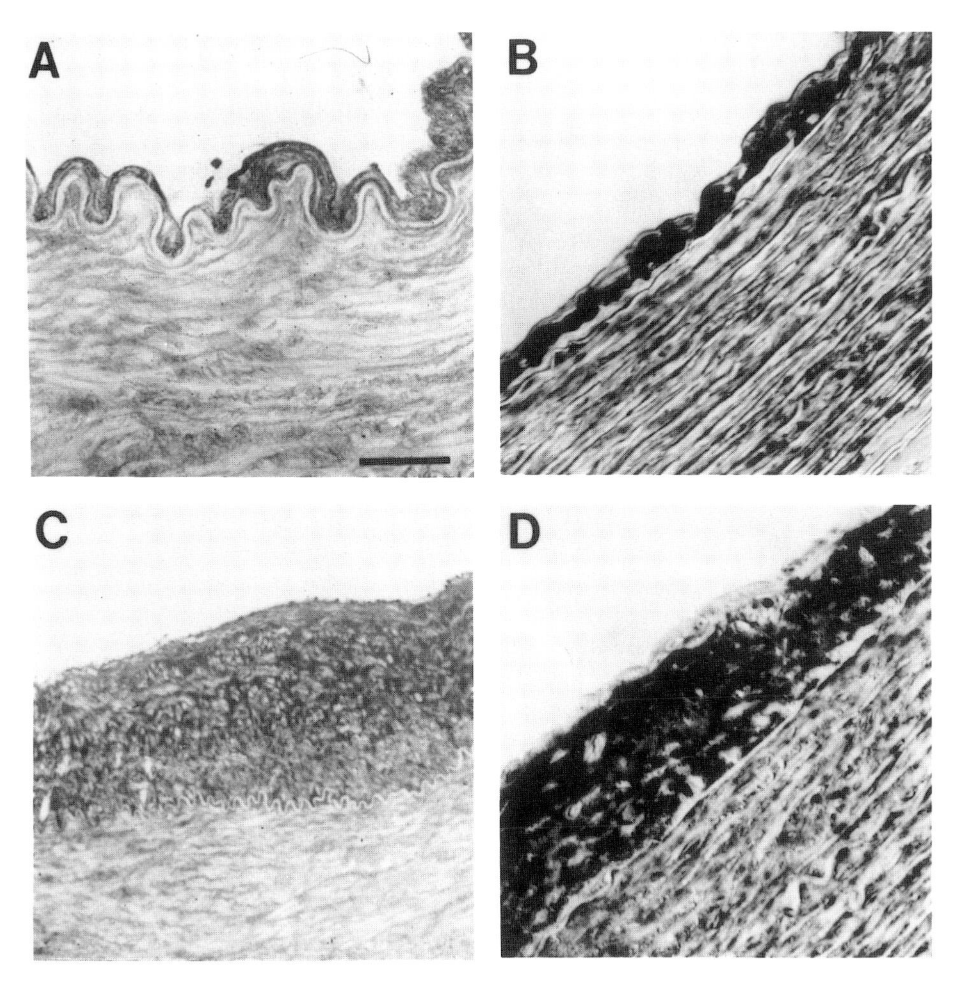

Figure 5-8. Light micrographs illustrating that the narrow intima of a normal blood vessel stains more intensely with (A) Alcian blue, and (B) a monoclonal antibody against aortic CSPG than the underlying medial layer. Vessels undergoing intimal hyperplasia (early atherosclerosis) have thickened intimas that also stain intensely with Alcian blue (C) and the antibody (D). Bar = 50 μm. (From Wight *et al.*, 1987.)

terial lesion formation, markedly stimulate synthesis of the large CSPG (Schönherr *et al.*, 1991). These results suggest that metabolism of this CSPG is a major factor in the progression of an atherosclerotic lesion.

4.3. Extracellular: Small Interstitial Proteoglycans

Several human diseases are characterized by abnormal processing of the small interstitial PGs. Skin fibroblasts from a patient exhibiting a progeroid-like disease with signs of the Ehlers–Danlos syndrome convert only about half of the core protein of decorin to the mature PG form; the remaining core protein is secreted normally, but without the DS chain (Kresse *et al.*, 1987; Quentin *et al.*, 1990). The patient exhibited delayed mental development and multiple abnormalities of connective tissue, including growth failure; osteopenia of all, and dysplasia of some bones; loose, but elastic skin; delayed wound healing; and hypermobile joints. Such results suggest that altered processing of decorin profoundly affects organization and function of the extracellular matrix in several connective tissues in this patient.

Marfan syndrome is a group of inherited connective tissue disorders characterized by musculoskeletal and cardiovascular abnormalities, lens dislocation, and autosomal dominant inheritance. By a median age of 45, ~ 85% of these patients die of cardiovascular abnormalities: aortic valve insufficiency, aortic dilation, and dissecting aneurysms. Skin fibroblasts cultured from a subset of these patients demonstrate decreased transcription and translation of decorin or biglycan (Pulkkinen *et al.*, 1990; T. Wight, personal communication). Decreases in these PGs could contribute to the disruption of the matrix and the loss of tissue elastic and tensile properties that characterize the cardiovascular system of these patients. It remains to be determined whether these alterations are related directly to defective PG genes or to changes induced in PG metabolism by other genes in affected patients (see also p. 83).

Turner syndrome patients have sex chromosome anomalies, usually monosomy X, but also X with structurally abnormal X or Y chromosomes. Invariably, the patients have short stature and a variety of skeletal dysplasias. Conversely, patients with higher numbers of sex chromosomes, such as in Klinefelters syndrome, have tall stature with limbs disproportionately long compared to the trunk. Fibroblast cultures derived from Turner syndrome patients contained only ~50% as much mRNA for the biglycan core protein as age-matched control cultures and only synthesized ~50% as much biglycan (Vetter *et al.*, 1991). Conversely, fibroblast cultures derived from patients with extra sex chromosomes consistently contained elevated (up to twofold) amounts of the mRNA and produced proportionally increased amounts of biglycan. Decorin and collagen synthesis in all cases did not differ significantly from controls. These results suggest that expression of the biglycan gene, which is on the long arm of the X chromosome, is critically involved in skeletal growth, particularly long-bone growth.

Most patients with corneal macular dystrophy synthesize an altered form of the KSPG in which the lactosaminoglycan chains are not sulfated (Nakazawa *et al.*, 1984; Midura *et al.*, 1990). As a consequence, precipitates form and accumulate with time in the collagen lattice of the stroma. This leads to increased corneal opacity and eventually requires a corneal transplant to restore vision, usually when the patient is in the teens. This suggests that continued metabolism of normal corneal KSPGs is required to maintain the highly ordered structure of the stroma necessary for corneal transparency. One patient with a subtype of corneal macular dystrophy synthesized normal KSPGs but the small, interstitial DSPGs had much shorter DS chains (Midura *et al.*, 1990).

This patient expressed the characteristic macular precipitates in the stroma at an earlier age than in the predominant form of the disease.

4.4. Extracellular: Basement Membrane Proteoglycans

Several studies have shown that basement membrane HSPGs are vital for normal functions of a variety of tissues (for review see Kanwar, 1984). The glomerular basement membrane loses its ability to exclude ferritin and albumin after perfusion with an enzyme which selectively degrades HS (Kanwar *et al.*, 1980). Nephrotic kidneys which allow serum proteins to filter into the urine (proteinuria) are very deficient in HSPG content (Vernier *et al.*, 1983). Kidneys in experimentally diabetic animals show reduced synthesis and content of HSPGs, and the HS chains are undersulfated (Rohrbach *et al.*, 1982; for review see Kanwar, 1984). It is not clear what factor(s) contribute to altered basement membrane HSPG metabolism in diabetes, but the end result appears to be significant alterations in the properties of these tissues.

Heparan sulfate PGs accumulate in amyloids, a generic term referring to specific extracellular deposits of insoluble fibrillar protein(s). Amyloids increase in organs and tissues until they lose their capacity to function normally. Basement membrane HSPGs have been identified in at least five different forms of amyloids (for review see Snow and Wight, 1989; Kisilevsky, 1990), including amyloid in neuronal plaques and blood vessels in Alzheimer's disease. The colocalization of HSPGs with amyloid in these tissues suggests that PGs may influence amyloid processing. It will be important to determine the mechanisms of this association and the role of PGs in this disease process.

4.5. Mucopolysaccharidoses

Mucopolysaccharidoses are forms of lysosomal storage diseases in which GAG catabolism is defective and abnormal quantities of GAGs accumulate in cells and are excreted in the urine (for reviews see Fluharty, 1987; Neufeld and Muenzer, 1989). Clinically, the hallmarks of the mucopolysaccharidoses are skeletal malformation, anomalies in tissues rich in connective tissue, and in many cases, severe mental retardation.

Normal catabolism of most of the GAGs metabolized in the body occurs in lysosomes and requires a cohort of exoglycosidases and sulfatases, which operate sequentially on nonreducing ends of the chains. For example, if the nonreducing end of a HS chain were 2-sulfated iduronic acid, a specific 2-sulfatase would be required to remove the sulfate, and then an α-iduronidase would remove the iduronic acid, thereby exposing a nonreducing terminal glucosamine residue (Fig. 5-9, 1 and 2). If this glucosamine contained an *N*-sulfate, a specific *N*-sulfatase would remove it. The exposed amino group on the glucosamine must then be *N*-acetylated, using acetyl CoA, by the only known enzyme

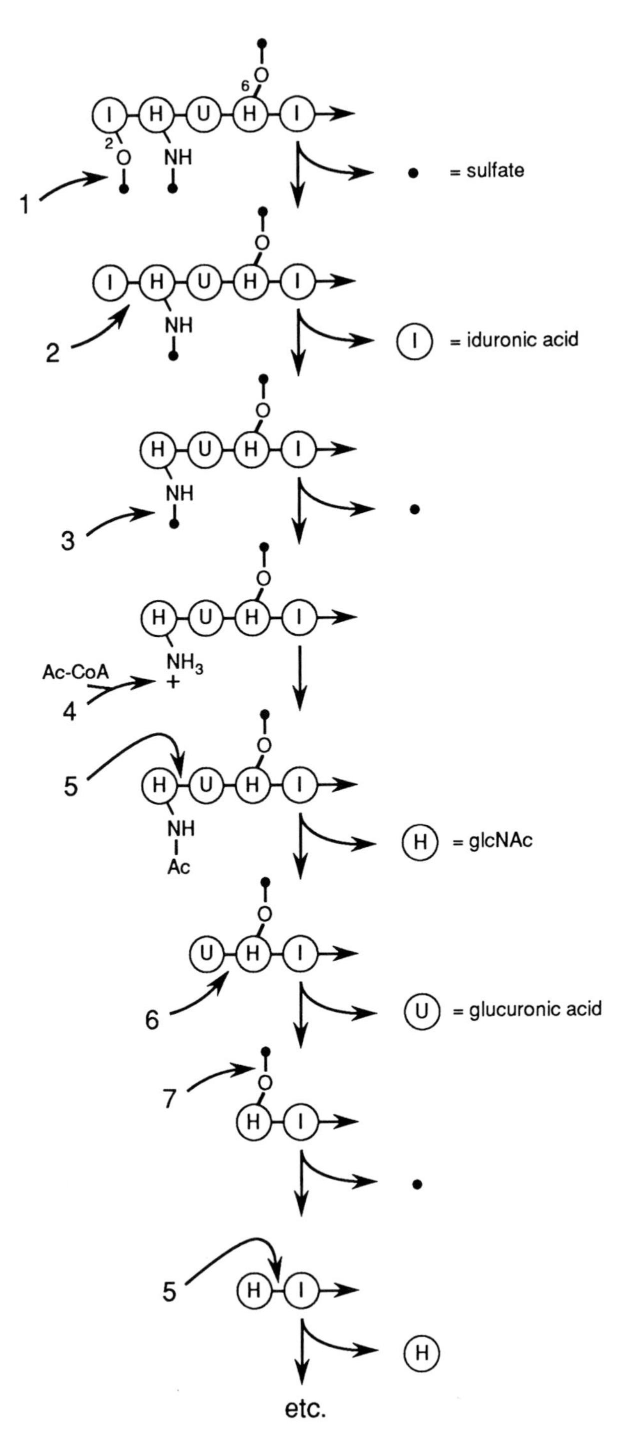

Figure 5-9. Pathway for catabolism of GAGs. See text for details.

in the catabolic sequence which is not a hydrolase (Fig. 5-9, 3). Only then can the α-N-acetylglucosaminidase remove the GlcNAc residue to expose the next hexuronic acid; and so on until the GAG is completely depolymerized.

Many of the enzymes required were first identified as a factor in a particular form of mucopolysaccharidoses in which a particular enzyme activity is absent. When the iduronic acid 2-sulfatase is absent, for example, DS and HS chains with nonreducing terminal 2-sulfated iduronic acid accumulate in lysosomes. A patient with this deficiency would be identified as having Hunter syndrome. If the α-iduronidase is defective or absent, nonsulfated iduronic acid residues accumulate on the nonreducing end of DS and HS chains (Hurler and Scheie syndromes). A deficiency in the N-acetyl synthetase or in the α-N-acetylglucosaminidase leads to accumulation of HS in lysosomes of Sanfillippo C and Sanfillippo B patients, respectively. Distinct syndromes have also been identified for each of the other enzymes known to be required for complete GAG degradation in the lysosomes.

5. Concluding Remarks

A large number of posttranslational modifications are involved in the transformation of a core protein to a mature PG. Some of these involve unique pathways, such as those for elongating and modifying GAG chains, while others utilize complex carbohydrate pathways common to other molecules, such as N-linked oligosaccharide synthesis on glycoproteins and O-linked oligosaccharide synthesis on mucins. For PGs with KS, nature has evolved pathways to use these oligosaccharides as linkages to proteins. In the case of aggrecan, KS chains were added to the large number of O-linked oligosaccharides already present on the same core protein with CS chains. This effectively increases the charge density within the domain of the PG and contributes to its inherent resistance to deformation. In the case of the corneal KSPG, KS was added to N-linked oligosaccharides of a structural glycoprotein, and this is now essential for tissue transparency. Most of these synthetic pathways are centered in the Golgi and its associated network. Because connective tissue cells are exceptionally active in synthesizing PGs, these cells should provide excellent models for cell biologists to explore the sorting and compartmentalization functions that are the hallmark of this cell organelle. PG metabolism is a dynamic process throughout the lifetime of most tissues. Thus, alterations in the normal balance of synthesis and catabolism can lead to disease states. While molecular biology can identify potential defects in core proteins, it is likely that many of the changes will be indirect and relate to one or more of the host of enzymes required to make a mature PG. Studies on the structure of PGs and the regulation of their metabolism are required to determine the manifold roles PGs have in cell biological processes and tissue functions.

ACKNOWLEDGMENT. The original research reported in this chapter was supported by NIH grants HL-18645 and DE-08229, and the National Institute of Dental Research, National Institutes of Health, Bethesda, Maryland.

References

Beaven, L. A., Davies, M., Couchman, J. R., Williams, M. A., and Mason, R. M., 1989, *In vivo* turnover of the basement membrane and other heparan sulfate proteoglycans of rat glomerulus, *Arch. Biochem. Biophys.* **269:**576–585.

Bourdon, M. A., 1990, Structure and role of cloned small proteoglycans, in: *Extracellular Matrix Genes* (L. Sandell and C. Boyd, eds.), pp. 157–174, Academic Press, New York.

Bourdon, M. A., Oldberg, Å., Pierschbacher, M., and Ruoslahti, E., 1985, Molecular cloning and sequence analysis of a chondroitin sulfate proteoglycan cDNA, *Proc. Natl. Acad. Sci. USA* **82:**1321–1325.

Camejo, G., 1982, The interaction of lipids and lipoproteins with the intercellular matrix of arterial tissue. Its possible role in atherogenesis, *Adv. Lipid Res.* **19:**1–53.

Caterson, B., Mahmoodian, F., Sorrell, J. M., Hardingham, T. E., Bayliss, M. T., Carney, S. L., Ratcliffe, A., and Muir, H., 1991, Modulation of native chondroitin sulphate structure in tissue development and in disease, *J. Cell Sci.* **97:**411–417.

Chin, J. R., Murphy, G., and Werb, Z., 1985, Stromelysin, a connective tissue-degrading metalloendopeptidase secreted by stimulated rabbit synovial fibroblasts in parallel with collagenase: biosynthesis, isolation, characterization and substrates, *J. Biol. Chem.* **260:**12367–12376.

Duskin, D., and Mahoney, W. C., 1982, Relationship of the structure and biological activity of the natural homologues of tunicamycin, *J. Biol. Chem.* **257:**3105–3109.

Engström-Laurent, A., 1989, Changes in hyaluronan concentration in tissues and body fluids in disease states, *Ciba Found. Symp.* **143:**233–247.

Esko, J. D., Elgavish, A., Prasthofer, T., Taylor, W. H., and Weinke, J. L., 1986, Sulfate transport-deficient mutants of Chinese hamster ovary cells. Sulfation of glycosaminoglycans dependent on cysteine, *J. Biol. Chem.* **261:**15725–15733.

Fluharty, A. L., 1987, Diseases of glycosaminoglycan and proteoglycan metabolism, in: *Connective Tissue Disease: Molecular Pathology of the Extracellular Matrix* (J. Uitto and A. Perejda, eds.), pp. 491–521, Dekker, New York.

Gillard, G. C., Reilly, H. C., Bell-Booth, P. G., and Flint, M. H., 1979, The influence of mechanical forces on the glycosaminoglycan content of the rabbit flexor digitorum profundus tendon, *Connect. Tissue Res.* **7:**37–46.

Hanover, J. A., Elting, J., Mintz, G. R., and Lennarz, W. J., 1982, Temporal aspects of the N- and O-glycosylation of human chorionic gonadotropin, *J. Biol. Chem.* **257:**10172–10177.

Hart, G. W., and Lennarz, W. J., 1978, Effects of tunicamycin on the biosynthesis of glycosaminoglycans by embryonic chick cornea, *J. Biol. Chem.* **253:**5795–5801.

Hascall, V. C., 1988, Proteoglycans: The chondroitin sulfate/keratan sulfate proteoglycan of cartilage, *ISI Atlas of Science Biochemistry* **1:**189–198.

Hascall, G. K., 1980, Ultrastructure of the chondrocytes and extracellular matrix of the Swarm rat chondrosarcoma, *Anat. Rec.* **198:**135–146.

Hascall, V. C., and Hascall, G. K., 1981, Proteoglycans, in: *Cell Biology of Extracellular Matrix* (E. D. Hay, ed.) pp. 39–63, Plenum Press, New York.

Hascall, V. C., Morales, T. I., Hascall, G. K., Handley, C. J., and McQuillan, D. J., 1983, Biosynthesis and turnover of proteoglycans in organ culture of bovine articular cartilage, *J. Rheumatol. Suppl.* **10:**45–52.

Hascall, V. C., Luyten, F. P., Plaas, A. H. K., and Sandy, J. D., 1990, Steady-state metabolism of proteoglycans in bovine articular cartilage explants, in: *Methods in Cartilage Research* (A. Maroudas and K. Kuettner, eds.), pp. 108–112, Academic Press, New York.

Hirschberg, C. B., and Snider, M. D., 1987, Topography of glycosylation in the rough endoplasmic reticulum and Golgi apparatus, *Annu. Rev. Biochem.* **56:**63–87.

Humphries, D. E., and Silbert, J. E., 1988, Chlorate: A reversible inhibitor of proteoglycan sulfation, *Biochem. Biophys. Res. Commun.* **154:**365–371.

Huttner, W. B., and Baeuerle, P. A., 1988, Protein sulfation on tyrosine, in: *Modern Cell Biology*, Vol. 6 (B. Satir, ed.), pp. 97–140, Liss, New York.

Inerot, S., Heinegård, D., Olsson, S.-E., Telhag, H., and Audell, L., 1991, Proteoglycan alterations during developing experimental osteoarthritis in a novel hip joint model, *J. Orthop. Res.* **9:** 658–673.

Ishihara, M., Fedarko, N. S., and Conrad, H. E., 1987, Involvement of phosphatidylinositol and insulin in the co-ordinate regulation of proteoheparan sulfate metabolism and hepatocyte growth, *J. Biol. Chem.* **262:**4708–4716.

Juul, S. W., Wight, T. N., and Hascall, V. C., 1991, Proteoglycans of the lung, in: *The Lung: Scientific Foundations* Vol. 1 (R. Crystal and J. West, eds.), pp. 413–420. Raven Press, New York.

Kanwar, Y. S., 1984, Biophysiology of glomerular filtration and proteinuria, *Lab. Invest.* **51:**7–21.

Kanwar, Y. S., Linker, A., and Farquhar, M. G., 1980, Increased permeability of the glomerular basement membrane to ferritin after removal of glycosaminoglycans (heparan sulfate) by enzyme digestion, *J. Cell Biol.* **86:**683–693.

Kimata, K., Barrach, H. J., Brown, K. S., and Pennypacker, J. P., 1981, Absence of proteoglycan core protein in cartilage from the cmd (cartilage matrix deficiency) mouse, *J. Biol. Chem.* **256:**6961–6968.

Kimura, J. H., Hardingham, T. E., and Hascall, V. C., 1980, Assembly of newly synthesized proteoglycan and link protein into aggregates in cultures of chondrosarcoma chondrocytes, *J. Biol. Chem.* **255:**7134–7143.

Kisilevsky, R., 1990, Heparan sulfate proteoglycans in amyloidogenesis: An epiphenomenon, a unique factor, or the tip of a more fundamental process, *Lab. Invest.* **63:**589–591.

Kornfeld, R., and Kornfeld, S., 1985, Assembly of asparagine-linked oligosaccharides, *Annu. Rev. Biochem.* **54:**631–664.

Kresse, H., Rosthoj. S., Quentin, E., Hollmann, J., Glössl, J., Okada, S., and Tonnesen, T., 1987, Glycosaminoglycan-free small proteoglycan core protein is secreted by fibroblasts from a patient with a syndrome resembling progeroid, *Am. J. Hum. Genet.* **41:**436–453.

Laurent, T. C., and Fraser, J. R. E., 1986, Properties and turnover of hyaluronan, *Ciba Found. Symp.* **124:**9–29.

Lindahl, U., 1989, Biosynthesis of heparin and related structures, in: *Heparin: Chemical and Biological Properties, Clinical Applications* (D. Lane and U. Lindahl, eds.), pp. 159–189, Arnold, London.

Lohmander, L. S., Hascall, V. C., and Caplan, A. I., 1979, Effects of 4-methyl umbelliferyl-β-D-xylopyranoside on chondrogenesis and proteoglycan synthesis in chick limb bud mesenchymal cell cultures, *J. Biol. Chem.* **254:**10551–10561.

Lohmander, L. S., Fellini, S. A., Kimura, J. H., Stevens, R. L., and Hascall, V. C., 1983, Formation of proteoglycan aggregates in rat chondrosarcoma chondrocyte cultures treated with tunicamycin, *J. Biol. Chem.* **258:**12280–12286.

Lohmander, L. S., Hascall, V. C., Yanagishita, M., Kuettner, K. E., and Kimura, J. H., 1986, Post-translational events in proteoglycan synthesis: Kinetics of synthesis of chondroitin sulfate and oligosaccharides on the core protein, *Arch. Biochem. Biophys.* **250:**211–227.

Lohmander, L. S., Dahlberg, L., Ryd, L., and Heinegård, D., 1989, Increased levels of proteoglycan fragments in knee joint fluid after injury, *Arthritis Rheum.* **32:**1434–1442.

Lugemwa, F. N., and Esko, J. D., 1991, Estradiol-β-D-xyloside, an efficient primer for heparan sulfate biosynthesis, *J. Biol. Chem.* **266:** 6674–6677.

Mason, R. M., Kimura, J. H., and Hascall, V. C., 1982, Biosynthesis of hyaluronic acid in cultures of chondrocytes from the Swarm rat chondrosarcoma, *J. Biol. Chem.* **257:**2236–2245.

Midura, R. J., and Hascall, V. C., 1989, Analysis of the proteoglycans synthesized by corneal explants from embryonic chicken: Structural characterization of the keratan sulfate and dermatan sulfate proteoglycans from corneal stroma, *J. Biol. Chem.* **264:**1423–1430.

Midura, R. J., Hascall, V. C., MacCallum, D. K., Meyer, R. F., Thonar, E. J.-M. A., Hassell, J. R., Smith, C. F., and Klintworth, G. K., 1990, Proteoglycan biosynthesis by human corneas from patients with types 1 and 2 macular corneal dystrophy, *J. Biol. Chem.* **265:**15947–15955.

Morales, T. I., and Hascall, V. C., 1988, Correlated metabolism of proteoglycans and hyaluronic acid in bovine cartilage organ cultures, *J. Biol. Chem.* **263:**3632–3638.

Mörgelin, M., Paulsson, M., Malmström, A., and Heinegård, D., 1989, Shared and distinct structural features of interstitial proteoglycans from different bovine tissues revealed by electron microscopy, *J. Biol. Chem.* **264:**12080–12090.

Nakazawa, K., Hassell, J. R., Hascall, V. C., Lohmander, L. S., Newsome, D. A., and Krachmer, J., 1984, Defective processing of keratan sulfate in macular corneal dystrophy, *J. Biol. Chem.* **259:**13751–13757.

Neame, P. J., Choi, H. U., and Rosenberg, L. C., 1989, The primary structure of the core protein of the small, leucine-rich proteoglycan (PG-I) from bovine articular cartilage, *J. Biol. Chem.* **264:**8653–8660.

Neufeld, E. F., and Muenzer, J., 1989, The mucopolysaccharidoses, in: *The Metabolic Basis of Inherited Disease,* Sixth ed. (C. Scriver, A. Beaudet, W. Sly, and D. Valle, eds.), pp. 1565–1587, McGraw-Hill, New York.

Ng, K. F., and Schwartz, N. B., 1989, Solubilization and partial purification of hyaluronate synthetase from oligodendroglioma cells, *J. Biol. Chem.* **264:**11776–11783.

Nilsson, B., De Luca, S., Lohmander, L. S., and Hascall, V. C., 1982, Structures of N-linked and O-linked oligosaccharides on proteoglycan monomer isolated from the Swarm rat chondrosarcoma, *J. Biol. Chem.* **257:**10920–10927.

Nilsson, B., Nakazawa, K., Hassell, J. R., Newsome, D. A., and Hascall, V. C., 1983, Structure of oligosaccharides and the linkage region between keratan sulfate and the core protein on proteoglycans from monkey cornea, *J. Biol. Chem.* **258:**6056–6063.

Nuwayhid, N., Glaser, J. H., Johnson, J. C., Conrad, H. E., Hauser, S. C., and Hirschberg, C. B., 1986, Xylosylation and glucuronosylation reactions in rat liver Golgi apparatus and endoplasmic reticulum, *J. Biol. Chem.* **261:**12936–12941.

Okayama, M., Kimata, K., and Suzuki, S., 1973, The influence of p-nitrophenyl beta-xyloside on the synthesis of proteochondroitin sulfate by slices of embryonic chick cartilage, *J. Biochem. (Tokyo)* **74:**1069–1073.

Orkin, R. W., Pratt, R. M., and Martin, G. R., 1976, Undersulfated glycosaminoglycans in the cartilage matrix of brachymorphic mice, *Dev. Biol.* **50:**82–94.

Orkin, R. W., Williams, B. R., Cranley, R. E., Poppke, D. C., and Brown, K. S., 1977, Defects in the cartilaginous growth plates of brachymorphic mice, *J. Cell Biol.* **73:**287–299.

Pennypacker, J. P., and Goetinck, P. F., 1976, Biochemical and ultrastructural studies of collagen and proteochondroitin sulfate in normal and nanomelic cartilage, *Dev. Biol.* **50:**35–47.

Philipson, L. H., and Schwartz, N. B., 1984, Subcellular localization of hyaluronate synthetase in oligodendroglioma cells, *J. Biol. Chem.* **259:**5017–5023.

Pratt, R. M., Yamada, K. M., Olden, K., Ohanian, S. H., and Hascall, V. C., 1979, Tunicamycin-induced alterations in the synthesis of sulfated proteoglycans and cell surface morphology in the chick embryo fibroblast, *Exp. Cell Res.* **118:**245–252.

Prehm, P., 1989, Identification and regulation of the eukaryotic hyaluronate synthase, *Ciba Found. Symp.* **143:**21–40.

Pulkkinen, L., Kainulainen, K., Krusius, T., Makinen, P., Schollin, J., Gustavsson, K. H., and Peltonen, L., 1990, Deficient expression of the gene coding for decorin in a lethal form of Marfan syndrome, *J. Biol. Chem.* **265:**17780–17785.

Quentin, E., Gladen, A., Rodén, L., and Kresse, H., 1990, A genetic defect in the biosynthesis of dermatan sulfate proteoglycan: galactosyltransferase I deficiency in fibroblasts from a patient with a progeroid syndrome, *Proc. Natl. Acad. Sci. USA* **87:**1342–1346.

Robinson, H. C., Brett, M. J., Tralaggan, P. J., Lowther, D., and Okayama, M., 1975, The effect of D-xylose, beta-D-xylosides and beta-D-galactosides on chondroitin sulphate biosynthesis in embryonic chicken cartilage, *Biochem. J.* **148:**25–34.

Rodén, L., 1980, Structure and metabolism of connective tissue proteoglycans, in: *The Biochemistry of Glycoproteins and Proteoglycans* (W. Lennarz, ed.), pp. 267–371, Plenum Press, New York.

Rohrbach, D. H., Hassell, J. R., Kleinman, H. K., and Martin G. R., 1982, Alterations in the basement membrane (heparan sulfate) proteoglycan in diabetic mice, *Diabetes* **31:**185–188.

Rohrmann, K., Niemann, R., and Buddecke, E., 1985, Two N-acetylgalactosaminyltransferases are involved in the biosynthesis of chondroitin sulfate, *Eur. J. Biochem.* **148:**463–469.

Saxne, T., Heinegård, D., Wollheim, F. A., and Pettersson, H., 1985, Difference in cartilage proteoglycan level in synovial fluid in early rheumatoid arthritis and reactive arthritis, *Lancet* **ii:**127–128.

Schachter, H., 1986, Biosynthetic controls that determine the branching and microheterogeneity of protein-bound oligosaccharides, *Biochem. Cell Biol.* **64:**163–181.

Schönherr, E., Järveläinen, H., Sandell, L. J., and Wight, T. N., 1991, Effects of PDGF and TGF-β on the synthesis of a large versican-like chondroitin sulfate proteoglycan by arterial smooth muscle cells, *J. Biol. Chem.* in press.

Schwartz, N. B., Galligani, L., Ho, P. L., and Dorfman, A., 1974, Stimulation of synthesis of free chondroitin sulfate chains by beta-D-xylosides in cultured cells, *Proc. Natl. Acad. Sci. USA* **71:**4047–4051.

Snow, A. D., and Wight, T. N., 1989, Proteoglycans in the pathogenesis of Alzheimer's disease and other amyloidoses, *Neurobiol. Aging* **10:**481–497.

Sommarin, Y., and Heinegård, D., 1983, Specific interaction between cartilage proteoglycans and hyaluronic acid at the chondrocyte cell surface, *Biochem. J.* **214:**777–784.

Stevens, R. L., Razin, E., Austen, K. F., Hein, A., Caulfield, J. P., Seno, N., Schmid, K., and Akiyama, F., 1983, Synthesis of chondroitin sulfate E glycosaminoglycan onto p-nitrophenyl-beta-D-xyloside and its localization to the secretory granules of rat serosal mast cells and mouse bone marrow-derived mast cells, *J. Biol. Chem.* **258:**5977–5984.

Sugahara, K., and Schwartz, N. B., 1979, Defect in 3-phosphoadenosine-5-phosphosulfate formation in brachymorphic mice, *Proc. Natl. Acad. Sci. USA* **76:**6615–6618.

Takeuchi, Y., Sakaguchi, K., Yanagishita, M., Aurbach, G. D., and Hascall, V. C. 1990, Extracellular calcium regulates distribution and transport of heparan sulfate proteoglycans in a rat parathyroid cell line, *J. Biol. Chem.* **265:**13661–13668.

Thonar, E. J.-M. A., Lohmander, L. S., Kimura, J. H., Fellini, S. A., Yanagishita, M., and Hascall, V. C., 1983, Biosynthesis of O-linked oligosaccharides on proteoglycans by chondrocytes from the Swarm rat chondrosarcoma, *J. Biol. Chem.* **258:**11564–11570.

Thonar, E. J.-M. A., Lenz, M. E., Klintworth, G. K., Caterson, B., Pachman, L. M., Glickman, P., Katz, R., Huff, J., and Kuettner, K. E., 1985, Quantification of keratan sulfate in blood as a marker of cartilage catabolism, *Arthritis Rheum.* **28:**1367–1376.

Vernier, R. L., Klein, D. J., Sisson, S. P., Mahan, J. D., Oegema, T. R., and Brown, D. M., 1983, Heparan sulfate-rich anionic sites in the human glomerular basement membrane, *N. Engl. J. Med.* **309:**1001–1009.

Vertel, B. M., Velasco, A., LaFrance, S., Walters, L., and Kaczman-Daniel, K., 1989, Precursors of chondroitin sulfate proteoglycan are segregated within a subcompartment of the chondrocyte endoplasmic reticulum, *J. Cell Biol.* **109:**1827–1836.

Vetter, U., Fedarko, N. S., Fisher, L. W., Young, M. F., Termine, J. D., Just, W., Vogel, W., and Gehron-Robey, P., 1991, Biglycan synthesis in fibroblasts of patients with Turner syndrome and other sex chromosome anomalies, *J. Clin. Invest.* in press.

Vogel, K. G., and Koob, T. J., 1989, Structural specialization in tendons under compression, *Int. Rev. Cytol.* **115:**267–293.

Wight, T. N., 1989, Cell biology of arterial proteoglycans, *Arteriosclerosis* **9:**1–20.

Wight, T. N., Lark, M. W., and Kinsella, M. G., 1987, Blood vessel proteoglycans, in: *Biology of Proteoglycans* (T. Wight and R. Mecham, eds.), pp. 267–300, Academic Press, New York.

Yanagishita, M., 1986, Tunicamycin inhibits proteoglycan synthesis in rat ovarian granulosa cells in culture, *Arch. Biochem. Biophys.* **251:**287–298.

Yanagishita, M., and Hascall, V. C., 1987, Proteoglycan metabolism by rat ovarian granulosa cells *in vitro*, in: *Biology of Proteoglycans* (T. Wight and R. Mecham, eds.), pp. 105–128, Academic Press, New York.

Yanagishita, M., and McQuillan, D. J., 1989, Two forms of plasma membrane-intercalated heparan sulfate proteoglycans in rat ovarian granulosa cells: Labeling of proteoglycans with a photoactivatable hydrophobic probe and effect of the membrane anchor-specific phospholipase C, *J. Biol. Chem.* **264:**17551–17558.

Yanagishita, M., Salustri, A., and Hascall, V. C., 1989, Determination of the specific activity of hexosamine precursors by analysis of double labeled disaccharides from chondroitinase digestion of chondroitin/dermatan sulfate, *Methods Enzymol.* **179:**435–455.

Chapter 6

Collagen Biosynthesis

BJORN REINO OLSEN

1. Introduction

The biosynthesis of the collagen components of connective tissues involves the regulated expression of members of the large collagen gene superfamily as well as complex interactions between primary collagen translation products and several cotranslational and posttranslational modifying enzymes (for review, see Vuorio and de Crombrugghe, 1990; Mayne and Burgeson, 1987; Ninomiya *et al.*, 1990; Fleischmajer *et al.*, 1990). Although the total number of genes that code for collagenous polypeptides has not yet been established, more that 25 genes have been cloned and characterized and the number is rapidly increasing. In addition, several proteins (Clq, acetylcholinesterase, conglutinin, macrophage scavenger receptor, surfactant proteins, mannose-binding lectins) that are not usually included among the collagens contain triple-helical sequence domains and are therefore also collagenous in nature.

As discussed in Chapter 1, the collagens fall into several groups or subclasses that are structurally and functionally distinct. This is also evident from analysis of the exon/intron structure of cloned collagen genes and from comparisons of the nucleotide and derived amino acid sequences of cloned cDNAs. In fact, examination of the gene structure of a collagen provides currently the best basis for classifying that particular collagen. Within each subclass different members have similar exon structure and highly conserved sequences, and must have evolved by a series of duplications of a parent gene. Between subclasses there are large differences in genomic structure. For example, while fibrillar collagens are encoded by large genes that contain more than 50 small exons (Vuorio and de Crombrugghe, 1990), the short-chain collagens—type VIII and X—are encoded by genes that contain 1 very large exon in addition to only 2 or 3 small exons (LuValle *et al.*, 1988; Yamaguchi *et al.*, 1991). Therefore,

Abbreviations used in this chapter: CAT, chloramphenicol acetyl transferase; FACIT, fibril-associated collagens with interrupted triple-helices; NF-1, nuclear factor-1; pN-collagen, NH_2-terminal peptide extension-collagen; RER, rough endoplasmic reticulum; SLS, segment-long-spacing.

BJORN REINO OLSEN • Department of Anatomy and Cellular Biology, Harvard Medical School, Boston, Massachusetts 02115.

Cell Biology of Extracellular Matrix, Second Edition, edited by Elizabeth D. Hay, Plenum Press, New York, 1991.

transcription of different collagen genes and associated RNA processing events vary enormously between different subclasses of collagens, and mechanisms of gene expression used by members of one subclass may be totally irrelevant to members of a different subclass.

In spite of this complexity, certain basic biosynthetic events are shared by all types of collagen polypeptides. This is because all collagens as components of the extracellular matrix are secretory proteins and contain triple-helical domains. Thus, the primary translation products of collagen mRNA molecules contain hydrophobic signal sequences found in most other secretory polypeptides. During or after ribosomal synthesis the polypeptide chains are extensively modified by cotranslational and posttranslational modifying enzymes. These modification reactions include removal of signal peptide sequences, as well as hydroxylation, glycosylation, and disulfide-bond formation. In the rough endoplasmic reticulum (RER) or in the Golgi complex the hydroxylated and glycosylated chains fold into characteristic triple-helices and globular domains. Following secretion from cells, the collagen molecules may be modified further by extracellular proteases and lysyl oxidase.

In the present chapter we will first examine the structure of the biosynthetic precursors of fibrillar collagens (procollagens) and the transcription of fibrillar collagen genes. We will then discuss genes that encode nonfibrillar types of collagen. Consideration will also be given to the questions of translation of mRNA molecules into collagen polypeptides, and the different posttranslational modification reactions that lead from primary translation products to functional collagen molecules. Finally, we will briefly discuss genetic defects in collagen biosynthesis.

2. Structure of Fibrillar Procollagens

Procollagens, the precursor of fibrillar collagens, represent the major intracellular forms of collagen. They are probably also the molecular forms of fibrillar collagens that are discharged into the extracellular space by cells synthesizing collagen (for review see Miller, 1985). Fibrillar procollagens are larger than fibrillar collagens because of peptide extensions, propeptides, at both the NH_2 and the COOH termini of the collagen α chains (Fig. 6-1). Most of the initial information about the structure of procollagen came from studies in type I procollagen, but subsequent investigations of procollagen types II, III, V, and XI indicate that all the fibrillar procollagens have similar domain structures (see Mayne and Burgeson, 1987).

The initial evidence of NH_2 and COOH propeptides in type I procollagen came from both chemical and morphological data. First, studies using vertebrate collagenases to cleave procollagen molecules into two fragments of unequal size indicated that both fragments contained extra peptides that were not present when collagen molecules were cleaved with the same type of enzyme (Tanzer *et al.*, 1974; Byers, *et al.*, 1975; Fessler, *et al.*, 1975; Olsen *et al.*, 1976). Second, immunological data showed that procollagen molecules contain two

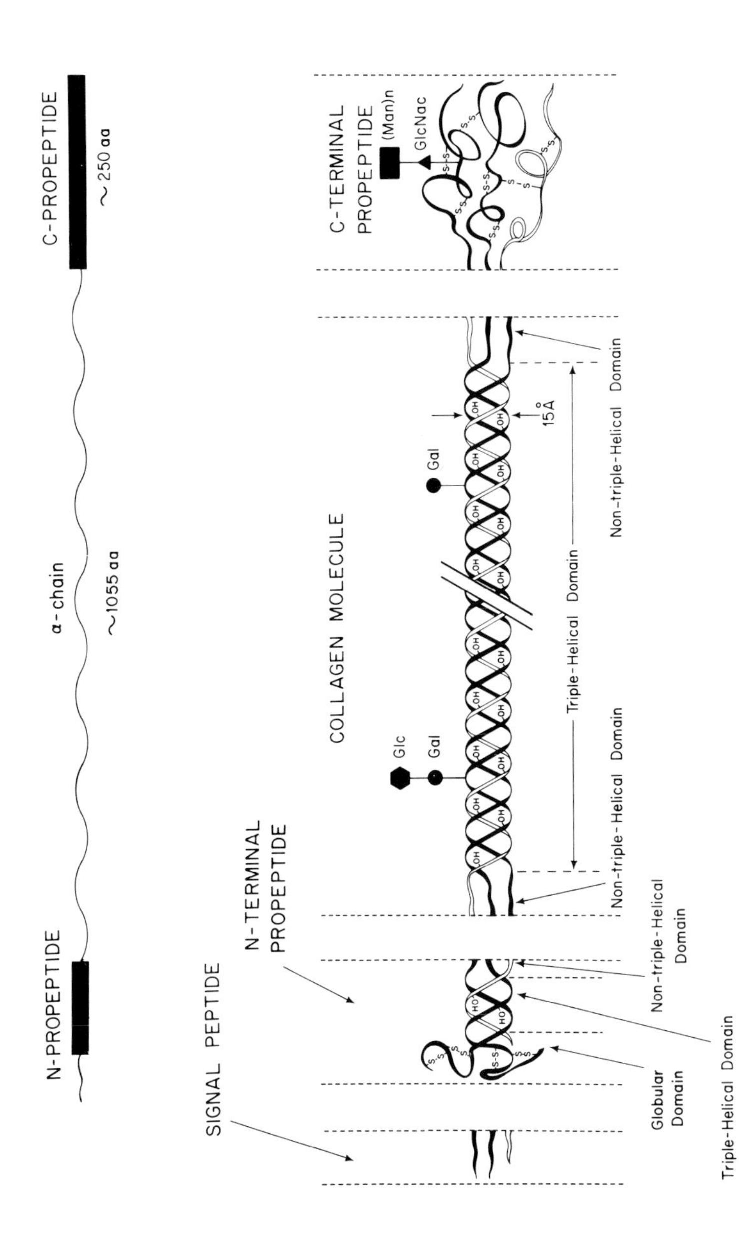

Figure 6-1. The domains of preproα chains and procollagen type I. During the synthesis of procollagen polypeptide chains on membrane-bound ribosomes, the chains are synthesized as preproα chains (top part of figure) with a hydrophobic signal peptide at their NH_2 terminal. The signal peptides are rapidly cleaved off the preproα chains as they enter the rough endoplasmic reticulum. (Modified from Prockop and Guzman, 1977.)

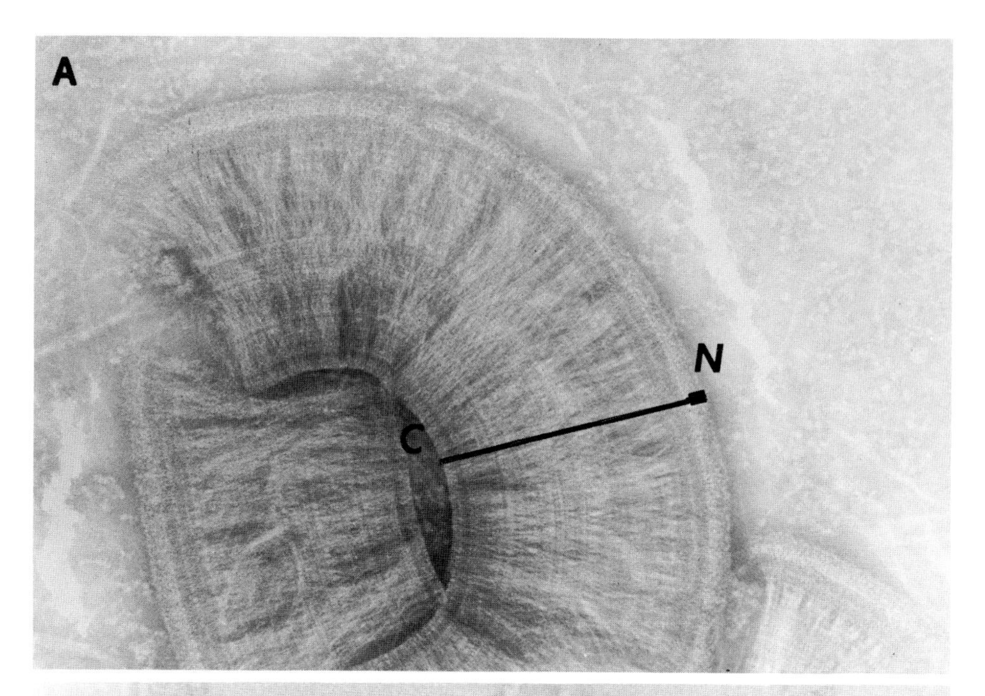
A
N
C

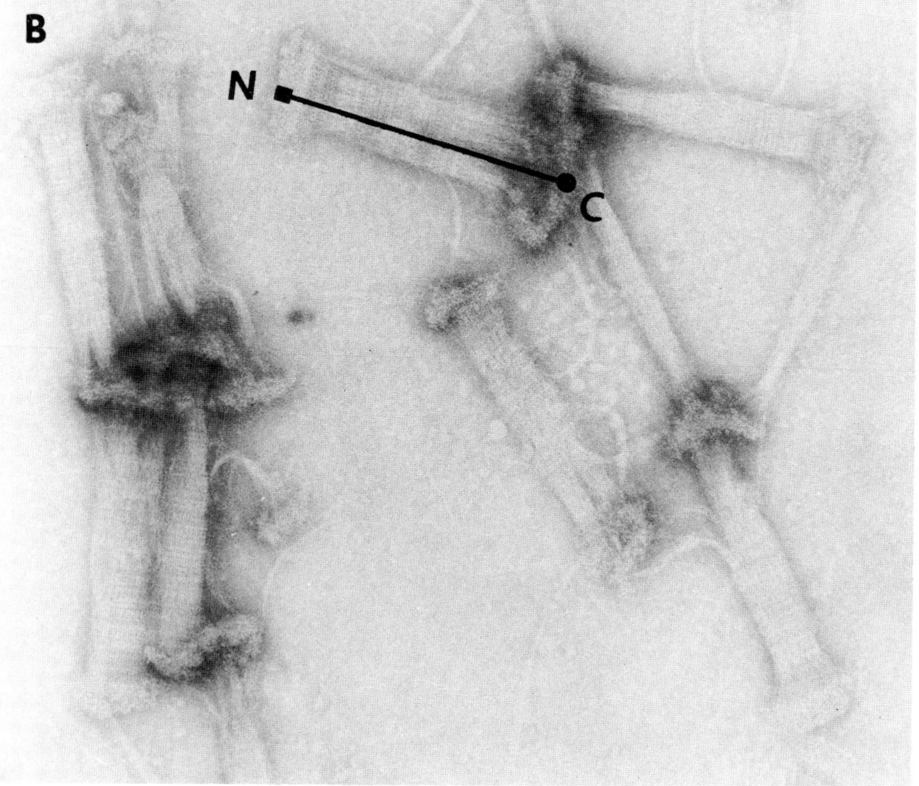
B
N
C

independent sets of antigenic determinants not present in collagen (Nist *et al.*, 1975). Third, electron microscopic examination of SLS crystallites made from procollagen showed extra pieces not found in collagen SLS at both ends of the aggregates (Tanzer *et al.*, 1974; Hoffmann *et al.*, 1976) (Fig. 6-2).

Chemical characterization of NH_2 propeptides (Becker *et al.*, 1976; Hörlein *et al.*, 1979) and amino acid sequences derived from cDNA sequences (for review see Vuorio and de Crombrugghe, 1990) have indicated that the NH_2 propeptide contains three structural domains in addition to a signal peptide sequence.* These domains are (1) a cysteine-rich globular domain, (2) a short triple-helical region, and (3) a short non-triple-helical region that connects the peptide with the major triple-helical domain. However, the size and sequence of these domains show considerable variation among different types of fibrillar procollagen chains, and even in the same chain from different animal species. For example, the cysteine-rich domain is 65–71 amino acid residues in the α1(I) and α1(III) collagen chains, but not present in the α2(I) chain. In α1(XI) procollagen chains the cysteine-rich domain is extremely long (383 amino acid residues in the human chain) and accounts for about 75% of the NH_2 propeptide (Yoshioka and Ramirez, 1990). The triple-helical domain varies also, from 39 amino acid residues in the α1(III) chain to 79 residues in the α1(II) and α2(V) chains. The domain shows a variable number of interruptions in the Gly-X-Y repeat structure among different types of NH_2 propeptides.

During the conversion of procollagen to fibrillar collagen, procollagen N-proteinases remove a portion or all of the NH_2 propeptide. The cleavage usually occurs at a Pro-Gln or Ala-Gln peptide bond in the connecting peptide portion of the NH_2 propeptide, leaving a short, non-triple-helical extension (telopeptide) at the NH_2 end of each collagen α chain. The enzymes require that the procollagen is in a native triple-helical conformation (Tuderman *et al.*, 1978; Hojima *et al.*, 1989). Thus, deletions within the major triple-helical domain of type I procollagen that cause misalignment of the NH_2 propeptides prevent cleavage of the propeptide and result in a connective tissue abnormality (Dombrowski *et al.*, 1989).

The COOH propeptide of fibrillar procollagens consists of three peptide chains that are linked by interchain disulfide bonds. These interchain bonds are formed by three or four (depending on the type of chain) cysteinyl residues

*Chemical characterization of the NH_2 propeptides of calf and sheep type I procollagen was made possible through the discovery of a genetic defect, dermatosparaxis, in these animals. As will be discussed later in this chapter, in dermatosparaxis there is a defect in the normal conversion of type I procollagen to collagen in skin so that precursor molecules with an NH_2-terminal peptide extension (pN-collagen) accumulate extracellularly in the skin.

Figure 6-2. SLS aggregates of type I pN-collagen (A) and procollagen (B) isolated from the medium of embryonic chick tendon fibroblasts incubated in suspension culture. The pN-collagen (see footnote on p. 177) reveals an NH_2-terminal segment (NH_2 propeptide) that is not present in control collagen SLS. The intact procollagen (B) reveals both NH_2- and COOH-terminal (COOH propeptide) segments. The aggregates were negatively stained with 1% potassium phosphotungstate. ×20,000.

located relatively close to the collagen triple-helix. Four additional, more COOH-terminal, cysteinyl residues in each chain form two intrachain disulfide bonds that stabilize the globular conformation of the COOH propeptide trimer.

A comparison of amino acid sequences derived from cDNA and genomic DNA sequences shows that the COOH propeptide domain exhibits the highest degree of sequence similarity between different fibrillar procollagen types and between animal species (Kimura *et al.*, 1989a; Vuorio and de Crombrugghe, 1990). The locations of the cysteinyl residues and the local sequences surrounding these residues are particularly well conserved. The only variation seen among the cysteines is that some chains [α1(I) α1(II), α1(III), α2(XI)] contain four cysteinyl residues of the interchain disulfide bonding type, while other chains [α2(I), α2(V), α1(XI)] contain only three such residues (Weil *et al.*, 1987; Bernard *et al.*, 1988). The COOH propeptides also contain mannose and *N*-acetylglucosamine, sugar residues that are not found within the collagen triple-helix (Olsen *et al.*, 1977; Clark, 1979). The function of this oligosaccharide is not known, but the sequence surrounding the carbohydrate attachment site is highly conserved among different procollagen types (Yamada *et al.*, 1983). During conversion of procollagen to fibrillar collagen, a procollagen C-proteinase cleaves an Ala-Asp or Glu-Asp peptide bond between the collagen triple-helix and the COOH propeptide so that a short non-triple-helical extension (telopeptide) is left at the COOH end of each collagen α chain. The length of this telopeptide shows considerable variation (11–27 amino acid residues) among different procollagen types. After cleavage the COOH propeptide is believed to be degraded, but the COOH propeptide of type II collagen has been found to accumulate in the extracellular matrix of mineralizing cartilage and may possibly play a role in the mineralization process (Poole *et al.*, 1984; van der Rest *et al.*, 1986).

Several functional roles have been proposed and partially demonstrated for the propeptides of fibrillar procollagens. These roles include: (1) chain association and triple-helix formation during molecular assembly; (2) intracellular transport and packaging of procollagen in secretion granules; (3) fibrillogenesis in the extracellular matrix; and (4) regulation of collagen synthesis. The COOH propeptide contains information required for polypeptide chain association in the RER. The folding of the COOH propeptide and formation of its intra- and interchain disulfide bonds precede triple-helix formation and folding of the rest of procollagen molecules. In fact, it is believed that the folding of the major triple-helical domain in procollagen molecules occurs in a zipperlike fashion starting at the COOH end. It is not surprising, therefore, that mutations in procollagen chains which affect the folding of the COOH propeptide interfere with the ability of mutant chains to be incorporated into triple-helical molecules (Pihlajaniemi *et al.*, 1984; Bateman *et al.*, 1989; Willing *et al.*, 1990). The NH_2 propeptide appears to have a role in regulating collagen fibril diameters (for review and discussion see Chapman, 1989). Experiments with cells in culture and in cell-free systems also suggest that the NH_2 propeptide or peptide fragments derived from it may function as a negative feedback inhibitor of collagen biosynthesis (Bornstein and Sage, 1989). The physiological significance of this inhibition is, however, not clear.

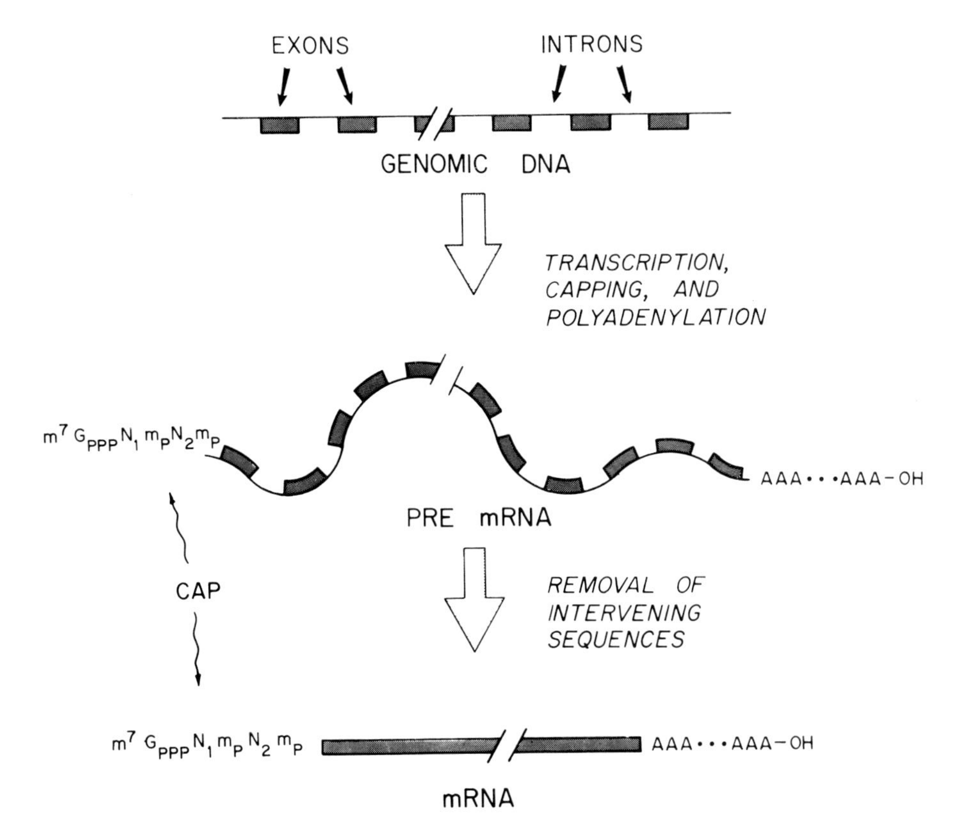

Figure 6-3. Most eukaryotic genes contain coding regions (exons) interrupted by noncoding intervening sequences (introns). The initial RNA transcripts of the genes are copies of both exons and introns. During maturation (processing) of the RNA, splicing enzymes remove the intron sequences, producing mature mRNA molecules. The maturation process also includes addition of methylated nucleotides at the 5′ end of the mRNA (the CAP structure) and of a poly A tail at the 3′ end.

3. Collagen Gene Structure and Regulation of mRNA Levels

Control of protein synthesis can be exerted through control of mRNA levels by regulation of gene transcription and RNA maturation (Fig. 6-3) or the stability of mRNA. In the case of collagen there is generally good correlation between the rate of collagen synthesis and the level of mRNA in the cell. For example, when chicken fibroblasts are transformed with Rous sarcoma virus, the proportion of type I procollagen synthesized by the cells drops about tenfold. At the same time, there is a four- to tenfold decrease in the amount of mRNA coding for type I collagen (Rowe *et al.*, 1978). A similar correlation between mRNA levels and protein synthesis is found for type X collagen in chondrocytes undergoing hypertrophy (LuValle *et al.*, 1989). As changes in mRNA levels are likely to involve changes in transcription of collagen genes, changes in posttranscriptional processing of mRNA precursor molecules, and/or changes in the stability of mRNA, it is clearly important to isolate and

Table I. Chromosomal Location of Fibrillar Collagen Genes[a]

Gene locus	Chain designation	Chromosomal location
COL1A1	$\alpha1$(I)	17q21.3-q22
COL1A2	$\alpha2$(I)	7q21.3-q22
COL2A1	$\alpha1$(II)	12q13-q14
COL3A1	$\alpha1$(III)	2q24.3-q31
COL5A1	$\alpha1$(V)	—
COL5A2	$\alpha2$(V)	2q24.3-q31
COL5A3	$\alpha3$(V)	—
COL11A1	$\alpha1$(XI)	1p21
COL11A2	$\alpha2$(XI)	6p212

[a]The symbols used for the human fibrillar collagens are those assigned by the Nomenclature Committee of the Human Gene Mapping Workshops. The type number is given after the letters COL (for collagen), followed by A (for α chain) and a number specifying the α chain. The chromosomal locations of the fibrillar collagen gene loci are indicated in the right-hand colum, using the standard description of G-banded chromosomes. Thus, the region indicated as 17q21.3-q22 represents the region defined by bands 21.3 and 22 on the q arm of chromosome 17.

characterize collagen cDNAs and genes and study the transcription of the genes in well-defined systems.

3.1. Fibrillar Collagen Genes

The subclass of fibrillar collagens, whose triple-helical molecules are the major building blocks of cross-striated collagen fibrils, is encoded by (at least) nine distinct genes. Most of these genes have been cloned and their chromosomal locations in the human genome determined. As can be seen in Table I, the genes are dispersed among different chromosomes, with the exception of $\alpha1$(III) and $\alpha2$(V), which are both on chromosome 2. Therefore, the tight coordinate control of fibrillar collagen gene transcription, for example of the $\alpha1$(I) and $\alpha2$(I) collagen genes, obviously must involve mechanisms that are not based on syntenic relationships.

The fibrillar collagen genes are large (15–50 kb) and have a complex exon structure (> 50 exons) (Fig. 6-4) (Boedtker *et al.*, 1985; Ramirez *et al.*, 1985; Upholt *et al.*, 1985; Ramirez, 1989; Upholt, 1989; Upholt and Olsen, 1991; Vuorio and de Crombrugghe, 1990). A small, but variable number of exons (< 10, exact number depending on chain type) encodes the NH_2 propeptide and the signal peptide. The COOH propeptide is encoded by four exons. The remaining exons (usually 44) contain the coding information for the major triple-helical domain. About half of these triple-helical exons are 54 bp long, some are 108 bp, while others are 45, 99, or 162 bp. The NH_2 and COOH ends of the triple-helix are encoded by junction exons which also code for telopeptides and part of propeptides. All the triple-helical exons start with a complete codon for glycine, and end with a complete codon for the amino acid residue in the Y position of the Gly-X-Y repeats. Thus, all triple-helical exons are multiples of 9 bp and code for an integral number of Gly-X-Y triplets. Since a majority of the triple-helical exons are 54 bp or a multiple of 54 (108 or 162), it has been proposed that the ancestral fibrillar collagen gene arose by amplification of a primordial gene containing a 54-bp exon unit (Yamada *et al.*, 1980).

3.2. Transcriptional Regulation of Fibrillar Collagen Genes

Several types of data collectively indicate that the regulation of fibrillar collagen synthesis by cells in culture or in developing organisms occurs primarily by regulation of the mRNA levels rather than by control of mRNA translation. An interesting exception is the presence of relatively high levels of α2(I) collagen mRNA in chondrocytes without synthesis of proα2(I) collagen chains. This is because chondrocytes utilize a transcription start site within the α2(I) collagen gene which is different from that used in fibroblasts. The consequence of this alternative transcription is that while the α2(I) mRNA is translated into proα2(I) chains in fibroblasts, the α2(I) mRNA in chondrocytes does not code for proα2(I) but potentially codes for a noncollagenous polypeptide (Bennet and Adams, 1990).

Our understanding of how the transcription of different fibrillar collagen genes is regulated is only fragmentary but studies of type I, II, and III collagen genes suggest that the genes are regulated by very complex and diverse mechanisms (for reviews see Ramirez and DiLiberto, 1990; Vuorio and de Crombrughe, 1990). These mechanisms ensure the precise tissue-specific transcription of the fibrillar procollagen genes during development and in adult organisms (Fig. 6-5), and allow for modulation of their activity by growth factors, hormones, and cytokines. The current efforts to understand these mechanisms are directed at (1) identifying DNA sequences in and around the genes, so-called *cis*-acting sequences, to which regulatory proteins, *trans*-acting factors, bind (2) determining the number and properties of such *trans*-acting factors, and (3) analyzing how the DNA/regulatory protein complex facilitates the interaction of RNA polymerase II with the genes.

For fibrillar collagen genes (as in most other eukaryotic genes) many of the important *cis*-acting regulatory elements are located in a relatively small region 5′ to the transcription start site (Fig. 6-6). For example, when a 2000-bp fragment 5′ to the transcription start site of the mouse α2(I) collagen gene was placed next to the gene for bacterial chloramphenicol acetyl transferase (CAT), and the construct introduced into the germline of mice, CAT expression was high in tissues that express high levels of type I collagen and low in tissues that are low in type I (Khillan *et al.*, 1986). It appears therefore that sequences important for tissue-specific expression are located within 2000 bp 5′ of the transcription start site in the mouse α2(I) collagen gene. Within this upstream promoter region several specific sites for DNA-binding factors have been identified in both the mouse α2(I) and α1(I) collagen genes. DNA-binding factors have also been purified and characterized. As a result, some of the details of DNA–protein interactions within the type I collagen gene promoters are now emerging.

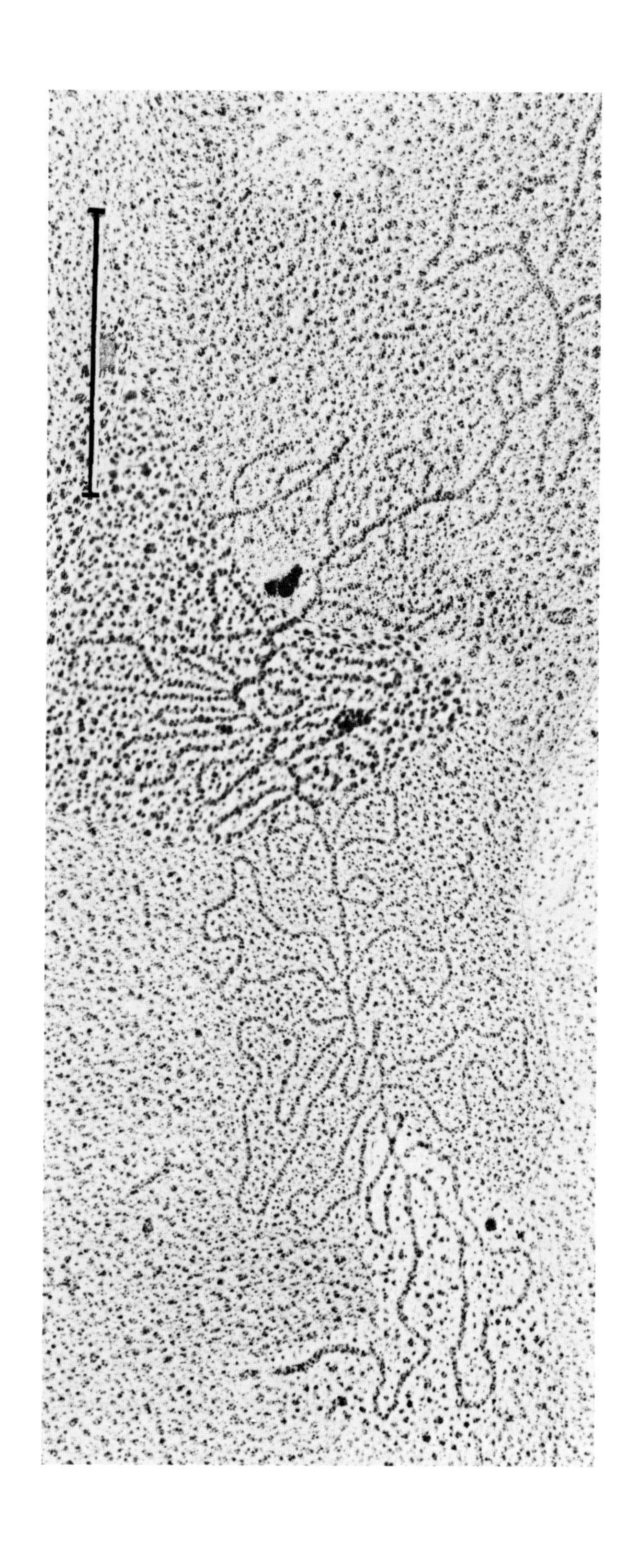

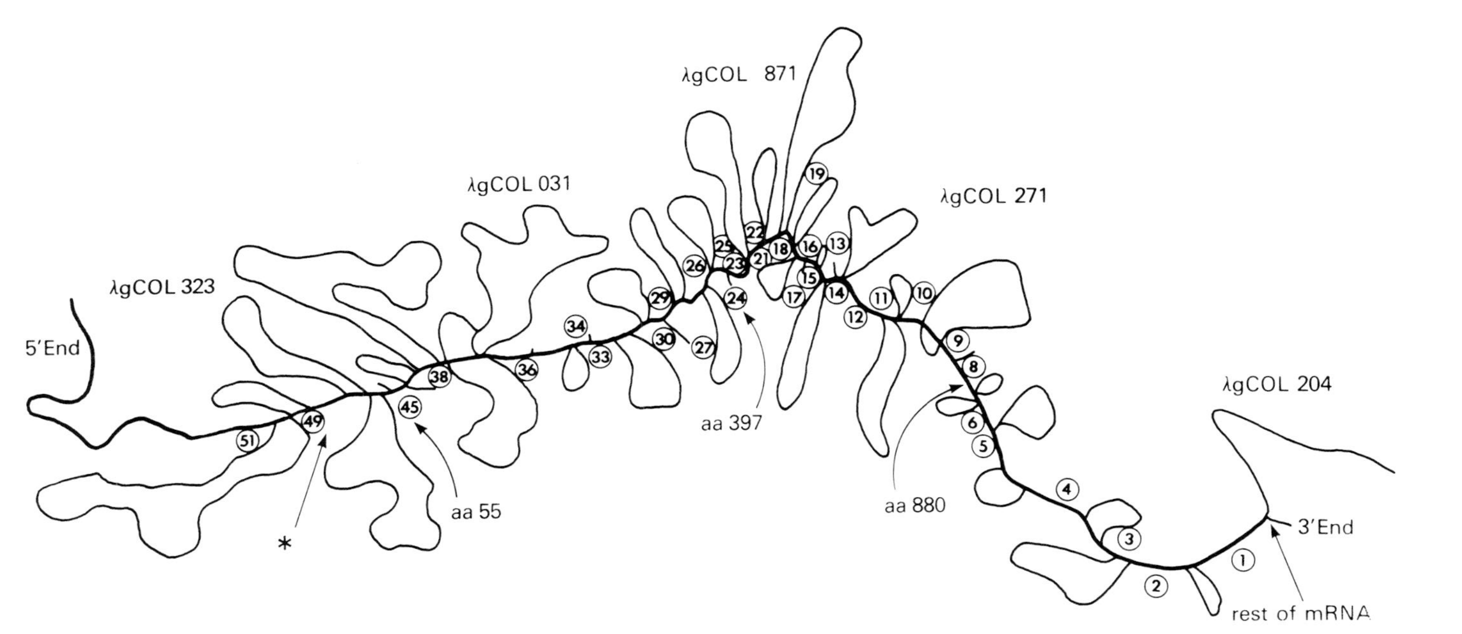

Figure 6-4. Electron micrograph and diagram showing the exon/intron structure of the chick proα2(I) collagen gene. The top part of the figure is a composite of several "R-loop" electron micrographs. The "R-loops" were prepared by incubating mRNA for type I collagen with one of several genomic clones coding for different portions of the proα2 chain under conditions that favor formation of stable DNA–RNA hybrids. As mRNA molecules contain sequences that are complementary to exon sequences but not to intron sequences, RNA and exons will form a double-stranded "core" in the "R-loop" structure with introns forming loops along this core. The bottom part of the figure is a tracing of the structure seen in the top part. The circled numbers indicate exons counted from the 3′ end of the mRNA, and the amino acids 55, 397, and 880 in the α2-chain of collagen are indicated. The genomic clones λgCOL204, λgCOL271, λgCOL871, λgCOL031, and λgCOL323 were used to obtain the "R-loop" structure. Bar = 1300 bp. (From Vogeli *et al.*, 1981.)

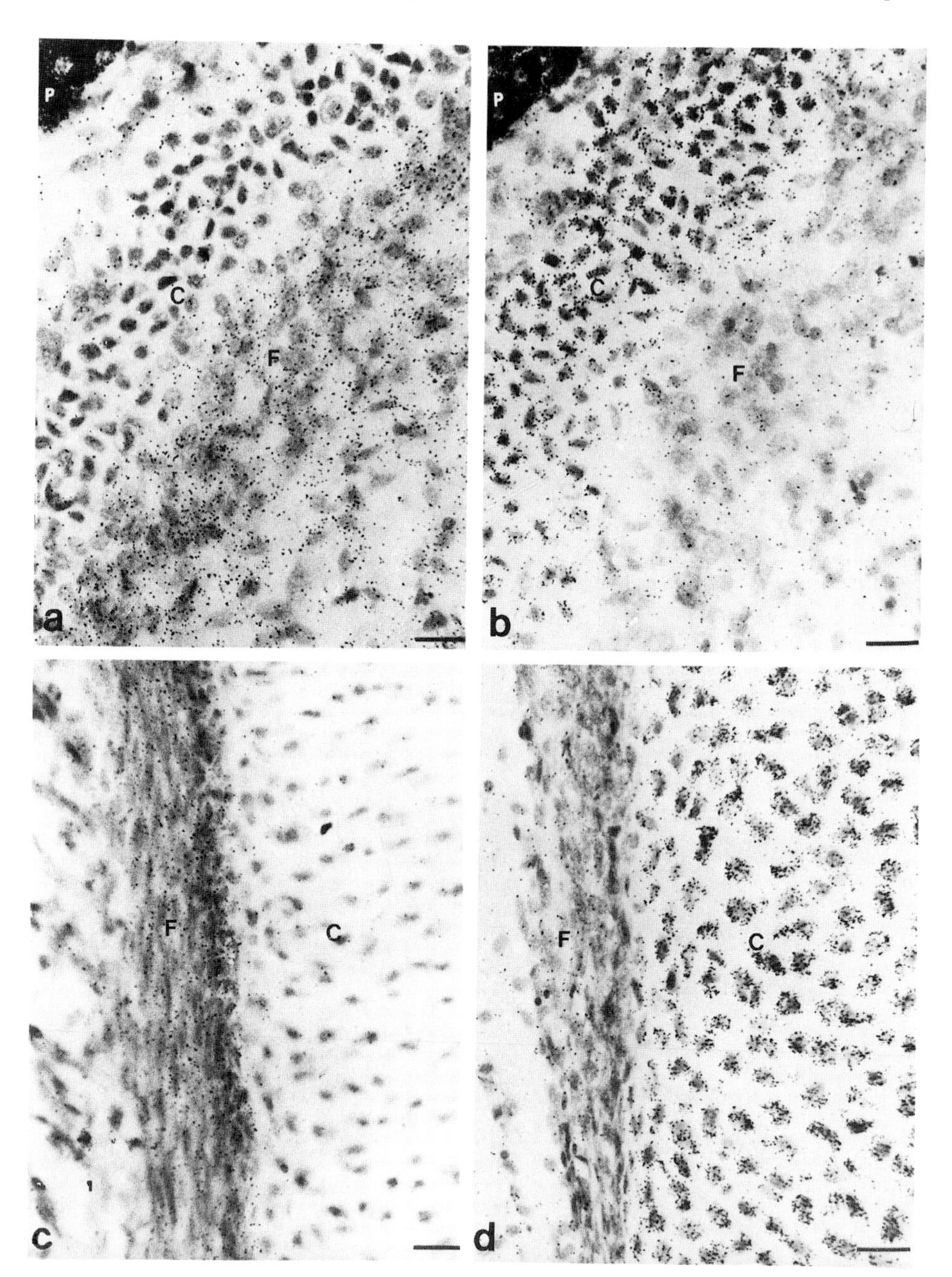

Figure 6-5. Localization of mRNAs encoding α2(I) and α1(II) collagen chains in fibroblasts and chrondrocytes by *in situ* hybridization. Paraffin sections from a 10-day-old whole chick embryo were processed for *in situ* hybridization with ^{3}H-labeled cDNA probes, followed by autoradiography. (a) Area including scleral cartilage (C) and surrounding fibroblast sheet (F) labeled with the α2(I) probe. P indicates a portion of the pigment epithelium. (b) Area of sclera as in a, labeled with the α1(II) probe. (c) Area of wing cartilage (C) and surrounding fibroblasts (F) labeled with α2(I) probe. (d) Same area as in c, labeled with the α1(II) probe. Bars = 20 μm. (From Hayashi *et al.*, 1986.)

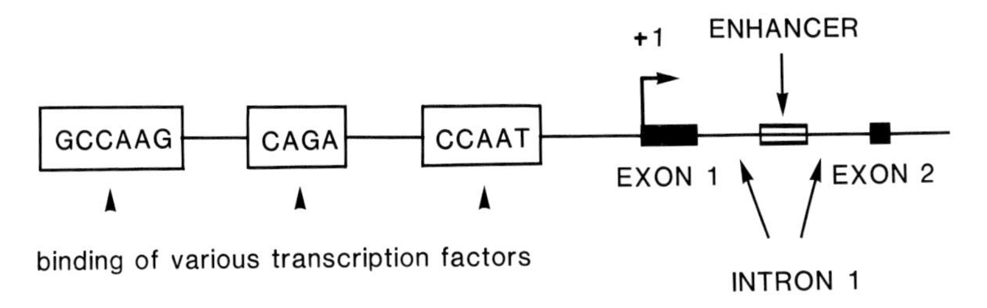

Figure 6-6. Diagram of the 5′ region of the mouse α2(I) collagen gene. Upstream of the transcription start site (+1) several transcription factor binding sites have been identified. Three such sites are indicated by the boxed-in sequences. Exons 1 and 2 are also shown, and an enhancer is located in intron 1.

Among regulator proteins that stimulate transcription, one factor binds to CCAAT sequences in both the mouse α1(I) and α2(I) genes. It is a heterodimer of two distinct subunits and belongs to a large group of CCAAT-binding proteins in eukaryotic cells (Hatamochi *et al.*, 1988; Maity *et al.*, 1988). An additional stimulating factor, NF-1, has been shown to bind to a binding site upstream of CCAAT in the mouse α2(I) gene (Oikarinen *et al.*, 1987), and there is evidence that the promoter in this gene interacts with other transcriptional stimulators as well (Ramirez and DiLiberto, 1990; Vuorio and de Crombrugghe, 1990). However, not all *trans*-acting transcription factors stimulate transcription. In the mouse α2(I) gene there is evidence that two negative factors bind close to the stimulatory CCAAT-binding protein (Vuorio and de Crombrugghe, 1990). Thus, although no mechanistic explanation can yet be offered for the precise tissue-specific transcriptional regulation of the fibrillar collagen genes (see Chapter 1), the continued analysis of the interaction between negatively and positively acting DNA-binding proteins is likely to provide such an explanation in the future.

For the type I and II collagen genes there is also evidence that sequences located in the first (most 5′) intron of the genes are important for the transcriptional regulation of the genes (Rossi and de Crombrugghe, 1987; Rossouw *et al.*, 1987; Bornstein *et al.*, 1987, 1988; Bornstein and McKay, 1988; Horton *et al.*, 1987). These enhancer elements appear to be quite complex and contain both positive and negative *cis*-acting sequences. It is believed that protein factors bind to the enhancer sequences, and at the same time interact with proteins in the 5′ promoter region of the genes. How proteins, bound to DNA sequences that are far apart in the gene, can interact is not entirely clear, but probably involves the looping-out of the DNA so that the DNA-binding sites are brought into close proximity.

An interesting case which illustrates the complexity of fibrillar collagen transcriptional regulation, and provides a useful animal model of a collagen genetic defect, is the Mov13 mutation in the mouse (Schnieke *et al.*, 1983). This mutation is produced by stable integration of the Moloney murine leukemia virus genome into the first intron of the α1(I) collagen gene. The presence of the proviral DNA in this location changes the chromatin conformation of the α1(I) gene in fibroblasts, and makes the gene transcriptionally inactive, possibly by disrupting the interaction between promoter and enhancer regulatory sites (see above) (Breindl *et al.*, 1984). Fibroblasts in culture and in embryos that are homozygous for the mutation do not produce any α1(I) collagen chains. Since α2(I) collagen chains cannot form triple-helices that are stable at 37°C, no type I collagen is therefore produced; the excess α2(I) chains are degraded. The lack of type I collagen causes homozygous embryos to die *in utero* between days 11 and 14 of gestation, because of rupture of large blood vessels (Jaenisch *et al.*, 1983). Surprisingly, osteoblasts and odontoblasts from homozygous Mov13 embryos produce large amounts of type I collagen, in spite of the presence of proviral sequences in the first intron of the α1(I) collagen gene. Thus, tooth rudiments isolated from 13-day-old homozygous embryos and grown as transplants produce normal dentin (Kratochwil *et al.*, 1989). This suggests that odontoblasts and osteoblasts regulate their transcription of the α1(I) collagen gene by a mechanism that is different from that used by fibroblasts, and it points out the necessity to consider both gene-specific and cell specific aspects of transcriptional regulation of collagen genes.

3.3. Nonfibrillar Collagen Genes

In addition to fibrillar collagens, the building blocks of cross-striated fibrils, extracellular matrices contain a variety of collagenous proteins that form different supramolecular structures (Chapter 1). These nonfibrillar collagens are clearly heterogeneous, both structurally and functionally, and based on the structure of their genes they fall into distinct subclasses.

One subclass contains molecules that are or are thought to be associated with collagen fibrils. Since the molecules in this subclass contain more than one triple-helical domain connected by non-triple-helical regions, the subclass has been named the FACIT group for *f*ibril-*a*ssociated *c*ollagens with *i*nterrupted *t*riple-helices (Ninomiya *et al.*, 1990). The genes in the FACIT group are currently five: The α1(IX), α2(IX), and α3(IX) genes encode the polypeptides of the heterotrimeric type IX collagen, the α1(XII) gene codes for the homotrimeric type XII collagen, and the α1(XIV) gene codes for the chains of homotrimeric type XIV collagen.

A partial exon structure has been determined for the α1(IX) and α1(XII) genes (Ninomiya *et al.*, 1990; Olsen *et al.*, 1989; Gordon *et al.*, 1989) and the α1(IX) gene has been localized to the long arm of human chromosome 6 (6q13) (Kimura *et al.*, 1989b). The complete exon structure of the chicken α2(IX) gene is known, and this gene thus serves as a prototype FACIT gene (Fig. 6-7) (Ninomiya *et al.*, 1990). The gene contains 32 exons spread over about 10 kb of DNA. Although type IX collagen differs from fibrillar collagens in that it contains three short triple-helical domains instead of one long uninterrupted triple-helix, about half of the exons that code for triple-helical sequences in the α2(IX) gene are 54 bp or some slight variation of that size, as in the fibrillar collagen genes. However, toward the 5′ and 3′ ends of the α2(IX) gene, the

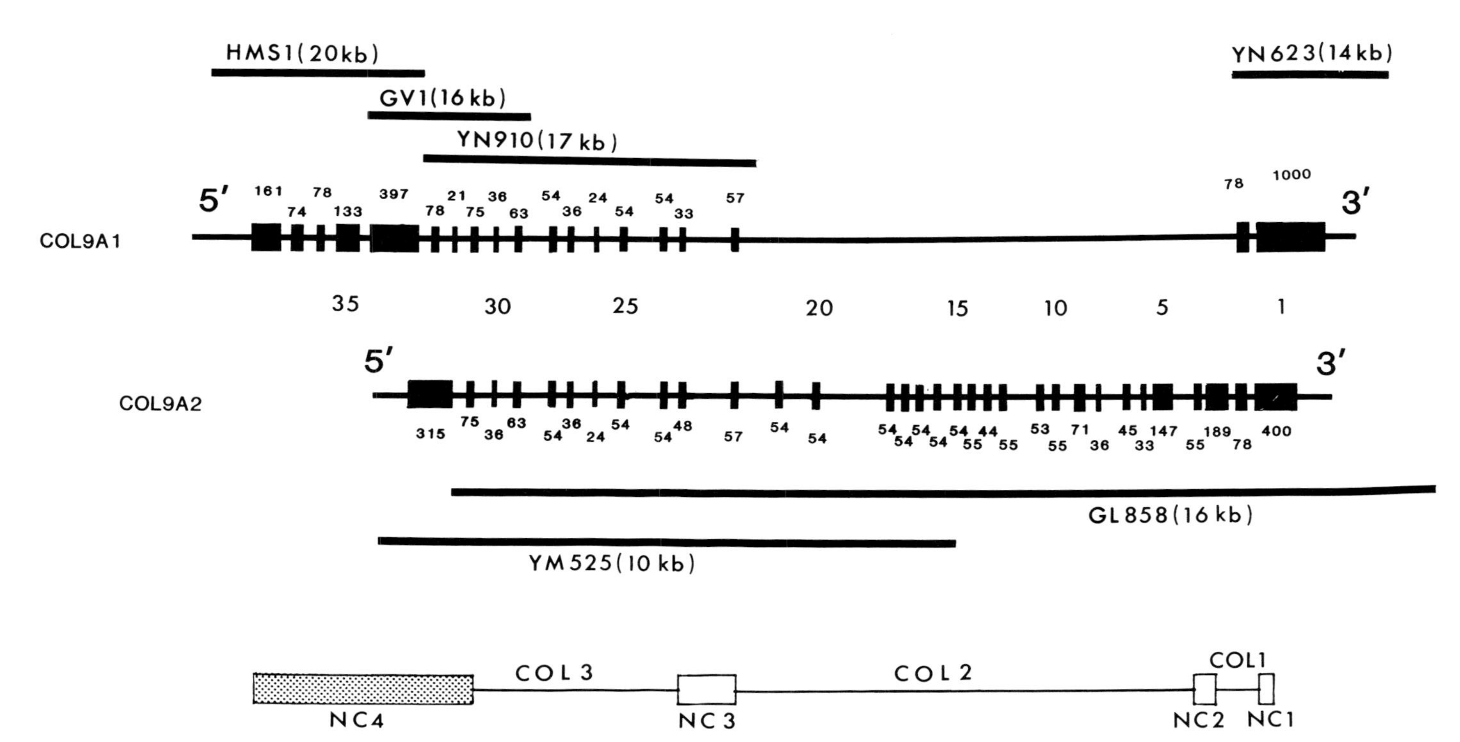

Figure 6-7. Diagram showing the exon structure of the α1(IX) and α2(IX) collagen genes (COL9A1 and COL9A2) and the chicken genomic clones that have been isolated for the two genes. The exons are numbered from the 3′ end; exon sizes (in bp) are given above and below the genes. At the bottom of the diagram, the positions (relative to the exons) of the various domains of the type IX collagen chains are indicated. Triple-helical domains (COL1, COL2, and COL3) are indicated by a line; the non-triple-helical domains (NC1, NC2, NC3, and NC4) are indicated by boxed-in areas. The large NC4 domain is only present in the α1(IX) chain. Note that clones covering the central portion of the α1(IX) gene have not yet been isolated. (From Ninomiya *et al.*, 1990.)

exons are quite different from what they are in fibrillar collagen genes. Therefore, although the α2(IX) gene must be related (because of the 54 bp motif) to the fibrillar collagen genes, it has followed a different evolutionary path. The complete exon structure of the α1(IX) gene is not yet established, but a substantial portion of the chicken gene has been characterized (Fig. 6-7). This partial structure shows that the α1(IX) and α2(IX) genes are very similar and probably arose by duplication from a common precursor, although their sizes are very different [100 kb for chicken α1(IX) versus 10 kb for chicken α2(IX)].

A second subclass consists of molecules that contain a short (about 450 amino acid residues versus about 1000 residues for type I collagen) triple-helical domain. This group is therefore classified as short-chain collagens and three genes are currently known to belong in this group: α1(VIII), α2(VIII), and α1(X) (Ninomiya *et al.*, 1986; LuValle *et al.*, 1988; Yamaguchi *et al.*, 1989, 1991). The two type VIII genes code for the polypeptides of the (probably) heterotrimeric type VIII collagen in Descemet's membrane and vascular walls (Yamaguchi *et al.*, 1989; Muragaki *et al.*, 1991), and the α1(X) gene codes for the subunits of type X collagen, a product of hypertrophic chondrocytes (Chapter 1). The structure of the short-chain collagen genes is completely different from all other collagen genes characterized to date. Instead of multiple, small exons, the short-chain genes contain a single, large exon (> 2000 bp) that encodes the entire triple-helix (about 460 amino acid residues) and a carboxyl globular domain (about 160 amino acid residues). In the 5′ portion of the genes, two [α1(X)] or three [α1(VIII)] small exons encode the 5′ untranslated region and an amino-terminal globular domain. How such a gene structure evolved is not clear, but one possibility is a homologous recombination event between a double-stranded cDNA and a previously existing intron-containing gene.

A third subclass includes the genes that code for collagenous polypeptides in basement membranes. Although the most common form of basement membrane collagen is composed of α1(IV) and α2(IV) chains, it is now known that there are at least five genes in this subclass: α1(IV), α2(IV), α3(IV), α4(IV), and α5(IV) (Burbelo *et al.*, 1989; Butkowski *et al*,. 1987; Saus *et al.*, 1988; Hostikka *et al.*, 1990). The exon structure of the type IV genes differs significantly from that of fibrillar collagen and FACIT genes in that triple-helical exons show considerable divergence from the 54 bp motif, but the genes do contain multiple short triple-helical exons in contrast to the short-chain collagen genes (see Vuorio and de Crombrugghe, 1990).

The α1(IV) and α2(IV) genes have been located to human chromosome 13 (13q34), where the two genes are arranged in a head-to-head orientation separated by a short (< 150 bp) intergenic DNA segment. The same situation has been demonstrated in the mouse. The two genes are therefore transcribed in opposite directions and on opposite strands of the DNA double-helix (Pöschl *et al.*, 1988; Soininen *et al.*, 1988; Burbelo *et al.*, 1988; Kaytes *et al.*, 1988).

The α5(IV) gene is X-linked in humans and is located at the locus for X-linked Alport's syndrome (Xq22) (Hostikka *et al.*, 1990). In fact, many patients with this hereditary kidney disease have been shown to have major defects in the α5(IV) collagen gene. Since the α5(IV) chain is located in the glomerular basement membrane, it is likely that the disease is directly caused by a defect in this gene (Barker *et al.*, 1990).

The remaining nonfibrillar collagen genes are difficult to classify based on current information. The gene encoding type VII collagen has not yet been isolated, and genes encoding type VI [α1(VI), α2(VI), α3(VI)] (Bonaldo *et al.*, 1990; Chu *et al.*, 1990) and XIII [α1(XIII)] collagens (Pihlajaniemi *et al.*, 1987a, 1990) are different from the subclasses described above. Undoubtedly, further characterization of collagen genes will provide a better understanding of where these collagens belong in the superfamily of collagen genes and how they evolved.

3.4. Transcriptional Regulation of Nonfibrillar Collagen Genes

As for the fibrillar collagens, available evidence suggests that transcriptional regulation of nonfibrillar collagen genes plays a very important part in their tissue- and cell-specific expression. A striking example is provided by the regulated expression of the short-chain collagen type X. This collagen is synthesized only in hypertrophic chondrocytes. Thus, during skeletal development the protein appears in areas of endochondral ossification (Schmid and Linsenmayer, 1985). The synthesis of the protein is due to a rapid increase in the level of α1(X) mRNA as chondrocytes become hypertrophic (Fig. 6-8). Assays of rates of transcription initiation with isolated cell nuclei suggest that this increased level of mRNA can be accounted for by a rapid increase in transcription rate (LuValle *et al.*, 1989). The mechanism for this increase is not known.

An interesting example of tissue-specific variations in collagen gene expression is provided by type IX collagen. As discussed in Chapter 1, this FACIT molecule is located on the surface of type II/type XI-containing fibrils (Vaughan *et al.*, 1988). The α1(IX) gene, encoding one of three distinct polypeptide subunits of type IX collagen, contains two alternative promoters and transcription start sites (Nishimura *et al.*, 1989; Muragaki *et al.*, 1991). In cartilage, most of the α1(IX) transcripts arise from an upstream start site and encode an amino-terminal globular domain of about 250 amino acid residues attached to the triple-helical part of the α1(IX) chain. In embryonic cornea (Svoboda *et al.*, 1988), and the vitreous body of the eye (Wright and Mayne, 1988; Yada *et al.*, 1990), most transcripts start from an alternate exon 1 located in the sixth intron of the gene. This gives rise to a short transcript that is identical in its triple-helical region to the transcript from the upstream start, but has a different 5′ end. Since the downstream start is 3′ to most of the exons encoding the amino-terminal globular domain, this domain is missing in the α1(IX) chains translated from the short mRNA (Fig. 6-9). This could therefore be a mechanism used by cells to modulate the surface structure of type II/type XI-containing collagen fibrils, and it could account for differences in the way fibrils are organized in different tissues.

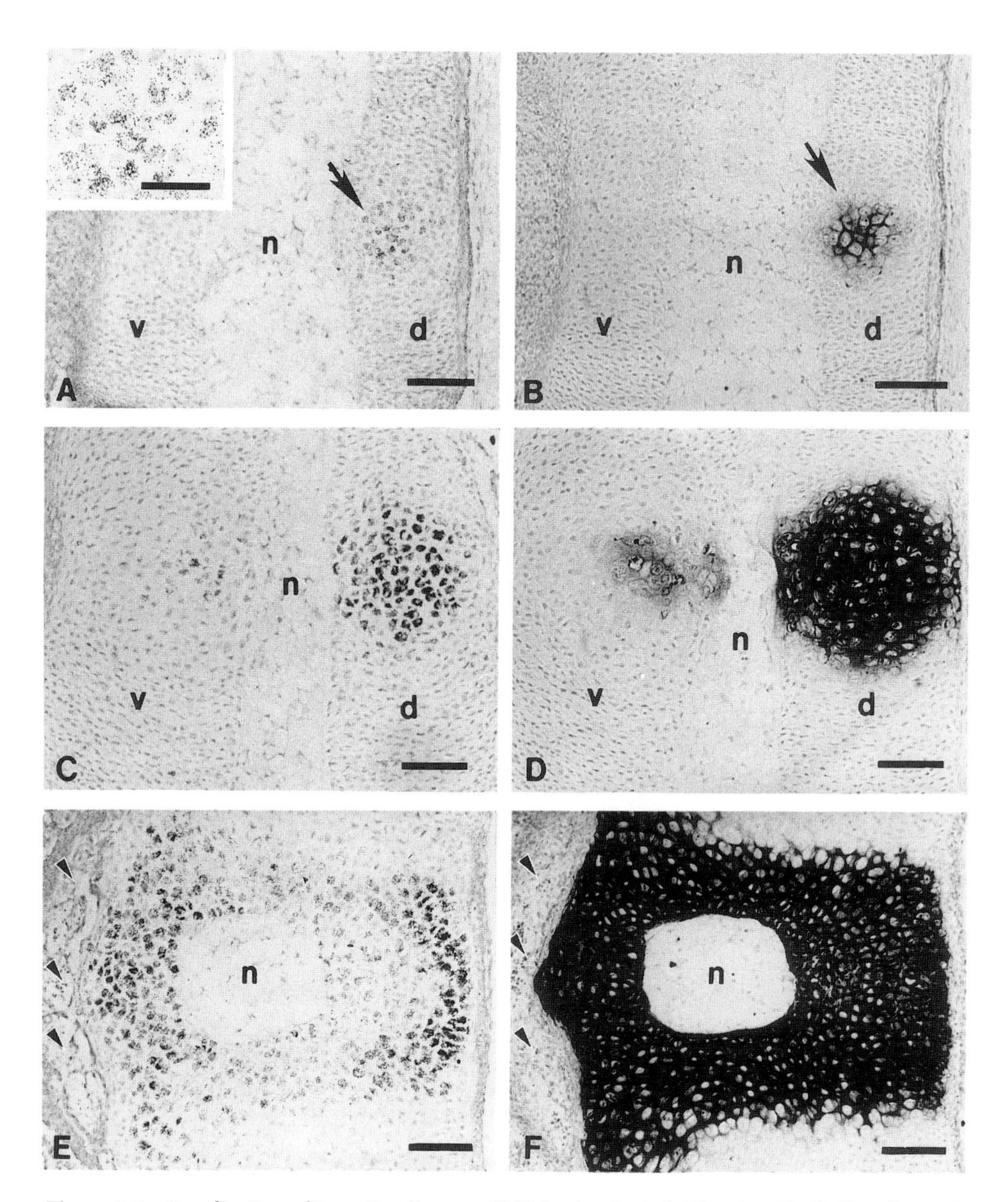

Figure 6-8. Localization of type X collagen mRNA by *in situ* hybridization (A, C, E) and type X protein by immunostaining (B, D, F). Sagittal sections (A–D) and cross sections (E, F) of vertebral body segments of chick embryos at stages 36 (A, B), 37 (C, D), and 38 (E, F) were processed for *in situ* hybridization with a ^{3}H-labeled cDNA for type X collagen, and for staining with a type X-specific monoclonal antibody and the peroxidase technique. The notochord (n) is in the middle of each figure. Dorsal (d) and ventral (v) halves of vertebrae are indicated. Insert in A shows at high magnification newly developed hypertrophic chrondrocytes positive for type X collagen mRNA. As indicated by arrows in A and B, hypertrophy and type X collagen expression starts in a small, localized area, and spreads rapidly through the cartilage. Arrowheads in E and F indicate dilated blood vessels. Bars = 100 μm; bar in inset-40 μm. (From Iyama *et al.*, 1991.)

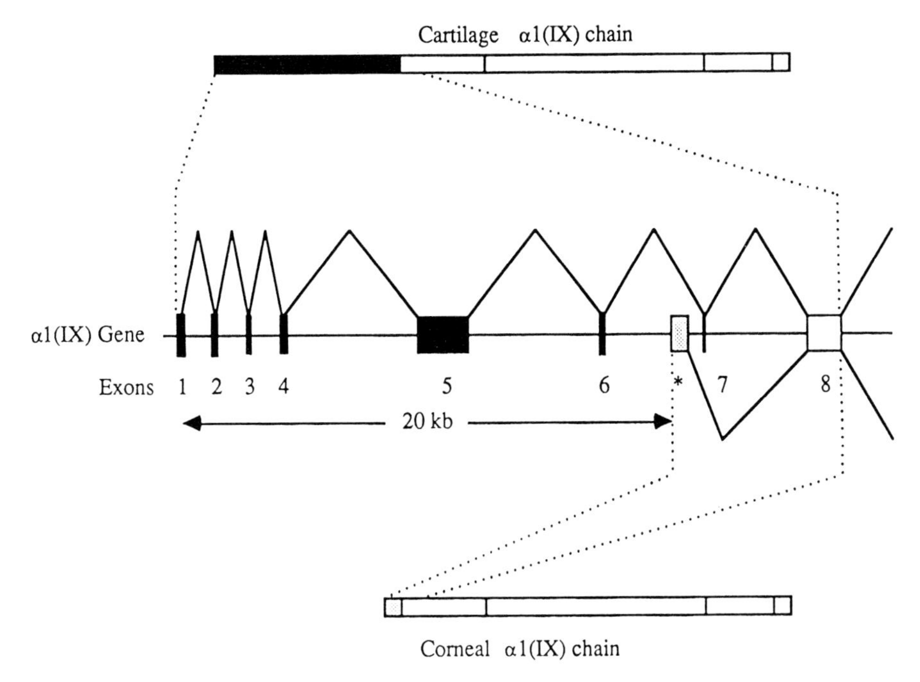

Figure 6-9. Diagram of the 5′ region of the α1(IX) collagen gene with two transcription start sites and two alternative first exons (exon 1 and exon *). The two exons are located about 20 kb apart. In cartilage most transcripts start at the upstream site and the mRNA contains the sequences of exons 1–8 as indicated by the splice pattern above the gene; in embryonic cornea, most transcripts start at the downstream site and the mRNA contains the sequences of exons * and 8 as indicated by the splice pattern below the gene. The α1(IX) collagen chains in cartilage and cornea differ therefore in their NH_2-terminal regions. (From Nishimura *et al.*, 1989.)

4. Cotranslational and Posttranslational Modification of Intracellular Procollagen

Numerous early studies with subcellular fractions of cells showed that when cells or tissues are incubated with radioactive proline, radioactive peptidyl-hydroxyproline or bacterial collagenase-sensitive peptides can be recovered in microsomes and polysomes. Such studies coupled with a demonstration of prolyl hydroxylase activity in microsomal fractions form the basis for the conclusion that collagen polypeptides are synthesized by membrane-bound polysomes and hydroxylated while still attached to ribosomes (for review see Prockop *et al.*, 1979).

All collagen polypeptides, whether they belong to fibrillar or nonfibrillar collagens, are synthesized with hydrophobic signal sequences at the NH_2 termini of the nascent chains, and these signal sequences are rapidly processed in the RER. As a result, collagen polypeptides that contain this NH_2-terminal hydrophobic signal sequence can only be observed by translating collagen mRNA in a cell-free system without microsomal membranes. The signal sequences serve as vectors for ensuring that the nascent protein chains will cross the membrane of the RER. As collagen chains isolated from intracellular compartments have about the same molecular size as chains of collagen molecules secreted into the medium by cells in culture, the removal of signal peptide segments must be a rapid process, occurring either during translation on ribosomes or shortly after completion of polypeptide synthesis in the RER.

The removal of the signal peptide sequence is only the first step in a series of modification steps that must be completed before a functional collagen molecule is formed. A critical next step is synthesis of hydroxyproline. The formation of a fibrillar collagen triple-helix that is stable at 37°C requires the presence of about 100 hydroxyproline residues per α chain (for review see Prockop *et al.*, 1979). The thermal stability of triple-helices in nonfibrillar collagens also depends on a critical number of hydroxyproline residues per chain. Hydroxyproline is synthesized from prolyl residues after their incorporation into nascent peptides.

The synthesis of hydroxyproline occurs within the RER (see below). In addition, some of the lysyl residues are hydroxylated and glycosylated and the COOH propeptides are glycosylated in this compartment. Subcellular fractionation studies suggested that posttranslational modification enzymes are distributed in several subcellular compartments (for review see Fessler and Fessler, 1979; Prockop *et al.*, 1979; Bornstein and Sage, 1980), but these studies are difficult to interpret because clean fractions, especially of the Golgi complex, were not obtained.

In the RER the newly synthesized collagen chains are disulfide linked into triple-stranded molecules, but the precise site of triple-helix formation is unknown. However, hydroxylation of prolyl and lysyl residues and glycosylation of hydroxylysyl residues occur only on unfolded polypeptide chains. Therefore, if glycosylation events do continue after collagen reaches the Golgi complex, the triple-helical domains cannot fold until after the molecules have reached this compartment.

4.1. Hydroxylation of Prolyl and Lysyl Residues

Hydroxylation of prolyl and lysyl residues by three hydroxylases located in the RER has been extensively studied (for a review see Kivirikko and Myllylä, 1984, 1985; Kivirikko *et al.*, 1989, 1990). Two of the hydroxylases convert some prolyl residues to 4-hydroxyproline or 3-hydroxyproline, and the third enzyme converts some lysyl residues to hydroxylysine. The three enzymes are mixed-function oxygenases. Their reactions involve the cofactors ferrous iron and ascorbate, and the cosubstrates oxygen and α-ketoglutarate. During the hydroxylation reaction, α-ketoglutarate is stoichiometrically decarboxylated to succinate as the peptide substrate is hydroxylated.

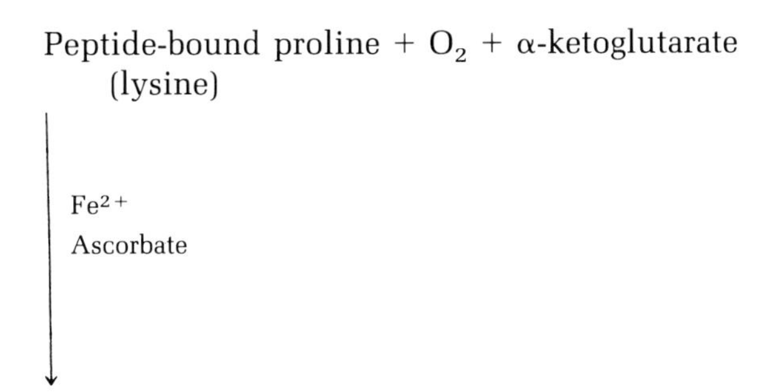

The requirement for iron as a cofactor makes it possible to inhibit the hydroxylases in cells by introducing an iron chelator such as α,α′-dipyridyl (for review see Prockop *et al.*, 1979).

The requirement for ascorbate is highly specific, but ascorbate is not consumed stoichiometrically in the enzyme reaction. It appears that ascorbate is needed for the decarboxylation of α-ketoglutarate, and this can occur even without hydroxylation of prolyl residues in a peptide substrate. In ascorbate deficiency (scurvy), proα chains of fibrillar procollagens are underhydroxylated and do not form stable triple-helices at 37°C. As discussed below, such unfolded proα chains are secreted from cells at a low rate and do not form functional collagen fibrils.

The three hydroxylases require a peptide substrate in a non-triple-helical conformation, and the susceptible prolyl and lysyl residues must be within specific amino acid sequences (for review see Kivirikko and Myllylä, 1984, 1985). Prolyl 4-hydroxylase will only hydroxylate prolyl residues in the Y position of peptides containing X-Y-Gly sequences. Prolyl 3-hydroxylase will only hydroxylate prolyl residues in the X position of peptides containing X-Hyp-Gly sequences. Within the collagen domain of proα chains, lysyl hydroxylase appears to have the same position specificity as prolyl 4-hydroxylase in that it only hydroxylates lysyl residues in the Y position of peptides containing X-Y-Gly sequences. However, within the telopeptide regions of fibrillar proα chains, the lysine before glycine rule does not hold (Chapter 1). The hydroxylysines synthesized within those regions are of special importance because they are converted to the corresponding aldehydes by lysyl oxidase and participate in cross-link formation in the extracellular matrix.

Lysyl hydroxylase (for review see Kivirikko and Myllylä, 1985) is a dimer of two subunits each about 90 kDa (Turpeenniemi *et al.*, 1977). It is a microsomal enzyme and its solubility properties suggest that it is tightly associated with the membrane of the RER. It reacts more efficiently with increasing lengths of peptide substrate. Prolyl 3-hydroxylase has only been partially characterized (Risteli *et al.*, 1977), but prolyl 4-hydroxylase has been isolated as a pure protein from several sources (Berg and Prockop, 1973; Kivirikko and Myllylä, 1984) and characterized in detail by analysis of cDNA and genomic DNA clones (see below).

The active prolyl 4-hydroxylase enzyme is a tetramer of 240 kDa with the subunit structure $\alpha_2\beta_2$, where α and β are two types of subunits of around 60 kDa. Both subunits are glycoproteins. The enzyme is located within the RER, as shown by electron microscopy using ferritin-labeled anti-prolyl hydroxylase antibodies (Olsen *et al.*, 1975) and by its presence in microsomal subcellular fractions (Diegelmann *et al.*, 1973; Harwood *et al.*, 1974; Peterkofsky and Assad, 1976). The enzyme does not appear to be an integral membrane protein or to be tightly associated with the RER membrane, for it is easily soluble after mild detergent treatment of microsomes to disrupt the microsomal membrane. Although the active prolyl 4-hydroxylase is a tetramer of 240 kDa, about 30–99% of the enzyme protein in cells is in an inactive form of about 60 kDa (see Prockop *et al.*, 1979). This inactive form consists of the β subunit (Chen-Kiang *et al.*, 1977). Comparisons of various cells and tissues indicate that the ratio of active tetramers to the inactive form is roughly proportional to the rate of collagen synthesis. Conversion of prolyl 4-hydroxylase from an inactive to an active form may therefore be important in regulating collagen synthesis during development or during fibrosis. Data on the assembly of active, tetrameric prolyl 4-hydroxylase in embryonic chick tendon fibroblasts indicate that the inactive form serves as a precursor of β subunits in the active form of the enzyme (Berg *et al.*, 1980). Changes in the amount of active prolyl 4-hydroxylase can thus be brought about by changes in the rate of synthesis of α subunits, and assembly of active enzyme involves the formation of tetramers by newly synthesized α subunits and preformed β subunits (Berg *et al.*, 1980).

The cloning of cDNA and genomic DNA for the β subunit of prolyl 4-hydroxylase revealed that the protein has multiple identities (Pihlajaniemi *et al.*, 1987b; Parkkonen *et al.*, 1988; Kao *et al.*, 1988). First, the β subunit of prolyl 4-hydroxylase and the enzyme protein disulfide isomerase in the RER (Edman *et al.*, 1985) are products of the same gene (for review see Bassuk and Berg, 1989). Protein disulfide isomerase catalyzes disulfide bond formation in a variety of secretory and cell surface proteins, by forming, reducing, or isomerizing disulfide bonds. Most of the disulfide isomerase in cells appears to be associated with free β protein in RER, and corresponds in this form to the previously described "inactive" β subunits of prolyl 4-hydroxylase. However, the disulfide isomerase activity is not lost when β protein and α subunits are assembled into active hydroxylase tetramers $\alpha_2\beta_2$. Thus, purified prolyl 4-hydroxylase has an associated protein disulfide isomerase activity (Koivu *et al.*, 1987).

Second, the β subunit also serves as a major thyroid hormone binding protein in the RER (Cheng *et al.*, 1987; Yamauchi *et al.*, 1987; Gong *et al.*, 1988), and it catalyzes the conversion of thyroxin to triiodothyronine (Boado *et al.*, 1988). Third, the β subunit is closely related, perhaps identical, to the glycosylation site binding component of oligosaccharyl transferase (Geetha-Habid *et al.*, 1988). This component recognizes the -Asn-X-Ser/Thr- acceptor site in nascent glycopeptides so that the catalytic part of the transferase can transfer the oligosaccharide chain from dolichol pyrophosphate to the asparagine residue (Kaplan *et al.*, 1988).

The β subunit of prolyl 4-hydroxylase is therefore a remarkable multifunctional protein residing in the RER. Among its many roles is also the function of retaining prolyl 4-hydroxylase in the RER, since it has the sequence -Lys-Asp-Glu-Leu, necessary and sufficient for the retention of polypeptides in the RER, at its carboxyl end (Kivirikko *et al.*, 1990; Munro and Pelham, 1987). The α subunit does not have such an RER retention signal. The α subunit has also been studied by molecular cloning, but unlike the β subunit its sequence appears to be unique (Helaakoski *et al.*, 1989). Interestingly, the RNA transcripts from the α subunit gene undergo alternative splicing, such that the mRNA contains one of two mutually exclusive exon sequences. Cells therefore contain two types of α subunit mRNA. The functional significance of this, however, is not known.

4.2. Disulfide-Bond Formation and Triple-Helix Formation

The formation of disulfide bonds between proα chains in fibrillar procollagens contributes to the rate of formation and stability of the triple-helical conformation of procollagen. The NH_2 propeptides were initially envisioned as registration peptides (Speakman, 1971), but these peptides usually contain only intrachain disulfide bonds, and it is now known that the COOH propeptides contain the information required for proper chain association (Rosenbloom *et al.*, 1976). The folding of the COOH propeptides, stabilized by disulfide bonds, therefore precedes the formation of the fibrillar collagen triple-helix (Dölz and Engel, 1990). In support of this is the finding that chain association and interchain disulfide bond formation (catalyzed by protein disulfide isomerase) are processes that can proceed in the absence of prolyl hydroxylation and triple-helix formation (see Prockop *et al.*, 1979).

A striking illustration of the importance of proper folding of the COOH propeptides for the incorporation of type I procollagen chains into a trimeric molecule, was provided by a case of autosomal recessive, nonlethal form of osteogenesis imperfecta. In this case, the son of consanguineous third-cousin parents inherited an α2(I) collagen gene with a 4-bp deletion toward the COOH end of the COOH propeptide (Pihlajaniemi *et al.*, 1984). This altered the reading frame and therefore the amino acid sequence of the COOH region of the proα2(I) propeptide, and eliminated the most distal cysteinyl residue in the COOH propeptide sequence. Because a new translation stop codon was present close to the normal stop codon, the size of the proα2(I) COOH propeptide did not change much; only the most distal sequence was altered. With the change in sequence, and with the distal cysteine eliminated, the COOH propeptide could not fold properly and was not incorporated into type I procollagen trimers (Nicholls *et al.*, 1979; Deak *et al.*, 1983). Thus, although proα2 (I) chains were synthesized and present in RER, they were rapidly degraded and did not appear in secreted procollagen. Instead of normal heterotrimeric type I procollagen, fibroblasts and osteoblasts in the patient secreted proα1(I) homotrimers. This did not lead to major abnormalities in skin, but caused severe defects in bone (Nicholls *et al.*, 1984).

The exact time or location for triple-helix formation is not known. Although some reports suggest that helix formation can occur while the polypeptide chains are still attached to ribosomes (Veis and Brownell, 1977), most available evidence favors a later stage of intracellular assembly. What is established, however, is that hydroxyproline plays a critical role in stabilizing the triple-helix under physiological conditions. Without any hydroxyproline residues, triple-helical type I collagen will melt around 25°C, whereas with about 100 hydroxyproline residues per α chain, the melting temperature is about 40°C (Berg and Prockop, 1973; Rosenbloom *et al.*, 1973). It is, therefore, clear that stable procollagen triple-helices cannot form in cells at 37°C unless about half the proline residues have been hydroxylated by prolyl 4-hydroxylase. Several types of experiments in which the formation of the triple-helix is prevented indicate that nonhelical procollagen is secreted by cells more slowly than helical procollagen. In fact, nonhelical procollagen is secreted with a rate constant that is only one-tenth of the rate constant for the secretion of helical molecules (Kao *et al.*, 1977, 1979). This finding probably explains the observed correlation between the rate of secretion of procollagen from cells and the level of hydroxylation of prolyl residues.

Nonfibrillar collagens do not contain sequence domains that are homologous to the COOH propeptides of fibrillar procollagens. How such collagens, frequently containing multiple triple-helical domains, are assembled and folded in cells is not known. Since some of the nonfibrillar collagens (such as types IV, VI, VII, VIII, X) contain relatively large non-triple-helical COOH domains, it is possible that these domains are functionally equivalent to the COOH propeptide of fibrillar collagens. Other nonfibrillar collagens (types IX, XII, XIV) have only short COOH peptide tails, and whether these tails contain sufficient information for initial chain association is an open question.

4.3. Glycosylation of Hydroxylysyl and Asparaginyl Residues in Collagens

O-Glycosylation of hydroxylysine in the collagen domains of both fibrillar and nonfibrillar collagens occurs by sequential action of galactosylhydroxylysyltransferase and glucosylgalactosylhydroxylysyltransferase (for review see Kivirikko and Myllylä, 1979). The enzymes require hydroxylysine in polypeptides with a non-triple-helical conformation as substrate. Galactosylhydroxylysyltransferase transfers galactose from UDP-galactose to peptidyl hydroxylysine, forming peptidyl *O*-α-D-galactopyranosylhydroxylysine. Glucosylgalactosylhydroxylysyltransferase transfers glucose from UDP-glucose to galactosylhydroxylysine, forming 2-*O*-α-D-glucosyl-*O*-β-D-galactopyranosylhydroxylysine (Fig. 6-10).

The extent of glycosylation of lysyl residues is variable for the different

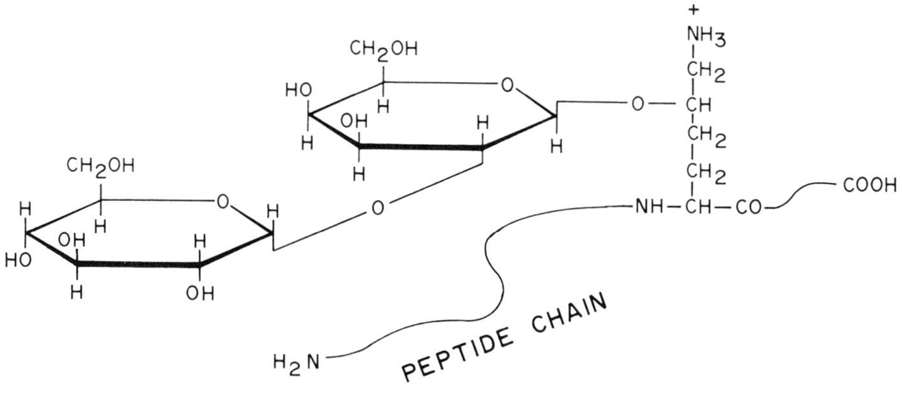

Figure 6-10. Structure of the disaccharide units attached to hydroxylysine in collagen.

genetic types of collagen and is also variable in the same type of collagen from different tissues in the same organism. As the glucosylgalactosylhydroxylysyltransferase does not act on triple-helical substrates, the degree of glycosylation of lysyl residues will depend on the rate of triple-helix formation in cells. The function of *O*-glycosylation in collagen is not known. It is clear, however, that it is not essential for secretion of the molecule from cells, as collagen secretion does not appear to be impaired in patients with type VI Ehlers–Danlos syndrome characterized by lack of hydroxylysine and glycosylated hydroxylysine residues (Quinn and Krane, 1976).

The COOH propeptides of fibrillar procollagens and some domains in nonfibrillar collagens are glycosylated in the RER by transfer of a mannose-rich oligosaccharide side chain from a dolichol phosphate intermediate to an asparaginyl residue in the polypeptide chain. The acceptor site for the oligosaccharide side chain contains the sequence Asn-X-Ser/Thr, and, as discussed above, the acceptor site recognition subunit of the glycosyltransferase is probably identical to the β subunit of prolyl hydroxylase (Geetha-Habid *et al.*, 1988). The synthesis of the dolichol phosphate intermediate is sensitive to tunicamycin. This drug, therefore, inhibits N-glycosylation of collagen (Duksin and Bornstein, 1977; Tanzer *et al.*, 1977; Hously *et al.*, 1980).

5. Intracellular Transport of Procollagen

Secretion of collagen involves transport from the RER to the Golgi complex, packaging in the Golgi region, and translocation of secretory vesicles. Some of these transport processes can be perturbed experimentally. For example, the secretion of procollagen by cultured cells is sensitive to local anesthetics, colchicine, vinblastine, cytochalasin B, uncouplers of oxidative phosphorylation, and the Na^+ ionophore monensin. This is not surprising since these agents have been shown to arrest the secretion of proteins in several other

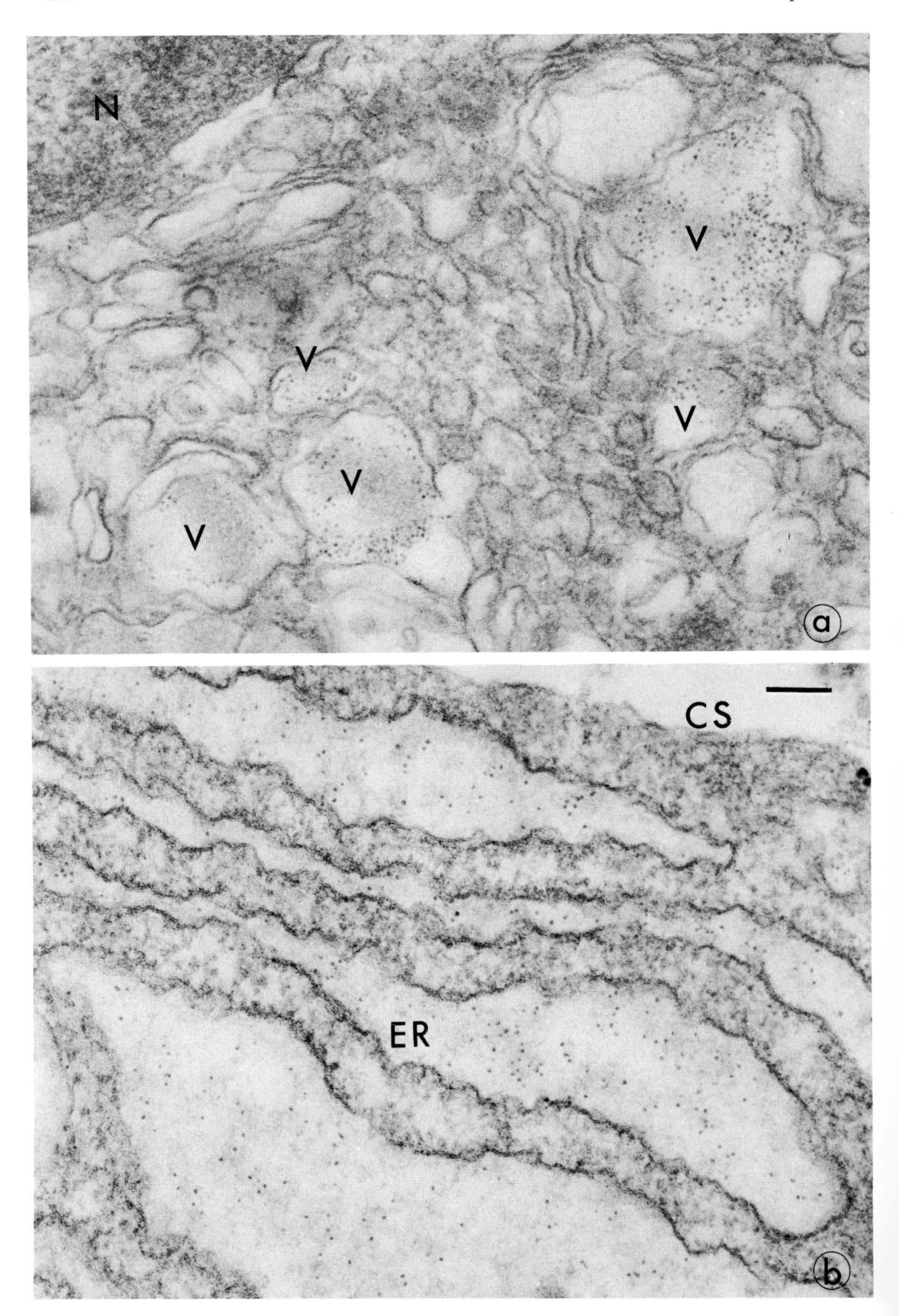
N
V
V
V
V
V
V
a
CS
ER
b

cellular systems, and collagen is transported by ion- and energy-dependent mechanisms common to the export of other secretory proteins.

5.1. The Role of the Golgi Complex in Procollagen Processing and Packaging

Early autoradiographic studies of collagen biosynthesis led to conflicting interpretations of the role of the Golgi complex in collagen secretion (for review see Ross, 1975). Autoradiographic studies by Revel and Hay (1963) and Weinstock and Leblond (1974) suggested that all the proteins synthesized and exported by chondroblasts and odontoblasts, including collagen, were routed through the Golgi complex and packaged there into secretory granules prior to secretion. In contrast, Ross and Benditt (1965) and Salpeter (1968) interpreted their autoradiographic data to indicate that fibroblasts and chondroblasts might use two different pathways for secretion of exportable proteins. One pathway would involve the RER–Golgi route; the other pathway, which might include collagen, could be a direct one from the RER to the extracellular space, circumventing the Golgi complex.

It is now clear, however, that Golgi vacuoles do contain collagen and are involved in the processing and secretion of collagen. Bacterial collagenase-sensitive ^{14}C-labeled polypeptides can be chased from a microsomal fraction through a Golgi-enriched fraction to the extracellular medium in embryonic chick tendon fibroblasts (Harwood *et al.*, 1976). Also, in Golgi-enriched subcellular fractions obtained from embryonic chick tendon fibroblasts, Harwood *et al.* (1976) could demonstrate the presence of disulfide-linked procollagen polypeptides. In addition, the presence of procollagen in Golgi vacuoles and cisternae of the endoplasmic reticulum has been demonstrated in chick tendon and cornea fibroblasts (Olsen and Prockop, 1974; Nist *et al.*, 1975) with ferritin-labeled antibodies against procollagen (Fig. 6-11). Finally, immunoperoxidase techniques have been successfully used to demonstrate procollagen in Golgi vacuoles of odontoblasts (Fig. 6-12).

5.2. The Packaging of Procollagen Aggregated in Secretory Granules

There is a considerable amount of morphological evidence for the conclusion that procollagen is packaged in the form of cylindrical aggregates in Golgi-derived secretory granules, at least in some cell types. The data are especially

←

Figure 6-11. Immunohistochemical localization of procollagen in tendon fibroblasts. The primary antibody [anti-proα1(I)] was labeled with ferritin using a second antibody against it. The secretory organelles were permeabilized to antibody by partially homogenizing isolated fibroblasts, thus breaking their membranes. (a) The contents of the Golgi vacuoles (V) are decorated with ferritin particles (small dots) indicating the presence of procollagen. N, nucleus. (b) The contents of the cisternae of the endoplasmic reticulum (ER) are also well decorated by the ferritin-labeled antibody, but the cell surface (CS) is not. Both ×120,000; bar = 80 nm. (From Nist *et al.*, 1975.)

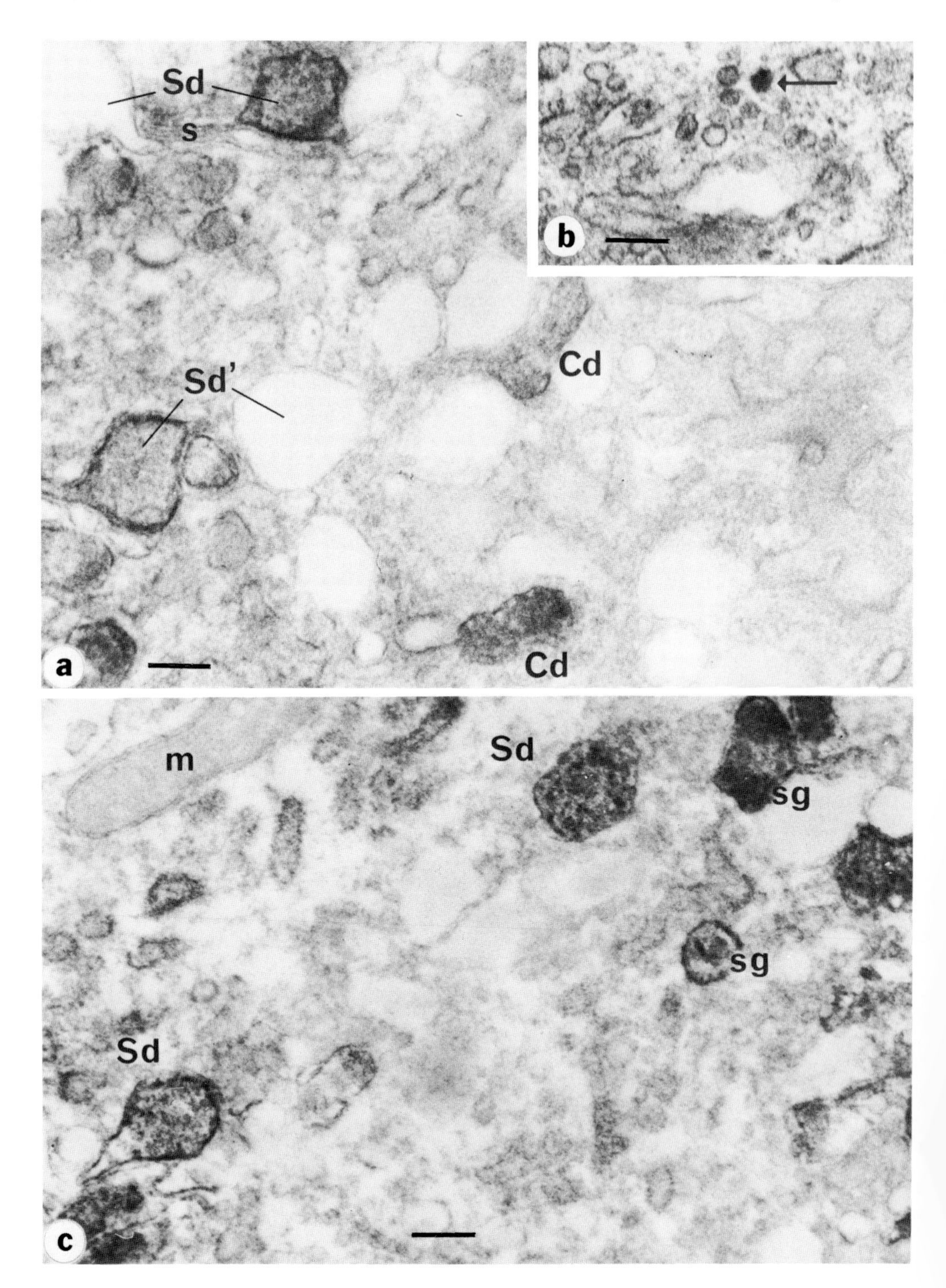
Sd
s
b
Sd'
Cd
a
Cd
m
Sd
sg
sg
Sd
c

convincing in the case of rat odontoblasts, for peroxidase-labeled anti-procollagen antibodies have been used to specifically label the aggregates in these cells (Karim *et al.*, 1979) (Fig. 6-13). It has been suggested that such aggregates could play the role of structural intermediates in fibril formation in the extracellular space (Bruns *et al.*, 1979), but the data available do not allow a definitive conclusion at this point. Considering the finding that collagen fibrils contain several collagenous components (Chapter 1), an interesting question is when the different components are brought together. Is it possible that multi-component building blocks of collagen fibrils are assembled in the Golgi complex and secretory granules, or are all assembly steps occurring after exocytosis (see Chapter 8 for further discussion)?

6. Proteolytic Processing of Procollagens

The propeptides of fibrillar procollagens must be removed before the molecules can form stable collagen fibrils with normal structure. Two classes of proteolytic enzymes are needed for this processing. One class of enzyme, the procollagen N-proteinase, is responsible for removal of the NH_2 propeptides. The second class of enzyme, the procollagen C-proteinase, removes the COOH propeptide from procollagen. It is conceivable that each class contains several enzymes that are specific for different genetic types of procollagens. Since cultured cells secrete procollagen into their culture medium, it is generally assumed that the proteinases cleave the propeptides extracellularly. However, whether this is actually the case *in vivo* and occurs in all tissues, has not been directly demonstrated.

Nonfibrillar collagens are usually not proteolytically processed like the fibrillar procollagens. However, all collagens are substrates for lysyl oxidase, which oxidizes susceptible lysyl side chains to lysyl aldehydes. Such aldehydes form covalent cross-links both between subunits within triple-helical molecules, as well as between molecules within polymeric structures (see Chapters 1 and 3).

6.1. Procollagen Proteinases

An enzyme that has a neutral pH optimum and removes the NH_2 propeptides from type I and II procollagen, but not type III procollagen, has been

Figure 6-12. Odontoblasts stained with peroxidase-labeled anti-procollagen type I antibodies. (a) Golgi complex showing spherical distension (Sd) and flattened portions of saccules (s). Some distensions are not stained; others are filled with reaction products. Two cylindrical distensions (Cd) are weakly stained. ×42,000; bar = 240 nm. (b) Here, a spherical distension is associated with transfer vesicles, one of which is strongly reactive (arrow). ×36,000; bar = 280 nm. (c) Another Golgi region showing two stained spherical distensions (Sd). The right part of the figure includes secretory granules (sg). Mitochondria (m) are unstained. ×36,000; bar = 280 nm. (From Karim *et al.*, 1979.)

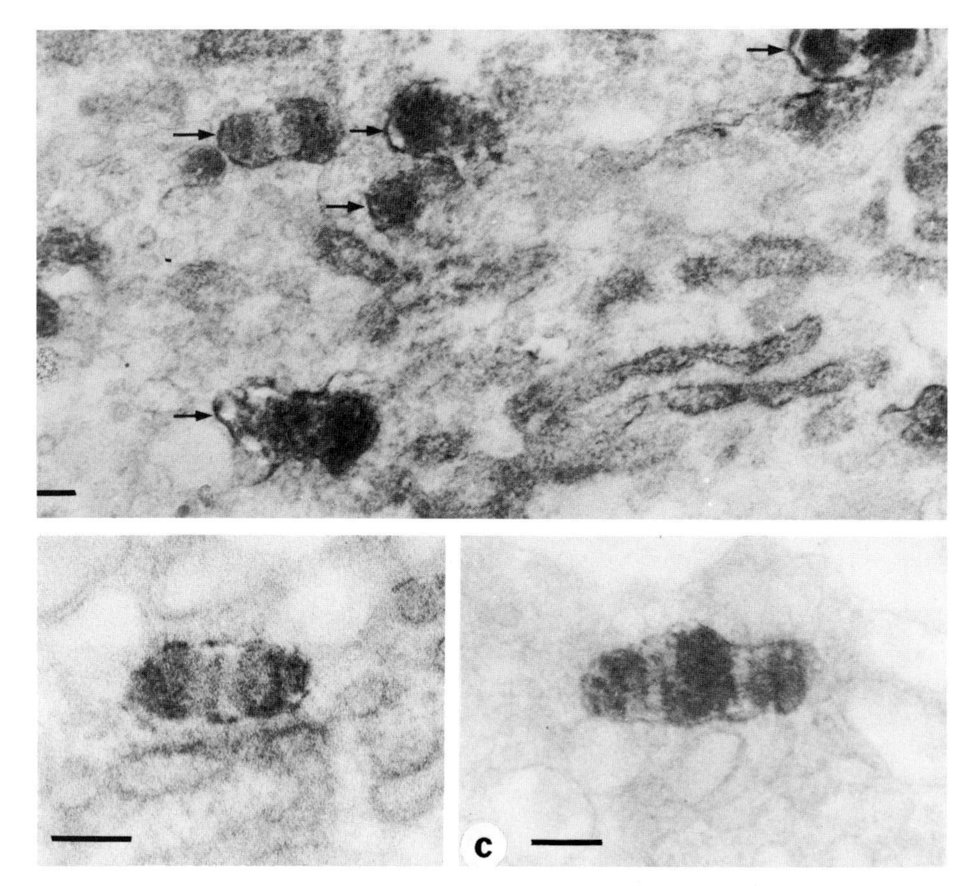

Figure 6-13. Apical cytoplasms of odontoblasts stained with peroxidase-labeled anti-procollagen type I antibodies. (a) In the upper right of this micrograph a typical secretory granule is cut longitudinally. Other granules are also reactive (arrows). Note the concentration of reaction product at the two extremities of the granules, giving them a bipolar appearance. (b) Higher magnification of such a granule. (c) This granule appears to be a double granule with a dilated dense area in the middle of the whole structure. Bars = 200 nm. (From Karim *et al.*, 1979.)

demonstrated in extracts from a variety of tissues (Kohn *et al.*, 1974; Tuderman *et al.*, 1977, 1978) and extensively purified (Hojima *et al.*, 1989). The molecular weight and subunit composition of this Ca^{2+}-dependent enzyme have been unclear because the enzyme is susceptible to degradation during purification (Tuderman *et al.*, 1978; Tanzawa *et al.*, 1985), but recently an apparently intact form of 500 kDa has been purified from embryonic chick tendons (Hojima *et al.*, 1989). The enzyme molecule contains four polypeptide subunits ranging from about 60 kDa to 160 kDa, and the catalytic activity is probably associated with the two largest subunits.

As discussed above, the type I/II procollagen proteinase requires the substrate to be in a native conformation. Synthetic peptides with sequences identical to the cleavage site in proα1(I) chains are not cleaved by the enzyme (Morikawa *et al.*, 1980). Mutations in humans that produce in-frame deletions of amino acid sequences in either proα1(I) or proα2(I) chains, make all three chains in the mutant procollagen molecules resistant to enzyme cleavage (Dombrowski *et al.*, 1989). That such mutations can affect the cleavage of the NH_2 propeptide even when they are located far downstream in the major collagen triple-helix, can best be understood by considering that the folding of the type I triple-helix occurs from the COOH end, and the effect of in-frame deletions anywhere in the triple-helix are therefore "transmitted" to the NH_2 propeptide region as well (Dombrowski *et al.*, 1989).

A different N-proteinase that cleaves type III procollagen has been purified from fibroblasts, aorta, and placenta. The molecular size and subunit composition of this enzyme are still unclear, since enzymes from different sources have been found to have different molecular weights (Nusgens *et al.*, 1980; Halila and Peltonen, 1984, 1986).

An enzyme that removes the COOH propeptide from type I procollagen has also been identified and purified from several tissue sources (Goldberg *et al.*, 1975; Duksin *et al.*, 1978; Leung *et al.*, 1979; Njieha *et al.*, 1982; Hojima *et al.*, 1985). It differs from the N-proteinase in that it cleaves type I, II, and III procollagens and shows the same activity with native as well as denatured substrates. Like the N-proteinase, it has a neutral pH optimum and requires Ca^{2+} for activity.

6.2. Defects in Collagen Biosynthesis

Dermatosparaxis, a genetic defect characterized by extreme fragility of skin, was first described in cattle (O'Hara *et al.*, 1970; Lenaers *et al.*, 1971) and later in sheep (Helle and Ness, 1972; Fjølstad and Helle, 1974). The disease is transmitted as an autosomal recessive defect. The striking clinical manifestations are cutaneous. Large wounds are generated by relatively minor traumas (Fig. 6-14). In cattle, death ultimately occurs due to wound infection and septicemia. In sheep, the animals are more seriously affected and lambs usually succumb within a few days following birth. Electron microscopy of tissues of affected animals shows strikingly abnormal collagen fibrils in skin (Fig. 6-15). The fibrils are multibranched, twisted ribbons, irregularly dispersed within an excessive amount of amorphous material (O'Hara *et al.*, 1970; Fjølstad and Helle, 1974). Other tissues that contain type I procollagen, such as tendon and bone, appear normal or only slightly affected. That tendon and skin are affected to a different extent has also been demonstrated by X-ray diffraction studies (Cassidy *et al.*, 1980). Analysis of the molecular packing in collagen fibrils by low-angle X-ray diffraction shows that in dermatosparactic tendons, most fibrils have a normal arrangement of collagen molecules, whereas in dermatosparactic skin, few, if any, normal fibrils are present.

A molecular explanation for the abnormal collagen fibrils in dermatosparactic skin is offered by the observation that skin extracts from calves with the genetic defect are deficient in collagen N-proteinase activity (Kohn *et*

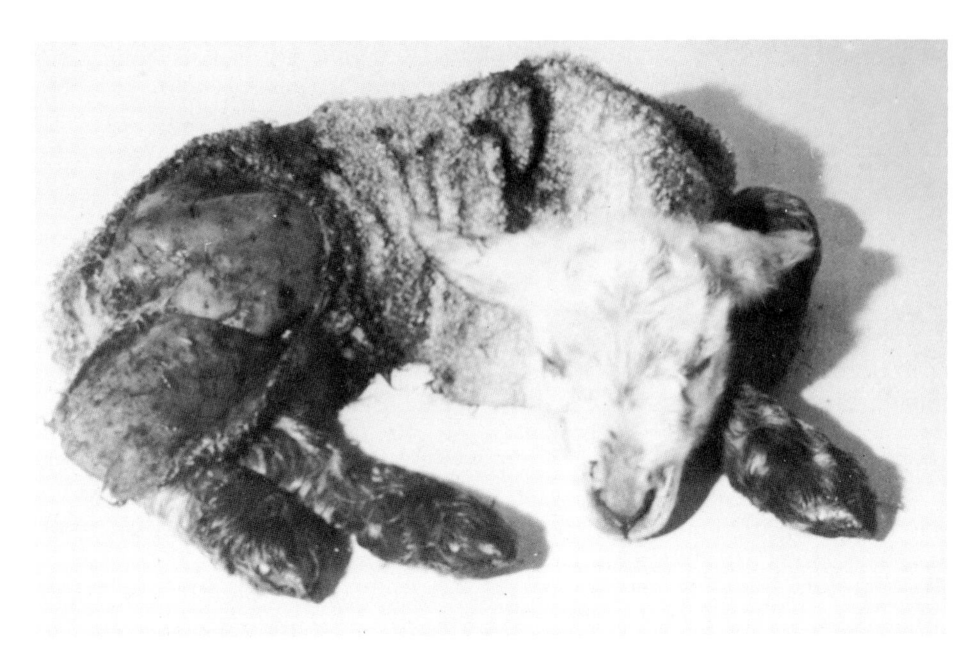

Figure 6-14. Dermatosparactic lamb 10 hr after birth showing lacerations of the skin. (From Fjølstad and Helle, 1974; courtesy of Dr. O. Helle.)

al., 1974). The lack of N-proteinase activity in the skin leads to accumulation of molecules that contain the NH_2 propeptides, pN-collagen. The abnormally twisted fibrils observed in dermatosparactic skin consist of pN-collagen molecules, as has been shown by direct immunostaining of the fibrils with ferritin-labeled antibodies directed against the NH_2 propeptide (Wick *et al.*, 1978) (Fig. 6-16). Apparently, the presence of the NH_2 propeptide prevents the normal packing of the collagen molecules into cylindrical fibrils.

The inability of pN-collagen to form cylindrical collagen fibrils has also been demonstrated *in vitro*. Human pN-collagen, generated by digesting type I procollagen with partially purified C-proteinase, forms thin sheets of variable widths, when incubated under conditions that favor collagen fibril formation (Hulmes *et al.*, 1989). With mixtures of collagen type I and pN-collagen, pleomorphic structures are formed; increasing amounts of pN-collagen cause fibril cross sections to become more distorted and more similar to the cross sections seen in dermatosparactic skin.

Electron microscopic examination of normal embryonic tissues with labeled antibodies against the NH_2 propeptides of type I procollagen shows preferential labeling along the surface of fibrils with small diameters (Fleischmajer *et al.*, 1983). It has therefore been suggested that pN-collagen molecules may limit the thickness of collagen fibrils even in normal tissues, and that the rate of proteolytic removal of the NH_2 propeptides may regulate collagen fibril growth (Chapman, 1989; Fleischmajer, *et al.*, 1983). The *in vitro* experiments with mixtures of collagen and pN-collagen described above tend to support this

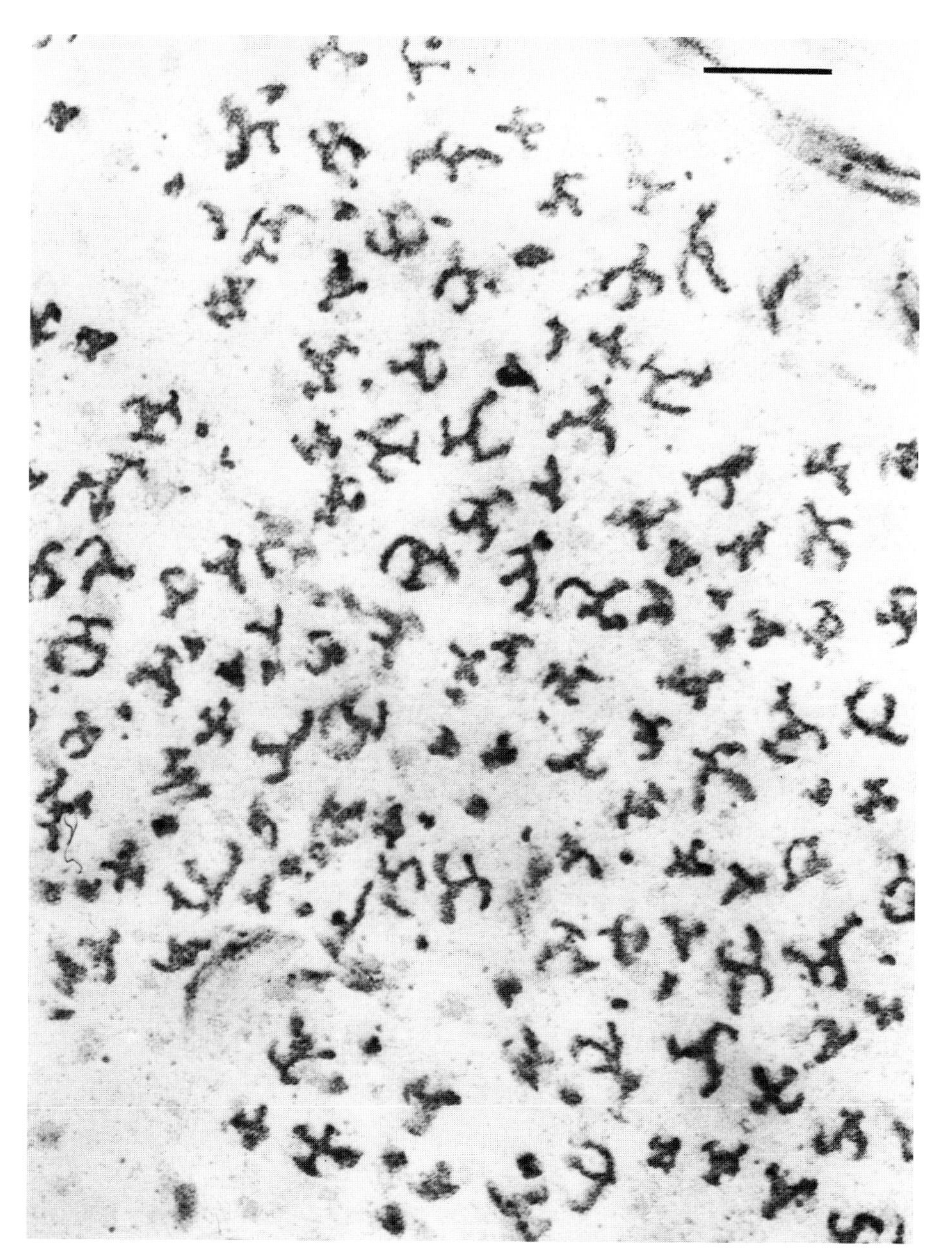

Figure 6-15. Cross section of collagen fibrils in the corium of dermatosparactic lamb showing the characteristic "hieroglyphic" appearance. ×90,000; bar = 200 μm. (From Fjølstad and Helle, 1974; courtesy of Dr. O. Helle.)

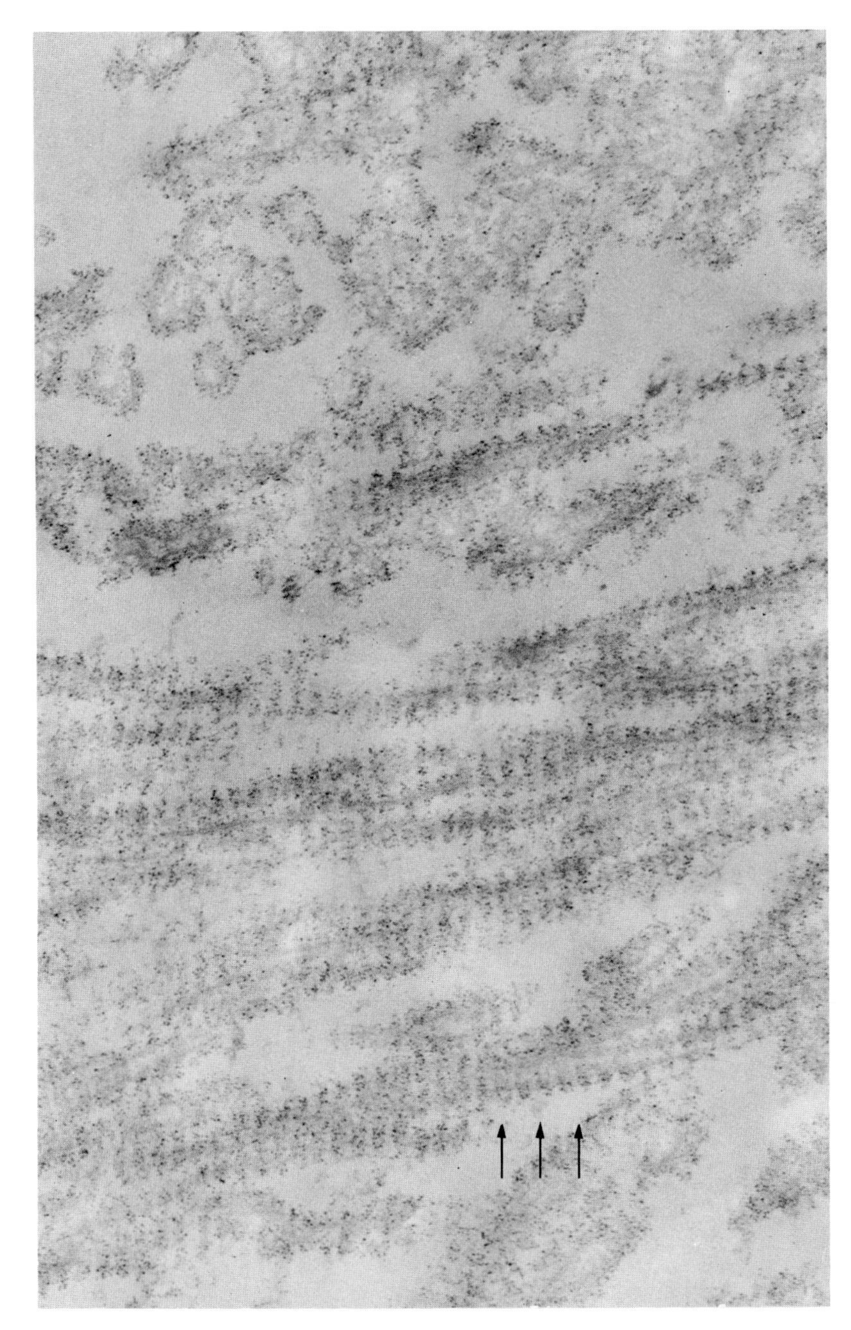

Figure 6-16. Dermatosparactic sheep skin with ferritin-labeled anti-pN-collagen type I. Note the localization of ferritin at regular intervals along the fibrils (arrows). ×60,000. (From Wick *et al.*, 1978.)

hypothesis in a general sense. However, collagen fibrils *in vivo* are generally cylindrical in shape, whereas dermatosparactic fibrils and polymers made with pN-collagen *in vitro* are thin banded sheets. Therefore, it is likely that additional factors are important for the precise control of fibril diameter and shape *in vivo* (Hulmes *et al.*, 1989) (see Chapters 1 and 8).

Several mutations in human procollagen genes that lead to abnormal biosynthesis of procollagen types I, II, and III have been identified (see Byers and Bonadio, 1989; Prockop *et al.*, 1990; Lee *et al.*, 1989; Vissing *et al.*, 1989; Tromp *et al.*, 1989; Superti-Furga *et al.*, 1989). The clinical consequences of the abnormalities are diseases belonging to osteogenesis imperfecta, Ehlers–Danlos syndrome, and various types of chondrodysplasias. The mutations include deletions, insertions, point mutations, and RNA splice mutations in the α1(I), α2(I), α1(II), and α1(III) collagen genes.

7. Concluding Remarks

The collagen gene superfamily comprises a large number of genes whose products constitute the major protein scaffold components of the extracellular matrix. Although we do not yet know the precise function of all those components, it is already evident that they play important roles in embryonic development and tissue morphogenesis.

During the last two decades a large amount of information has been accumulated about the posttranslational events involved in collagen biosynthesis. We have perhaps more detailed knowledge about the posttranslational modifications of collagen than of any other secretory protein, with the possible exception of proteoglycans (Chapter 5). The functional role of some of these modifications is known: 4-hydroxyproline allows the formation of stable triple-helices at body temperature, hydroxylysine participates in the formation of stable cross-links. The functional importance of other modifications is still obscure; the significance of *O*-glycosylation of hydroxylysine residues within collagen domains and *N*-glycosylation of asparagine residues is not understood.

Our understanding of the molecular basis for the modulation of collagen types during development and morphogenesis and the regulation of collagen production during connective tissue repair is rapidly becoming more sophisticated. As in many other areas of biochemistry and cell biology, the use of recombinant DNA approaches has contributed enormously to that increased sophistication. It is interesting to note that some of the most recently described collagens (types XII and XIII) were first discovered by molecular cloning of DNA (Gordon *et al.*, 1989; Pihlajaniemi *et al.*, 1987a). The continued use of such techniques to study collagens coupled with modern methods of protein chemistry and cell and developmental biology should rapidly lead to a better understanding of tissue and organ assembly principles that extend beyond the realm of collagen cell and molecular biology. For the real understanding of how tissues, organs, and organisms are assembled, maintained, and repaired, detailed insights into the nature of matrix scaffolds formed by collagenous proteins will undoubtedly be essential.

ACKNOWLEDGMENTS. Original contributions from this laboratory were supported in part by research grants AR36819, AR36820, and HL33014 from the National Institutes of Health. Expert secretarial assistance was provided by Ms. M. Rodin.

References

Barker, D. F., Hostikka, S. L., Zhou, J., Chow, L. T., Oliphant, A. R., Gerken, S. C., Gregory, M. C., Skolnick, M. H., Atkin, C. L., and Tryggvason, K., 1990, Identification of mutations in the COL4A5 collagen gene in Alport syndrome, *Science* **248**:1224–1227.

Bassuk, J. A., and Berg, R. A., 1989, Protein disulphide isomerase, a multifunctional endoplasmic reticulum protein, *Matrix* **9**:244–258.

Bateman, J. F., Lamande, S. R., Dahl, H.-H. M., Chan, D., and Mascara, T., 1989, A frameshift mutation results in a truncated nonfunctional carboxyl-terminal proα1(I) propeptide of type I collagen in osteogenesis imperfecta, *J. Biol. Chem.* **264**:10960–10964.

Becker, V., Timpl, R., Helle, O., and Prockop, D. J., 1976, NH_2-terminal extensions on skin collagen from sheep with a genetic defect in conversion of procollagen into collagen *Biochemistry* **15**:2853–2862.

Bennet, V. D., and Adams, S. L., 1990, Identification of a cartilage-specific promoter within intron 2 of the chick α2(I) collagen gene, *J. Biol. Chem.* **265**:2223–2230.

Berg, R. A., and Prockop, D. J., 1973, Affinity column purification of protocollagen proline hydroxylase from chick embryos and further characterization of the enzyme, *J. Biol. Chem.* **248**:1175–1182.

Berg, R. A., Kao, W. Y., and Kedersha, L. L., 1980, The assembly of tetrameric prolyl hydroxylase in tendon fibroblasts from newly synthesized α-subunits and from preformed cross-reacting protein, *Biochem. J.* **189**:491–499.

Bernard, M., Yoshioka, H., Rodriguez, E., van der Rest, M., Kimura, T., Ninomiya, Y., and Olsen, B. R., 1988, Cloning and sequencing of proα1(XI) collagen cDNA demonstrates that type XI belongs to the fibrillar class of collagens and reveals that the expression of the gene is not restricted to cartilaginous tissue, *J. Biol. Chem.* **263**:17159–17166.

Boado, R. J., Campbell, D. A., and Chopra, I. J., 1988, Nucleotide sequence of rat liver iodothyronine 5′-monodeiodinase (5′MD): Its identity with the protein disulfide isomerase, *Biochem. Biophys. Res. Commun.* **155**:1287–1304.

Boedtker, H., Finer, M., and Aho, S., 1985, The structure of the chicken α2 collagen gene, *Ann. N.Y. Acad. Sci.* **470**:85–116.

Bonaldo, P., Russo, V., Bucciotti, F., Doliana, R., and Colombatti, A., 1990, Structural and functional features of the α3 chain indicate a bridging role for chicken collagen VI in connective tissues, *Biochemistry* **29**:1245–1254.

Bornstein, P., and McKay, J., 1988, The first intron of the α1(I) collagen gene contains several transcriptional regulatory elements, *J. Biol. Chem.* **263**:1603–1606.

Bornstein, P., and Sage, H., 1980, Structurally distinct collagen types, *Annu. Rev. Biochem.* **49**:957–1003.

Bornstein, P., and Sage, H., 1989, Regulation of collagen gene expression, *Prog. Nucleic Acid Res. Mol. Biol.* **37**:67–106.

Bornstein, P., McKay, J., Morishima, J. K., Devarayalu, S., and Gelinas, R. E., 1987, Regulatory elements of the first intron contribute to transcriptional control of the human α1(I) collagen gene, *Proc. Natl. Acad. Sci. USA* **84**:8869–8873.

Bornstein, P., McKay, J., Liska, D. J., Apone, S., and Devarayalu, S., 1988, Interactions between the promoter and first intron are involved in transcriptional control of α1(I) collagen gene expression, *Mol. Cell. Biol.* **8**:4851–4857.

Breindl, M., Harbers, K., and Jaenisch, R., 1984, Retrovirus-induced lethal mutation in collagen I gene of mice is associated with an altered chromatin structure, *Cell* **38**:9–16.

Bruns, R. R., Hulmes, D. J. S., Therrien, S. F., and Gross, J., 1979, Procollagen segment-long-spacing crystallites: Their role in collagen fibrillogenesis, *Proc. Natl. Acad. Sci. USA* **76**:313–317.

Burbelo, P., Martin, G., and Yamada, Y., 1988, α1(IV) and α2(IV) collagen genes regulated by a bidirectional promoter and a shared enhancer, *Proc. Natl. Acad. Sci. USA* **85**:9674–9682.

Burbelo, P., Killen, P. D., Ebihara, I., Sakurai, Y., and Yamada, Y., 1989, Structure and expression of collagen IV genes, in: *Collagen—Molecular Biology*, Vol. IV (B. R. Olsen and M. E. Nimni, eds.), pp. 51–64, CRC Press, Boca Raton, Fla.

Butkowski, R. J., Langeveld, J. P. M., Wieslander, J., Hamilton, J., and Hudson, B. G., 1987, Localization of the Goodpasture epitope to a novel chain of basement membrane collagen, *J. Biol. Chem.* **262**:7874–7877.

Byers, P. H., and Bonadio, J. F., 1989, The nature, characterization and phenotypic effects of mutations that affect collagen structure and processing, in: *Collagen—Molecular Biology*, Vol. IV (B. R. Olsen and M. E. Nimni, eds.), pp. 125–139, CRC Press, Boca Raton, Fla.

Byers, P. H., Click, E. M., Harper, E., and Bornstein, P., 1975, Interchain disulfide bonds in procollagen are located in a large non-triple-helical COOH-terminal domain, *Proc. Natl. Acad. Sci. USA* **72**:3009–3013.

Cassidy, K., Eikenberry, E. F., Olsen, B. R., and Brodsky, B., 1980, X-ray diffraction investigations of collagen fibril structure in dermatosparactic lamb tissues, *Lab. Invest.* **43**:542–546.

Chapman, J. A., 1989, The regulation of size and form in assembly of collagen fibrils *in vivo*, *Biopolymers* **28**:1367–1382.

Cheng, S.-Y., Gong, Q.-H., Parkinson, C., Robinson, E. A., Appella, E., Merlino, G. T., and Pastan, I., 1987, The nucleotide sequence of human cellular thyroid hormone binding protein present in endoplasmic reticulum, *J. Biol. Chem.* **257**:11221–11227.

Chen-Kiang, S., Cardinale, G. J., and Udenfriend, S., 1977, Homology between a prolyl hydroxylase subunit and a tissue protein that crossreacts immunologically with the enzyme, *Proc. Natl. Acad. Sci. USA* **74**:4420–4424.

Chu, M.-L., Pan, T.-C., Conway, D., Saitta, B., Stokes, D., Kuo, H.-J., Glanville, R. W., Timpl, R., Mann, K., and Deutzmann, R., 1990, The structure of type VI collagen, *Ann. N.Y. Acad. Sci.* **580**:55–63.

Clark, C. C., 1979, The distribution and initial characterization of oligosaccharide units on the COOH-terminal propeptide extensions of the pro-α1 and pro-α2 chains of type I procollagen, *J. Biol. Chem.* **254**:10798–10802.

Deak, S. B. M., Pope, F. M., and Prockop, D. J., 1983, The molecular defect in a nonlethal variant of osteogenesis imperfecta, *J. Biol. Chem.* **258**:15192–15197.

Diegelmann, R. F., Bernstein, L., and Peterkofsky, B., 1973, Cell-free synthesis on membrane-bound polysomes of chick embryo connective tissue and the localization of prolyl hydroxylase on the polysome–membrane complex, *J. Biol. Chem.* **248**:6514–6521.

Dölz, E., and Engel, J., 1990, Nucleation, propagation, and direction of triple-helix formation in collagens, I, III, and IV and in gelatin as monitored by electron microscopy, *Ann. N.Y. Acad. Sci.* **580**:421–424.

Dombrowski, K. E., Vogel, B. E., and Prockop, D. J., 1989, Mutations that alter the primary structure of type I procollagen have long-range effects on its cleavage by procollagen N-proteinase, *Biochemistry* **28**:7107–7112.

Duksin, D., and Bornstein, P., 1977, Impaired conversion of procollagen to collagen by fibroblasts and bone treated with tunicamycin, an inhibitor of protein glycosylation, *J. Biol. Chem.* **252**:955–962.

Duksin, D., Davidson, M. M., and Bornstein, P., 1978, The role of glycosylation in the enzymatic conversion of procollagen to collagen: Studies using tunicamycin and concanavalin A, *Arch. Biochem. Biophys.* **185**:326–332.

Edman, J. C., Ellis, L., Blacher, R. W., Roth, R. A., and Rutter, W. J., 1985, Sequence of protein disulphide isomerase and implications of its relationship to thioredoxin, *Nature* **317**:267–270.

Fessler, L. I., and Fessler, J. H., 1979, Characterization of type III procollagen from chick embryo blood vessels, *J. Biol. Chem.* **254:**233–239.

Fessler, L. I., Morris, N. P., and Fessler, J. H., 1975, Procollagen: Biological scission of amino and carboxyl extension peptides, *Proc. Natl. Acad. Sci. USA* **72:**4905–4909.

Fjølstad, M., and Helle, O., 1974, A hereditary dysplasia of collagen tissues in sheep, *J. Pathol.* **112:**183–188.

Fleischmajer, R., Olsen, B. R., Timpl, R., Perlish, J. S., and Lovelace, O., 1983, Collagen fibril formation during embryogenesis, *Proc. Natl. Acad. Sci. USA* **80:**3354–3358.

Fleischmajer, R., Kühn, K., and Olsen, B. R. (eds), 1990, *Structure, Molecular Biology, and Pathology of Collagen, Ann. N.Y. Acad. Sci.* **580.**

Geetha-Habid, M., Noiva, R., Kaplan, H. A., and Lennarz, W. J., 1988, Glycosylation site binding protein, A component of oligosaccharyl transferase, is highly similar to three other 57 kd luminal proteins of the ER, *Cell* **54:**1053–1060.

Goldberg, B., Taubman, M. B., and Radin, B., 1975, Procollagen peptidase: Its mode of action on the native substrate, *Cell* **4:**45–50.

Gong, Q.-H., Fukuda, T., Parkinson, C., and Cheng, S.-Y., 1988, Nucleotide sequence of a full-length cDNA clone encoding a mouse cellular thyroid hormone binding protein (p55) that is homologous to protein disulphide isomerase and the β-subunit of prolyl 4-hydroxylase, *Nucleic Acids Res.* **16:**1203.

Gordon, M. K., Gerecke, D. R., Dublet, B., van der Rest, M., and Olsen, B. R., 1989, Type XII collagen—A large multidomain molecule with partial homology to type IX collagen, *J. Biol. Chem.* **264:**19772–19778.

Halila, R., and Peltonen, L., 1984, Neutral protease cleaving the N-terminal propeptide of type III procollagen: Partial purification and characterization of the enzyme from smooth muscle cells of bovine aorta, *Biochemistry* **23:**1254–1256.

Halila, R., and Peltonen, L., 1986, Purification of human procollagen type III N-proteinase from placenta and preparation of antiserum, *Biochem. J.* **239:**47–52.

Harwood, R., Grant, M. E., and Jackson, D. S., 1974, Collagen biosynthesis: Characterization of subcellular fractions from embryonic chick fibroblasts and the intracellular localization of protocollagen prolyl and protocollagen lysyl hydroxylases, *Biochem. J.* **176:**283–294.

Harwood, R., Grant, M. E., and Jackson, D. S., 1976, The influence of α,α′-bipyridyl, colchicine and antimycin A on the secretory process in embryonic chick tendon and cartilage cells, *Biochem. J.* **176:**283–294.

Hatamochi, A., Golumbeck, P. T., van Schaftinger, E., and de Crombrugghe, B., 1988, A CCAAT DNA binding factor consisting of two different components that are both required for DNA binding, *J. Biol. Chem.* **263:**5940–5947.

Hayashi, M., Ninomiya, Y., Parsons, T., Hayashi, K., Olsen, B. R., and Trelstad, R. L., 1986, Differential localization of mRNAs of collagen types I and II in chick fibroblasts, chondrocytes, and corneal cells by *in situ* hybridization using cDNA probes, *J. Cell Biol.* **102:**2302–2309.

Helaakoski, T., Vuori, K., Myllylä, R., and Kivirikko, K. I., 1989, Molecular cloning of the α-subunit of human prolyl 4-hydroxylase: The complete cDNA-derived amino acid sequence and evidence for alternative splicing of RNA transcripts, *Proc. Natl. Acad. Sci. USA* **86:**4392–4396.

Helle, O., and Ness, N. N., 1972, A hereditary skin defect in sheep, *Acta Vet. Scand.* **13:**443–445.

Hoffmann, H.-P., Olsen, B. R., Chen, H.-T., and Prockop, D. J., 1976, Segment-long-spacing aggregates and isolation of COOH-terminal peptides from type I procollagen, *Proc. Natl. Acad. Sci. USA* **73:**4304–4308.

Hojima, Y., van der Rest, M., and Prockop, D. J., 1985, Type I procollagen carboxyl-terminal proteinase from chick embryo tendons, *J. Biol. Chem.* **260:**15996–16003.

Hojima, Y., McKenzie, J. A., van der Rest, M., and Prockop, D. J., 1989, Type I procollagen N-proteinase from chick embryo tendons. Purification of a new 500-kDa form of the enzyme and identification of the catalytically active polypeptides *J. Biol. Chem.* **264:**11336–11345.

Hörlein, D., Fietzek, P. P., Wachter, E., Lapière, C. M., and Kühn, K., 1979, Amino acid sequence of the aminoterminal segment of dermatosparactic calf skin procollagen type I, *Eur. J. Biochem.* **99:**31–38.

Horton, W., Miyashita, T., Kohno, K., Hassell, J. R., and Yamada, Y., 1987, Identification of a phenotype-specific enhancer in the first intron of the rat collagen II gene, *Proc. Natl. Acad. Sci. USA* **84:**8864–8868.

Hostikka, S. L., Eddy, R. L., Byers, M. G., Höyhtyä, M., Schows, T. B., and Tryggvason, K., 1990, Identification of a distinct type IV collagen α chain with restricted kidney distribution and assignment of its gene to the locus of X chromosome-linked Alport syndrome, *Proc. Natl. Acad. Sci. USA* **87:**1606–1610.

Hously, T. J., Rowland, F. N., Ledger, P. W., Kaplan, J., and Tanzer, M. L., 1980, Effects of tunicamycin on the biosynthesis of procollagen by human fibroblasts, *J. Biol. Chem.* **255:**121–128.

Hulmes, D. J. S., Kadler, K. E., Mould, A. P., Hojima, Y., Holmes, D. F., Cummings, C., Chapman, J. A., and Prockop, D. J., 1989, Pleomorphism in type I collagen fibrils produced by persistence of the procollagen N-propeptide, *J. Mol. Biol.* **210:**337–345.

Iyama, K., Ninomiya, Y., Olsen, B. R., Linsenmayer, T. F., Trelstad, R. L., and Hayashi, M., 1991, Spatio-temporal pattern of type X collagen gene expression and collagen deposition in embryonic chick vertebrae undergoing endochondral ossification, *Anat. Rec.* **229:** 462–472.

Jaenisch, R., Harbers, K., Schnieke, A., Lohler, J., and Chumakov, I., 1983, Germline integration of Moloney murine leukemia virus at the Mov 13 locus leads to recessive lethal mutations and early embryonic death, *Cell* **32:**209–216.

Kao, W. W.-Y., Berg, R. A., and Prockop, D. J., 1977, Kinetics for the secretion of procollagen by freshly isolated tendon cells, *J. Biol. Chem.* **252:**8391–8397.

Kao, W. W.-Y., Prockop, D. J., and Berg, R. A., 1979, Kinetics for the secretion of non-helical procollagen by freshly isolated tendon cells, *J. Biol. Chem.* **254:**2234–2243.

Kao, W. W.-Y., Nakazawa, M., Aida, T., Everson, W. V., Kao, C. W.-C., Seyer, J., and Hughes, S. H., 1988, Isolation of cDNA clones and genomic DNA clones of β-subunit of chicken prolyl 4-hydroxylase, *Connect. Tissue Res.* **18:**157–174.

Kaplan, H. A., Naider, F., and Lennarz, W. J., 1988, Partial characterization and purification of the glycosylation site recognition component of oligosaccharyl-transferase, *J. Biol. Chem.* **263:**7814–7820.

Karim, A., Cournil, I., and Leblond, C. P., 1979, Immunochemical localization of procollagens, *J. Histochem. Cytochem.* **27:**1070.

Kaytes, P., Wood, L., Theriault, N., Kurkinen, M., and Vogeli, G., 1988, Head-to-head arrangement of murine type IV collagen genes, *J. Biol. Chem.* **263:**19274–19277.

Khillan, J. S., Schmidt, A., Overbeek, P. A., de Crombrugghe, B., and Westphal, H., 1986, Developmental and tissue-specific expression directed by the α2 type I collagen promoter in transgenic mice, *Proc. Natl. Acad. Sci. USA* **83:**725–729.

Kimura, T., Cheah, K. S. E., Chan, S. D. H., Lui, V. C. H., Mattei, M.-G., van der Rest, M., Ono, K., Solomon, E., Ninomiya, Y., and Olsen, B. R., 1989a, The human α2(XI) collagen (COL11A2) chain, *J. Biol. Chem.* **264:**13910–13916.

Kimura, T., Mattei, M.-G., Stevens, J. W., Goldring, M. B., Ninomiya, Y., and Olsen, B. R., 1989b, Molecular cloning of rat and human type IX collagen cDNA and localization of the α1(IX) gene on the human chromosome 6, *Eur. J. Biochem.* **179:**71–78.

Kivirikko, K. I., and Myllylä, R., 1979, Collagen glycosyltransferases, *Int. Rev. Connect. Tissue Res.* **8:**23–72.

Kivirikko, K. I., and Myllylä, R., 1984, Biosynthesis of collagens, in: *Extracellular Matrix Biochemistry,* (K. A. Piez and A. H. Reddi, eds.), pp. 83–118, Elsevier, New York.

Kivirikko, K. I., and Myllylä, R., 1985, Post-translational processing of procollagens, *Ann. N.Y. Acad. Sci.* **460:**187–201.

Kivirikko, K. I., Myllylä, R., and Pihlajaniemi, T., 1989, Protein hydroxylation: Prolyl 4-hydroxylase, an enzyme with four cosubstrates and a multifunctional subunit, *FASEB J.* **3:**1609–1617.

Kivirikko, K. I., Helaakoski, T., Tasanen, K., Vuori, K., Myllylä, R., Parkkonen, T., and Pihlajaniemi, T., 1990, Molecular biology of prolyl 4-hydroxylase, *Ann. N.Y. Acad. Sci.* **580:**132–142.

Kohn, L. D., Isersky, C., Zupnik, J., Lenaers, A., Lee, G., and Lapiere, C. M., 1974, Calf tendon procollagen peptidase: Its purification and endopeptidase mode of action, *Proc. Natl. Acad. Sci. USA* **71:**40–44.

Koivu, J., Myllylä, R., Helaakoski, T., Pihlajaniemi, T., Tasanen, K., and Kivirikko, K. I., 1987, A single polypeptide acts both as the β subunit of prolyl 4-hydroxylase and as a protein disulphide-isomerase, *J. Biol. Chem.* **262**:6447–6449.

Kratochwil, K., von der Mark, K., Kollar, E. J., Jaenisch, R., Mooslehner, K., Schwarz, M., Haase, K., Gmachl, I., and Harbers, K., 1989, Retrovirus-induced insertional mutation of Mov 13 mice affects collagen I expression in a tissue-specific manner, *Cell* **57**:807–816.

Lee, B., Vissing, H., Ramirez, F., Rogers, D., and Rimoin, D., 1989, Identification of the molecular defect in a family with spondyloepiphyseal dysplasia, *Science* **244**:978–980.

Lenaers, A., Ansay, M., Nusgens, B. V., and Lapiere, C. M., 1971, Collagen made of extended α-chains, procollagen, in genetically-defective dermatosparaxic calves, *Eur. J. Biochem.* **23**:533–543.

Leung, M. K., Fessler, L. I., Greenberg, D. B., and Fessler, J. H., 1979, Separate amino and carboxyl procollagen peptidases in chick embryo tendon, *J. Biol. Chem.* **254**:224–232.

LuValle, P., Ninomiya, Y., Rosenblum, N. D., and Olsen, B. R., 1988, The type X collagen gene: Intron sequences split the 5′ untranslated region and separate the coding regions for the non-collagenous amino-terminal and triple-helical domains, *J. Biol. Chem.* **263**: 18378–18385.

LuValle, P., Hayashi, M., and Olsen, B. R., 1989, Transcriptional regulation of type X collagen during chondrocyte maturation, *Dev. Biol.* **133**:613–616.

Maity, S. N., Golumbeck, P. T., Karsenty, G., and de Crombrugghe, B., 1988, Selective activation of transcription by a novel CCAAT binding factor, *Science*, **241**:582–585.

Mayne, R., and Burgeson, R. E. (eds.), 1987, *Structure and Function of Collagen Types*, Academic Press, New York.

Miller, E. J., 1985, The structure of fibril-forming collagens, *Ann. N.Y. Acad. Sci.* **460**:1–13.

Morikawa, T., Tuderman, L., and Prockop, D. J., 1980, Inhibitors of procollagen N-protease. Synthetic peptides with sequences similar to the cleavage site in the proα1(I) chain, *Biochemistry* **19**:2646–2650.

Munro, S., and Pelham, H. R. B., 1987, A C-terminal signal prevents secretion of luminal ER proteins, *Cell* **48**:899–907.

Muragaki, Y., Jacenko, O., Apte, S., Mattei, M.-G., Ninomiya, Y., and Olsen, B. R., 1991, The α2(VIII) collagen gene—a novel member of the short-chain collagen family located on the human chromosome 1, *J. Biol. Chem.* **266**: 7721–7727.

Nicholls, A. C., Pope, F. M., and Schloon, H., 1979, Biochemical heterogeneity of osteogenesis imperfecta: New variant, *Lancet*,**1**:1193.

Nicholls, A. C., Osse, G., Schloon, H. G., Lenard, H. G., Deak, S., Myers, J. C., Prockop, D. J., Weigel, W. R. F., Fryer, P., and Pope, F. M., 1984, The clinical features of homozygous α2(I) deficient osteogenesis imperfecta, *J. Med. Genet.* **21**:257–262.

Ninomiya, Y., Gordon, M., van der Rest, M., Schmid, T., Linsenmayer, T., and Olsen, B. R., 1986, The developmentally regulated type X collagen gene contains a long open reading frame without introns, *J. Biol. Chem.* **261**:5041–5050.

Ninomiya, Y., Castagnola, P., Gerecke, D., Gordon, M., Jacenko, O., LuValle, P., McCarthy, M., Muragaki, Y., Nishimura, I., Oh, S., Rosenblum, N., Sato, N., Sugrue, S., Taylor, R., Vasios, G., Yamaguchi, N., and Olsen, B. R., 1990, The molecular biology of collagens with short triple-helical domains, in: *Collagen Genes: Extracellular Matrix Genes* (L. J. Sandell and C. D. Boyd, eds.), pp. 79–114, Academic Press, New York.

Nishimura, I., Muragaki, Y., and Olsen, B. R., 1989, Tissue-specific forms of type IX collagen-proteoglycan arise from the use of two widely separated promoters, *J. Biol. Chem.* **264**:20033–20041.

Nist, C., von der Mark, K., Hay, E. D., Olsen, B. R., Bornstein, P., Ross, R., and Dehm, P., 1975, Location of procollagen in chick corneal and tendon fibroblasts with ferritin-conjugated antibodies, *J. Cell Biol.* **65**:75–87.

Njieha, F. K., Morikawa, T., Tuderman, L., and Prockop, D. J., 1982, Partial purification of a procollagen C-proteinase. Inhibition by synthetic peptides and sequential cleavage of type I procollagen, *Biochemistry* **21**:757–764.

Nusgens, B., Goebels, Y., Shinkai, H., and Lapiere, C. M., 1980, Procollagen type III N-terminal endopeptidase in fibroblast culture, *Biochem. J.* **191**:699–706.

O'Hara, P. J., Read, W. K., Romane, W. M., and Bridges, C. H., 1970, A collagenous tissue dysplasia of calves, *Lab. Invest.* **23**:307–314.

Oikarinen, J., Hatamochi, A., and de Crombrugghe, B., 1987, Separate binding sites for nuclear factor 1 and a CCAAT DNA binding factor in the mouse α2(I) collagen promoter, *J. Biol. Chem.* **262**:11064–11070.

Olsen, B. R., and Prockop, D. J., 1974, Ferritin-conjugated antibodies used for labeling of organelles involved in the cellular synthesis and transport of procollagen, *Proc. Natl. Acad. Sci. USA* **71**:2033–2037.

Olsen, B. R., Berg, R. A., Kishida, Y., Prockop, D. J., 1975, Further characterization of embryonic tendon fibroblasts and the use of immuno-ferritin techniques to study collagen biosynthesis, *J. Cell Biol.* **64**:340–355.

Olsen, B. R., Hoffman, H.-P., and Prockop, D. J., 1976, Interchain disulfide bonds at the COOH-terminal end of procollagen synthesized by matrix-free cells from chick embryonic tendon and cartilage, *Arch. Biochem. Biophys.* **175**:341–350.

Olsen, B. R., Guzman, N. A., Engel, J., Condit, C., and Aase, S., 1977, Purification and characterization of a peptide from the carboxy-terminal region of chick tendon procollagen type I, *Biochemistry* **16**:3030–3036.

Olsen, B. R., Gerecke, D., Gordon, M., Green, G., Kimura, T., Konomi, H., Muragaki, Y., Ninomiya, Y., Nishimura, I., and Sugrue, S., 1989, A new dimension in extracellular matrix, in: *Collagen: Biochemistry, Biotechnology, and Molecular Biology,* Vol. IV, (B. R. Olsen, and M. E. Nimni, eds.), pp. 2–19, CRC Press, Boca Raton, Florida.

Parkkonen, R., Kivirikko, K. I., and Pihlajaniemi, T., 1988, Molecular cloning of a multifunctional chicken protein acting as the prolyl 4-hydroxylase β-subunit, protein disulphide isomerase and a cellular thyroid binding protein. Comparison of cDNA-deduced amino acid sequences with those in other species, *Biochem. J.* **256**:1005–1011.

Peterkofsky, B., and Assad, R., 1976, Submicrosomal localization of prolyl hydroxylase from chick embryo limb bone, *J. Biol. Chem.* **251**:4771–4777.

Pihlajaniemi, T., Dickson, L. A., Pope, F. M., Korhonen, V. R., Nicholls, A., Prockop, D. J., and Myers, J. C., 1984, Osteogenesis imperfecta: Cloning of a proα2(I) collagen gene with a frame-shift mutation, *J. Biol. Chem.* **259**:12941–12944.

Pihlajaniemi, T., Myllylä, R., Seyer, J., Kurkinen, M., and Prockop, D. J., 1987a, Partial characterization of a low molecular weight human collagen that undergoes alternative splicing, *Proc. Natl. Acad. Sci. USA* **84**:940–944.

Pihlajaniemi, T., Helaakoshi, T., Tasanen, K., Myllylä, R., Huhtala, M.-L., Koivu, J., and Kivirikko, K. I., 1987b, Molecular cloning of the β subunit of human prolyl 4-hydroxylase. This subunit and protein disulphide isomerase are products of the same gene, *EMBO J.* **6**:643–649.

Pihlajaniemi, T., Tamminen, M., Sandberg, M., Hirvonen, H., and Vuorio, E., 1990, The α1 chain of type XIII collagen: Polypeptide structure, alternative splicing, and tissue distribution, *Ann. N.Y. Acad. Sci.* **580**:440–443.

Poole, A. R., Pidoux, I., Reiner, A., Choi, H., and Rosenberg, L. C., 1984, Association of an extracellular protein (chondrocalcin) with the calcification of cartilage in endochondral bone formation, *J. Cell Biol.* **98**:54–65.

Pöschl, E., Pollner, R., and Kuhn, K., 1988, The genes for the α1(IV) and α2(IV) chains of human basement membrane collagen type IV are arranged head-to-head and separated by a bi-directional promoter of unique structure, *EMBO J.* **7**:2687–2695.

Prockop, D. J., and Guzman, N. A., 1977, Collagen diseases and the biosynthesis of collagen, *Hosp. Pract.* **December**:61–68.

Prockop, D. J., Kivirikko, K. I., Tuderman, L., and Guzman, N. A., 1979, The biosynthesis of collagen and its disorders, *N. Engl. J. Med.* **301**(July 5 and July 12):13–24, 77–85.

Prockop, D. J., Olsen, A., Kontusari, S., Hyland, J., Ala-Kokko, L., Vasan, N. S., Barton, E., Buck, D., Harrison, K., and Brent, R. L., 1990, Mutations in human procollagen genes. Consequences of the mutations in man and in transgenic mice, *Ann. N.Y. Acad. Sci.* **580**:330–339.

Quinn, R. A., and Krane, S. M., 1976, Abnormal properties of collagen lysyl hydroxylase from skin fibroblasts of siblings with hydroxylysine-deficient collagen, *J. Clin. Invest.* **57**:83–93.

Ramirez, F., 1989, Organization and evolution of the fibrillar collagen genes, in: *Collagen—Molecular Biology*, Vol. IV (B. R. Olsen and M. E. Nimni, eds.), pp. 21–30, CRC Press, Boca Raton, Fla.

Ramirez, F., and DiLiberto, M., 1990, Complex and diversified regulatory programs control the expression of vertebrate collagen genes, *FASEB J.* **4:**1616–1623.

Ramirez, F., Bernard, M., Chu, M.-L., Dickson, L., Sangiorgi, F., Weil, D., de Wet, W., Junien, C., and Sobel, M., 1985, Isolation and characterization of the human fibrillar collagen genes, *Ann. N.Y. Acad. Sci.* **460:**117–129.

Revel, J.-P., and Hay, E. D., 1963, An autoradiographic and electron microscopic study of collagen synthesis in differentiating cartilage, *Z. Zellforsch. Mikrosk. Anat.* **61:**110–114.

Risteli, J., Tryggvason, K., and Kivirikko, K. I., 1977, Prolyl 3-hydroxylase: Partial characterization of the enzyme from rat kidney cortex, *Eur. J. Biochem.* **73:**485–492.

Rosenbloom, H., Harsch, M., and Jimenez, S. A., 1973, Hydroxyproline content determines the denaturation temperature of chick tendon procollagen, *Arch. Biochem. Biophys.* **158:**478–484.

Rosenbloom, J., Endo, R., and Harsch, M., 1976, Termination of procollagen chain synthesis by puromycin, *J. Biol. Chem.* **251:**2070–2076.

Ross, R., 1975, Connective tissue cells, cell proliferation and synthesis of extracellular matrix—A review, *Philos. Trans. R. Soc. London Ser. B* **271:**247–259.

Ross, D. W., and Benditt, E. P., 1965, Wound healing and collagen formation. V. Quantitative electron microscope autoradiographic observations of proline ^3H utilization by fibroblasts, *J. Cell Biol.* **27:**83–106.

Rossi, P., and de Crombrugghe, B., 1987, Identification of a cell-specific transcriptional enhancer in the first intron of the mouse alpha 2 (type I) collagen gene, *Proc. Natl. Acad. Sci. USA* **84:**5590–5594.

Rossouw, C. M. S., Vergeer, W. P., du Plooy, S. J., Bernard, M. P., Ramirez, F., and de Wet, W. J., 1987, DNA sequences in the first intron of the human pro-α1(I) collagen gene enhance transcription, *J. Biol. Chem.* **262:**15151–15157.

Rowe, D. W., Moen, R. C., Davidson, J. M., Byers, P. H., Bornstein, P., and Palmiter, R. D., 1978, Correlation of procollagen mRNA levels in normal and transformed chick embryo fibroblasts with different rates of procollagen synthesis, *Biochemistry* **17:**1581–1590.

Salpeter, M. M., 1968, H^3-proline incorporation into cartilage: Electron microscope autoradiographic observations, *J. Morphol.* **124:**387–421.

Saus, J., Wieslander, J., Langeveld, J. P. M., Quinones, S., and Hudson, B. G., 1988, Identification of the Goodpasture antigen as the α3(IV) chain of collagen IV, *J. Biol. Chem.* **263:**13374–13380.

Schmid, T. M., and Linsenmayer, T. F., 1985, Immunohistochemical localization of short chain cartilage collagen (type X) in avian tissues, *J. Cell Biol.* **100:**598–605.

Schnieke, A., Harbers, K., and Jaenisch, R., 1983, Embryonic lethal mutation in mice induced by retrovirus insertion in the α1(I) collagen gene, *Nature* **304:**315–320.

Soininen, R., Huotari, M., Hostikka, S. L., Prockop, D. J., and Tryggvason, K., 1988, The structural genes for α1 and α2 chains of human type IV collagen are divergently encoded on opposite DNA strands and have an overlapping promoter region, *J. Biol. Chem.* **263:**17217–17220.

Speakman, P. T., 1971, Proposed mechanism for the biological assembly of collagen triple-helix, *Nature* **229:**241–243.

Superti-Furga, A., Steinmann, B., Ramirez, F., and Byers, P. H., 1989, Molecular defects of type III procollagen in Ehlers–Danlos syndrome type IV, *Hum. Genet.* **82:**104–108.

Svoboda, K. K., Nishimura, I., Sugrue, S. P., Ninomiya, Y., and Olsen, B. R., 1988, Embryonic chicken cornea and cartilage synthesize type IX collagen molecules with different amino-terminal domains, *Proc. Natl. Acad. Sci. USA* **85:**7496–7500.

Tanzawa, K., Berger, J., and Prockop, D. J., 1985, Type I procollagen N-proteinase from whole chick embryos, *J. Biol. Chem.* **260:**1120–1126.

Tanzer, M. L., Church, R. L., Yaeger, J. A., Wampler, D. E., and Park, E.-D., 1974, Procollagen: Intermediate forms containing several types of peptide chains and noncollagen peptide extensions at NH_2 and COOH ends, *Proc. Natl. Acad. Sci. USA* **71:**3009–3019.

Tanzer, M. L., Rowland, F. N., Murray, L. W., and Kaplan, J., 1977, Inhibitory effects of tunicamycin on procollagen biosynthesis and secretion, *Biochim. Biophys. Acta* **500:**187–196.

Tromp, G., Kuivaniemi, H., Shikata, H., and Prockop, D. J., 1989, A single base mutation that substitutes serine for glycine 790 of the α1(III) chain of type III procollagen exposes an arginine and causes Ehlers–Danlos syndrome IV, *J. Biol. Chem.* **264:**1349–1352.

Tuderman, L., Puistola, U., Anttinen, H., and Kivirikko, K. I., 1977, Partial purification and characterization of a neutral protease which cleaves the N-terminal propeptides from procollagen, *Biochemistry* **17:**2948–2954.

Tuderman, L., Kivirikko, K. I., and Prockop, D. J., 1978, Partial purification and characterization of a neutral protease which cleaves the N-terminal propeptides from procollagen, *Biochemistry* **17:**2948–2954.

Turpeenniemi, T. M., Puistola, U., Anttinen, H., and Kivirikko, K. I., 1977, Affinity chromatography of lysyl hydroxylase on concanavalin A-agarose, *Biochim. Biophys. Acta* **483:**215–219.

Upholt, W. B., 1989, The type II collagen gene, in: *Collagen–Molecular Biology*, Vol. IV (B. R. Olsen and M. E. Nimni, eds.), pp. 31–49, CRC Press, Boca Raton, Fla.

Upholt, W. B., and Olsen, B. R., 1991, The active genes of cartilage, in: *Cartilage: Molecular Aspects* (B. Hall and S. A. Newman, eds.), pp. 1–57, CRC Press, Boca Raton, Fla.

Upholt, W. B., Strom, C. M., and Sandell, L. J., 1985, Structure of the type II collagen gene, *Ann. N.Y. Acad. Sci.* **460:**130–140.

van der Rest, M., Rosenberg, L. C., Olsen, B. R., and Poole, A. R., 1986, Chondrocalcin is identical with the C-propeptide of type II procollagen, *Biochem. J.* **237:**923–925.

Vaughan, L., Mendler, M., Huber, J., Bruckner, P., Winterhalter, K. H., Irwin, M. H., and Mayne, R., 1988, D-periodic distribution of collagen type IX along cartilage fibrils, *J. Cell Biol.* **106:**991–997.

Veis, A., and Brownell, A. G., 1977, Triple-helix formation on ribosome-bound nascent chains of procollagen: Deuterium–hydrogen exchange studies, *Proc. Natl. Acad. Sci. USA* **74:**215–219.

Vissing, H., D'Alessio, M., Lee, B., Ramirez, F., Godfrey, M., and Hollister, D. W., 1989, Glycine to serine substitution in the triple-helical domain of proα1(II) collagen results in a lethal perinatal form of short-limbed dwarfism, *J. Biol. Chem.* **264:**18265–18267.

Vogeli, G., Ohkubo, H., Avvedimento, V. E., Sullivan, M., Yamada, Y., Mudryj, M., Pastan, I., and de Crombrugghe, B., 1981, A repetitive structure in the chick α2 collagen gene, *Cold Spring Harbor Symp. Quant. Biol.* **45:**777–783.

Vuorio, E., and de Crombrugghe, B., 1990, The family of collagen genes, *Annu. Rev. Biochem.* **59:**837–872.

Weil, D., Bernard, M., Gargano, S., and Ramirez, F., 1987, The proα2(V) collagen gene is evolutionarily related to the major fibrillar-forming collagens, *Nucleic Acids Res.* **15:**181–198.

Weinstock, M., and Leblond, C. P., 1974, Synthesis, migration and release of precursor collagen by odontoblasts as visualized by radioautography after [^{3}H]proline administration, *J. Cell Biol.* **60:**92–127.

Wick, G., Olsen, B. R., and Timpl, R., 1978, Immunohistologic analysis of fetal and dermatosparactic calf and sheep skin with antisera to procollagen and collagen type I, *Lab. Invest.* **39:**151–156.

Willing, M. C., Cohn, D. H., and Byers, P. H., 1990, Frameshift mutation near the 3′ end of the COL1A1 gene of type I collagen predicts an elongated proα1(I) chain and results in osteogenesis imperfecta type I, *J. Clin. Invest.* **85:**282–290.

Wright, D. W., and Mayne, R., 1988, Vitreous humor of chicken contains two fibrillar systems: An analysis of their structure, *J. Ultrastruct. Mol. Struct. Res.* **100:**224–234.

Yada, T., Suzuki, S., Kobayashi, K., Kobayashi, M., Hoshino, T., Horie, K., and Kimata, K., 1990, Occurence in chick embryo vitreous humor of a type IX collagen proteoglycan with an extraordinary large chondroitin sulfate chain and short α1 polypeptide, *J. Biol. Chem.* **265:**6992–6999.

Yamada, Y., Avvedimento, V. E., Mudryj, M., Ohkubo, H., Vogeli, G., Irani, M., Pastan, I., and de Crombrugghe, B., 1980, The collagen gene: Evidence for its evolutionary assembly by amplification of a DNA segment containing an exon of 54 bp, *Cell* **22:**877–892.

Yamada, Y., Kühn, K., and de Crombrugghe, B., 1983, A conserved nucleotide sequence, coding for a segment of the C-propeptide, is found at the same location in different collagen genes, *Nucleic Acids Res.* **11:**2733–2744.

Yamaguchi, N., Benya, P. D., van der Rest, M., and Ninomiya, Y., 1989, The cloning and sequencing of α1(VIII) collagen cDNAs demonstrate that type VIII collagen is a short chain collagen and contains triple-helical and carboxyl-terminal non-triple-helical domains similar to those of type X collagen, *J. Biol. Chem.* **264**:16022–16029.

Yamaguchi, N., Mayne, R., and Ninomiya, Y., 1991, The α1(VIII) collagen gene is homologous to the α1(X) collagen gene and contains a large exon encoding the entire triple-helical and carboxyl-terminal non-triple-helical domain of the α1(VIII) polypeptide *J. Biol. Chem.* **266**:4508–4513.

Yamauchi, K., Yamamoto, T., Hayashi, H., Koya, S., Takikawa, H., Toyoshima, K., and Horiuchi, R., 1987, Sequence of membrane-associated thyroid hormone binding protein from bovine liver: Its identity with protein disulphide isomerase, *Biochem. Biophys. Res. Commun.* **146**:1485–1492.

Yoshioka, H., and Ramirez, F., 1990, Pro-α1(XI) collagen, *J. Biol. Chem.* **265**:6423–6426.

Chapter 7

Matrix Assembly

DAVID E. BIRK, FREDERICK H. SILVER, and ROBERT L. TRELSTAD

1. Introduction

Matrix assembly, disassembly, and reassembly of collagenous matrices are dynamic processes which occur when an organism undergoes morphogenesis, growth, repair, or regeneration. They are complex events which ascend the biological hierarchy from molecular interactions among adjacent molecules to macroscopic interactions among adjacent tissues. The multiple levels of organization involved in matrix assembly from molecules to final structure, while not understood, can be appreciated by consideration of your hand: over 90% of the organic matter in the skin, tendons, ligaments, and bones is type I collagen. The genetic and chemical details of type I collagen are relatively well understood and discussed in detail in other chapters; however, the assembly algorithms which partition and position these molecules in three-dimensional space into a functional grasping and holding device are not. *In vitro* studies of isolated type I collagen and procollagen have indicated that much of the assembly process of a collagen fibril can be explained by physicochemical forces, often described as "self-assembly." Light microscopic studies of type I collagen-rich tissues reveal a variety of architectures ranging from near-complete molecular alignment in tendons and ligaments; a tight, orthogonal network in bone; and a loose network arrangement in skin (Fig. 7-1). Ultrastructural studies of cells producing type I collagen indicate that fibrils form in association with the cell's external surface in a complex topographic relationship and that the initial elements formed by the cells are relatively short, discontinuous fibril segments. The relationships of the cell-defined extracellular domains with matrix assembly permit cellular regulation of the self-assembly process by the manipulation of the microenvironment and temporal and spatial regulation by determination of the time and place of matrix deposition.

DAVID E. BIRK, FREDERICK H. SILVER, and ROBERT L. TRELSTAD • Department of Pathology, Robert Wood Johnson Medical School, Piscataway, New Jersey 08854.

Cell Biology of Extracellular Matrix, Second Edition, edited by Elizabeth D. Hay, Plenum Press, New York, 1991.

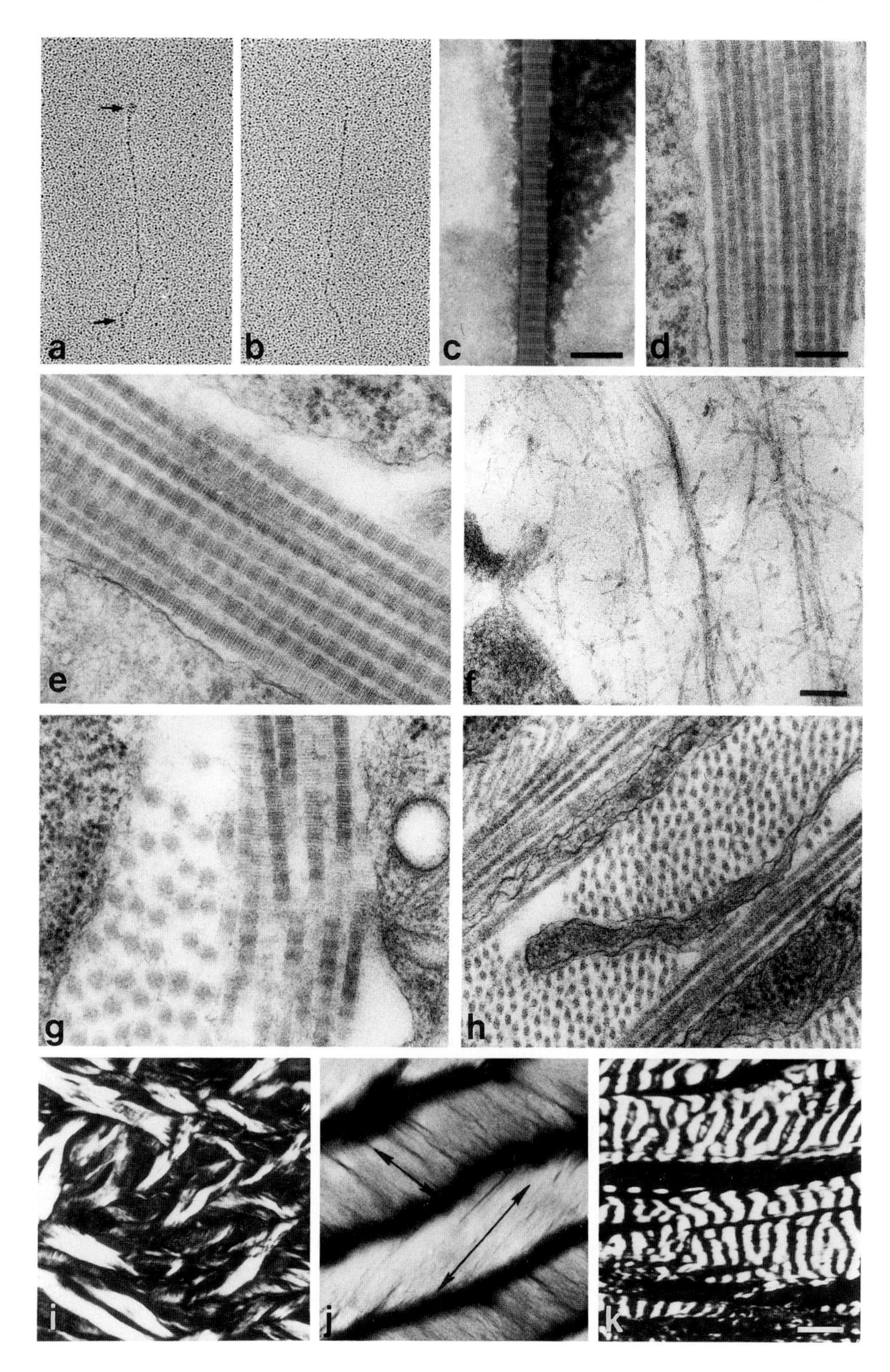
a
b
c
d
e
f
g
h
i
j
k

1.1. From Collagen Molecules to Matrix Architecture: Hierarchies of Order

Adequate description of matrix assembly requires knowledge of the stratum in the hierarchy of biological structure at which attention is being directed and recognition that attributes of the assembled elements change as the hierarchy is ascended. The shape and size of your hand are primarily defined by the type I collagen that comprise it, yet it is apparent that the biological forces which partitioned the collagens in that unique shape are not resident in the collagens themselves and that your hand did not self-assemble. The collagen fibrils which lie in the skin of your hand on a proximodistal axis are not, to our knowledge, chemically different from those which lie mediolaterally. An upper boundary, therefore, exists on the influence of primary structure on tissue architecture, yet as we will see, molecular attributes may be reflected as distantly as macroscopic tissue form. The supramolecular forces and transient "architectural elements" which govern morphogenesis and repair reactions are just beginning to be understood. In this chapter we intend to focus on the variety of forces and factors which operate in the assembly of collagenous matrices beginning at the molecular level and ending at the tissue level. Our subject will be the fibril-forming collagens with major emphasis on type I collagen and on the assembly of tendons.

1.2. Type I Collagen: A Prototype for Fibril Assembly

Type I collagen was the first macromolecule to be shown to undergo spontaneous assembly *in vitro* into a structure which resembles that found in native tissue (Gross *et al.*, 1954). Collagen is now joined by many examples, e.g., actin and tubulin, in which the assembly to higher-ordered forms is driven by phys-

Figure 7-1. Elements of the matrix hierarchy beginning at the molecular level and ending at the macroscopic, illustrating stages of assembly. Type I procollagen (a) and collagen (b) molecules are seen in rotary-shadowed transmission electron micrographs. These are rod-shaped, semiflexible molecules approximately 300 nm long which assemble to form the collagen fibril (c). The collagen fibril in c was formed *in vitro* and is negatively stained with 67-nm periodic light/dark zones representing overlap/hole zones. In tissues the fibril is usually collected into bundles which in turn form larger aggregates or fibers (d). Tissues have different collagen molecules, fibrils, fibril bundles, and architectures. Transmission electron micrographs from 14-day chick embryo tendon (e), sternal cartilage (f), dermis (g), and corneal stroma (h) illustrating differences in fibril size and organization. Light micrographs of adult chick dermis (i), intervertebral disc (j) and tendon (k) taken with polarizing optics illustrating differences in the organization, packing, and orientation of the fibrillar matrix. Dermal collagen fibrils are organized into a wickerwork of cablelike structures. This woven structure gives the skin a pliable quality. The fibrils in the intervertebral disc are organized into sheets within which all fibrils course in the same direction (arrows) and between which there is a shift in orientation. The fibrils in tendon are oriented along one axis, but the fibrils contain a crimp seen as an alternation in the birefringence; this crimp serves a biomechanical function (see text). Bars: (c, d) 300 nm; (e–h) 100 nm; (i–k) 25 μm. (i–k, from Trelstad and Silver, 1981.)

icochemical forces derived from the primary sequence (Baker *et al.*, 1990). Attempts to understand collagen assembly at the cellular level, in homopolymeric and heteropolymeric fibrils, and in the interstices of tissues indicate that a full understanding of matrix assembly must: address regulatory events which begin with the self-assembly of molecules driven by enzymatic processing and side-chain interactions; explain the macroscopic conditions of cells and tissues, growing in size, but often maintaining a constancy of shape; and consider the requirement that many of these multimolecular aggregates resist the force of gravity and provide a functional biomaterial from which to craft an organism.

The multimerism of the matrix poses major problems for the study of matrix assembly. For example, it cannot be assumed that all components of a multimeric aggregate are equally codistributed at all stages of the assembly process. Partitioning of products clearly begins within the cell and the pathway to the outside of the cell is not necessarily going to be the same for all components. The highly glycosylated proteoglycans will have different pathways through the intracellular compartments from the less glycosylated type I collagen. The kinetics and sites of mixing will accordingly play a role in the assembly of the matrix. It is also likely that new elements will add to the matrix at some time following initial deposition. It is known, for example, that proteoglycan in cartilage matrix has a half-life of 3 to 4 weeks (Chapter 5), whereas its collagenous network is much more stable (Chapter 8). Such postdepositional steps may require enzymatic processing of the surface of an aggregate; formation and/or breakage of covalent cross-links; possible disruption of the aggregates to achieve intercalation; or some other form of disassembly and reassembly.

With these caveats in mind, a consideration of the mechanisms of collagen fibril assembly *in vitro* could be regarded an academic argument because most such data are generated using purified, individual components. Nonetheless, a great deal has been learned by this approach which represents, in a sense, a limit or boundary condition on the assembly process as well as a guide to potential mechanisms operative at subsequent stages of assembly.

2. Structural Elements: Monomers, Multimers, Fibrils, Tissues

2.1. Monomer Structure: Amino Acids, Carbohydrates, Extension Peptides

The primary structure of type I collagen and other fibril-forming collagens are reviewed in Chapter 1. The profile of charged and hydrophobic amino acids along the triple-helix is responsible for the lateral packing of molecules in the fibril, described as a D-staggered arrangement (Wallace, 1985). The genetically determined sequence of amino acids dictates many of the assembly properties of type I collagen, a relationship emphasized in type I collagen mutations such as osteogenesis imperfecta in which single amino acid substitutions and/or various deletions lead to altered fibrillar forms (Prockop *et al.*, 1989). At the same time, it is apparent that only portions of the primary structure are responsible for the D packing in that types I, II, and III will, in pure form, make fibrils with similar 67-nm patterns (Lapiere *et al.*, 1977; Birk and Silver, 1984; Burgeson, 1988) and will form heterotypic fibrils *in vivo* (Chapter 1).

Fibril-forming collagens contain bound carbohydrates discussed elsewhere (Chapters 1, 6). These relatively bulky carbohydrates influence the assembly processes at several levels: within the cell during packaging, transport, and early assembly; and in extracellular compartments during final deposition. The covalently linked glycosaminoglycan chain on type IX collagen creates a large charged domain which certainly influences the character of its various tissue forms (Bruckner *et al.*, 1985, 1988; Chapter 1).

Procollagen peptides are transient or permanent structural elements depending on collagen type, sites of deposition, and conditions of deposition. The non-triple-helical portions of type I collagen at either or both amino and carboxy termini significantly influence assembly. If these regions of the molecules are present as fully unprocessed procollagen extensions, then significant alteration of assembly occurs (Capaldi and Chapman, 1982; Miyahara *et al.*, 1984; Hulmes *et al.*, 1989; Mould *et al.*, 1990; Fleischmajer *et al.*, 1981, 1988). *In vitro,* if proteases are provided to process the procollagen extensions, fibrillar aggregates form into 67-nm striated fibrils, which have a blunted end and a pointed end, and in which the addition of new material occurs at the pointed end (Kadler *et al.*, 1987, 1990).

2.2. Multimers

2.2.1. From 0-D to 4-D Overlap

When purified solutions of collagens are incubated under physiological conditions, fibrils with the native periodicity spontaneously form. When this process is followed spectrophotometrically, the turbidity–time curve shows a lag phase, followed by an exponential growth phase, ending in a plateau (Fig. 7-2). The increase in turbidity is a measure of the diameter of the scattering objects and is discussed in detail in the first edition of this volume (Trelstad and Silver, 1981). Examination by electron microscopy of samples removed during the growth phase reveals a variety of multimeric aggregate forms. Samples examined from the plateau region show a tangle of native-looking fibrils (Fig. 7-1c). Examination of the lag phase has led to a variety of observations, using both microscopic and spectrophotometric techniques (Fig. 7-3). Some groups have identified intermediate structures which have D-periodic characteristics; others have failed to detect such aggregates (Trelstad *et al.*, 1976; Silver *et al.*, 1979; Silver and Trelstad, 1980; Suarez *et al.*, 1985; Ward *et al.*, 1986; Na, 1989).

When hydrophobic interactions among type I collagen molecules are predominant, the molecules align in parallel without any stagger, or in 0-D stagger, to form segment-long-spacing (SLS) crystallites. Such aggregates, by them-

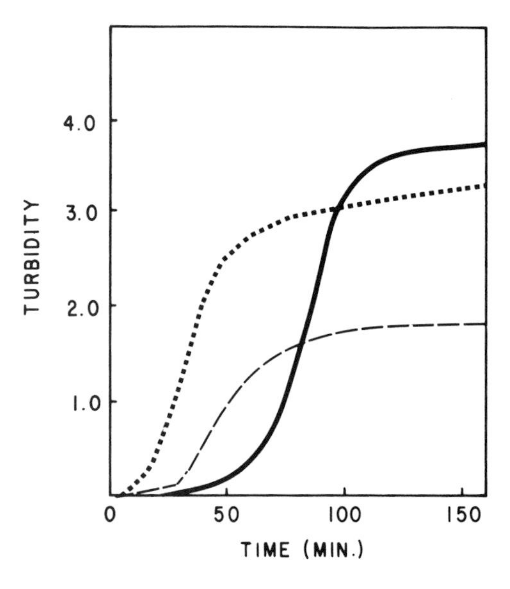

Figure 7-2. Turbidity–time curves for collagen type I (solid), type II (dashed), and type III (dotted) shown as percentage final turbidity versus time. Each curve demonstrates a lag phase, growth phase, and plateau phase. The turbidity at the plateau phase is different for each type, indicating differences in the diameter of the fibrils formed. (From Birk and Silver, 1984.)

selves, cannot form long, thin polymers or fibrils because there is no overlap or stagger between them. In the 1-D staggered arrangements, the adjacent molecules are displaced along the molecular axis by 234 amino acids. In the 4-D stagger, there are interactions among sets of charged amino acid pairs located at positions 53, 54, and 56, near the amino terminus, and positions 990 and 992, near the carboxy terminus, as well as charged residues in the nonhelical ends. While all D-staggered arrangements are limitations to linear packing of the molecules, rotations of the backbone of the triple-helix suggest that lateral packing of collagen molecules into fibrils can be quite variable and thereby potentially responsible for polymorphic fibril forms (Silver, 1982).

Recognition of the existence of aggregates in the lag phase of the assembly reaction is central to understanding the characteristics of the self-assembly

Figure 7-3. Self-assembly of type I collagen. A schematic series of the assembly stages of type I collagen derived from studies using light scattering and electron microscopy to identify intermediate forms. The individual monomers interact in a 4-D stagger to form dimers and trimers (a). In the 4-D trimer at the upper right and elsewhere, the numbers refer to molecular segments of length D along the 4.4-D-long molecular axis. Number 1 is at the NH_2 terminal, 2 is at the interface of the first D segment with the second, and so on. A 4-D staggered configuration is the same as an overlap of 0.4-D. The 4-D trimers laterally associate to form an aggregate consisting of approximately 15 molecules. If the trimers overlap in a 1-D pattern, a relatively compact structure is formed (middle right, a), which can add to like structures in a linear and/or lateral manner to build a long subfibril as illustrated by the large central structure. The electron micrographs are from a collagen solution sampled at the end of the lag phase of the assembly sequence. The collagen molecules have aggregated into tactoidal structures that contain a 1-D stagger (b, c). In b, three aggregates, each with a 1-D stagger, are aligned end to end. These tactoidal aggregates appear to add to one another yielding subfibrils. The narrow subfibrils laterally entwine at a later stage to form wider fibrils (d). Bars = 67 nm (b), 300 nm (c, d). (From Trelstad and Silver, 1981.)

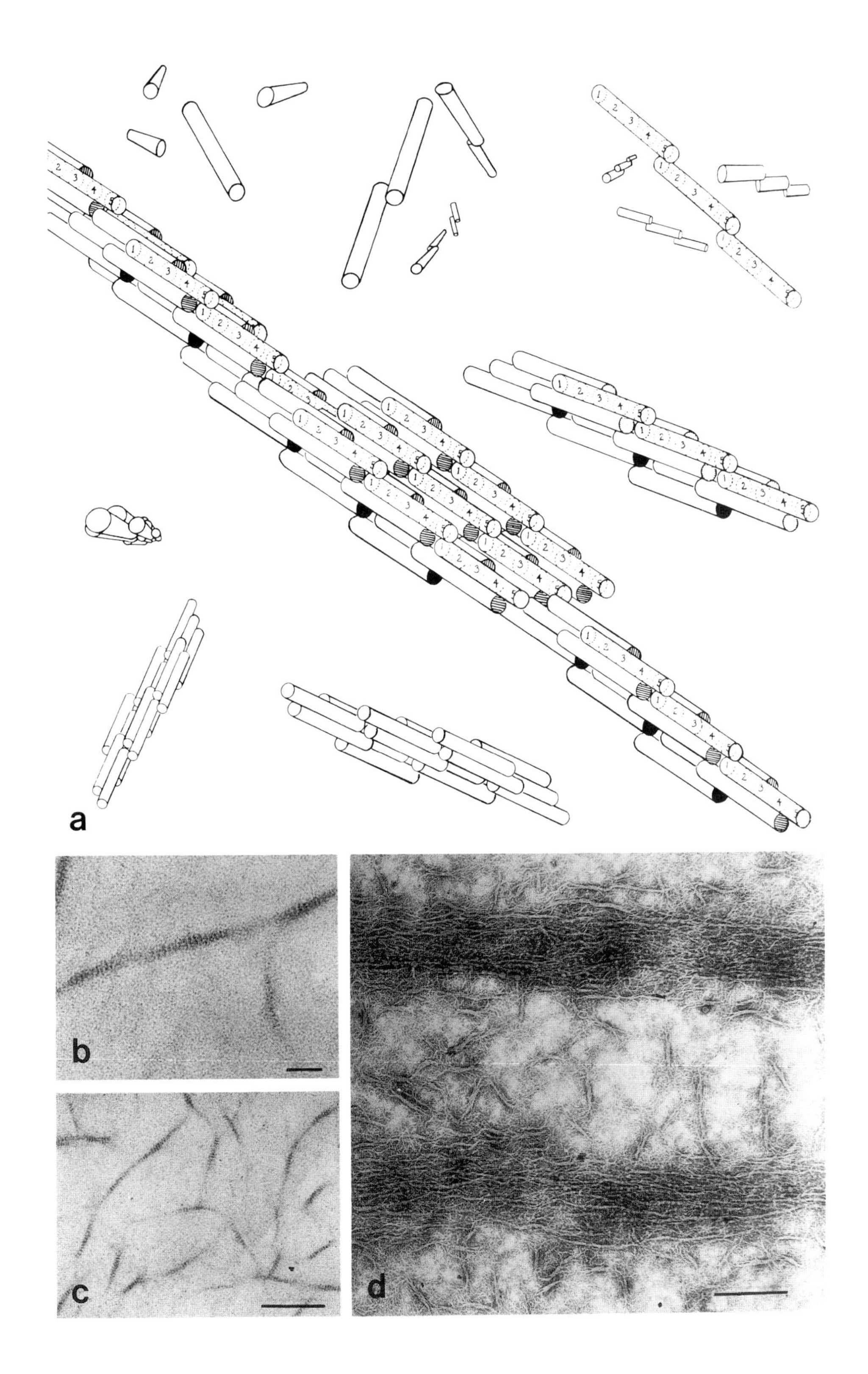
a
b
c
d

reaction. If the reaction is a typical nucleation-growth process, the lag phase is typically the period of formation of "nuclei" onto which monomers then add in a rapid fashion with the formation of the nuclei a rate-limiting step (Na, 1989). If the lag phase is a period of formation of intermediate subassemblies, such aggregates would then be expected to add to each other to form the structures identified in the growth phase. The nucleation-growth hypothesis suggests that collagen fibril formation is a diffusion-mediated process in which monomeric collagen adds to a multimer, initially a nucleus of a few molecules which have attained some kind of thermodynamically unfavored arrangement, but having reached this configuration are then capable of promoting monomeric additions (Kadler *et al.*, 1987; Na, 1989). The intermediate assembly hypothesis suggests that collagen forms intermediate aggregates which then add to each other to form higher multiples of macromolecules culminating ultimately in a fibril (Trelstad *et al.*, 1976; Bruns *et al.*, 1979; Hulmes *et al.*, 1983; Suarez *et al.*, 1985; Silver and Trelstad, 1979; Ward *et al.*, 1986). The literature concerning collagen fibril assembly contains support both for the arguments of monomeric addition through nucleation-growth and for intermediate addition through multistep assembly. Resolution of this issue is beyond the scope of this chapter, although there are several obvious challenges that must be addressed. While the nucleation and growth process is sufficient to explain the physicochemical events in solution, it has not been accompanied by any demonstration or identification of the "nuclei." Since it is now possible to visualize individual molecules by microscopy and/or spectroscopy, it would seem essential that such "nuclei" be rigorously characterized. Conversely, those who have identified "intermediate" subassemblies must develop suitable theoretical treatments to describe the kinetics of the multistep nature of the process and the annealing of intermediates into larger aggregates.

2.3. Fibrils

2.3.1. Heteropolymerism

Proper understanding of matrix assembly can only be achieved in a circumstance where the final product of assembly can be fully defined, structurally and chemically. New information as to the composition of collagen fibrils gained from both chemical analyses and immunohistochemistry makes it likely that most, if not all, fibrils are heteropolymeric (see heterotypic fibrils, Chapter 1).

Heteropolymerism results from mixing of different collagen species within a single fibril. Immunolocalization and/or chemical cross-link studies have shown that collagen types I and V (Fitch *et al.*, 1984, 1988; Birk *et al.*, 1988), types I and II (Hendrix *et al.*, 1982; Linsenmayer *et al.*, 1990), types II and IX (Vaughan *et al.*, 1988; van der Rest and Mayne, 1988), types II and XI (Mendler *et al.*, 1989), types I and III (Henkel and Glanville, 1982; Keene *et al.*, 1987) can be present as heterotypic fibrils. Immunoelectron microscopic studies of the

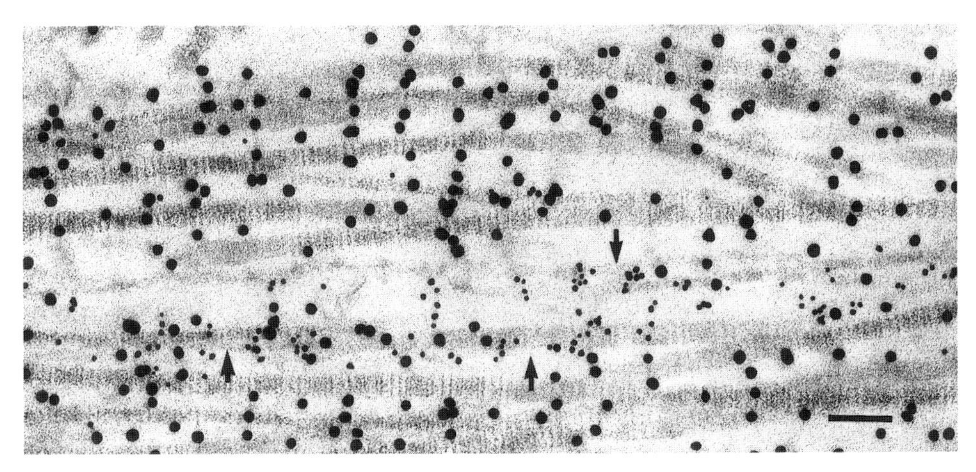

Figure 7-4. Colocalization of collagen types I and V in the chick embryo cornea. Sections from lathyritic corneas were incubated in PBS for 15 hr at 4°C, a condition where fibril structure is partially disrupted, fixed in paraformaldehyde, and incubated simultaneously with monoclonal antibodies against type I collagen coupled to 10-nm gold and against type V collagen coupled to 5-nm gold. Both collagen types I and V are found within a single fibril. Most of the fibrils are labeled with antibodies against type I collagen while only those fibrils whose structure has been partially disrupted by the pretreatment with cold PBS are labeled with the antibodies against type V collagen (arrows). In some cases the fibril is so dissociated that the labeled structure is difficult to identify. The compact fibrils are not decorated with 5-nm gold particles coupled to anti-type V collagen. Those fibrils formed in the presence of βAPN, and therefore not fully cross-linked, have been partially dissociated by the cold treatment and are labeled with antibodies specific for type V collagen. Bar = 100 nm. (From Birk *et al.*, 1988.)

corneal stroma demonstrate that corneal fibrils are heterotypic and composed of 80% collagen type I and 20% collagen type V (Birk *et al.*, 1988) (Fig. 7-4). Type V collagen also copolymerizes with type I *in vitro*, producing fibrils of smaller diameters with small increases in type V content (Fig. 7-5). It is noteworthy that the terminal globular domain of type V collagen is important for the full measure of this diameter-regulating effect; the influence of type V may also stem from its slightly longer triple-helix. It is also possible that a heteropolymer could be comprised of two different forms of the same molecule, e.g., type I procollagen and fully processed type I collagen in varying proportions. Further exploration of heteromolecular fibrils comprised of the multiple collagens will be important in our effort to understand the exact composition and hence character of specific tissues.

In addition to the multiplicity of collagens which might be constituents of the same fibril, other matrix macromolecules also have been shown to be closely associated with the fibril, particularly proteoglycans, and influence fibril formation *in vitro* (Scott and Haigh, 1988; Birk and Lande, 1981; Vogel *et al.*, 1984; Chandrasekhar *et al.*, 1984; Amudeswari *et al.*, 1987; Ruoslahti, 1988; Garg *et al.*, 1989; Brown and Vogel, 1990). Small dermatan sulfate proteoglycans from tendon have a high affinity for type I collagen and their speci-

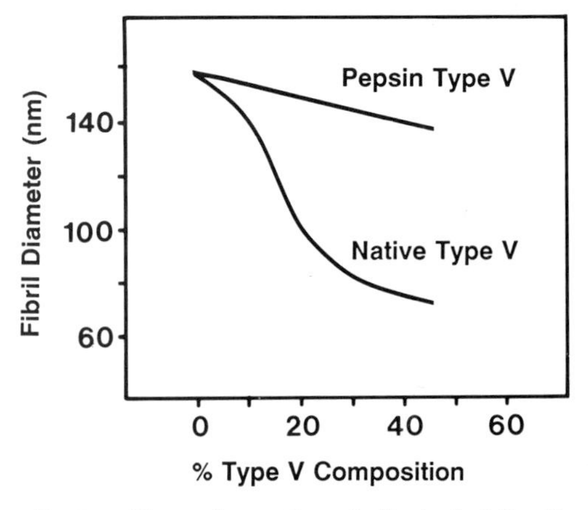

Figure 7-5. Regulation of collagen fibril diameter: type I/V heterotypic fibrils *in vitro*. The effect of type V collagen copolymerization on the regulation of fibril diameter is illustrated for native type V collagen and pepsin-treated type V collagen. Pepsin treatment removes an amino-terminal, noncollagenous peptide. When type I and V collagen are mixed and allowed to self-assemble *in vitro*, the native form of the molecule markedly reduces mean fibril diameter at concentrations between 10 and 30% type V collagen. However, the pepsin-treated form is much less effective. These observations indicate that the diameter-regulating effect of type V resides in the amino-terminal, nonhelical portion of the molecule. This nonhelical portion may sterically alter the addition of type I collagen, thus limiting lateral growth of fibrils.

ficity of binding lies in the protein core (see Chapter 2). Scott has proposed that small proteoglycans bound to the fibril surface are major regulators of fibril growth and has further suggested that lateral fusion of fibrils might occur when such proteoglycan sheaths are removed from the fibril surface (Scott, 1986). A major correlation exists among fibril diameter and mechanical properties and the content and composition of glycosaminoglycans (Flint *et al.*, 1984; Merrilees *et al.*, 1987; Parry and Craig, 1988).

2.3.2. X-Ray Diffraction

From nearly 30 years of study, a consensus about the packing of the molecules along the axis of a collagen fibril has been reached which states that there is a stagger of approximately 67 nm along the fibril axis. While there is close correlation of this dimension with the 234-amino-acid overlap, it is apparent that factors such as heteropolymerism may alter the D period in type I collagen-rich structures from different tissues such as skin or tendon (Hulmes and Miller, 1979; Eikenberry *et al.*, 1982; Brodsky *et al.*, 1982; Hulmes *et al.*, 1985). More detailed studies of type I collagen by Fraser and colleagues (Fraser *et al.*, 1987) using high- and low-angle X-ray diffraction suggest that the molecular segments in the gap region are more mobile than those in the overlap region and that the segments in the gap region are kinked. The highly ordered overlap region of the collagen fibrils is thought to consist of a crystalline array of molecular segments inclined by a small angle with respect to the fibril axis whereas the gap region is less well ordered and includes a D-periodic molecular crimp. Crimps are periodic folds or waves in structures and, interestingly, in tendons there is a macroscopic crimp in the fibril/bundle structure. Biomechanical studies indicate that this crimp plays an important role in early load bearing (Silver and Doillon, 1989; Kato *et al.*, 1991). Crimping is an example where molecular attributes confer properties upon the multimer which are reflected in the macroscopic structure.

2.3.3. Electron Microscopy

The first evidence for a periodic structure of the collagen fibril was obtained nearly one half century ago by transmission electron microscopy (see Gross, 1974, for review). The current view of the overlap of adjacent type I collagen molecules in a fibril is shown in Figs. 1-5 and 7-6. The molecules are in a staggered arrangement of charged residues and the summation of this charge profile leads to the classical striated pattern of the fibril. The packing arrangement thus produces a surface charge profile on the fibril which is derived from, but different from, the charge profile of the primary amino acid sequence. The surface features of the molecule, including charge profile and bulk energy, are thus attributes of the formed structure, attributes which do not exist in the isolated molecules.

The "gap" or "hole" regions in the fibril result from the fact that the overlap distance (1-D) is not integrally related to the length of the molecule (4.4-D). The spaces are sites within which other molecules, or parts of molecules, might reside, for example, a procollagen extension peptide, a cross-linked telopeptide, or the side chain of type IX collagen. The cross-linkages can be between molecules of the same type (homotypic cross-links) or between collagens of different types (heterotypic cross-links) (Henkel and Glanville, 1982; van der Rest and Mayne, 1988). During bone mineralization the initial formation of hydroxyapatite crystals seems to occur within the hole region (Glimcher, 1989).

2.4. Tissues

2.4.1. Ligaments and Tendons

Ligaments and tendons are cablelike structures that cyclically transmit force without permanent changes in dimension. Tendons are composed of fascicles which in turn are composed of bundles which in turn are composed of crimped, aligned collagen fibrils (Kastelic *et al.*, 1978; McBride *et al.*, 1985; Mosler *et al.*, 1985; Folkhard *et al.*, 1987; Baer *et al.*, 1988; Rowe, 1988). The collagen fibrils are deposited in the bundles and fascicles as discontinuous structures (McBride *et al.*, 1985; Birk and Trelstad, 1986) and are buckled into planes of wavy parallel fibrils which branch and rotate and so are not an ideal parallel aligned collagen network (Birk *et al.*, 1989a). With polarizing microscopy, tendons demonstrate a characteristic series of light and dark bands which Diamant *et al.* (1972) interpreted as arising from the periodic arrangement of collagen fibers along the tendon axis (Fig. 7-1). Ligaments such as the anterior cruciate connecting the femur and tibia have additional levels of structural complexity in that the collagen network is twisted by about 180° from the femoral to tibial attachment sites.

2.4.2. Cornea

The cornea is a transparent structure comprised of collagen fibrils with uniform diameters and uniform interfibrillar spacings. The fibrils are grouped as bundles and the bundles are organized into orthogonal lamellae (Fig. 7-1). In submammals, the corneal stroma consists of orthogonal lamellae that describe a right-handed spiral resembling a cholesteric liquid crystal (Trelstad, 1982). Remarkably, this pattern is identical in both eyes and thus bilaterally asymmetrical (Trelstad and Coulombre, 1971). In mammals, the macroscopic organization of the stroma, while orthogonal, is less regular than that found in submammals.

2.4.3. Skin

The type I collagen fibrils in the dermis form a nonwoven wavy network (Fig. 7-1) that maintains its shape in response to applied mechanical loads by alignment of the fibers with the stress direction. Skin is able to bear loads in any direction within its plane and the collagen network is loose enough to be able to rearrange and orient freely.

3. Morphogenesis via Cell-Mediated Assembly

3.1. Intracellular Compartments

The intracellular synthesis and processing of the collagens and proteoglycans are detailed in Chapters 5 and 6. Because many matrix structures are heteropolymeric, the question of when and where mixing of components occurs is important and it would seem likely that it begins within the cell in one of the intracellular compartments. The stoichiometry of such intracellular mixing will have impact on the character of the products which can assemble as well as their thermal stability (Bruckner and Eikenberry, 1984). During intracellular processing, events other than heteropolymeric mixing, of importance for the matrix assembly, would include addition of proteoglycans and of enzymes for procollagen processing and enzymes for cross-linking.

The time, site, and control of procollagen processing will have considerable influence on the assembly of the matrix. The sites of procollagen processing and the specific portions of the molecules processed appear to be highly variable. The possible sites of processing include intracellular and extracellular compartments and in the latter circumstance, the processing may occur at some period in time after discharge from the cell (Chapter 6). At present the regulation of processing within the context of *in vivo* assembly is not well understood.

3.2. Vectorial Discharge

The vectorial discharge of matrix constituents into the extracellular space is another way by which cells exercise influence over the matrix assembly process. The movement of secretory vacuoles to the cell surface for discharge does not result in their random accretion around the entire perimeter of the cell, but usually results in their preferred localization to one region or pole of the cell. The vectorial or polarized discharge of secretory products is a well-recognized phenomenon in exocrine and some endocrine glands and is generally associated with the coordinate, polarized positioning of the Golgi apparatus in the secretory pole of each epithelial cell.

Similar coordinate, polarized patterns of intracellular organelles have been described in fibroblasts, osteoblasts, odontoblasts, chondroblasts, and matrix-producing epithelia (Trelstad, 1970, 1971, 1977, 1980; Weinstock and Leblond, 1974; Trelstad and Hayashi, 1979; Trelstad and Birk, 1984, 1985). The polarized discharge of matrix components from odontoblasts, osteoblasts, and the corneal epithelium are good examples of how cells influence the organization of the architecture of the matrix by polarized discharge.

3.3. Extracellular Compartments

Extracellular compartments have been recognized for a number of years, but their unique attributes have only recently begun to be examined. As a generalization, it should be understood that the space outside the cell is not a homogeneous domain with a random distribution of diffusing materials. Rather the space outside the cell is highly structured. The extracellular compartments over which the cell has extensive control exist at the immediate interface between the matrix and the cell surface. Associated with osteoclasts is an extracellular compartment which has a unique pH and enzymatic composition (Baron *et al.*, 1985). Associated with a variety of fibroblasts are a series of at least three extracellular compartments within which type I collagen assembles into fibrils, bundles, and macroaggregates (Trelstad and Hayashi, 1979; Birk and Trelstad, 1986; Birk *et al.*, 1989b) (Figs. 7-6, 7-7, 7-8). In the embryonic chick tendon fibroblast, the first of these three extracellular compartments is a narrow channel containing one to three forming fibril segments. The three-dimensional course of a forming fibril segment within a narrow channel can be traced in serial sections and shown to extend from the perinuclear Golgi region to an opening at the cell surface or a micrometer into the adjacent bundle (Fig. 7-9). The finding of both ends of the fibril segments was the first *in situ* demonstration that an intermediate of large size was involved in fibril assembly (Birk *et*

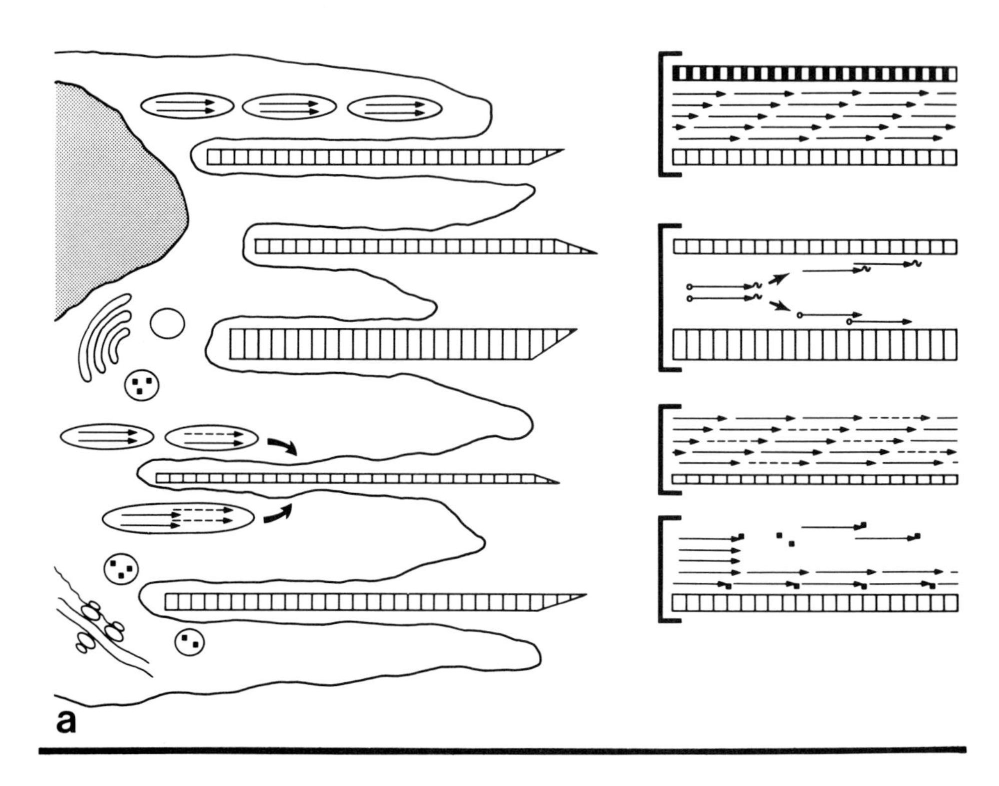
a

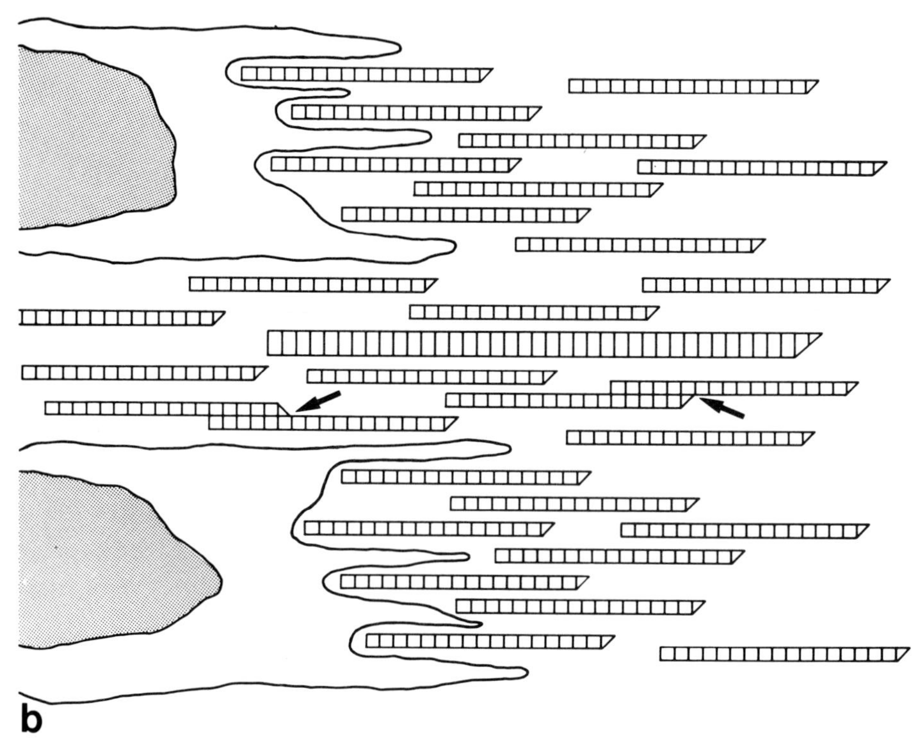
b

al., 1989b). It is suspected that many of the forces of self-assembly discussed earlier operate within this compartment as well as other enzymatic steps in procollagen processing.

The second extracellular compartment associated with the fibroblasts derives from fusions of the first with the result that larger, bundle-forming compartments are formed. The bundle-forming compartment is usually defined by the plasma membrane of a single cell, but may involve the surfaces of other cells. The fibril segments, formed in the first compartment, are collected in this second compartment where they laterally associate and form the basic bundle structure of the macroscopic collagen fiber. The fibril segments persist as discontinuous elements and can be readily detected in serial sections (Fig. 7-10) until their apparent fusion during later stages of maturation. Manipulation of individual fibril segments in the bundle-forming compartment is probably effected by interactions of the cell's surface with the collagen fibrils, the traction and pushing process (Harris *et al.*, 1981; Stopak *et al.*, 1985) possibly mediated by matrix receptors (Gullberg *et al.*, 1989).

The third extracellular compartment associated with the fibroblasts is derived from the second and consists of laterally associated bundles which form tissue-specific macroaggregates. For example, in the tendon the macroscopic organization of the molecules–segments–fibrils–bundles is uniaxial and suit-

---

Figure 7-6. In a, a number of possible mechanisms controlling fibril diameter during assembly are illustrated. At the top left, three elongated vacuoles, containing collagen molecules fuse in tandem to form the long, narrow fibril-forming compartments. The extracellular channel generated in this manner is open to the extracellular space and extends deep into the cell. It is within this compartment that collagen assembles to form fibril segments approximately 10 μm long. This partitioning of the extracellular space provides a mechanism for the fibroblast to exert control over the extracellular steps in matrix assembly. In the upper right, the traditional two-dimensional presentation of collagen packing into native fibrils is presented. These models were derived from electron microscopic and X-ray diffraction studies of the fibril. For the beginning student, the 1-D stagger of adjacent molecules is most readily seen in this presentation. The second grouping illustrates the influence of procollagen processing on fibril diameter. The differences in final fibril diameter depend on the sequence and extent of the procollagen processing. If the amino propetide is removed, followed by the carboxy propeptide, large-diameter fibrils form. If the sequence is reversed smaller diameter fibrils result. In the third grouping, the influence of collagen–collagen interactions on fibril diameter is illustrated. The interaction of type V collagen with type I collagen in the corneal stroma has been shown to result in small-diameter fibrils. Numerous other heterotypic interactions also have been implicated in diameter regulation. Finally, a model of a fibril is presented in which proteoglycans, fibril-associated collagens, or some other component associates with the fibril surface. A variety of fibril growth-inhibiting interactions have been postulated. In b, a model of the post-depositional events in extracellular matrix assembly is summarized. The channels fuse laterally to form a second compartment within which fibril bundles form. Within this compartment fibril segments coalesce to form discontinuous fibril bundles. Finally, the bundle-forming compartments continue to aggregate laterally, the intervening projections of cytoplasm retract, and a compartment is generated within which broader bundles are generated. Within this compartment lateral (arrows) and/or linear fusions of fibril segments occur. This may be the mechanism responsible for the genesis of longer fibrils with larger diameters that are seen as development proceeds. The fibrils are presented with the same orientation for simplicity.

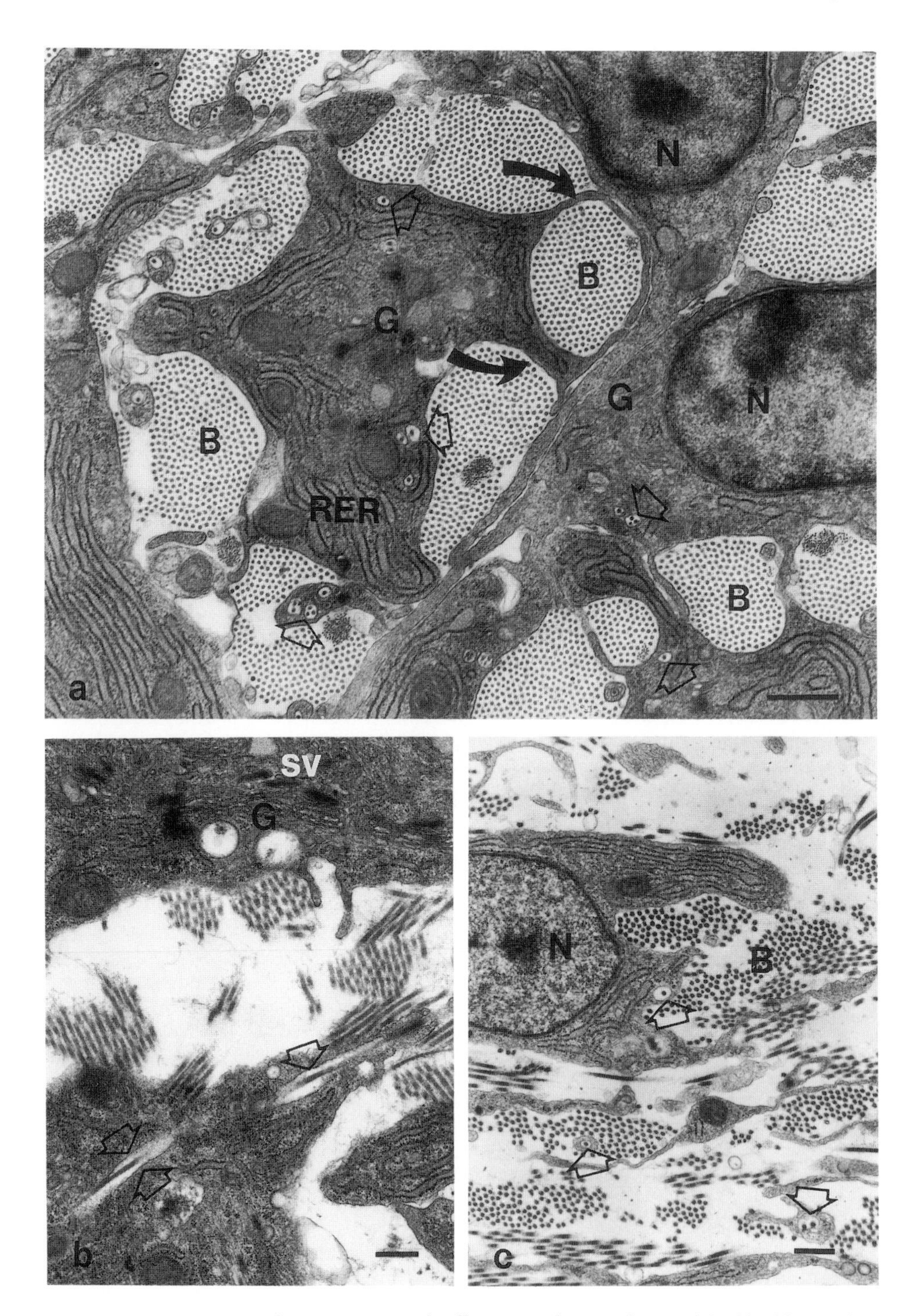

Figure 7-7. Transmission electron micrographs illustrating the complexity of the fibroblast surface and the partitioning of the extracellular space with formation of extracellular compartments in

able for tension bearing in only one direction whereas in skin the macroscopic organization of the molecules–segments–fibrils–bundles is biaxial and suitable for tension bearing in all directions. The third compartment is usually defined by the apposition of a number of adjacent cell bodies and/or extending cell processes.

Fibroblasts, in general, compartmentalize the extracellular space and although differences exist among tissues such as the tendon, cornea, and skin, the same basic pattern of extracellular compartments is seen throughout and assumed to be generalized to most connective tissues (Birk and Trelstad, 1984, 1985; Birk *et al.*, 1990b; Ploetz *et al.*, 1991). The recognition of the existence of extracellular compartments and our beginning understanding of the ways by which events within these compartments might be regulated, opens a new area of study in matrix cell biology. Certainly the matters outside the cell are under cellular control. Some events that might be managed by simple diffusion-mediated processes might not require the formation or presence of an extracellular compartment; others might. The lytic activities of the osteoclast are best kept restricted to the exact surfaces of the bone at which resorption is taking place (Baron *et al.*, 1985). Similarly, matrix deposition requires active cellular participation to produce a matrix which is of the proper size, shape, place, and orientation. The shape of the collagen fibril is primarily determined by self-assembly forces which operate within the fibril formation compartments; the size or diameter of the fibril is influenced by a number of factors discussed below, including mixing of various products into heteropolymers during synthesis, packaging, transport, and excretion; and finally, the place and orientation of the fibril segment is dependent on its release by the cell in a particular orientation and by its subsequent repositioning. The compartments formed by the cell serve these purposes and others and further characterization of the nonhomogeneous nature of the extracellular space will be essential to continue to understand completely the multistep processes that operate in morphogenesis.

---

tendon (a) and dermis (b, c). Within these extracellular compartments the assembly of the matrix from fibrils to fibril bundles to collagen macroaggregates occurs. A section cut perpendicular to the tendon axis in a 14-day chick embryo (a) shows the fibril-forming channels (open arrowhead); the bundle-forming spaces (B); the macroaggregate-forming spaces (curved arrows); and bundles coalescing into macroaggregates. Analogous compartments are seen in 15-day chick embryo dermis (b, c). Secretory vacuoles containing procollagen are aligned in tandem in a region containing an extensive Golgi complex in b. During fibrillogenesis, these vacuoles fuse with the cell surface and one another giving rise to the long narrow channels, where fibril assembly occurs (arrowheads). In c, cross sections of fibril formation channels are illustrated (arrowheads). In the dermis, the fibril bundles (B) are closely associated with the cell surface, being enveloped by a single fibroblast. The continuous fusion of extracytoplasmic channels (arrowheads) with the cell surface and the retraction of the cell with membrane recycling, presumably is responsible for the fibril bundles of increasing size which then become delimited by adjacent fibroblasts. Sections 200–250 nm thick. N, nucleus; G, Golgi complex; RER, rough endoplasmic reticulum; SV, secretory vesicles. Bars = 1 μm (a), 500 nm (b, c). (b and c modified from Ploetz *et al.*, 1991.)

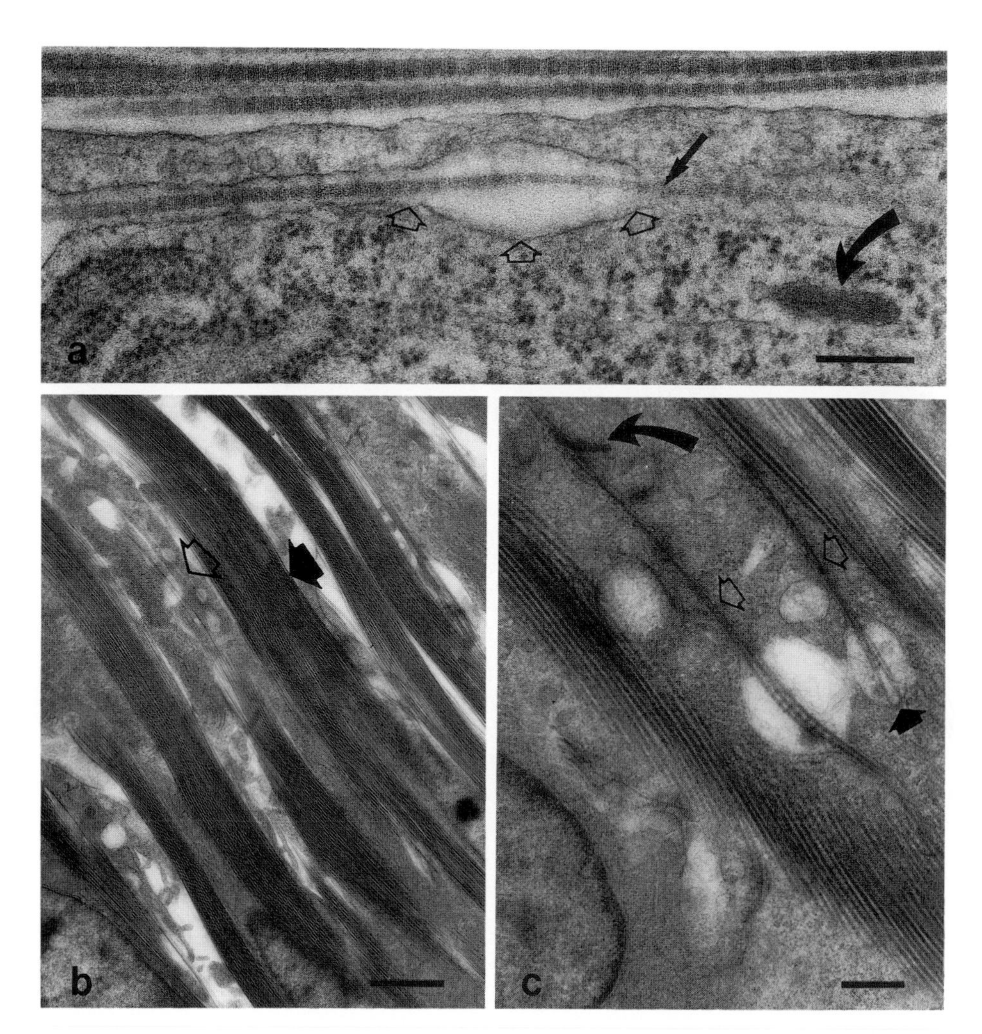

Figure 7-8. Transmission electron micrographs of sections cut parallel to the axes of 14-day chick embryo tendons illustrating extracellular compartments. (a) A portion of a fibroblast with a narrow cytoplasmic channel containing a focal dilatation and a collagen fibril (solid arrow). Secretory vacuoles containing collagen aggregates (curved arrow) are often observed associated with these dilatations. Bar = 300 nm. (b) High-voltage electron micrograph showing portions of four adjacent fibroblasts. The cells are intimately associated with the matrix and channels containing fibrils are present. At the open arrowhead, one such channel is associated with numerous secretory vacuoles. Larger bundles of fibrils are also present in close association with the cell surface (closed arrowhead). Bar = 2 μm. (c) High-voltage electron micrograph illustrating several narrow, deep cytoplasmic channels (open arrowheads). Vacuoles are seen associated with, and sometimes continuous with these channels. The solid arrowhead indicates a less-ordered region containing a hairpin loop in the collagen fibril. These less-ordered regions are commonly observed and may be sites of fibril growth. The curved arrow indicates a long, dense secretory vacuole containing electron-dense material. This vacuole length is sufficient to contain 0-D staggered aggregates. Bar = 500 nm. (Modified from Birk and Trelstad, 1986.)

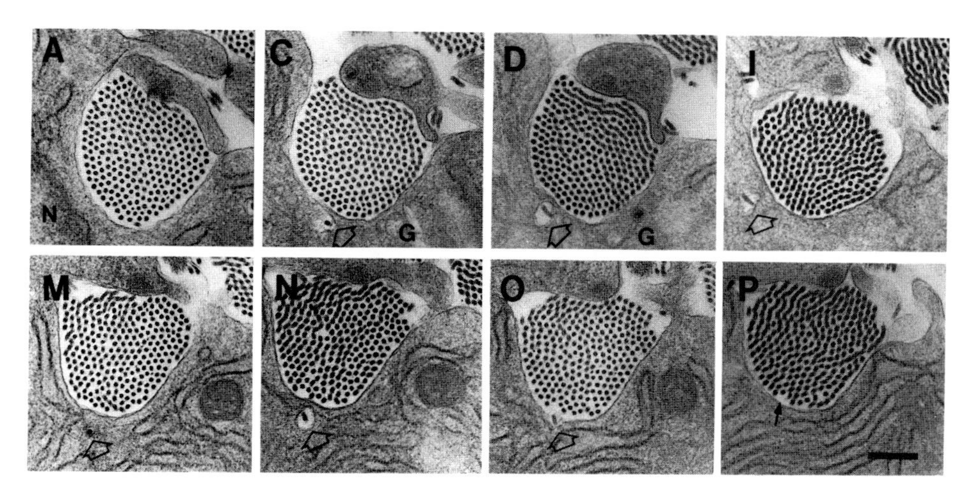

Figure 7-9. Fibril segments are assembled within fibril-forming channels. These high-voltage electron micrographs are from a serial set of thick sections (0.5 μm) cut perpendicular to the axis of 14-day chick embryo tendons (Birk and Trelstad, 1986). In this selected set from the series (in which the original set was labeled from A to P), a collagen fibril segment, within a narrow channel (open arrowhead), is followed from its initial appearance (B) in a perinuclear region surrounded by elements of the Golgi to its opening to the extracellular space (0). At this point the fibril is seen to terminate as it joins a fibril bundle (small arrow). This fibril segment is approximately 8 μm long. N, nucleus; G, Golgi complex. Bar = 500 nm.

4. Morphogenesis via Postdepositional Fusions, Rearrangements, and Remodeling

The matrix which is deposited in the embryo is not the same matrix which is present in the adult. Major revisions occur during maturation in the size of an organism, but often not in its shape, which require new deposition, fusions, rearrangements, and remodeling. That new materials are required is understood; that remodeling, usually meaning degradation of existing structures followed by new deposition, is required has also been understood. What has not been given much attention are fusions and/or rearrangements of existing structures.

4.1. Collagen Fibril Segments: Fusions and Rearrangements

Fusions and rearrangements of macroaggregates are newly recognized processes by which matrix architecture is altered following deposition. By fusion is meant the coalescence of macromolecular aggregates into larger, ordered macromolecular aggregates in a regulated and specific manner. By rearrangement is meant the physical repositioning of macroscopic aggregates.

The evidence for physical rearrangement of the microenvironment by cells

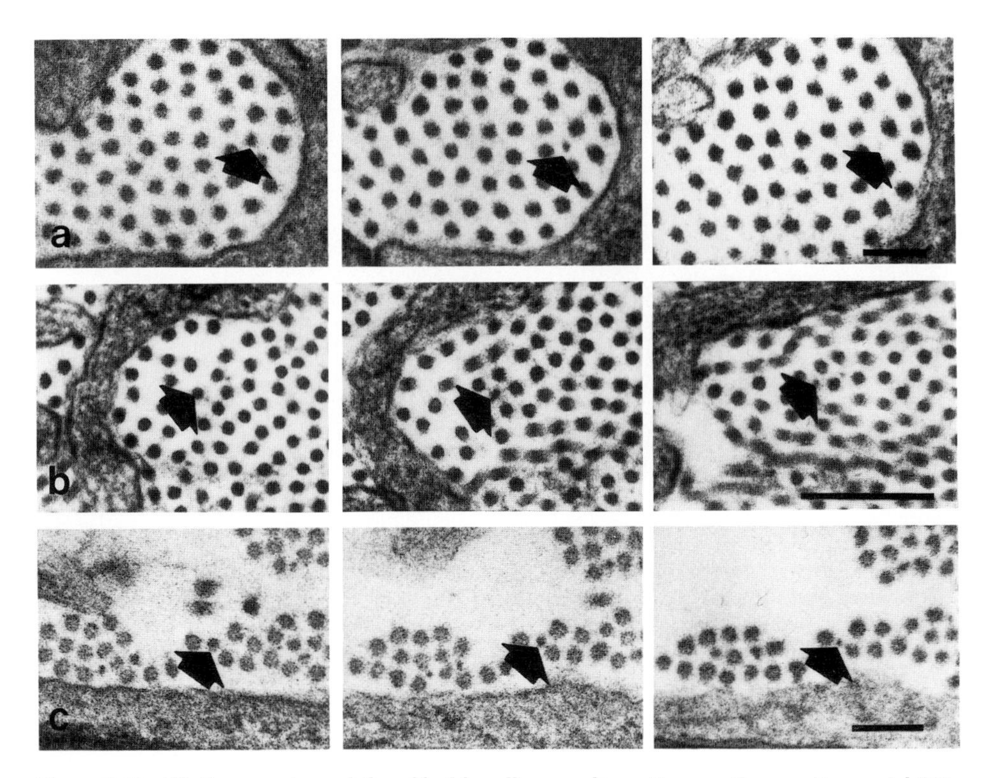

Figure 7-10. Fibril segments, and thus fibril bundles, are discontinuous. Consecutive, serial 200-nm-thick sections were cut perpendicular to collagen bundles in the chick embryo (a) 14-day tendon; (b) the 14-day corneal stroma; and (c) the 15-day dermis. The serial transmission electron micrographs demonstrate the ends of collagen fibrils within fibril bundles (arrows). Not all plates are presented as consecutive sections, but in no case is more than 0.4 μm missing between the sections presented. Bars = 300 nm.

Figure 7-11. Collagen fibril segments: three-dimensional computer-assisted reconstruction. Representative digitized electron micrographs from a 14-day chick embryo tendon showing a portion of a fibril bundle (box). The fibril profiles within the selected area were contoured for 55 consecutive 150-nm-thick sections. Fibril profiles were numbered and followed throughout this 8.5-μm data set. The ten fibril profiles which remained within this box were selected for image reconstruction. Three of the ten fibrils have at least one end within the 8.5 μm of serial data (#1, 6, 9). The fibril segment #1 has both ends present and is approximately 8.0 μm long. The other nine fibrils were followed over 12 μm and about 70% of the fibril ends were identified. This suggests that most, if not all, of the collagen fibrils are present as fibril segments at this stage of matrix development. These contours were transferred into a subroutine of MOVIE.BYU for the next steps in reconstruction. For presentation, not all of the fibrils seen in this group in the micrographs are present in this reconstruction and therefore this series of fibrils appears more open than is the case in the actual bundle. In this stero pair, the front fibrils have been removed. It is evident that the fibrils are discontinuous within the fibril bundles of a 14-day chick embryo tendon. The beginning and end of a fibril segment (red) are shown in relationship to other fibrils. Also shown is the termination of one neighboring fibril (cyan) and the beginning of another adjacent fibril (green). The fibrils shift in the bundle in relation to one another such that nearest neighbors constantly are changing and a woven composite structure results. This reconstruction is presented with the Z scale at 10% of actual and the arrows indicate the level of the two sections presented. (Modified from Birk *et al.*, 1989b.)

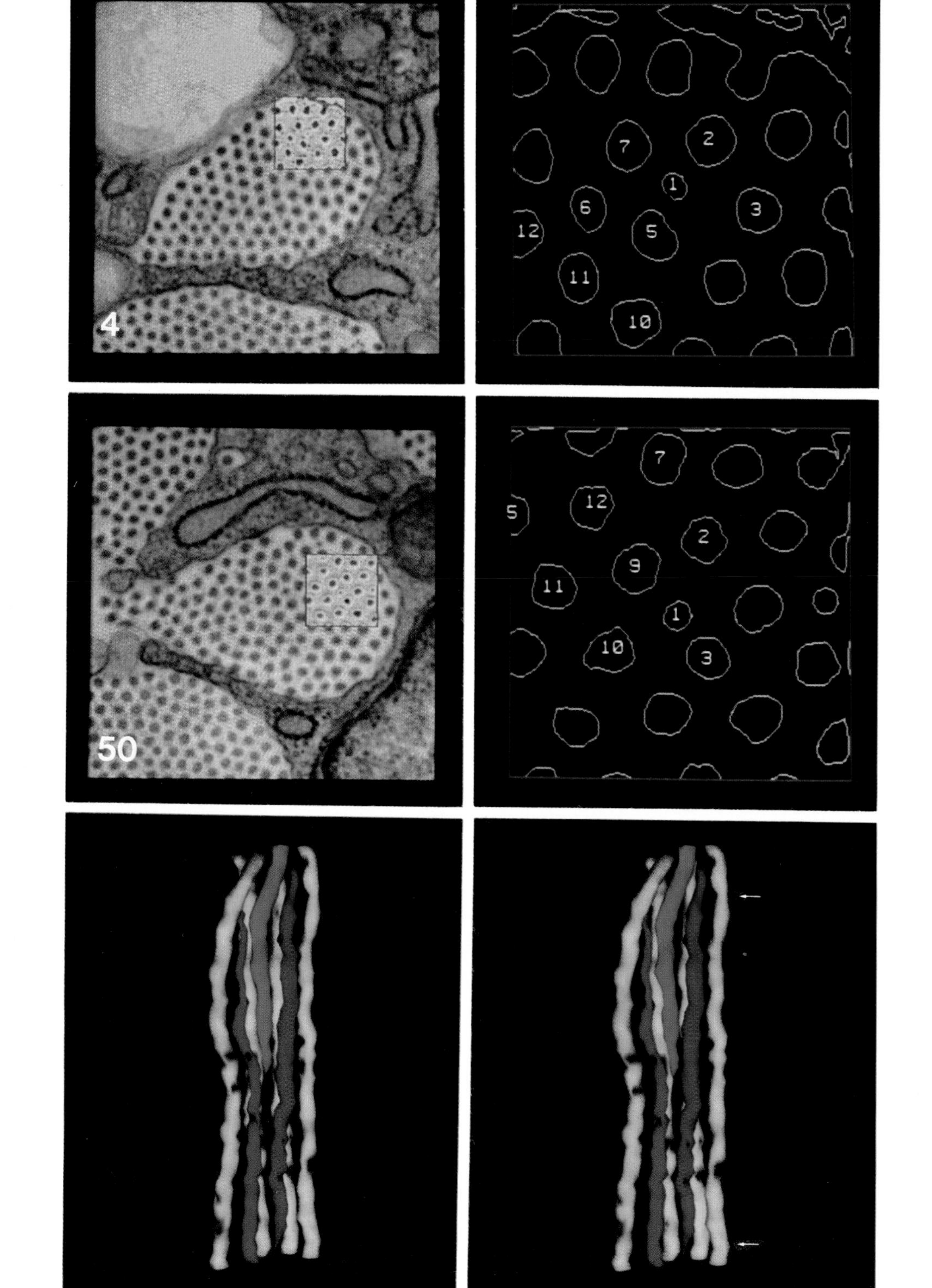
4
7
2
1
6
3
12
5
11
10
50
7
12
5
2
9
11
1
10
3

is extensive and has a relatively long history. Weiss (1934) showed that fibroblasts can align a plasma clot; Grillo and Gross (1967) studied the contraction of cutaneous wounds in guinea pigs; Stopak *et al.* (1985) and Harris *et al.* (1981) showed that collagen injected into tissues or placed in culture can be aligned by fibroblasts; and Bell *et al.* (1979) and others (Guidry and Grinnell, 1987) showed that fibroblast-populated collagen lattices undergo a significant decrease in size owing to the presence of the fibroblasts.

Only recently has evidence been obtained that macromolecular collagen aggregates undergo fusion. The identification of fibril segments *in situ* (Fig. 7-11) (Birk *et al.*, 1989b, 1990b) and their successful isolation (Fig. 7-12) offer

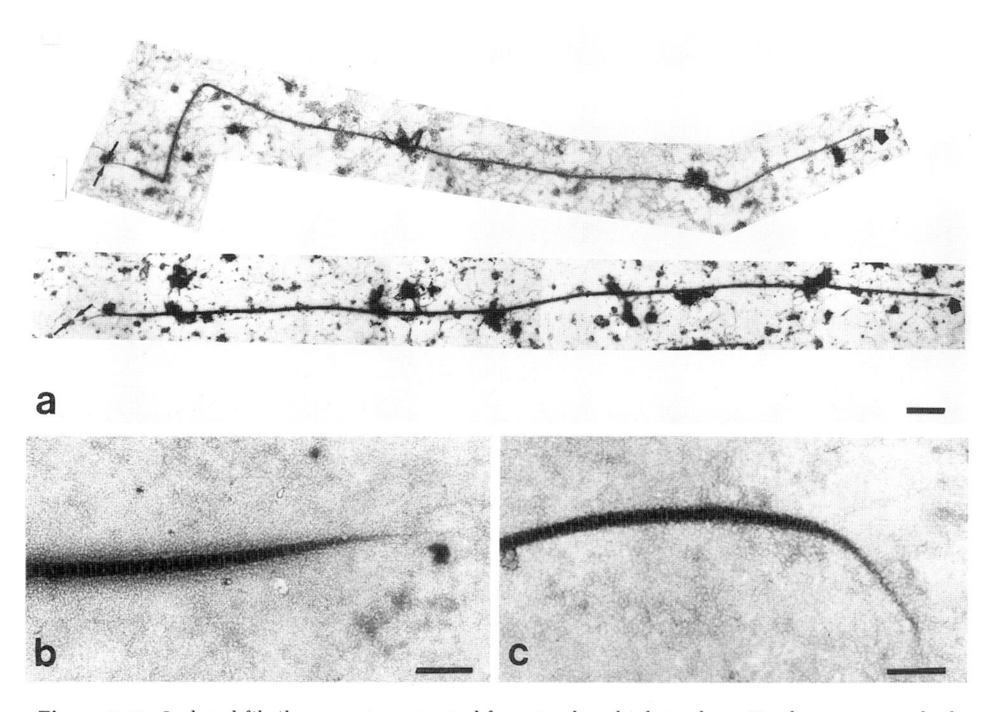

Figure 7-12. Isolated fibril segments, extracted from 14-day chick tendons. Tendons were washed, swollen, and homogenized at 4°C in PBS pH 7.3 with protease inhibitors. This procedure almost completely disrupted the 14-day tendons. The suspension was negatively stained and observed by transmission electron microscopy. Intact structures of discrete lengths were observed and two examples are presented in a. We have measured 25 such structures from the 14- to 15-day chick embryo tendon and found a mean length of about 30 μm (range 12–82 μm), most (16) in the 17 to 35 μm range. Only structures with clearly identified, intact ends were measured. The segments shown in a are 23–24 μm. The lengths of extracted segments are longer than those measured from serial reconstructions, a difference which might be accounted for by: (1) the reconstructions done to date are from limited data sets which may select for shorter segments; (2) the reconstructions have been done from small bundles of fibrils which may be newly deposited and less mature; (3) we have used a mixture of 14- to 15-day tendons for the extractions which may be more mature. The ends of extracted fibril segments were asymmetric as described *in situ*, with short (b) and long (c) tapers.

new avenues to pursue in this area of macroscopic matrix assembly. That structures very similar to fibril segments can form *in vitro* adds to the opportunities to explore these new structures (Kadler *et al.*, 1987, 1990).

The collagen fibril segments which form *in situ* in the developing 14-day chick tendon are approximately 10 μm in length and 40 nm in diameter. The segments form within one of the extracellular compartments described above and range in length from 5 to 20 μm in the chick tendon. The fibril segments which form in these compartments are relatively blunt at one end, tapering over 2 μm, and relatively sharp at the other, tapering over 5 μm. These asymmetric fibril segments identified by serial section reconstructions (Fig. 7-11) (Birk *et al.*, 1989b) and extracted from developing tissues (Fig. 7-12) are consistent with the segmentlike structures formed *in vitro* (Haworth and Chapman, 1977; Holmes and Chapman, 1979; Kadler *et al.*, 1987, 1990).

Individual fibril segments from 14-day chick embryo tendon (Fig. 7-12), dermis and cornea have been isolated by gentle homogenization with physiologic buffers. The isolated segments from the tendons range in length from 12 to 82 μm, with the majority in the 17 to 35 μm range (mean 30 μm). The segment ends are asymmetric as described *in situ*, with a short and long taper (Fig. 7-12). Only a few segments from 17- to 18-day chick embryo tendons have been isolated, a finding consistent with the proposal that there is lateral and/or linear fusion of fibril segments to form thicker, more continuous fibrils and with the observations that there is a very rapid increase in tendon fibril diameter from 40 nm at day 14 to 60 nm at day 19 (Birk *et al.*, 1990). Scott (1986) has already proposed that a fusion of fibrils occurs, with possible regulation via a surface coat of dermatan sulfate proteoglycan. The data available do not directly address the mechanism(s) involved in the fusion, but would be consistent with the idea that the process occurs with some varying degree of overlap between adjacent segments.

Before leaving a consideration of the fusion of macroaggregates, it should be noted that similar phenomena have been reported for actin and tubulin, two filamentous intracytoplasmic macroaggregates which also may assemble, in part, by a nucleation and growth mechanism (Caplow *et al.*, 1986; Rothwell *et al.*, 1986; Murphy *et al.*, 1988; Williams and Rone, 1989).

4.2. Collagen Fibril Diameter Distributions

The diameter of a collagen fibril is an important parameter for its function. Since the lengths of fibrils in tissues are generally much beyond a critical length (*vide infra*), the major variable that determines the strength of a tendon is based on fibril diameter and uniaxial alignment. In the cornea, transparency is determined in part by the diameter and spacing of the constituent collagen fibrils. The factors contributing to the regulation of collagen fibril diameter are several and include: (1) primary structural or sequence determinants in the collagen molecule; (2) molar ratios and spatial distributions of molecular types of collagens within a fibril; (3) molar ratios and spatial distributions of noncollagenous macromolecules associated with a fibril; (4) rates and extents of processing of amino- and carboxy-terminal non-triple-helical sequences; (5) assembly conditions.

Detailed measurements of collagen fibril diameters in a number of tissues have shown that fibril diameter distributions are not continuous and this finding has led to a number of suggestions pertaining to the regulation of fibril growth (Parry and Craig, 1988). Parry and Craig (1979) proposed that a subunit of a diameter initially estimated at 8 nm, later revised to approximately 12 nm, is an integral element that adds to existing fibrils to increase their diameter by an integral factor of the subfibril. Another explanation for limitation of fibril diameter is that fibrils grow by accretion of monomeric procollagen onto their surfaces to a particular limit dictated by the species of collagen and the extent of procollagen processing (Hulmes, 1983; Fleischmajer *et al.*, 1990). A variant on this model has been proposed by Chapman (1989), who posits the existence of a growth inhibitor associated with the collagen monomer which is restricted to the surface of the growing fibril whose growth ceases when the inhibitors cover the fibril surface. Enzymatic or cell-mediated removal of the inhibitors would then allow lateral growth to proceed again and cycles of this process would effect uniform and synchronous growth over large populations of fibrils. This particular model requires axial order (D-periodicity) in the fibrils and fluidity in lateral packing prior to cross-linking. Another suggestion is that the heteropolymeric character of the fibril regulates diameters by unique interactions between molecules of similar primary structure and/or among molecules of differing primary structure. Molecular mixing of collagen types I and V within a single fibril has considerable consequences for the diameter (Fig. 7-4) (Adachi and Hayashi, 1986; Birk *et al.*, 1990a). And finally, yet another mechanism for fibril diameter regulation is that fibrils fuse laterally and linearly, like liquid crystals, to form thicker, longer structures (Fig. 7-6) (Birk *et al.*, 1989a, 1990b). This model, like that proposed by Chapman (1989), requires that intermediate aggregates of collagen, such as fibril segments, remain D-periodic, but in a liquid-crystalline state such that the molecules are free to undergo rearrangements.

Each of these hypothetical mechanisms for the regulation of fibril growth is likely to contain elements that are operative *in vivo*. It seems likely that more than one mechanism will operate in a particular tissue, depending on the chemistry of the constituents, the nature of the biomechanical environment, and the explicit function that has evolved at that site. For example, the cornea is simultaneously continuous with the skin, part of the wall of the eye, and transparent; the sclera shares two of these attributes, but not the third. Accordingly, the regulation of the corneal collagen fibril diameter is likely to be a very local process, which nonetheless reflects general mechanisms used by the organism, but to differing degrees, elsewhere.

4.3. Collagen Fibril Ends and Lengths

The identification of fibril segments as distinct entities within the forming tendon, skin, and cornea also indicates that fibrils have distinct ends and lengths within tissues (Figs. 7-10 and 7-11). The seeming endlessness of collagen fibrils has been questioned by a number of investigators, particularly Nemetschek and colleagues in their studies of the biomechanical properties of tissues (Mosler *et al.*, 1985; Folkhard *et al.*, 1987). These investigators proposed that the matrix was comprised of discontinuous elements based on combined stress and ultrastructural studies. From estimates of the mean collagen fibril length in rat tail tendon, Craig and colleagues (1989) have concluded that the fibril length is at least equal to the critical length required to maintain the appropriate tensile properties and in mature tissues, in excess of 100 times the critical length. McBride *et al.* (1988) have calculated that the critical length of the collagen fibril in chick tendon is between 30 and 100 μm. The three-dimensional reconstructions of Birk *et al.* (1989b) have conclusively demonstrated the existence of discontinuous segments in the forming tendon.

5. Macroscopic Properties of Type I Collagen-Rich Tissues

Biomechanical studies of type I collagen fibrils, formed either *in vitro* or *in vivo*, indicate that fibrils behave under mechanical stress as viscoelastic elements that are cross-linked into continuous networks and that noncovalently associated macromolecules are not major determinants of tissue strength.

5.1. Basic Biomechanics

Extracellular matrix acts to transmit force, maintain shape under gravitational forces, and absorb applied mechanical energy (Wainwright *et al.*, 1976). To accomplish these tasks, macromolecules, primarily the collagens, are assembled into continuous networks that have appropriate geometrics. The term *fiber* is used to describe the macroscopic, fibrous elements of these networks; fibers are comprised of bundles of collagen fibrils. Tissues that transmit forces such as the tendon contain dense, aligned collagen fibers in fasciculated structures that deform minimally under physiological loads. In contrast, tissues that act to absorb impact forces such as the skin are formed into structures that contain wavy collagen fibers that are able to align during loading. In both of these cases, networks of cross-linked collagen fibers bear the loads and prevent premature tissue failure. In most soft tissues, removal of proteoglycans and elastin leaves an intact collagen network which prevents failure. While aggregation of proteoglycans restricts the movement of proteoglycan monomers and affects local concentrations within tissues, it does not significantly increase the shear stiffness of the macroaggregate. Owing to the large difference in stiffness of proteoglycans and collagen fibers, the proteoglycans do not mediate stress transfer between neighboring fibers in tissues; however, they probably are important in fibrillar deformation within a single fiber (Hardingham *et al.*, 1987).

One way to analyze the response of extracellular matrix to loads is to study the load deformation or the stress–strain behavior. A specimen of known length is stretched at a constant rate and the force of resistance generated by the specimen is measured. The force, measured in kilograms, is usually divided by the original cross-sectional area of the specimen, measured in square centimeters, to yield a normalized quantity termed the stress, which has units of kg/cm^2 (Fig. 7-13). The strain is merely the change in specimen length divided by the original length. The slope of a plot of stress versus strain is referred to as the modulus and reflects the resistance offered by a material to deformation. A typical stress–strain or load-deformation curve is shown in Fig. 7-13.

The stress–strain curve of tendon is characterized by a nonlinear toe region, a linear region, and a nonlinear yield and failure region. The toe region is believed to reflect the uncrimping of collagen fibers which are straightened during the linear region and begin to fail by fibril defibrillation during the yield and failure region (McBride *et al.*, 1988). The stress–strain curve for wet tendons tested in uniaxial tension is almost linear after an initial nonlinear region and is characterized by ultimate tensile strength and strain of as high as 1200

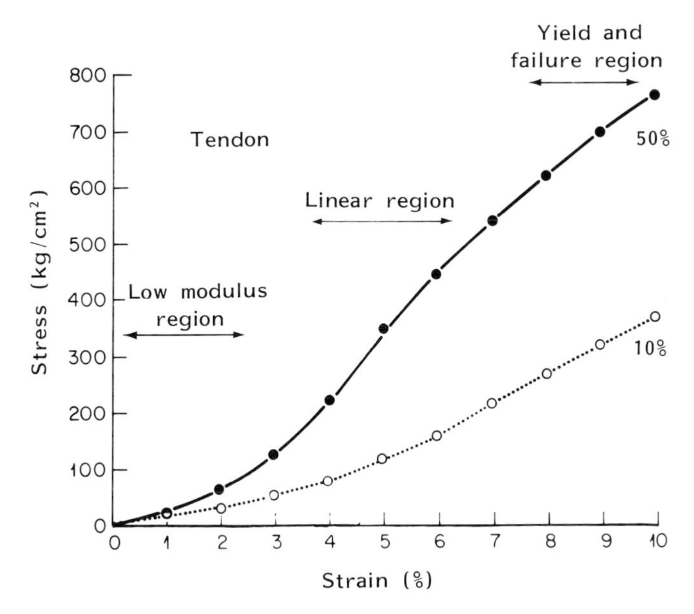

Figure 7-13. Stress–strain curve for tendon. Stress–strain curves for wet rat tail tendon in tension at strain rates of 10% and 50% per min. The strain rate dependence is a result of viscoelasticity which is prominent in connective tissues. The early part of each curve, or toe region, represents straightening of the crimp pattern in collagen fibers while the fibers are stretched during the linear region. Fibril disintegration occurs via a shear-slip mechanism in the yield and failure region. The low-modulus region extends to strains as high as 50% in skin and aorta due to the straightening of wavy collagen fibers that become aligned by the applied stress. The stress of failure reflects the number of collagen fibers that are aligned along the direction or applied stress before failure. Load-deformation curves have similar shapes and reflect the same physical phenomena. Stress is equal to load/cross-sectional area. (From Trelstad and Silver, 1981.)

kg/cm² and 10% and 1470 kg/cm² and 70% for tendon and ligament, respectively.

The low-modulus region varies from a strain of less than 1% for bone to a strain in excess of 50% for skin and cardiovascular tissue and is correlated with the density of aligned fibers and the weave of the collagen network. Dense aligned collagen fibers in tendon deform only a few percent while wavy nonwoven networks in skin and cardiovascular tissue require much larger deformations before they become taut. Alignment of crimped (tendon) and wavy (skin, cardiovascular tissues) collagen fibers requires very low force levels (Silver and Doillon, 1989; Kato et al., 1991).

Once collagen fibers are directly brought under tension, they offer a high resistance to further deformation as reflected in the linear region of the stress–strain curve. During this phase of deformation, stretching of the triple-helix and interfibrillar slippage are the mechanisms by which the applied loads are borne. Mechanical properties of individual collagen fibers 50 to 100 μm in diameter have been measured in uniaxial tension and ultimate tensile stresses as high as 800 kg/cm², strains up to 10% and moduli in excess of 5000 kg/cm² have been found (Kato *et al.*, 1989). These parameters reflect the stiffness of the collagen triple-helix which has been reported to be approximately 40,000 kg/cm² (Nestler *et al.*, 1983). Once collagen fibers are maximally stretched, mechanical failure in the terminal nonlinear region occurs by defibrillation.

In order for dense aligned and nonwoven wavy collagen networks to effectively bear stress, they must contain intra- and intermolecular cross-links. This allows applied loads to be evenly distributed throughout the collagen network. Inhibition of collagen cross-linking results in tissues that are unable to support even moderate mechanical loads without premature mechanical failure.

5.2. Viscoelastic Properties

The mechanical properties of different tissues described above, while reflecting inherent differences in the composition and geometries of the collagen fibers, are also dependent on their viscoelastic behavior. The viscoelastic behavior of extracellular matrix is reflected in the time dependence of the modulus and the strain rate dependence of the stress–strain curve and can lead to premature failure of a tissue. Viscoelastic behavior is of primary importance for tissues such as tendons, ligaments, and aorta that experience millions of repetitive cycles during a normal lifetime. A single cycle consists of application of a load which is followed by rapid unloading. Although some tissues continuously deform, they ultimately reach a set of equilibrium dimensions. Complicating the viscoelastic behavior of tissues is the cellular feedback that occurs resulting in synthesis of additional connective tissue components in high-stress areas. These factors are responsible for the increased wall thickness and diameter seen in the aorta of older individuals which may lead to dilatation and aneurysm formation (Silver *et al.*, 1991).

5.3. Reconstituted Matrices

Studies of the assembly and properties of reconstituted collagen fibers underscore the complexities of geometrical patterns found in nature and the remarkable variabilities of natural mechanical properties. After cross-linking with exogenous chemicals, reconstituted fibrils assembled from type I collagen under optimal *in vitro* conditions have ultimate tensile strengths and moduli quite similar to those of native fibers. Structures formed from several hundred aligned type I collagen fibers in a matrix that lacks other components have a stress–strain curve that is similar in shape to that observed for tendon from animals. Therefore, in the case of aligned collagen networks, the most important design parameter required for mechanical stability is cross-linking and fibrillar alignment. In the absence of cross-links the mechanical behavior of reconstituted collagen fibers is dramatically different from the properties of native collagen. However, while native tendon contains only about four stress-bearing cross-links per molecule, these reconstituted collagen networks with equivalent mechanical stability require several hundred cross-links per molecule. This indicates that the proper placement of intra- and intermolecular cross-links is critical to formation of networks that prevent premature mechanical failure and further indicates that during morphogenesis the collagen fibril is effectively assembled in a manner not achieved *in vitro*. The differences presumably derive from many of the characteristics discussed in this chapter, including the heteropolymeric nature of the fibrils, the multistep process of assembly, the possible enzymatic processing of intermediate forms, and the newly recognized fusion of segments. In addition, the close relationship that exists between normal collagen processing, assembly, and cross-linking and the primary structure of the type I collagen molecule illustrate that a single genetic or acquired mutation may lead to profound changes in mechanical properties (Kato and Silver, 1990; Kato *et al.*, 1991; Silver *et al.*, 1991).

6. Morphogenesis: A Synthesis

Full understanding of morphogenesis will require knowledge of components and their interactions at different levels in the overall architecture of the fully formed object. And it will require knowledge of the transient spatiotemporal presence of regulators not found in the final product. In short, we will need to know the architect, builder, and materials.

For type I collagen-rich tissues, we are close to knowing the materials. A chemical description of all the components of your hand is likely within the decade. The physicochemical attributes of many of these components will be further clarified by a variety of present and future techniques. These studies will clearly show that interactions among the components lead to multimeric structures which have unique shapes and properties. The extent to which molecular attributes can affect multimolecular assemblages will be thoroughly explored. Self-assembly will be better understood for its power and its limitations.

The identification of tissue architects has only begun. We have much to learn about the manner by which transient influences operate during matrix deposition, whether by large molecules or small, whether by forces operating at large or small distance. These factors, recognized by earlier developmental biologists as "fields," "gradients," "inductors," are now being explored and shown to be chemicals such as retinoids, regulators such as homeotic genes, and physical forces which range from ion currents to gravity (Gilbert, 1988).

The builder of the matrix has undergone extensive dissection and inspection during the past 30 years. Our understanding of cells, and particularly the fibroblast, now allows for consideration of epicellular as well as subcellular phenomena. In these considerations, we should uncover new patterns and attributes of groups of cells which are unknown and unexpected. The coordinate behavior of groups of cells will lead us to new understanding of the "social" character of tissue structure and function.

Progress in all three of these areas has impacted on our understanding of matrix assembly. In the matter of materials, there has been the identification of the components of the matrix and the establishment of self-assembly as a necessary and/or sufficient condition for the D-stagger of the type I molecules and a constraint on fibril diameters. The recognition that heteropolymeric interactions are operative in nearly all fibrils only serves to enhance the importance of intermolecular interactions. In the matter of architects, there has been attention to the site and timing of procollagen processing, the possible episodic degradation of the matrix by proteinases, the interesting effects of retinoids on matrix stability, and the likelihood that gradients of diffusible substances are present in and/or bound to the matrix. And in the matter of the builder, there has been the characterization of intra- and extracellular compartments in the progression of the collagens from a newly synthesized to a newly polymerized structure.

As progress is made in uncovering the basic reactants and principles of the assembly process, it is likely that descriptors of the forms generated will also improve. At present the language of the biologist is limited when attempting formal descriptions of size and shape. Furthermore, formal relationships between objects of different sizes and shapes are usually stated in informal language. There is no formal language of morphogenesis. Attempts to provide a more formal language have been made by a variety of authors including Thompson (1969), Thom (1975), Stevens (1974), Prigogine (1980), Meinhardt (1982), Mandelbrot (1983), Malacinski and Bryant (1984), and Lord and Wilson (1984).

A characteristic of developing systems which reflects features of other systems in which there is the emergence of order is that they are iterative (May, 1987). That is, the output from the last step serves as the input for the next. Such iterative processes generate forms which are self-similar at successive levels of organization; such structures are called fractals (Mandelbrot, 1983). The branching-upon-branching character of the collagen in the tendon (Birk *et al.*, 1989b) raises the possibility that these collagen aggregates resemble fractal-like structures (Trelstad *et al.*, 1987; Barnsley *et al.*, 1987). That fact, coupled with the observations that purified collagens will assemble into branched structures under a variety of circumstances (Miyahara *et al.*, 1984; Hulmes *et al.*, 1989), raises the interesting possibility that the hierarchy of structure within some of the type I collagen-rich tissues can formally be considered fractal-like and be explored for rules of assembly which might operate within the rules of fractal generation.

ACKNOWLEDGMENTS. This work was supported by grants EY05129 and AR37003 from the National Institutes of Health. The authors wish to thank Joanne Babiarz, Rita Hahn, Emanuel Zycband, Kimiko Hayashi, and Daniel Casper for assistance.

References

Adachi, E., and Hayashi, T., 1986, In vitro formation of hybrid fibrils of type V collagen and type I collagen. Limited growth of type I collagen into thick fibrils by type V collagen, *Connect. Tissue Res.* **14:**257–266.

Amudeswari, S., Liang, J. N., and Chakrabarti, B., 1987, Polar–apolar characteristics and fibrillogenesis of glycosylated collagen, *Collagen Relat. Res.* **7:**215–223.

Baer, E., Cassidy, J. J., and Hiltner, A., 1988, Hierarchical structure of collagen and its relationship to the physical properties of tendon, in: *Collagen*, Vol. II (M. E. Nimni, eds.), pp. 177–200, CRC Press, Boca Raton.

Baker, H. N., Rothwell, S. W. Grasser, W. A., Wallis, K. T., and Murphy, D. B., 1990, Copolymerization of two distinct tubulin isotypes during microtubule assembly in vitro, *J. Cell Biol.* **110:**97–104.

Barnsley, M. F., Massopust, P., Strickland, H., and Sloan, A. D., 1987, Fractal modeling of biological structures, *Ann. N.Y. Acad. Sci.* **504:**179–194.

Baron, R., Neff, L., Louvard, D., and Courtoy, P. J., 1985, Cell-mediated extracellular acidification and bone resorption: Evidence for a low pH in resorbing lacunae and localization of a 100-kD lysosomal membrane protein at the osteoclast ruffled border, *J. Cell Biol.* **101:**2210–2222.

Bell, E., Ivarsson, B., and Merrill, C., 1979, Production of a tissue-like structure by contraction of collagen lattices by human fibroblasts of different proliferative potential in vitro, *Proc. Natl. Acad. Sci. USA* **76:**1274–1278.

Birk, D. E., and Lande, M. A., 1981, Corneal and scleral collagen fiber formation in vitro, *Biochim. Biophys. Acta* **670:**362–369.

Birk, D. E., and Silver, F. H., 1984, Collagen fibrillogenesis in vitro: Comparison of types I, II, and III, *Arch. Biochem. Biophys.* **235:**178–185.

Birk, D. E., and Trelstad, R. L., 1984, Extracellular compartments in matrix morphogenesis: Collagen fibril, bundle, and lamellar formation by corneal fibroblasts, *J. Cell Biol.* **99:**2024–2033.

Birk, D. E., and Trelstad, R. L., 1985, Fibroblasts create compartments in the extracellular space where collagen polymerizes into fibrils and fibrils associate into bundles, *Ann. N.Y. Acad. Sci.* **460:**258–266.

Birk, D. E., and Trelstad, R. L., 1986, Extracellular compartments in tendon morphogenesis: Collagen fibril, bundle, and macroaggregate formation, *J. Cell Biol.* 103:231–240.

Birk, D. E., Fitch, J. M., Babiarz, J. P., and Linsenmayer, T. F., 1988, Collagen type I and type V are present in the same fibril in the avian corneal stroma, *J. Cell Biol.* **106:**999–1008.

Birk, D. E., Southern, J. F., Zycband, E. I., Fallon, J. T., and Trelstad, R. L., 1989a, Collagen fibril bundles: A branching assembly unit in tendon morphogenesis, *Development* **107:**437–443.

Birk, D. E., Zycband, E. I., Winkelmann, D. A., and Trelstad, R. L., 1989b, Collagen fibrillogenesis in situ: Fibril segments are intermediates in matrix assembly, *Proc. Natl. Acad. Sci. USA* **86:**4549–4553.

Birk, D. E., Fitch, J. M., Babiarz, J. P., Doane, K. J., and Linsenmayer, T. F., 1990a, Collagen fibrillogenesis in vitro: Interaction of types I and V collagen regulates fibril diameter, *J. Cell Sci.* **95:**649–657.

Birk, D. E., Zycband, E. I., Winkelmann, D. A., and Trelstad, R. L., 1990b, Collagen fibrillogenesis in situ. Discontinuous segmental assembly in extracellular compartments, *Ann. N.Y. Acad. Sci.* **580:**176–194.

Brodsky, B., Eikenberry, E. F., Belbruno, K. C., and Sterling, K., 1982, Variations in collagen fibril structure in tendons, *Biopolymers* **21:**935–951.

Brown, D. C., and Vogel, K. G., 1990, Characteristics of the in vitro interaction of a small proteoglycan (PG II) of bovine tendon with type I collagen, *Matrix* **9:**468–478.

Bruckner, P., and Eikenberry, E. F., 1984, Procollagen is more stable in cellulo than in vitro, *Eur. J. Biochem.* **140:**397–399.

Bruckner, P., Vaughan, L., and Winterhalter, K. H., 1985, Type IX collagen from sternal cartilage of chicken embryo contains covalently bound glycosaminoglycans, *Proc. Natl. Acad. Sci. USA* **82:**2608–2612.

Bruckner, P., Mendler, M., Steinmann, B., Huber, S., and Winterhalter, K. H., 1988, The structure of human collagen type IX and its organization in fetal and infant cartilage fibrils, *J. Biol Chem.* **263:**16911–16917.

Bruns, R. R., Hulmes, D. J., Therrien, S. F., and Gross, J., 1979, Procollagen segment-long-spacing crystallites: Their role in collagen fibrillogenesis, *Proc. Natl. Acad. Sci. USA* **76:**313–317.

Burgeson, R. E., 1988, New collagens, new concepts, *Annu. Rev. Cell Biol.* **4:**551–577.

Capaldi, M. J., and Chapman, J. A., 1982, The C-terminal extrahelical peptide of type I collagen and its role in fibrillogenesis in vitro, *Biopolymers* **21:**2291–2313.

Caplow, M., Shanks, J., and Brylawski, B. P., 1986, Differentiation between dynamic instability and end-to-end annealing models for length changes of steady-state microtubules, *J. Biol. Chem.* **261:**16233–16240.

Chandrasekhar, S., Kleinman, H. K., Hassell, J. R., Martin, G. R., Termine, J. D., and Trelstad, R. L., 1984, Regulation of type I collagen fibril assembly by link protein and proteoglycans, *Collagen Relat. Res.* **4:**323–337.

Chapman, J. A., 1989, The regulation of size and form in the assembly of collagen fibrils in vivo, *Biopolymers* **28:**1367–1382.

Craig, A. S., Britles, M. J., Conway, J. F., and Parry, D. A. D., 1989, An estimate of the mean length of collagen fibrils in rat tail-tendon as a function of age, *Connect. Tissue Res.* **19:**51–62.

Diamant, J., Keller, A., Baer, E., Lith, N., and Arridge, R. G. C., 1972, Collagen; ultrastructure and its relationship to mechanical properties as a function of aging, *Proc. Roy. Soc. Lond.* **B180:**293–315.

Eikenberry, E. F., Brodsky, B., and Parry, D. A. D., 1982, Collagen fibril morphology in developing chick metatarsal tendons: 1. X-ray diffraction studies, *Int. J. Biol. Macromol.* **4:**322–328.

Fitch, J. M., Gross, J., Mayne, R., Johnson Wint, B., and Linsenmayer, T. F., 1984, Organization of collagen types I and V in the embryonic chicken cornea: Monoclonal antibody studies, *Proc. Natl. Acad. Sci. USA* **81:**2791–2795.

Fitch, J. M., Birk, D. E., Mentzer, A., Hasty, K. A., Mainardi, C., and Linsenmayer, T. F., 1988, Corneal collagen fibrils: Dissection with specific collagenases and monoclonal antibodies, *Invest. Ophthalmol. Vis. Sci.* **29:**1125–1136.

Fleischmajer, R., Timpl, R., Tuderman, L., Raisher, L., Wiestner, M., Perlish, J. S., and Graves, P. N., 1981, Ultrastructural identification of extension aminopropeptides of type I and III collagens in human skin, *Proc. Natl. Acad. Sci. USA* **78:**7360–7364.

Fleischmajer, R., Perlish, J. S., Timpl, R., and Olsen, B. R., 1988, Procollagen intermediates during tendon fibrillogenesis, *J. Histochem. Cytochem.* **36:**1425–1432.

Fleischmajer, R., Perlish, J. S., Burgeson, R. E., Shaikh Bahai, F., and Timpl, R., 1990, Type I and type III collagen interactions during fibrillogenesis, *Ann. N.Y. Acad. Sci.* **580:**161–175.

Flint, M. H., Craig, A. S., Reilly, H. C., Gillard, G. C., and Parry, D. A., 1984, Collagen fibril diameters and glycosaminoglycan content of skins—Indices of tissue maturity and function, *Connect. Tissue Res.* **13:**69–81.

Folkhard, W., Mosler, E., Geerckern, W., Knorzer, E., Nemetschek-Gansler, H., Nemetschek, T., and Koch, M. H. J., 1987, Quantitative analysis of the molecular sliding mechanism in native tendon collagen: Time resolved dynamic studies using synchrotron radiation, *Int. J. Biol. Macromol.* **9**:169–175.

Fraser, R. D., MacRae, T. P., and Miller, A., 1987, Molecular packing in type I collagen fibrils, *J. Mol. Biol.* **193**:115–125.

Garg, A. K., Berg, R. A., Silver, F. H., and Garg, H. G., 1989, Effect of proteoglycans on type I collagen fibre formation, *Biomaterials* **10**:413–419.

Gilbert, S. F., 1988, *Developmental Biology*, Sinauer, Sunderland, Mass.

Glimcher, M. J., 1989, Mechanism of calcification: Role of collagen fibrils and collagen–phosphoprotein complexes in vitro and in vivo, *Anat. Rec.* **224**:139–153.

Grillo, H. C., and Gross, J., 1967, Collagenolytic activity during mammalian wound repair, *Dev. Biol.* **15**:300–317.

Gross, J., 1974, Collagen biology: Structure, degradation, and disease, *Harvey Lect.* **68**:351–432.

Gross, J., Highberger, J., and Schmitt, F. O., 1954, Collagen structures considered as states of aggregation of a kinetic unit. The tropocollagen particle, *Proc. Natl. Acad. Sci. USA* **40**:679–688.

Guidry, C., and Grinnell, F., 1987, Contraction of hydrated collagen gels by fibroblasts: Evidence for two mechanisms by which collagen fibrils are stabilized, *Collagen Relat. Res.* **6**:515–529.

Gullberg, D., Terracio, L., Borg, T. K., and Rubin, K., 1989, Identification of integrin-like matrix receptors with affinity for interstitial collagens, *J. Biol. Chem.* **264**:12686–12694.

Hardingham, T. E., Muir, H., Kwan, M. K., Lai, W. M., and Mow, V. C., 1987, Viscoelastic properties of proteoglycan solutions with varying proportions present as aggregates, *J. Orthopaedic Res.* **5**:36–46.

Harris, A. K., Stopak, D., and Wild, P., 1981, Fibroblast traction as a mechanism for collagen morphogenesis, *Nature* **290**:249–251.

Haworth, R. A., and Chapman, J. A., 1977, A study of the growth of normal and iodinated collagen fibrils in vitro using electron microscope autoradiography, *Biopolymers* **16**:1895–1906.

Hendrix, M. J., Hay, E. D., von der Mark, K., and Linsenmayer, T. F., 1982, Immunohistochemical localization of collagen types I and II in the developing chick cornea and tibia by electron microscopy, *Invest. Ophthalmol. Vis. Sci.* **22**:359–375.

Henkel, W., and Glanville, R. W., 1982, Covalent crosslinking between molecules of type I and type III collagen, *Eur. J. Biochem.* **122**:205–213.

Holmes, D. F., and Chapman, J. A., 1979, Axial mass distributions of collagen fibrils grown in vitro: Results for the end regions of early fibrils, *Biochem. Biophys. Res. Commun.* **87**:993–999.

Hulmes, D. J., 1983, A possible mechanism for the regulation of collagen fibril diameter in vivo, *Collagen Relat. Res.* **3**:317–321.

Hulmes, D. J., and Miller, A., 1979, Quasi-hexagonal molecular packing in collagen fibrils, *Nature* **282**:878–880.

Hulmes, D. J., Bruns, R. R., and Gross, J., 1983, On the state of aggregation of newly secreted procollagen, *Proc. Natl. Acad. Sci. USA* **80**:388–392.

Hulmes, D. J., Holmes, D. F., and Cummings, C., 1985, Crystalline regions in collagen fibrils, *J. Mol. Biol.* **184**:473–477.

Hulmes, D. J., Kadler, K. E., Mould, A. P., Hojima, Y., Holmes, D. F., Cummings, C., Chapman, J. A., and Prockop, D. J., 1989, Pleomorphism in type I collagen fibrils produced by persistence of the procollagen N-propeptide, *J. Mol. Biol.* **210**:337–345.

Kadler, K. E., Hojima, Y., and Prockop, D. J., 1987, Assembly of collagen fibrils de novo by cleavage of the type I pC-collagen with procollagen C-proteinase. Assay of critical concentration demonstrates that collagen self-assembly is a classical example of an entropy-driven process, *J. Biol. Chem.* **262**:15696–15701.

Kadler, K. E., Hojima, Y., and Prockop, D. J., 1990, Collagen fibrils in vitro grow from pointed tips in the C- to N-terminal direction, *Biochem. J.* **268**:339–343.

Kastelic, J., Galeski, A., and Baer, E., 1978, The multicomposite structure of tendon, *Connect. Tissue Res.* **6**:11–23.

Kato, Y. P., and Silver, F. H., 1990, Formation of continuous collagen fibres: Evaluation of biocompatibility and mechanical properties, *Biomaterials* **11**:169–175.

Kato, Y. P., Christiansen, D. L., Hahn, R. A., Shieh, S.-J., Goldstein, J. D., and Silver, F. H., 1989, Mechanical properties of collagen fibers: A comparison of reconstituted and rat tail tendon fibers, *Biomaterials* **10:**38–42.

Kato, Y. P., Dunn, M. G., Zawadsky, J. P., Tria, A. J., and Silver, F. H., 1991, Regeneration of achilles tendon using a collagen prosthesis: Results of a year long implantation study, *J. Bone J. Surg.* **73**(A): 561–574.

Keene, D. R., Sakai, L. Y., Bachinger, H. P., and Burgeson, R. E., 1987, Type III collagen can be present on banded collagen fibrils regardless of fibril diameter, *J. Cell Biol.* **105:**2393–2402.

Lapiere, C. M., Nusgens, B., and Pierard, G. E., 1977, Interaction between collagen type I and type III in conditioning bundles organization, *Connect. Tissue Res.* **5:**21–29.

Linsenmayer, T. F., Fitch, J. M., and Birk, D. E., 1990, Heterotypic collagen fibrils and stabilizing collagens. Controlling elements in corneal morphogenesis? *Ann. N.Y. Acad. Sci.* **580:**143–160.

Lord, E. A., and Wilson, C. B., 1984, *The Mathematical Description of Shape and Form,* Horwood/Wiley, New York.

Malacinski, G. M., and Bryant, S. V., 1984, *Pattern Formation: A Primer in Developmental Biology,* Macmillan Co., New York.

Mandelbrot, B. B., 1983, *The Fractal Geometry of Nature,* Freeman, San Francisco.

May, R. M., 1987, Nonlinearities and complex behavior in simple ecological and epidemiological models, *Ann. N.Y. Acad. Sci.* **504:**1–15.

McBride, D. J., Hahn, R. A., and Silver, F. H., 1985, Morphological characterization of tendon development during chick embryogenesis, *Int. J. Biol. Macromol.* **7:**71–76.

McBride, D. J., Jr., Trelstad, R. L., and Silver, F. H., 1988, Structural and mechanical assessment of developing chick tendon, *Int. J. Biol. Macromol.* **10:**194–200.

Meinhardt, H., 1982, *Models of Biological Pattern Formation,* Academic Press, New York.

Mendler, M., Eich Bender, S. G., Vaughan, L., Winterhalter, K. H., and Bruckner, P., 1989, Cartilage contains mixed fibrils of collagen types II, IX, and XI, *J. Cell Biol.* **108:**191–197.

Merrilees, M. J., Tiang, K. M., and Scott, L., 1987, Changes in collagen fibril diameters across artery walls including a correlation with glycosaminoglycan content, *Connect. Tissue Res.* **16:**237–257.

Miyahara, M., Hayashi, K., Berger, J., Tanzawa, F. K., Trelstad, R. L., and Prockop, D. J., 1984, Formation of collagen fibrils by enzymatic cleavage of precursors of type I collagen in vitro, *J. Biol. Chem.* **259:**9891–9898.

Mosler, E., Folkhard, W., Knorzer, E., Nemetschek-Gansler, H., Nemetschek, T., and Koch, M. H. J., 1985, Stress induced molecular rearrangement in tendon collagen, *J. Mol. Biol.* **182:**589–596.

Mould, A. P., Hulmes, D. J., Holmes, D. F., Cummings, C., Sear, C. H., and Chapman, J. A., 1990, D-periodic assemblies of type I procollagen, *J. Mol. Biol.* **211:**581–594.

Murphy, D. B., Gray, R. O., Grasser, W. A., and Pollard, T. D., 1988, Direct demonstration of actin filament annealing in vitro, *J. Cell Biol.* **106:**1947–1954.

Na, G. C., 1989, Monomer and oligomer of type I collagen: Molecular properties and fibril assembly, *Biochemistry* **28:**7161–7167.

Nestler, F. H., Hvidt, S., Ferry, J. D., and Veis, A., 1983, Flexibility of collagen determined from dilute solution viscoelastic measurements, *Biopolymers* **22:**1747–1758.

Parry, D. A., and Craig, A. S., 1979, Electron microscope evidence for an 80 A unit in collagen fibrils, *Nature* **282:**213–215.

Parry, D. A. D., and Craig, A. S., 1988, Collagen fibrils during development and maturation and their contribution to the mechanical attributes of connective tissue, in: *Collagen,* Vol. 2 (M. E. Nimni, eds.), pp. 1–24, CRC Press, Boca Raton.

Ploetz, C., Zycband, E. I., and Birk, D. E., 1991, Collagen fibril assembly and deposition in the developing dermis: Segmental deposition in extracellular compartments, *J. Struct. Biol.* **106:** 73–81.

Prigogine, L., 1980, *From Being to Becoming: Time and Complexity in Physical Sciences,* Freeman, San Francisco.

Prockop, D. J., Kadler, K. E., Kuivaniemi, H., Tromp, G., and Vogel., B. E., 1989, Type I procollagen: The gene–protein system that harbors most of the mutations causing osteogenesis imperfecta and probably more common heritable disorders of connective tissue, *Am. J. Med. Genet.* **34:**60–67.

Rothwell, S. W., Grasser, W. A., and Murphy, D. B., 1986, End-to-end annealing of microtubules in vitro, *J. Cell Biol.* **102:**619–627.

Rowe, R. W. D., 1988, The structure of rat tail tendon, *Connect. Tissue Res.* **14:**9–20.

Ruoslahti, E., 1988, Structure and biology of proteoglycans, *Annu. Rev. Cell Biol.* **4:**229–255.

Scott, J. E., 1986, Proteoglycan–collagen interactions, *Ciba Found. Symp.* **124:**104–124.

Scott, J. E., and Haigh, M., 1988, Identification of specific binding sites for keratan sulphate proteoglycans and chondroitin–dermatan sulphate proteoglycans on collagen fibrils in cornea by the use of cupromeronic blue in 'critical-electrolyte-concentration' techniques, *Biochem. J.* **253:**607–610.

Silver, F. H., 1982, A molecular model for linear and lateral growth of type I collagen fibrils, *Collagen Relat. Res.* **2:**219–229.

Silver, F. H., and Doillon, C. J., 1989, *Biocompatibility: Interactions of Biological and Implantable Materials,* Vol. 1, Polymers, VCH Publishers, New York.

Silver, F. H., and Trelstad, R. L., 1979, Linear aggregation and the turbidimetric lag phase: Type I collagen fibrillogenesis in vitro, *J. Theor. Biol.* **81:**515–526.

Silver, F. H., and Trelstad, R. L., 1980, Type I collagen in solution. Structure and properties of fibril fragments, *J. Biol. Chem.* **255:**9427–9433.

Silver, F. H., Langley, K. H., and Trelstad, R. L., 1979, Type I collagen fibrillogenesis: Initiation via reversible linear and lateral growth steps, *Biopolymers* **18:**2523–2535.

Silver, F. H., Kato, Y. P., Ohno, M., and Wasserman, A. J., 1992, Analysis of soft tissues as composites, in: *Biomedical Applications of Composites* (C. Migliarese and J. L. Kardos, eds.), CRC Press, Boca Raton, Fla., in press.

Stevens, P. S., 1974, *Patterns in Nature,* Atlantic/Little, Brown, Boston.

Stopak, D., Wessels, N. K., and Harris, A. K., 1985, Morphogenetic rearrangement of injected collagen in developing chicken limb buds, *Proc. Natl. Acad. Sci. USA* **82:**2804–2808.

Suarez, G., Oronsky, A. L., Bordas, J., and Koch, M. H., 1985, Synchrotron radiation x-ray scattering in the early stages of in vitro collagen fibril formation, *Proc. Natl. Acad. Sci. USA* **82:**4693–4696.

Thom, R., 1975, *Structural Stability and Morphogenesis,* Benjamin, Reading, Mass.

Thompson, D., 1969, *On Growth and Form, Abridged Edition* (J. T. Bonner, ed.), Cambridge University Press, London.

Trelstad, R. L., 1970, The Golgi apparatus in chick corneal epithelium: Changes in intracellular position during development, *J. Cell Biol.* **45:**34–42.

Trelstad, R. L., 1971, Vacuoles in the embryonic chick corneal epithelium, an epithelium which produces collagen, *J. Cell Biol.* **48:**689–694.

Trelstad, R. L., 1977, Mesenchymal cell polarity and morphogenesis of chick cartilage, *Dev. Biol.* **59:**153–163.

Trelstad, R. L., 1982, The bilaterally asymmetrical architecture of the submammalian corneal stroma resembles a cholesteric liquid crystal, *Dev. Biol.* **92:**133–134.

Trelstad, R. L., and Birk, D. E., 1984, Collagen fibril assembly at the surface of polarized cells, in: *Extracellular Matrix in Development* (R. L. Trelstad, ed.), pp. 513–543, Liss, New York.

Trelstad, R. L., and Birk, D. E., 1985, The fibroblast in morphogenesis and fibrosis: Cell topography and surface-related functions, *Ciba Found. Symp.* **114:**4–19.

Trelstad, R. L., and Coulombre, A. J., 1971, Morphogenesis of the collagenous stroma in the chick cornea, *J. Cell Biol.* **50:**840–858.

Trelstad, R. L., and Hayashi, K., 1979, Tendon collagen fibrillogenesis: Intracellular subassemblies and cell surface changes associated with fibril growth, *Dev. Biol.* **71:**228–242.

Trelstad, R. L., and Silver, F. H., 1981, Matrix assembly, in: *Cell Biology of Extracellular Matrix* (E. D. Hay, ed.), pp. 179–215, Plenum Press, New York.

Trelstad, R. L., Hayashi, K., and Gross, J., 1976, Collagen fibrillogenesis: Intermediate aggregates and suprafibrillar order, *Proc. Natl. Acad. Sci. USA* **73:**4027–4031.

Trelstad, R. L., Hayashi, M., Hayashi, K., and Birk, D. E., 1987, Fibrils, Fibonacci and fractals: Searching for rules and rulers of morphogenesis in the orthogonal stroma of the chick cornea,

in: *The Microenvironment and Vision* (J. B. Sheffield and S. R. Hilfer, eds.), pp. 1–27, Springer-Verlag, Berlin.
van der Rest, M., and Mayne, R., 1988, Type IX collagen proteoglycan from cartilage is covalently crosslinked to type II collagen, *J. Biol. Chem.* **263:**1615–1618.
Vaughan, L., Mendler, M., Huber, S., Bruckner, P., Winterhalter, K. H., Irwin, M. I., and Mayne, R., 1988, D-periodic distribution of collagen type IX along cartilage fibrils, *J. Cell Biol.* **106:**991–997.
Vogel, K. G., Paulsson, M., and Heinegard, D., 1984, Specific inhibition of type I and type II collagen fibrillogenesis by small proteoglycan of tendon, *Biochem. J.* **223:**587–597.
Wainwright, S. A., Biggs, W. D., Currey, J. D., and Gosline, J. M., 1976, *Mechanical Design in Organisms*, Princeton University Press, Princeton, N.J.
Wallace, D., 1985, The role of hydrophobic bonding in collagen fibril formation: A quantitative model, *Biopolymers* **24:**1705–1720.
Ward, N. P., Hulmes, D. J., and Chapman, J. A., 1986, Collagen self-assembly in vitro: Electron microscopy of initial aggregates formed during the lag phase, *J. Mol. Biol.* **190:**107–112.
Weinstock, M., and Leblond, C. P., 1974, Synthesis, migration and release of precursor collagen by odontoblasts as visualized by radioautography after ^{3}H proline administration, *J. Cell Biol.* **60:**92–127.
Weiss, P., 1934, In vitro experiments on the factors determining the course of the outgrowing nerve fiber, *J. Exp. Zool.* **68:**393–448.
Williams, R. C., Jr., and Rone, L. A., 1989, End-to-end joining of taxol-stabilized GDP-containing microtubules, *J. Biol. Chem.* **264:**1663–1670.

Chapter 8

Extracellular Matrix Degradation

CAROLINE M. ALEXANDER and ZENA WERB

1. Introduction

Many processes are included within the scope of the term *matrix degradation*. Specialized matrices such as bone and cartilage, which provide structural support for the animal, are dynamic structures that resorb or expand in response to hormonal stimuli. The enzymatic machinery for the degradation of large quantities of collagen and other matrix components must be under the appropriate transcriptional regulation in the controlling cell type. Similarly, if a whole organ or tissue is removed, such as during uterine and mammary involution, enzymes are required that are capable of lysing basement membranes and interstitial matrix and processing cellular debris so that all the components can be absorbed by scavenging cell types. Matrix degradation is also a part of invasive cellular migration, when cells displace gels of interstitial molecules or cross basement membranes, the natural barriers to cell migration. In contrast to the involution reaction, lysis associated with cell motility is a limited reaction that leaves the bulk of the matrix intact and is directed only at the "structural kingpins," those molecules that constitute a barrier to cell displacement. Cell growth, division, and expansion also demand flexibility in the growth matrix, such as that associated with tissue regeneration and during development. Matrix is also a reservoir of growth factors and possible morphogens, and rather specific cleavage reactions directed against carrier molecules like vitronectin

Abbreviations used in this chapter: bFGF, basic fibroblast growth factor; FACIT, fibril-associated collagens with interrupted triple-helices; IL, interleukin; MDCK, Madin Darby canine kidney cell; MEP, major excreted protein; PA, plasminogen activator; PAI, PA inhibitor; PMN, polymorphonuclear leukocytes; PN, protease inhibitor; PTH, parathyroid stimulating hormone; PUMP, putative metalloproteinase; RSV, Rous sarcoma virus; SPARC, secreted protein acidic and rich in cysteine; TGF-β, transforming growth factor-beta; TIMP, tissue inhibitor of metalloproteinases; TPA, tissue-type PA; UPA, urokinase-type PA; ZP, zona pellucida.

CAROLINE M. ALEXANDER and ZENA WERB • Laboratory of Radiobiology and Environmental Health, Department of Anatomy, and Programs in Cell and Developmental Biology, University of California, San Francisco, California 94143-0750.

Cell Biology of Extracellular Matrix, Second Edition, edited by Elizabeth D. Hay, Plenum Press, New York, 1991.

and heparan sulfate may have a role in determining the course of further cell growth and proteolysis.

As expected from the complex demands of these various processes, many enzymes from several different classes are involved. Two principal methods exist for matrix degradation: the extracellular secretion of catabolic enzymes or the internalization of matrix molecules followed by lysosomal degradation. Secreted enzymes important for extracellular matrix destruction tend to have neutral pH optima. The most important class are the metalloproteinases, a classic example of an enzyme family of related activities that vary from highly specific (such as collagenase) to broadly reactive (such as stromelysin). The pattern of molecular lysis may start with a primary denaturing cleavage by a highly specific enzyme followed by lysis to small soluble products by another family member. Hence, the cleavage of a single bond in collagen type I/III by collagenase renders the collagen fragments susceptible to stromelysin. Although the metalloproteinases tend to be the enzymes directly responsible for matrix degradation, the inhibition of other enzyme classes can slow the process. This is due at least in part to the activation requirement of metalloproteinases. The second type of reaction is controlled mainly by the discrimination of target proteins at the cell surface and activation of the internalization machinery. Proteins may be targeted for this pathway by limited proteolytic cleavage by the secreted enzymes. Thus, the whole process is a very typical biological cascade, in which key regulatory events exist at many levels. The characteristics and control of various enzyme activities are discussed in more detail below.

2. Enzymes and Inhibitors Involved in Matrix Degradation

2.1. Metalloproteinases and Inhibitors

The metalloproteinases are named for their catalytically active chelated zinc and calcium dependence, and constitute an increasing number of family members with various and partly overlapping substrate specificities (Table I). The identification of additional members of the metalloproteinase family may depend only upon the choice of substrate for assay and tissue source. So far, the metalloproteinases have been grouped into the same family on the basis of extracellular matrix substrate specificity, zinc dependence, inhibition by the tissue inhibitor of metalloproteinases (TIMP), secretion as a zymogen, and, more recently, sequence relatedness. The domainal structure of these proteinases is illustrated in Fig. 8-1. All contain the zinc-chelating motif (VAAHExGH), which follows the pattern of proteins that contain zinc, with the fourth site of the zinc tetrahedron available for catalysis (Van Wart and Birkedal-Hansen, 1990). All enzymes also contain the 5′ sequence (PRCGxPDV) critical to activation of the proenzyme, just upstream of the N-terminus of the active protein. Amino acid homologies show that there is 71–85% identity for homologous enzymes in different species and 40–50% identify between family

Table I. Substrate Specificity of Extracellular Matrix-Degrading Proteinases[a]

ECM protein	Enzyme								
	Collagenase	PMN collagenase	Stomelysin I, II/transin	Matrilysin (gelatinase, 72kDa)	Invadolysin (gelatinase, 92kDa)	PUMP	UPA	TPA	Plasmin
Collagen type									
I	+	+	–	–	–	–	–	–	–
II	+	+	–	–	–				–
III	+	+	–	–	–	+			+
IV	–	–	+	+	+	+	–	–	+
V	–		+	+	+				+
VI	–		–	–					
VII	–		+	+	+				
VIII									
IX	–		+	–					
X	+			+					
Gelatin	+	+	+	+	+	+	–	–	+
Fibronectin	–		+	+	+	+	+	–	+
Laminin	–		+	+ (?)	+ (?)				+
Proteoglycans	–		+		+	+			+
Elastin	–	–	+	+	+	–	–	–	–
Plasminogen	–		–				+	+	–

[a]PMN, polymorphonuclear leukocyte; –, no activity; +, activity; blank, not determined. Modified from Alexander and Werb (1989).

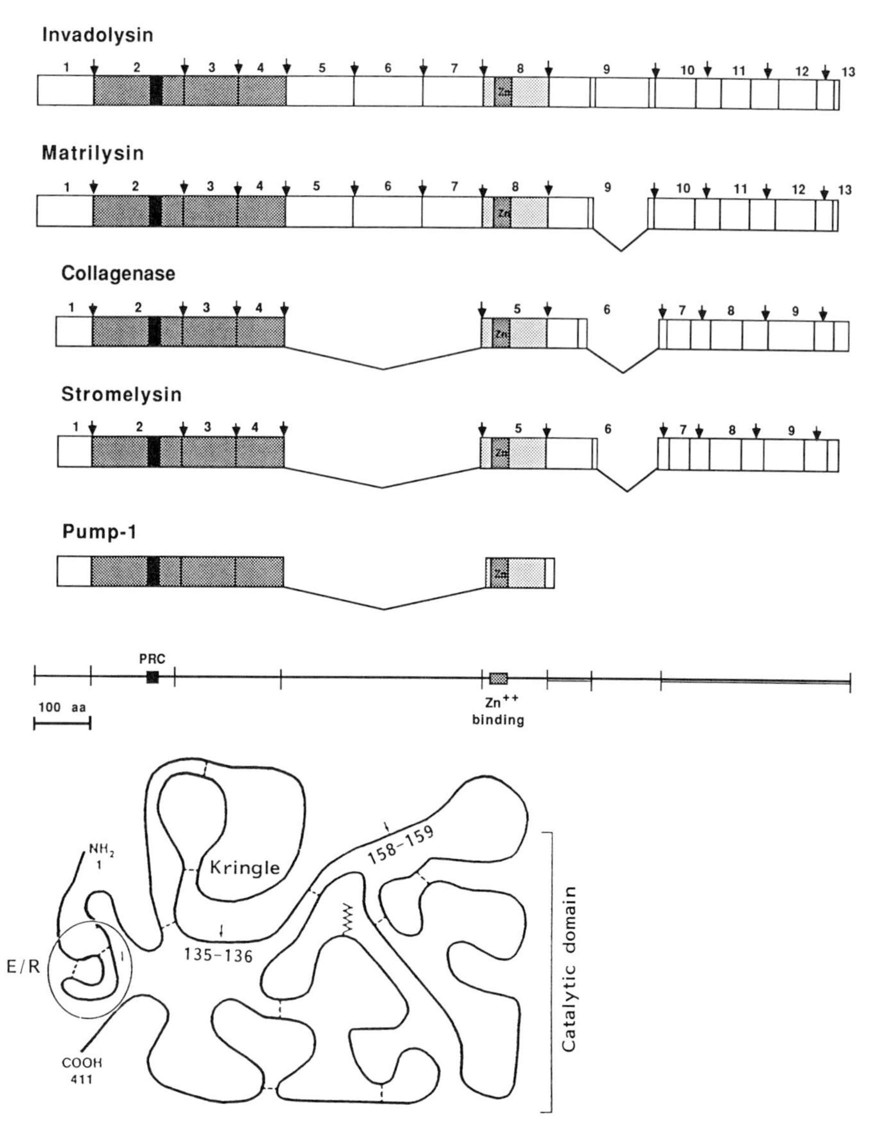

Figure 8-1. Structural and functional domains of metalloproteinases and the serine proteinase UPA. A illustrates the metalloproteinase consensus sequence, and includes the activation sequence (PRC motif) and the zinc-binding peptide. Along the top of each sequence are numbered the introns, together with the intron–exon boundaries. The N-terminal blank boxes represent signal sequences; C-terminal blank boxes illustrate the hemopexin and vitronectin homology domains. Lightly stippled boxes indicate the sequences inserted into the larger members of the family, and the darker stippling illustrates the consensus domains common to all members so far identified. The UPA sequence (B) includes the N-terminal EGF-like repeat with receptor binding sequence (E/R), a single kringle consensus domain, and the remainder of the sequence which includes a classic serine proteinase catalytic domain. Disulfide bonds are shown by dotted lines. Cleavage between residues 158 and 159 produces the two-chain active UPA molecule, and a cleavage at 135–136 generates the attenuated molecule described in Section 3.4. B is modified from Haber *et al.*, 1989, *Science* **243:**51–56.

members. The extent of sequence conservation is different for the various domains. For example, the N-terminal domain up to the insertion of the collagen type V homology unit in the 92-kDa gelatinase invadolysin is 54% homologous to the 72-kDa gelatinase matrilysin; after the insertion point, homology tails off to 27% (Wilhelm *et al.*, 1989). Rather precisely inserted sequences (see Fig. 8-1) account for the increase in size of the larger family members, and these sequences have homology to other extracellular matrix proteins (reviewed in Alexander and Werb, 1989), namely, collagen type V and the collagen-binding domain of fibronectin.

This class of enzymes has confusing systems of nomenclature. The enzymes known as collagenases (interstitial of fibroblast type and granulocyte type) have single highly specific cleavage sites in interstitial collagen types I and III and degrade gelatin only poorly. (Gelatin is denatured collagen that has lost the typical triple left-handed helix that protects collagen from proteolytic attack.) The collagen fragments denature at physiological temperatures and are then a substrate for other proteinases, the so-called gelatinases, a class that includes other metalloproteinase family members and serine proteinases such as plasmin. As a denatured protein, gelatin is a good target for many proteinases: the ability to degrade gelatin *in vitro* may not, however, be relevant to the physiological substrate of any particular gelatinase. Thus, matrilysin is a gelatinase but also a specific collagenase that can cleave native collagen types IV, V, VII, and X. Type IV collagen is one of the major structural networks of the basement membrane, and is one of the most important target molecules to implicate an enzyme in a cellular invasive process. In contrast to these rather specific reactions, enzymes such as stromelysin and putative metalloproteinase (PUMP) can degrade many of the components of the matrix, including proteoglycans, gelatins, glycoproteins such as fibronectin and laminin, and elastin (reviewed by Alexander and Werb, 1989): these enzymes fit the definition of truly lytic enzymes, degrading macromolecules to many small soluble products. The tertiary structure of these enzymes is not yet available, and so the molecular determinants of the substrate specificity can only be extrapolated from the conserved and divergent areas of sequence. The reason for the existence of several variants of the same enzyme (granulocyte and fibroblast collagenases, stromelysins I and II) has not been established, but they have very different promoter sequences (Breathnach *et al.*, 1987) and are known to be independently regulated in various cell types. Metalloproteinase expression has been observed in most cell types, endothelial, mesenchymal, and epithelial.

There is a family of specific metalloproteinase inhibitors which form high-affinity complexes with the proteinases. The first and best characterized is TIMP (28.5 kDa), which is effective against all enzymes so far isolated. A new family member was recently described (TIMP-2, 21 kDa) (De Clerck *et al.*, 1989; Boone *et al.*, 1990; Stetler-Stevenson *et al.*, 1989) with a similar but not identical inhibitor profile for the enzymes tested: the slightly different K_i values for various enzymes may indicate different physiological roles. TIMP-2 will select matrilysin from a mixture of other enzymes *in vitro* (Stetler-Stevenson *et al.*,

1989); the partner *in vivo* might be expected to depend on the local complement of other inhibitors and enzymes. Certainly expression of the two inhibitors is differently regulated, both in a wide range of tumor cell lines (Stetler-Stevenson *et al.*, 1990) and in embryonic stem cells (our observations). Some data suggest that there are more members in this class of inhibitors (Herron *et al.*, 1986; Apodaca *et al.*, 1990; Cawston *et al.*, 1990); these are usually identified either on reverse zymograms or as inhibitors of the release of radioactivity from labeled matrix molecules by purified enzyme preparations. Significant properties of TIMP and TIMP-2 include six internal cystine bonds that create a very stable molecule resistant to heat treatment or partial trypsin proteolysis but extremely sensitive to reducing agents (see Fig. 8-2; Okada *et al.*, 1988). There appear to be no autonomous domains or fragments that have inhibitor activity (Coulombe and Skup, 1988). Despite the conservation of sequence (66%

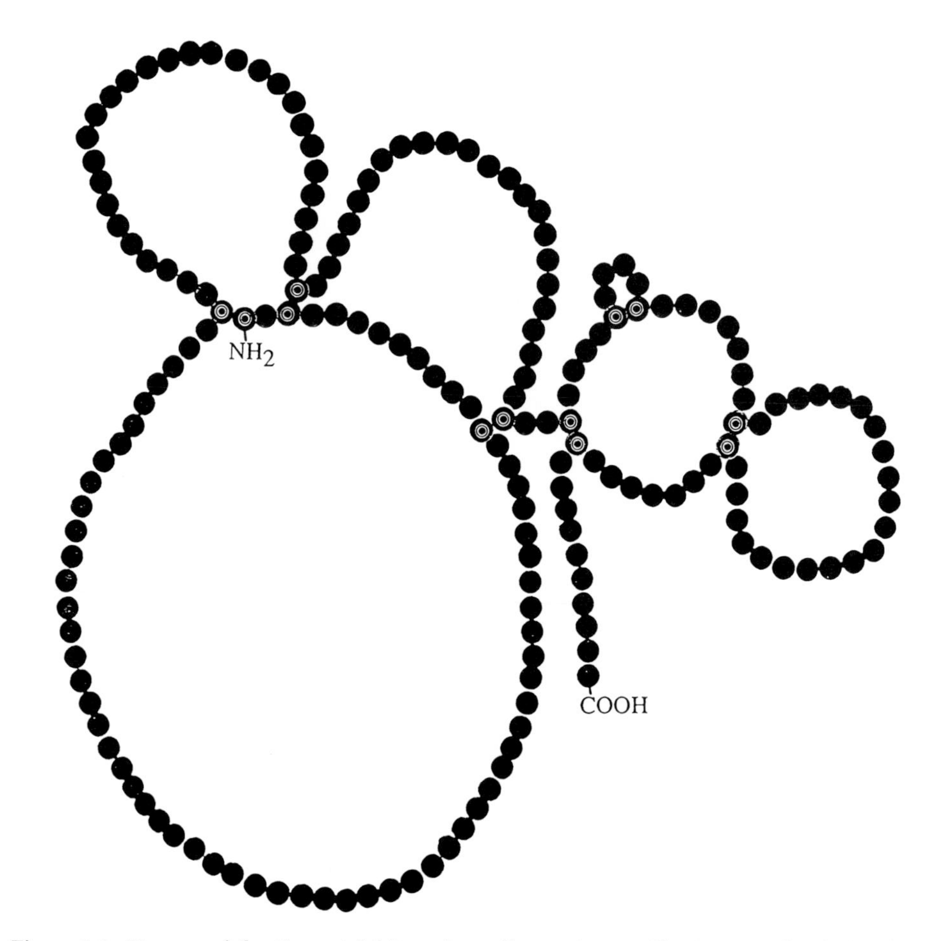

Figure 8-2. Diagram of the tissue inhibitor of metalloproteinases. The TIMP molecule has been drawn to show the extensive internal disulfide bonding; cysteine residues are marked with a white ring. The TIMP-2 molecule is similarly knotted together and shares the resistance to proteinases and heat denaturation.

Table II. Polypeptide Inhibitors of Extracellular Matrix-Degrading Proteinases

Inhibitor	Specificity	Regulation
α_2-macroglobulin	All classes	Made in liver, also locally; also binds growth factors
Pregnancy zone protein	All classes	Made in placenta
Ovostatin	Metalloproteinases	Found in avian eggs
α_1-proteinase inhibitor	Leukocyte elastase	Made in liver and macrophages; can be inactivated by metalloproteinases
α_2-antiplasmin	Plasmin	Made in liver; cannot inhibit receptor-bound plasmin
Aprotinin	Plasmin, trypsin	Found in mast cells; inhibits receptor-bound plasmin
Protease nexin-1	Plasmin, thrombin, UPA	Made by fibroblasts, endothelial cells; matrix-bound PN-1 loses capacity to inhibit plasmin and UPA
PAI-1	UPA, TPA	Upregulated by growth factors, cytokines, glucocorticoids
PAI-2	UPA	Made by macrophages; mostly intracellular
TIMP-1	Metalloproteinases	Upregulated by growth factors, cytokines, retinoids; binds to proinvasin
TIMP-2	Metalloproteinases	Appears to be constitutive; promatrilysin binds 1 mole

homology) between TIMP and TIMP-2, neither antisera raised to these proteins nor cDNA probes tend to cross-react. The enzyme binding site for inhibitor has not been defined. The TIMP gene is single copy and X-linked in both mouse and man (Huebner *et al.*, 1986). Other proteinase inhibitors of rather broad specificity (including serine proteinases and metalloproteinases) that exist at high concentration in serum (e.g., α2-macroglobulin, see Table II) probably do not have a significant role in controlling physiological remodeling processes but would serve to absorb excess proteolytic activity that escapes into the interstitium.

2.2. Serine Proteinases and Inhibitors

The codependence of matrix destruction on both metalloproteinase and serine proteinase activities has now been demonstrated for several model systems, and it has long been suggested that the two classes are linked in an activation cascade (Werb *et al.*, 1977; Saksela and Rifkin, 1988). That the metalloproteinases are downstream of the serine proteinases is suggested by several sets of experiments, notably those of Reich *et al.* (1988), who studied the invasive properties of melanoma cells. Serine proteinase inhibitors inhibited amnion invasion by these cells, but this block could be circumvented by the addition of alternative sources of metalloproteinase activator. Similarly, bovine aortic endothelial cells have been found to rely on a cascade of interaction in the presence of transforming growth factor-β (TGF-β) (Mignatti *et al.*, 1989). A mammary carcinoma cell line has also been shown to use UPA/collagenase chain of lysis. The conclusion is that the serine proteinases are probably not directly responsible for matrix catabolism and are useful principally as activators for the workhorse metalloproteinases.

The serine proteinases so far implicated are the plasminogen/plasminogen activator (PA) system. The abundant serum protein plasminogen (90 kDa) is converted to the broad-specificity proteinase plasmin by cleavage of a single bond by the PAs. The lytic potential of plasmin is thus held in check by the PAs, serving as rate-determining enzymes. There are two distinct PA molecules with different purposes. Tissue-type PA (TPA, 70 kDa) is involved in regulating the clotting process, and urokinase-type PA (UPA, 50 kDa) is active in fibrinolysis as well as in biological remodeling and invasive processes. The structure and function of the PAs are discussed in detail in a recent review (Saksela and Rifkin, 1988). UPA is secreted as an inactive single-chain proform, which is cleaved to an active two-chain disulfide-linked form. The physiologic activator of UPA has not been identified, but may be either plasmin or kallikrein: both can act as activators *in vitro*. Urokinase is a single copy gene, and the protein is organized on a pattern of rather autonomous domains (diagrammed in Fig. 8-1) consisting of an N-terminal epidermal growth factor-like receptor-binding domain, followed by a cross-linked kringle structure and a typical serine proteinase catalytic domain. Immunolocalization studies have shown that UPA is widely distributed in mouse tissues, particularly prominent in fibroblastlike cells in connective tissues of the gastrointestinal tract, pancreas, trachea, and bronchi, in epithelial cells of kidney and lung, and in decidual cells of mouse placenta (Larsson *et al.*, 1984).

There are several inhibitors of the serpin class of PA and plasmin, including PA inhibitor (PAI)-1 (45 kDa), PAI-2 (60 kDa), protease nexin I (PN-1, 45 kDa), α_2-antiplasmin, and aprotinin (7 kDa) (Saksela and Rifkin, 1988; Laiho and Keski-Oja, 1989). PAI-1 is present at significant concentration in serum; it is synthesized by endothelium and is present in many other cell types including platelets. Unlike the metalloproteinase inhibitors, it has no internal disulfide bonds. During the interaction with UPA, PAI-I is cleaved, and dissociation of the urokinase/PAI-1 complex *in vitro* releases active enzyme and inactive inhibitor. PAI-2 is produced by macrophages in placenta and is probably associated with inflammatory reactions; PN-1 is made by many cell types, especially by fibroblasts.

2.3. Intracellular Enzymes

Enzymes that are typically associated with intracellular compartments and internal processing reactions can still contribute significantly to the process of matrix degradation. Cells may either internalize matrix components, or release enzymes sequestered in specialized internal granules or vacuoles after a particular stimulus.

Internalization of matrix molecules invokes a delivery mechanism, either a specific cell surface receptor-directed process or generalized bulk flow, to deliver proteins to the catabolic machinery of the acidic lysosomal compartment, which includes aspartic proteinases, serine proteinases, metalloproteinases, and many nonproteinase enzymes. Certainly, there tends to be an aspect of extracellular matrix lysis which is resistant to the addition of inhibitors (e.g., see Behrendtsen *et al.*, 1991). This may represent the contribution of the internalization process, but its significance is not known. Lysosomal glycosidases that catabolize proteoglycans are clearly critical to matrix degradation: deficiencies of particular enzyme activities result in the pathological accumulation of proteins targeted for catabolism and eventual dysfunction of the whole cell. The processes that coordinate the cytoskeletal organization, the discharge of vacuoles, and the directional clearing of matrix during migration are not yet established.

Lysosomal enzymes can be secreted, notably in pathological circumstances, but at a neutral extracellular pH it is not clear what significance this secretion has for general matrix remodeling processes unless the cells have the ability to compartmentalize extracellular spaces like osteoclasts (Werb, 1989; Zucker, 1988). For example, the major excreted protein (MEP) of transformed mouse fibroblasts (compared with their normal counterparts) has been identified as cathepsin L, and considerable effort has been invested in finding out whether the secretion of this protein determines the invasive, transformed phenotype typical of this cell type. However, overexpression of MEP (by amplification of a transfected plasmid in 3T3 cells) does not confer any of the characteristics typical of transformation, such as change of cell morphology, increased growth rate, growth in soft agar, growth in low serum, or tumorigenicity in nude mice (reviewed in Alexander and Werb, 1989). We should not exclude a possible synergy of other changes in the tumor cells that might make this enzyme a significant player in the invasive process; however, the overproduction and secretion of an enzyme usually targeted to a particular subcellular compartment represents a considerable upset of transcriptional control and/or intracellular trafficking, and may be just a marker of the deranged expression typical of tumor cells.

Several cell types derived from the hematopoietic lineage are specialized to secrete lytic vacuoles or granules as part of their physiological role. Osteoclasts, macrophages, neutrophils, eosinophils, killer T cells, and mast cells are all capable of adhering to a matrix or to a target organism or cell and discharging vacuolar contents into microenvironments delimited by membrane adhesions. Osteoclasts are typically highly polarized with respect to their substrate of bone. The membrane apposed to the bone matrix is termed a ruffled border, clearly interdigitated by clear zones of lysis. Lacunae, or excavations in the bone surface, rapidly appear as the enzymes of the acid pH lysosomes are discharged. Neutrophils contain at least two populations of vesicles or granules: the so-called azurophilic primary granules, which contain all the acid hydrolases and peroxidase as well as elastase and cathepsin G, and the secondary granules, which contain enzymes of neutral pH optima like collagenase and invasin. The collagenase and elastase contained in these granules are isozymes specific to polymorphonuclear leukocytes (PMN): other enzymes such as invadolysin or cathepsin G are apparently identical to their counterparts in other cell types. Both killer T cells and mast cells secrete granules loaded with digestive enzymes (such as the serine proteinases granzymes and mast cell chymase and tryptase) that could function as metalloproteinase activators at sites of inflammation.

Zones of contact between cells like neutrophils and their target are also rather insulated, and in the presence of proteinase inhibitors, the process of matrix degradation by PMNs is confined to the marginated zones subjacent to the cells (Campbell and Campbell, 1988). This may allow proteinases to function in the presence of high serum concentrations of various proteinase inhibitors.

The contribution of particular enzymes or classes of enzyme to any process can be examined by using generic chemical inhibitors, more specific protein inhibitors, or antibodies that interfere with enzyme activity (Librach *et al.*, 1991; Behrendtsen *et al.*, 1991; Sullivan and Quigley, 1986; Ossowski and Reich, 1983). Preferably, information should come from more than one source, since chemical inhibitors tend to have nonspecific side effects but do not have the steric constraints of larger protein molecules.

3. Control of the Process of Matrix Degradation

3.1. Transcriptional Regulation

The genetic control of expression of various components of the matrix mobilization cascade is a finely tuned and highly responsive process that responds to many physiological signals. This is obvious from the result of several cloning strategies: for example, TIMP was isolated from subtracted cDNA libraries of fibroblasts as an mRNA induced by serum feeding (Edwards *et al.*, 1986) and by virus infection in mouse (Gewert *et al.*, 1987). The expression of collagenase and stromelysin is induced by many, very diverse stimuli (reviewed in Werb, 1989; see Table III). Environmental stresses and insults such as ultraviolet irradiation (Stein *et al.*, 1989), heat shock (Vance *et al.*, 1989), and hyperoxic shock (Horowitz *et al.*, 1989) can be eliciting stimuli for enzyme expression. The rounding of cells that is induced by treatment of fibroblasts with antiactin drugs or proteinases or by growth in retractable collagen gels induces the secretion of stromelysin and collagenase, and this induction may serve to reinforce the change in cell–matrix interaction (Werb *et al.*, 1989). Physical stimuli like these may be relevant to the remodeling processes associated with wound repair. For example, urokinase is expressed at the edges of wounds scraped into confluent endothelial cell cultures (Pepper *et al.*, 1987): this activity is linked to the processes of motility and invasion in endothelial

Table III. Some Factors Regulating Expression of ECM-Degrading Proteinases and Their Inhibitors by Fibroblasts[a]

Regulator	Effect			
	Collagenase	TIMP	UPA	PAI/ protease nexin
Cytokines (interleukin-1, tumor necrosis factor-α)	S	S	S	S
Phorbol esters	S	S	S	S
Growth factors	S	S	S	S
TGF-β	I	S	I	S
Glucocorticoids	I	None	I	S
Retinoids	I	S	I	ND
Transformation	S	I	S	I
Matrix interactions	S	None	S	ND

[a]I, inhibits; S, stimulates; ND, not determined. Modified from Alexander and Werb (1989).

cells, and the levels of urokinase return to normal when the wound has closed. The inductive stimulus in operation here may be the release from confluence-induced quiescence into the cell cycle (G_0 to G_1 transition).

Proteinase expression is specifically regulated by changes in the interaction of receptors with matrix molecules. For example, collagenase, stromelysin, and invadolysin can be induced in rabbit synovial fibroblasts by the binding of fibronectin fragments or specific antibody to the α5β1 integrin fibronectin receptor on the synovial fibroblast cell surface (reviewed in Werb *et al.*, 1991). Collagenase and stromelysin are frequently coordinately expressed but control of their expression can be dissociated depending on cell type, growth conditions, and inductive stimulus (MacNaul *et al.*, 1990).

The flexibility of the metalloproteinase profile introduces a variable that is usually not appreciated in the description and interpretation of the effects of a variety of reagents (notably the RGD adhesion peptide) on cellular properties like adhesion, cell shape, and migration. These reagents may produce changes indirectly by altering the secretion of remodeling enzymes.

The expression of proteinases and inhibitors responds to phorbol esters and to many growth factors (see Table III). The complex transcriptional response of these proteins to growth factors has also been described in other reviews (Laiho and Keski-Oja, 1989; Matrisian and Hogan, 1990). TGF-β appears to be the principal controller of matrix remodeling (reviewed by Roberts *et al.*, 1990). In fact, this growth factor was originally isolated on the basis of its ability to induce fibrosis, i.e., an excess deposition of matrix at the site of application. Not only is the production of matrix components increased, but the remodeling enzyme equilibrium is shifted to increase the production of inhibitor and decrease the production of lytic enzymes (reviewed by Alexander and Werb, 1989; Matrisian and Hogan, 1990). The response of cells to TGF-β *in vitro* is reinforced by the correlation of the localization patterns for TGF-β, the metalloproteinase inhibitor TIMP, and several matrix components in the developing embryo: all are prominent in bone (during ossification) and other remodeling tissues (Heine *et al.*, 1990; Flenniken and Williams, 1990; Nomura *et al.*, 1989).

Net degradation of matrix correlates with enzyme activity and is inversely related to inhibitor production. Transformation tends to increase enzyme synthesis and decrease inhibitor synthesis, giving a net lytic profile for transformed cell types. Inhibitor and proteinase are not always controlled in opposite directions; some growth factors induce both. This may not be as futile as it would seem: enzyme and inhibitor may be generated at different cell aspects and be related to directional movement of the cell. Coexpression may also serve to localize and control the process of degradation (see Section 4). Alternatively, the expression of each may be temporally regulated and may be produced in bursts of one or the other. Certainly, populations of fibroblasts (Fig. 8-3) producing collagenase have been demonstrated to be heterogeneous in this respect, but no complete immunolocalization studies of pairs of enzymes and inhibitors are available to assess this hypothesis. In summary, the expression of members of this interactive cascade is highly labile and would be expected to respond to many physiological situations, adjusting the motility of cells and altering the components of the matrix and the overall tissue architecture.

The collagenase and stromelysin promoters have been used for sophisticated studies of interactions with transcriptional activators and inhibitors, e.g., the modulation of the Fos/Jun interaction at the TPA-responsive element (reviewed in Matrisian and Hogan, 1990) and the response of elements in the promoter sequences to various intracellular signaling pathways (Gutman and Wasylyk, 1990; Chiu *et al.*, 1989). These analyses should reveal the basis for the observations of synergy, inhibition, and dominance of various classes of the stimuli listed above. It should also explain why no one growth factor is guaranteed to induce either a proteinase or an inhibitor: this depends upon cell type, growth stage, and the environment of other effectors (Kumkumian *et al.*, 1989; Sottile *et al.*, 1989; Alexander and Werb, 1989). Induction of enzymes may be partly a result of changes of mRNA stability [e.g., the phorbol ester induction of collagenase in synovial fibroblasts (Brinckerhoff *et al.*, 1986)] although proteinase mRNA species tend to be rather stable.

3.2. Activation of Latent Proenzymes

The control of the catalytic activity of matrix-degrading enzymes constitutes the finest level of control. Many cell types consistently produce a number of latent metalloproteinases but do not express any lytic activity until an activator, such as PA/plasmin, becomes available. Urokinase, plasmin, and the metalloproteinases are all secreted as zymogens, but only the activation of plasminogen has been well characterized. Plasmin can activate urokinase, and plasmin, human neutrophil elastase, and many other proteinases can activate the metalloproteinases *in vitro*. However, it is not at all clear that any of these activators are relevant *in vivo*.

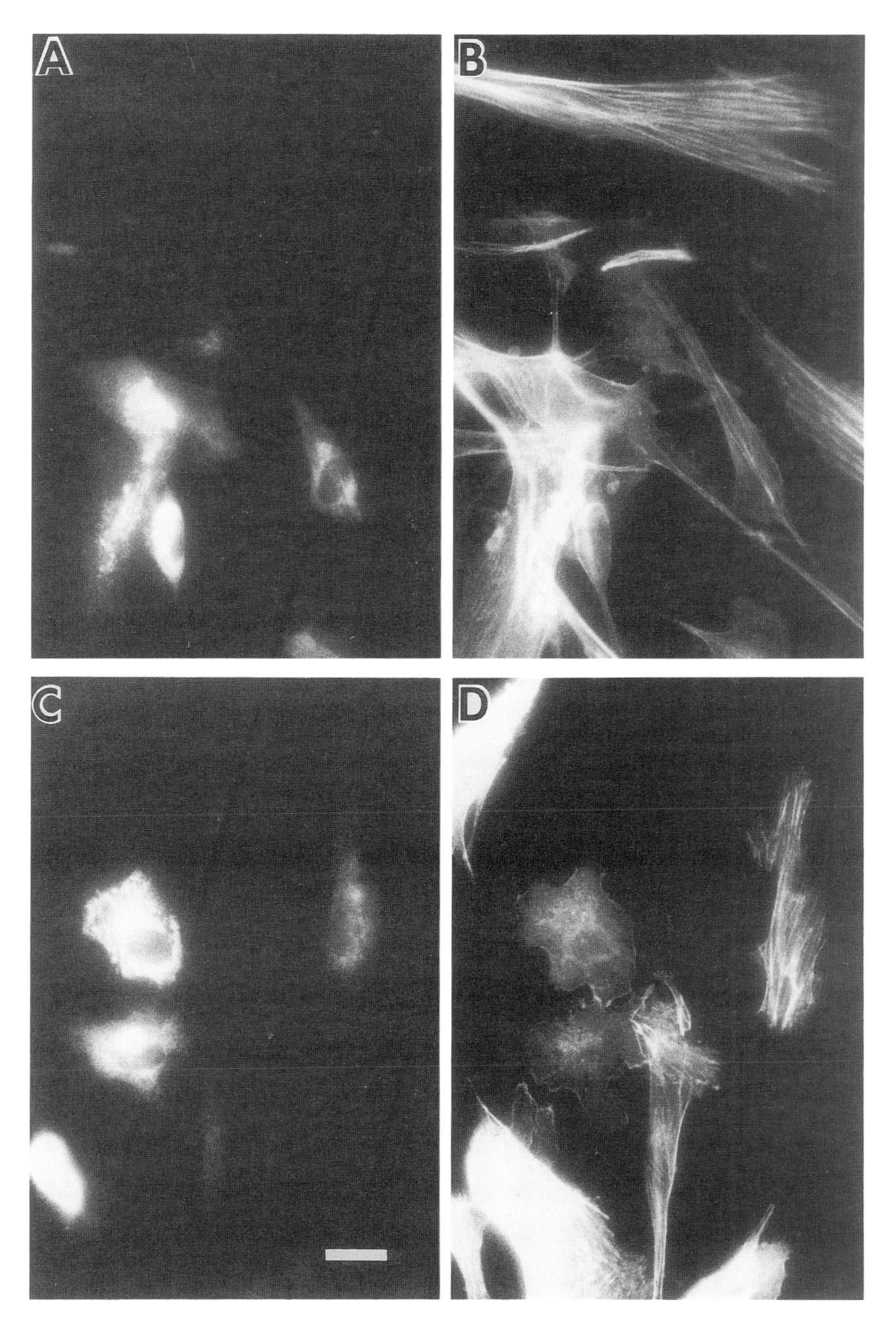

Figure 8-3. Heterogeneity of collagenase expression in fibroblast populations. Cultured rabbit synovial fibroblasts stained with an anticollagenase antiserum (panels A and C) reveal a subpopulation of the cells that actively synthesize collagenase. The total cell population is stained with antiactin antibodies (panels B and D). Bar = 20 μm.

Metalloproteinase activity is not expressed until the interaction of the cysteine in the pro-sequence (PRCGxPDV) with the catalytically active zinc is perturbed (Van Wart and Birkedal-Hansen, 1990). In general, this is accompanied by an autolytic reaction which removes about 8 kDa of N-terminal sequence. Many disparate reagents can act as activators in the test tube, from chaotropes (such as sodium dodecyl sulfate) to organomercurial compounds to the proteinases described above. The common feature of all these reagents is their ability to destabilize the propeptide–zinc interaction (Springman *et al.*, 1990). Activating proteinases chew into the propeptide sequence, promoting an autolytic cleavage seven residues beyond the cysteine-chelating zinc. The propeptide sequence is apparently critical to the stability of the proteinase: if the residues in the propeptide sequence R^{89} and C^{90} (numbered according to the sequence of rat stromelysin) are altered by site-directed mutagenesis, the enzyme is autolytic and is secreted as fragments. Mutation of surrounding residues enables the enzyme to sidestep the need for an activator, and it is secreted in autoactivated form (Park *et al.*, 1991; Sanchez-Lopez *et al.*, 1988). The details of mechanism and the intermediates generated by each activating proteinase have been described for human stromelysin (Nagase *et al.*, 1990), and will hopefully lead to the identification of the physiological activator.

There are antibodies against collagenase or matrilysin that distinguish active and proenzyme forms of the enzymes (Birkedal-Hansen *et al.*, 1988; Monteagudo *et al.*, 1990): otherwise, immunolocalization studies will identify both. Other antibodies can distinguish free enzyme from enzyme complexed with inhibitor (Schleef *et al.*, 1986). The presence of SDS activates latent metalloproteinases on zymograms so that both the high-molecular-weight latent and cleaved active forms will clear the substrate; however, the release of the propeptide is not absolutely required for enzyme activity; this should be kept in mind when lower gel mobility is used to identify the inactive form.

It has been shown that collagenase can activate stromelysin and that TIMP can be inactivated by neutrophil elastase and other serine proteinases (Okada *et al.*, 1988) [the network of interaction has been reviewed by Werb (1989) and Moscatelli and Rifkin (1988)]. Enzymes can inactivate inhibitors; for example, several metalloproteinases including macrophage elastase can degrade the serpin α_1-proteinase inhibitor, presumably amplifying serine proteinase activity (Banda *et al.*, 1988). Whether any of these reactions ever operate in a natural setting is not clear.

Different cell types may depend upon different activators. For example, the Ehrlich ascites tumor cell line expresses a surface-bound trypsinlike serine proteinase, a melanoma cell line expresses cathepsin G at the cell membrane, and serum kallikrein was identified as a potential activator for many cell types (reviewed by Moscatelli and Rifkin, 1988). Triggered neutrophils and eosinophils are highly lytic cell types and have a specialized method for metalloproteinase activation. These cells generate hypochlorous or hypobromous acid from oxygen during a respiratory burst and these chlorinated oxidants can activate latent gelatinases (Peppin and Weiss, 1986). Reactive oxygen metabolites have also been associated with the oxidation and inactivation of proteinase inhibitors such as α_1-proteinase inhibitor (Desrochers and Weiss, 1988).

3.3. Inhibitors

3.3.1. Antiinvasive Properties of Proteinase Inhibitors

The inhibitors of the metalloproteinases have been shown to be powerful antiinvasive agents when added to various model systems, both *in vitro* and *in vivo*. Thus, adding TIMP to cultures of melanoma cells in an amnion adhesion assay can totally inhibit their invasion (Schultz *et al.*, 1988). Furthermore, coinjecting TIMP with melanoma cells significantly inhibits their metastasis to the lung. Reducing the concentration of TIMP produced by 3T3 cells with an antisense construct increases the metastatic potential and tumorigenicity of these cells when injected back into nude mice (Khokha *et al.*, 1989). It is important to know more about why TIMP is so effective in order to assess its therapeutic use in preventing metastasis of various cancer cell types.

3.3.2. Proenzyme Susceptibility to Inhibitor Interaction

There may be two roles for the metalloproteinase inhibitors, since there appear to be two distinct binding interactions for some of the enzymes. The TIMP inhibitor binding site is unavailable in the proenzyme of either collagenase or stromelysin, but TIMP binds to both proenzyme and active forms of invadolysin (Wilhelm *et al.*, 1989): similarly, TIMP-2 can interact with both forms of matrilysin (Stetler-Stevenson *et al.*, 1989). Suppression of the catalytic activity of these enzymes appears to require two molecules of inhibitor.

3.3.3. Clearance of Enzyme–Inhibitor Complexes

Enzyme–inhibitor complexes are bound together tightly or irreversibly and are targeted for destruction. For example, once receptor-bound urokinase is inhibited by PAI-2, the whole complex is rapidly cleared from the monocyte surface, internalized, and degraded (Estreicher *et al.*, 1990). A similar process has been described for the PAI-1/UPA complex (Cubellis *et al.*, 1990). PN-1, when bound to some factor at the fibroblast cell surface or fibroblast matrix, narrows its specificity of interaction from urokinase, thrombin, and plasminogen to thrombin alone (Wagner *et al.*, 1989), and the protease nexin I/serine proteinase complex is rapidly endocytosed (reviewed in Werb, 1989). Similarly, α_2-macroglobulin proteinase complexes are cleared by a receptor-mediated mechanism. For the cell to reestablish proteolysis, not only does it need to synthesize more active enzyme, but the receptor needs to be regenerated at the cell surface. Very few data are available concerning the control of proteinase receptor expression.

3.4. Specificity of Attenuated Enzymes

Several reports of a 21- to 26-kDa N-terminal proteolytic cleavage product of stromelysin suggest that this protein (missing the C-terminal hemopexin-like domain) retains its substrate specificity. A 22-kDa (25-kDa glycosylated form) N-terminal fragment of collagenase, on the other hand, loses the ability to cleave native collagen, and acquires specificity for many additional matrix components, converging toward the specificity profile of the enzymes stromelysin and PUMP (Clark and Cawston, 1989). Whether these autolytic cleavage products, or the autolytic cleavage products of other family members, exist *in vivo* has not been established. However, these observations raise the possibility of a succession of enzyme derivatives modulating total substrate specificity. Plasmin proteolysis of UPA generates a 33-kDa proteinase that lacks the receptor-binding and kringle domain (see Fig. 8-1). This attenuated molecule has indeed been shown to exist *in vivo:* what the role of this non-surface-bound enzyme is, and whether it is inhibitable, has not been established (Stump *et al.*, 1986).

4. Localization of Reactions

Extracellular proteolysis, or the process of self-digestion, is obviously a critically controlled reaction. In the event of poor transcriptional control allowing excessive proteolytic activity, there is yet another level of control in place that has the effect of limiting the damage that could result from this interstitial lysis. Most reactions are clearly localized and are catalyzed by solid surfaces. Probably little proteolysis occurs *in vivo* in the soluble phase. A number of strategies have so far been identified that serve to localize proteolysis. The first is the immobilization of enzyme activities by binding to receptors on the cell surface or matrix. This may be accompanied by the predisposition of the enzyme or inhibitor to denature in free solution. The second is the protection of matrix with high concentrations of bound inhibitors, presumably left behind by the cellular architect.

4.1. Cell-Associated Proteolysis

It is clear that proteolytic reactions are effective only in close proximity to the cell membrane. This has been demonstrated for a variety of cell types and reactions (for reviews see Zucker, 1988; Moscatelli and Rifkin, 1988). For example, trophoblast outgrowths clear areas of matrix immediately subjacent to the cells, in a process that is TIMP-inhibitable and probably depends upon invadolysin (Librach *et al.*, 1991; Behrendtsen *et al.*, 1991). In RSV-transformed chick fibroblasts and 3T3 mouse fibroblasts, areas of fibronectin degradation are localized to cell contact points, the so-called rosette contacts that concentrate cytoskeletal molecules like α-actinin and vinculin and signaling molecules like *src*. Two cell membrane-associated enzymes have been identified as fibronectinases: a 120-kDa serine proteinase and a 150-kDa metalloproteinase, of which the metalloproteinase appeared to account for most activity. Similarly, the restriction of degradation was demonstrated for endothelial cells on a collagen type IV substrate, tumor cells growing on purified matrix molecules, on reconstituted matrices, and on matrices secreted by endothelial cells (reviewed by Moscatelli and Rifkin, 1988). Adding solubilized matrix to the cultured cells resulted in very little degradation, as did the incubation of substrate with medium conditioned by these cell types. This clearly illustrates that unless the molecules are in close proximity to the basal aspect of these many cell types, they are resistant to cleavage. Furthermore, it has been shown that the lytic zone delimited by tightly apposed membrane and substrate can exclude macromolecular intervention. For example, B16 melanoma cells pretreated with anticollagenase antiserum show a reduced invasive reaction, but this antiserum is ineffective after the adhesive interaction of these cells and the amnion substrate has been established (Mignatti *et al.*, 1986).

In common with many other proteins, the metalloproteinases show a directional preference for secretion in polarized cell types. For example, in primary cultures of mouse mammary cells, matrilysin is secreted at the basal side. Very little enzyme is directed toward the apical lumen (Talhouk *et al.*, 1991). Endothelium also secretes matrix components and metalloproteinases basally (Fig. 8-4), whereas TIMP shows little selectivity (Unemori *et al.*, 1990). MDCK cells are typically used as the tissue culture model of a polarized cell: these cells secrete invadolysin apically (Fig. 8-4). Fibroblasts, an apolar cell type, that were grown on similar floating substrates showed no such specificity. These results fit well with the lack of degradation of matrix molecules typically observed when the molecules are added in solution to the apical side of a cell culture.

Both normal trophoblast and tumor cells form specialized pseudopodia ("invadopodia," see Fig. 8-9) or membranous fingers that extend the cell surface area considerably. Presumably, these pseudopodia concentrate enzyme secretion or the cell surface adsorption of enzyme activity.

4.2. Plasminogen Activator and Plasmin Cell Surface Receptors

Theoretically, one of the principal ways to control extracellular lysis would be to bind the enzymes to the cell surface via a specific receptor, and to denature or inactivate free enzyme. This is certainly a feature of the PA and plasmin enzymes (reviewed by Blasi *et al.*, 1987; Saksela and Rifkin, 1988) and may well be important to the metalloproteinase family, although no metalloproteinase receptor has yet been demonstrated. It is true that metalloproteinases are frequently purified not from the soluble supernatant fraction, but from the insoluble cell-associated pellet that remains after homogenization of bone explant, rheumatoid synovium, uterus, or tadpole tail and from plasma membranes of tumor cell lines (reviewed by Zucker, 1988; Zucker *et al.*, 1990).

The UPA receptor is expressed by a wide variety of cell types and has a high-affinity interaction with UPA (Blasi *et al.*, 1987). Inhomogeneity of the distribution of urokinase activity was first observed as a fibrinolytic activity associated with the growth cones of neurons and with the soma or cell bodies (Krystosek and Seeds, 1981). The UPA receptor is an integral membrane protein

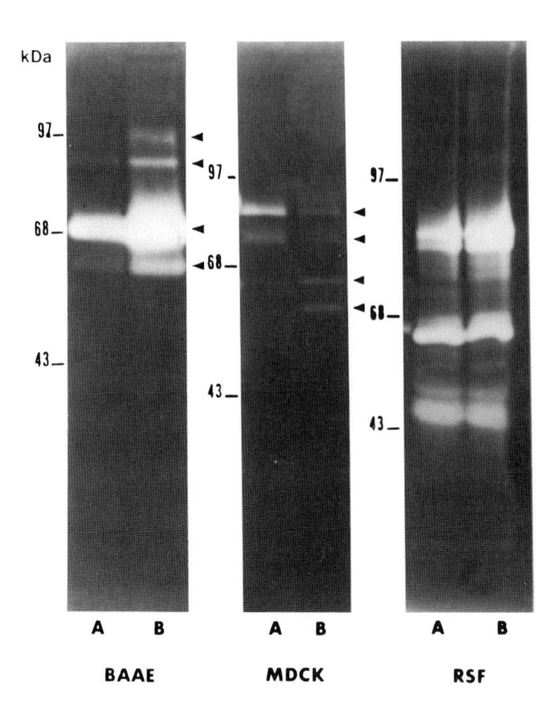

Figure 8-4. Polarization of secretion of the metalloproteinases. Metalloproteinases are secreted asymmetrically in polarized cell types (Unemori *et al.*, 1990). Conditioned medium was collected from the apical (A) and basal (B) compartments of monolayers of bovine aortic arch endothelial cells (BAAE A and B), and of Madin Darby canine kidney cells (MDCK A and B) and analyzed by gelatin substrate gel zymography, a technique that visualizes enzyme activity by clearing of substrate enmeshed in the matrix of a polyacrylamide gel. The metalloproteinases are considerably enriched in the basal compartment of endothelial cells, but in the apical compartment of MDCK cells. Fibroblasts, an apolar cell type, are included as a control (RSF A and B). Polarization phenomena are significant to remodeling reactions that are initiated by these cell types.

that responds to the internal cytoskeletal organization of fibroblasts so as to localize to focal attachments (Pollanen *et al.*, 1988; Hebert and Baker, 1988). Furthermore, localization of the PAI-1 inhibitor shows that whereas the enzyme is focused at sites of cell–substratum interaction, the inhibitor is distributed as a homogeneous carpet under the cells (reviewed by Saksela and Rifkin, 1988). This illustrates again that the production of inhibitor/enzyme pairs by the same cell is not necessarily a futile process: here the pair are separated by location. Surface-bound UPA has been observed to be polarized to the leading edge to migrating monocytes, as would be predicted for an enzyme that may contribute to the invasive process (Fig. 8-5; Estreicher *et al.*, 1990).

Binding of urokinase to the cell surface receptor is critical to its activity: secreted urokinase does not significantly contribute to the invasive process without being surface-bound. Thus, a human cell line expressing human UPA

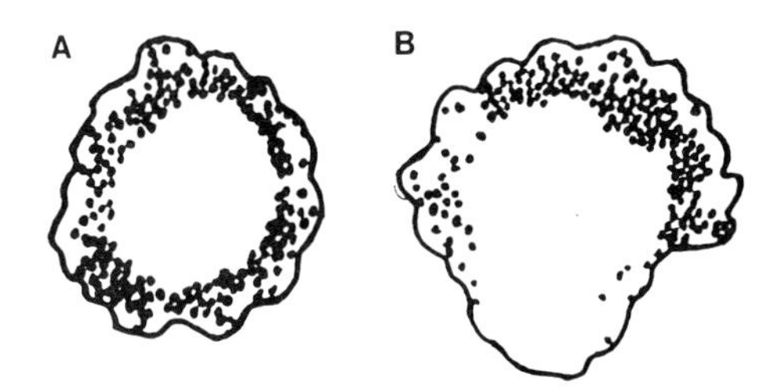

Figure 8-5. Polarization of UPA/UPA receptor complex to the leading edge of migrating cells. U937, a human monocytic cell line, was used to demonstrate the polarization of the UPA receptor in a gradient of the chemotactic peptide N-formyl-L-methionyl-L-leucyl-L-phenylalanine (fMLP). The unoccupied UPA receptor was visualized with iodinated UPA (bound for 30 min at 4°C) and autoradiography. In cells placed in a uniform concentration of fMLP, the receptor is evenly distributed along the lamellae of the cell (A). In a gradient of fMLP (source at top of picture), the UPA receptor concentrates along the ruffled leading edge of the cell (B). (Redrawn from Estreicher *et al.*, 1990.)

receptor and transfected with mouse UPA under the control of an inducible promoter shows no increase in invasiveness. The mouse UPA cannot interact with the human receptor. On the other hand, the addition of human UPA does enhance invasion (Ossowski, 1988).

Expression of the TPA receptor is more restricted and is typically expressed only on endothelial cell types. There also appears to be a cell surface binding site for plasmin in cells of the hematopoietic lineage, and plasmin thus immobilized is resistant to inhibition by α_2-antiplasmin but not aprotinin (described by Moscatelli and Rifkin, 1988). The adsorption of plasmin to cell surfaces is significant in studies that exclude a role for plasmin/plasminogen in cell-associated processes by using plasminogen depletion of serum as the negative control. The presence of proteinase cell surface receptors introduces another level of control for the process of proteolysis. Their synthesis, affinity, and internalization would be predicted to respond to hormones and growth factors. An example is the down-regulation of plasmin receptors on the surface of sarcoma cells in response to glucocorticoids (Pollanen, 1989).

4.3. Matrix-Bound Enzymes and Inhibitors

Some matrices are highly resistant to cell invasion, e.g., that of the central nervous system, and of cirrhotic liver (see Section 8.1). Cartilage has been shown to contain inhibitors of serine proteinases, of collagenase, and of thiol proteinases that can be extracted with guanidinium thiocyanate and remain active even after this treatment (see Moscatelli and Rifkin, 1988; Alexander and Werb, 1989). This is not just a passive copolymerization process; there are specific binding sites on particular matrix components for these inhibitors. Recently, vitronectin was identified as the matrix receptor for PAI-1 (Wiman *et al.*, 1988); unless this inhibitor is rapidly assimilated to the matrix it is denatured to an inactive form. This has been proposed to protect capillary walls from TPA-catalyzed breakdown. PAI-2 does not show this extreme instability. Protease nexin-1 also locates primarily to the matrix (Farrell *et al.*, 1988), probably bound by heparan sulfate proteoglycan, which potentiates its activity (Hiramoto and Cunningham, 1988). TIMP is usually purified by heparin affinity chromatography, perhaps an indication of a physiological proteoglycan interaction.

Plasminogen/plasmin binds to many matrix molecules in a lysine-dependent process (reviewed by Moscatelli and Rifkin, 1988), and the immobilized enzyme is less susceptible to inhibition by α_2-antiplasmin. Plasminogen activators have also been identified in the matrix secreted by bovine aortic endothelial cells and other cell types, including neurons (Krystosek and Seeds, 1986). The metalloproteinases invadolysin and matrilysin have domains with homology to the collagen-binding domain of fibronectin, and these enzymes are typically purified on gelatin affinity columns (Apodaca *et al.*, 1990). The inclusion of conserved domains may only reflect the evolutionary origins of the enzymes (Patthy, 1985). Not only does the matrix bind enzymes and inhibitors, but it is a reservoir of cytokines and growth factors. Local destruction of matrix could be expected to release these factors to areas of cell contact, thereby affecting growth and differentiation of the local cell population. The immunolocalization of bFGF to basement membranes and mesenchymal matrices of diverse tissues (Gonzalez *et al.*, 1990) is founded in the affinity of this growth factor for high-molecular-weight heparan sulfate proteoglycans. bFGF can be released from these complexes by UPA (Saksela and Rifkin, 1990) and is capable of eliciting further transcription of enzymes and inhibitors to propagate the effect. TGF-β1 has also been immunolocalized to connective tissue, specifically around myotubes, notochord, and endocardial cushions in the developing mouse embryo (Heine *et al.*, 1987). The growth factors TGF-β_1, -β_2, and bone morphogenetic proteins 1-, -2A, and -3 have all been isolated from demineralized bone; neuronal matrices too are laced with factors and cytokines.

Not only does the binding of degradative enzymes serve to localize and restrict the process of degradation, but it may also concentrate various components of the cascade to bring them to a critical mass capable of effective catalysis.

5. Physiological Processes of Degradation and Remodeling of Extracellular Matrix

Matrix degradation, measured as collagen turnover, is a very slow process in structures like skin and tendon of adult animals. However, there are various sites of degradation, remodeling, and migration that are obvious physiological niches for the enzyme activities we have described. Very little data are available to describe their expression and control *in vivo*: however, a brief description of classic cases may serve to illuminate the extent of each process and the types of enzyme activity that would be required. Other areas of extensive remodeling that are not discussed below because of the paucity of useful information are the processes of nerve regeneration and wound healing. Both are associated with involutive, inflammatory, and regenerative phases that involve a complex interplay of cell types and growth factor elicitations.

A novel approach that has the capability of identifying remodeling processes has been described recently by Wu *et al.* (1990). These investigators constructed a substrate (the collagen $\alpha 1$ chain) resistant to cleavage by the degrading enzyme collagenase (by mutating the G^{775}–I/L^{776} cleavage site by site-directed mutagenesis of cloned sequence). This substrate is incorporated into trimers to form mixed wild-type/mutant triple-helices, conferring immunity to cleavage upon the whole molecule. The recognition site for collagenase must therefore involve all three strands. Conventional transgenesis using this mutated gene would be expected to identify sites of collagen type I turnover in mice. Another novel experimental approach to identifying sites of enzyme activity by immunolocalization is to visualize the degraded matrix target. For example, it has been shown that an antibody can discriminate between denatured and native collagen type II, which is useful in the study of activation of chondrocytes by IL-1 (Dodge and Poole, 1989).

5.1. Bone Deposition and Remodeling

Bone remodeling is a constant process, both in the adult and during the developmental expansion of the fetal skeleton. The degradation of this matrix is delicately balanced against synthesis, and dysfunctional aspects of the degradative machinery are revealed as failure of bone rudiments to grow (osteopetrosis) or pathologic removal of compact bone by inappropriate stimulation of osteoclasts (osteoporosis). The dynamic equilibrium of bone matrix is regulated by several hormonal stimuli (reviewed by Huffer, 1988; Vaes, 1988), notably parathyroid hormone (PTH, stimulating resorption) and calcitonin (stimulating assimilation). In adult bone, two main cell populations are thought to be responsible for maintaining the equilibrium: osteoblasts, a highly biosynthetic cell type capable of complex processes of matrix fibrillogenesis, and osteoclasts, a cell type originating in the macrophage lineage. Degradation of a microscopic domain (such as a region of a Haversian osteon), and its subsequent replacement, appears to be a cyclical process of activation of one cell type or the other. Both cell types are highly polarized with respect to matrix. Osteoblasts line canals that contain capillaries; they secrete and assemble matrix basally to generate spicules or tubes. Underneath the osteoblast cell layer is a layer of unmineralized bone or osteoid which apparently consists of all the protein connective tissue elements without calcium phosphate in the form of hydroxyapatite. Osteoclasts resorb matrix by using the sequestered battery of enzymes described in Section 2.3.

Osteoid is not a permissive substrate for osteoclast attachment and activity, and requires preparation in some fashion by osteoblasts. In fact, osteoblasts direct many of the activities of osteoclasts (Chambers, 1988). For example, osteoclasts, although they are the principal PTH-responsive cell type, do not even express the PTH receptor: instead, PTH interacts with osteoblasts and the signal is translated into a stimulatory activity for osteoclasts. PTH-stimulated osteoblasts round up and make an attachment surface available to osteoclasts. Osteoblasts stimulated *in vitro* by 1,25-vitamin D_3 respond by degrading collagen films, in a process that depends upon the PA/plasmin/metalloproteinase cascade (Thomson *et al.*, 1989). In the absence of serum, there is no degradation in response to vitamin D_3; collagenolysis can be restored in these cultures by the addition of exogenous plasminogen. All degradation is inhibited by the addition of TIMP. Proteinases and inhibitors collected in the medium do not reflect the cellular collagenolytic activities accurately: conditioned medium is not collagenolytic in soluble fibril assays and does not contain measurable collagenase (Thomson *et al.*, 1989; Rifas *et al.*, 1989), but does contain matrilysin and high levels of TIMP (the inhibitor presumably obscuring any collagenolytic activity of conditioned medium). Cultured osteoblasts express cell-associated PA and high levels of evenly staining cell surface-associated PAI-1 (Thomson *et al.*, 1989), in common with the fibrosarcoma cells described in Section 4.1.

Matrix is degraded during several phases of bone development. Bone is a totally nonelastic medium that relies totally on breakdown and resynthesis for expansion. There are two main routes of bone development: endochondral ossification and intramembranous ossification. The first route employs a cartilage template developed from a mesenchymal anlage and is typical of long bones and mandible. Around the center of the cartilage model, a bony collar (the periosteum) is laid down by a process of intramembranous ossification within the perichondrium (a connective tissue sheath encompassing the cartilage template). Cells including osteoblasts and other precursors such as fibroblasts, break out of the periosteum, traverse layers of connective tissue, and invade the hypertrophic cartilage in the reactive center of the cartilage model. This invasion is diagrammed in Fig. 8-6. The calcified cartilage template is degraded by the incoming cells, and osteoblasts aggregate to synthesize and deposit bone. The matrix degraded during this process is based on cartilage collagen type II, together with an alternatively spliced form of collagen type IX (a FACIT collagen), which is associated with the transition to hypertrophy. Type X collagen is also typical of hypertrophic cartilage, and is a substrate for several gelatinases. The wave of ossification proceeds bidirectionally outwards, generating a characteristic pattern of zones of resting, growing and hypertrophic cartilage ahead of the ossifying tissue. Meanwhile, in the wake of the synthetic process, another wave of invasion by hematopoietic cells removes the core of newly synthesized bone to generate a hollow marrow populated by developing hematopoietic lineages. The head of the bone is excavated by a second group of periosteal buds. Intramembranous bone such as calvaria is laid down within a fibrous layer of connective tissue; this process is probably not associated with the massive resorption and replacement typical of endochondral bone formation.

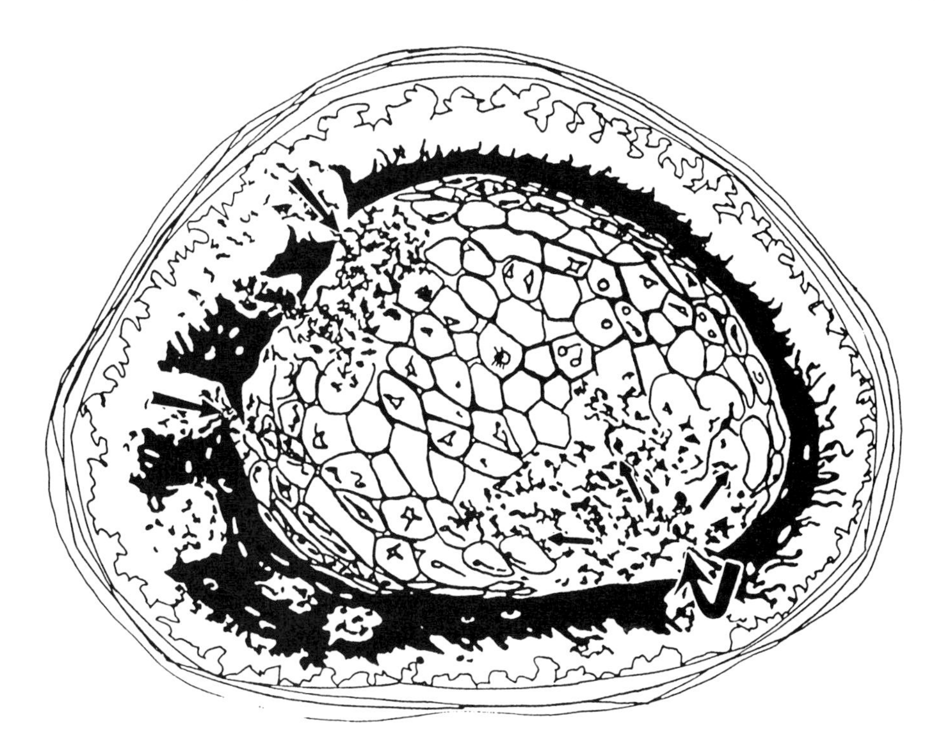

Figure 8-6. The invasive process associated with ossification of cartilage rudiments. A transverse section of human second metacarpal (11 week old) shows that the dense sheath of connective tissue (the periosteum) is degraded and removed at the points of invasion of periosteal buds (arrows). Ossification begins as a collar around the center of the cartilage precursor and then spreads outwards. Calcified cartilage is destroyed by the invading cells; then blood vessels enter the matrix and osteoblasts develop and initiate ossification. (Redrawn from Gray *et al.*, 1957.)

These are complicated developmental processes, and little information is so far available concerning the enzymes involved in the invasive and resorptive phases. Chick limbs from various developmental stages have been dissected into distinct zones and the proteinases secreted by these zones in organ culture have been analyzed (Mikuni-Takagaki and Cheng, 1987). The resting, proliferating, and hypertrophic cartilage regions together with periosteal bone do not show net proteolytic activity (measured as the ability to degrade proteoglycans or gelatin), but several activities have been identified in marrow, marrow cavity surround, and tarsus. These activities fall into the typical TIMP-inhibitable latent gelatinase class and into a second class of spontaneously active metalloproteinases active at neutral pH and resistant to inhibition by TIMP.

The expression of TIMP is directly correlated with the ossification of tissues, where it may be acting as a safeguard protecting the newly deposited matrix from the plentiful proteolytic machinery. TIMP appears at the first sites of ossification in the embryo, namely, mandible and clavicle (intramembranous

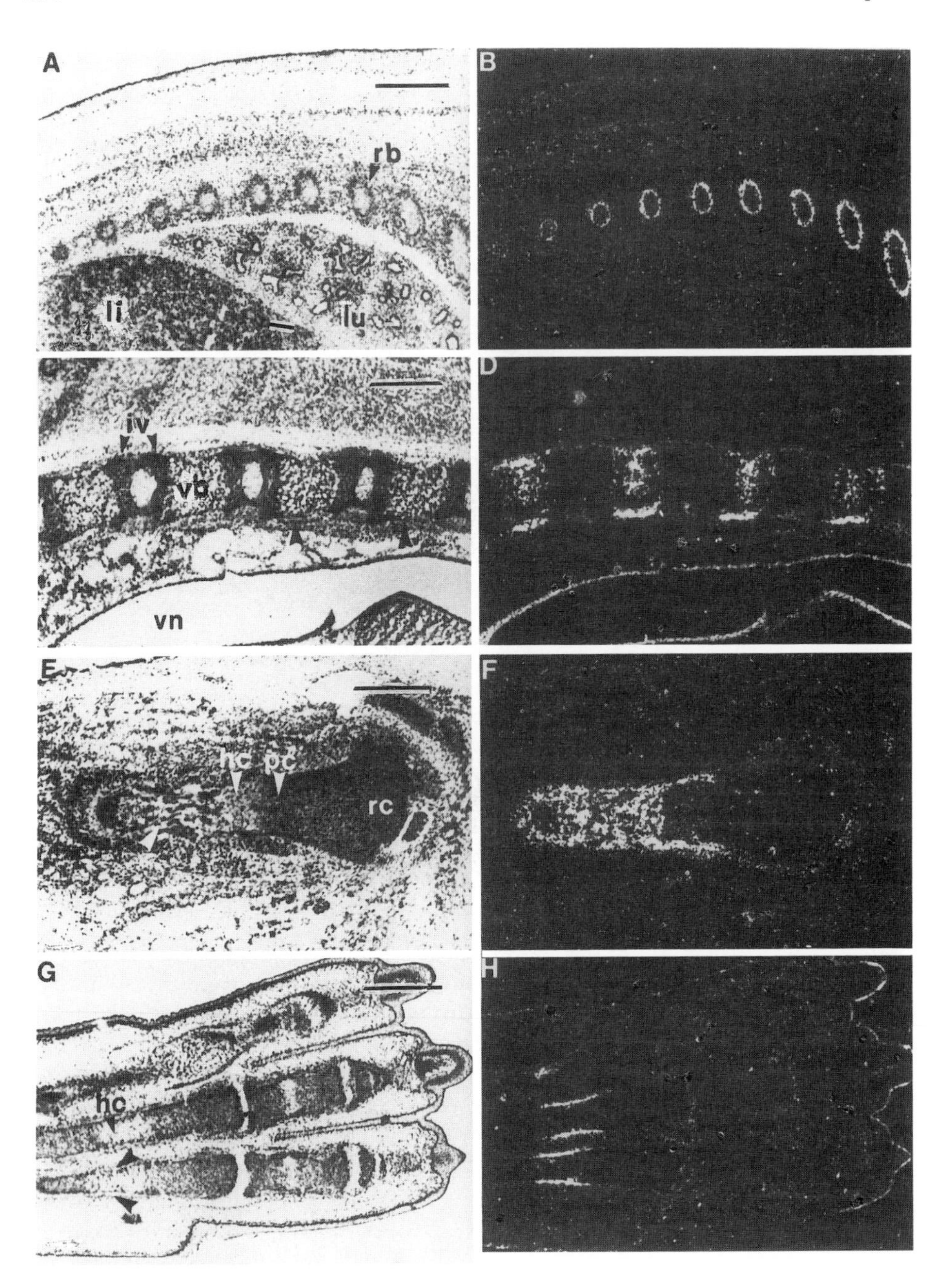

Figure 8-7. Localization of mRNA for TIMP in mouse embryos. *In situ* hybridization of TIMP mRNA reveals the expression pattern of this inhibitor in the bones of a developing mouse embryo. Panels A, C, E, and G correspond to the sections in panels B, D, F, and H, respectively, under brightfield illumination. Panels A and B show sagittal sections of rib in sections from a 13.5-day p.c. embryo. Panels C and D show the vertebral discs of a 17.5-day p.c. embryo, where TIMP expression complements the pattern of the intervertebral discs (iv) and also stains the endothelium of the vein (vn). Panels E and F show a frontal section of the 15.5-day p.c. humerus: TIMP is

ossification), followed by ribs and long bones (see Fig. 8-7). High-resolution data are not yet available to establish the cellular source of TIMP, but the *in vitro* data described above would suggest it to be osteoblasts.

Cartilage operates in regions of wear and tear to absorb mechanical stresses and is therefore also continuously remodeling. Immunolocalization of TIMP, collagenase, stromelysin, and gelatinase in rabbit femoral growth plate shows that all three enzymes and the inhibitor are made by the growth plate chondrocytes (Brown *et al.*, 1989). Little other information is available that describes the dynamic processing of this tissue, but it is known that cartilage proteoglycans turn over quite rapidly (Chapter 5), and that articular cartilage is a rich source of stromelysin (Gunja-Smith *et al.*, 1989).

5.2. Mammary Gland Involution

During pregnancy the mammary gland consists of a dense branching alveolar system of epithelial cells. After the cessation of lactation, the gland regresses and the epithelial component, together with basement membranes and underlying matrix, is destroyed and replaced by fat tissue. The composition of the matrix is critical to the expression of differentiated mammary gland function.

A careful study of the gelatinases secreted by mammary cells has shown that active matrilysin is secreted basally, toward the basement membrane and underlying matrix, at the time of active involution (i.e., about 3 days after pup removal). The active enzyme is also associated with the remodeling phases of the pregnant gland, whereas only inactive matrilysin is secreted in the fully expanded lactating gland. The expression of TIMP is maximal during the early stages of involution and may prevent a premature onset of the breakdown process (Talhouk *et al.*, 1991). Immunostaining of human mammary gland tissue has revealed that the myoepithelial cell population is principally responsible for the synthesis of matrilysin (Monteagudo *et al.*, 1990).

A similar pattern has been observed for UPA. During lactation, UPA activity is very low and increases slowly during involution (Ossowski *et al.*, 1979; Busso *et al.*, 1989). However, this enzyme appears to immunostain to the apical side of mammary epithelial cells (Larsson *et al.*, 1984), and evidence is beginning to accumulate that suggests that UPA may have a role in establishing and maintaining the morphology of branching ducts. This enzyme is also apically located in angiogenic endothelium and the ducts of the vas deferens.

In later stages of involution the proteinase profile may be dominated by macrophage infiltration.

---

expressed in zones of endochondral bone deposition, and excluded from cartilage. Panels G and H show a frontal section of a 17.5-day p.c. hindfoot. Arrowheads indicate the primary ossification collar of the forming metatarsal, which is clearly coincident with TIMP expression. hc, hypertrophic cartilage; li, liver; lu, lung; pc, proliferating cartilage; rb, ribs; rc, resting cartilage; vb, vertebral body. Bars = 200 μm. (From Flenniken and Williams, 1990.)

5.3. Uterine Involution

The postpartum involution of uterus involves one of the most rapid processes of extracellular matrix known: in rodent species, 90% of wet weight of tissue resorbs in 5–6 days. At this stage the uterus consists of maternal decidual tissue and muscle intertwined in a dense mass of structural tissue and basement membranes. Two enzyme activities have been correlated with the onset of involution (Woessner and Taplin, 1988): collagenase and an enzyme of 24 kDa (probably PUMP). Immunohistochemical staining for collagenase shows the enzyme to be present at relatively high levels even in nonpregnant uterus, so the amount of procollagenase is not likely to be rate determining. Collagenase is secreted by smooth muscle cells of myometrium, and also probably by the many highly reactive cell types invading the uterus at parturition (including macrophages and eosinophils). These invading cells may regulate activation of the preexisting zymogens. The source of PUMP is not known. The total collagen content of the uterus is under hormonal control: thus, estradiol maintains the amount of uterine collagen by inhibiting the breakdown process, presumably via control of the proteolytic activity of the endogenous histiocyte population.

5.4. Ovulation and Fertilization

Ovulation is another example of a process that apparently depends on the PA–metalloproteinase cascade. Inhibitors of both serine proteinases and collagenase can prevent follicular rupture (Brannstrom *et al.*, 1988; Beers, 1975). Ovulation starts with the cleavage of the proteoglycan that is the main component of follicular fluid to liquify this matrix, followed by an asymmetric rupture of the thecal wall. PAs may also help to avoid the formation of blood clots as blood vessels break. Granulosa cells, thecal cells, and the egg itself all produce PAs; the source of the collagenase is not clear. Considering the intensity of the proteolytic processes, it is interesting that as the follicle continues to develop into the corpus luteum, the expression of TIMP soars to one of the highest observed for any tissue [detected by *in situ* hybridization of mRNA by Nomura *et al.* (1989) and by protein immunostaining by our group (unpublished observation)]. This again may be a protective backlash to prevent any further tissue dissolution.

The mammalian egg is surrounded by a protective matrix called the zona pellucida, which consists mainly of three glycoproteins (ZP1, 2, and 3). These proteins are cleaved by specific proteinases during the processes of coat hardening, penetration of the zona by the sperm, and hatching of the blastocyst before implantation. Secretion of the ZP2 proteinase is correlated with hardening of the zona pellucida coincident with activation of the egg. This proteinase is secreted within 5 min of activation and diffuses into the zona, making a single cut within ZP2 (Moller and Wassarman, 1989). Another serine proteinase, acrosin, is carried by the sperm head to facilitate the acrosomal reaction (entry of sperm into the egg). This enzyme cleaves two of the three ZP proteins (Hardy *et al.*, 1989; Adham *et al.*, 1989). The hatching enzyme, strypsin, is secreted only by the mural trophectoderm (the distal end of the embryo, responsible for implantation) and has a trypsinlike specificity that generates sufficient localized ZP lysis to allow the embryo to squeeze out (Perona and Wassarman, 1986).

Recently, hatching of the sea urchin blastula from the hard coat that protects it during early development has been found to depend on an enzyme closely resembling stromelysin (Lepage and Gache, 1990). There is a remarkable conservation of sequence from the sea urchin to mammalian counterparts (PRCgVPDV, urchin activation sequence; VAAHEfGH, urchin zinc binding region).

5.5. Preimplantation Development

At the blastocyst stage, the embryo develops structurally to delineate cell compartments and support further growth. Very early cell types express the PAs and various members of the metalloproteinase and inhibitor classes. The mRNA transcripts for TIMP, stromelysin, and collagenase are present in the egg and increase at the blastocyst stage. This is associated with the appearance of the proteinase activities, identified by substrate gel electrophoresis (Brenner *et al.*, 1989). Parietal endoderm synthesizes TPA, whereas the other cell types make UPA in the early embryo. Two early cell lineages that are active in remodeling processes are the parietal endoderm and the trophoblast. Trophoblast is responsible for the invasion of the mammalian blastocyst and is one of the most highly invasive normal cell types. This property is strictly temporally limited (for further discussion see Section 6.4). Parietal endoderm migrates out from the inner cell mass of the early blastocyst, along the inner aspect of the trophoblast cell layer. These are highly motile cells responsible for the synthesis and deposition of Reichert's membrane, the first basement membrane. Parietal endoderm cells are an example of a cell type that can deposit a highly complex matrix in the wake of a directed migration. Outgrowth of these cells is influenced by adding TIMP to cultures of blastocysts, implicating metalloproteinases in this migratory process (Behrendtsen *et al.*, 1991).

5.6. Postimplantation Development

Very little is known about the expression of metalloproteinases in the gastrulating embryo. With the development of the third germ layer, the mesoderm, comes the first connective tissue surrounding motile fibroblast cell types. There are very active processes of remodeling during notochord induction, somitogenesis, heart formation, limb bud morphogenesis, neural outgrowth, Wallerian degeneration, and many others. In sea urchin, there is rapid collagen degradation during gastrulation. One of the few developmental analy-

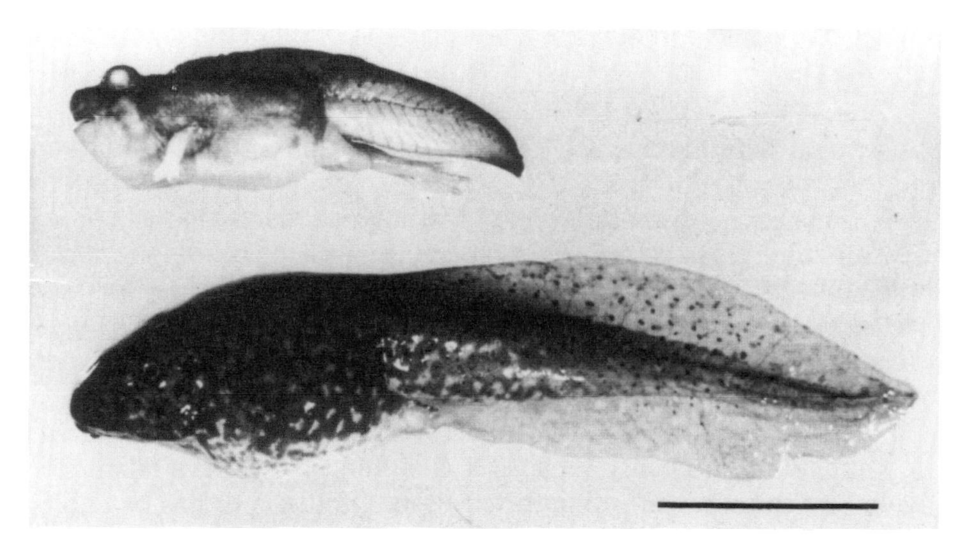

Figure 8-8. Tadpole tail resorption. At the top is a tadpole (*Rana catesbeiana*) treated with thyroxine for 9 days. Below is a resting tadpole. The collagenous substructure of the tail undergoes a hormone-dependent resorption. Bar = 1 inch. (From Gross, 1981.)

ses of expression of any of the components involved in matrix degradation describes the distribution of TIMP mRNA in the developing embryo by using *in situ* hybridization and RNase protection (Nomura *et al.*, 1989; Flenniken and Williams, 1990). TIMP can be detected by RNase protection in embryos only 7.5 days old, and even earlier when the reverse transcription-polymerase chain reaction is used (Brenner *et al.*, 1989). TIMP is highly expressed in bone from 15.5 days onwards (see Section 5.1 and Fig. 8-7), and is clearly regulated during morphogenesis of kidney, first branchial arch, lung, gut, and extraembryonic tissues. Higher-resolution studies may help to elucidate the role of TIMP in these processes.

5.7. Tadpole Tail

The first description of matrix-degrading metalloproteinases came from Gross and Lapiere in 1962 studying *Rana* tadpole tail involution (reviewed by Gross, 1966). They found an enzyme, collagenase, that was capable of degrading triple-helical collagen and was actively produced by fragments of tissue from the tails of tadpoles induced to undergo morphogenesis by thyroxine (Fig. 8-8). Adding fragments of separated tail fin tissues to radioactive collagen gels showed that it is the epidermis and not the mesenchyme that is responsible for collagenase production. In contrast, the mesenchyme is the source of hyaluronidase. There is a balance between the degradation of the collagen fibers of the tail and the new synthesis of collagen which maintains the animal's function during the water-to-land transition; new collagen is completely protected from the degradative process. Morphologically, the collagen fibrils appear to be frayed and disorganized during the involution process. Mesenchymal cells invade the collagenous lamellae and contain collagen fibrils in phagocytotic vesicles. No more details are available to describe this involutive mechanism.

6. Lysis of Basement Membranes

The basement membrane compartmentalizes various cell types so that they are not in surface contact or junctional communication. This lamina is a network of intertwined macromolecules with properties that differ from those of the stromal, or interstitial, milieu, mainly in that it constitutes an effective barrier to cell migration. Some normal cells, however, are able to cross basement membranes. The invasion processes may have either of two characteristics: the cells may invade in columns or aggregates (like endothelium or trophoblast) or as single cells (such as PMNs or macrophages).

6.1. Endothelium: The Process of Angiogenesis

The outgrowth of capillaries starts with the localized lysis of an established vessel wall and basement membrane. An activated population of endothelial cells forms a bud and grows out as a tube or cord of cells in response to angiogenic factors. Endothelial cells migrate as a community of cells, comprising lytic apical cells supported by a more proximal proliferative population. This communication can be demonstrated in wounded endothelial cell monolayers *in vitro:* groups of about six or seven of the migrating cells are linked by gap junctions (Pepper *et al.*, 1989). The outgrowth of endothelium and invasion through the vascular basement membrane into the interstitial matrix, and migration into the stromal matrix underneath, is dependent on the action of proteinases (Moscatelli and Rifkin, 1988; Mignatti *et al.*, 1989: Folkman and Klagsbrun, 1987). Endothelial cells respond *in vitro* to the addition of angiogenic factors (e.g., bFGF) with a surge of PA and metalloproteinase synthesis (Saksela *et al.*, 1987; Saksela and Rifkin, 1988). Endothelial cell invasion is best modeled in three-dimensional matrices, since it has been clearly shown that these cells respond differently with respect to differentiation and mitogenesis depending on cell interactions and the tensile properties of matrix (Ingber and Folkman, 1989; Merwin *et al.*, 1990). The polarity that is established in gel cultures is important to remodeling reactions that depend on directional secretion of enzymes. Endothelial cells seeded onto a collagen gel, fibrin clot, or amniotic membrane, respond to various stimuli by invading the matrix and forming tubules. The invasion correlates with an increase in proteolytic activity, and is totally dependent on the activity of PA, plasmin, cell contact, and metalloproteinases (Montesano and Orci, 1985; Mignatti *et al.*, 1989). Anticatalytic antibodies to collagenase and to matrilysin abolish the invasive process (Mignatti *et al.*, 1989). Stimuli that can induce the invasive phenotype include angiogenic factors that are likely to be relevant *in vivo*, such as bFGF, and reagents like phorbol esters and vanadate (an inhibitor of tyrosine phosphatases), all of which probably stimulate protein tyrosylation (Montesano *et al.*, 1988). It is important that these experiments use a physiologically relevant matrix: for example, if endothelial cells are grown on artificial fibrin clots in the presence of phorbol esters they lyse the matrix entirely in the absence of serine proteinase inhibitors, probably exaggerating the contribution of UPA to the outgrowth process. On collagen gels, the stimulation of PA activity induced by phorbol esters is not sufficient to dissolve the gel (Montesano and Orci, 1985). The morphology of the sprouting capillary is the result of a rather subtle interplay of proteinases and inhibitors, which is discussed in Section 7.6.

6.2. Neutrophils: The Process of Extravasation

In response to inflammatory signals originating in the interstitium, circulating neutrophils cross the basement membrane of the postcapillary venule and migrate toward the stimulus. The adhesive interactions that direct this process are relatively well characterized and are followed by the intrusion of cellular processes from the neutrophils into the endothelium and displacement of this cell layer. The cells then appear to pause at the basement membrane before completing the extravasation; the process of membrane dissolution is still a puzzle. Matrilysin is rapidly secreted from the specific granules of neutrophils as the cells contact chemoattractants. No obvious gaps in the membrane can be identified by microscopic examination, but the integrity of a basement membrane (reconstructed *in vitro*) is transiently affected (Huber and Weiss, 1989). In this model system, migration appears to be resistant to inhibitors of elastase, cathepsin G, metalloproteinases, endoglycosidases, and oxygen radical formation, and the disruption of barrier function is restored rapidly by the endothelium after transmigration.

6.3. Macrophages: The Response to Injury

Macrophages play four major roles in matrix remodeling: they degrade matrix directly by secreting UPA and metalloproteinases; they participate in the activation of proteinases made by other cells; they produce cytokines (e.g., IL-1 and tumor necrosis factor-α) that can stimulate fibroblasts and endothelial cells to express metalloproteinases; and they are phagocytic, ingesting and digesting fragments of matrix produced by proteolytic attack or by injury. The central role of macrophages in the repair of injury has been inferred from experiments in which animals were treated with antimacrophage serum or with glucocorticoids prior to wounding (Leibovich and Ross, 1975). If inflammatory macrophages are thus prevented from entering the sites of injury, the resulting wounds repair very poorly. Macrophages participate in many phases of the injury response, from directing restorative cellular proliferation to restructuring of the extracellular matrix. The process of matrix degradation by macrophages begins as the cell homes along a trail of chemoattractant toward a site of injury. During chemotaxis the macrophage becomes polarized and concentrates the UPA–UPA receptor complex to the leading edge of the cell (see Section 4.2 and Fig. 8-5). This activator and downstream proteinase activity are thus focused to the cellular margin responsible for breaching matrix barriers. Invadolysin, the metalloproteinase most strongly implicated in the invasive behavior of other cell types, is also the major secreted metalloproteinase of macrophages (Behrendtsen *et al.*, 1991). The role of macrophages in matrix degradation may be particularly significant because of their ability not only to secrete relevant proteinases but to activate them even in the tissue culture dish.

6.4. Trophoblast: The Mammalian Implantation Process

The trophoblast is one of the first terminally differentiated cell types in mammalian embryos and the first step toward establishing the placenta responsible for nutritional support of the fetus. The limited time frame of invasiveness observed for trophoblast *in vivo* (4.5–7.5 days for mouse, first trimester for human; Fisher *et al.*, 1989) can be modeled *in vitro* by plating whole embryos or isolated trophoblast cells on suitable complex matrices. Invadolysin is prominent in the lytic profiles of both the mouse and human models and has been directly implicated in the process by inhibition with an anticatalytic antibody to invadolysin (Fig. 8-9; Librach *et al.*, 1991; Behrendtsen *et al.*, 1991). The significance of PA in this process is unclear. UPA is produced by trophoblast during the time frame of uterine invasion in mouse (Strickland *et al.*, 1976; Sappino *et al.*, 1989): in fact, PA isozyme expression is carefully controlled in early mouse embryo and can be used as a marker of cell type. PAI-2, the PA inhibitor prominent in placenta, is immunolocalized to the outermost layer of syncytiotrophoblast in the placental villus (Astedt *et al.*, 1986). However, inhibitors of PAs affect only partially, or do not affect at all, the implantation process modeled *in vitro* (Librach *et al.*, 1991).

7. Cell–Cell Interaction: The Net Remodeling Process

Cell types interact to generate a remodeling environment, and it is a well-established principle that mesenchymal cells can direct the epithelium during morphogenetic tissue development. One of the first examples of interactive proteolysis was described by Grillo and Gross (1967), who observed that a wound edge in guinea pig skin was highly collagenolytic, whereas if the epithelium and stromal cells were separated, the activity was greatly reduced. Combinatorial effects like this can arise in many ways: growth factors produced in one cell type stimulate production of proteinases (or inhibit synthesis of inhibitors) in another cell type; an activator is produced in the first cell type

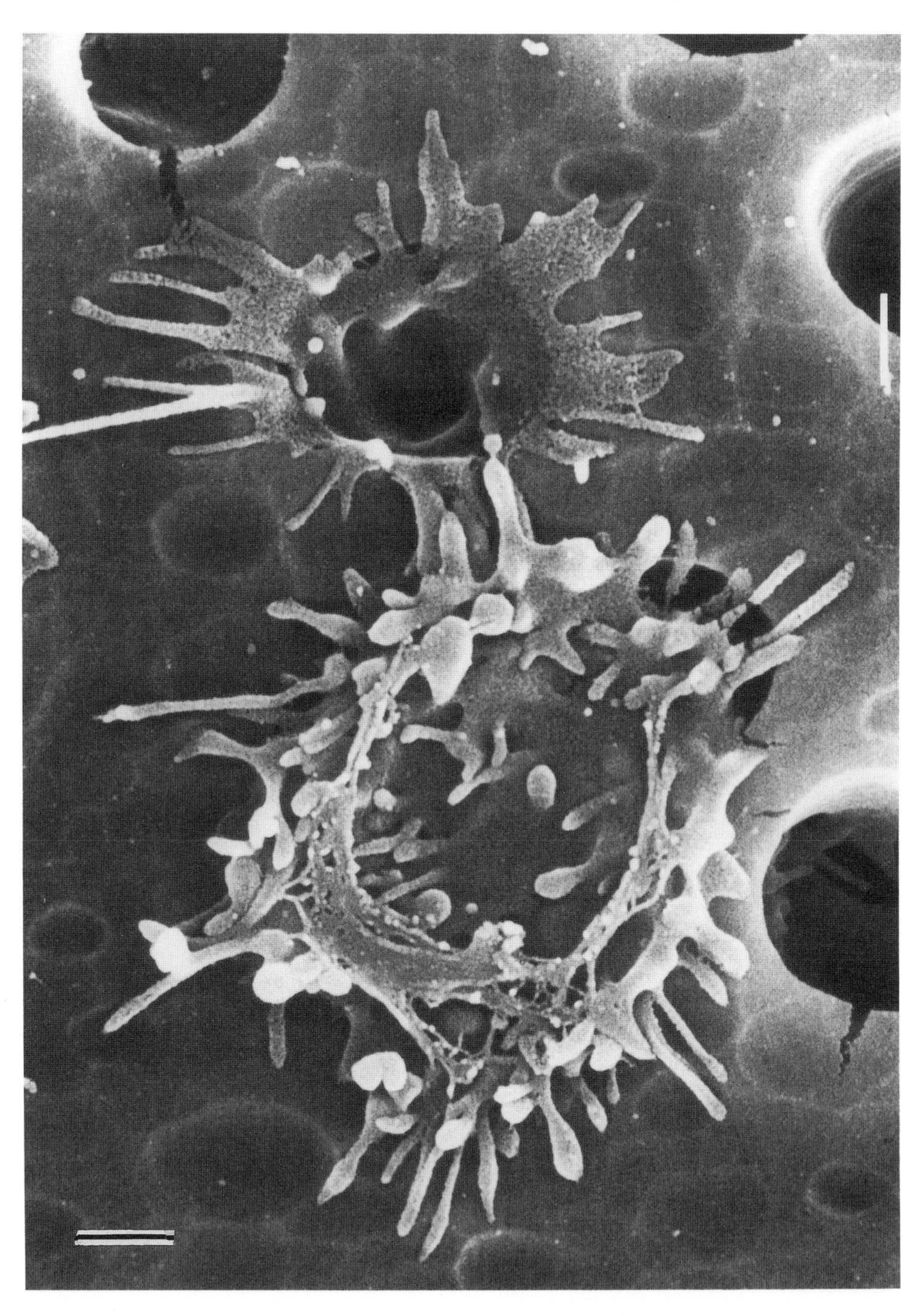

Figure 8-9. Scanning electron micrographs of human cytotrophoblast invasion. Human cytotrophoblast cells from first-trimester pregnancies were isolated and cultured on Matrigel-coated polycarbonate filters as described by Librach *et al.* (1991). During the invasive period of trophoblast development, these cells invade the matrix as columns or aggregates and extend fingerlike invadopodia with highly invasive properties. On polycarbonate filters with a 3-μm pore diameter, the cells cross the matrix and invadopodia appear at the basal side of the Millipore filter. These have been visualized by scanning electron microscopy. Bar = 1 μm. (From C. Librach, UCSF.)

and latent effector enzymes in the other; one cell may have the receptor for a proteinase and the other may produce the proteinase itself. The interaction of one cell can also be expressed through secretion of a particular matrix molecule, such as tenascin, SPARC, or fibronectin, which induces enzyme expression (Werb *et al.*, 1991), or via a particular cell–cell receptor interaction that induces some secondary signal. Examples are described below.

7.1. Cell Interaction-Dependent Collagenase Induction

Collagenase induction is one of the results of many different types of cell interactions, e.g., lymphocyte-stimulated macrophages, macrophage-stimulated synovial fibroblasts, corneal epithelial cells and corneal fibroblasts, epidermal keratinocytes and fibroblasts, carcinoma cells and macrophages, and carcinoma cells and fibroblasts (reviewed by Lyons *et al.*, 1989). If induction is mediated by growth factor release, the recent availability of many purified growth factors may establish these responses in serum-free culture. However, there are also more complicated cell–cell communication loops such as that described for mammary carcinoma cells by Lyons *et al.* (1989). An epithelial-like subclone (which typically arises as a natural variant during culture of mammary carcinoma cells) produces a growth factor that sustains a myoepithelial-like subtype, which in turn responds to this factor with release of another factor which induces collagenase in the epithelial cell type. The myoepithelial stimulus may not be a secreted growth factor: a 58-kDa integral membrane protein on the surface of a tumor cell line can induce collagenase in contacting fibroblasts (Ellis *et al.*, 1989).

7.2. Separation of UPA and UPA Receptor Expression

UPA is a good candidate for the cell type-specific expression of receptor and enzyme; indeed, this separation has been demonstrated for spermatocytes, which express urokinase receptor and bind urokinase, which is secreted into the lumen of the ductus deferens by the epithelium (Huarte *et al.*, 1987).

7.3. Modulation of Endothelial Cell Migration by Accessory Cells

The migratory reactions that are inhibited in endothelial cells by TGF-β1 and bFGF (discussed in Section 6.1) can be mimicked *in vitro* by coculture with pericytes or smooth muscle cells. Pericytes exist on the abluminal side of endothelial cells in capillaries, and smooth muscle cells are separated from endothelium by a basement membrane in the larger blood vessels. After wounding of an endothelial monolayer, the outgrowth of endothelial cells is reduced in the presence of either of these cell types. Inhibition depends on cell–cell contact and on the presence of plasmin and either TGF-β1 or bFGF (Antonelli-Orlidge *et al.*, 1989; Sato *et al.*, 1990). All of these cells produce latent TGF-β1 and urokinase, and so the source of the cellular synergy is puzzling. It is also known that bovine endothelial cells inhibit collagenolysis catalyzed by fibrosarcoma and melanoma cells, and this has been traced to the production of two proteinase inhibitors, TIMP and TIMP-2 (De Clerck *et al.*, 1989).

7.4. Communication between Neural Cell Types by Means of Proteinase Mediators

Glial cells produce a factor that stimulates neurite outgrowth and stabilizes growth cone adhesion: this has been identified as PN-1 (Gloor *et al.*, 1986). The target for this inhibitor has not been formally identified, but it is known that neuronal growth cones express a localized fibrinolytic activity and deposit urokinase in the matrix (Krystosek and Seeds, 1981, 1986). As granule cells migrate, UPA activity increases; migration is slowed by glial-derived PN-1 (Lindner *et al.*, 1986). Hence, these two cell processes show opposite responses to the inhibitor: one is stimulated, the other is inhibited. This may reflect a rather subtle equilibrium that allows the adhesion and extension of growth cones.

Three-dimensional cultures of astrocytes are an impenetrable barrier to outgrowth of axons from postnatal neurons; however, embryonic dorsal root ganglia can extend axons through these cultures for considerable distances. The growth of the embryonic ganglia was much reduced in the presence of serine proteinase inhibitors, but not in the presence of TIMP (Fawcett and Housden, 1990).

7.5. Tumorigenicity Resulting from a Shift in the Proteinase/Inhibitor Equilibrium

3T3 fibroblasts in culture synthesize a number of proteinases and inhibitors (including TIMP-2). If the synthesis of TIMP is decreased in these cells with antisense TIMP mRNA, they become invasive and metastatic (Khokha *et al.*, 1989). This suggests that the invasive capability of cells can change with a shift in the proteinase/inhibitor equilibrium.

7.6. Dependence of Vasculogenesis on the Balance of Proteinases and Inhibitors

The process of angiogenesis has been described in Section 6.1 as a controlled physiological invasive reaction dependent on the PA/plasmin/ metalloproteinase cascade. Manipulation of the proteolytic profiles using growth factors has been used to correlate the endothelial tubular morphology with a particular balance of proteinases and inhibitors. Thus, the angiogenic factor bFGF can induce the synthesis of more PA than PAI-1, with the result that the

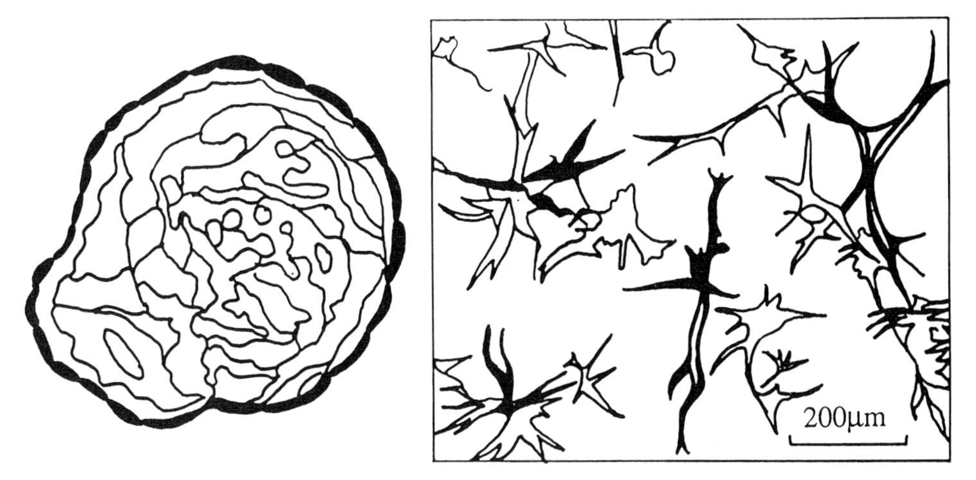

Figure 8-10. The effect of serine proteinase inhibitors on vasculogenesis *in vitro*. Endothelioma cells, derived from hemangiomas induced in mice by expression of polyoma virus middle T antigen, form cystlike structures grown in fibrin gels *in vitro* that resemble their parent hemangioma (A). Cysts form after 6–12 days of culture from dispersed single cells. In the presence of the serine proteinase inhibitor Trasylol (1000 kIU/ml), the morphology of cultures is altered to thin branching cords of endothelium split by narrow spaces (B). The expression of middle T antigen in the endothelioma cells is associated with very high UPA expression (in the absence of any PAI-1). Normal endothelium in culture forms thin branching cords similar to those of the endothelioma cells in the presence of serine proteinase inhibitor, suggesting that the cystic phenotype is the result of excessive fibrinolysis. (Redrawn from Montesano *et al.*, 1990.)

endothelium buds out into the matrix. TGF-β1 also induces PA and PAI-1, but production of the inhibitor exceeds that of the proteinase and the induction is on a different time scale, producing net antiproteolysis and no vasculogenesis (Pepper *et al.*, 1990). Adding both bFGF and TGF-β produces a compromise, and the pattern of endothelial invasion is limited to solid cords of cells in the upper layers of a fibrin clot. The antagonistic effects of bFGF and TGF-β1 are also observed on amniotic membrane substrates (Mignatti *et al.*, 1989). Actually, both of these reagents are angiogenic *in vivo*, probably owing to a mediation of the TGF-β1 response by fibroblasts (Yang and Moses, 1990). Endothelial cells transformed with the polyoma virus middle T antigen form hemangiomas *in vivo*, and when grown *in vitro* on a fibrin clot they form a rather similar cystlike structure. These cells also express high UPA activity and low PAI-1, and the architecture changes to a typical tubular morphology in the presence of serine proteinase inhibitors (Fig. 8-10; Montesano *et al.*, 1990).

7.7. Effect of the Collagenase–TIMP Equilibrium on Salivary Gland Morphogenesis

The salivary gland can develop *in vitro* from a bud excised at 12 days p.c. to a more mature gland 24 hr later, with a consistent branching morphology.

The addition of collagenase or collagenase inhibitor to the culture decreases or increases the cleft formation, respectively (reviewed by Alexander and Werb, 1989). Hence, the presence of more inhibitor (and more intact collagen) is correlated with a more convoluted structure.

8. The Pathological Invasive Phenotype

8.1. Tumorigenic Cell Types: Metastasis and Invasion

Metastatic tumor cells have the ability to cross basement membranes and displace most tissue types. Over the years, the atypical invasive properties of tumorigenic cell types have been correlated with the usurped expression of various enzymes and with no one activity in particular. Many reviews deal specifically with this area (Zucker, 1988; Moscatelli and Rifkin, 1988; Dano *et al.*, 1985). The variety of enzymes identified is not surprising, since the cascade is sufficiently complex that it can be altered at many levels to produce a lytic phenotype. Purely correlative studies that look at just one component can be misleading to the identification of critical factors that confer metastatic and invasive properties upon tumor cells. PA was originally isolated as a fibrinolytic activity associated with the tumorigenic phenotype. However, it is usually present in catalytic quantities and does not have the correct substrate specificity to be lytic in its own right. It depends upon target enzymes such as plasminogen or metalloproteinases, and the resulting properties of cells depend completely on the downstream cascade. Metastatic invasion requires the expression of an active enzyme that can degrade the cross-linked basement membrane: of the metalloproteinases identified so far, stromelysin, invadolysin, matrilysin, and PUMP can all degrade type IV collagen and possibly other critical structural elements. Pulmonary metastasis has been clearly correlated with the ability of melanoma cell variants to degrade type IV collagen (Liotta *et al.*, 1980), but again this correlation is not absolute (Starkey *et al.*, 1987). Which of the enzymes is most useful to the invasive process may depend on which activators are available to any particular transformed cell type. The presence of a lytic cascade of PAs and metalloproteinases was demonstrated for melanoma cells (Reich *et al.*, 1988) using an amniotic membrane invasion assay, for RSV-transformed chick embryo fibroblasts and for tumor cells (reviewed by Saksela and Rifkin, 1988). Amniotic membrane is one of several complex membranes that can be used to test invasive potential; others include the chick chorioallantoic membrane, mouse urinary bladder, blood vessel walls, and lens capsule (described by Zucker, 1988). Transformation of a wide variety of cells with the H-*ras* oncogene results in highly gelatinolytic cells: the invasive phenotype has been attributed to invadolysin for fibroblasts and colonic carcinoma cells, and to matrilysin for bronchial epithelial cells (Bernhard *et al.*, 1990; reviewed in Alexander and Werb, 1989). Stromelysin expression is linked to the transition from a benign premalignant papilloma to the aggressive squamous cell carcinoma in the two-step mouse skin model of carcinogenesis

(Ostrowski *et al.*, 1988). It is possible that deletion of gain-of-function mutations in the inhibitor or proteinase genes may be responsible for later stage hyperplastic–neoplastic transitions.

The expression of PAs has been associated with loss of contact inhibition, rounding, and enhanced migration in transformed cells (Sullivan and Quigley, 1986). The analysis of the contribution of PA to invasive reactions is not a simple one, since many factors are involved, including expression and occupancy of the urokinase receptor and the coexpression of inhibitors.

We have already mentioned that the matrix of the central nervous system is highly nonpermissive for invasion by foreign cell types (see Section 4.3): this is attributed to the presence of at least two inhibitors secreted by the oligodendrocytes and present in the myelin sheath (Caroni and Schwab, 1988). The C6 glioblastoma cell line can invade this tissue by expressing a cell-associated metalloproteinase that is not inhibitable by TIMP (Paganetti *et al.*, 1988). Other highly invasive cell types, like B16 melanoma, that express enzymes that allow invasion into most other matrices still cannot infiltrate optic nerve explants. This does suggest that there is a normal role for the glioblastoma metalloproteinase during development of the nervous system and raises the possibility of the existence of other, as yet unidentified, highly specific enzymes tailored for invasive reactions in particular types of matrix.

8.2. The Invasion of Parasites

Any invasive parasite is capable of penetrating a tissue layer that would normally constitute a barrier, e.g., the mucosal epidermal linings of lung or bowel and the cornified epidermis of skin. Matrix-degrading enzymes of at least two classes have been recruited by parasites to use during invasion, and they have been developed to have very broad proteolytic capabilities (reviewed by McKerrow *et al.*, 1989). The parasite cercariae (a schistosome) invades elastin-rich skin by secreting a serine proteinase elastase from acetabular glands during the invasive phase of its life cycle. This enzyme can also degrade fibronectin and collagen type IV. Other organisms like *Ascaris* that cross lung and bowel are not elastinolytic but collagenolytic (McKerrow and Doenhoff, 1988). The transformation of parasites, like cercariae to schistosomula, involves shedding of the glycocalyx. These remodeling processes depend on proteolysis, and elucidation of the mechanism could lead to a disruptive intervention that would be valuable in eliminating the parasites at particular phases of the life cycle.

9. Disease States Associated with an Imbalance of Lytic Enzymes and Inhibitors

All inflammatory conditions are associated with lysis of connective tissue to a greater or lesser extent, and various cell types are recruited to sites of inflammation that have highly reactive proteolytic profiles. An increase of stromelysin and elastase has been observed at sites of gingivitis, chronic bronchitis, abscesses, and peritonitis. Myocardial infarction is associated with a decrease of collagen and elevated proteinase activity.

Excessive deposition or removal of collagen-based matrices is typical of many pathological conditions, from scleroderma to ulcers. Some of these are due to malfunctions of collagen biosynthesis, and some are due to inappropriate metabolism by the matrix-degrading cascade. Only two examples will be discussed further to try to show where a balance of the enzymes and inhibitors is critical to normal tissue function.

9.1. Rheumatoid Arthritis

The synovium is a dense membrane of connective tissue that secretes lubricating fluids for joint and tendon sheaths. Inappropriate activation and inflammation of the synovial membrane is characteristic of the disease of rheumatoid arthritis, producing localized loss of bone and tendon. The inflammatory cell mass of the rheumatoid synovium (pannus) contains many different types, including synovial cells, lymphocytes, monocytes, and endothelium, which produce a cascade of inflammatory cytokines, of which the primary one is probably IL-1 (reviewed by Krane *et al.*, 1990). This cascade results in the erosion of bone via activation of osteoclasts, and degradation of soft connective tissue, joint capsule, tendons, and cartilage by the highly proteolytic effusions of the activated synovium. Synovial and articular cartilage cell types, when transferred to primary culture, produce a variety of proteinases that are elevated in cells from a diseased source (reviewed by Werb, 1989) but the cell types actually responsible for the pathology can be less ambiguously identified if enzyme activities are studied *in vivo*. Using *in situ* hybridization, collagenase has been localized to type A macrophage-like synovial cells and not to fibroblasts to endothelium or to lymphocytes (McCachren *et al.*, 1990). Immunostaining for stromelysin in samples from patients with rheumatoid arthritis shows that this enzyme is present both in fibroblasts and in endothelial cells (Case *et al.*, 1989). Because TGF-β dominantly inhibits the induction of proteolytic enzymes by cytokines (including IL-1) at inflammatory sites, this may become useful in controlling conditions such as rheumatoid arthritis (Andrews *et al.*, 1989).

9.2. Emphysema

The purpose of the multiplicity of proteinase inhibitors, and their high concentration in serum, seems rather vague and unspecific. However, without the full complement of proteinase inhibitors, matrix destruction at sites of localized inflammation and reaction can go out of control. The destruction of pulmonary tissue characteristic of emphysema is a dramatic example of the effect of a dysfunctional circulatory inhibitor. Emphysema is discussed in many reviews (e.g., Perlmutter and Pierce, 1989). Elastase is secreted by neutrophils during inflammation of the bronchial tree and normally is inhibited by circulatory α_1-proteinase inhibitor (also known rather ambiguously as α_1-antitrypsin). This inhibitor is highly pleiomorphic, existing as at least 75 natural variants, which range from normal to deficient, dysfunctional, or absent. The common disease-associated allele depresses production of inhibitor to 10–20% of normal. This results in the pathological loss of the elastin meshwork characteristic of alveolar connective tissue and results in the premature onset of emphysema. α_1-Proteinase inhibitor is made in the liver, and the most common deficiency variant of the inhibitor is denatured and insoluble and accumulates in the reticulum of hepatocytes, generating a hepatic pathology too. In smokers, α_1-proteinase inhibitor in the lung is destroyed by oxidants in cigarette smoke, or by metalloproteinases produced by macrophages and neutrophils attracted by the particulate deposits. This results in excess neutrophil elastase activity, elastin destruction, and emphysema.

10. Future Directions

The process of invasion, and the destruction of the barrier to cell motility, is tempered by the need of the invading cells to have an adhesive matrix that provides the cell–matrix interactions critical to motility and intracellular signaling. Intuitively, we would expect that the proteinases required to specifically cleave molecules that impede the cell's motility would not destroy other molecules that provide directional information or traction. Whether fragments of collagen, the result of collagenolysis, are sufficient for an adhesive interaction is not known. However, a peptide that spans the G^{775}–I/L^{776} cleavage site of collagenase competes for binding of cells to collagen type I (Qian *et al.*, 1990). Whether collagenase predigestion enhances cell adhesion to collagen or destroys it is not clear. A recent report suggests that metalloproteinase treatment of collagenous substrates enhances the motility of tumor cells (Terranova *et al.*, 1989): it has also been observed that adding urokinase to occupy the UPA receptor of keratinocytes enhances the mobility of these cells in a Boyden chamber (Del Rosso *et al.*, 1990). It is not known whether the products of stromelysin degradation of basement membranes characterized by Bejarano *et al.* (1988) can interact with cells. Lytic products of matrix macromolecules may actually direct and enhance migration. Fibronectin fragments are capable of interacting at the fibronectin receptor to induce collagenase and stromelysin expression (Werb *et al.*, 1989), which would presumably propagate the fragmentation process. Intact fibronectin reacts with cells at the classic RGD cell adhesion site; however, denatured fibronectin has other cryptic binding sites on the β1 chain that may interact with various receptors and induce alternative signaling responses. Fragments of the matrix macromolecules induce different motile properties in cells: for example, a substrate of the noncollagenous, nonhelical fragment of collagen type IV allows melanoma cells to adhere, but the triple-helical portion of the molecule allows the cells to adhere and be motile (Chelberg *et al.*, 1989).

The concept that migration and cell adhesion represent an equilibrium between establishing the contacts and breaking them is suggested by a number of observations. Receptor-bound urokinase is localized to the focal adhesions of cells (see Section 4), and so an activator for the enzymes that degrade substrate molecules is concentrated to the actual point of cell attachment. The balance of enzyme and inhibitor required to support neuronal extension has been discussed in Section 7.4. Urokinase is localized to the growth cones of neurons; however, the addition of protease nexin-1 stimulates their outgrowth, suggesting that the making and breaking of growth cone adhesions is a delicate balance.

It is not clear how structures such as basement membrane or tendon, which are continuously functional structures, can grow and remodel. Presumably, the process must involve limited site-specific nicking and intercalation of new macromolecules. Data derived by morphological observation suggest that degradation of collagen fibrils may differ grossly between the pathological destruction of matrix and the tightly regulated physiological processes. A detailed analysis of the proteolytic mechanisms involved would be worthwhile.

The discovery of the FACIT collagens (see Chapter 6) has changed ideas about the degradation of collagenous matrices. These collagens may well coat collagen fibrillar bundles and protect the internal fibrils from attack by the type I and III collagenases. Degradation may be multiphasic, involving the initial attack of the FACIT collagens by enzymes like stromelysin.

There is still considerable doubt about the physiological activation of the metalloproteinases. What are the rate-limiting steps for the initiation of matrix degradation? Is the apparent redundancy of enzymes and inhibitors tailored to cell type-specific transcription, or do the proteins have properties and specificities we have not yet identified? As more information becomes available from studies of the degradative machinery *in vivo*, some of these questions should be answered. The application of molecular and genetic analyses to these problems, both *in vivo* and *in vitro*, should rapidly yield new insights into the mechanisms of matrix remodeling.

ACKNOWLEDGMENTS. We thank Mary McKenney for editorial assistance with the manuscript, Ole Behrendtsen for figure preparation, and E. Unemori for unpublished data. This work was supported by a grant from the National Institutes of Health (HD 23539) and a contract (DE-AC03-76-SF01012) from the Office of Health and Environmental Research, U.S. Department of Energy.

References

Adham, I. M., Klemm, U., Maier, W. M., Hoyer-Fender, S., Tsaousidou, S., and Engel, W., 1989, Molecular cloning of preproacrosin and analysis of its expression pattern in spermatogenesis, *Eur. J. Biochem.* **182:**563–568.

Alexander, C. M., and Werb, Z., 1989, Proteinases and extracellular matrix remodeling, *Curr. Opin. Cell Biol.* **1**:974–982.
Andrews, H. J., Edwards, T. A., Cawston, T. E., and Hazleman, B. L., 1989, Transforming growth factor-beta causes partial inhibition of interleukin 1-stimulated cartilage degradation *in vitro*, *Biochem. Biophys. Res. Commun.* **162**:144–150.
Antonelli-Orlidge, A., Saunders, K. B., Smith, S. R., and D'Amore, P. A., 1989, An activated form of transforming growth factor β is produced by cocultures of endothelial cells and pericytes, *Proc. Natl. Acad. Sci. USA* **86**:4544–4548.
Apodaca, G., Rutka, J. T., Bouhana, K., Berens, M. E., Giblin, J. R., Rosenblum, M. L., McKerrow, J. H., and Banda, M. J., 1990, Expression of metalloproteinases and metalloproteinase inhibitors by fetal astrocytes and glioma cells, *Cancer Res.* **50**:2322–2329.
Astedt, B., Hagerstrand, I., and Lecander, I., 1986, Cellular localization in placenta of placental type plasminogen activator inhibitor, *Thromb. Haemost.* **56**:63–65.
Banda, M. J., Rice, A. G., Griffin, G. L., and Senior, R. M., 1988, Alpha-1 proteinase inhibitor is a neutrophil chemoattractant after proteolytic inactivation by macrophage elastase, *J. Biol. Chem.* **263**:4481–4484.
Beers, W. H., 1975, Follicular plasminogen and plasminogen activator and the effect of plasmin on ovarian follicle wall, *Cell* **6**:379–386.
Behrendtsen, O., Alexander, C. M., and Werb, Z., 1991, Metalloproteinases mediate extracellular matrix degradation by cells from mouse blastocyst outgrowths, *Development* submitted.
Bejarano, P. A., Noelken, M. E., Suzuki, K., Hudson, B. G., and Nagase, H., 1988, Degradation of basement membranes by human matrix metalloproteinase-3 (stromelysin), *Biochem. J.* **256**:413–419.
Bernhard, E. J., Muschel, R. J., and Hughes, E. N., 1990, M_r 92,000 gelatinase release correlates with the metastatic phenotype in transformed rat embryo cells, *Cancer Res.* **50**:3872–3877.
Birkedal-Hansen, B., Moore, W. G., Taylor, R. E., Bhown, A. S., and Birkedal-Hansen, H., 1988, Monoclonal antibodies to human fibroblast procollagenase. Inhibition of enzymatic activity, affinity purification of the enzyme, and evidence for clustering of epitopes in the NH_2-terminal end of the activated enzyme, *Biochemistry* **27**:6751–6758.
Blasi, F., Vassalli, J.-D., and Dano, K., 1987, Urokinase-type plasminogen activator: Proenzyme, receptor, and inhibitors, *J. Cell Biol.* **104**:801–804.
Boone, T. C., Johnson, M. J., De Clerck, Y. A., and Langley, K. E., 1990, cDNA cloning and expression of a metalloproteinase inhibitor related to tissue inhibitor of metalloproteinases, *Proc. Natl. Acad. Sci. USA* **87**:2800–2804.
Brannstrom, M., Woessner, J. F., Jr., Koos, R. D., Sear, C. H. J., and LeMaire, W. J., 1988, Inhibitors of mammalian tissue collagenase and metalloproteinases suppress ovulation in the perfused rat ovary, *Endocrinology* **122**:1715–1721.
Breathnach, R., Matrisian, L. M., Gesnel, M. C., Staub, A., and Leroy, P., 1987, Sequences coding for part of oncogene-induced transin are highly conserved in a related rat gene, *Nucleic Acids Res.* **15**:1139–1151.
Brenner, C. A., Adler, R. R., Rappolee, D. A., Pedersen, R. A., and Werb, Z., 1989, Genes for extracellular matrix-degrading metalloproteinases and their inhibitor, TIMP, are expressed during early mammalian development, *Genes Dev.* **3**:848–859.
Brinckerhoff, C. E., Plucinska, I. M., Sheldon, L. A., and O'Connor, G. T., 1986, Half-life of synovial cell collagenase mRNA is modulated by phorbol myristate acetate but not by all-trans-retonoic acid or dexamethasone, *Biochemistry* **25**:6378–6384.
Brown, C. C., Hembry, R. M., and Reynolds, J. J., 1989, Immunolocalisation of metalloproteinases and their inhibitor in the rabbit growth plate, *J. Bone J. Surg.* **71**:580–593.
Busso, N., Huarte, J., Vassalli, J.-D., Sappino, A.-P., and Belin, D., 1989, Plasminogen activators in the mouse mammary gland: Decreased expression during lactation, *J. Biol. Chem.* **264**:7455–7457.
Campbell, E. J., and Campbell, M. A., 1988, Pericellular proteolysis by neutrophils in the presence of proteinase inhibitors: Effects of substrate opsonization, *J. Cell Biol.* **106**:667–676.
Caroni, P., and Schwab, M. E., 1988, Antibody against myelin-associated inhibitor of neurite

growth neutralizes nonpermissive substrate properties of CNS white matter, *Neuron* **1**:85–96.

Case, J. P., Lafyatis, R., Remmers, E. F., Kumkumian, G. K., and Wilder, R. L., 1989, Transin/stromelysin expression in rheumatoid synovium. A transformation-associated metalloproteinase secreted by phenotypically invasive synoviocytes, *Am. J. Pathol.* **135**:1055–1064.

Cawston, T. E., Curry, V. A., Clark, I. M., and Hazleman, B. L., 1990, Identification of a new metalloproteinase inhibitor that forms tight-binding complexes with collagenase, *Biochem. J.* **269**:183–187.

Chambers, T. J., 1988, The regulation of osteoclastic development and function, *Ciba Found. Symp.* **136**:92–107.

Chelberg, M. K., Tsilibary, E. C., Hauser, A. R., and McCarthy, J. B., 1989, Type IV collagen-mediated melanoma cell adhesion and migration: Involvement of multiple, distinct domains of the collagen molecule, *Cancer Res.* **49**:4796–4802.

Chiu, R., Angel, P., and Karin, M., 1989, Jun-B differs in its biological properties from, and is a negative regulator of, c-Jun, *Cell* **59**:979–986.

Clark, I. M., and Cawston, T. E., 1989, Fragments of human fibroblast collagenase. Purification and characterization, *Biochem. J.* **263**:201–206.

Coulombe, B., and Skup, D., 1988, *In vitro* synthesis of the active tissue inhibitor of metalloproteinases encoded by a complementary DNA from virus-infected murine fibroblasts, *J. Biol. Chem.* **263**:1439–1443.

Cubellis, M. V., Wun, T. C., and Blasi, F., 1990, Receptor-mediated internalization and degradation of urokinase is caused by its specific inhibitor PAI-1, *EMBO J.* **9**:1079–1085.

Dano, K., Andreasen, P. A., Grondahl-Hansen, J., Kristensen, P., Nielsen, L. S., and Skriver, L., 1985, Plasminogen activators, tissue degradation, and cancer, *Adv. Cancer Res.* **44**:139–266.

De Clerck, Y. A., Yean, T.-D., Ratskin, B. J., Lu, H. S., and Langley, K. E., 1989, Purification and characterisation of two related but distinct metalloproteinase inhibitors secreted by bovine aortic endothelial cells, *J. Biol. Chem.* **264**:17445–17453.

Del Rosso, M., Fibbi, G., Dini, G., Grappone, C., Pucci, M., Caldini, R., Magnelli, L., Fimiani, M., Lotti, T., and Panconsei, E., 1990, Role of specific membrane receptors in urokinase-dependent migration of human keratinocytes, *J. Invest. Dermatol.* **94**:310–316.

Desrochers, P. E., and Weiss, S. J., 1988, Proteolytic inactivation of alpha-1-proteinase inhibitor by a neutrophil metalloproteinase, *J. Clin. Invest.* **81**:1646–1650.

Dodge, G. R., and Poole, A. R., 1989, Immunohistochemical detection and immunochemical analysis of type II collagen degradation in human normal, rheumatoid, and osteoarthritic articular cartilages and in explants of bovine articular cartilage cultured with interleukin-1, *J. Clin. Invest.* **83**:647–661.

Edwards, D. R., Waterhouse, P., Holman, M. L., and Denhardt, D. T., 1986, A growth-responsive gene (16C8) in normal mouse fibroblasts homologous to a human collagenase inhibitor with erythroid-potentiating activity: Evidence for inducible and constitutive transcripts, *Nucleic Acids Res.* **14**:8863–8877.

Ellis, S. M., Nabeshima, K., and Biswas, C., 1989, Monoclonal antibody preparation and purification of a tumor cell collagenase-stimulatory factor, *Cancer Res.* **49**:3385–3391.

Estreicher, A., Muhlhauser, J., Carpentier, J.-L., Orci, L., and Vassalli, J.-D., 1990, The receptor for urokinase type plasminogen activator polarises expression of the protease to the leading edge of migrating monocytes and promotes degradation of enzyme inhibitor complexes, *J. Cell Biol.* **111**:783–792.

Farrell, D. H., Wagner, S. L., Yuan, R. H., and Cunningham, D. D., 1988, Localisation of protease nexin-1 on the fibroblast extracellular matrix, *J. Cell. Physiol.* **134**:179–188.

Fawcett, J. W., and Housden, E., 1990, The effects of protease inhibitors on axon growth through astrocytes, *Development* **109**:59–66.

Fisher, S. J., Cui, T. Y., Zhang, L., Hartman, L., Grahl, K., Zhang, G. Y., Tarpey, J., and Damsky, C. H., 1989, Adhesive and degradative properties of human placental cytotrophoblast cells in vitro, *J. Cell Biol.* **109**:891–902.

Flenniken, A. M., and Williams, B. R. G., 1990, Developmental expression of the endogenous TIMP gene and a TIMP-lacZ fusion gene in transgenic mice, *Genes Dev.* **4**:1094–1106.

Folkman, J., and Klagsbrun, M., 1987, Angiogenic factors, *Science* **235**:442–447.
Gewert, D. R., Coulombe, B., Castelino, M., Skup, D., and Williams, B. R. G., 1987, Characterization and expression of a murine gene homologous to human EPA/TIMP: A virus-induced gene in the mouse, *EMBO J.* **6**:651–657.
Gloor, S., Odink, K., Guenther, J., Nick, H., and Monard, D., 1986, A glia-derived neurite promoting factor with protease inhibitory activity belongs to the protease nexins, *Cell* **47**:687–693.
Gonzalez, A.-M., Buscaglia, M., Ong, M., and Baird, A., 1990, Distribution of basic fibroblast growth factor in the 18-day rat fetus: Localisation in the basement membranes of diverse tissues, *J. Cell Biol.* **110**:753–765.
Gray, D. J., Gardner, E., and O'Rahilly, R., 1957, The prenatal development of the skeleton and joints of the human hand, *Am. J. Anat.* **101**:169–223.
Grillo, H. C., and Gross, J., 1967, Collagenolytic activity during mammalian wound repair, *Dev. Biol.* **15**:300–317.
Gross, J., 1966, How tadpoles lose their tails, *J. Invest. Dermatol.* **47**:274–277.
Gross, J., 1981, An essay on biological degradation of collagen, in: *Cell Biology of Extracellular Matrix* (E. D. Hay, ed.), pp. 217–258, Plenum Press, New York.
Gunja-Smith, Z., Nagase, H., and Woessner, J. F., Jr., 1989, Purification of the neutral proteoglycan-degrading metalloproteinase from human articular cartilage tissue and its identification as stromelysin matrix metalloproteinase-3, *Biochem. J.* **258**:115–119.
Gutman, A., and Wasylyk, B., 1990, The collagenase gene promoter contains a TPA and oncogene-responsive unit encompassing the PEA3 and AP-1 binding sites, *EMBO J.* **9**:2241–2246.
Hardy, D. M., Schoots, A. F., and Hedrick, J. L., 1989, Caprine acrosin. Purification, characterization and proteolysis of the porcine zona pellucida, *Biochem. J.* **257**:447–453.
Hebert, C. A., and Baker, J. B., 1988, Linkage of extracellular plasminogen activator to the fibroblast cytoskeleton: Colocalization of cell surface urokinase with vinculin, *J. Cell Biol.* **106**:1241–1247.
Heine, U., Munoz, E. F., Flanders, K. C., Ellingsworth, L. R., Lam, H. Y., Thompson, N. L., Roberts, A. B., and Sporn, M. B., 1987, Role of transforming growth factor-beta in the development of the mouse embryo, *J. Cell Biol.* **105**:2861–2876.
Heine, U. I., Munoz, E. F., Flanders, K. C., Roberts, A. B., and Sporn, M. B., 1990, Colocalization of TGF beta-1 and collagen I and collagen III, fibronectin and glycosaminoglycans during lung branching morphogenesis, *Development* **109**:29–36.
Herron, G. S., Banda, M. J., Clark, E. J., Gavrilovic, J., and Werb, Z., 1986, Secretion of metalloproteinases by stimulated capillary endothelial cells. II. Expression of collagenase and stromelysin activities is regulated by endogenous inhibitors, *J. Biol Chem.* **261**:2814–2818.
Hiramoto, S. A., and Cunningham, D. D., 1988, Effects of fibroblasts and endothelial cells on inactivation of target proteases by protease nexin-1, heparin cofactor II and C1-inhibitor, *J. Cell. Biochem.* **36**:199–207.
Horowitz, S., Dafni, N., Shapiro, D. L., Holm, B. A., Notter, R. H., and Quible, D. J., 1989, Hyperoxic exposure alters gene expression in the lung: Induction of the tissue inhibitor of metalloproteinases mRNA and other RNAs, *J. Biol. Chem.* **264**:7092–7095.
Huarte, J., Belin, D., Bosco, D., Sappino, A.-P., and Vassalli, J.-D., 1987, Plasminogen activator and mouse spermatozoa: Urokinase synthesis in the male genital tract and binding of the enzyme to the sperm cell surface, *J. Cell Biol.* **104**:1281–1289.
Huber, A. R., and Weiss, S. J., 1989, Disruption of the subendothelial basement membrane during neutrophil diapedesis in an in vitro construct of a blood vessel wall, *J. Clin. Invest.* **83**:1122–1136.
Huebner, K., Isobe, M., Gasson, J. C., Golde, D. W., and Croce, C. M., 1986, Localization of the gene encoding human erythroid-potentiating activity to chromosome region Xp11.1–Xp11.4, *Am. J. Hum. Genet.* **38**:819–826.
Huffer, W. E., 1988, Morphology and biochemistry of bone remodeling: Possible control by vitamin D, parathyroid hormone, and other substances, *Lab. Invest.* **59**:418–442.
Ingber, D. E., and Folkman, J., 1989, Mechanochemical switching between growth and differentiation during fibroblast growth factor-stimulated angiogenesis in vitro: Role of extracellular matrix, *J. Cell Biol.* **109**:317–330.

Khokha, R., Waterhouse, P., Yagel, S., Lala, P. K., Overall, C. M., Norton, G., and Denhardt, D. T., 1989, Antisense RNA-induced reduction in murine TIMP levels confers oncogenicity on Swiss 3T3 cells, *Science* **243:**947–950.

Krane, S. M., Conca, W., Stephenson, M. L., Amento, E. P., and Goldring, M. B., 1990, Mechanisms of matrix degradation in rheumatoid arthritis, *Ann. N.Y. Acad. Sci.* **580:**340–354.

Krystosek, A., and Seeds, N. W., 1981, Plasminogen activator release at the neuronal growth cone, *Science* **213:**1532–1534.

Krystosek, A., and Seeds, N. W., 1986, Normal and malignant cells, including neurons, deposit plasminogen activator on the growth substrata, *Exp. Cell Res.* **166:**31–46.

Kumkumian, G. K., Lafyatis, R., Remmers, E. F., Case, J. P., Kim, S. J., and Wilder, R. L., 1989, Platelet-derived growth factor and IL-1 interactions in rheumatoid arthritis. Regulation of synoviocyte proliferation, prostaglandin production, and collagenase transcription, *J. Immunol.* **143:**833–837.

Laiho, M., and Keski-Oja, J., 1989, Growth factors in the regulation of pericellular proteolysis: A review, *Cancer Res.* **49:**2533–2553.

Larsson, L.-I., Skriver, L., Nielsen, L. S., Grondahl-Hansen, J., Kristensen, P., and Dano, K., 1984, Distribution of urokinase-type plasminogen activator immunoreactivity in the mouse, *J. Cell Biol.* **98:**894–903.

Leibovich, S. J., and Ross, R., 1975, The role of the macrophage in wound repair: A study with hydrocortisone and antimacrophage serum, *Am. J. Pathol.* **78:**71–100.

Lepage, T., and Gache, C., 1990, Early expression of a collagenase-like hatching enzyme gene in the sea urchin embryo, *EMBO J.* **9:**3003–3012.

Librach, C. L., Werb, Z., Fitzgerald, M. L., Chiu, K., Corwin, N. M., Esteves, R. A., Grobelny, D., Galardy, R., Damsky, C. H., and Fisher, S. J., 1991, 92 kD type IV collagenase mediates invasion of human cytotrophoblasts, *J. Cell Biol.* **113:**437–449.

Lindner, J., Guenther, J., Nick, H., Zinser, G., Antonicek, H., Schachner, M., and Monard, D., 1986, Modulation of granule cell migration by a glia-derived protein, *Proc. Natl. Acad. Sci. USA* **83:**4568–4571.

Liotta, L. A., Tryggvason, K., Garbisa, S., Hart, I., Foltz, C. M., and Shafie, S., 1980, Metastatic potential correlates with enzymatic degradation of basement membrane collagen, *Nature* **284:**67–68.

Lyons, J. G., Siew, K., and O'Grady, R. L., 1989, Cellular interactions determining the production of collagenase by a rat mammary carcinoma cell line, *Int. J. Cancer* **43:**119–125.

MacNaul, K. L., Chartrain, N., Lark, M., Tocci, M. J., and Hutchinson, N. I., 1990, Discoordinate expression of stromelysin, collagenase, and tissue inhibitor of metalloproteinase-1 in rheumatoid human synovial fibroblasts, *J. Biol Chem.* **265:**17238–17245.

Matrisian, L. M., and Hogan, B. L. M., 1990, Growth factor-regulated proteases and extracellular matrix remodeling during mammalian development, *Curr. Top. Dev. Biol.* **24:**219–259.

McCachren, S. S., Haynes, B. F., and Niedel, J. E., 1990, Localization of collagenase mRNA in rheumatoid arthritis synovium by *in situ* hybridization histochemistry, *J. Clin. Immunol.* **10:**19–27.

McKerrow, J. H., and Doenhoff, M. J., 1988, Schistosome proteases, *Parasitol. Today* **4:**334–340.

McKerrow, J. H., Skanari, J., Brown, M., Brindley, P., Railey, J., Weiss, N., and Resnick, S. D., 1989, Proteolytic enzymes as mediators of parasite invasion of skin, in: *Models in Dermatology,* Vol. 4 (H. I. Maibach and N. J. Lowe, eds.), p. 276–284 Karger, Basel.

Merwin, J. R., Anderson, J. M., Kocher, O., van Itallie, C. M. and Madri, J. A., 1990, Transforming growth factor β1 modulates extracellular matrix organization and cell–cell junctional complex formation during in vitro angiogenesis, *J. Cell. Physiol.* **142:**117–128.

Mignatti, P., Robbins, E., and Rifkin, D. B., 1986, Tumor invasion through the human amniotic membrane: Requirement for a proteinase cascade, *Cell* **47:**487–498.

Mignatti, P., Tsuboi, R., Robbins, E., and Rifkin, D. B., 1989, In vitro angiogenesis on the human amniotic membrane: Requirement for basic fibroblast growth factor-induced proteinases, *J. Cell Biol.* **108:**671–682.

Mikuni-Takagaki, Y., and Cheng, Y.-S., 1987, Metalloproteinases in endochondral bone formation:

Appearance of tissue inhibitor-resistant metalloproteinases, *Arch. Biochem. Biophys.* **259**:576–588.

Moller, C. C., and Wassarman, P. M., 1989, Characterization of a proteinase that cleaves zona pellucida glycoprotein ZP2 following activation of mouse eggs, *Dev. Biol.* **132**:103–112.

Monteagudo, C., Merino, M. J., San-Juan, J., Liotta, L. A., and Stetler-Stevenson, W. G., 1990, Immunohistochemical distribution of type IV collagenase in normal, benign, and malignant breast tissue, *Am. J. Pathol.* **136**:585–592.

Montesano, R., and Orci, L., 1985, Tumor-promoting phorbol esters induce angiogenesis in vitro, *Cell* **42**:469–477.

Montesano, R., Pepper, M. S., Belin, D., Vassalli, J.-D., and Orci, L., 1988, Induction of angiogenesis in vitro by vanadate, an inhibitor of phosphotyrosine phosphatases, *J. Cell. Physiol.* **134**:460–466.

Montesano, R., Pepper, M. S., Mohle-Steinlein, U., Risau, W., Wagner, E. F., and Orci, L., 1990, Increased proteolytic activity is responsible for the aberrant morphogenetic behavior of endothelial cells expressing the middle T oncogene, *Cell* **62**:435–445.

Moscatelli, D., and Rifkin, D. B., 1988, Membrane and matrix localization of proteinases: A common theme in tumor cell invasion and angiogenesis, *Biochim. Biophys. Acta* **948**:67–85.

Nagase, H., Enghild, J. J., Suzuki, K., and Salvesen, G., 1990, Stepwise activation mechanisms of the precursor of matrix metalloproteinase 3 (stromelysin) by proteinases and (4-aminophenyl)-mercuric acetate, *Biochemistry* **29**:5783–5789.

Nomura, S., Hogan, B. L. M., Wills, A. J., Heath, J. K., and Edwards, D. R., 1989, Developmental expression of tissue inhibitor of metalloproteinase (TIMP) RNA, *Development* **105**:575–583.

Okada, Y., Watanabe, S., Nakanishi, I., Kishi, J.-I., Hayakawa, T., Watorek, W., Travis, J., and Nagase, H., 1988, Inactivation of tissue inhibitor of metalloproteinase by neutrophil elastase and other serine proteinases, *FEBS Lett.* **229**:157–160.

Ossowski, L., 1988, *In vivo* invasion of modified chorioallantoic membrane by tumor cells: The role of cell surface-bound urokinase, *J. Cell Biol.* **107**:2437–2445.

Ossowski, L., and Reich, E., 1983, Antibodies to plasminogen activator inhibit human tumor metastasis, *Cell* **35**:611–619.

Ossowski, L., Biegel, D., and Reich, E., 1979, Mammary plasminogen activator: Correlation with involution, hormonal modulation and comparison between normal and neoplastic tissue, *Cell* **16**:929–940.

Ostrowski, L. E., Finch, J., Krieg, P., Matrisian, L., Patskan, G., O'Connell, J. F., Phillips, J., Slaga, T. J., Breathnach, R., and Bowden, G. T., 1988, Expression pattern of a gene for a secreted metalloproteinase during late stages of tumor progression, *Mol. Carcinogenesis* **1**:13–19.

Paganetti, P. A., Caroni, P., and Schwab, M. E., 1988, Glioblastoma infiltration into central nervous system tissue *in vitro:* Involvement of a metalloprotease, *J. Cell Biol.* **107**:2281–2291.

Park, A. J., Matrisian, L. M., Kells, A. F., Pearson, R., Yuan, Z., and Navre, M., 1991, Mutational analysis of the transin (rat stromelysin) autoinhibitor region demonstrates a role for residues surrounding the cysteine switch, *J. Biol. Chem.* **266**:1584–1590.

Patthy, L., 1985, Evolution of the proteases of blood coagulation and fibrinolysis by assembly from modules, *Cell* **41**:657–663.

Pepper, M. S., Vassalli, J.-D., Montesano, R., and Orci, L., 1987, Urokinase-type plasminogen activator is induced in migrating capillary endothelial cells, *J. Cell Biol.* **105**:2535–2541.

Pepper, M. S., Spray, D. C., Chanson, M., Montesano, R., Orci, L., and Meda, P., 1989, Junctional communication is induced in migrating capillary endothelial cells, *J. Cell Biol.* **109**:3027–3038.

Pepper, M. S., Belin, D., Montesano, R., Orci, L., and Vassalli, J.-D., 1990, Transforming growth factor-beta 1 modulates basic fibroblast growth factor-induced proteolytic and angiogenic properties of endothelial cells in vitro, *J. Cell Biol.* **111**:743–755.

Peppin, G. J., and Weiss, S. J., 1986, Activation of the endogenous metalloproteinase, gelatinase, by triggered human neutrophils, *Proc. Natl. Acad. Sci. USA* **83**:4322–4326.

Perlmutter, D. H., and Pierce, J. A., 1989, The alpha-1-antitrypsin gene and emphysema, *Am. J. Physiol.* **257**:L146–162.

Perona, R. M., and Wassarman, P. M., 1986, Mouse blastocysts hatch *in vitro* by using a trypsin-like proteinase associated with cells of mural trophectoderm, *Dev. Biol.* **114:**42–52.

Pollanen, J., 1989, Down-regulation of plasmin receptors on human sarcoma cells by glucocorticoids, *J. Biol. Chem.* **264:**5628–5632.

Pollanen, J., Hedman, K., Nielsen, L. S., Dano, K., and Vaheri, A., 1988, Ultrastructural localization of plasma membrane-associated urokinase-type plasminogen activator at focal contacts, *J. Cell Biol.* **106:**87–95.

Qian, J. J., Liu, X., Zha, J., Sorensen, K. R., Scaria, P. V., and Bhatnagar, R. S., 1990, How do cells see collagen? *J. Cell Biol.* **111:**24a.

Reich, R., Thompson, E. W., Iwamoto, Y., Martin, G. R., Deason, J. R., Fuller, G. C., and Miskin, R., 1988, Effects of inhibitors of plasminogen activator, serine proteinases, and collagenase IV on the invasion of basement membranes by metastatic cells, *Cancer Res.* **48:**3307–3312.

Rifas, L., Halstead, L. R., Peck, W. A., Avioli, L. V., and Welgus, H. G., 1989, Human osteoblasts in vitro secrete tissue inhibitor of metalloproteinases and gelatinase but not interstitial collagenase as major cellular products, *J. Clin. Invest.* **84:**686–694.

Roberts, A. B., Heine, U. I., Flanders, K. C., and Sporn, M. B., 1990, Transforming growth factor-beta. Major role in regulation of extracellular matrix, *Ann. N.Y. Acad. Sci.* **580:**225–232.

Saksela, O., and Rifkin, D. B., 1988, Cell-associated plasminogen activation: Regulation and physiological functions, *Annu. Rev. Cell Biol.* **4:**93–126.

Saksela, O., and Rifkin, D. B., 1990, Release of basic fibroblast growth factor–heparan sulfate complexes from endothelial cells by plasminogen activator-mediated proteolytic activity, *J. Cell Biol.* **110:**767–775.

Saksela, O., Moscatelli, D., and Rifkin, D. B., 1987, The opposing effects of basic fibroblast growth factor and transforming growth factor beta on the regulation of plasminogen activator activity in capillary endothelial cells, *J. Cell Biol.* **105:**957–963.

Sanchez-Lopez, R., Nicholson, R., Gesnel, M.-C., Matrisian, L. M., and Breathnach, R., 1988, Structure–function relationships in the collagenase family member transin, *J. Biol. Chem.* **263:**11892–11899.

Sappino, A. P., Huarte, J., Belin, D., and Vassalli, J. D., 1989, Plasminogen activators in tissue remodeling and invasion: mRNA localisation in mouse ovaries and implanting embryos, *J. Cell Biol.* **109:**2471–2479.

Sato, Y., Tsuboi, R., Lyons, R., Moses, H., and Rifkin, D. B., 1990, Characterisation of the activation of latent TGFβ by co-cultures of endothelial cells and pericytes or smooth muscle cells: A self-regulating system, *J. Cell Biol.* **111:**757–763.

Schleef, R. R., Wagner, N. V., Sinha, M., and Loskutoff, D. J., 1986, A monoclonal antibody that does not recognize tissue-type plasminogen activator bound to its naturally occurring inhibitor, *Thromb. Haemost.* **56:**328–332.

Schultz, R. M., Silberman, S., Persky, B., Bajkowski, A. S., and Carmichael, D. F., 1988, Inhibition by human recombinant tissue inhibitor of metalloproteinases of human amnion invasion and lung colonization by murine B16-F10 melanoma cells, *Cancer Res.* **48:**5539–5545.

Sottile, J., Mann, D. M., Diemer, V., and Millis, A. J. T., 1989, Regulation of collagenase and collagenase mRNA production in early- and late-passage human diploid fibroblasts, *J. Cell. Physiol.* **138:**281–290.

Springman, E. B., Angleton, E. L., Birkedal-Hansen, H., and Van Wart, H. E., 1990, Multiple modes of activation of latent human fibroblast collagenase: evidence for the role of a Cys^{73} active-site zinc complex in latency and a "cysteine switch" mechanism for activation, *Proc. Natl. Acad. Sci. USA* **87:**364–368.

Starkey, J. R., Stanford, D. R., Magnuson, J. A., Hamner, S., Robertson, N. P., and Gasic, G. J., 1987, Comparison of basement membrane matrix degradation by purified proteases and by metastatic tumor cells, *J. Cell. Biochem.* **35:**31–49.

Stein, B., Rahmsdorf, H. J., Steffen, A., Litfin, M., and Herrlich, P., 1989, UV-induced DNA damage is an intermediate step in UV-induced expression of human immunodeficiency virus type 1, collagenase, c-fos, and metallothionein, *Mol. Cell. Biol.* **9:**5169–5181.

Stetler-Stevenson, W. G., Krutzsch, H. C., and Liotta, L. A., 1989, Tissue inhibitor of metalloproteinase (TIMP-2): A new member of the metalloproteinase inhibitor family, *J. Biol. Chem.* **264:**17374–17378.

Stetler-Stevenson, W. G., Brown, P. D., Onisto, M., Levy, A. T., and Liotta, L. A., 1990, Tissue inhibitor of metalloproteinases-2 (TIMP-2) mRNA expression in tumour cell lines and human tumor tissues, *J. Biol. Chem.* **265:**13933–13938.

Strickland, S., Reich, E., and Sherman, M. I., 1976, Plasminogen activator in early embryogenesis: Enzyme production by trophoblast and parietal endoderm, *Cell* **9:**231–240.

Stump, D. C., Lijnen, H. R., and Collen, D., 1986, Purification and characterization of a novel low molecular weight form of single-chain urokinase-type plasminogen activator, *J. Biol. Chem.* **261:**17120–17126.

Sullivan, L. M., and Quigley, J. P., 1986, An anticatalytic monoclonal antibody to avian plasminogen activator: Its effect on behavior of RSV-transformed chick fibroblasts, *Cell* **45:**905–915.

Talhouk, R. S., Chin, J. R., Unemori, E. N., Werb, Z., and Bissell, M. J., 1991, Proteinases of the mammary gland: Developmental regulation *in vivo* and vectorial secretion in culture, *Development* **112:**439–449.

Terranova, V. P., Maslow, D., and Markus, G., 1989, Directed migration of murine and human tumor cells to collagenases and other proteases, *Cancer Res.* **49:**4835–4841.

Thomson, B. M., Atkinson, S. J., McGarrity, A. M., Hembry, R. M., Reynolds, J. J., and Meikle, M. C., 1989, Type I collagen degradation by mouse calvarial osteoblasts stimulated with 1,25-dihydroxyvitamin D-3: Evidence for a plasminogen–plasmin–metalloproteinase activation cascade, *Biochim. Biophys. Acta* **1014:**125–132.

Unemori, E. N., Bouhana, K. S., and Werb, Z., 1990, Vectorial secretion of extracellular matrix proteins, matrix-degrading proteinases, and tissue inhibitor of metalloproteinases by endothelial cells, *J. Biol. Chem.* **265:**445–451.

Vaes, G., 1988, Cellular biology and biochemical mechanism of bone resorption. A review of recent developments on the formation, activation, and mode of action of osteoclasts, *Clin. Orthop. Relat. Res.* **231:**239–271.

Vance, B. A., Kowalski, C. G., and Brinckerhoff, C. E., 1989, Heat shock of rabbit synovial fibroblasts increases expression of mRNAs for two metalloproteinases, collagenase and stromelysin, *J. Cell Biol.* **108:**2037–2043.

Van Wart, H. E., and Birkedal-Hansen, H., 1990, The cysteine switch: A principle of regulation of metalloproteinase activity with potential applicability to the entire matrix metalloproteinase gene family, *Proc. Natl. Acad. Sci. USA* **87:**5578–5582.

Wagner, S. L., Lau, A. L., and Cunningham, D. D., 1989, Binding of protease nexin-1 to the fibroblast surface alters its target proteinase specificity, *J. Biol. Chem.* **264:**611–615.

Werb, Z., 1989, Proteinases and matrix degradation, in: *Textbook of Rheumatology,* 3rd ed. (W. N. Kelley, E. D. Harris, Jr., S. Ruddy, and C. B. Sledge, eds.), pp. 300–321, W. B. Saunders, Philadelphia.

Werb, Z., Mainardi, C. L., Vater, C. A., and Harris, E. D., Jr., 1977, Endogenous activation of latent collagenase by rheumatoid synovial cells: Evidence for a role of plasminogen activator, *N. Engl. J. Med.* **296:**1017–1023.

Werb, Z., Tremble, P. M., Behrendtsen, O., Crowley, E., and Damsky, C. H., 1989, Signal transduction through the fibronectin receptor induces collagenase and stromelysin gene expression, *J. Cell Biol.* **109:**877–889.

Werb, Z., Tremble, P., and Damsky, C. H., 1990, Regulation of extracellular matrix degradation by cell–extracellular matrix interactions, *Cell Differ. Dev.* **32:**299–306.

Wilhelm, S. M., Collier, I. E., Marmer, B. L., Eisen, A. Z., Grant, G. A., and Goldberg, G. I., 1989, SV-40 transformed human lung fibroblasts secrete a 92-kDa type IV collagenase which is identical to that secreted by normal human macrophages, *J. Biol. Chem.* **264:**17213–17221.

Wiman, B., Almquist, A., Sigurdardottir, O., and Lindahl, T., 1988, Plasminogen activator inhibitor 1 (PAI) is bound to vitronectin in plasma, *FEBS Lett.* 242:125–128.

Woessner, J. F., Jr., and Taplin, C. J., 1988, Purification and properties of a small latent matrix metalloproteinase of the rat uterus, *J. Biol. Chem.* **263:**16918–16925.

Wu, H., Byrne, M. H., Stacey, A., Goldring, M. B., Birkhead, J. R., Jaenisch, R., and Krane, S. M., 1990, Generation of collagenase-resistant collagen by site-directed mutagenesis of murine pro alpha 1(I) collagen gene, *Proc. Natl. Acad. Sci. USA* **87**:5888–5892.

Yang, E. Y., and Moses, H. L., 1990, Transforming growth factor β1-induced changes in cell migration, proliferation, and angiogenesis in the chicken chorioallantoic membrane, *J. Cell Biol.* **111**:731–741.

Zucker, S., 1988, A critical appraisal of the role of proteolytic enzymes in cancer invasion: Emphasis on tumor surface proteinases, *Cancer Invest.* **6**:219–231.

Zucker, S., Moll, U. M., Lysik, R. M., DiMassimo, E. I., Stetler-Stevenson, W. G., Liotta, L. A., and Schwedes, J. W., 1990, Extraction of type IV collagenase/gelatinase from plasma membranes of human cancer cells, *Int. J. Cancer* **46**:1137–1142.

Part III

What Does Matrix Do for Cells?

Chapter 9

Proteoglycans and Hyaluronan in Morphogenesis and Differentiation

BRYAN P. TOOLE

1. Introduction

The objective of this chapter is to discuss the roles of hyaluronan (HA) and proteoglycans (PGs) in developmental processes and tissue remodeling. In the first section I have summarized recent progress on the molecular basis of HA and PG interactions with the surface of cells, since this is the most likely means whereby HA and PGs would influence cell behavior during development. This is followed by a review of cellular studies, mainly performed *in vitro*, that implicate HA and PGs in specific aspects of cell behavior that would be relevant to development *in vivo*. Finally, I have chosen three developmental systems to illustrate ways in which HA or PGs are involved in actual morphogenetic and differentiative processes.

2. Mechanisms of Hyaluronan– and Proteoglycan–Cell Interaction

2.1. Hyaluronan-Binding Proteins: The Hyaladherins

Early studies suggested that there were at least two classes of HA-binding proteins (HABPs). This classification was based both on functional roles and on binding characteristics, with the first class requiring an HA decasaccharide for

Abbreviations used in this chapter: PG, proteoglycan; HA, hyaluronan (hyaluronate, hyaluronic acid); HABP, HA-binding protein; HS, heparan sulfate; CS, chondroitin sulfate; DS, dermatan sulfate; GAG, glycosaminoglycan; HSPG, heparan sulfate proteoglycan; CSPG, chondroitin sulfate proteoglycan; DSPG, dermatan sulfate proteoglycan; FGF, fibroblast growth factor; TGF-β, transforming growth factor-β; NCAM, neural cell adhesion molecule.

BRYAN P. TOOLE • Department of Anatomy and Cellular Biology, Tufts University Health Science Schools, Boston, Massachusetts 02111.

Cell Biology of Extracellular Matrix, Second Edition, edited by Elizabeth D. Hay, Plenum Press, New York, 1991.

efficient binding and consisting mainly of structural HA-binding molecules, e.g., the link proteins, hyaluronectin, and HA-binding PGs such as aggrecan, versican, and PG-M (Tengblad, 1981; Bertrand and Delpech, 1985; Yamagata *et al.*, 1986). The second class consisted of cell surface-associated, putative HA receptors that require only an HA hexasaccharide for binding (Underhill *et al.*, 1983; Laurent *et al.*, 1986; Nemec *et al.*, 1987). However, these distinctions have now blurred and it is likely that all of these HABPs belong to a single family with homologous HA-binding sequence motifs. For this reason I suggest that the HABPs be given a family name. My personal preference would be "the HAHAgrins" since it nicely depicts the repetitive structure of HA and its multivalent interactions with binding proteins, a parallel connotation to the integrins, and, most of all, the amusing intensity of the naming game in biology today. One of my colleagues suggested "the hyaloscams"; I trust this was due to the HABPs being substrate and cell adhesive molecules rather than his skeptical attitude toward them. Third prize goes to "the gripagags," but this was suggested by collagen people and therefore is automatically ineligible. Alas, I will turn my back on these wonderful suggestions and reluctantly acknowledge the seriousness of our scientific endeavors by suggesting that this family be termed "the hyaladherins."

2.1.1. Types of Hyaladherins

The most thoroughly studied HA-binding macromolecules are the link proteins and the large PG, termed aggrecan, of cartilage (see Chapter 2). Tandemly repeated domains in link protein appear to constitute the HA-binding region (Goetinck *et al.*, 1987). Four different peptides, derived from two corresponding regions of each of the tandem repeats, have been found to block binding of HA to link protein, implying that all four regions are involved in the binding. Since each of these peptides contains a cluster of positively charged amino acids and since binding was found to decrease with increased ionic strength, it is believed that binding of link protein to HA is ionic in nature (Goetinck *et al.*, 1987). The amino-terminal region of the core protein of aggrecan contains an HA-binding domain homologous to the tandem repeats in link protein (Neame *et al.*, 1987). Recently, related homologies have been described in several other macromolecules: (1) near the amino-terminus of the HA-binding PG produced by human fibroblasts, named versican (Zimmermann and Ruoslahti, 1989); (2) within an HABP produced by human glial cells and probably closely related to hyaluronectin (Perides *et al.*, 1989) and versican (Zimmermann and Ruoslahti, 1989); and (3) at the N-terminus of the adhesive protein(s) variously known as CD44, Pgp-1, Hermes antigen, ECMRIII collagen receptor, H-CAM, etc.—herein termed CD44 (Stamenkovic *et al.*, 1989; Goldstein *et al.*, 1989).

It is generally accepted that the link proteins and aggrecan of cartilage are structural macromolecules that, through interaction with HA, form gigantic ternary complexes that contribute to the physical properties of cartilage. Similar structural complexes of HA, link protein, and PGs such as versican presumably participate in building the matrices of several other connective tissues. The structure and distribution of these complexes are reviewed in Chapter 2. However, there is also a group of HABPs that apparently serve as cell surface receptors for HA and mediate the effects of HA on cell behavior that are described later in this chapter.

Putative HA receptors were originally described as HA-binding sites of high affinity on the surface of transformed 3T3 cells (Underhill and Toole, 1979) and fibroblasts (Angello and Hauschka, 1980). The HA receptor has now been identified in 3T3 and BHK cells as an 85-kDa glycoprotein recognized by the K3 monoclonal antibody (Underhill *et al.*, 1987) (herein termed the 85-kDa HA receptor). Another 85-kDa HABP is present at the surface of cultured fibroblasts, but this HABP is derived from serum and has properties distinct from the BHK receptor (Yoneda *et al.*, 1990). A putative HA receptor was also isolated from the culture medium of 3T3 cells and chick embryo fibroblasts as a mixture of 56- to 70-kDa proteins that were subsequently shown to form large aggregates at the cell surface (Turley, 1989). The HABPs above interact directly or indirectly with the cytoskeleton (Lacy and Underhill, 1987; Turley *et al.*, 1990) and mediate various effects of HA on cell behavior (see Sections 3 and 4), but their biochemical interrelationships are not yet known. We have recently obtained monoclonal antibodies to HABPs that are widely distributed on the surfaces of embryonic, transformed, and tumor cells. One of these antibodies (MAb IVd4) recognizes proteins of 93, 90, and 69 kDa that are probably related to the 85-kDa HA receptor since this antibody blocks binding of HA to transformed 3T3 cells (Banerjee and Toole, 1991).

A cell surface HABP is also involved in uptake of HA from the circulation by liver endothelial cells. Binding, internalization, and degradation of HA by liver endothelial cells have been demonstrated in culture, and the HA-binding sites involved are recycled in a fashion suggesting a coated-pit pathway of endocytosis of HA (Laurent *et al.*, 1986; McGary *et al.*, 1989). However, these sites recognize chondroitin sulfate (CS) and heparin as well as HA (Laurent *et al.*, 1986; Raja *et al.*, 1988). The HABP that mediates endocytosis has not been identified, but it is probably different from the 85-kDa HA receptor (Raja *et al.*, 1988).

Fibrinogen from some species also binds HA and a role for this binding in wound healing has been proposed (Frost and Weigel, 1990).

2.1.2. CD44 Is Related to the 85-kDa Hyaluronan Receptor

Recently, it has become apparent that the 85-kDa HA receptor is the same as or very similar to CD44 (reviewed in Toole, 1990). CD44 is a cell surface glycoprotein involved in lymphocyte activation and adhesion, and homing of circulating lymphocytes via interaction with high endothelial venules, which are the entry sites to organized lymphoid tissues (reviewed in Haynes *et al.*, 1989). The evidence suggesting a relationship of the 85 kDa HABP to CD44 includes expression of the HA receptor on the surface of COS cells transfected with cDNAs for CD44, inhibition of CD44 binding to tissues by treatment with hyaluronidase or competition with HA, inhibition of HA binding to lymphocytes by antibodies to CD44, and inhibition of HA-mediated lymphocyte–lymphocyte and lymphocyte–stromal cell interactions with antibodies to CD44 (Aruffo *et al.*, 1990; Culty *et al.*, 1990; Lesley *et al.*, 1990; Miyake *et al.*, 1990).

Since CD44 contains a domain homologous to the HA-binding regions of link protein and cartilage PG (Stamenkovic *et al.*, 1989; Goldstein *et al.*, 1989), and since CD44 and the 85-kDa HA receptor are very similar, if not identical, then the HA receptor would be expected to exhibit homologies with the HA-binding domains of structural HABPs such as aggrecan, versican, and link protein. However, other data distinguish the HA-binding properties of these structural molecules and the HA receptor. The first and most marked distinction is in the length of HA oligosaccharide required for recognition by the HABPs. Link protein, aggrecan, hyaluronectin, and PG-M of chick embryo fibroblasts require an HA decasaccharide for binding (Tengblad, 1981; Bertrand and Delpech, 1985; Yamagata *et al.*, 1986). Oligosaccharides of smaller lengths exhibit negligible binding, and the affinities of binding of HA decasaccharide and HA polymer do not differ greatly (Tengblad, 1981). However, cell surface HA receptors recognize HA hexasaccharide, and affinity of binding increases considerably with the length of the ligand (Underhill *et al.*, 1983; Laurent *et al.*, 1986; Nemec *et al.*, 1987). Second, the HA receptor is not as highly specific for HA as link protein and the PGs (Underhill *et al.*, 1983). Third, increases in ionic strength enhance the affinity of interaction of HA with the HA receptor (Underhill and Toole, 1980; Nemec *et al.*, 1987), but decrease affinity for link protein and aggrecan (Goetinck *et al.*, 1987; Tengblad, 1981). Presumably, sequence and conformational differences in the HA-binding regions of the structural HABPs and CD44/HA receptor allow for these differences in ligand recognition. Despite the differences in binding behavior, it is clear that these macromolecules comprise a family that share related HA-binding motifs. Therefore, they deserve a family name: the hyaladherins.

2.2. Hyaluronan-Dependent Pericellular Coats

Several types of cells exhibit large pericellular coats that exclude particles, such as fixed erythrocytes or latex beads, and are destroyed by treatment with hyaluronidases. The structure of these coats depends on interaction of pericellular HA with an HABP, since antibody to this HABP (the MAb IVd4 mentioned in Section 2.1.1) blocks coat formation by rat fibrosarcoma cells and chondrocytes (Fig. 9-1; Q. Yu, S. Banerjee, and B. Toole, unpublished data). The involvement of HA–HABP interaction is supported by the finding that HA hexasaccharides also inhibit coat formation. Large coats can be reconstructed by addition of HA and aggrecan to cells stripped of their coats by pretreatment with hyaluronidase; these coats, as well as those formed metabolically, are inhibited by addition of HA hexasaccharides (Knudson and Knudson, 1991). As discussed in Section 4.3, similar HA-dependent coats may be important in

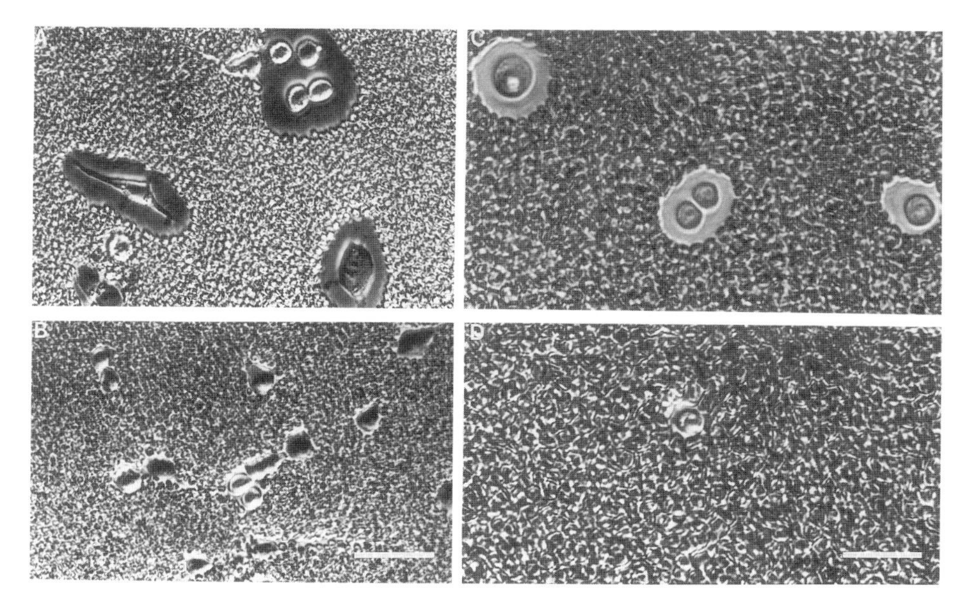

Figure 9-1. HA-dependent pericellular coats. A and B show rat fibrosarcoma cells; C and D show embryonic chick chondrocytes. A and C are controls; B and D were treated with MAb IVd4, a monoclonal antibody raised against chick embryo HABP. The coats (A, C) are visualized by exclusion of particles, here fixed red blood cells. In B and D, the antibody has blocked coat formation (several chondrocytes in D are covered by the red blood cells). Results similar to those shown in A and C have been obtained for several other cell types, e.g., early embryonic mesodermal cells (Knudson and Toole, 1985) and myoblasts (Orkin *et al.*, 1985). For all these cell types, treatment with HA-specific hyaluronidase removes the coats. Bars = 100 μm (A, B), 40 μm (C, D).

regulating the interaction of mesodermal cells and the initial formation of cartilage in early development of the limb.

2.3. Cell Surface Binding Proteins for Proteoglycans

Retention of PGs at the cell surface has been shown to occur by at least four direct mechanisms (see Fig. 9-2A–D): (1) noncovalent interaction of GAG side chains with cell surface binding sites; (2) noncovalent interaction of core protein with cell surface sites; (3) intercalation of core protein into the plasma membrane; (4) insertion into the plasma membrane via a phosphatidylinositol moiety covalently attached to the core protein. In addition, there are many indirect possibilities that involve binding to another extracellular macromolecule that is in turn bound to an integrin (Fig. 9-2E). Evidence has been obtained suggesting that each of these possibilities may be important in various cellular events and some of these are discussed in later sections. In this section, I will briefly discuss the first two mechanisms, those that involve cell surface

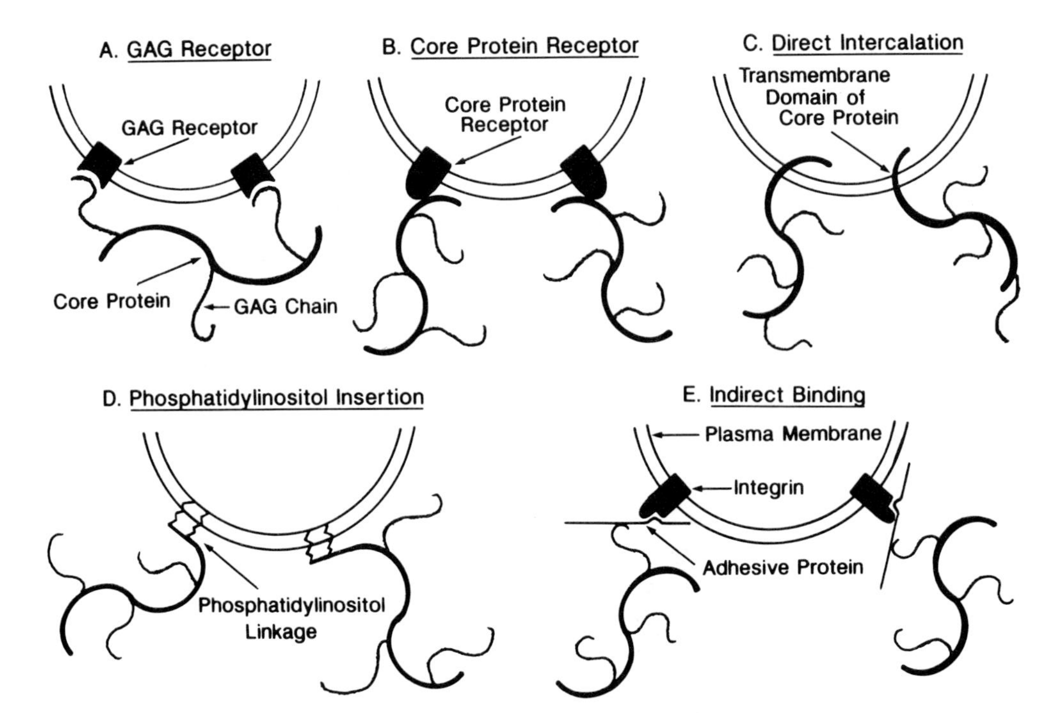

Figure 9-2. Modes of interaction of PGs with the cell surface. A and B show receptor-mediated interactions as discussed in Section 2.3. In the case of GAG receptors, multivalent interactions are likely and could occur with more than one chain of a single PG, as shown, or with more than one repeating sequence within a chain. C and D show two means by which the core protein of a PG can be directly bound to the plasma membrane—via a transmembrane sequence within the core protein, as occurs with syndecan (Saunders *et al.*, 1989), or via a covalently attached phosphatidylinositol group, as occurs in hepatocyte and Schwann cell PGs (Ishihara *et al.*, 1987; Carey and Stahl, 1990). E represents several possible indirect interactions where a PG is bound to another matrix macromolecule, such as fibronectin or laminin, which is in turn bound to a cell surface receptor such as an integrin.

binding proteins, i.e., putative receptors, for the GAG or core protein components of PGs.

2.3.1. Heparin-Binding Proteins

Heparan sulfate (HS)-PGs have been implicated in a variety of cellular events (for examples, see Sections 3 and 4), pointing to the likely involvement of cell surface receptors recognizing these molecules. Several investigators have demonstrated binding of HS or its close relative, heparin, to the surface of a wide variety of cell types, e.g., hepatocytes (Kjellen *et al.*, 1980), smooth muscle cells (Castellot *et al.*, 1985), tumor cells (Bilozur and Biswas, 1990), and spermatozoa (Miller *et al.*, 1990). The strength of binding appears to depend mainly on degree of sulfation (Kjellen *et al.*, 1980), but other factors may also be important (Wright *et al.*, 1989). Molecular modeling studies have pointed to consensus sequences containing clusters of basic amino acids as likely sites for interaction of heparin-related GAG with proteins (Cardin & Weintraub, 1989); thus, ionic interactions between these sequences and the sulfate groups of heparin/HS probably mediate binding.

Heparin/HS-binding proteins have been isolated from tumor cell membranes and identified as two doublets of ~16 and ~32 kDa that bind to HSPG and heparin with very high affinity (~ 10^{-10} M) (Bilozur and Biswas, 1990). Proteins of similar sizes to these 16- and 32-kDa doublets have been isolated from membranes derived from vascular smooth muscle cells (M. E. Bilozur, J. J. Castellot, and C. Biswas, personal communication) and spermatozoa (Miller *et al.*, 1990). In the latter case, 15- to 17-kDa heparin-binding proteins were shown to be absorbed from seminal plasma and to be important for sperm capacitation (Miller *et al.*, 1990). A widely distributed, 78-kDa, heparin-binding protein has also been isolated; antibody to this protein inhibits HSPG binding to smooth muscle cells (Lankes *et al.*, 1988). Although it is clear that binding proteins for heparin/HS are present on the surface of many cell types, the molecular characteristics, interrelationships, and functions of these proteins are not yet clear.

2.3.2. Core Protein-Binding Proteins

A 38-kDa membrane protein that binds to the core protein of basement membrane HSPG has been identified in several cell types (Clement *et al.*, 1989). Again, however, the function and relationship of this binding protein to other matrix receptors are unknown.

The core protein of the DSPG, decorin, has been implicated in receptor-mediated endocytosis of this PG. Recently, overlay experiments probing interaction of decorin core protein with the endosomal proteins of fibroblasts and osteosarcoma cells have led to identification of 51- and 26-kDa proteins that are candidates for mediating recognition and internalization of decorin (Hausser *et al.*, 1989).

3. Influence of Proteoglycans and Hyaluronan on Cell Behavior

3.1. Cell–Cell Adhesion

Until recently it was thought that PGs play secondary roles in cell–cell adhesion, e.g., as modulators of homophilic binding of neural cell adhesion molecules (NCAMs). However, it now appears that PGs may play a more direct role (Bernfield and Sanderson, 1990; Reyes *et al.*, 1990). This topic, which is still relatively undeveloped, will be discussed in the context of epithelial differentiation (Section 4.1) and neural development (Section 4.2), in which systems most of the relevant studies have been performed.

HA can act as a direct mediator of cell–cell adhesion, both homotypic and heterotypic. A well-documented example is the divalent cation-independent

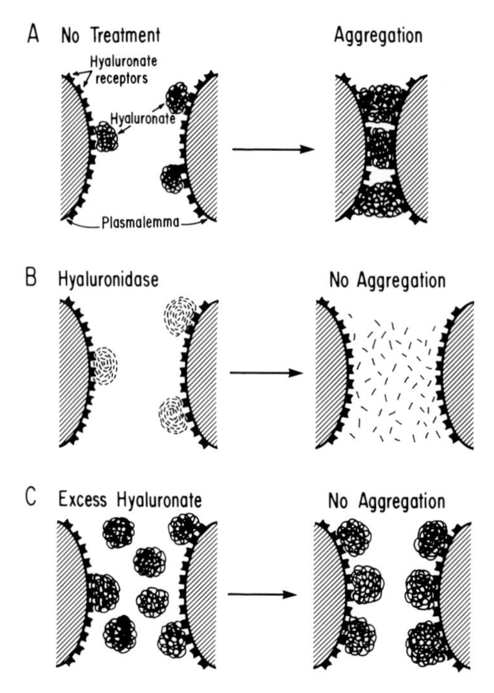

Figure 9-3. HA-mediated aggregation of cells. HA can cross-bridge cells via a calcium-independent, multivalent interaction with receptors on adjacent cells. This aggregation is inhibited by treatment of cells with hyaluronidase (B), or excess HA (C); the latter causes saturation of the receptors, inhibiting cross-bridging (see Section 3.1). If HA receptors are present but there is little or no HA on the surface of the cells (see Fig. 4C), as occurs with some lymphocytes and macrophages, then addition of HA will induce aggregation (Green *et al.*, 1988; Lesley *et al.*, 1990). If the receptors are partially occupied (see Fig. 4A), as occurs with some virally transformed cells, then the cells will spontaneously aggregate (Underhill and Toole, 1981; Green *et al.*, 1988). If the cells have large HA-dependent coats, as occurs with early embryonic limb mesodermal cells, aggregation will not occur (Knudson, 1991). (From Toole, 1981.)

aggregation of several transformed cell lines, which is inhibited by treatment with hyaluronidases, high concentrations of HA, or the K3 antibody to the 85-kDa HA receptor (Underhill and Toole, 1981; Green *et al.*, 1988). The mechanism underlying this aggregation phenomenon is multivalent cross-bridging by HA of HA receptor sites on adjacent cells; inhibition of cross-bridging at high HA concentrations is due to saturation of these receptors (see Fig. 9-3).

Addition of low concentrations of HA to lymphocyte lines or macrophages induces their aggregation (Figs. 9-3 and 9-4), and this aggregation is inhibited by antibody to the 85-kDa HA receptor (Green *et al.*,1988) or to CD44 (Lesley *et al.*, 1990), a molecule now known to be related to the HA receptor (see Section 2.1). Also, interaction of some lymphocyte lines with marrow-derived stromal cell lines is mediated by interaction of HA on the stromal cell surface with CD44 on the lymphocyte surface (Miyake *et al.*, 1990). However, the mediation of homotypic lymphocyte aggregation and lymphocyte–stromal cell adhesion by HA and CD44 is restricted to a subpopulation of lymphocyte lines; many lines that express CD44 do not exhibit HA-dependent interactions. Nevertheless, these findings may be indicative of a role for HA–CD44 interactions in lymphocyte maturation (Miyake *et al.*, 1990). Although CD44 is thought to be involved in the homing of lymphocytes to high endothelial venules in the gut (Haynes *et al.*, 1989), it is not yet clear to what extent the HA-binding function of CD44 is important. Binding of soluble CD44–immunoglobulin fusion proteins to high endothelial cells is blocked by treatment of the cells with hyaluronidase or by competition with HA (Aruffo *et al.*, 1990), suggesting a role

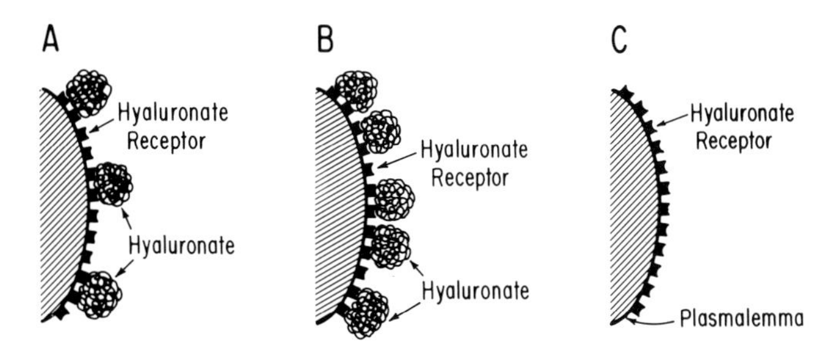

Figure 9-4. Three configurations of HA and HA receptors at the cell surface. (A) The HA receptors are partially occupied, as occurs in some virally transformed cells (Underhill and Toole, 1981; Green *et al.*, 1988). (B) The HA receptors are fully or almost fully occupied, as occurs in some cells with HA-dependent coats such as embryonic chondrocytes (see Fig. 9-1; Knudson and Toole, 1985; 1987). (C) The HA receptors are unoccupied, as occurs in some lymphocytes and macrophages (Green *et al.*, 1988; Lesley *et al.*, 1990). As explained in Fig. 9-3, these different arrangements lead to differences in aggregation properties. (From Toole, 1981.)

for HA in homing; however, binding of intact lymphocytes is not inhibited by these treatments (Culty *et al.*, 1990), indicating that more complex events than HA–CD44 interaction are required. Another function of importance *in vivo* that might be related to the above is HA-mediated adhesion of macrophages to the peritoneal wall following antigen injection (Shannon *et al.*, 1980).

HA-mediated cell adhesion is also of importance in events leading to cartilage differentiation during limb development; this is discussed in Section 4.3.

3.2. Cell–Matrix Adhesion

PGs are important regulators of cell interactions with the extracellular matrix. They have been shown to participate in formation of the transmembrane complexes that result from adhesion of various cell types to extracellular matrices and to modulate initial attachment to these matrices (Fig. 9-5).

3.2.1. Cell Surface Heparan Sulfate Proteoglycan as a Matrix Receptor

Many cell types have been shown to attach to fibronectin via interaction of an Arg-Gly-Asp (RGD)-containing cell-binding domain within fibronectin to a receptor of the integrin class at their cell surface. For most of these cells, especially fibroblasts and related cell lines, this interaction is sufficient for both attachment and spreading of the cells on a fibronectin substratum (see Chapter 10). However, subsequent formation of focal adhesions and other attachment specializations requires both the RGD-containing sequence and a heparin-binding domain within fibronectin. For example, Chinese hamster ovary cell mutants that are deficient in PG synthesis attach and spread on a fibronectin substratum, but do not form stress fibers or focal adhesions (LeBaron *et al.*, 1988). Cell surface HSPG, possibly intercalated directly into the plasma mem-

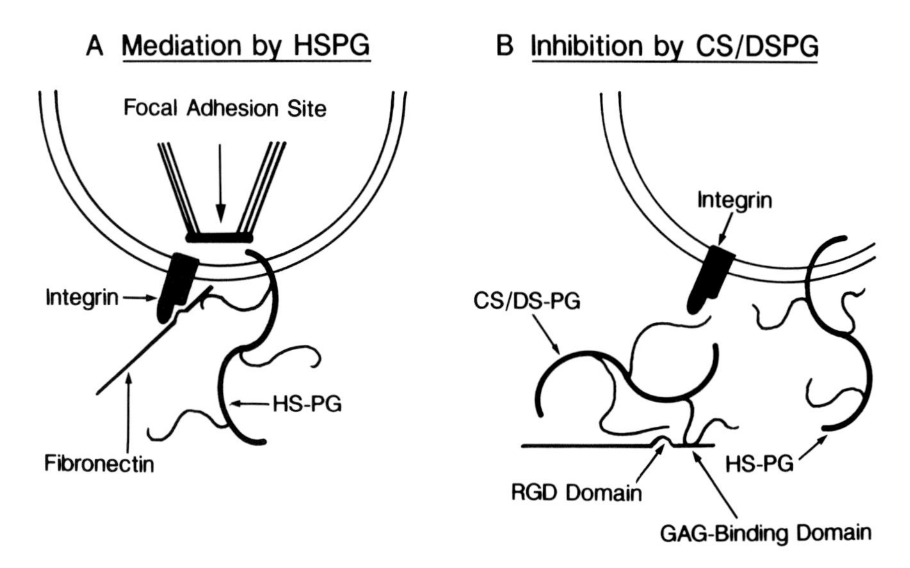

Figure 9-5. Modulation of cell–matrix adhesion by PGs. (A) Cooperative interactions of different regions of fibronectin with its integrin receptor and with cell surface HSPG have been shown to participate in formation of attachment specializations such as focal adhesion sites. (B) Some soluble PGs inhibit interaction of cells with fibronectin by steric interference with its binding to integrin. The molecular size of the PGs in these models has deliberately been exaggerated to emphasize the large hydrodynamic volume that they occupy. The models are discussed further in Section 3.2.

brane, appears to be responsible for interaction of the cell with the heparin-binding domain of fibronectin (Woods *et al.*, 1985) (see Fig. 9-5A). In contrast, thrombospondin, which also interacts with cell surface HSPG, antagonizes the formation of focal adhesions (Murphy-Ullrich and Hook, 1989).

The relationship of epithelial cells to their extracellular matrices is quite different than that of fibroblasts, yet interaction of epithelial cells with fibronectin also involves both the RGD-containing and heparin-binding domains; again, the latter domain interacts with an HS-containing PG, in this case syndecan, at the epithelial surface (Saunders and Bernfield, 1988). The core protein of syndecan has been shown to contain a transmembrane domain that directly intercalates the PG into the epithelial cell membrane and an external domain to which is attached varying proportions of CS and HS chains (see Fig. 9-2C) (Saunders *et al.*, 1989) that mediate the attachment to fibronectin (Koda *et al.*, 1985; Saunders and Bernfield, 1988).

In addition to the interactions with fibronectin described above, cell surface HSPGs of both fibroblastic and epithelial cells mediate interaction of these cells with collagens. Heparin and HS bind more strongly to type V collagen than to fibronectin, laminin, or other collagen types, and the attachment of CHO cells to type V collagen is mediated by cell surface HSPG (LeBaron *et al.*, 1989). Mammary epithelial cells attach readily to type I, III, or V collagens via cell surface syndecan (Koda *et al.*, 1985).

Binding of syndecan to collagen or fibronectin is mainly mediated by its HS chains; thus, changes in ratio of HS to CS in this hybrid PG, either within different epithelial structures (Sanderson and Bernfield, 1988) or in response to regulatory factors such as TGF-β (Rasmussen and Rapraeger, 1988), may be an important functional feature of syndecan. Another potential role of syndecan is in the attachment of B lymphocytes to extracellular matrices, since syndecan is expressed in B-lymphocyte precursors in the marrow and in differentiated plasma cells, but not in circulating B lymphocytes (Sanderson *et al.*, 1989). Syndecan may mediate interaction of the precursor cells with extracellular matrix in the marrow stroma while maturing and in connective tissue after differentiation, but would not be required while the cells are in circulation and thus not attached to a matrix.

3.2.2. Inhibition of Attachment by Soluble Proteoglycans

Whereas cell surface PG participates positively in cell–fibronectin and cell–collagen interactions, soluble PGs usually have an inhibitory effect. For example, attachment of L2 rat yolk sac tumor cells to fibronectin and type I collagen is inhibited by serglycin, a PG derived from the L2 tumor (Brennan *et al.*, 1983). Inhibition is dependent on the ability of the substratum to bind the PG rather than direct interaction of PG with the cells. Since serglycin does not inhibit attachment of L2 cells to fragments of fibronectin lacking the heparin-binding site (Brennan *et al.*, 1983), but inhibits binding of intact fibronectin to its integrin receptor (Hautanen *et al.*, 1989), it has been proposed that inhibition of attachment to fibronectin results from steric interference with the cell attachment domain of this molecule (Fig. 9-5B).

In the case of fibroblasts, a mixture of the DS/CSPGs, decorin and biglycan, has been shown to inhibit attachment to fibronectin (Lewandowska *et al.*, 1987). Again, the evidence obtained points to binding of PG to fibronectin, leading to steric inhibition of fibronectin–integrin interaction, as the mechanism of inhibition of cell attachment. However, these PGs do inhibit fibroblast attachment to fibronectin fragments lacking the heparin-binding site; evidently this is due to a second, "cryptic," GAG-binding site in these fragments (Lewandowska *et al.*, 1987). Inhibition of fibroblast attachment to a wide variety of substrata, including fibronectin, has also been obtained with a large CSPG, termed PG-M, derived from chick embryo fibroblasts. This inhibitory effect, however, appears to be independent of binding of PG-M to the adhesion proteins used as substrata and to arise from adsorption of PG-M to uncoated sites on the culture surface and consequent prevention of cell attachment by repulsion or steric effects (Yamagata *et al.*, 1989). Previous work had shown that substrata of sulfated GAGs can mediate attachment, but this attachment is probably relatively weak (Erickson and Turley, 1983).

Further work is required to clarify this area of research but it is apparent that the amounts, composition, and supramolecular organization of cell surface-associated and extracellular PGs will have profound regulatory effects on attachment of cells to their extracellular matrices, and therefore on the consequences of such attachment, e.g., cell movement.

3.3. Cell Movement

Migratory cells usually lack organized stress fibers and focal adhesions, whereas these structures are prominent in sessile cells on planar substrata *in vitro* (Chapter 12). Thus, the effects that cell surface-associated and extracellular PGs exert on formation of these structures (see Section 3.2) are likely to be important in regulating cell migration on planar substrata. This conclusion is supported by the findings that migrating endothelial cells exhibit increased CS/DSPG synthesis as compared to sessile cells *in vitro* (Kinsella and Wight, 1986) and removal of CS/DSPG from early rat embryos inhibits emigration of cranial neural crest cells *in vivo* (Morriss-Kay and Tuckett, 1989). Other studies, however, indicate that CSPGs inhibit migration of embryonic mesenchymal cells, but their effect depends on interactions with other macromolecules. For example, inhibition by cartilage aggrecan of neural crest cell emigration from explants of amphibian neural tube is due to its interaction with cell surface HA (Perris and Johannson, 1990). This particular mechanism does not explain the effects of cell surface HSPG on focal adhesions or the inhibition of attachment by serglycin, decorin, and biglycan, since these PGs do not bind HA.

Many studies have suggested a role for HA in cell migration and these studies point to several ways that HA might exert an effect. The first possibility relates to the need for a translocating cell to detach from its substratum, at least partially, as well as to attach. Whereas the "footpads" deposited by attaching cells are enriched in HS, those deposited by migrating cells contain more HA and undersulfated CS (Rollins and Culp, 1979). Also, weakly adhesive variants of Chinese hamster ovary cells produce more cell surface HA than tightly adhering variants (Barnhart *et al.*, 1979), and embryonic mesenchymal cells do not attach and spread efficiently on an HA substratum (Erickson and Turley, 1983). These experiments suggest that endogenously produced HA might weaken attachment to adhesive substrata, facilitating the partial detachment that is believed necessary for cells to migrate (but see Chapter 12).

The second possible role of HA is to create hydrated pathways that may facilitate invasion of tissues by separating cellular or fibrous barriers (Fig. 9-6). Hydration of tissues has long been known to depend on the concentration and physical state of HA within tissue compartments (see Comper and Laurent, 1978; Toole, 1981). During embryonic development, regeneration, and tumorigenesis, invasion of cells usually occurs into highly hydrated, HA-rich matrices (reviewed in Toole, 1981; Biswas and Toole, 1987). However, treatment with HA-specific hyaluronidase does not inhibit migration of neural crest cells or primary mesenchyme in the embryo (Anderson and Meier, 1982; Morriss-Kay *et al.*, 1986), making the above postulate unlikely unless the structural cues for emigration significantly precede actual emigration or small but sufficient amounts of HA are still produced during the course of the hyaluronidase-

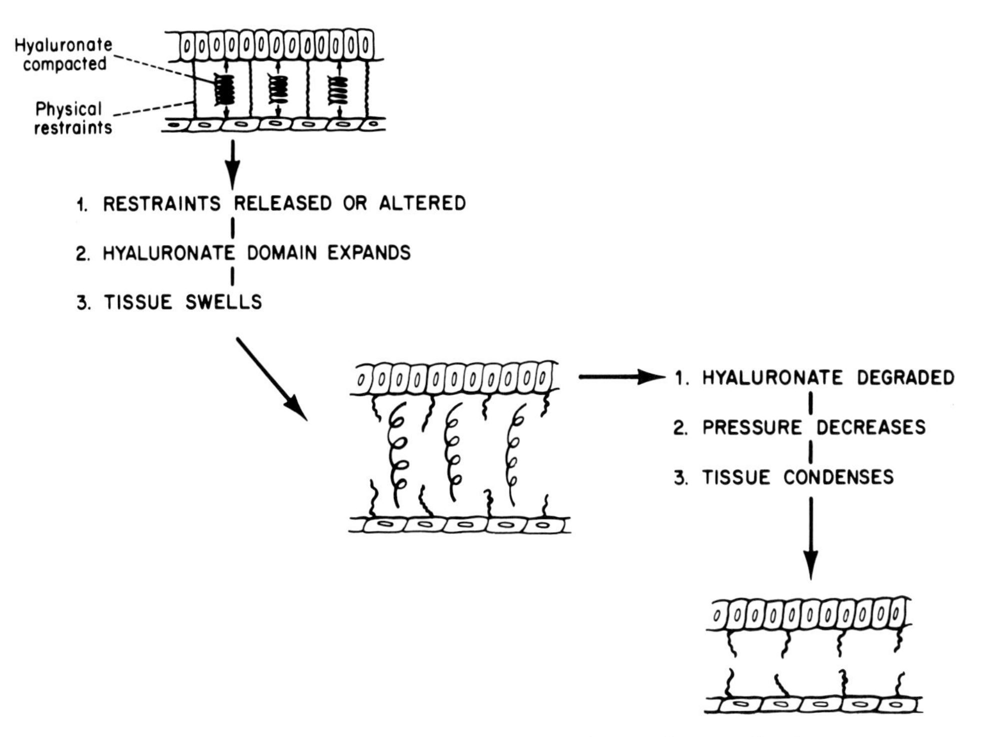

Figure 9-6. HA-mediated tissue expansion. Accumulation of HA within a confined tissue compartment leads to increased swelling pressure due to compaction of the HA network within the compartment (represented by the tightly coiled springs). On removal of constraints on the given tissue compartment, the tissue will swell via hydration of the HA network (expansion could also occur due to increases in HA concentration that are sufficient to overcome the restraints). Expansion would lead to separation of cell or fiber layers, creating spaces that may act as avenues for cell migration (see Section 3.3); regionalized restrictions to expansion could also lead to tissue deformations such as occur during neural tube formation (see Section 4.2). Degradation of HA would lead to reduced swelling pressure, allowing recondensation as occurs in the embryonic cornea (see Toole, 1981). (From Toole, 1981.)

treatment experiments. With respect to the latter possibility, it should be noted that hyaluronidase treatment of cells in culture stimulates HA synthesis (Philipson *et al.*, 1985).

Recent evidence suggests a direct role for cell surface HA in regulation of migration. An HABP is concentrated in the ruffling lamellae of actively locomoting cells (Turley and Auersperg, 1989) and addition of further HABP plus HA to fibroblasts stimulates cell movement (Turley *et al.*, 1985). The endogenous cell surface HABP is associated with protein kinase activity that is stimulated by addition of HA to the cells or to the isolated HABP complex, suggesting the possibility that instructive signal transduction results from interaction of HA with cell surface HABPs (Turley, 1989). This might occur by the apparent interaction of cell surface HABPs with the cytoskeleton (Lacy and Underhill, 1987; Turley *et al.*, 1990).

A 70-kDa factor associated with fetal fibroblasts and fibroblasts derived from cancer patients stimulates confluent cells to penetrate collagen gels by an HA-dependent mechanism. Both factor-induced stimulation of confluent adult fibroblasts and the spontaneously elevated ability of fetal and cancer patient fibroblasts to invade the collagen gels are neutralized by treating the cells with HA-specific hyaluronidase (Schor *et al.*, 1989). Also, the relative abilities of confluent cells to penetrate collagen gels correlate with the level of HA synthesis and size of HA produced at confluence (Chen *et al.*, 1989). Possibly related to these phenomena are the presence of an HA-stimulatory factor in cancer patient sera (Decker *et al.*, 1989) and the finding that several types of tumor cells produce a factor that stimulates HA production in adult fibroblasts (Knudson *et al.*, 1984) and embryonic mesodermal cells (Knudson and Knudson, 1990). However, these potential relationships are not yet established.

3.4. Cell Proliferation

PGs have significant influences on cell proliferation. The available evidence has led to two categories of proposals to explain the effects of PGs: direct effects on cell metabolism that lead to changes in proliferation and indirect effects whereby PGs interact with protein growth factors, altering their efficacy in a positive or negative manner (Fig. 9-7).

3.4.1. Inhibition of Smooth Muscle Cell Proliferation by Heparin-Related Compounds

Numerous studies have demonstrated an inhibitory effect of heparin and HS on proliferation of smooth muscle and related cells *in vitro* and *in vivo*, a finding that has generated great interest, because of its potential importance in vascular wall homeostasis and in clinical applications (reviewed in Castellot *et al.*, 1987). This inhibition is exerted by HS species secreted by confluent, but not proliferating, vascular endothelial (Castellot *et al.*, 1981) or smooth muscle cells (Fritze *et al.*, 1985), pointing to the possibility of either paracrine or autocrine pathways of action. These observations, as well as direct analyses of heparin fragments of varying composition (Wright *et al.*, 1989) and the finding that HS from lung endothelium exhibits much more potent antiproliferative activity than commercial heparin (Benitz *et al.*, 1990), indicate that specific sequences may be important in the inhibitory action of heparin/HS. Inhibition by heparin leads to a block in the G_0 to S phase transition (Castellot *et al.*, 1987; Pukac *et al.*, 1990) and induces specific changes in production of several proteins (e.g., see Majack and Bornstein, 1985; Cochran *et al.*, 1988).

The molecular events that result in inhibition of smooth muscle cell proliferation by heparin/HS are not yet clear. Direct binding and internalization of heparin by smooth muscle cells have been demonstrated (Castellot *et al.*, 1985) and putative cell surface binding proteins have been partially characterized (see Section 2.3). However, although the number of binding sites correlates

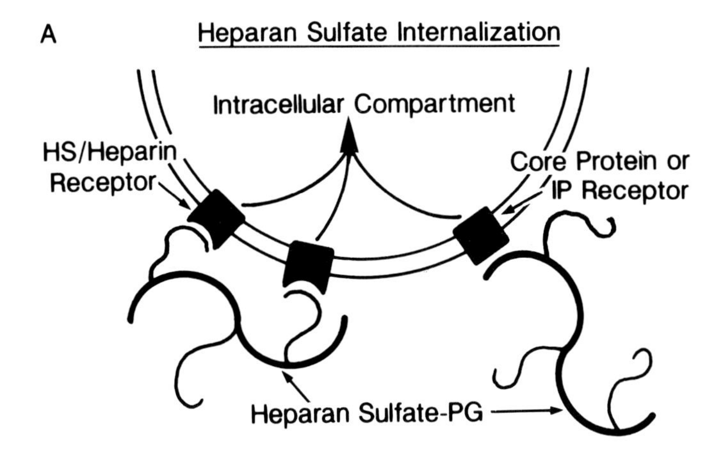

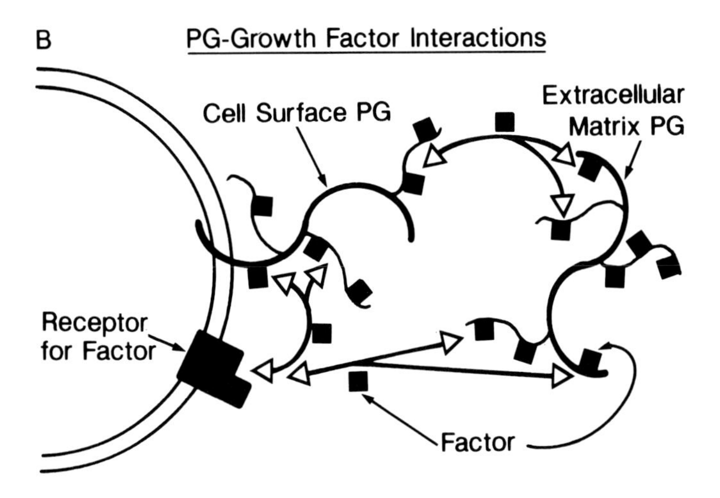

Figure 9-7. Mechanisms of action of PGs in regulation of cell proliferation. Two possible ways by which PGs might influence proliferation are shown and further discussed in Section 3.4. (A) HSPG interacts with three types of receptors on the cell surface, via GAG receptors (see Section 2.3), core protein receptors (Section 2.3), or inositol phosphate (IP) receptors (Section 3.4), any of which might mediate endocytosis of the PG. HS chains or fragments thereof would then be transported by an unknown pathway to an intracellular compartment, thought to be the nucleus (Fedarko *et al.*, 1989; Ishihara and Conrad, 1989), where they would influence events that control proliferation. (B) Many factors that regulate growth bind to HSPG and other PGs, via their GAG chains or their core protein. Interaction of a factor with cell surface or matrix PG could lead to: (1) potentiation of the action of the factor, e.g., acidic FGF (Damon *et al.*, 1989); (2) storage and slow release allowing prolonged action, e.g., basic FGF (Rogelj *et al.*, 1989; Presta *et al.*, 1989; Flaumenhaft *et al.*, 1989); (3) sequestration or neutralization such that the factor cannot interact with its receptor, e.g., thrombospondin (Majack *et al.*, 1988), FGF (Kardami *et al.*, 1988), and TGF-β (Yamaguchi *et al.*, 1990).

with susceptibility to heparin inhibition (Castellot *et al.*, 1985), there is no evidence for direct coupling of cell surface binding to the intracellular events leading to inhibition.

Several indirect mechanisms whereby heparin might exert its inhibitory effect are possible. In the light of the fact that proliferation of most normal cells is dependent on attachment to their substratum, competitive effects of heparin-related compounds on the interaction of cell surface HSPG with fibronectin, collagen, or thrombospondin (see Section 3.2) may be important for proliferation. Of relevance to this possibility are the effects of heparin on the composition of extracellular matrix produced by smooth muscle cells, especially with respect to collagen (Majack and Bornstein, 1985) and thrombospondin (Majack *et al.*, 1988). Reciprocal effects may also be involved in these interactions, since the nature of matrix upon which the cells are cultured influences the production of antiproliferative species of HS by endothelial cells and the responsiveness of smooth muscle cells to these species (Herman and Castellot, 1987).

Other indirect mechanisms of potential importance include prevention of binding of thrombospondin to the smooth muscle cell surface, release of TGF-β from association with α_2-macroglobulin, and sequestration of heparin-binding growth factors. Thrombospondin is required for smooth muscle cell proliferation, and its mitogenic effect and binding to the smooth muscle cell surface are blocked by heparin (Majack *et al.*, 1988). Binding of thrombospondin to the cell surface has been shown, in other cell types, to be mediated by cell surface HSPG (e.g., see Sun *et al.*, 1989), and thus heparin would presumably act by competitive inhibition of binding to this PG. TGF-β is an inhibitor of smooth muscle proliferation, but is partially sequestered in inactive form by serum α_2-macroglobulin. Heparin may cause dissociation of the complex of TGF-β and α_2-macroglobulin due to binding of TGF-β to heparin; consequently, TGF-β inhibition of proliferation would be increased (McCaffrey *et al.*, 1989). Although these mechanisms are compelling, they contradict previous studies demonstrating that heparin inhibition occurs in medium containing serum depleted of heparin-binding factors (Reilly *et al.*, 1986). This contradiction would be resolved if the action of the heparin-binding factor is autocrine, i.e., derived from the cell itself rather than from serum, or if multiple mechanisms of action of heparin/HS are involved.

3.4.2. Binding of Growth Factors to Proteoglycans

It is not yet clear whether the inhibitory effect of heparin-related compounds on smooth muscle cell proliferation is due to direct binding at the cell surface or to interactions with other factors that stimulate or inhibit proliferation. However, it is clear in other cellular systems that many of the effects of PGs and GAGs, including heparin-related compounds, are due to their ability to bind growth factors in culture media, at the cell surface or in the extracellular matrix. These interactions can lead to positive or negative effects on proliferation. For example, inhibition of proliferation of skeletal myoblasts by heparin is due to sequestration of a heparin-binding factor in serum, probably FGF (Kardami *et al.*, 1988). Since basic FGF is present in the extracellular matrix of muscle *in vivo* (DiMario *et al.*, 1989), interaction of HS and FGF may be of regulatory importance in muscle growth and regeneration.

Basic FGF is also present in extracellular matrices produced by endothelial cells, in linkage with HSPG (Baird and Ling, 1987; Folkman *et al.*, 1988). Proliferation of vascular endothelial cells will take place on cornea endothelial extracellular matrix in the absence of basic FGF added to the medium, but is dependent on basic FGF sequestered in this matrix (Rogelj *et al.*, 1989). Basic FGF bound to HS in the extracellular matrix of vascular endothelium is released slowly in active form allowing long-term action (Presta *et al.*, 1989; Flaumenhaft *et al.*, 1989). However, it is not clear by what mechanism the FGF is released, e.g., via competitive binding with secreted HS or heparanase-mediated degradation of the matrix-bound HSPG.

In addition to its role in binding basic FGF to extracellular matrices, HSPG may protect FGF from inactivation (Saksela *et al.*, 1988). Heparin also potentiates the activity of acidic FGF by protection from proteolysis (Damon *et al.*, 1989). These results contrast with the inhibition by sequestration of FGF that occurs with skeletal myoblasts, as mentioned above (Kardami *et al.*, 1988); the reason for this difference is not clear.

Interaction with HSPG also leads to potentiation of a Schwann cell mitogen unrelated to FGF. This mitogen is bound to the surface of neurons as a complex with the PG and stimulates Schwann cell division only after contact between the two cell types (Ratner *et al.*, 1988). Hemopoiesis is promoted by interactions between precursor cells and the marrow stroma that also involve HSPG. HSPG in the extra- or pericellular matrices of marrow stromal cells retains hemopoietic growth factors in active form, and removal of HS reverses the stimulatory effect of stroma on hemopoiesis (Luikart *et al.*, 1990).

Finally, modulation of cell proliferation by the DSPG, decorin, can be explained by its ability to bind TGF-β and neutralize its effects (Yamaguchi *et al.*, 1990). In this case, however, binding of the PG is via its core protein rather than its GAG chains. Since TGF-β also stimulates the synthesis of decorin, decorin may be a feedback regulator of TGF-β action. The TGF-β type III receptor is a cell surface PG that also binds TGF-β to its core protein (Cheifetz and Massague, 1989). This "receptor" is not responsible for direct mediation of TGF-β action (Boyd and Massague, 1989) and thus may also serve as a storage site or regulator of TGF-β action.

3.4.3. Nuclear Heparan Sulfate

Conrad and colleagues have described the growth-related localization of a specific subclass of HS chains, enriched in glucuronate-2-sulfate, in the nucleus of hepatoma cells. The concentration of these HS chains in the nucleus of logarithmically growing cells remains within a narrow range, but rises approximately threefold as the cells reach confluence. Conditions that prevent this rise lead to loss of contact inhibition (Ishihara and Conrad, 1989). The core protein of HSPG, which is initially attached to the plasma membrane by a phosphatidylinositol moiety (see Fig. 9-2D), may be cleaved from the plasma membrane and the PG endocytosed via interaction with an inositol phosphate receptor (Ishihara *et al.*, 1987). After release of HS chains from the internalized PG, a small proportion of these chains, enriched in glucuronate-2-sulfate, is transferred to the nucleus by an unknown pathway. Cell surface HSPGs prepared from log-phase and confluent hepatoma cells have opposing effects on growth when added back to these cells (Fedarko *et al.*, 1989). Addition of PG from confluent cells leads to transient appearance of HS chains in the nucleus and inhibition of growth in the G_1 phase of the cell cycle. In contrast, PG from log-phase cells does not result in significant targeting to the nucleus and has no effect on the growth of log-phase cells; however, it increases the growth rate of cells whose growth has slowed due to depletion of their medium of serum or insulin.

The opposing effects of HSPG from growing and confluent hepatoma cells are reminiscent of the effects on smooth muscle cells of HS species secreted by growing and confluent smooth muscle (Fritze *et al.*, 1985) or endothelial cells (Castellot *et al.*, 1981). Heparin inhibition of smooth muscle growth is not dependent on the presence of glucuronate-2-sulfate, suggesting a different mechanism from that in hepatoma cells (Wright *et al.*, 1989). However, the enrichment of this group in nuclear HS remains a correlation observed only in hepatoma cells; it has not yet been shown that this moiety is a necessary sequence for the action of HS. In fact, no mechanism of action of nuclear HS is established, but two possibilities have been suggested. First, it is known that heparin has significant effects on transcription and DNA replication in cell-free assays (see Fedarko and Conrad, 1986, for discussion) and thus may act directly in the nucleus. Second, HS may act as a carrier for heparin-binding growth factors into the nucleus where they may exert effects on transcription (see Fedarko *et al.*, 1989; Ishihara and Conrad, 1989, for discussion).

In summary, the observations described in the above sections point strongly to the possibility that HSPG exerts its effects on proliferation at multiple sites within the complex extracellular and intracellular signaling pathways that regulate growth. The specific response obtained will depend on sequences present within the HS chains of the particular HSPG present, its supramolecular organization at the cell surface or within the extracellular matrix, and the signaling pathways operating within a given cell type under the particular physiological conditions.

3.4.4. Hyaluronan–Cell Interaction

A strong, positive correlation has been demonstrated between HA synthesis, HA synthetase activity, and cell proliferation in culture (e.g., see Brecht *et al.*, 1986), and cells *in vivo* usually proliferate within an HA-rich extracellular matrix (see Toole, 1981). Recently, the 85-kDa HA receptor (see Section 2.1) has been localized to the surface of a variety of proliferating epithelial cells, e.g., in the basal layers of the epidermis and the base of intestinal crypts, but the areas rich in receptor appeared to vary in their occupancy by HA (Alho and Underhill, 1989). Although HA and the HA receptor are usually enriched in the proliferating zones, they are not restricted to these areas (Alho and Underhill, 1989; Tammi *et al.*, 1989).

Inhibition of HA synthesis in fibroblasts leads to arrest of these cells in mitosis just prior to rounding, possibly since HA-mediated detachment from the substratum is required (Brecht *et al.*, 1986) (also see Section 3.3). Endothelial cell proliferation is stimulated by HA-derived oligosaccharides but inhibited by polymeric HA (West and Kumar, 1989). We have also found that polymeric HA inhibits proliferation of fibroblasts, but only at high concentrations of high-molecular-weight polymer. Low concentrations of high-molecular-weight HA or high concentrations of low-molecular-weight HA either had no effect or were somewhat stimulatory (Goldberg and Toole, 1987). Clearly, much further work is required to establish the precise relationship of HA to proliferation.

4. Function of Proteoglycans and Hyaluronan in Development

4.1. Epithelial Branching and Differentiation

Interaction of epithelia with extracellular matrices is essential for their morphogenesis, their differentiation, and the stability of their phenotype, a theme that recurs throughout this volume (e.g., see Chapter 12). A role for PGs is especially well established with respect to epithelial–mesenchymal influences in epithelial morphogenesis.

4.1.1. Basement Membrane Proteoglycans in Epithelial Branching

Branching morphogenesis is an event common to the development of many organs, e.g. lung, mammary gland, kidney, and salivary gland, and is dependent on interaction of epithelium and mesenchyme (Spooner *et al.*, 1986). Treatment of embryonic salivary gland epithelium with hyaluronidase prior to recombination with mesenchyme (Banerjee *et al.*, 1977), or inhibition of PG synthesis in salivary gland or kidney primordia with β-xyloside (Spooner *et al.*, 1985; Lelongt *et al.*, 1988), blocks branching morphogenesis. Bernfield and co-workers (reviewed in Bernfield *et al.*, 1984) have shown that during branching of salivary gland primordia, GAGs accumulate preferentially in the quiescent clefts and are turned over more rapidly in the lobules at the active sites of branching (Fig. 9-8). The basal lamina in the clefts is stable and of typical composition, i.e., enriched in HSPG, laminin, and type IV collagen. Type I and III collagen also accumulate in the clefts and may be responsible for stabilization of these regions (Bernfield *et al.*, 1984; Spooner *et al.*, 1986; Fukuda *et al.*, 1988). The basal lamina of the lobules, however, is depleted of HSPG and type IV collagen, and enriched in HA and CS, which turn over rapidly due to interaction with the mesenchyme. A neutral hyaluronidase pro-

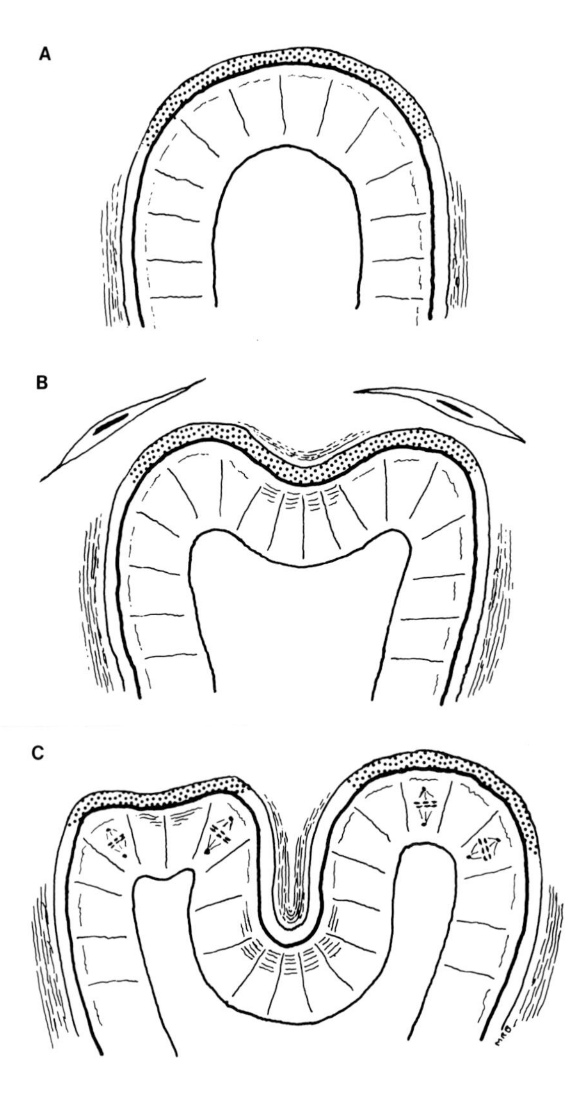

Figure 9-8. The role of extracellular matrix components in branching morphogenesis. (A) In the primary lobule, HA and CSPG are synthesized actively and rapidly turned over at the distal end due to action of a mesenchymal hyaluronidase (strippling represents the sites of active GAG turnover). (B) Mature basal lamina components and interstitial collagen fibers accumulate at the site of cleft formation; HA and CSPG continue to turn over rapidly in the lobules. (C) Turnover of HA and CSPG becomes restricted to the distal tips of the secondary lobules, and collagen fibers accumulate further in the cleft. (From Bernfield and Banerjee, 1972.)

duced by the mesenchyme may mediate turnover of the lobule GAGs *in vivo* (Bernfield *et al.*, 1984). The differential distribution and dynamic state of these epithelial and mesenchymal elements would allow some areas of the epithelium to proliferate and undergo the shape changes necessary for branching, while other areas become stabilized as clefts.

4.1.2. Syndecan in Epithelial Morphogenesis and Differentiation

Interaction with extracellular matrices has a profound influence on epithelial differentiation (e.g., see Chapter 12) but, until recently, little was known about the role of PGs in this regard. Syndecan is a hybrid HS/CSPG intercalated into the plasma membrane of many types of cells (see Section 3.2.1), most but not all of which are epithelial (Hayashi *et al.*, 1987; Bernfield and Sanderson, 1990). Syndecan varies in the ratio and amount of HS and CS chain substitution and in its distribution on the cell surface in simple versus stratified epithelia. These differences in composition and distribution are reproduced when the organization of epithelia changes from simple to stratified and vice versa. These observations, and the finding that epithelial cells transfected with antisense cDNA to syndecan adopt fibroblastic shape, organization, and behavior, suggest strongly that syndecan is required for maintenance of epithelial phenotype (Sanderson and Bernfield, 1988; Bernfield and Sanderson, 1990). This idea is supported further by the fact that syndecan is lost prior to epithelial–mesenchymal transformation in the embryo and appears during mesenchymal–epithelial transformation (Chapter 12).

Syndecan not only binds extracellular matrix, but also binds basic FGF, another function that could alter the state of differentiation of the cell (Bernfield and Sanderson, 1990). Treatment of dedifferentiated hepatocytes in culture with various GAGs or PGs restores gap junction formation and transcription of liver-specific mRNAs, and alters cell shape (Spray *et al.*, 1987). Possibly, these exogenous GAGs and PGs displace or modify heparin-binding regulatory factors bound to cell surface syndecan or related PG (Kjellen *et al.*, 1981), thus influencing differentiation.

Epithelial–mesenchymal interactions are essential for differentiation of most organs and close correlations have been noted between expression of several matrix or cell surface macromolecules and the events that follow such interactions (e.g., see Ekblom *et al.*, 1986; Dauca *et al.*, 1990). Most of these observations suggest important functions for these molecules in histogenesis, but little is known concerning their specific mechanistic involvement in the interactions themselves or the onset of differentiation. Syndecan expression in the transformation of nephrogenic mesenchyme to epithelium (see Chapter 12) appears to be induced by a tissue interaction. Embryonic oral epithelium induces condensation of adjacent mesenchyme and this condensed mesenchyme in turn induces formation of the enamel organ in the oral epithelium. The condensed mesenchyme and the oral epithelium express syndecan transiently during induction (Vainio *et al.*, 1989). As discussed above and in Section 3.2, syndecan is thought to participate in the induction and stabilization of epithelial morphology by acting as a mediator of cell–matrix adhesion but, on the basis of the above observations, it may also participate in mesenchymal cell condensation, epithelial cell adhesion, and growth factor sequestration (Bernfield and Sanderson, 1990).

4.2. Neural Development

One of the most rapidly growing areas of research on the role of PGs and HA in development is their relationship to the many facets of neural development. The nervous system, both adult and developing, contains a bewildering array of PGs; most of these PGs have not been well characterized and it is not known how many are related, polymorphic forms with various combinations of GAG sizes and types attached to the same or homologous core proteins. Also, distribution of the PGs and HA ranges from extracellular to cell surface, cytoplasmic, and nuclear; this cellular distribution changes radically in different regions of the nervous system and with developmental stage (see Margolis and Margolis, 1989; Hockfield, 1990; Herndon and Lander, 1990). These data imply multiple functions for PGs and HA in the nervous system and the preliminary mechanistic studies performed to date support this conclusion.

4.2.1. Neural Tube Closure

Several observations suggest that HA may play a role in the tissue elevation and folding that occur during formation of the neural tube from the neural plate. As stated in Section 3.3, accumulation of HA within confined compartments gives rise to internal tissue pressures that, on reduction of restrictive forces or increases in HA concentration, can lead to increased hydration and tissue expansion (see Fig. 9-6). Such forces could in turn contribute to tissue deformations involved in events such as neural fold elevation and closure (see Toole, 1981). HA accumulates in the mesenchyme of cranial neural folds with a distribution that suggests it may play such a role (Morris-Wiman and Brinkley, 1990). Schoenwolf and Fisher (1983) demonstrated that treatment of early chick embryos with HA-specific hyaluronidase gives rise to large numbers of open neural tube defects, but Morriss-Kay *et al.* (1986) did not obtain this result with early rat embryos. HA also accumulates at the posterior neuropore of embryonic mice and is less abundant at sites where closure is already complete, while HA accumulation is reduced at the neuropore of curly tail mutant mice that exhibit neural closure defects (Copp and Bernfield, 1988). Although additional work is required to substantiate the above, it seems likely that physical forces resulting from HA accumulation may contribute to early neural morphogenesis.

4.2.2. Neural Cell Interactions

One of the major mediators of cell–cell adhesion in the nervous system is NCAM; homophilic interaction between NCAM molecules on adjacent cells and modulation of this interaction due to varying content of polysialic acid are thought to underlie the mechanism of action of NCAM in neuronal adhesion (reviewed in Rutishauser *et al.*, 1988). However, NCAM also contains a heparin-binding domain that is essential for neuronal adhesion (Reyes *et al.*, 1990). Although it is clear that HSPG is involved in NCAM-mediated cell adhesion, the relative importance of HSPG-induced conformation changes that influence homophilic NCAM binding versus heterophilic binding of cell surface HSPG to NCAM is not clear.

Other cell interactions in the nervous system may also be influenced by PGs. For example, interaction between cytotactin (tenascin) and a neuronally produced CSPG has been implicated in neuron–glial cell adhesion (Hoffman and Edelman, 1987). PGs are also present within neuromuscular junctions and surrounding synapses in the central nervous system but their precise role therein is not known (for discussion see Sanes, 1989; Hockfield, 1990). As with fibroblastic and epithelial cells (see Section 3.2), HSPGs are likely to be involved in adhesion of neurons and Schwann cells to fibronectin or laminin; this in turn would be important for promoting neurite outgrowth (Chernoff, 1988; Carey and Stahl, 1990; Hockfield, 1990).

4.2.3. Neurite Outgrowth

An area of considerable interest is the importance of extracellular macromolecules in regulating neurite outgrowth. Much of the work in this area has focused on laminin since it is an excellent substratum for neuron outgrowth (Sanes, 1989; Lander, 1989). However, the stimulatory effect on neurite growth of conditioned media from a variety of cell types appears to be due to a complex of laminin and HSPG, the activity of which is dependent on both components (e.g., see Chiu *et al.*, 1986; Dow *et al.*, 1988). HSPG may regulate the length of neurites, while laminin influences initiation and branching (Hantaz-Ambroise *et al.*, 1987). One likely mechanism of action of HSPG would be to mediate or modulate binding of laminin to the neuronal cell surface; however, the heparin-binding domain of laminin does not appear to be involved (Edgar *et al.*, 1988).

CS and keratan sulfate PGs may act as barriers to the growth of axons into specific areas of the brain since CS and keratan sulfate are enriched in such areas *in vivo* and neurite growth is inhibited on substrata of PGs containing these GAGs. Inhibition is strongest with hybrid CS/keratan sulfate PG (Snow *et al.*, 1990).

4.3. Limb Development

The embryonic limb is one of the most extensively studied developmental systems with respect to the role of extracellular matrix macromolecules. The stage dependency and distribution of numerous matrix constituents have been described (reviewed in Solursh, 1990). Some of the reasons for this concentration of effort are ease of dissection, the depth of detailed biological knowledge that has accumulated over many decades, the well-defined steps in differentiation that occur therein, and the availability of cell culture systems that mimic major events in limb development, especially chondrogenesis and myogenesis.

Embryonic limb development begins by proliferation and immigration of mesodermal cells, which is followed by formation of "condensations" in the chondrogenic and myogenic regions prior to overt differentiation. The subectodermal mesoderm, i.e., the future dermis, remains noncondensed during premyogenic and prechondrogenic condensation, retaining mesenchymal morphology and large HA-rich spaces between cells (Singley and Solursh, 1981). Condensation of the mesoderm is an important step in early formation of cartilage and muscle.

4.3.1. Proteoglycans in Limb Development

Several types of PG participate in limb development. Of obvious importance are the PG components of differentiated cartilage, muscle, and fibrous connective tissue; the nature and functions of these PGs are discussed elsewhere in this volume (Chapters 2 and 5). Basement membrane HSPG (Solursh and Jensen, 1988) and syndecan (Solursh *et al.*, 1990) are present transiently in early mesoderm prior to differentiation, but their functional significance in the mesoderm is not known. The large PG, PG-M, of fibroblasts is also produced by limb mesoderm and becomes concentrated in precartilage condensations prior to overt differentiation (Kimata *et al.*, 1986). This PG is of particular interest because of its restricted localization and its ability to bind both HA and fibronectin (Yamagata *et al.*, 1986), each of which is probably involved in early chondrogenesis. Fibronectin promotes precartilage condensation in culture via interactions involving its heparin-binding domain (Frenz *et al.*, 1989). Most of the above-mentioned PGs are able to interact with this domain, but it is not known whether one of these PGs is a natural ligand and whether this interaction is critical *in vivo*.

4.3.2. Hyaluronan–Cell Interactions during Mesodermal Condensation

HA is a ubiquitous component of the extracellular matrices in which cells migrate and proliferate during embryonic development (reviewed in Toole, 1981). Mesodermal cells in the early chick embryo limb bud are separated by extensive HA-rich spaces (Singley and Solursh, 1981) and these cells in culture produce large HA-dependent pericellular coats (Knudson and Toole, 1985; see Fig. 9-1). When the chondrogenic and myogenic areas of the limb bud become condensed *in vivo*, i.e., separated by less matrix, the mesodermal cells do not exhibit visible coats in culture. During differentiation of chondrocytes, which are again separated by extensive spaces *in vivo*, large pericellular coats are reexpressed in culture. Chondrocyte coat structure is still dependent on HA even though PG is now a quantitatively more prominent component (Goldberg and Toole, 1984; Knudson and Toole, 1985). In contrast to chondrogenesis, differentiation of myoblasts is accompanied by loss of pericellular coats (Orkin *et al.*, 1985), apparently a necessary step since myoblasts cultured on an HA substratum fail to fuse (Kujawa *et al.*, 1986). Thus, there is a close correlation between the presence of large intercellular spaces *in vivo* and the expression of HA-dependent coats *in vitro* (see Fig. 9-9).

In parallel to the loss of ability to express HA-dependent coats at the time of condensation, there is a large decrease in the ratio of HA to CSPG produced by the condensing mesoderm as compared to precondensation mesoderm (Knudson and Toole, 1985). This is accompanied by expression of cell surface, membrane-associated binding sites for HA (Knudson and Toole, 1987). HA-binding sites are known, in other systems, to be involved in endocytosis en route to degradation of HA (Laurent *et al.*, 1986; McGary *et al.*, 1989), and so it is reasonable to suppose that the appearance of HA-binding sites at the time of

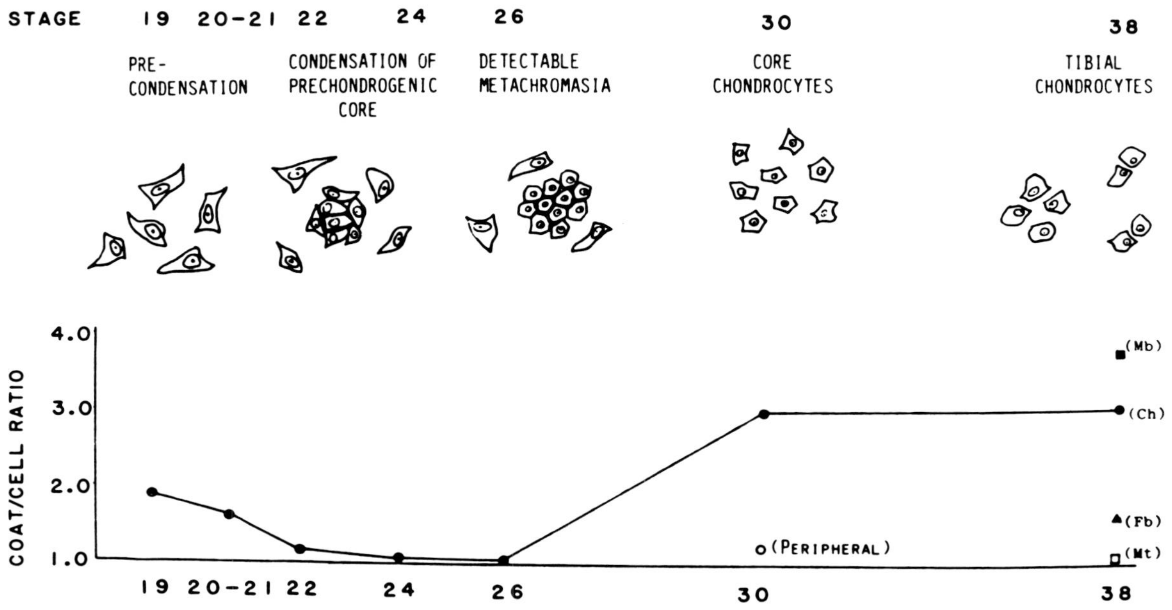

Figure 9-9. Comparison of pericellular coats *in vitro* with cell organization *in vivo*. The sizes of pericellular coats are represented as mean coat-to-cell ratios (ratios of coat perimeter to cell perimeter) obtained with cells cultured from limbs at the stages indicated. Prior to condensation, the mesodermal cells *in vivo* are highly separated by HA-rich matrix and *in vitro* they exhibit large HA-dependent coats. When the mesoderm becomes condensed *in vivo*, the cells lack coats *in vitro*. On differentiating, chondrocytes again elaborate large HA-dependent coats *in vitro* and an extensive matrix *in vivo*. Myoblasts exhibit coats prior to fusion but lose them on fusion (see Section 4.3). Ch, chondrocytes; Fb, fibroblasts; Mb, myoblasts; Mt, myotubes. (From Knudson and Toole, 1985.)

condensation represents an increase in receptor-mediated endocytosis of HA. Decreased HA synthesis and coat production together with increased endocytosis and degradation of HA would lead to a dramatic reduction in volume of matrix between the mesodermal cells, thus allowing them to "condense" (Fig. 9-10A versus B).

In addition to the events leading to reduction in matrix volume, condensation may also involve direct cell interactions. Multiple adhesive steps are probably involved in myogenesis and chondrogenesis, but an early step is likely to be cross-bridging of cells by multivalent interaction of HA with the cell surface HA-binding sites that become expressed at this stage (see Section 3.1 and Fig. 9-3). Mesodermal cells from condensation-stage limbs, but not from precondensation limbs, aggregate *in vitro* via such HA cross-bridging (Knudson, 1991). Thus, condensation may be largely explained by a combination of receptor-mediated removal of preexisting pericellular HA, decreased HA synthesis and coat assembly, and HA-mediated cross-bridging of the mesodermal cells (Fig. 9-10B).

4.3.3. Hyaluronan–Cell Interactions during Cartilage and Muscle Differentiation

Recent studies in our laboratory (R. Turner, S. Banerjee, and B. Toole, unpublished data) have shown that interference with binding of HA to the surface of limb mesodermal cells blocks chondrogenesis in culture. In these studies, chondrogenesis was blocked by treatment of the cell either with monoclonal antibody to HABP (MAb IVd4 discussed in Section 2.1) or with HA hexasaccharide, both of which reagents interfere with binding of polymeric HA to HABP. It is likely that the HA–HABP interaction blocked in these experiments is involved in matrix assembly.

Many studies indicate that HA plays an important role in matrix assembly by chondrocytes. For example, the well-documented interaction of HA with link protein and aggrecan is central to the structure of cartilage matrix (see Chapter 2). Recent evidence indicates that retention of HA–PG aggregates in the pericellular matrix by binding to other HABPs is also important. Study of chondrocytes in culture has revealed the following information about the pericellular matrix of chondrocytes. It has been shown, first, that chondrocyte coats are destroyed by treatment with HA-specific hyaluronidase. The loss of coats is accompanied by loss of much of the aggrecan associated with the chondrocyte cell layer, indicating that it was retained by an interaction with HA (Goldberg and Toole, 1984). Second, production of these pericellular coats by chondrocytes is inhibited by HA hexasaccharides (Knudson and Knudson, 1991) or the MAb IVd4 antibody to HABP (see Fig. 9-1). The IVd4 antibody reacts strongly with embryonic cartilage, but only after treatment of the tissue with HA-specific hyaluronidase (Fig. 9-11), implying that the HABP is occupied by HA *in vivo*. The relationship of the proteins recognized by MAb IVd4 (see Section 2.1) with other HABPs in cartilage (e.g., Crossman and Mason, 1990) is not known. Third, pericellular coats can be rebuilt by addition of HA

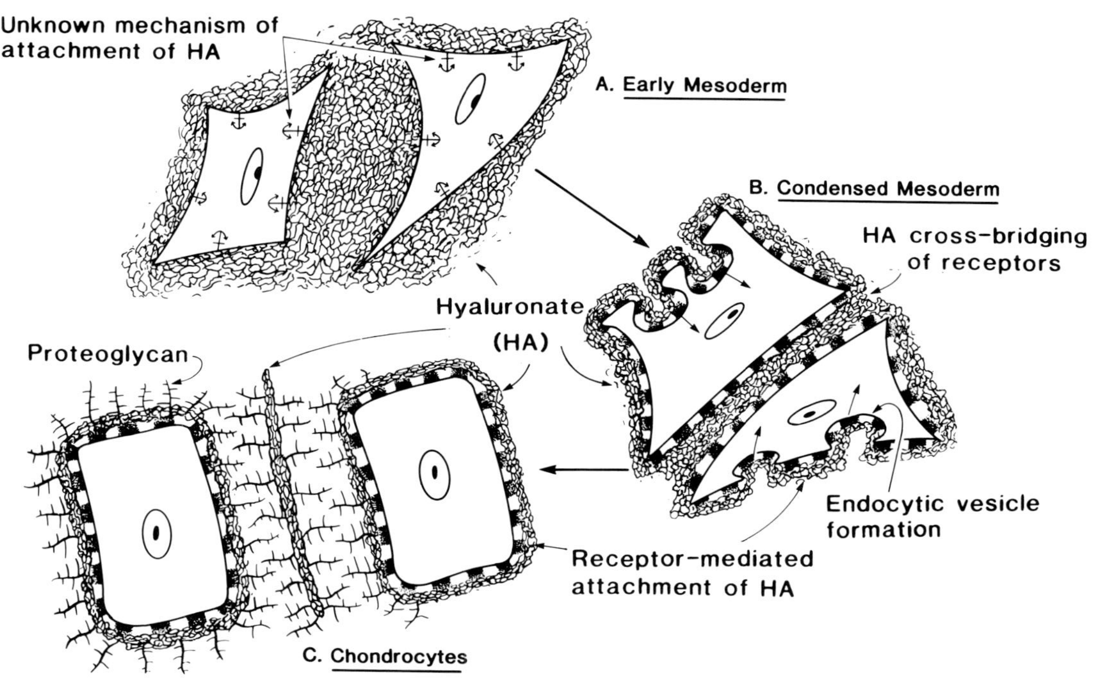

Figure 9-10. The hypothesized role of HA–cell interactions in limb mesoderm condensation and chondrogenesis. The mechanism of interaction of HA with the precondensation mesodermal cells is not known since HA-binding sites have not been detected (Knudson and Toole, 1987). HA-binding sites are detected in condensation stage mesoderm and may mediate endocytosis and cross-bridging of cells in the condensate. Binding sites are also found in chondrocytes and may mediate retention of HA–PG aggregates in the pericellular matrix (see Section 4.3). (From Toole *et al.*, 1989.)

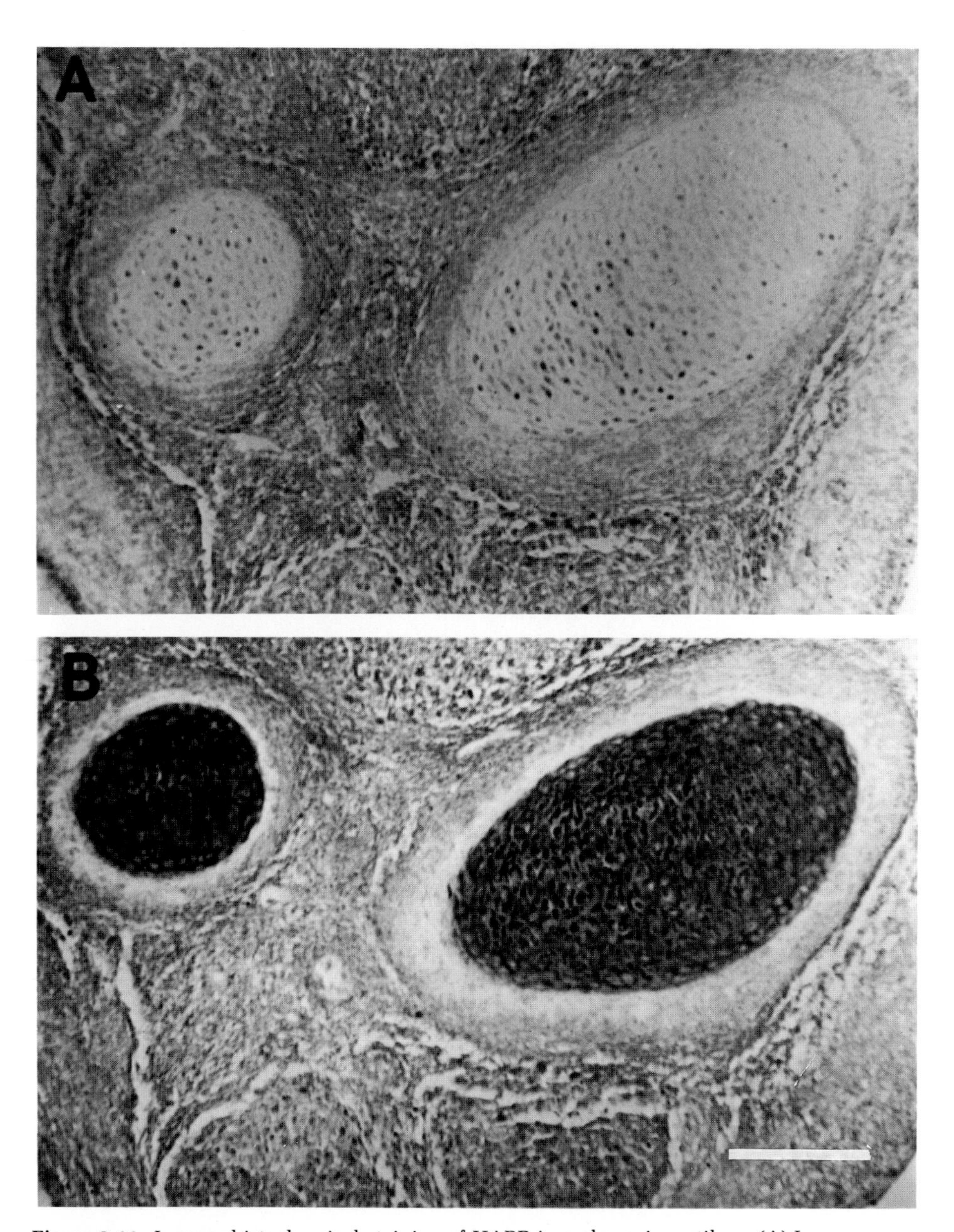

Figure 9-11. Immunohistochemical staining of HABP in embryonic cartilage. (A) Immunoperoxidase staining of 7-day chick embryo limb tissues with MAb IVd4 to chick embryo HABP (see Section 2.1). Muscle and connective tissue, except perichondrium and subectodermal mesoderm, stain moderately; cartilage stains weakly. (B) Staining of a section similar to A after treatment of the section with HA-specific hyaluronidase. Cartilage now stains very strongly, implying that a high concentration of HA-occupied HABP was present. Bar = 0.2 mm.

and aggrecan to cells stripped of their coats by treatment with hyaluronidase, but this process is inhibited in the presence of HA hexasaccharides (Knudson and Knudson, 1991). Since HA hexasaccharides do not inhibit HA–link or HA–aggrecan binding (Tengblad, 1981), but do inhibit HA binding to other HABPs, we conclude that HA–PG aggregates are retained in the pericellular matrix by interaction with other HABPs, presumably those recognized by MAb IVd4 (see Fig. 9-10C).

Whereas HA is clearly an important structural component of differentiated cartilage, there is little evidence that HA plays a central role in the structure of differentiated muscle. However, HA may be important in early events prior to cytodifferentiation. Exposure of myoblasts to a substratum to which HA is conjugated inhibits their differentiation and maintains the cells in a proliferative state (Kujawa *et al.*, 1986). Myoblasts themselves, but not myotubes, exhibit HA-dependent coats (Orkin *et al.*, 1985) and membrane-associated HA-binding sites (Knudson and Toole, 1987), characteristics that may be important in their response to cellular interactions prior to differentiation.

4.3.4. Regulation of Hyaluronan Synthesis

As discussed in Section 3.4, a positive relationship may exist between HA synthesis and cell proliferation, two prominent activities during the earliest stages of limb development. Basic FGF is a major growth factor in the early developing limb; the amount of this factor is highest in precondensation-stage limb buds, when HA synthesis and cell proliferation are maximal, and decreases during subsequent condensation and differentiation of the mesoderm (Munaim *et al.*, 1988; Seed *et al.*, 1988). Purified basic FGF stimulates HA synthesis and HA-dependent coat formation in cultures of limb mesoderm (Munaim *et al.*, 1991). Thus, FGF may be important in the coordinated regulation of cell proliferation, HA synthesis, and pericellular matrix assembly in the precondensation limb bud mesoderm. Diffusible growth factors, especially FGF, have been implicated in the growth regulatory effects of the apical ectodermal ridge and zone of polarizing activity regions of the early limb bud (Bell and McLachlan, 1985; Aono and Ide, 1988). Therefore, it will be of interest to determine whether or not the apical ridge and polarizing zone influence HA-rich matrix assembly; such activity could contribute to their morphogenetic roles.

Separate to the effects of the apical ridge, the ectoderm in general produces a factor or combination of factors that prevent differentiation of nearby mesoderm to cartilage and cause the retention of mesenchymal characteristics, including HA-rich intercellular spaces (Solursh *et al.*, 1981; Hurle *et al.*, 1989; Solursh, 1990). One of the factors produced by the ectoderm stimulates HA synthesis and formation of HA-dependent coats in condensation-stage mesodermal cells (Knudson and Toole, 1988). Antibody raised against TGF-β inhibits these effects of the ectodermal factor (Toole *et al.*, 1989). TGF-β itself stimulates HA-dependent coat formation and HA synthesis (Munaim *et al.*, 1991) but, unlike the ectodermal factor, TGF-β also stimulates CS synthesis (Kulyk *et al.*, 1989; Munaim *et al.*, 1991). Carcinoma cells, but not normal adult epithelial cells, also produce a factor that stimulates HA synthesis in limb mesodermal cells. It seems likely that this factor would be similar to that produced by the limb ectoderm but its effect is not blocked by neutralizing antibody to TGF-β (Knudson and Knudson, 1990). We conclude from these studies that the ectoderm produces a factor that causes subjacent mesoderm to maintain a high rate of HA synthesis relative to the central condensed mesoderm; however, its relationship to TGF-β requires further investigation.

Thus, a combination of regulatory factors, including factors related to FGF and TGF-β, may regulate regional matrix production during condensation and other morphogenetic events in the limb bud. Cooperative effects between factors related to FGF and TGF-β have been demonstrated in other morphogenetic systems (see Smith, 1989; Whitman and Melton, 1989), and these agents interact with and influence the composition of extracellular matrices (e.g., see Section 3.4). It is anticipated, then, that future research will continue to focus on their important role in regulation of early morphogenetic events in the limb.

ACKNOWLEDGMENTS. I thank Drs. Shib Banerjee, Merton Bernfield, Chitra Biswas, John Castellot, Cheryl Knudson, Warren Knudson, and Charles Underhill for sharing unpublished data and for useful discussions concerning the material in this review. The work from my laboratory was supported by grants (DE05838 and HD23681) from the National Institutes of Health.

References

Alho, A. M., and Underhill, C. B., 1989, The hyaluronate receptor is preferentially expressed on proliferating epithelial cells, *J. Cell Biol.* **108:**1557–1565.

Anderson, C. B., and Meier, S., 1982, Effect of hyaluronidase treatment on the distribution of cranial neural crest cells in the chick embryo, *J. Exp. Zool.* **221:**329–335.

Angello, J. C., and Hauschka, S. D., 1980, Hyaluronate–cell interaction: Effects of exogenous hyaluronate on muscle fibroblast cell surface composition, *Exp. Cell Res.* **125:**389–400.

Aono, H., and Ide, H., 1988, A gradient of responsiveness to the growth-promoting activity of ZPA (zone of polarizing activity) in the chick limb bud, *Dev. Biol.* **128:**136–141.

Aruffo, A., Stamenkovic, I., Melnick, M., Underhill, C. B., and Seed, B., 1990, CD44 is the principal cell surface receptor for hyaluronate, *Cell* **61:**1303–1313.

Baird, A., and Ling, N., 1987, Fibroblast growth factors are present in the extracellular matrix produced by endothelial cells *in vitro:* Implications for a role of heparinase-like enzymes in the neovascular response, *Biochem. Biophys. Res. Commun.* **142:**428–435.

Banerjee, S. D., and Toole, B. P., 1991, Monoclonal antibody to chick embryo hyaluronan-binding protein: Changes in distribution of binding protein during early brain development, *Dev. Biol.* **146:**186–197.

Banerjee, S. D., Cohn, R. H., and Bernfield, M. R., 1977, Basal lamina of embryonic salivary epithelia. Production by the epithelium and role in maintaining lobular morphology, *J. Cell Biol.* **73:**445–463.

Barnhart, B. J., Cox, S. H., and Kraemer, P. M., 1979, Detachment variants of Chinese hamster cells. Hyaluronic acid as a modulator of cell detachment, *Exp. Cell Res.* **119:**327–332.

Bell, K. M., and McLachlan, J. C., 1985, Stimulation of division in mouse 3T3 cells by coculture with embryonic chick limb tissue, *J. Embryol. Exp. Morphol.* **86:**219–226.

Benitz, W. E., Kelley, R. T., Anderson, C. M., Lorant, D. E., and Bernfield, M., 1990, Endothelial heparan sulfate proteoglycan. I. Inhibitory effects on smooth muscle cell proliferation, *Am. J. Respir. Cell Mol. Biol.* **2**:13–24.

Bernfield, M. R., and Banerjee, S. D., 1972, Acid mucopolysaccharide (glycosaminoglycan) at the epithelial–mesenchymal interface of mouse embryo salivary glands, *J. Cell Biol.* **52**:664–673.

Bernfield, M., and Sanderson, R. D., 1990, Syndecan, a developmentally regulated cell surface proteoglycan that binds extracellular matrix and growth factors, *Philos. Trans. R. Soc. London* **327**:171–186.

Bernfield, M., Banerjee, S. D., Koda, J. E., and Rapraeger, A. C., 1984, Remodeling of the basement membrane as a mechanism of morphogenetic tissue interaction, in: *The Role of Extracellular Matrix in Development* (R. L. Trelstad, ed.), pp. 545–572, Liss, New York.

Bertrand, P., and Delpech, B., 1985, Interaction of hyaluronectin with hyaluronic acid oligosaccharides, *J. Neurochem.* **45**:434–439.

Bilozur, M. E., and Biswas, C., 1990, Identification and characterization of heparan sulfate-binding proteins from human lung carcinoma cells, *J. Biol. Chem.* **265**:19697–19703.

Biswas, C., and Toole, B. P., 1987, Modulation of the extracellular matrix by tumor cell–fibroblast interactions, in: *Cell Membranes* (E. Elson, W. Frazier and L. Glaser, eds.), pp. 341–363, Plenum Press, New York.

Boyd, F. T., and Massague, J., 1989, Transforming growth factor-β inhibition of epithelial cell proliferation linked to the expression of a 53-kDa membrane receptor, *J. Biol. Chem.* **264**:2272–2278.

Brecht, M., Mayer, U., Schlosser, E., and Prehm, P., 1986, Increased hyaluronate synthesis is required for fibroblast detachment and mitosis, *Biochem. J.* **239**:445-450.

Brennan, M. J., Oldberg, A., Hayman, E. G., and Ruoslahti, E., 1983, Effect of a proteoglycan produced by rat tumor cells on their adhesion to fibronectin–collagen substrata, *Cancer Res.* **43**:4302–4307.

Cardin, A. D., and Weintraub, H. J. R., 1989, Molecular modeling of protein–glycosaminoglycan interactions, *Arteriosclerosis* **9**:21–32.

Carey, D. J., and Stahl, R. C., 1990, Identification of a lipid-anchored heparan sulfate proteoglycan in Schwann cells, *J. Cell Biol.* **111**:2053–2062.

Castellot, J. J., Addonizio, M. L., Rosenberg, R., and Karnovsky, M. J., 1981, Cultured endothelial cells produce a heparinlike inhibitor of smooth muscle cell growth, *J. Cell Biol.* **90**:372–379.

Castellot, J. J., Wong, K., Herman, B., Hoover, R. L., Albertini, D. F., Wright, T. C., Caleb, B. L., and Karnovsky, M. J., 1985, Binding and internalization of heparin by vascular smooth muscle cells, *J. Cell. Physiol.* **124**:13–20.

Castellot, J. J., Wright, T. C., and Karnovsky, M. J., 1987, Regulation of vascular smooth muscle cell growth by heparin and heparan sulfates, *Semin. Thromb. Hemost.* **13**:489–503.

Cheifetz, S., and Massague, J., 1989, Transforming growth factor-β (TGF-β) receptor proteoglycan. Cell surface expression and ligand binding in the absence of glycosaminoglycan chains. *J. Biol. Chem.* **264**:12025–12028.

Chen, W. Y. J., Grant, M. E., Schor, A. M., and Schor, S. L., 1989, Differences between adult and foetal fibroblasts in the regulation of hyaluronate synthesis: Correlation with migratory activity, *J. Cell Sci.* **94**:577–584.

Chernoff, E. A., 1988, The role of endogenous heparan sulfate proteoglycan in adhesion and neurite outgrowth from dorsal root ganglia, *Tissue Cell* **20**:165–178.

Chiu, A. Y., Matthew, W. D., and Patterson, P. H., 1986, A monoclonal antibody that blocks the activity of a neurite regeneration-promoting factor: Studies on the binding site and its localization in vivo, *J. Cell Biol.* **103**:1383–1398.

Clement, B., Segui-Real, B., Hassell, J. R., Martin, G. R., and Yamada, Y., 1989, Identification of a cell surface-binding protein for the core protein of the basement membrane proteoglycan, *J. Biol. Chem.* **264**:12467–12471.

Cochran, D. L., Castellot, J. J., Robinson, J. M., and Karnovsky, M. J., 1988, Heparin modulates the secretion of a major excreted protein-like molecule by vascular smooth muscle cells, *Biochim. Biophys. Acta* **967**:289–295.

Comper, W. D., and Laurent, T. C., 1978, Physiological function of connective tissue polysaccharides, *Physiol. Rev.* **58**:255–315.

Copp, A. J., and Bernfield, M., 1988, Accumulation of basement membrane-associated hyaluronate is reduced in the posterior neuropore region of mutant (curly tail) mouse embryos developing spinal neural tube defects, *Dev. Biol.* **130:**583–590.

Crossman, M. V., and Mason, R. M., 1990, Purification and characterization of a hyaluronan-binding protein from rat chondrosarcoma, *Biochem. J.* **266:** 399–406.

Culty, M., Miyake, K., Kincade, P. W., Sikorski, E., Butcher, E. C., and Underhill, C., 1990, The hyaluronate receptor is a member of the CD44 family of cell surface glycoproteins, *J. Cell Biol.* **111:**2765–2774.

Damon, D. H., Lobb, R. R., D'Amore, P. A., and Wagner, J. A., 1989, Heparin potentiates the action of acidic fibroblast growth factor by prolonging its biological half-life, *J. Cell. Physiol.* **138:**221–226.

Dauca, M., Bouziges, F., Colin, S., Kedinger, M., Keller, J. M., Schilt, J., Simon-Assmann, P., and Haffen, K., 1990, Development of the vertebrate small intestine and mechanisms of cell differentiation, *Int. J. Dev. Biol.* **34:**205–218.

Decker, M., Chiu, E. S., Dollbaum, C., Moiin, A., Hall, J., Spendlove, R., Longaker, M. T., and Stern, R., 1989, Hyaluronic acid-stimulating activity in sera from the bovine fetus and from breast cancer patients, *Cancer Res.* **49:**3499–3505.

DiMario, J., Buffinger, N., Yamada, S., and Strohman, R. C., 1989, Fibroblast growth factor in the extracellular matrix of dystrophic (mdx) mouse muscle, *Science* **244:**688–690.

Dow, K. E., Mirski, S. E. L., Roder, J. C., and Riopelle, R. J., 1988, Neuronal proteoglycans: Biosynthesis and functional interaction with neurons in vitro, *J. Neurosci.* **8:**3278–3289.

Edgar, D., Timpl, R., and Thoenen, H., 1988, Structural requirements for the stimulation of neurite outgrowth by two variants of laminin and their inhibition by antibodies, *J. Cell Biol.* **106:**1299–1306.

Ekblom, P., Vestweber, D., and Kemler, R., 1986, Cell–matrix interactions and cell adhesion during development, *Annu. Rev. Cell Biol.* **2:**27–47.

Erickson, C. A., and Turley, E. A., 1983, Substrata formed by combinations of extracellular matrix components alter neural crest cell motility in vitro, *J. Cell Sci.* **61:**299–323.

Fedarko, N. S., and Conrad, H. E., 1986, A unique heparan sulfate in the nuclei of hepatocytes: Structural changes with the growth state of the cells, *J. Cell Biol.* **102:**587–599.

Fekardo, N. S., Ishihara, M., and Conrad, H. E., 1989, Control of cell division in hepatoma cells by exogenous heparan sulfate proteoglycan, *J. Cell Physiol.* **139:**287–294.

Flaumenhaft, R., Moscatelli, D., Saksela, O., and Rifkin, D. B., 1989, Role of extracellular matrix in the action of basic fibroblast growth factor: Matrix as a source of growth factor for long-term stimulation of plasminogen activator production and DNA synthesis, *J. Cell. Physiol.* **140:**75–81.

Folkman, J., Klagsbrun, M., Sasse, J., Wadzinski, M., Ingber, D., and Vlodavsky, I., 1988, A heparin-binding angiogenic protein—basic fibroblast growth factor—is stored within basement membrane, *Am. J. Pathol.* **130:**393–400.

Frenz, D. A., Akiyama, S. K., Paulsen, D. F., and Newman, S. A., 1989, Latex beads as probes of cell surface–extracellular matrix interactions during chondrogenesis: Evidence for a role for amino-terminal heparin-binding domain of fibronectin, *Dev. Biol.* **136:**87–96.

Fritze, L. M. S., Reilley, C. F., and Rosenberg, R. D., 1985, An antiproliferative heparan sulfate species produced by postconfluent smooth muscle cells, *J. Cell Biol.* **100:**1041–1049.

Frost, S. J., and Weigel, P. H., 1990, Binding of hyaluronic acid to mammalian fibrinogens, *Biochim. Biophys. Acta* **1034:**39–45.

Fukuda, Y., Masuda, Y., Kishi, J.-I., Hashimoto, Y., Hayakawa, T., Nogawa, H., and Nakanishi, Y., 1988, The role of interstitial collagens in cleft formation of mouse embryonic submandibular gland during initial branching, *Development* **103:**259–267.

Goetinck, P. F., Stirpe, N. S., Tsonis, P. A., and Carlone, D., 1987, The tandemly repeated sequences of cartilage link protein contain the sites for interaction with hyaluronic acid, *J. Cell Biol.* **105:**2403–2408.

Goldberg, R. L., and Toole, B. P., 1984, Pericellular coat of chick embryo chondrocytes: Structural role of hyaluronate, *J. Cell Biol.* **99:**2114–2122.

Goldberg, R. L., and Toole, B. P., 1987, Hyaluronate inhibition of cell proliferation, *Arthritis Rheum.* **30:**769–778.

Goldstein, L. A., Zhou, D. F. H., Picker, L. J., Minty, C. N., Bargatze, R. F., Ding, J. F., and Butcher, E. C., 1989, A human lymphocyte homing receptor, the Hermes antigen, is related to cartilage proteoglycan core and link proteins, *Cell* **56:**1063–1072.
Green, S. J., Tarone, G., and Underhill, C. B., 1988, Aggregation of macrophages and fibroblasts is inhibited by a monoclonal antibody to the hyaluronate receptor, *Exp. Cell Res.* **178:**224–232.
Hantaz-Ambroise, D., Vigny, M., and Koenig, J., 1987, Heparan sulfate proteoglycan and laminin mediate two different types of neurite outgrowth, *J. Neurosci.* **7:**2293–2304.
Hausser, H., Hoppe, W., Rauch, U., and Kresse, H., 1989, Endocytosis of a small dermatan sulphate proteoglycan. Identification of binding proteins, *Biochem. J.* **263:**137–142.
Hautanen, A., Gailit, J., Mann, D. M., and Ruoslahti, E., 1989, Effects of modifications of the RGD sequence and its context on recognition by the fibronectin receptor, *J. Biol. Chem.* **264:**1437–1442.
Hayashi, K., Hayashi, M., Jalkanen, M., Firestone, J. H., Trelstad, R. L., and Bernfield, M., 1987, Immunocytochemistry of cell surface heparan sulfate proteoglycan in mouse tissues. A light and electron microscopic study, *J. Histochem. Cytochem.* **35:**1079–1088.
Haynes, B. F., Telen, M. J., Hale, L. P., and Denning, S. M., 1989, CD44—A molecule involved in leukocyte adherence and T-cell activation, *Immunol. Today* **10:**423–428.
Herman, I. M., and Castellot, J. J., 1987, Regulation of vascular smooth muscle cell growth by endothelial-synthesized extracellular matrices, *Arteriosclerosis* **7:**463–469.
Herndon, M. E., and Lander, A. D., 1990, A diverse set of developmentally regulated proteoglycans is expressed in the rat central nervous system, *Neuron* **4:**949–961.
Hockfield, S., 1990, Proteoglycans in neural development, *Semin. Dev. Biol.* **1:**55–63.
Hoffman, S., and Edelman, G. M., 1987, A proteoglycan with HNK-1 antigenic determinants is a neuron-associated ligand for cytotactin, *Proc. Natl. Acad. Sci. USA* **84:**2523–2527.
Hurle, J. M., Ganan, Y., and Macias, D., 1989, Experimental analysis of the in vivo chondrogenic potential of the interdigital mesenchyme of the chick leg bud subjected to local ectodermal removal, *Dev. Biol.* **132:**368–374.
Ishihara, M., and Conrad, H. E., 1989, Correlations between heparan sulfate metabolism and hepatoma growth, *J. Cell Physiol.* **138:**467–476.
Ishihara, M., Fedarko, N. S., and Conrad, H. E., 1987, Involvement of phosphatidylinositol and insulin in the coordinate regulation of proteoheparan sulfate metabolism and hepatocyte growth, *J. Biol. Chem.* **262:**4708–4716.
Kardami, E., Spector, D., and Strohman, R. C., 1988, Heparin inhibits skeletal muscle growth in vitro, *Dev. Biol.* **126:**19–28.
Kimata, K., Oike, Y., Tani, K., Shinomura, T., Yamagata, M., Uritani, M., and Suzuki, S., 1986, A large chondroitin sulfate proteoglycan (PG-M) synthesized before chondrogenesis in the limb bud of chick embryo, *J. Biol. Chem.* **261:**13517–13525.
Kinsella, M. G., and Wight, T. N., 1986, Modulation of sulfated proteoglycan synthesis by bovine aortic endothelial cells during migration, *J. Cell Biol.* **102:**679–687.
Kjellen, L., Oldberg, A., and Hook, M., 1980, Cell-surface heparan sulfate: Mechanisms of proteoglycan–cell association, *J. Biol. Chem.* **255:**10407–10413.
Kjellen, L., Pettersson, I., and Hook, M., 1981, Cell-surface heparan sulfate: An intercalated membrane proteoglycan, *Proc. Natl. Acad. Sci. USA* **78:**5371–5375.
Knudson, C. B., 1991, Cell–cell adhesion of limb bud mesoderm mediated by hyaluronan, submitted.
Knudson, C. B., and Knudson, W., 1990, Similar epithelial–stromal interactions in the regulation of hyaluronate production during limb morphogenesis and tumor invasion, *Cancer Lett.* **52:**113–122.
Knudson, W., and Knudson, C. B., 1991, Assembly of a chondrocyte-like pericellular matrix on non-chondrogenic cells. Role of the cell surface hyaluronan receptors in the assembly of a pericellular matrix, *J. Cell Sci.* **99:**227–235.
Knudson, C. B., and Toole, B. P., 1985, Changes in the pericellular matrix during differentiation of limb bud mesoderm, *Dev. Biol.* **112:**308–318.
Knudson, C. B., and Toole, B. P., 1987, Hyaluronate–cell interactions during differentiation of chick embryo limb mesoderm, *Dev. Biol.* **124:**82–90.

Knudson, C. B., and Toole, B. P., 1988, Epithelial–mesenchymal interaction in the regulation of hyaluronate production during limb development, *Biochem. Int.* **17:**735–745.

Knudson, W., Biswas, C., and Toole, B. P., 1984, Interactions between human tumor cells and fibroblasts stimulate hyaluronate synthesis, *Proc. Natl. Acad. Sci. USA* **81:**6767–6771.

Koda, J. E., Rapraeger, A., and Bernfield, M., 1985, Heparan sulfate proteoglycans from mouse mammary epithelial cells. Cell surface proteoglycan as a receptor for interstitial collagens, *J. Biol. Chem.* **260:**8157–8162.

Kujawa, M. J., Pechak, D. G., Fiszman, M. Y., and Caplan, A. I., 1986, Hyaluronic acid bonded to cell culture surfaces inhibits the program of myogenesis, *Dev. Biol.* **113:**10–16.

Kulyk, W. M., Rodgers, B. J., Greer, K., and Kosher, R. A., 1989, Promotion of embryonic chick limb cartilage differentiation by transforming growth factor-beta, *Dev. Biol.* **135:**424–430.

Lacy, B. E., and Underhill, C. B., 1987, The hyaluronate receptor is associated with actin filaments, *J. Cell Biol.* **105:**1395–1404.

Lander, A. D., 1989, Understanding the molecules of neural cell contacts: Emerging patterns of structure and function, *Trends Neurosci.* **12:**189–195.

Lankes, W., Griesmacher, A., Grunwald, J., Schwartz-Albiez, R., and Keller, R., 1988, A heparin-binding protein involved in inhibition of smooth-muscle cell proliferation, *Biochem. J.* **251:**831–842.

Laurent, T. C., Fraser, J. R. E., Pertoft, H., and Smedsrod, B., 1986, Binding of hyaluronate and chondroitin sulphate to liver endothelial cells, *Biochem. J.* **234:**653–658.

LeBaron, R. G., Esko, J. D., Woods, A., Johansson, S., and Hook, M., 1988, Adhesion of glycosaminoglycan-deficient Chinese hamster ovary cell mutants to fibronectin substrata, *J. Cell Biol.* **106:**945–952.

LeBaron, R. G., Hook, A., Esko, J. D., Gay, S., and Hook, M., 1989, Binding of heparan sulfate to type V collagen. A mechanism of cell–substrate adhesion, *J. Biol. Chem.* **264:**7950–7956.

Lelongt, B., Makino, H., Dalecki, T. M., and Kanwar, Y. S., 1988, Role of proteoglycans in renal development, *Dev. Biol.* **128:**256–276.

Lesley, J., Schulte, R., and Hyman, R., 1990, Binding of hyaluronic acid to lymphoid cell lines is inhibited by monoclonal antibodies against Pgp-1, *Exp. Cell Res.* **187:**224–233.

Lewandowska, K., Choi, H. U., Rosenberg, L. C., Zardi, L., and Culp, L. A., 1987, Fibronectin-mediated adhesion of fibroblasts: Inhibition by dermatan sulfate proteoglycan and evidence for a cryptic glycosaminoglycan-binding domain, *J. Cell Biol.* **105:**1443–1454.

Luikart, S. D., Maniglia, C. A., Furcht, L. T., McCarthy, J. B., and Oegema, T. R., 1990, A heparin sulfate-containing fraction of bone marrow stroma induces maturation of HL-60 cells in vitro, *Cancer Res.* **50:**3781–3785.

Majack, R. A., and Bornstein, P., 1985, Heparin regulates the collagen phenotype of vascular smooth muscle cells: Induced synthesis of an M_r 60,000 collagen, *J. Cell Biol.* **100:**613–619.

Majack, R. A., Goodman, L. V., and Dixit, V. M., 1988, Cell surface thrombospondin is functionally essential for vascular smooth muscle cell proliferation, *J. Cell Biol.* **106:**415–422.

Margolis, R. U., and Margolis, R. K., 1989, Nervous tissue proteoglycans, *Dev. Neurosci.* **11:**276–288.

McCaffrey, T. A., Falcone, D. J., Brayton, C. F., Agarwal, L. A. Welt, F. G. P., and Weksler, B. B., 1989, Transforming growth factor-β activity is potentiated by heparin via dissociation of the transforming growth factor-β/α_2-macroglobulin inactive complex, *J. Cell Biol.* **109:**441–448.

McGary, C. T., Raja, R. H., and Weigel, P. H., 1989, Endocytosis of hyaluronic acid by rat liver endothelial cells. Evidence for receptor recycling, *Biochem. J.* **257:**875–884.

Miller, D. J., Winer, M. A., and Ax, R. L., 1990, Heparin-binding proteins from seminal plasma bind to bovine spermatozoa and modulate capacitation by heparin, *Biol. Reprod.* **42:**899–915.

Miyake, K., Underhill, C. B., Lesley, J., and Kincade, P. W., 1990, Hyaluronate can function as a cell adhesion molecule and CD44 participates in hyaluronate recognition, *J. Exp. Med.* **172:**69–75.

Morriss-Kay, G., and Tuckett, F., 1989, Immunohistochemical localisation of chondroitin sulphate proteoglycans and the effects of chondroitinase ABC in 9-to 11-day rat embryos, *Development* **106:**787–798.

Morriss-Kay, G. M., Tuckett, F., and Solursh, M., 1986, The effects of Streptomyces hyaluronidase on tissue organization and cell cycle time in rat embryos, *J. Embryol. Exp. Morphol.* **98:**59–70.

Morris-Wiman, J., and Brinkley, L. L., 1990, Changes in mesenchymal cell and hyaluronate distribution correlate with in vivo elevation of the mouse mesencephalic neural folds, *Anat. Rec.* **226:**383–395.

Munaim, S. I., Klagsbrun, M., and Toole, B. P., 1988, Developmental changes in fibroblast growth factor in the chicken embryo limb bud, *Proc. Natl. Acad. Sci. USA* **85:**8091–8093.

Munaim, S. I., Klagsbrun, M., and Toole, B. P., 1991, Hyaluronan-dependent pericellular coats of chick embryo limb mesodermal cells: induction by basic fibroblast growth factor, *Dev. Biol.* **143:**297–302.

Murphy-Ullrich, J. E., and Hook, M., 1989, Thrombospondin modulates focal adhesions in endothelial cells, *J. Cell Biol.* **109:**1309–1319.

Neame, P. J., Christner, J. E., and Baker, J. R., 1987, Cartilage proteoglycan aggregates. The link protein and proteoglycan amino-terminal globular domains have similar structures, *J. Biol. Chem.* **262:**17768–17778.

Nemec, R. E., Toole, B. P., and Knudson, W., 1987, The cell surface hyaluronate binding sites of invasive human bladder carcinoma cells, *Biochem. Biophys. Res. Commun.* **149:**249–257.

Orkin, R. W., Knudson, W., and Toole, B. P., 1985, Loss of hyaluronate-dependent coat during myoblast fusion, *Dev. Biol.* **107:**527–530.

Perides, G., Lane, W. S., Andrews, D., Dahl, D., and Bignami, A., 1989, Isolation and partial characterization of a glial hyaluronate-binding protein, *J. Biol. Chem.* **264:**5981–5987.

Perris, R., and Johansson, S., 1990, Inhibition of neural crest cell migration by aggregating chondroitin sulfate proteoglycans is mediated by their hyaluronan-binding region, *Dev. Biol.* **137:**1–12.

Philipson, L. H., Westley, J., and Schwartz, N. B., 1985, Effect of hyaluronidase treatment of intact cells on hyaluronate synthetase activity, *Biochemistry* **24:**7899–7906.

Presta, M., Maier, J. A. M., Rusnati, M., and Ragnotti, G., 1989, Basic fibroblast growth factor is released from endothelial extracellular matrix in a biologically active form, *J. Cell Physiol.* **140:**68–74.

Pukac, L. A., Castellot, J. J., Wright, T. C., Caleb, B. L., and Karnovsky, M. J., 1990, Heparin inhibits c-fos and c-myc mRNA expression in vascular smooth muscle cells, *Cell Regul.* **1:**435–443.

Raja, R. H., McGary, C. T., and Weigel, P. H., 1988, Affinity and distribution of surface and intracellular hyaluronic acid receptors in isolated rat liver endothelial cells, *J. Biol. Chem.* **263:**16661–16668.

Rasmussen, S., and Rapraeger, A., 1988, Altered structure of the hybrid cell surface proteoglycan of mammary epithelial cells in response to transforming growth factor-β, *J. Cell Biol.* **107:**1959–1967.

Ratner, N., Hong, D., Lieberman, M. A., Bunge, R. P., and Glaser, L., 1988, The neuronal cell-surface molecule mitogenic for Schwann cells is a heparin-binding protein, *Proc. Natl. Acad. Sci. USA* **85:**6992–6996.

Reilly, C. F., Fritze, L. M. S., and Rosenberg, R. D., 1986, Heparin inhibition of smooth muscle cell proliferation: A cellular site of action, *J. Cell. Physiol.* **129:**11–19.

Reyes, A. A., Akeson, R., Brezina, L., and Cole, G. J., 1990, Structural requirements for neural cell adhesion molecule–heparin interaction, *Cell Regul.* **1:**567–576.

Rogelj, S., Klagsbrun, M., Atzmon, R., Kurokawa, M., Haimovitz, A., Fuks, Z., and Vlodavsky, I., 1989, Basic fibroblast growth factor is an extracellular matrix component required for supporting the proliferation of vascular endothelial cells and the differentiation of PC12 cells, *J. Cell Biol.* **109:**823–831.

Rollins, B. J., and Culp, L. A., 1979, Glycosaminoglycans in the substrate adhesion sites of normal and virus-transformed murine cells, *Biochemistry* **18:**141–148.

Rutishauser, U., Acheson, A., Hall, A. K., Mann, D. M., and Sunshine, J., 1988, The neural cell adhesion molecule (NCAM) as a regulator of cell–cell interactions, *Science* **240:**53–57.

Saksela, O., Moscatelli, D., Sommer, A., and Rifkin, D. B., 1988, Endothelial cell-derived heparan sulfate binds basic fibroblast growth factor and protects it from proteolytic degradation, *J. Cell Biol.* **107:**743–751.

Sanderson, R. D., and Bernfield, M., 1988, Molecular polymorphism of a cell surface proteoglycan: Distinct structures on simple and stratified epithelia, *Proc. Natl. Acad. Sci. USA* **85:**9562–9566.

Sanderson, R. D., Lalor, P., and Bernfield, M., 1989, B lymphocytes express and lose syndecan at specific stages of differentiation, *Cell Regul.* **1**:27–35.

Sanes, J. R., 1989, Extracellular matrix molecules that influence neural development, *Annu. Rev. Neurosci.* **12**:491–516.

Saunders, S., and Bernfield, M., 1988, Cell surface proteoglycan binds mouse mammary epithelial cells to fibronectin and behaves as a receptor for interstitial matrix, *J. Cell Biol.* **106**:423–430.

Saunders, S., Jalkanen, M., O'Farrell, S., and Bernfield, M., 1989, Molecular cloning of syndecan, an integral membrane proteoglycan, *J. Cell Biol.* **108**:1547–1556.

Schoenwolf, G. C., and Fisher, M., 1983, Analysis of the effects of Streptomyces hyaluronidase on formation of the neural tube, *J. Embryol. Exp. Morphol.* **73**:1–15.

Schor, S. L., Schor, A. M., Grey, A. M., Chen, J., Rushton, G., Grant, M. E., and Ellis, I., 1989, Mechanism of action of the migration stimulating factor produced by fetal and cancer patient fibroblasts: Effect on hyaluronic acid synthesis, *In Vitro Cell Dev. Biol.* **25**:737–746.

Seed, J., Olwin, B. B., and Hauschka, S. D., 1988, Fibroblast growth factor levels in the whole embryo and limb bud during chick development, *Dev. Biol.* **128**:50–57.

Shannon, B. T., Love, S. H., and Myrvik, Q. N., 1980, Participation of hyaluronic acid in the macrophage disappearance reaction, *Immunol. Commun.* **9**:357–370.

Singley, C. T., and Solursh, M., 1981, The spatial distribution of hyaluronic acid and mesenchymal condensation in the embryonic chick wing, *Dev. Biol.* **84**:102–120.

Smith, J. C., 1989, Mesoderm induction and mesoderm-inducing factors in early amphibian development, *Development* **105**:665–677.

Snow, D. M., Lemmon, V., Carrino, D. A., Caplan, A. I., and Silver, J., 1990, Sulfated proteoglycans in astroglial barriers inhibit neurite outgrowth in vitro, *Exp. Neurol.* **109**:111–130.

Solursh, M., 1990, The role of extracellular matrix molecules in early limb development, *Semin. Dev. Biol.* **1**:45–53.

Solursh, M., and Jensen, K. L., 1988, The accumulation of basement membrane components during the onset of chondrogenesis and myogenesis in the chick wing bud, *Development* **104**:41–49.

Solursh, M., Singley, C. T., and Reiter, R. S., 1981, The influence of epithelia on cartilage and loose connective tissue formation by limb mesenchyme cultures, *Dev. Biol.* **86**:471–482.

Solursh, M., Reiter, R. S., Jensen, K. L., Kato, M., and Bernfield, M., 1990, Transient expression of a cell surface heparan sulfate proteoglycan (syndecan) during limb development, *Dev. Biol.* **140**:83–92.

Spooner, B. S., Bassett, K., and Stokes, B., 1985, Sulfated glycosaminoglycan deposition and processing at the basal epithelial surface in branching and β-D-xyloside-inhibited embryonic salivary glands, *Dev. Biol.* **109**:177–183.

Spooner, B. S., Thompson-Pletscher, H. A., Stokes, B., and Bassett, K. E., 1986, Extracellular matrix involvement in epithelial branching morphogenesis, in: *Developmental Biology: A Comprehensive Synthesis* (M. S. Steinberg, ed.), pp. 225–260, Plenum Press, New York.

Spray, D. C., Fujita, M., Saez, J. C., Choi, H., Watanabe, T., Hertzberg, E., Rosenberg, L. C., and Reid, L. M., 1987, Proteoglycans and glycosaminoglycans induce gap junction synthesis and function in primary liver cultures, *J. Cell Biol.* **105**:541–551.

Stamenkovic, I., Amiot, M., Pesando, J. M., and Seed, B., 1989, A lymphocyte molecule implicated in lymph node homing is a member of the cartilage link protein family, *Cell* **56**:1057–1062.

Sun, X., Mosher, D. F., and Rapraeger, A., 1989, Heparan sulfate-mediated binding of epithelial cell surface proteoglycan to thrombospondin, *J. Biol. Chem.* **264**:2885–2889.

Tammi, R., Ripellino, J. A., Margolis, R. U., Maibach, H. I., and Tammi, M., 1989, Hyaluronate accumulation in human epidermis treated with retinoic acid in skin organ culture, *J. Invest. Dermatol.* **92**:326–332.

Tengblad, A., 1981, A comparative study of the binding of cartilage link protein and the hyaluronate-binding region of the cartilage proteoglycan to hyaluronate-substituted Sepharose gel, *Biochem. J.* **199**:297–305.

Toole, B. P., 1981, Glycosaminoglycans in morphogenesis, in: *Cell Biology of Extracellular Matrix* (E. D. Hay, ed.), pp. 259–294, Plenum Press, New York.

Toole, B. P., 1990, Hyaluronan and its binding proteins, the hyaladherins, *Curr. Opin. Cell Biol.* **2**:839–844.

Toole, B. P., Munaim, S. I., Welles, S., and Knudson, C. B., 1989, Hyaluronate–cell interactions and growth factor regulation of hyaluronate synthesis during limb development, *Ciba Found. Symp.* **143**:138–149.

Turley, E. A., 1989, Hyaluronic acid stimulates protein kinase activity in intact cells and in an isolated protein complex, *J. Biol. Chem.* **264**:8951–8955.

Turley, E., and Auersperg, N., 1989, A hyaluronate binding protein transiently codistributes with $p21^{k\text{-}ras}$ in cultured cell lines, *Exp. Cell Res.* **182**:340–348.

Turley, E. A., Bowman, P., and Kytryk, M. A., 1985, Effects of hyaluronate and hyaluronate binding proteins on cell motile and contact behavior, *J. Cell Sci.* **78**:133–145.

Turley, E. A., Brassel, P., and Moore, D., 1990, A hyaluronan-binding protein shows a partial and temporally regulated codistribution with actin on locomoting chick heart fibroblasts, *Exp. Cell Res.* **187**:243–249.

Underhill, C. B., and Toole, B. P., 1979, Binding of hyaluronate to the surface of cultured cells, *J. Cell Biol.* **82**:475–484.

Underhill, C. B., and Toole, B. P., 1980, Physical characteristics of hyaluronate binding to the surface of simian virus 40-transformed 3T3 cells, *J. Biol. Chem.* **255**:4544–4549.

Underhill, C. B., and Toole, B. P., 1981, Receptors for hyaluronate on the surface of parent and virus-transformed cell lines—Binding and aggregation studies, *Exp. Cell Res.* **131**:419–424.

Underhill, C. B., Chi-Rosso, G., and Toole, B. P., 1983, Effects of detergent solubilization on the hyaluronate-binding protein from membranes of simian virus 40-transformed 3T3 cells, *J. Biol. Chem.* **258**:8086–8091.

Underhill, C. B., Green, S. J., Comoglio, P. M., and Tarone, G., 1987, The hyaluronate receptor is identical to a glycoprotein of M_r 85,000 (gp85) as shown by a monoclonal antibody that interferes with binding activity, *J. Biol. Chem.* **262**:13142–13146.

Vainio, S., Jalkanen, M., and Thesleff, I., 1989, Syndecan and tenascin expression is induced by epithelial–mesenchymal interactions in embryonic tooth mesenchyme, *J. Cell Biol.* **108**:1945–1954.

West, D. C., and Kumar, S., 1989, The effect of hyaluronate and its oligosaccharides on endothelial cell proliferation and monolayer integrity, *Exp. Cell Res.* **183**:179–196.

Whitman, M., and Melton, D. A., 1989, Growth factors in early embryogenesis, *Annu. Rev. Cell Biol.* **5**:93–117.

Woods, A., Couchman, J. R., and Hook, M., 1985, Heparan sulfate of rat embryo fibroblasts. A hydrophobic form may link cytoskeleton and matrix components, *J. Biol. Chem.* **260**:10872–10879.

Wright, T. C., Castellot, J. J., Petitou, M., Lormeau, J. C., Choay, J., and Karnovsky, M. J., 1989, Structural determinants of heparin's growth inhibitory activity. Interdependence of oligosaccharide size and charge, *J. Biol. Chem.* **264**:1534–1542.

Yamagata, M., Yamada, K. M., Yoneda, M., Suzuki, S., and Kimata, K., 1986, Chondroitin sulfate proteoglycan (PG-M-like proteoglycan) is involved in the binding of hyaluronic acid to cellular fibronectin, *J. Biol. Chem.* **261**:13526–13535.

Yamagata, M., Suzuki, S., Akiyama, S. K., Yamada, K. M., and Kimata, K., 1989, Regulation of cell–substrate adhesion by proteoglycans immobilized on extracellular substrates, *J. Biol. Chem.* **264**:8012–8018.

Yamaguchi, Y., Mann, D. M., and Ruoslahti, E., 1990, Negative regulation of transforming growth factor-β by the proteoglycan decorin, *Nature* **346**:281–284.

Yoneda, M., Suzuki, S., and Kimata, K., 1990, Hyaluronic acid associated with the surfaces of cultured fibroblasts is linked to a serum-derived 85-kDa protein, *J. Biol. Chem.* **265**:5247–5257.

Zimmermann, D. R., and Ruoslahti, E., 1989, Multiple domains of the large fibroblast proteoglycan, versican, *EMBO J.* **8**:2975–2981.

Chapter 10

Integrins as Receptors for Extracellular Matrix

ERKKI RUOSLAHTI

1. Introduction

Cells bind to extracellular matrices as a means of anchoring themselves and to derive traction for migration. Matrices may also impart signals for growth and differentiation. The binding of cells to extracellular matrices is mediated by cell surface receptors. The primary class of these receptors is a family of transmembrane proteins known as integrins. The integrins bind to extracellular matrix proteins at specialized cell attachment sites that often have the tripeptide sequence Arg-Gly-Asp or RGD as the target sequence for the integrin binding. While the specificity of the cell–matrix interactions seems to come from the integrin binding, most adhesive matrix molecules also contain sites that can interact with the glycosaminoglycan component of proteoglycans. The binding of cell surface proteoglycans to such sites is likely to play an augmenting role in cell adhesion. The proteoglycan interactions are discussed elsewhere in this volume. This chapter summarizes the current status of integrins, focusing on their role as extracellular matrix receptors.

2. Structural Features of Integrins

Integrins are a family of membrane glycoproteins consisting of two subunits, α and β. The primary structure of many of these subunits has been deduced from sequencing of complementary DNA (reviewed in Hemler, 1990). This sequence information is the basis of the structural models for the various

Abbreviations used in this chapter: CHO cells, Chinese hamster ovary cells; CSAT, cell surface attachment molecule; ICAM, intercellular adhesion molecule; LFA, leukocyte function antigen; NMR, nuclear magnetic resonance; VCAM, vascular cell adhesion molecule.

ERKKI RUOSLAHTI • Cancer Research Center, La Jolla Cancer Research Foundation, La Jolla, California 92037.

Cell Biology of Extracellular Matrix, Second Edition, edited by Elizabeth D. Hay, Plenum Press, New York, 1991.

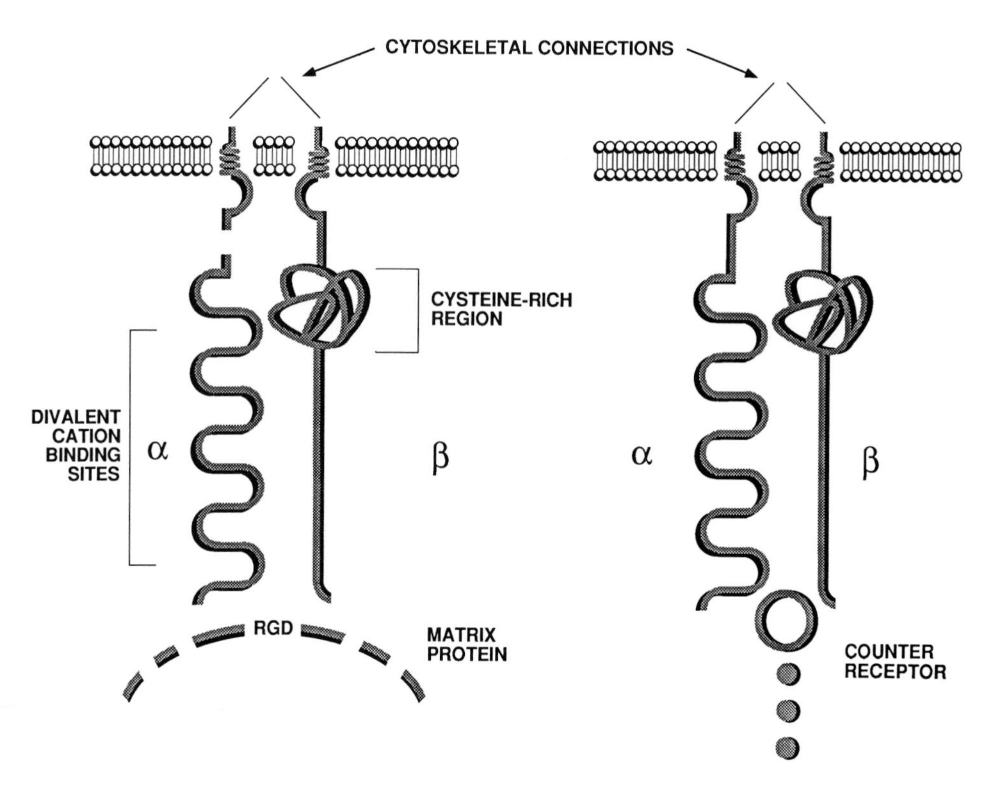

Figure 10-1. Schematic representation of the general structure of integrins. Two types of integrins are shown, one with an α subunit which is proteolytically processed into two disulfide-linked fragments at a cleavage site (shown as a gap in the structure on the left). The $\alpha_5\beta_1$ fibronectin receptor which binds to an RGD sequence is an example of this type of an integrin. Other integrins have α subunits that are not processed. An example is the $\alpha_L\beta_2$ integrin which binds to its counter-receptors, ICAM-1 and ICAM-2.

integrins depicted in Fig. 10-1. The α subunits are homologous to one another but not to the β subunits, which form their own homologous group. The extent of similarity at the amino acid sequence level within the α and β subunit groups is 40–50%. Both integrin subunits have a large extracellular domain, a transmembrane segment, and a cytoplasmic tail. The cytoplasmic portion is quite short (30 to 50 amino acids) in all the known subunits, except in the recently sequenced β_4 subunit which has a 1000-amino-acid cytoplasmic domain (Hogervorst *et al.*, 1990; Suzuki and Naitoh, 1990). Some of the α subunits (α_1, α_2, α_L, α_M, and α_X) contain an additional 180-amino-acid segment that has homologies with certain collagen-binding proteins (Pytela, 1988). The sizes of the integrin subunits vary from about 90,000 to 200,000 daltons for β_4. The extracellular domains of the α subunits contain several calmodulin-type divalent cation-binding sites, while the β subunits may have one such site (Loftus *et al.*, 1990); the divalent cation dependency of integrin function is likely to derive from the presence of these structures. The divalent cation dependency is very useful in the isolation of integrins by affinity chromatography on their insolubilized ligands, because some cations enhance the binding of integrins to their ligands and because EDTA can reverse the binding (Gailit and Ruoslahti, 1988).

The α and β subunits are noncovalently bound to one another, and this association is promoted by divalent cations. Cross-linking studies and analysis of the ligand specificities of individual integrins suggest that the ligand-binding site is formed by sequences from both subunits (Loftus *et al.*, 1990; Vogel *et al.*, 1990). The cytoplasmic domains of the integrins are thought to interact with components of the cytoskeleton. Such interactions may be regulated by phosphorylation of the β subunit cytoplasmic tail (Tapley *et al.*, 1989; Buck and Horwitz, 1987). These properties endow integrins with the ability to serve as a link between the cytoskeleton and the extracellular matrix (Hynes, 1987).

3. Integrin Diversity

There are 11 α subunits and six β subunits known at this time. All these subunits have been at least partially sequenced and thereby shown to be distinct (Hemler, 1990; Suzuki and Naitoh, 1990; Ignatius *et al.*, 1990; Sheppard *et al.*, 1990). Some additional α and β subunits that may be distinct from the sequenced ones have also been described. The α and β subunits, in various combinations, form at least 16 integrins (Fig. 10-2). It is likely that more will be discovered.

It was initially thought that integrins could be grouped in subfamilies, each characterized by a distinct β subunit combining with its own set of α subunits. It has since become clear that other combinations also exist, because a single α subunit can become paired with more than one β. The α_v subunit appears to be particularly versatile; it forms a subfamily of as many as four integrins by combining with different β subunits (see Vogel *et al.*, 1990). This diversity of the integrins provides cells with varied capabilities to recognize adhesive substrates, extracellular matrices in particular.

4. Three Modes of Integrin-Mediated Adhesion

4.1. Adhesion of Cells to Extracellular Matrix

Many integrins bind to extracellular matrix proteins, and thereby mediate cell–extracellular matrix interactions. Among the extracellular matrix ligands (see Chapter 4) for integrins are fibronectin, laminin, various collagens, entactin, tenascin, thrombospondin, von Willebrand factor, and vitronectin (Hemler, 1990; Ruoslahti and Pierschbacher, 1987). These matrix interactions are discussed in detail below.

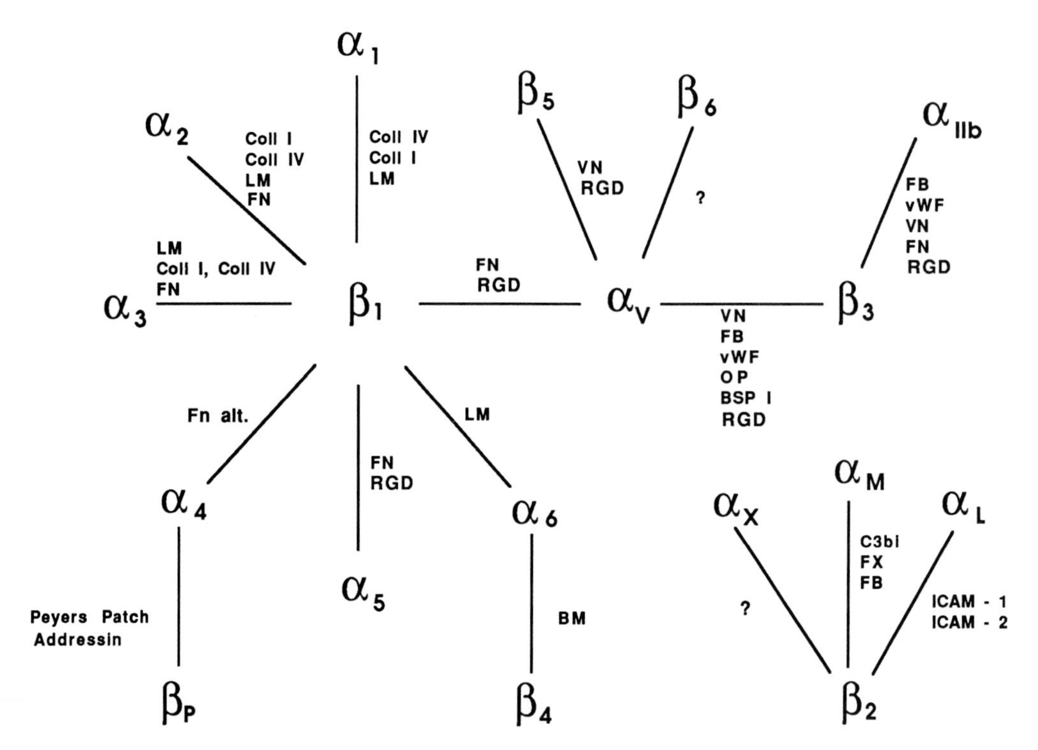

Figure 10-2. Integrin family. The known subunits, the subunit combinations that form the known integrins, and the known ligands for these integrins are shown. Also shown is the RGD specificity of those integrins that bind to this sequence. FN, fibronectin; Fn alt., fibronectin alternatively spliced domain; LM, laminin; VN, vitronectin; Coll, collagen; vWF, von Willebrand factor; FB, fibrinogen; OP, osteopontin; BSP 1, bone sialoprotein 1; ICAM-1, ICAM-2, intercellular adhesion molecules; FX, factor X; BM, basement membrane; C3bi, complement component C3bi.

4.2. Cell–Cell Adhesion

Other integrins bind to cell membrane proteins on the surface of another cell ("counterreceptors"), in order to mediate cell–cell adhesion. The intercellular adhesion proteins, ICAM-1 and ICAM-2, have been identified as "counterreceptors" for the leukocyte integrin LFA-1 (also known as CD11A/CD18 or $\alpha_L\beta_2$) (Springer, 1990), and the counterreceptor for the $\alpha_4\beta_1$ integrin is VCAM-1 (Elices *et al.*, 1990). ICAM-1, ICAM-2, and VCAM-1 are members of the immunoglobulin superfamily.

The $\alpha_4\beta_1$ integrin has a dual specificity; it not only binds to the VCAM-1 counterreceptor, but can also bind to fibronectin (Mould *et al.*, 1990; Guan and Hynes, 1990). Similar to the $\alpha_4\beta_1$ integrin, the $\alpha_2\beta_1$ and $\alpha_3\beta_1$ integrins, which recognize extracellular matrix molecules, have also recently been suggested to function in cell–cell adhesion. This suggestion is based on their appearance in cell–cell contact sites in keratinocytes (DeLuca *et al.*, 1990; Larjava *et al.*, 1990).

The β_1 subunit is implicated in cell–cell adhesion also by experiments using antibodies to disrupt the adhesions (Larjava *et al.*, 1990).

4.3. Cell Aggregation through Soluble Adhesion Proteins

In a third mode of interaction, the major integrin in platelets, $\alpha_{IIb}/\beta_{IIIa}$, or gp IIb/IIIa, promotes the binding of platelets to one another through soluble, multivalent mediator molecules such as fibrinogen and von Willebrand factor. These molecules have multiple binding sites for α_{IIb}/β_3 and as a result can bridge the α_{IIb}/β_3 integrin of adjacent platelets, causing aggregation. Such aggregation is a key in blood clotting (reviewed in Ginsberg *et al.*, 1988, and Phillips *et al.*, 1988).

5. The RGD Cell Attachment Site

The recognition site for many of the integrins that bind to extracellular matrix and platelet adhesion proteins is the tripeptide RGD (Ruoslahti and Pierschbacher, 1987). The role of the RGD sequence as the recognition site was demonstrated by making progressively smaller fragments of fibronectin and by assaying for the cell attachment-promoting activity in the fragments and in synthetic peptides reproducing the amino acid sequences of such fragments (Pierschbacher and Ruoslahti, 1984a). When coated onto a surface, the fragments and synthetic peptides containing the RGD sequence promote cell attachment, whereas in solution they inhibit the attachment of cells to a surface coated with fibronectin or the peptides themselves. Changes in the peptides as small as the exchange of alanine for the glycine or glutamic acid for the aspartic acid, which constitute the addition of a single methyl or methylene group to the RGD tripeptide, or the replacement of the arginine with a lysine residue, eliminate these activities of the peptides (Pierschbacher and Ruoslahti, 1984b). The RGD sequence is also the cell recognition site of a surprising number of other extracellular matrix and platelet adhesion proteins. In addition to fibronectin, these include vitronectin, collagens, fibrinogen, von Willebrand factor, osteopontin, bone sialoprotein I, thrombospondin, tenascin, laminin, and entactin (Ruoslahti and Pierschbacher, 1987; Ruoslahti, 1991).

That the various integrins have different ligand specificities despite their shared RGD target sequence requires an explanation. The sequences flanking the RGD sequence in fibronectin, vitronectin, and C3bi show some similarity to one another, and it has been suggested that the binding sites for the receptors would encompass this similarity region of about 30 amino acids (Wright *et al.*, 1987). Because the flanking sequences are only similar, not identical, specific recognition of a single protein would become possible.

While the flanking sequences may well be of some importance, several facts argue against an extended binding site. First, with the exception of some similarity on the NH_2-terminal side of the RGD in fibronectin and vitronectin, the RGD flanking sequences in various RGD adhesion proteins have little or no similarity, yet several of them can bind to the same receptor. Second, peptides with the fibronectin flanking sequences bind better to the vitronectin receptor than to the fibronectin receptor (Pytela *et al.*, 1985b). Finally, a number of recent observations support the alternative notion that the specificity could reside in the conformation of the RGD tripeptide, and that the role of the surrounding sequences would be to force the RGD determinant into an appropriate conformation. Thus, peptides containing altered RGD conformations brought about by the use of D-amino acids or cyclization of the peptide have different receptor specificities and affinities than their L-amino acid or linear counterparts (Pierschbacher and Ruoslahti, 1987). Moreover, it is known that the RGD sequence can take very different conformations in different proteins (Ruoslahti and Pierschbacher, 1987). Thus, it may be that the RGD sequences of proteins with incidental cell attachment activity happen to be in the conformation of one of the adhesion protein sequences. An inactive RGD sequence, on the other hand, may either not be available at the surface of the molecule containing it or, if available, its conformation may not fit any of the receptors.

A conformational change occurring in the fibronectin fragments may also be the explanation for the observation that the affinity of the cell attachment fragments spanning the RGD site for the $\alpha_5\beta_1$ fibronectin receptor decreases precipitously when the fragments are made smaller than about 40,000 daltons (Pytela *et al.*, 1985b; Obara *et al.*, 1988). A second "affinity site" has been postulated to reside on the N-terminal side of the RGD sequence to account for this phenomenon (Obara *et al.*, 1988). However, as discussed above, the shortening of the fragments is accompanied by an increase in the affinity of the shorter fragments for the $\alpha_v\beta_3$ vitronectin receptor, suggesting that a conformational change in the RGD site has taken place. The affinity site may, therefore, not be a binding site in the sense that the RGD is, but may rather be an extended polypeptide stretch, the presence of which is needed to maintain the RGD site in the conformation most suitable for the binding of the $\alpha_5\beta_1$ integrin. Crystallographic and NMR analyses under way in our laboratory and elsewhere (Baron *et al.*, 1990) will provide answers to these questions.

A peptide sequence entirely different from RGD has been identified as the target sequence of the $\alpha_4\beta_1$ integrin in fibronectin (Mould *et al.*, 1990; Guan and Hynes, 1990). This sequence, GPEILDVPST, is present in one of the alternatively spliced segments of fibronectin (Hynes, 1990) and it is obviously not only different from RGD, but also does not resemble the fibrinogen-derived sequence, KQAGDV, which binds to the α_{IIb}/β_3 integrin interchangeably with RGD (Ginsberg *et al.*, 1988). Peptides containing the RGD and other integrin recognition sequences provide excellent tools for probing cell adhesion functions as discussed later in this chapter.

6. Regulation of Integrin Activity and Specificity

While the matrix binding integrins of adherent cells appear to be constitutively activated, the main platelet integrin $\alpha_{IIb}\beta_3$ requires activation to bind to its ligands. This integrin is present at the surface of resting platelets, but no aggregation results, even though fibrinogen and other ligands are available (Ginsberg *et al.*, 1988; Phillips *et al.*, 1988). It is not known how activation of the platelets triggers the binding activity of $\alpha_{IIb}\beta_3$.

The leukocyte integrin LFA-1 also needs to be activated. The ligation of the T-cell receptor causes the activation of this integrin in lymphocytes (Dustin and Springer, 1989). Other integrins such as the $\alpha_5\beta_1$ fibronectin and $\alpha_6\beta_1$ laminin receptor, which are constitutively activated in many types of cells, are also controlled by activation and inactivation in leukocytes (Shimizu *et al.*, 1990). It may be important for blood cells to control their activation in this manner, so that their circulation through the body is not impeded until they become stimulated at the site of an injury or by some other activating event. Elucidation of the molecular mechanisms of integrin activation is one of the most important goals of research on these proteins.

Cells may also regulate integrin specificity. The $\alpha_2\beta_1$ integrin in platelets is a collagen receptor, but in some other cells, such as endothelial cells, it binds to laminin and fibronectin in addition to collagen (Kirchhofer *et al.*, 1990). The molecular basis of this phenomenon is not understood, but an alternative splicing, a posttranslational modification such as phosphorylation, or the modulation of the integrin by an additional component are some of the possibilities. The regulation of integrin binding activity and ligand specificity greatly enhances the functional diversity of integrins.

7. Receptors for Extracellular Matrix Proteins

As can be seen in Fig. 10-2, there is much overlap and redundancy in the binding of integrins to various ligands. Thus, one integrin can bind to more than one adhesive protein. On the other hand, one adhesive protein can have more than one integrin-type receptor. In addition, some of the adhesive proteins are recognized both by integrins and by other types of receptor molecules. It seems worthwhile to examine the types of receptors that recognize the various matrix proteins.

7.1. Receptors for Fibronectin

Fibronectin is recognized by at least six different integrins. The integrin now known as $\alpha_5\beta_1$ was the first one isolated and characterized (Pytela *et al.*, 1985a) and is now often referred to as the "classical" fibronectin receptor. It appears to be the main integrin attaching cells to fibronectin and, as discussed below, it at least partially controls the assembly of a fibronectin matrix around cells. The $\alpha_5\beta_1$ integrin binds to the RGD cell attachment site of fibronectin (Pytela *et al.*, 1985a; Fig. 10-3A). At least two other integrins bind to this same site: the recently identified $\alpha_v\beta_1$ integrin (Vogel *et al.*, 1990) and $\alpha_{IIb}\beta_3$ (Ginsberg *et al.*, 1988). The $\alpha_v\beta_1$ integrin has also been reported to recognize vitronectin (Bodary and McLean, 1990) and type I collagen (Dedhar and Gray,

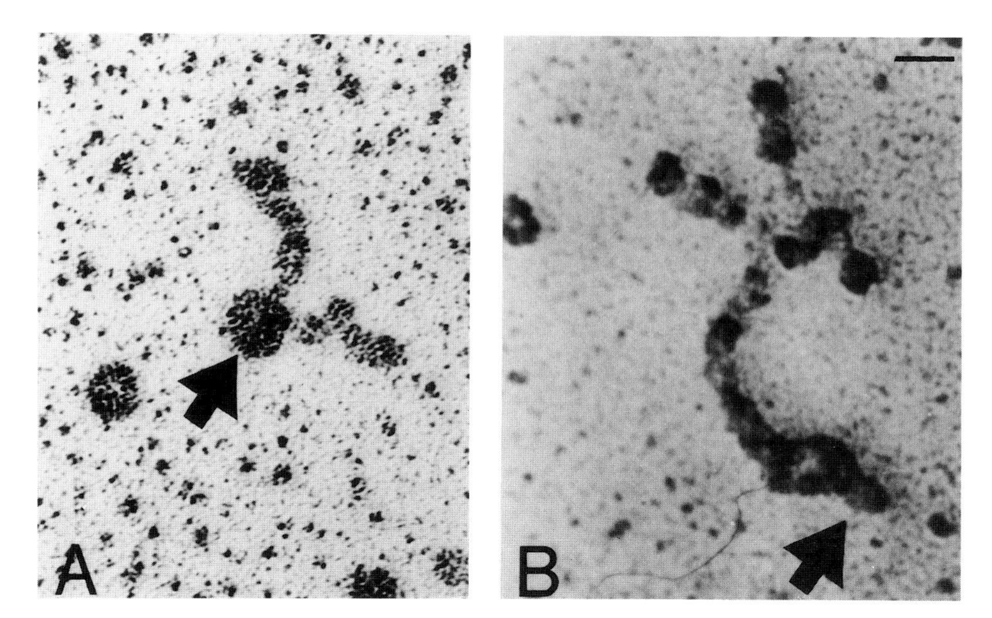

Figure 10-3. Electron microscopic localization of the binding of integrins to fibronectin and laminin. The $\alpha_5\beta_1$ integrin binds to fibronectin at the RGD site (A) and the $\alpha_3\beta_1$ recognizes a site in or near the globular region at the end of the long arm of laminin (B). The arrows point to the bound integrin molecules. Bar = 25 nm.

1990). However, we have not seen these activities in $\alpha_v\beta_1$ and find only fibronectin binding. The specificity of the $\alpha_v\beta_1$ integrin differs from $\alpha_5\beta_1$ in that it can be isolated on an RGD peptide column, whereas $\alpha_5\beta_1$ does not have sufficiently high affinity for the short RGD-containing peptides to bind to such an affinity matrix (Pytela *et al.*, 1985a). The differing specificities of the $\alpha_5\beta_1$, $\alpha_v\beta_1$, and of the $\alpha_v\beta_3$ vitronectin binding integrins clearly show that the β subunit contributes to the specificity of an integrin.

The $\alpha_4\beta_1$ integrin binds to fibronectin at a different site than the integrins discussed above. This integrin also binds to a cell surface counterreceptor VCAM-1 (Elices *et al.*, 1990). Fibronectin and VCAM-1 appear to bind to different sites on the integrin, because monoclonal antibodies have been found that inhibit one of these binding activities, but not the other (Elices *et al.*, 1990). The $\alpha_4\beta_1$ binding site in fibronectin resides in one of the alternatively spliced segments of fibronectin, the connecting segment. This segment undergoes a complex series of splicings. As far as is known, all types of fibronectins have the portion containing the $\alpha_4\beta_1$ recognition site in at least one of the subunits (Hynes, 1990). Plasma fibronectin, produced by liver cells, lacks the site in half of the subunits, whereas fibronectin from other types of cells usually has it in both subunits. Whether this higher density of binding sites in such fibronectin is functionally significant remains to be determined.

The $\alpha_3\beta_1$ integrin, which is primarily known as a collagen and laminin receptor, can also bind to fibronectin (Wayner and Carter, 1987; Hemler, 1990). I discuss the role of this integrin in the section on integrins in cancer. The $\alpha_2\beta_1$, which is primarily a collagen receptor, shows a cell-type-dependent ability to recognize laminin and fibronectin (Elices and Hemler, 1989; Kirchhofer *et al.*, 1990; Languino *et al.*, 1989). The binding of the $\alpha_2\beta_1$ and $\alpha_3\beta_1$ integrins to their ligands cannot be inhibited with the RGD-containing peptides; it will be interesting to see where in the fibronectin molecule the binding site is located. One of the important questions in the integrin field is to determine the specific roles of the multiple receptors for fibronectin and other matrix molecules. Such work should also determine the relative importance of proteoglycans as receptors for matrix molecules (see Chapter 4).

7.2. Receptors for Laminin

Like fibronectin, laminin has multiple integrin-type receptors, although not quite as many. The $\alpha_1\beta_1$, $\alpha_2\beta_1$, $\alpha_3\beta_1$, $\alpha_6\beta_1$, and $\alpha_v\beta_3$ integrins, as well as a novel integrin, are known to recognize laminin. The first three of these integrins bind to various degrees also to collagen and/or fibronectin (Wayner and Carter, 1987; Sonnenberg *et al.*, 1988; Gehlsen *et al.*, 1989; Languino *et al.*, 1989; Elices and Hemler, 1989; Kramer *et al.*, 1989, 1990; Hall *et al.*, 1990; Ignatius *et al.*, 1990). The chicken cell substrate attachment antigen (CSAT) is a mixture of β_1 integrins that appears to have the laminin-binding integrins as its major component (Horwitz *et al.*, 1985; Buck and Horwitz, 1987). In contrast, the $\alpha_6\beta_1$ integrin appears to be specific for laminin, and may be the main laminin receptor in tissues (Sorokin *et al.*, 1990). The α_6 subunit can also combine with β_4 to form a receptor that localizes at the junction between epithelial cells and their basement membrane (DeLuca *et al.*, 1990) [in hemidesmosomes when present (Stepp *et al.*, 1990)]. The recognition specificity of the $\alpha_6\beta_4$ integrin is not known, but it may also be a laminin receptor (Lotz *et al.*, 1990). Figure 10-4 shows examples of the varying tissue distribution of the integrin subunits.

An important question regarding the attachment of various types of cells to laminin is the relative importance of integrins and other types of laminin-binding proteins described in the literature. Antibody inhibition experiments conducted by many laboratories have shown that the adhesion of cells to laminin can be inhibited with antibodies against integrin subunits. While anti-β inhibits in all cell types, various α subunit-specific antibodies inhibit in some cells and not in others (see references above).

The early results on inhibition of cell attachment to laminin by antibodies against the 67-kDa laminin receptor (Liotta *et al.*, 1986) have not been reproduced (Basson *et al.*, 1990). The 67-kDa receptor is thought to recognize the sequence YIGSR, which comes from near the NH_2-terminus of the β_1 subunit, and peptides containing this sequence have been reported to inhibit the attachment of cells to laminin (Graf *et al.*, 1987). While this peptide has been found to be active in more complex assays (Grant *et al.*, 1989), its ability to inhibit cell

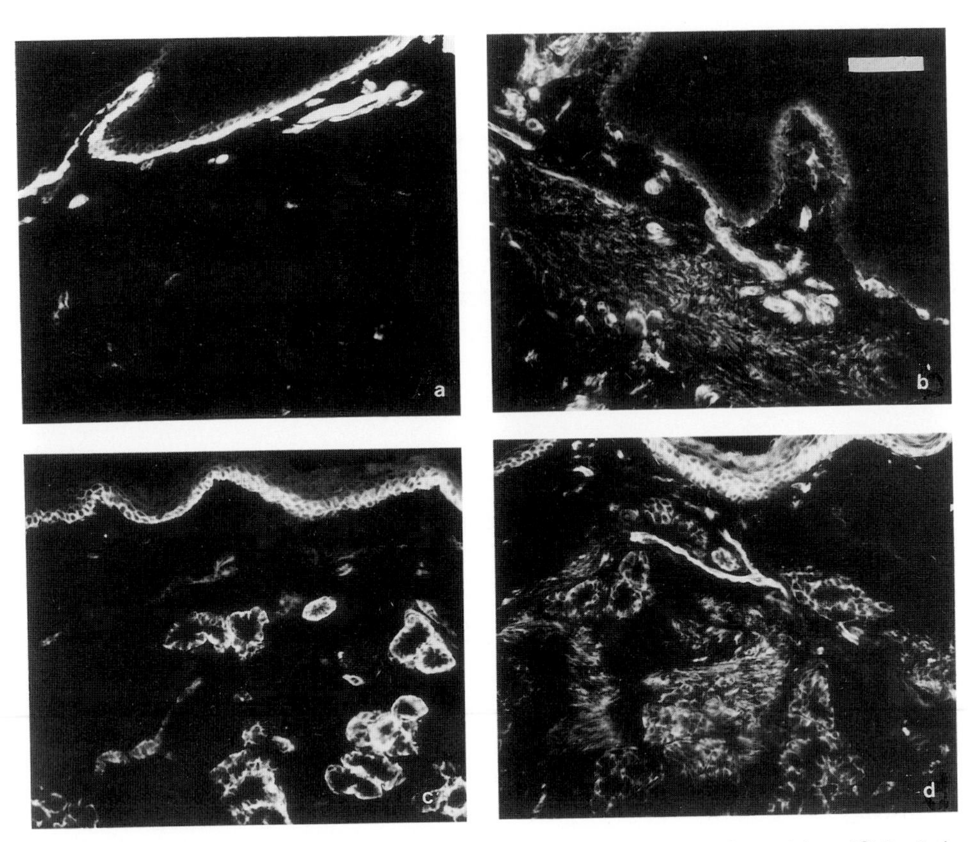

Figure 10-4. Distribution of various integrin subunits in the skin. (a) β_4; (b) α_3; (c) α_2; (d) β_1. β_4 is primarily basal and β_1, α_2, and α_3 are mainly on lateral epithelial cell surfaces. α_3 and β_1 are also present in the connective tissue. Bar = 25 μm. (Courtesy of Dr. Eva Engvall, La Jolla Cancer Research Foundation.)

attachment to laminin is controversial (Deutzmann *et al.*, 1990; but see also Chapter 12). The sequence of the 67-kDa receptor lacks a signal sequence (Makrides *et al.*, 1989), indicating that it is a cytoplasmic protein. There appears to be a consensus developing in the field that the 67-kDa receptor is likely to be a chaperone protein, an intracellular protein that guides the folding of nascent polypeptides by binding to them in the unfolded state. Whether the effects of the YIGSR peptide or cell adhesion are pharmacological effects on cells, or mediated by binding to a cell surface receptor is not clear.

Like fibronectin, laminin has more than one cell attachment site. Most of the integrins bind to an area near the globular region at the foot of the cross-shaped laminin molecule (Gehlsen *et al.*, 1989; Hall *et al.*, 1990; Deutzmann *et al.*, 1990; Fig. 10-3B). In addition, mouse laminin has a cell attachment site at the opposite end of the molecule (Goodman *et al.*, 1987). This site behaves as a cryptic site; it is activated in proteolytically cleaved laminin. This site is presumed to be located at an RGD sequence close to the N-terminal end of the A chain (Sasaki *et al.*, 1988), because its function can be inhibited with RGD-containing peptides (Aumailley *et al.*, 1987). However, this sequence is not an RGD in the human A chain (Haaparanta *et al.*, 1991). This lack of evolutionary conservation and the cryptic nature of the site cast doubt on the significance of this site. A YIGSR sequence has been proposed as an adhesive site in the β_1 chain (Graf *et al.*, 1987). Clearly, most of the cell attachment activity as well as the neurite-promoting activity (Edgar *et al.*, 1984; Engvall *et al.*, 1986) are mediated by the site located near the junction between the globular region formed by the C-terminal of the A chain and the rod-shaped part formed by all three of the subunits (Deutzmann *et al.*, 1990).

A complicating factor in the understanding of the role of laminin is that laminin is not a single protein. Of the three laminin subunits, A, B1, and B2, there is an alternative subunit at least for the A chain (Ehrig *et al.*, 1990) and the B1 chain (Hunter *et al.*, 1989). Moreover, all combinations of the subunits seem to exist, giving four different laminins (Engvall *et al.*, 1990). Much work will have to be done to elucidate the receptor specificities of these different laminins.

7.3. Receptors for Collagen

The collagen-binding integrins are $\alpha_1\beta_1$, $\alpha_2\beta_1$, and $\alpha_3\beta_1$ (Wayner and Carter, 1987; Ignatius *et al.*, 1990; Staatz *et al.*, 1989). There are many RGD sequences in collagens (Ruoslahti and Pierschbacher, 1987), but judging from lack of inhibition by RGD-containing peptides of the integrin binding to collagen in many cell types and integrin-binding activity in collagen fragments that do not contain any RGD sequence (Gullberg *et al.*, 1988), at least most of the collagen-mediated cell attachment is not RGD dependent. However, an RGD-containing peptide with a threonine residue on the C-terminal side of the aspartic acid of RGD has been found to partially inhibit the attachment of some cells to type I collagen (Dedhar *et al.*, 1987; Gehlsen *et al.*, 1988; Pignatelli and Bodmer, 1989). One possible explanation for these divergent results is that the RGD sequences of collagen exist in many conformations and that some of these may be recognized by one or more of the RGD-dependent integrins. Especially the α_v integrins, discussed below, are candidates for that role.

7.4. Receptors for Vitronectin

Several of the integrins belonging to the $\alpha_v\beta_3$ group of integrins bind to vitronectin. These include the classical vitronectin receptor, $\alpha_v\beta_3$ (Pytela *et al.*, 1985b), the $\alpha_v\beta_5$ integrin (Smith *et al.*, 1990), which may be the same as $\alpha_v\beta_s$ (Freed *et al.*, 1989), and $\alpha_{IIb}\beta_3$ (Pytela *et al.*, 1986). The newly described β_6 subunit may also combine with α_v to form a vitronectin receptor, because the amino acid sequence β_6 is more homologous to that of the β_3 subunit than those of the other subunits (Sheppard *et al.*, 1990).

The α_v receptors have multiple ligands; the $\alpha_v\beta_3$ integrin has been particularly studied in this regard. In addition to vitronectin, it binds to osteopontin, bone sialoprotein I, fibrinogen, von Willebrand factor, and thrombospondin (Cheresh, 1987; Reinholt *et al.*, 1990), each of which is an RGD protein. In addition, it binds well to small, RGD-containing fragments of fibronectin, and may even show some affinity for fibronectin (Pytela *et al.*, 1985b; see Smith *et al.*, 1990).

Considering the wide specificity of the $\alpha_v\beta_3$ integrin, it is surprising that individuals who lack the β_3 subunit essentially only show symptoms attributable to the lack of the $\alpha_{IIb}\beta_3$ integrin, i.e., anomalies in platelet function (Ginsberg *et al.*, 1988). One would expect changes in bone function, in particular, because that is where the integrin $\alpha_v\beta_3$ is abundantly expressed (J. Gailit and E. Ruoslahti, unpublished results). Perhaps the other α_v integrins can substitute for $\alpha_v\beta_3$.

8. The Role of Integrins in Matrix Assembly

The interaction of the integrins with the extracellular matrix is a two-way street; the integrins not only bind cells to the matrix, they also help shape the matrix. Most of the information in this regard comes from work on fibronectin, and that work is reviewed below. However, the deposition of many of the other matrix components appears to parallel that of fibronectin. Thus, the perturbation in the deposition of fibronectin that is common in malignantly transformed fibroblasts is accompanied by a lack of deposition of laminin, perlecan, and various collagens (see Ruoslahti, 1988). The matrix components need a cell to assemble them into the matrix and integrins play a role in it (Roman *et al.*, 1989). A possible exception is the interstitial collagens which can form fibrils spontaneously. However, a cellular contribution may be necessary in the assembly of a collagen matrix, because cell lines have been found that make type I procollagen but do not have it in their matrix (Vaheri *et al.*, 1978).

The matrix assembly has been particularly thoroughly explored with fibronectin. The primary functional form of fibronectin is the insoluble protein. The insoluble tissue fibronectin can interact with cells in a cooperative manner effecting cell attachment. This makes the process of fibronectin insolubilization an important aspect of cell adhesion.

The precise molecular mechanisms of the matrix assembly are not known, but three events appear important in it: the binding of the fibronectin to the $\alpha_5\beta_1$ integrin and to a matrix assembly receptor, fibronectin–fibronectin binding, and fibronectin–fibronectin cross-linking.

Matrix assembly by cultured cells is inhibited in the presence of a monoclonal antibody that prevents the binding of the $\alpha_5\beta_1$ integrin to fibronectin (Roman *et al.*, 1989), suggesting a role for this receptor. Moreover, gene transfer experiments have shown that the level of $\alpha_5\beta_1$ integrin in cells correlates with the level of matrix deposition (Giancotti and Ruoslahti, 1990; Fig. 10-5). On the other hand, the main argument in favor of a separate matrix assembly receptor

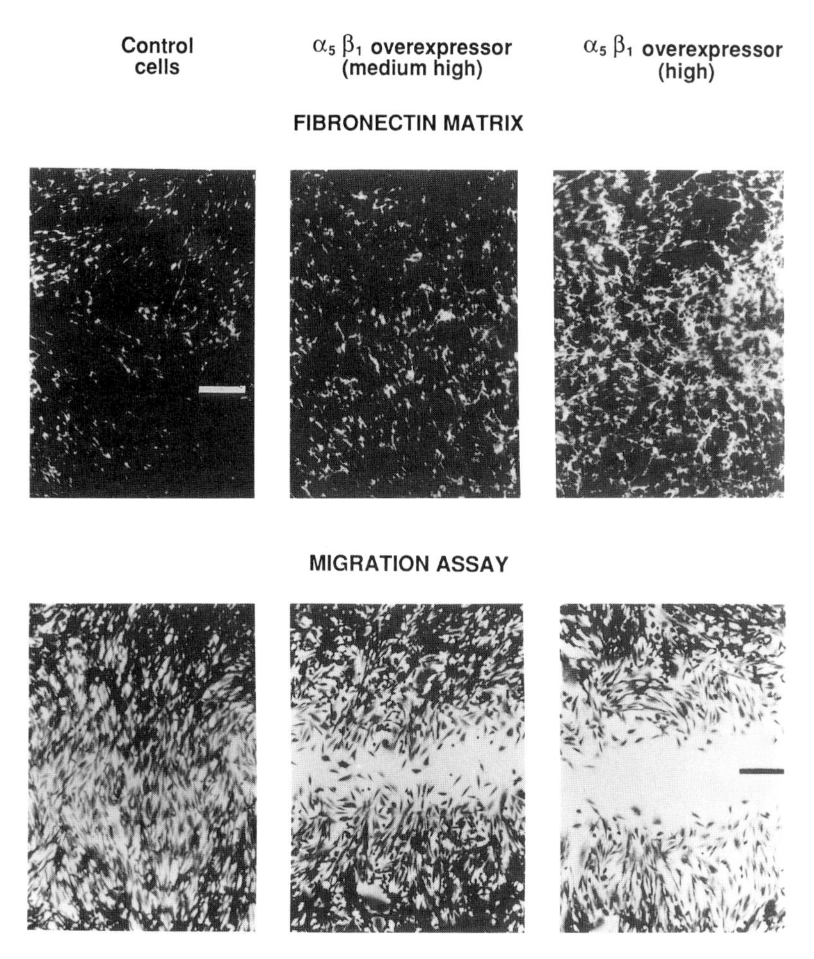

Figure 10-5. Phenotype changes in cells expressing elevated levels of the $\alpha_5\beta_1$ integrin from transfected cDNA. The deposition of fibronectin matrix is increased (upper row of panels), and the ability to migrate across a "wound" in culture is reduced (lower row) in the $\alpha_5\beta_1$ overexpressor cells compared to control-transfected cells. Bars = 60 μm (upper panel), 110 μm (lower panel.)

is that the NH_2-terminal domain of fibronectin interferes with the binding of fibronectin to a cell layer (see Peters *et al.*, 1990).

It may be that the RGD receptor initiates the deposition of fibronectin into fibrils by binding fibronectin to the cell surface. Additional steps are then needed to elongate the fibrils. These steps consist of fibronectin–fibronectin interaction and the cross-linking of adjacent fibronectin molecules to one another by disulfide bonding. A fibronectin–fibronectin interaction site has been found near the collagen-binding domain, and the cross-linking occurs through rearrangement of intramolecular disulfide bonds in the NH_2-terminal domains of adjacent fibronectin molecules into intermolecular bonds (see Roman *et al.*, 1989, and Peters *et al.*, 1990, for discussion of the assembly process).

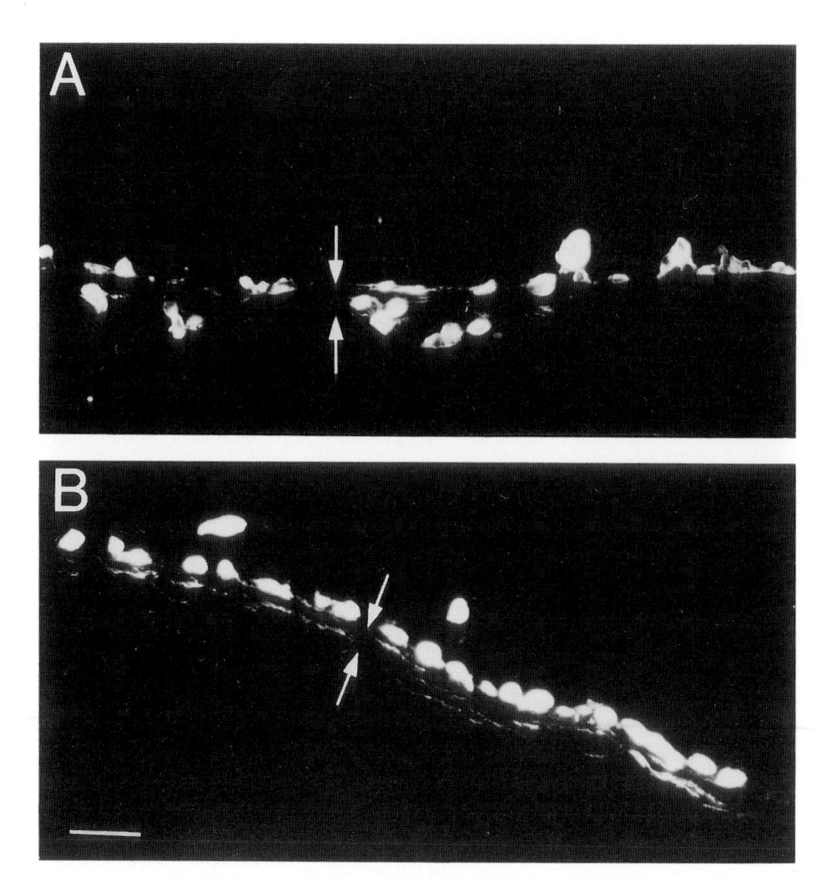

Figure 10-6. Inhibition of tumor cell invasion through amniotic membrane by RGD peptides. A375M human melanoma cells were labeled with the fluorescent lipophilic dye, DiI, and allowed to invade the amniotic membrane for 72 hr in the presence of either GRGESP (A) or GRGDTP (B). The cells were visualized by fluorescence microscopy of the sectioned amniotic membrane. Basement membrane surface (upper arrow) and the lower stromal surface (lower arrow) of the amniotic membrane are shown. Bar = 15 μm. (From Gehlsen *et al.*, 1988, with permission.)

Gangliosides may also play a role in matrix assembly, since it has been found that matrix formation can be restored by adding gangliosides (sialic acid-containing glycolipids) to a cell line that lacks both gangliosides and a matrix (Yamada *et al.*, 1983). Moreover, soluble gangliosides can inhibit attachment of cells to fibronectin. This effect may be through the RGD-directed receptor for fibronectin, because gangliosides have been found to interact with these receptors (Cheresh *et al.*, 1987). Further work on the ganglioside connection will be needed.

The presence of an extracellular matrix around a cell profoundly affects the biological behavior of the cell (Chapters 9, 12). A good example of this is the effects of a lack of extracellular matrix, especially fibronectin, in malignant cells.

9. Integrins in Cancer

While stationary adult cells use the integrin–extracellular matrix interactions for anchorage, embryonal cells, migrating adult cells, and tumor cells seem to use these interactions for traction to migrate and possibly as a stimulus to proliferate.

As discussed above, normal cells deposit fibronectin, laminin, collagens, and other extracellular matrix components around themselves as an adhesion network of insoluble protein. The reasons for the failure of many tumorigenic cells in culture to deposit such a matrix or do so to a lesser degree than normal cells are only partially understood. The expression of the $\alpha_5\beta_1$ integrin which is needed for the matrix deposition (Roman *et al.*, 1989; Giancotti and Ruoslahti, 1990) is often reduced in tumor cells (Plantefaber and Hynes, 1989). However, at least one other factor, a "matrix assembly receptor," is also needed for fibronectin deposition (see Peters *et al.*, 1990, and Fogarty *et al.*, 1990). This factor, which is absent in matrix-deficient cells (McKeown-Longo and Etzler, 1987), has not been characterized, but will obviously be an important object of future studies.

Modulation of the binding affinities of the integrins for their ligands could also explain the defective matrix deposition by transformed cells. Phosphorylation is one possible mechanism by which integrin affinity could be modulated *in vivo*. In transformed chicken cells, a fibronectin receptor complex becomes phosphorylated on a cytoplasmic tyrosine residue; this event has been reported to be accompanied by a reduction in fibronectin binding activity and by a loss of binding to the cytoskeletal protein talin (Tapley *et al.*, 1989).

A consequence of the lack of matrix deposition is that the tumor cells have an added degree of freedom; their mobility is not limited by adhesion to their own matrix. The importance of this constraint in normal cellular behavior is suggested by the $\alpha_5\beta_1$ gene transfer experiment mentioned above (Giancotti and Ruoslahti, 1990). The CHO cells expressing high levels of this integrin from the transfected genes not only deposited more fibronectin matrix, but had become less migratory than the control cells and grew less well in soft agar (Fig. 10-5). Moreover, unlike the original recipient cells and various control-transfected cell lines which are tumorigenic, the $\alpha_5\beta_1$ overexpressors failed to form tumors in nude mice. The low expression of the $\alpha_5\beta_1$ fibronectin receptors and the reduced assembly of a fibronectin matrix often seen in malignant cells may therefore be very closely associated with the expression of the tumorigenic phenotype.

There is another aspect to fibronectin in malignancy, however. It appears that fibronectin (and extracellular matrix in general) plays a dual role in malignancy. As discussed above, a tumor cell should lack its own extracellular matrix to be able to proliferate fast and migrate optimally. However, such a cell needs some matrix receptors to be able to derive traction for migration from the matrices of other cells. This is suggested by the fact that tumor cells and other migratory cells preferentially migrate on surfaces coated with adhesive extracellular matrix proteins. Moreover, the RGD peptides can inhibit migration of tumor cells through tissue in invasion assays (Fig. 10-6) and reduce tumor cell dissemination in tissues when injected together with tumor cells into mice (Humphries *et al.*, 1986; reviewed in Ruoslahti and Giancotti, 1989).

Changes in integrins may offer an explanation for these seemingly paradoxical observations. Transformed cells express reduced levels of the $\alpha_5\beta_1$ fibronectin receptor. However, they tend to also express higher levels of the $\alpha_3\beta_1$ integrin, which can serve as a receptor for fibronectin as well as for collagens and laminin (Plantefaber and Hynes, 1989).

The $\alpha_3\beta_1$ integrin binds to intact fibronectin or its cell-adhesive fragment at low salt concentrations (Wayner and Carter, 1987) and shows some affinity for fibronectin in receptor binding assays (Gehlsen *et al.*, 1989). It also seems capable of mediating cell adhesion to fibronectin as judged by antibody inhibition results (Wayner and Carter, 1987). It may be that $\alpha_3\beta_1$ has a lower affinity for fibronectin than $\alpha_5\beta_1$, and this feature precludes its participation in matrix assembly without affecting its activity as an adhesion receptor. It may also be important for the malignant properties of the cells expressing the $\alpha_3\beta_1$ integrin that this integrin is a laminin receptor.

Attachment of cells to laminin has been linked to increased tumorigenicity. For example, cells grown on laminin have been found to be more tumorigenic than cells grown on fibronectin (Terranova *et al.*, 1984). These observations suggest the following generalizations and testable hypotheses. The $\alpha_5\beta_1$ integrin contributes to the assembly of a fibronectin matrix and its expression tends to immobilize a cell and suppress its migration. Expression of the $\alpha_3\beta_1$ integrins allows binding to fibronectin for migration purposes but does not support matrix assembly. Moreover, the interaction of this integrin and possibly other laminin-binding integrins, increases the malignant potential of cells. This latter effect, as well as at least some of the antitumor effects of the $\alpha_5\beta_1$ integrin expression, may depend on the cell receiving a signal through the integrin rather than the integrin action being purely an adhesive phenomenon. This aspect of integrin function undoubtedly will be among the most widely studied subjects of future studies on integrins.

10. Future Prospects

As discussed above, there are indications that integrins can transmit signals from the extracellular matrix to the interior of the cell. It will be important to determine the consequences to the cell from the binding of the various integrins to their ligand proteins. Expression of individual integrins in tumor cells by gene transfer, followed by analysis of the various cellular properties that have changed, will provide direct information on such roles of integrins. *In vivo* studies can be conducted by disrupting integrin genes and by effecting temporally and spatially incorrect expression of integrins in transgenic mice. Comparison of integrins with the same ligand specificity in such studies should be particularly illuminating. In fact, the need for different types of cells to derive different messages from matrix interactions may well be the reason for the existence of some of the integrin diversity.

In the past, most of the studies on integrins have focused on their ability to mediate cell attachment. Like most other biological phenomena, cell adhesion is likely to be governed by both positive and negative signals. The negative signals, those that cause a cell to detach and/or migrate rather than attach, will probably become a much more actively studied area than has been the case in the past.

In parallel with the observation of the consequences of integrin expression and ligation, it will be necessary to work out the cytoplasmic connections of integrins to intracellular signal transduction pathways and to the cytoskeleton. Such connections may not only transmit signals from the exterior of the cell to its interior, but are likely to affect the affinity, specificity, and distribution of integrins making integrins transmitters of signals in both directions through the plasma membrane.

Finally, the RGD peptides show promise as modulators of activities with both the experimental and therapeutic applications. Better understanding of the binding requirements of those integrins that do not recognize RGD should lead to the development of specific low-molecular-weight inhibitors for these integrins also.

ACKNOWLEDGMENTS. I thank Drs. Eva Engvall and Lucia Languino for comments on the manuscript. The writing of this review and the author's original work are supported by grants CA 42507, CA 28896, and Cancer Center support grant CA 30199 from the National Cancer Institute and grant HL 26838 from the National Heart, Lung and Blood Institute, DHHS.

References

Aumailley, M., Nurcombe, V., Edgar, D., Paulsson, M., and Timpl, R., 1987, The cellular interactions of laminin fragments, *J. Biol. Chem.* **262:**11532–11538.

Baron, M., Norman, D., Willis, A., and Campbell, I. D., 1990, Structure of the fibronectin type 1 module, *Nature* **345:**642–646.

Basson, C. T., Knowles, W. J., Bell, L., Albelda, S. M., Castronovo, V., Liotta, L. A., and Madri, J. A., 1990, Spatiotemporal segregation of endothelial cell integrin and nonintegrin extracellular matrix-binding proteins during adhesion events, *J. Cell Biol.* **110:**789–801.

Bodary, S. C., and McLean, J. W., 1990, The integrin β_1 subunit associates with the vitronectin receptor α_v subunit to form a novel vitronectin receptor in a human embryonic kidney cell line, *J. Biol. Chem.* **265:**5938–5941.

Buck, C. A., and Horwitz, A. F., 1987, Cell surface receptors for extracellular matrix molecules, *Annu. Rev. Cell Biol.* **3:**179–205.

Cheresh, D. A., 1987, Human endothelial cells synthesize and express an Arg-Gly-Asp-directed adhesion receptor involved in attachment to fibrinogen and von Willebrand factor, *Proc. Natl. Acad. Sci. USA* **84:**6471–6475.

Cheresh, D., Pytela, R., Pierschbacher, M., Klier, F., Ruoslahti, E., and Reisfeld, R., 1987, An Arg-Gly-Asp-directed receptor on the surface of human melanoma cells exists in a divalent cation-dependent functional complex with the disialoganglioside GD2, *J. Cell Biol.* **105:**1163–1173.

Dedhar, S., and Gray, V., 1990, Isolation of a novel integrin receptor mediating ARG-GLY-ASP-directed cell adhesion to fibronectin and type I collagen from human neuroblastoma cells. Association of a novel β_1-related subunit with α_v, *J. Cell Biol.* **110:**2185–2193.

Dedhar, S., Ruoslahti, E., and Pierschbacher, M. D., 1987, A cell surface receptor complex for collagen type I recognizes the Arg-Gly-Asp sequence, *J. Cell Biol.* **104:**585–593.

DeLuca, M., Tamura, R. N., Kajiji, S., Bondanza, S., Rossino, P., Cancedda, R., Marchisio, P. C., and Quaranta, V., 1990, Polarized integrin mediates human keratinocyte adhesion to basal lamina, *Proc. Natl. Acad. Sci. USA* **87**:6888–6892.

Deutzmann, R., Aumailley, M., Wiedemann, H., Pysny, W., Timpl, R., and Edgar, D., 1990, Cell adhesion, spreading and neurite stimulation by laminin fragment E8 depends on maintenance of secondary and tertiary structure in its rod and globular domain, *Eur. J. Biochem.* **191**:513–522.

Dustin, M. L., and Springer, T. A., 1989, T-cell receptor cross-linking transiently stimulates adhesiveness through LFA-1, *Nature* **341**:619–624.

Edgar, D., Timpl, R., and Thoenen, H., 1984, The heparin-binding domain of laminin is responsible for its effects on neurite outgrowth and neuronal survival, *EMBO J.* **3**:1463–1468.

Ehrig, K., Leivo, I., Argraves, W. S., Ruoslahti, E., and Engvall, E., 1990, Merosin, a tissue-specific basement membrane protein, is a laminin-like protein, *Proc. Natl. Acad. Sci. USA* **87**:3264–3268.

Elices, M. J., and Hemler, M. E., 1989, The human integrin VLA-2 is a collagen receptor on some cells and a collagen/laminin receptor on others, *Proc. Natl. Acad. Sci. USA* **86**:9906–9910.

Elices, M. J., Osborn, L., Takada, Y., Crouse, C., Luhowsky, S., Hemler, M. E., and Lobb, R. R., 1990, VCAM-1 on activated endothelium interacts with the leukocyte integrin VLA-4 at a site distinct from the VLA-4/fibronectin binding site, *Cell* **60**:577–584.

Engvall, E., Davis, G. E., Dickerson, K., Ruoslahti, E., Varon, S., and Manthorpe, M., 1986, Mapping of domains in human laminin using monoclonal antibodies: Localization of the neurite-promoting site, *J. Cell Biol.* **103**:2457–2465.

Engvall, E., Earwicker, D., Haaparanta, T., Ruoslahti, E., and Sanes, J. R., 1990, Distribution and isolation of four laminin variants; tissue restricted distribution of heterodimers assembled from five different subunits, *Cell Regul.* **1**:731–740.

Fogarty, F. J., Akiyama, S. K., Yamada, K. M., and Mosher, D. F., 1990, Inhibition of binding of fibronectin to matrix assembly sites by anti-integrin ($\alpha_5\beta_1$) antibodies, *J. Cell Biol.* **111**:699–708.

Freed, E., Gailit, J., van der Geer, P., Ruoslahti, E., and Hunter, T., 1989, A novel integrin β subunit is associated with the vitronectin receptor α subunit (α_v) in a human osteosarcoma cell line and is a substrate for protein kinase, *EMBO J.* **8**:2955–2965.

Gailit, J., and Ruoslahti, E., 1988, Regulation of the fibronectin receptor affinity by divalent cations, *J. Biol. Chem.* **263**:12927–12932.

Gehlsen, K. R., Argraves, W. S., Pierschbacher, M. D., and Ruoslahti, E., 1988, Inhibition of *in vitro* tumor cell invasion by Arg-Gly-Asp-containing synthetic peptides, *J. Cell Biol.* **106**:925–930.

Gehlsen, K. R., Dickerson, K., Argraves, W. S., Engvall, E., and Ruoslahti, E., 1989, Subunit of a laminin-binding integrin and localization of its binding site on laminin, *J. Biol. Chem.* **264**:19034–19038.

Giancotti, F. G., and Ruoslahti, E., 1990, Elevated levels of the $\alpha_5\beta_1$ fibronectin receptor suppress the transformed phenotype of Chinese hamster ovary cells, *Cell* **60**:849–859.

Ginsberg, M. H., Loftus, J. C., and Plow, E. F., 1988, Cytoadhesins, integrins, and platelets, *Thromb. Haemos.* **59**:1–6.

Goodman, S. L., Deutzmann, R., and von der Mark, K., 1987, Two distinct cell-binding domains in laminin can independently promote nonneuronal cell adhesion and spreading, *J. Cell Biol.* **105**:589–598.

Graf, J., Iwamoto, Y., Sasaki, M., Martin, G. R., Kleinman, H. K., Robey, F. A., and Yamada, Y., 1987, Identification of an amino acid sequence in laminin mediating cell attachment, chemotaxis, and receptor binding, *Cell* **48**:989–996.

Grant, D. S., Tashiro, K.-I., Segui-Real, B., Yamada, Y., Martin, G. R., and Kleinman, H. K., 1989, Two different laminin domains mediate the differentiation of human endothelial cells into capillary-like structures in vitro, *Cell* **58**:933–943.

Guan, J.-L., and Hynes, R. O., 1990, Lymphoid cells recognize an alternatively spliced segment of fibronectin via the integrin receptor $\alpha_4\beta_1$, *Cell* **60**:53–61.

Gullberg, D., Terracio, L., Borg, T. K., and Rubin, K., 1988, Identification of integrin-like matrix receptors with affinity for interstitial collagens, *J. Biol. Chem.* **263**:12686–12694.

Haaparanta, T., Uitto, J., Ruoslahti, E., and Engvall, E., 1991, Molecular cloning of the cDNA encoding human laminin A chain, *Matrix* **11**:151–160.
Hall, D. E., Reichardt, L. F., Crowley, E., Holley, B., Moezzi, H., Sonnenberg, A., and Damsky, C. H., 1990, The α_1/β_1 and α_6/β_1 integrin heterodimers mediate cell attachment to distinct sites on laminin, *J. Cell Biol.* **110**:2175–2184.
Hemler, M. E., 1990, VLA proteins in the integrin family: Structures, functions, and their role on leukocytes, *Annu. Rev. Immunol.* **8**:365–400.
Hogervorst, F., Kuikman, I., von dem Borne, A. E. G. K., and Sonnenberg, A., 1990, Cloning and sequence analysis of beta-4 cDNA: An integrin subunit that contains a unique 118 kd cytoplasmic domain, *EMBO J.* 9:765–770.
Horwitz, A., Duggan, K., Greggs, R., Decker, C., and Buck, C., 1985, The cell substrate attachment (CSAT) antigen has properties of a receptor for laminin and fibronectin, *J. Cell Biol.* **101**:2134–2144.
Humphries, M. J., Olden, K., and Yamada, K. M., 1986, A synthetic peptide from fibronectin inhibits experimental metastasis of murine melanoma cells, *Science* **233**:467–470.
Hunter, D. D. Shah, V., Merlie, J. P., and Sanes, J. R., 1989, A laminin-like adhesive protein concentrated in the synaptic cleft of the neuromuscular junction, *Nature* **338**:229–234.
Hynes, R. O., 1987, Integrins: A family of cell surface receptors, *Cell* **48**:549–554.
Hynes, R. O., 1990, *Fibronectins*, Springer-Verlag, Berlin.
Ignatius, M. J., Large, T. H., Houde, M., Tawil, J. W., Barton, A., Esch, F., Carbonetto, S., and Reichardt, L. F., 1990, Molecular cloning of the rat integrin α_1-subunit: A receptor for laminin and collagen, *J. Cell Biol.* **111**:709–720.
Kirchhofer, D., Languino, L. R., Ruoslahti, E., and Pierschbacher, M. D., 1990, $\alpha_2\beta_1$ integrins from different cell types show different binding specificities, *J. Biol. Chem.* **265**:615–618.
Kramer, R. H., McDonald, K. A., and Vu, M. P., 1989, Human melanoma cells express a novel integrin receptor for laminin, *J. Biol. Chem.* **264**:15642–15649.
Kramer, R. H., Cheng, Y.-F., and Clyman, R., 1990, Human microvascular endothelial cells use β_1 and β_3 integrin receptor complexes to attach to laminin, *J. Cell Biol.* **111**:1233–1243.
Languino, L. R., Gehlsen, K. R., Wayner, E., Carter, W. G., Engvall, E., and Ruoslahti, E., 1989, Endothelial cells use $\alpha_2\beta_1$ integrin as a laminin receptor, *J. Cell Biol.* **109**:2455–2462.
Larjava, H., Peltonen, J., Akiyama, S. K., Yamada, S. S., Gralnick, H. R., Uitto, J., and Yamada, K. M., 1990, Novel function for β_1 integrins in keratinocyte cell–cell interactions, *J. Cell Biol.* **110**:803–815.
Liotta, L. A., Rao, C. N., and Wewer, U. M., 1986, Biochemical interactions of tumor cells with the basement membrane, *Annu. Rev. Biochem.* **55**:1037–1057.
Loftus, J. C., Plow, E. F., O'Toole, T. E., Glass, A., Frelinger, A. L., and Ginsberg, M. H., 1990, A β_3 integrin mutation abolishes ligand binding and alters divalent cation-dependent conformation, *Science* **249**:915–918.
Lotz, M. M., Korzelius, C. A., and Mercurio, A. M., 1990, Human colon carcinoma cells use multiple receptors to adhere to laminin: Involvement of $\alpha_6\beta_4$ and $\alpha_2\beta_1$ integrins, *Cell Regul.* **1**:249–257.
Makrides, S., Chitpatima, S. T., Bandyopadhyay, R., and Brawerman, G., 1988, Nucleotide sequence for a major messenger RNA for a 40 kilodalton polypeptide that is under translational control in mouse tumor cells, *Nucleic Acids Res.* **16**:2349.
McKeown-Longo, P. J., and Etzler, C. A., 1987, Induction of fibronectin matrix assembly in human fibrosarcoma cells by dexamethasone, *J. Cell Biol.* **104**:601–610.
Mould, A. P., Wheldon, L. A., Komoriya, A., Wayner, E. A., Yamada, K. M., and Humphries, M. J., 1990, Affinity chromatographic isolation of the melanoma adhesion receptor for the IIICS region of fibronectin and its identification as the integrin $\alpha_4\beta_1$, *J. Biol. Chem.* **265**:4020–4024.
Obara, M., Kang, M. S., and Yamada, K. M., 1988, Site-directed mutagenesis of the cell-binding domain of human fibronectin: Separable, synergistic sites mediate adhesive function, *Cell* **53**:649–657.
Peters, D. M. P., Portz, L. M., Fullerwider, J., and Mosher, D. F., 1990, Co-assembly of plasma and cellular fibronectins into fibrils in human fibroblast cultures, *J. Cell Biol.* **111**:249–256.

Phillips, D. R., Charo, I. F., Parise, L., and Fitzgerald, L. A., 1988, The platelet membrane glycoprotein IIb–IIIa complex, *Blood* **71**:831–843.
Pierschbacher, M. D., and Ruoslahti, E., 1984a, The cell attachment activity of fibronectin can be duplicated by small fragments of the molecule, *Nature* **309**:30–33.
Pierschbacher, M. D., and Ruoslahti, E., 1984b, Modifications of the cell attachment recognition site in fibronectin which retain activity, *Proc. Natl. Acad. Sci. USA* **81**:5985–5988.
Pierschbacher, M. D., and Ruoslahti, E., 1987, Influence of stereochemistry of the sequence Arg-Gly-Asp-Xxx on binding specificity in cell adhesion, *J. Biol. Chem.* **262**:17294–17298.
Pignatelli, M., and Bodmer, W. F., 1989, Integrin-receptor-mediated differentiation and growth inhibition are enhanced by transforming growth factor-β in colorectal tumour cells grown in collagen gel, *Int. J. Cancer* **44**:518–523.
Plantefaber, L. C., and Hynes, R. O., 1989, Changes in integrin receptors on oncogenically transformed cells, *Cell* **56**:281–290.
Pytela, R., 1988, Amino acid sequence of the murine Mac-1 α chain reveals homology with the integrin family and an additional domain related to von Willebrand factor, *EMBO J.* **7**:1371–1378.
Pytela, R., Pierschbacher, M. D., and Ruoslahti, E., 1985a, Identification and isolation of a 140 kilodalton cell surface glycoprotein with properties of a fibronectin receptor, *Cell* **40**:191–198.
Pytela, R., Pierschbacher, M. D., and Ruoslahti, E., 1985b, A 125/115 kd cell surface receptor specific for vitronectin interacts with the Arg-Gly-Asp adhesion sequence derived from fibronectin, *Proc. Natl. Acad. Sci. USA* **82**:5766–5770.
Pytela, R., Pierschbacher, M. D., Ginsberg, M. H., Plow, E. F., and Ruoslahti, E., 1986, Platelet membrane glycoprotein IIb/IIIa is a member of a family of arg-gly-asp-specific adhesion receptors, *Science* **231**:1559–1562.
Reinholt, F. P., Hultenby, K., Oldberg, Å., and Heinegård, D., 1990, Osteopontin—A possible anchor of osteoclasts to bone, *Proc. Natl. Acad. Sci. USA* **87**:4473–4475.
Roman, J., LaChance, R. M., Broekelmann, T. J., Kennedy, C. J. R., Wayner, E. A., Carter, W. G., and McDonald, J. A., 1989, The fibronectin receptor is organized by extracellular matrix fibronectin: Implications for oncogenic transformation and for cell recognition of fibronectin matrices, *J. Cell Biol.* **108**:2529–2543.
Ruoslahti, E., 1988, Fibronectin and its receptors, *Annu. Rev. Biochem.* **57**:375–413.
Ruoslahti, E., 1991, Integrins, *J. Clin. Invest.* **87**:1–5.
Ruoslahti, E., and Giancotti, F. G., 1989, Integrins and tumor cell dissemination, *Cancer Cells* **1**:119–126.
Ruoslahti, E., and Pierschbacher, M. D., 1987, New perspectives in cell adhesion: RGD and integrins, *Science* **238**:491–497.
Sasaki, M., Kleinman, H. K., Huber, H., Deutzmann, R., and Yamada, Y., 1988, Laminin, a multidomain protein, *J. Biol. Chem.* **263**:16536–16544.
Sheppard, D., Rozzo, C., Starr, L., Quaranta, V., Erie, D. J., and Pytela, R., 1990, Complete amino acid sequence of a novel integrin β subunit (β_6) identified in epithelial cells using the polymerase chain reaction, *J. Biol. Chem.* **265**:11502–11507.
Shimizu, Y., Van Seventer, G. A., Horgan, K. J., and Shaw, S., 1990, Regulated expression and binding of three VLA (β_1) integrin receptors on T cells, *Nature* **345**:250–253.
Smith, J. W., Vestal, D. J., Irwin, S. V., Burke, T. A., and Cheresh, D. A., 1990, Purification and functional characterization of integrin $\alpha_v\beta_5$. An adhesion receptor for vitronectin, *J. Biol. Chem.* **265**:11008–11013.
Sonnenberg, A., Modderman, P. W., and Hogervorst, F., 1988, Laminin receptor on platelets is the integrin VLA-6, *Nature* **336**:487–489.
Sorokin, L., Sonnenberg, A., Aumailley, M., Timpl, R., and Ekblom, P., 1990, Recognition of the laminin E8 cell-binding site by an integrin possessing the α_6 subunit is essential for epithelial polarization in developing kidney tubules, *J. Cell Biol.* **111**:1265–1273.
Springer, T. A., 1990, Adhesion receptors of the immune system, *Nature* **346**:425–433.
Staatz, W. D., Rajpara, S. M., Wayner, E. A., Carter, W. G., and Santoro, S. A., 1989, The membrane glycoprotein Ia-IIIa (VLA-2) complex mediates the Mg^{++}-dependent adhesion of platelets to collagen, *J. Cell Biol.* **108**:1917–1924.

Stepp, M. A., Spurr-Michaud, S., Tisdale, A., Elwell, J., and Gipson, I. K., 1990, $\alpha_6\beta_4$ integrin heterodimer is a component of hemidesmosomes, *Proc. Natl. Acad. Sci. USA* **87**:8970–8974.

Suzuki, S., and Naitoh, Y., 1990, Amino acid sequence of a novel integrin β_4 subunit and primary expression of the mRNA in epithelial cells, *EMBO J.* **9**:757–763.

Tapley, P., Horwitz, A., Buck, C., Duggan, K., and Rohrschneider, L., 1989, Integrins isolated from Rous sarcoma virus-transformed chicken embryo fibroblasts, *Oncogene* **4**:325–333.

Terranova, V. P., Williams, J. E., Liotta, L. A., and Martin, G. R., 1984, Modulation of metastatic activity of melanoma cells by laminin and fibronectin, *Science* **226**:982–985.

Vaheri, A., Kurkinen, M., Lehto, V.-P., Linder, E., and Timpl, R., 1978, Codistribution of pericellular matrix proteins in cultured fibroblasts and loss in transformation: Fibronectin and procollagen, *Proc. Natl. Acad. Sci. USA* **75**:4944–4948.

Vogel, B. E., Tarone, G., Giancotti, F. G., Gailit, J., and Ruoslahti, E., 1990, A novel fibronectin receptor with an unexpected subunit composition ($\alpha_v\beta_1$), *J. Biol. Chem.* **265**:5934–5937.

Wayner, E. A., and Carter, W. G., 1987, Identification of multiple cell adhesion receptors for collagen and fibronectin in human fibrosarcoma cells possessing unique α and common β subunits, *J. Cell Biol.* **105**:1873–1884.

Wright, S. D., Reddy, P. A., Jong, M. T. C., and Erickson, B. W., 1987, C3bi receptor (complement receptor type 3) recognizes a region of complement protein C3 containing the sequence Arg-Gly-Asp, *Proc. Natl. Acad. Sci. USA* **84**:1965–1968.

Yamada, K. M., Critchley, D. R., Fishman, P. H., and Moss, J., 1983, Exogenous gangliosides enhance the interaction of fibronectin with ganglioside-deficient cells, *Exp. Cell Res.* **143**:295–302.

Chapter 11

The Glomerular Basement Membrane
A Selective Macromolecular Filter

MARILYN GIST FARQUHAR

1. Introduction

The basement membrane of the glomerular capillaries of the mammalian kidney has been a favorite object for studies on the structure, function, and composition of basement membranes because of its important physiological role in glomerular filtration of macromolecules and because it is often involved in renal disease. This chapter will focus on summarizing what is known about the characteristics of the glomerular basement membrane (GBM) and will compare its properties to those of basement membranes from other sources. For additional information the reader can refer to several excellent recent reviews that deal with specific aspects of basement membrane biology (LeBlond and Inoue, 1989; Martin *et al.*, 1988; Paulsson, 1987; Timpl, 1989; Timpl and Dziadek, 1986; Yurchenco and Schittny, 1990).

1.1. What Are Basement Membranes and Where Are They Found?

To begin with, one should define the term *basement membrane*. Basement membranes, also called basal laminae, are continuous sheets of specialized extracellular matrix material composed of collagenous and noncollagenous glycoproteins and proteoglycans that are found wherever cells (other than connective tissue cells) meet connective tissue. This means that basement membranes are found in all of the following locations: at the dermal–epidermal

Abbreviations used in this chapter: GBM, glomerular basement membrane; EHS, Engelbreth–Holm sarcoma; CSPG, chondroitin/dermatan sulfate proteoglycan; HSPG, heparan sulfate proteoglycan; GAG, glycosaminoglycans.

MARILYN GIST FARQUHAR • Division of Cellular and Molecular Medicine, University of California, San Diego, La Jolla, California 92093-0651.

Cell Biology of Extracellular Matrix, Second Edition, edited by Elizabeth D. Hay, Plenum Press, New York, 1991.

junction of the skin; at the base of all lumen-lining epithelia throughout the digestive, respiratory, reproductive, and urinary tracts; underlying endothelia of capillaries and venules; around Schwann cells, adipocytes, skeletal and cardiac muscle cells; and at the base of parenchymatous cells of exocrine (pancreas, salivary glands) and endocrine (pituitary, thyroid, adrenal) glands. In effect, they are the natural substrates on which all cells (except connective tissue cells and blood cells) grow. They closely adjoin and are the products of the overlying cell layers (e.g., endothelium, epithelium, or muscle cells), and they serve to delimit the domain of connective tissue and provide a barrier between it and the domain of non-connective-tissue elements.

In all the locations mentioned, the basement membrane consists of a lamina densa, 20–50 nm in thickness, which runs parallel to the basal cell membranes of the cell layer in question (e.g., epithelium, endothelium) and is separated from the latter by a lighter, approximately 10-nm layer—referred to as the lamina lucida or lamina rara. Thus, the basement membrane usually faces cell membranes on one surface and extracellular matrix components on the other; however, there are a few locations (renal glomerulus, smooth muscle, lung alveoli, neuromuscular junction) where basement membranes face cell layers on both surfaces. The best studied example is the renal glomerulus of mammals where the GBM faces vascular endothelium on one surface and epithelium on the other.

1.2. What Is the Structure and Function of the Kidney Glomerulus?

In order to understand the functional role of the GBM, some knowledge of glomerular organization and physiology is needed. Therefore, anticipating that not all readers will have this background, a brief overview will be given here and a more detailed discussion of glomerular physiology appears in a later section.

The function of the renal glomerulus is to filter the blood, producing a protein-free filtrate of the blood plasma (the glomerular filtrate, the first step in urine formation), while retaining the cellular elements of the blood and the plasma proteins in the circulation. Each glomerulus consists of a tuft of anastomosing capillaries (Fig. 11-1) where filtration takes place. Physiologic studies (Brenner *et al.*, 1978) indicate that the glomerular filter has both size-selective and charge-selective properties, i.e., there is increasing restriction to passage of macromolecules across these capillaries with increasing size and increasing net negative charge of the molecules. To carry out this unique filtration function, the structure of glomerular capillaries is highly specialized. To begin with, they are divided into peripheral regions facing Bowman's capsule (see Fig. 11-1) which constitute the main filtration surface and axial regions buried deeper in the glomerular tuft (Figs. 11-2 and 11-3). The glomerular capillary wall is constructed of three cell types—endothelial, epithelial,* and mesangial cells—and the extracellular basement membrane (Figs. 11-2 and 11-3).

*By convention, the epithelium, because of its direct attachment to the GBM, is considered part of the wall of these specialized capillaries.

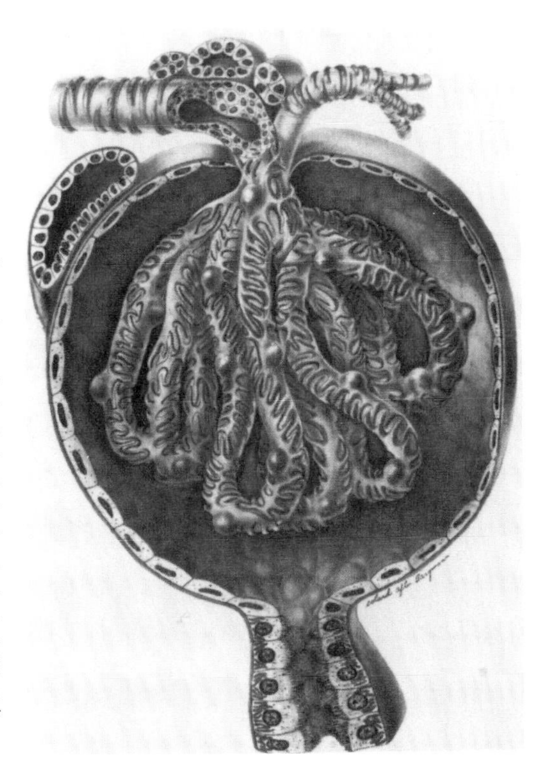

Figure 11-1. Schematic representation of the renal glomerulus, a tuft of capillaries surrounded by a capsule (Bowman's capsule). The latter consists of an inner, squamous epithelial layer surrounded by a basement membrane-like layer of extracellular matrix material. The blood enters the glomerulus through the afferent arteriole (top), which immediately branches into the capillary network. The glomerular filtrate is forced out of the capillaries by the hydrostatic pressure of the blood and collected in the capsular space between the capillaries and Bowman's capsule. From here it is funneled into the renal tubule (below) where readsorption of most of the fluid and salts takes place. (From Bloom and Fawcett, 1975, *A Textbook of Histology*, 10th ed.)

Filtration along peripheral regions is facilitated by the fact that the walls of these capillaries are very thin. Here the wall consists of the GBM, an attenuated endothelial layer found along its luminal surface, and the characteristic hoof-shaped, interdigitating foot processes of the epithelium that lie along its abluminal surface facing the urinary spaces (Figs. 11-4 and 11-5). The wall is streamlined because the endothelial layer is interrupted by porous openings or fenestrae and the epithelium is interrupted by the slits between the foot processes, with the GBM being the only continuous layer. Therefore, on morphological grounds, the GBM represents the logical candidate for the structure that represents the molecular filter. Permeability studies carried out with electron-dense tracers have verified that such is indeed the case (Farquhar, 1975, 1978, 1980; Karnovsky, 1979; Venkatachalam and Rennke, 1978).

The glomerular capillaries are unique in a number of respects:

1. Their basement membranes are thicker (350 nm in humans) and more compact than basement membranes in most other locations. This reflects the fact that they are formed during development by fusion of the endothelial and epithelial basement membranes (Reeves *et al.*, 1980; Sariola *et al.*, 1984).
2. The GBM faces cell layers (the endothelium and epithelium) on both surfaces, whereas, as already mentioned, in most other situations, the

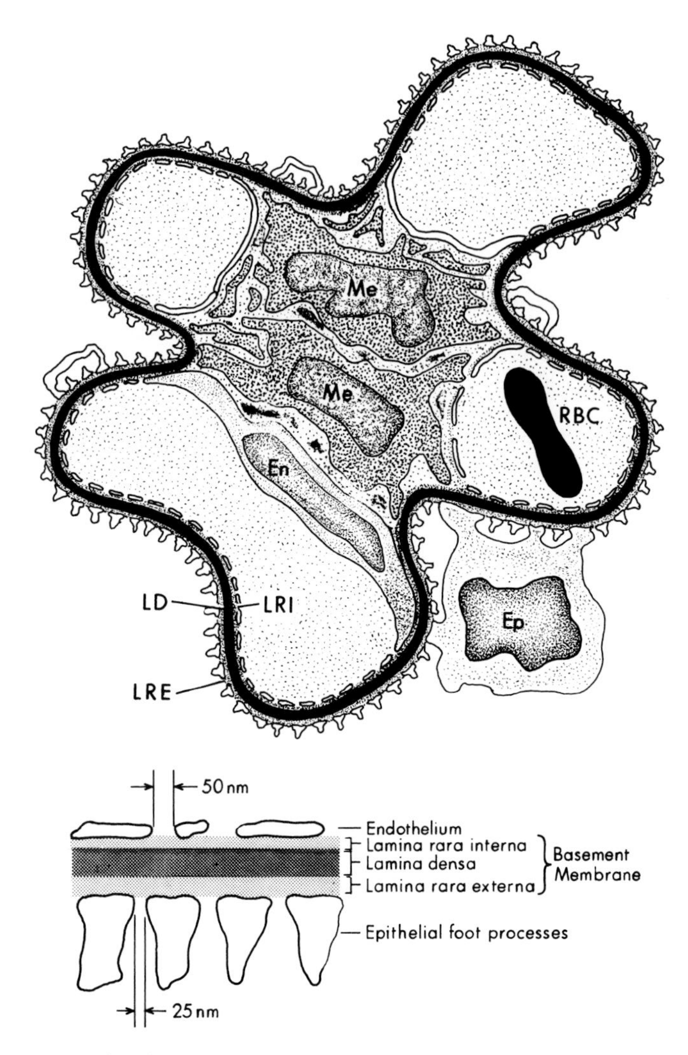

Figure 11-2. Diagram of a glomerular lobule showing the peripheral regions where the walls are very thin and the central or axial regions where the cell bodies of the endothelial (En) and mesangial (Me) cells are concentrated. Normally, filtration occurs mainly along the peripheral regions of these capillaries (enlarged below). The extracellular spaces between adjoining mesangial cells, between endothelial and mesangial cells, and between mesangial cells and the GBM contain a basement membrane-like material known as the mesangial matrix. It has a lower density and is looser in texture than the GBM, and it often contains fibrillar elements. (Redrawn after Latta *et al.*, 1960.)

Figure 11-3. Low-power electron micrograph of a portion of a renal glomerulus from a rat kidney showing portions of two glomerular capillaries (Cap), one of which contains a red blood cell (RBC). In the peripheral regions of these capillaries, the walls are very thin and consist only of an attenuated layer of endothelial cytoplasm interrupted by fenestrae (f), the GBM (B), and the foot processes (fp) of the epithelium (Ep). Part of a mesangial cell (Me) and the cell body of an endothelial cell (En) are present in the axial regions (bottom). Mesangial matrix is found in the spaces

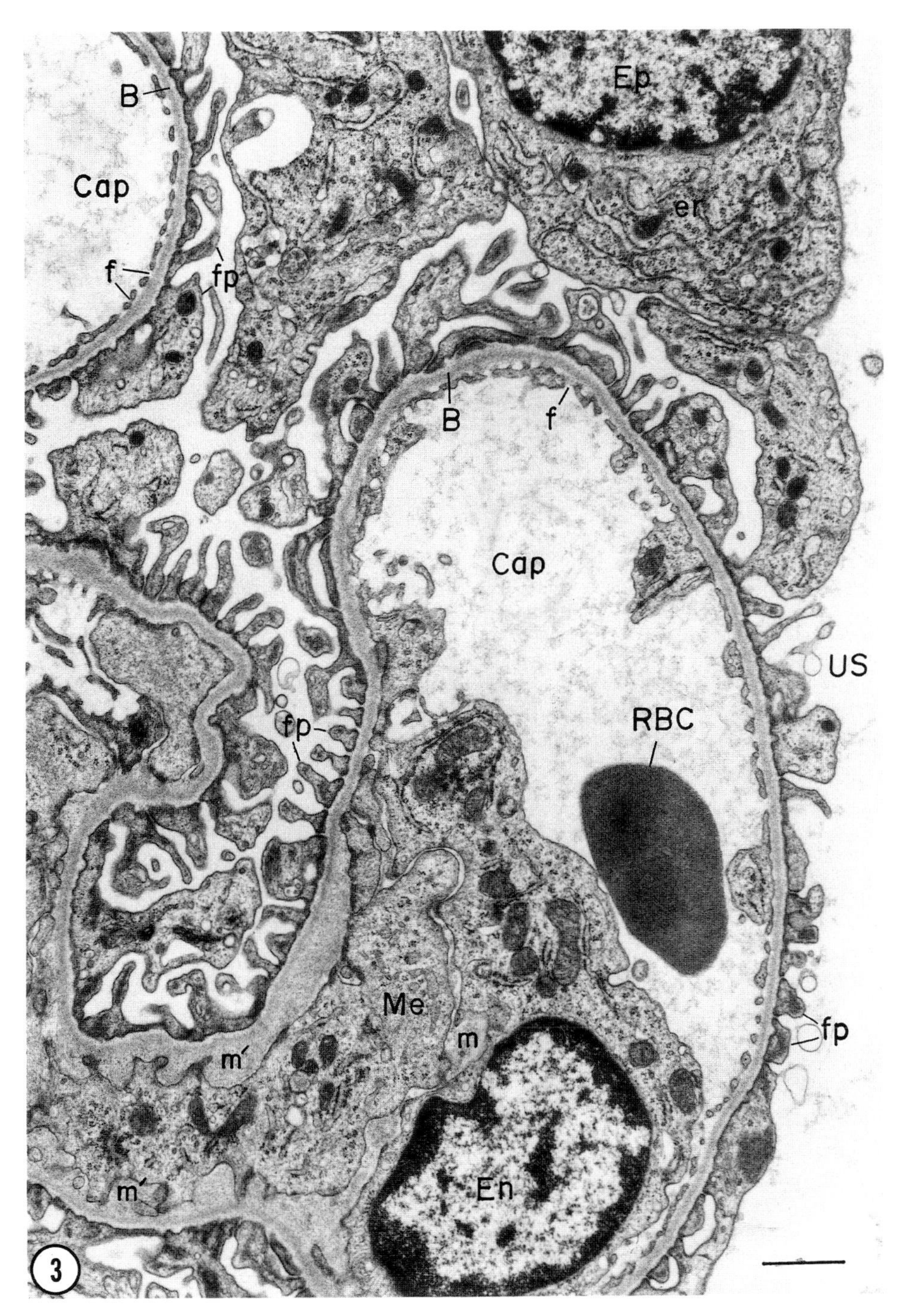

between endothelial and mesangial cells (m) and between the GBM and the pointed processes of the mesangial cells (m′). The cell body of an epithelial cell with abundant rough endoplasmic reticulum (er) is seen above. US, urinary spaces. Bar = 1 μm.

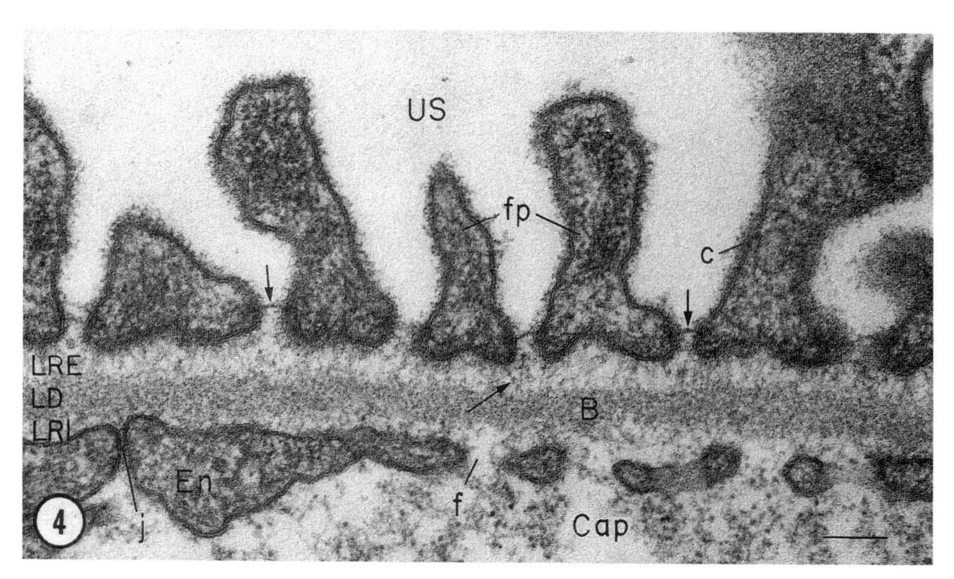

Figure 11-4. Enlargement of a peripheral region of a glomerular capillary (Cap) where filtration takes place. The filtration surface consists of the attenuated endothelium (En) interrupted by fenestrae (f) the GBM (B), and the epithelial foot processes (fp). The latter are separated by ~25-nm spaces known as the filtration slits which are bridged at their base by the so-called filtration slit diaphragms (short arrows). A thick cell coat (c) is visible on the cell membrane of the foot processes above the level of the slit diaphragms. Note that the endothelial fenestrae are open, i.e., they lack the diaphragms that are present in most other fenestrated capillaries (see Fig. 11-40) so that at their level the GBM is directly exposed to the blood plasma. The GBM consists of three layers: a central dense layer, the lamina densa (LD), and two adjoining layers of lower density, the lamina rara interna (LRI) and externa (LRE), adjoining the endothelium and epithelium, respectively. The lamina densa is composed of a fine (~3 nm) filamentous meshwork, and wispy filaments are seen extending from the lamina densa to the membranes of the foot processes (long arrow) and endothelium on either side. The glomerular filtrate passes through the endothelial fenestrae, permeates the GBM, and passes through the filtration slits to reach the urinary spaces (US), which are in continuity with Bowman's space. j, junction between two endothelial cells. Bar = 0.1 μm. (From Farquhar and Kanwar, 1980.)

basement membrane normally faces a cell layer on one side and extracellular matrix on the other.

3. The endothelial fenestrae are larger (50–100 nm) than those of other fenestrated capillaries and they are open, i.e., they lack the usual diaphragms that are invariably present in other fenestrated capillaries (see Fig. 11-40). As a consequence of this arrangement, the GBM is directly exposed to the blood plasma on its endothelial surface. This is a highly efficient arrangement to facilitate the filtration function, but it has the disadvantage that it renders the GBM vulnerable to toxic and immune injury (see below).
4. The epithelial cells (or podocytes) with their elaborate interdigitating foot processes which form the filtration slits are unique to these vessels. The integrity of the foot processes and slit arrangement is dependent on

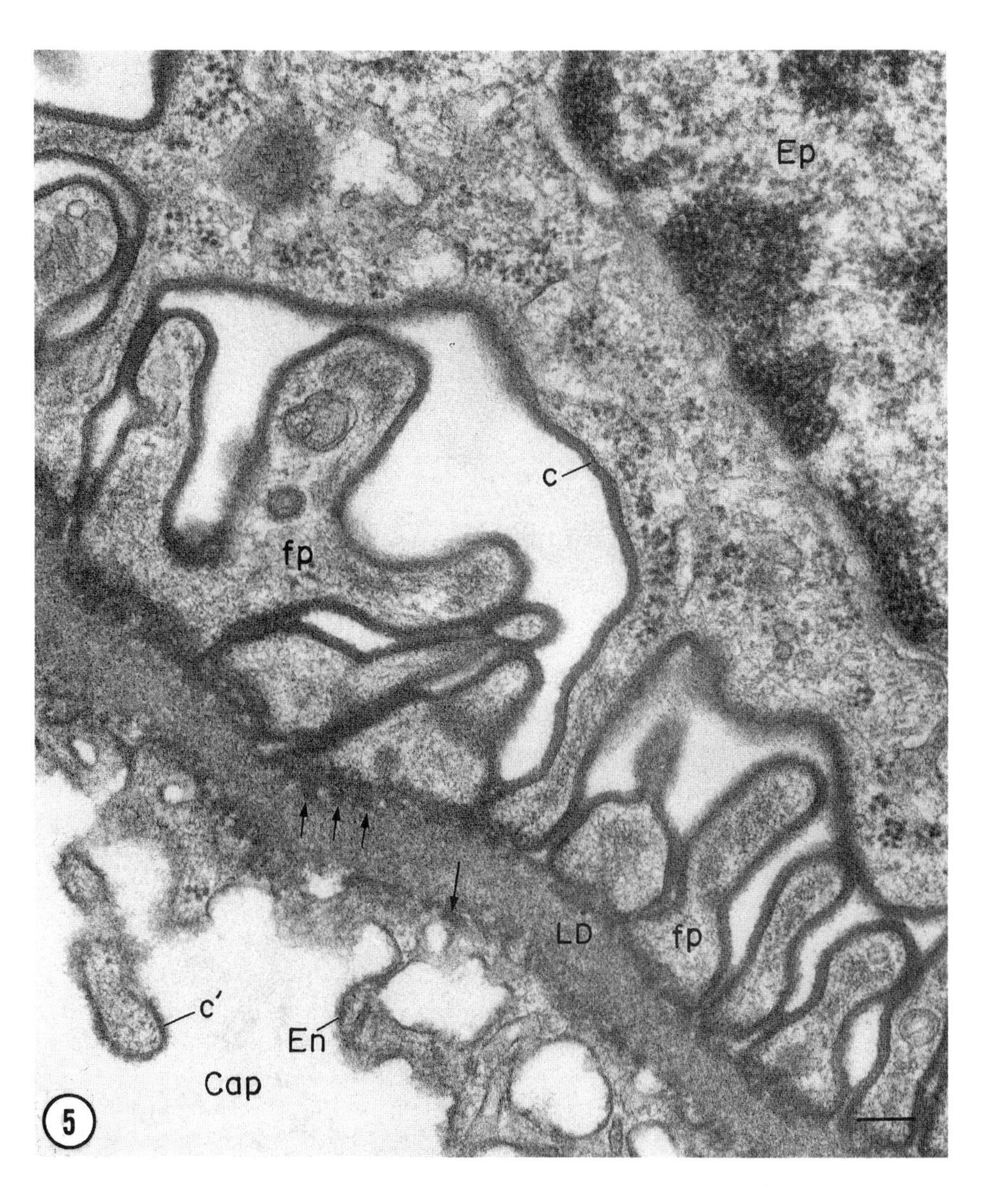

Figure 11-5. Field from the glomerulus of a normal rat kidney perfused with lysozyme. Lysozyme, which is a small cationic protein (pI 11.0), binds to negatively charged groups and acts as a stain for anionic groups. Here it binds in a thick layer to the conspicuous sialic acid-rich, polyanionic cell surface coat (c) on the epithelial cell body (Ep) and its foot processes (fp). A similar but thinner cell coat is seen on the endothelium (c′). Lysozyme also binds to the anionic sites in the laminae rarae interna (↓) and externa (↑) of the GBM, thereby rendering them more dense than the lamina densa (LD). Bar = 0.1 μm. (From Caulfield and Farquhar, 1978.)

the presence of an unusually thick, sialic acid-rich, cell surface coat or glycocalyx (Fig. 11-4), which can be stained with cationic stains (Fig. 11-5). It is now known to consist predominantly of oligosaccharide moieties of podocalyxin (Kerjaschki *et al.*, 1984) which are both sialylated (Kerjaschki *et al.*, 1985) and sulfated (Dekan *et al.*, 1991).

5. The mesangial cells, likewise, are specialized cellular elements. They resemble in some respects (e.g., presence of contractile proteins and contractability) the pericytes (modified smooth muscle cells) in other capillaries, but differ in being located on the luminal (rather than on the adventitial) side of the GBM and are concentrated in the axial regions.
6. The spaces between adjoining mesangial cells, between mesangial cells and endothelial cells, and between mesangial cells and the GBM are filled with a special extracellular matrix material known as the mesangial matrix (Figs. 11-2 and 11-3), which differs in morphology and composition from the GBM.

2. Organization of the GBM and Mesangial Matrix

The extracellular matrix of the glomerulus consists of the GBM which extends around the entire capillary wall and the mesangial matrix which is concentrated in the axial or mesangial regions. Our main attention in this chapter will be focused on the GBM because of its primary role in glomerular filtration; however, information on the structure and composition of the mesangial matrix will be compared and contrasted to the GBM wherever possible, as the mesangial matrix is often altered in disease states and thus is of considerable importance.

The GBM is organized into three distinct layers: a dense middle layer, the lamina densa, and two adjoining layers of lower electron density, known as the lamina rara interna and externa,* which are located between the lamina densa and the membranes of the adjoining endothelium and epithelium, respectively (Fig. 11-4). The lamina densa is composed of a tightly woven meshwork of fine (3 nm) filaments. The laminae rarae, as their name implies, have a lighter background density, but they are by no means empty spaces or retraction artifacts as sometimes claimed.

Several distinct types of fibrous elements can be recognized in the GBM (Farquhar, 1978): (1) the 3-nm filaments that comprise the lamina densa (Fig. 11-4); (2) large fibrils (10-nm diameter) located in the lamina rara interna (Fig. 11-6) and in the mesangial matrix; and (3) filaments of variable size seen in both of the laminae rarae extending between and apparently connecting the lamina densa and the adjoining cell membranes (Fig. 11-4). When sections are stained with cationic stains, such as ruthenium red, a network of proteoglycan granules (Kanwar and Farquhar, 1979a,b) can be detected in the laminae rarae (Figs. 11-7 to 11-9). They are arranged in a quasi-regular, latticelike array (Figs. 11-7 and 11-9), resembling the corneal basal lamina (Trelstad *et al.*, 1974).

*In the glomerulus, the lamina rara interna faces endothelium, whereas in other areas the lamina rara "interna" occupies the side of the basement membrane facing ECM.

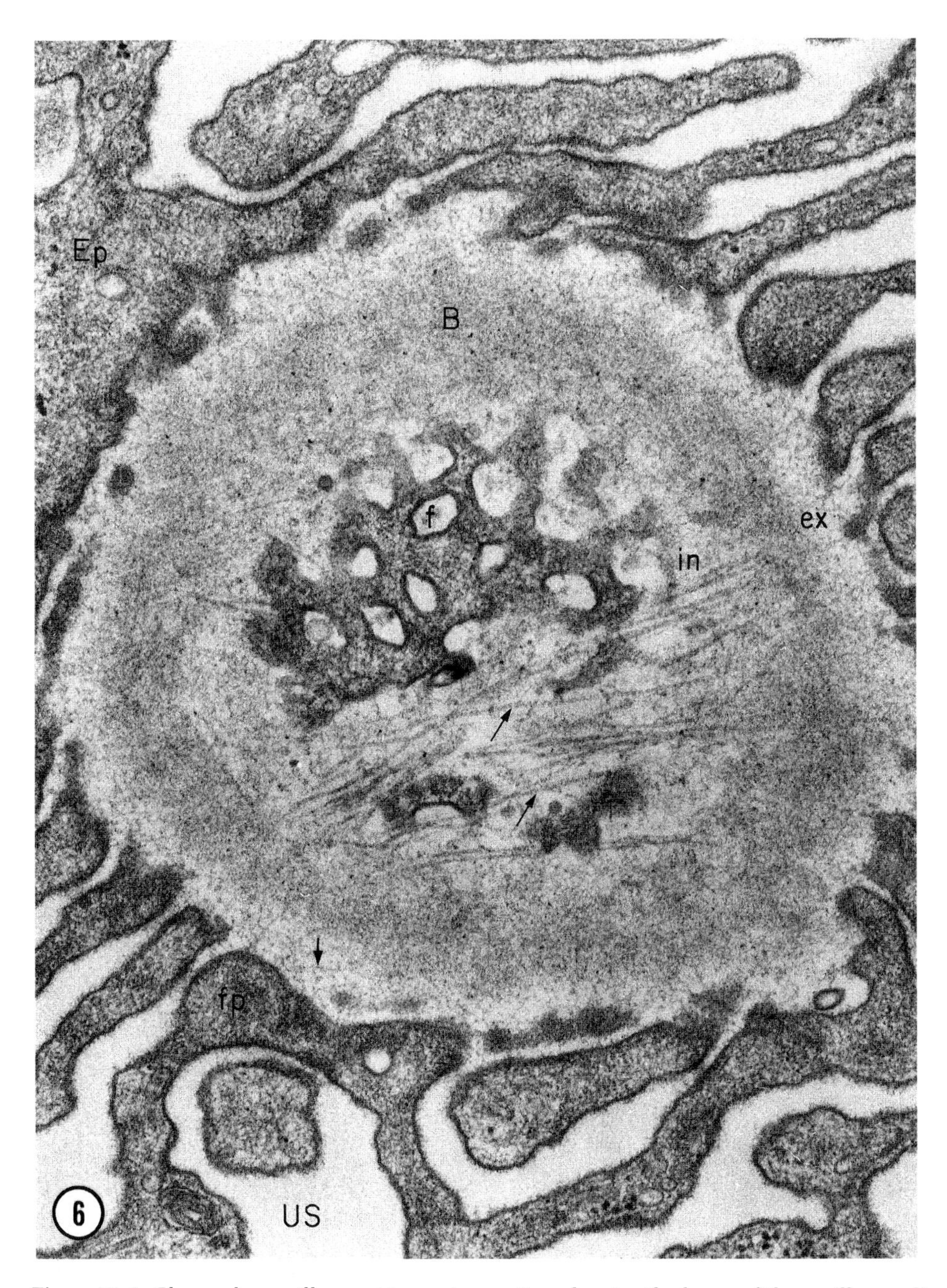

Figure 11-6. Glomerular capillary cut in grazing section, showing the layers of the capillary wall partially en face. The endothelial fenestrae (f) appear as irregularly circular, open portholes, and the three layers of the GBM—the lamina densa (B), the lamina rara interna (in), and the lamina rara externa (ex)—are cut broadly. Two types of fibrils are visible in the GBM: (1) a fine (~3 nm) fibrillar meshwork, which makes up the lamina densa and extends across the lamina rara externa to the epithelial cell membrane (short arrow), and (2) large (10 nm) double-walled tubular fibrils (long arrows), which are located between the endothelium and the GBM. ×50,400. (From Farquhar and Kanwar, 1982.)

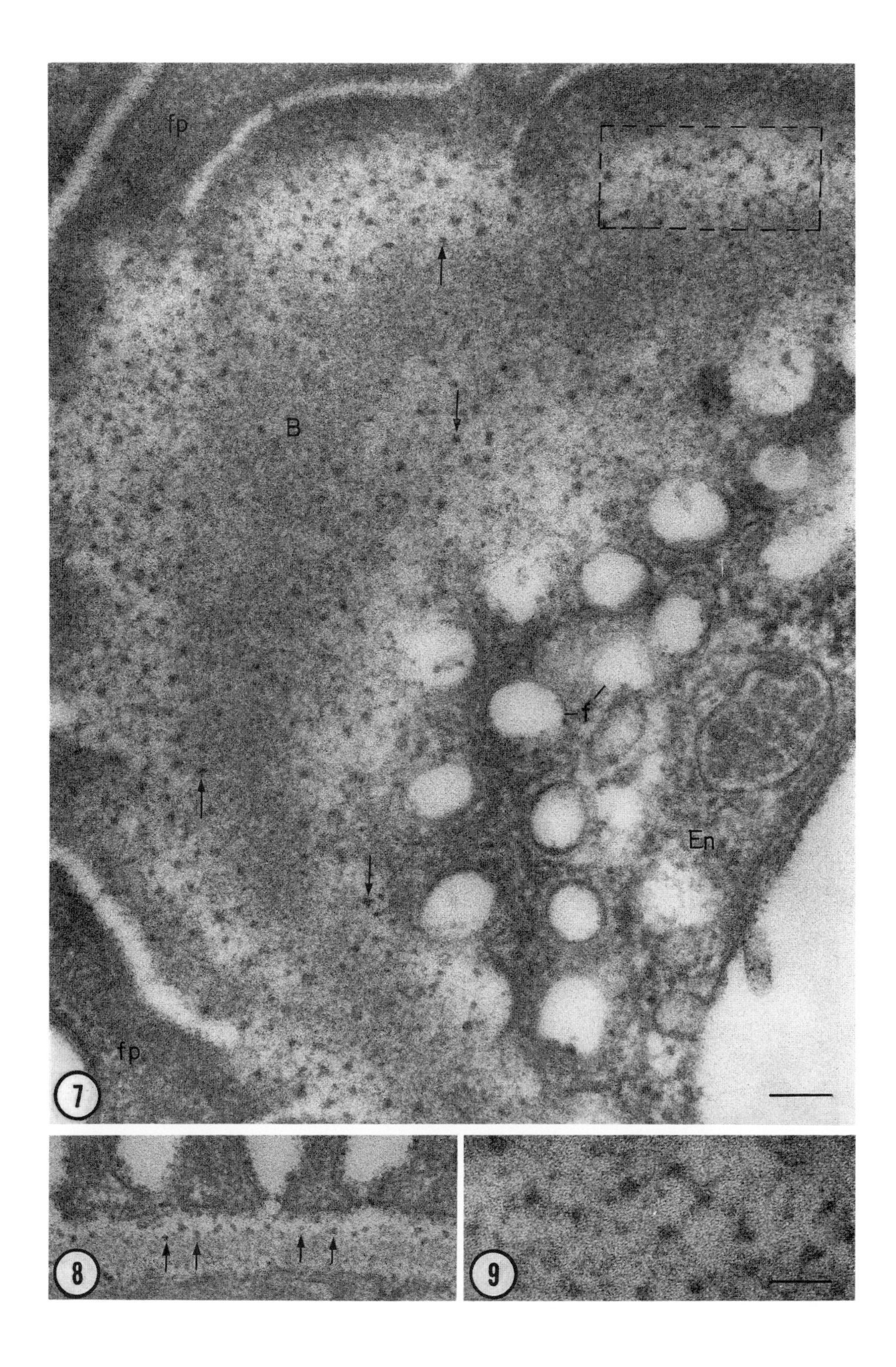
fp
B
f
En
fp
7
8
9

Another class of fine (2-nm diameter) filaments can be seen connecting the proteoglycan granules to each other, to the lamina densa, and to the adjoining cell layers (Fig. 11-7). Ideally, one would like to identify by immunocytochemical staining the chemical nature of each of the classes of microfibrils found in the GBM; however, to date this goal has not been achieved.

3. Glomerular Development

To obtain an understanding of glomerular organization and the origin of its basement membrane, it is helpful to know the essential steps in glomerular development. The kidney develops from two sources—the metanephric blastema which gives rise to the glomerulus and the proximal tubule, and the ureteric bud which gives rise to the distal and collecting segments of the tubule. The terminal ampulla of the ureteric bud induces formation of a renal vesicle, the earliest recognizable stage in glomerular differentiation. A cleft forms in the vesicle and it is invaded by mesenchyme, forming an S-shaped body (see Fig. 11-28) from which the glomerular and proximal tubule epithelia develop. Both begin as typical polarized epithelia with apical occluding junctions and basal basement membranes. Podocalyxin is present very early on the apical domain of the presumptive glomerular epithelium for which it represents a differentiation marker (Schnabel *et al.*, 1989). The glomerular endothelium and mesangium develop from the mesenchyme in the cleft of the S-shaped body to form capillary loops during which period the basement membrane of the presumptive glomerular epithelium comes to lie adjacent to the basement membrane of the presumptive glomerular endothelium. Initially the two are separated by a layer of extracellular matrix containing CSPG (Fig. 11-10), but gradually this matrix disappears, the two basement membranes fuse (Figs. 11-11 and 11-12), and the endothelial fenestrae open (Reeves *et al.*, 1980). At the same time, the apical epithelial junctions which are authentic tight junctions (Schnabel *et al.*, 1990) migrate as a wave down the lateral cell membrane toward the basal cell surface, podocalyxin extends laterally down to the tight junction (Schnabel *et al.*, 1989), basal interdigitation and elimination of the tight junction begins creating the elaborate foot process arrangement. With further maturation, additional basement membrane is added to the GBM by a splicing-type event (Abrahamson, 1985). Initially the GBM is permeable to anionic ferritin, but it becomes increasingly compact and impermeable to the

←

Figures 11-7 to 11-9. Portions of glomerular capillaries stained with ruthenium red to demonstrate anionic sites in the GBM. A network of polygonal, HSPG granules can be seen in the laminae rarae interna (↓) and externa (↑) of the GBM, which is cut in grazing section in Fig. 11-7 and in cross section in Fig. 11-8. Figure 11-9 is an enlargement of part of the lamina rara externa in Fig. 11-7 showing the quasi-regular, latticelike arrangement of the particles, which are located at regular (60 nm) intervals. Fine (2 nm) filaments connect the particles to one another and to the cell membranes of the epithelium and endothelium (En). Bars = 0.1 μm. (From Kanwar and Farquhar, 1979a.)

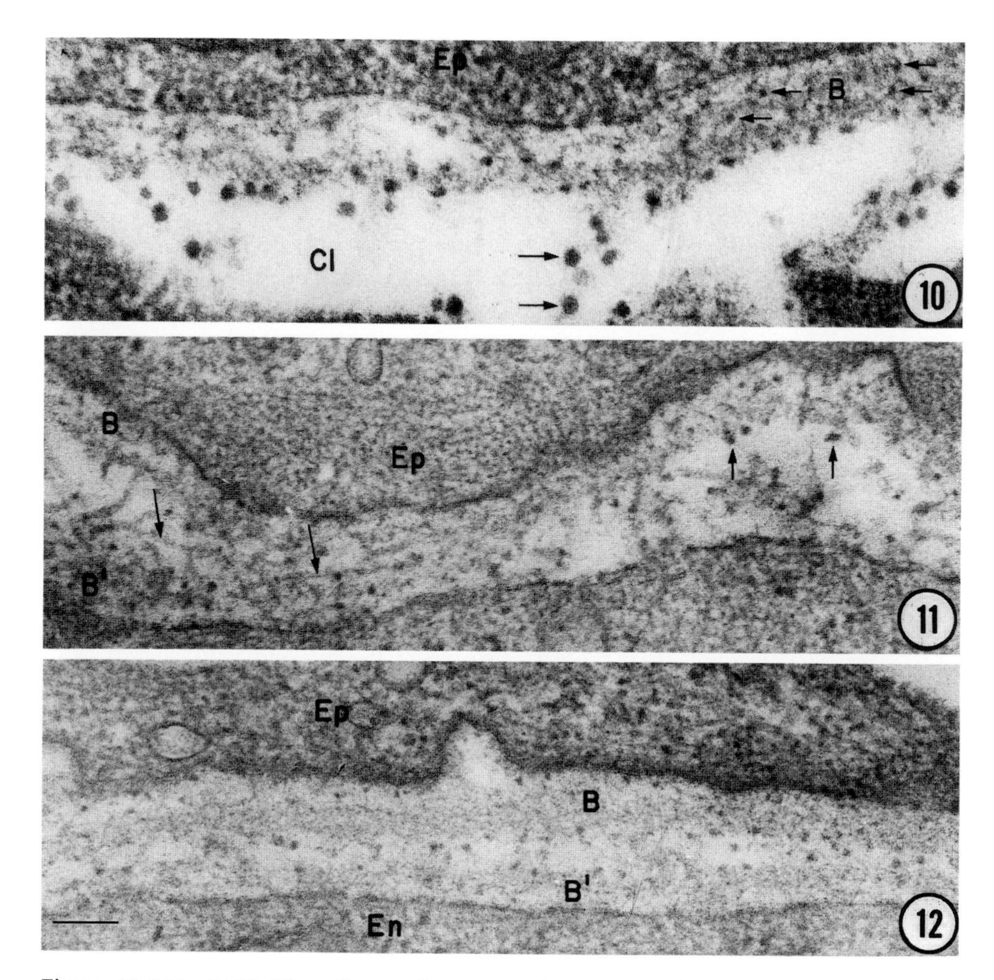

Figures 11-10 to 11-12. These figures show successive stages in the maturation of the GBM in preparations stained with ruthenium red. Figure 11-10, from the S-shaped body stage, illustrates the early epithelial basement membrane which is associated with mesenchyme in the cleft. No endothelium is present at this stage. Two types of ruthenium red-stained proteoglycan particles are seen: large (30 nm) CSPG particles associated with the mesenchyme located within the cleft (Cl) (long arrows) and smaller (10–15 nm) HSPG particles (short arrows) located within the epithelial basement membrane (B). Figure 11-11, from the early capillary loop stage, shows beginning fusion between the epithelial basement membrane (above) and the endothelial basement membrane (below). Vestiges of the large CSPG particles (short arrows) interconnected by thicker (10 nm) strands (long arrows) are seen in the cleft. These thick fibrils form a meshwork within the spaces destined to become the lamina densa. Smaller (10–15 nm) HSPG particles are also seen within the epithelial (B) and endothelial (B′) basement membranes. Figure 11-12, from the middle of the capillary loop stage, shows the epithelial (B) and endothelial (B′) basement membranes coming closer together, separated by only a thin layer of residual mesenchyme. Small (10–15 nm) HSPG particles are seen throughout both basement membranes at this stage. Bar = 0.1 μm. (From Reeves *et al.*, 1980.) (Reproduced from the *Journal of Cell Biology*, 1980, **85**:735–753, by copyright permission of the Rockefeller University Press.)

tracer as development progresses (Reeves *et al.*, 1980). The opening of the endothelial fenestrae and the opening of the epithelial slits serves to streamline the surface and facilitate the filtration process.

4. Composition of the GBM

4.1. Background Information

The inventory of GBM components, like that of other basement membranes (Table I), is now known to include type IV collagen, laminin, HSPG, nidogen or entactin, and BM40 (also known as osteonectin and SPARC). In fact, it is now known that these proteins are ubiquitous components of basement membranes (see reviews by Martin *et al.*, 1988, and Timpl *et al.*, 1979).

Information on the composition of the GBM has come from three main approaches: preparation and biochemical analysis of purified GBM fractions, immunocytochemistry carried out at the light or electron microscopic level, and cytochemical studies employing selective extraction *in situ*. The collective findings from all of these approaches are summarized below.

Until about 10 years ago, the bulk of the available information on the composition of the GBM was derived from biochemical studies on purified GBM fractions, studies in which Kefalides (Kefalides, 1973, 1978, 1979) and Spiro (Spiro, 1972, 1978) and their collaborators pioneered. This approach was made possible by the introduction in the early 1950s of a technique for the preparation of GBM fractions (Krakower and Greenspon, 1951) which involves isolation (by collection on a graded series of sieves) of glomeruli free from contamination by other kidney structures, especially the kidney tubules with their basement membranes and associated interstitial collagen and other matrix elements. Purified GBM fractions were then prepared therefrom by sonication (with or without TCA treatment), a procedure that undoubtedly resulted in

Table I. Intrinsic Basement Membrane Components[a]

Component	Chain structure	Major function
Collagen IV	α1(IV), α2(IV)	Structural support
Laminin	B1, B2, A	Cell attachment
Heparan sulfate proteoglycans		
Low density	(M_r, 500–650k); 3–4 heparan sulfate chains	Filtration
High density	(M_r, 130–250k); 4 heparan sulfate chains	Filtration
Chondroitin sulfate proteoglycans	3–4 chondroitin sulfate chains	
Nidogen/entactin	Single chain (M_r, 150k)	Laminin binding
BM-40/osteonectin/SPARC	Single chain (M_r, 35k)	Calcium binding

[a]Modified from Martin *et al.* (1988).

removing much of the laminae rarae and the mesangial matrix and thereby in concentrating the lamina densa. Since 15% of the glomeruli retain their capsules during the glomerular isolation procedure, the main contaminant of such fractions is the extracellular matrix layer of Bowman's capsule. Later on, a method was introduced to prepare GBM by detergent treatment (Meezan *et al.*, 1978) that yields intact, unbroken loops of GBM (see Figs. 11-15 to 11-17) with the mesangial matrix and laminae rarae apparently intact.

Biochemical characterization of GBM components has been rendered difficult by the fact that the GBM is extensively cross-linked: by disulfide bonds, by collagen-type cross-links (hydroxylysine), and by other unknown interactions, so that it is quite insoluble. Therefore, harsh procedures (such as pepsin digestion, or treatment with SDS after reduction and alkylation) must be used to solubilize it, and a wide variety of heterogeneous fragments result. An additional complication is that proteolysis occurs both *in vivo* and *in vitro*, a factor that further increases the heterogeneity of the fragments.

The majority of the biochemical studies on the GBM have concentrated on an analysis of its collagenous components and proteoglycans. Characterization of the glycoproteins present in the GBM has been very difficult due to its aforementioned highly cross-linked nature which has hindered isolation of intact glycoproteins from the GBM. To date, documentation of the presence of specific glycoproteins in the GBM has come largely from immunocytochemical studies in which cross-reactivity of the GBM to glycoproteins isolated from other sources (primarily the EHS sarcoma, a basement membrane-secreting tumor) has been demonstrated (see below).

4.2. Heparan Sulfate Proteoglycans

It is now clear that basement membranes contain as ubiquitous components a distinctive class of HSPG (Martin *et al.*, 1988; Timpl, 1989). The presence of proteoglycans in the GBM was first detected in the laminae rarae by using cationic stains or probe molecules, i.e., lysozyme (Fig. 11-5), ruthenium red (Figs. 11-7 to 11-12), cationized ferritin (Figs. 11-13 and 11-14), that bind at low pH specifically to sulfate groups in GAG. These probes demonstrated the presence of a network of particles or anionic sites in the laminae rarae of the GBM. The identity of the anionic sites as composed of heparan sulfate was initially established by cytochemical studies (Kanwar and Farquhar, 1979b) in which specific enzymes were perfused into the kidney or used to digest isolated GBM (Fig. 11-15 to 11-18). The anionic sites were no longer demonstrable after digestion with heparitinase (which specifically degrades heparan sulfate and heparin but not other GAG) but were unaffected by digestion with neuraminidase, chondroitnase ABC, or hyaluronidases.

The presence of HSPG was subsequently verified by extraction and biochemical analysis of the GAG moieties (Kanwar and Farquhar, 1979c; Parthasarathy and Spiro, 1981) and, later on, of the intact proteoglycans (Kanwar *et al.*, 1981; Parthasarathy and Spiro, 1984). Proteoglycans isolated from rat

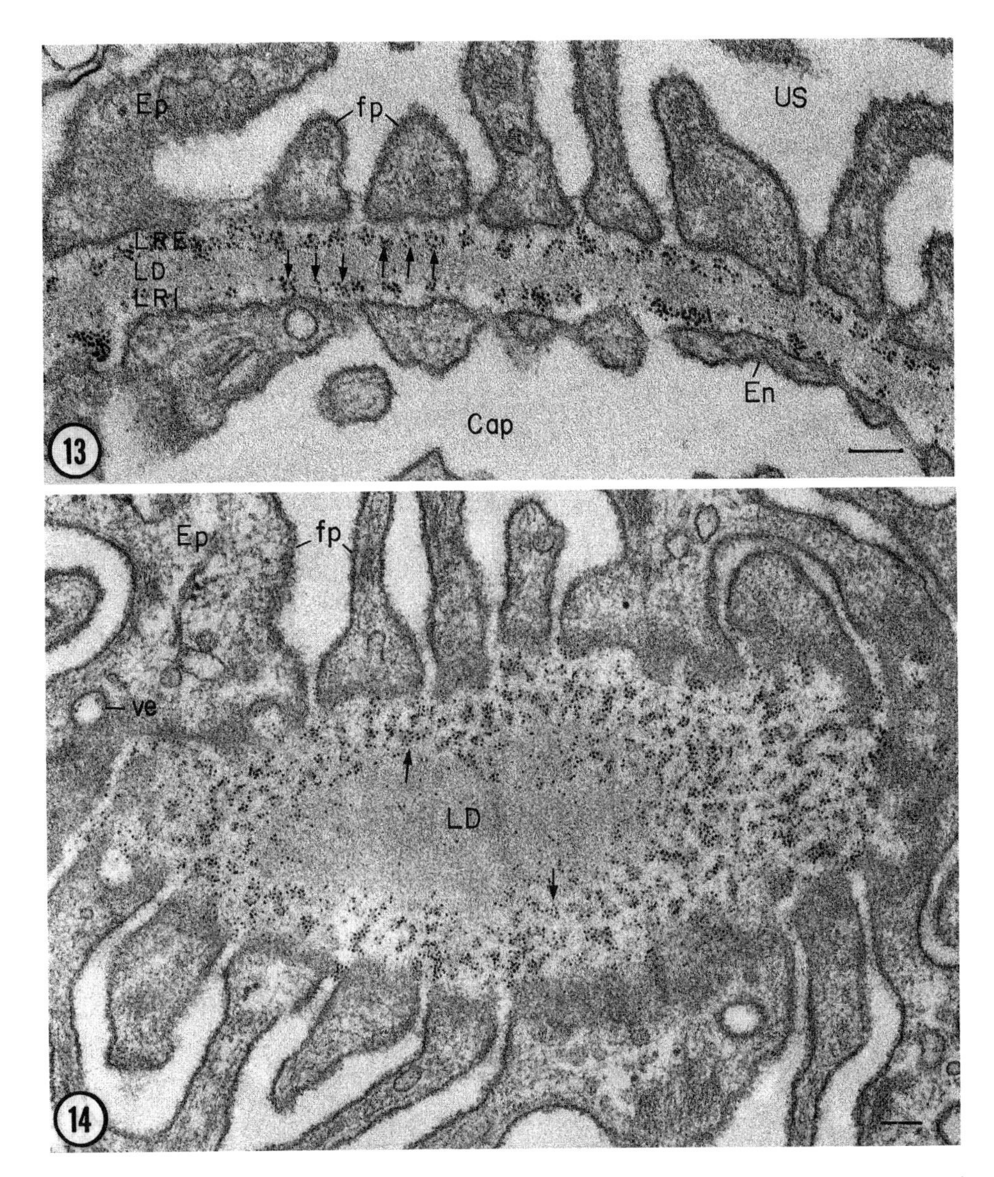

Figures 11-13 and 11-14. Portions of glomeruli of a rat given cationized ferritin (CF) (pI 7.3–7.5) *in vivo* by intravenous injection, to label anionic sites in the GBM. CF binds to anionic sites in the laminae rarae interna (LRI) and externa (LRE) of the GBM. Figure 11-13 shows that CF is distributed in discrete clusters located at regular ~60-nm intervals (arrows). Figure 11-14, which is from a grazing section of the same glomerulus as that in Fig. 11-13, illustrates the reticular pattern of CF binding in the lamina rara externa (arrows). Note that CF with this neutral pI binds only to the GAG moieties of the HSPG in the GBM and that there is no binding to the sialoglycoproteins on the endothelial (En) or epithelial (Ep) cell surfaces. Bars = 0.1 μm. (From Kanwar and Farquhar, 1979a.) (Reproduced from the *Journal of Cell Biology*, 1979, **81**:137–153, by copyright permission of the Rockefeller University Press.)

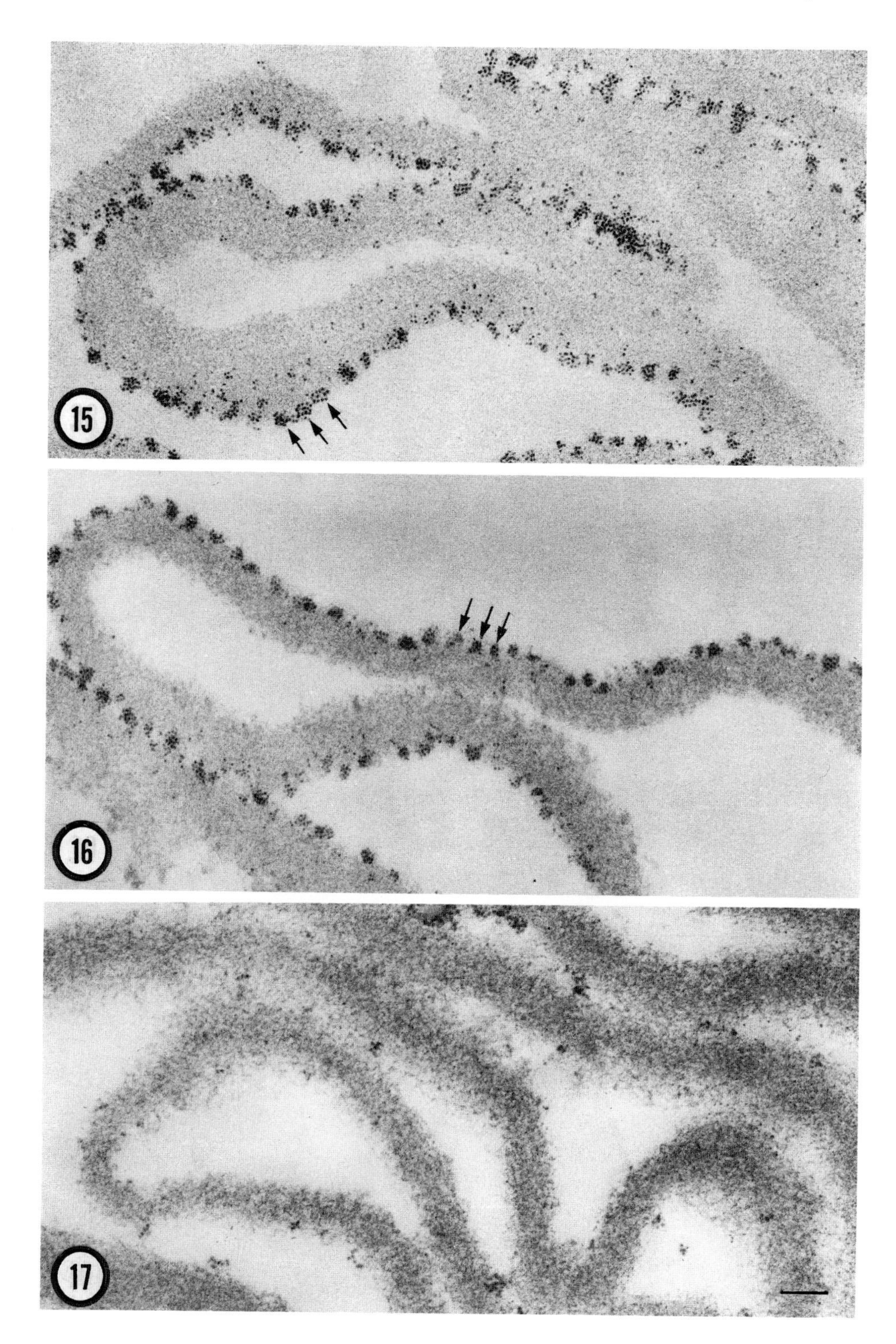
15
16
17

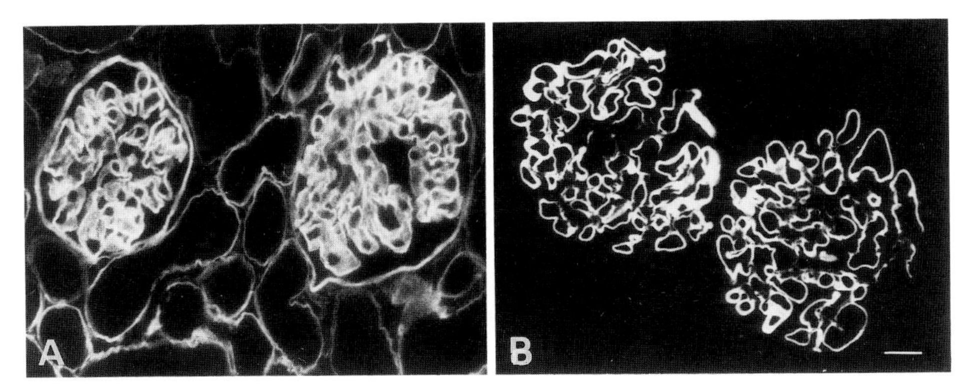

Figure 11-18. Fluorescence micrographs of semithin (0.5 μm) frozen sections of perfusion-fixed rat kidneys demonstrating the distribution of HSPG. (A) Section from a normal rat kidney stained by indirect immunofluorescence with anti-HSPG (GBM), an antibody that recognizes the core protein of basement membrane HSPG from rat kidney. The antibody stains all renal basement membranes, i.e., the GBM, Bowman's capsule, the basement membranes of renal tubules, and those of peritubular capillaries. (B) Direct immunofluorescence staining of the kidney of a rat at 2 hr after intravenous injection of anti-HSPG (GBM) IgG. The antibody binds only to the HSPG in the GBM and not to HSPG in other basement membranes, as only the GBM is accessible to circulating antibodies. (From Miettinen *et al.*, 1986).

GBM consisted predominantly (85%) of HSPG. Smaller amounts of CSPG as well as hyaluronic acid were also detected (Lemkin and Farquhar, 1981). Approximately 45% of the total GBM proteoglycans cannot be extracted by the usual procedures used for proteoglycans (involving use of 4 M guanidine chloride), as they are covalently bound to other GBM constituents. Comparative analyses revealed no differences in the GAG chains released from the extractable and nonextractable proteoglycans (Kanwar *et al.*, 1981).

Considerable diversity in the size of the core proteins of the HSPG isolated from basement membranes from different tissues and cultured cells has been reported. The HSPG from the EHS sarcoma matrix have been most thoroughly studied. They consist of two (high and low density) HSPG, 650 and 130 kDa, with core proteins of 450 and 21–34 kDa, respectively (Martin *et al.*, 1988). The large HSPG is believed to represent the precursor which is converted into the small HSPG by proteolytic processing (Ledbetter *et al.*, 1985). Antibodies raised against these two populations of HSPG cross-react with each other and with HSPG in the GBM and in all other basement membranes.

Figures 11-15 to 11-17. Loops of isolated GBM subjected to treatment with specific enzymes followed by incubation with cationized ferritin (CF) to label the GBM anionic sites. CF binding (arrows) is not affected by treatment with chondroitinase ABC (Fig. 11-15) or neuraminidase (Fig. 11-16), but is abolished by heparitinase (Fig. 11-17). CF molecules bind only to the outer or exposed side of the isolated GBM loops because the latter consist of intact, closed tubes, and the tracer does not have access to the inner or unexposed side of the GBM. Bar = 0.1 μm. (From Kanwar and Farquhar, 1979b.)

The overall size reported for the core protein of the HSPG from the GBM has varied considerably from 18 kDa (Kanwar *et al.*, 1984) up to 400 kDa (Klein *et al.*, 1988), probably due in part to proteolysis, to differences in antibody specificities, and to differences in deglycosylation procedures. Most recently, the core protein size of HSPG from the bovine (Parthasarathy and Spiro, 1984), rat (Pietromonaco and Farquhar, 1988), and human (Van Den Heuvel *et al.*, 1989) GBM has been shown to be ~130–160 kDa after complete chemical deglycosylation with trifluoromethanesulfonic acid (TFMS). This is considerably smaller than the core protein of the large, low-density HSPG from the EHS sarcoma and considerably larger than the small, high-density proteoglycan from the same source. The differences in size of the core proteins from different basement membrane sources coupled with the heterogeneity in staining of the same basement membrane with different antibodies (Couchman, 1987) suggest the existence of a family of basement membrane HSPG of related but nonidentical structure.

When antibodies raised against rat glomerular HSPG were used to localize the HSPG in the kidney, all kidney basement membranes (Fig. 11-18A) as well as basement membranes of other organs reacted with the antibody by immunofluorescence. When the HSPG was localized by immunoelectron microscopy within the GBM, they were found primarily in the lamina rara interna and externa regardless of whether the antigen was localized after *in vivo* binding (Miettinen *et al.*, 1986; Stow *et al.*, 1985) or after immunogold labeling of ultrathin cryosections (Kerjaschki *et al.*, 1986; Stow *et al.*, 1985) (see Figs. 11-19 and 11-45A). If, however, antibodies raised against the EHS sarcoma proteoglycan were used for labeling, all three layers of the GBM–lamina densa, lamina rara interna, and lamina rara externa—were stained by both immunoperoxidase (Laurie *et al.*, 1984; Mynderse *et al.*, 1983) and immunogold (Grant and LeBlond, 1988) methods. The basis for these differences in staining with different antibodies is not understood, but it has been proposed (Klein *et al.*, 1988) that the glomerular HSPG is synthesized as a larger molecular weight precursor which is subsequently proteolytically processed to a smaller form. If so, the immunocytochemical findings suggest that the processed form is concentrated in the laminae rarae.

The HSPG from the EHS sarcoma has been partially cloned and sequenced and found to consist of laminin-, EGF-, and N-CAM-like domains (Noonan *et al.*, 1988) (see Fig. 11-20). Partial clones for the rat (Pietromonaco and Farquhar, 1988) and human (Kallunki and Tryggvason, 1990) kidney basement membrane HSPG have also been sequenced and shown to contain LDL receptor-like repeating domains. It is not clear at present whether the size differences of basement membrane HSPG derived from different tissues reflect the fact that these HSPG represent separate gene products, arise from differential splicing of mRNA from the same gene, or are formed by proteolytic cleavage of a low-density proteoglycan (Martin *et al.*, 1988). The answer will become evident only when the complete amino acid sequence of basement membrane HSPG from a variety of sources is available.

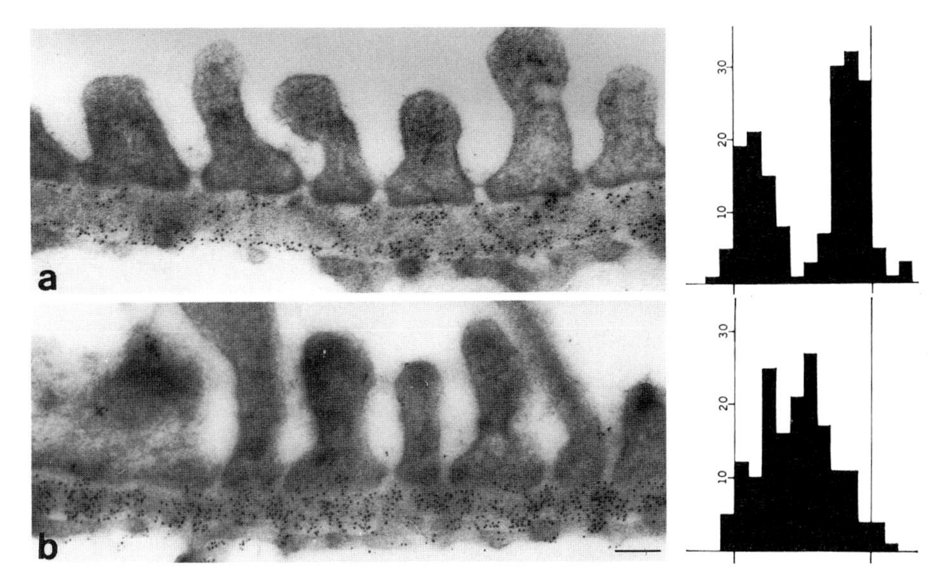

Figure 11-19. Immunogold labeling of the GBM in ultrathin frozen sections. In a, the section was incubated with anti-HSPG (GBM), an antibody that recognizes the core protein of basement membrane HSPG. Gold label is concentrated in both laminae rarae. In b, from a section incubated with antilaminin antibodies, gold particles are found in all layers of the GBM but are most concentrated in the lamina densa. Pieces of rat kidney were aldehyde-fixed by perfusion, frozen in liquid N_2, and ultrathin sections were cut on a Reichert Ultracut Ultramicrotome equipped with a cryo-attachment. The sections were mounted on electron microscope grids and incubated with a specific antibody followed by goat anti-rabbit IgG coupled to 10-nm colloidal gold. The number of gold particles in the layers of the GBM are expressed in the histograms accompanying each of the immunolocalizations. The right and left vertical lines on the histogram represent the positions of the endothelial and epithelial cell membranes, respectively, and the numbers indicate the number of gold particles. Bar = 0.2 μm. (From Kerjaschki *et al.*, 1986.)

4.3. Type IV Collagen

It has been known for some time that the GBM contains type IV collagen or basement membrane collagen (see Chapter 1) which was originally purified from the GBM and lens capsule (Kefalides, 1978). Collagen IV differs from other collagens in that it does not form fibers, it is secreted and assembles as a procollagen-like molecule, and it contains a unique highly cross-linked 7 S domain. It is known to self-assemble into stabile latticelike three-dimensional networks (Timpl, 1989; Yurchenco and Schittny, 1990) through interactions at its C- and N-terminal noncollagenous domains and by lateral associations (see Fig. 11-43). It functions to provide structural stability to basement membranes.

The GBM as well as the mesangial matrix were shown some time ago to react with antibodies prepared against type IV collagen (Roll *et al.*, 1980; Scheinman *et al.*, 1980a), (see Figs. 11-23 and 11-24). Originally, only two type

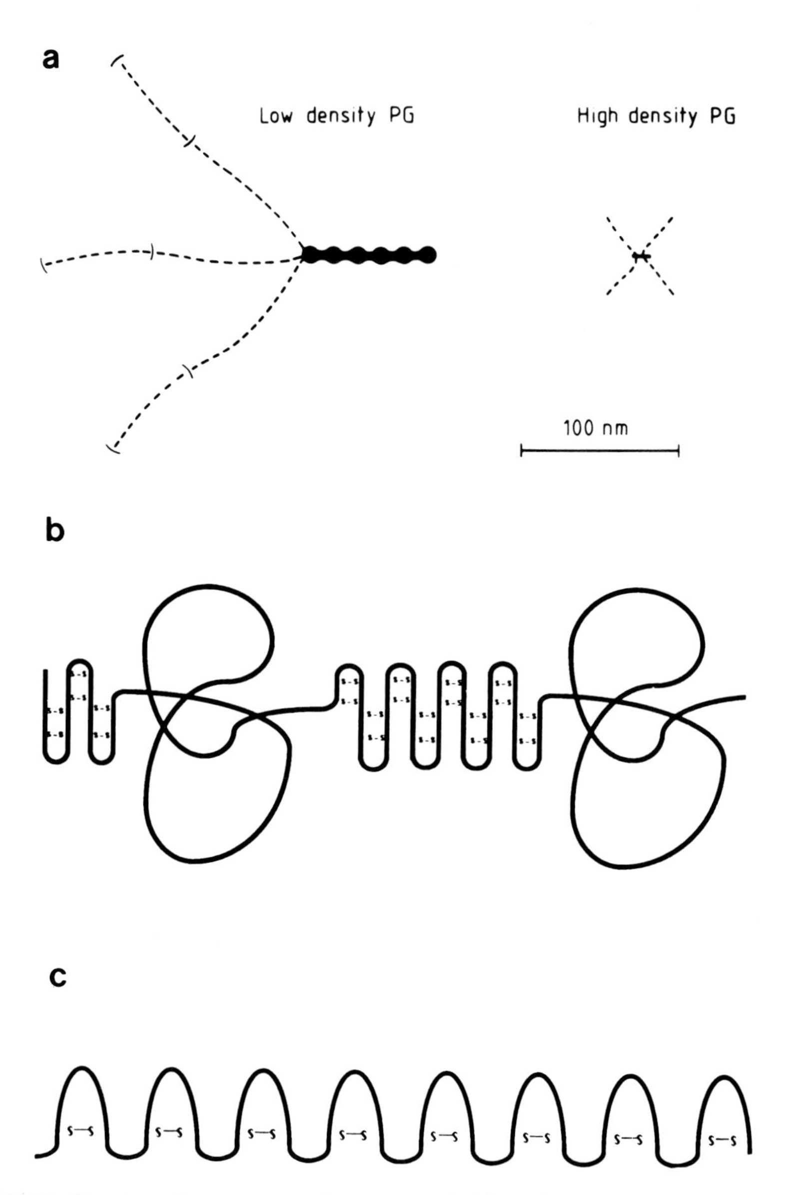

Figure 11-20. Structure of basement membrane-type HSPG from the EHS sarcoma matrix. (a) Shapes of high- and low-density HSPG. Protein cores are shown in black, and heparan sulfate chains are indicated by the dotted lines. (b, c) Protein domains present in the low-density HSPG predicted from the protein sequence. They include EGF repeats alternating with globular domains (b), and repeats of disulfide-bonded loops similar to N-CAM (c). (From Noonan *et al.*, 1988, and Timpl, 1989.)

IV collagen chains, α1(IV) and α2(IV), were identified. More recently, two new chains, α3(IV) and α4(IV), have been purified form kidney basement membranes (Butkowski *et al.*, 1989). These two new chains are distinguished by structural differences in a 27-kDa so-called "M region" of their NC1 domains, with M1a and M1b present in the NC1 region of the α1(IV) and α2(IV) chains, and M2 and M3 in the α3(IV) and α4(IV) chains. It has been demonstrated that the antigen involved in an anti-GBM glomerulonephritis known as Goodpasture's syndrome, resides in the NC1 region of the α3(IV) chain (Hudson *et al.*, 1989).

Immunoelectron microscopic studies using antibodies raised against the entire molecule (see Fig. 11-27) have localized type IV collagen to the lamina densa of the GBM (Courtoy *et al.*, 1982; Desjardins *et al.*, 1990; Grant *et al.*, 1989; Roll *et al.*, 1980).

When antibodies raised against the globular NC1 domain were used for immunolabeling, heterogeneity in the distribution of the M fragments was observed in the glomerulus: Labeling for M1 was most intense over the mesangial matrix whereas labeling for M2 and M3 were higher in the GBM. Within the GBM itself there were also differences: M1 was preferentially concentrated in the subendothelial regions of the GBM, whereas M2 and M3 were found throughout the entire thickness of the GBM. This heterogeneity indicates that there are subtle differences in the relative composition of collagen chains within the GBM and the mesangial matrix and within the GBM itself. This is not surprising in view of the dual (perhaps triple) cellular origin of the GBM and mesangial matrix. These differences probably reflect subtle differences in the structure and function of these matrices.

4.4. Chondroitin Sulfate Proteoglycans

As mentioned above, biochemical analysis of GBM fractions (which also contain the mesangial matrix) demonstrated the presence of chondroitin sulfate (Klein *et al.*, 1986; Lemkin and Farquhar, 1981) and CSPG (Kanwar *et al.*, 1981) which were similar in overall size to that of the HSPG. It could not be determined whether or not the chondroitin sulfate/dermatan sulfate and heparan sulfate GAG were on the same or different core proteins. Enzyme digestion studies suggested that the chondroitin sulfate GAG was present only in the mesangial matrix and not in the GBM (Kanwar *et al.*, 1983). Recently, an immunologically distinct CSPG (originally isolated from Reichert's membrane) has been identified that appears to be a widespread basement membrane component (McCarthy *et al.*, 1989). Using monoclonal antibodies directed against its core protein, it was possible to demonstrate the presence of this CSPG in basement membranes of nearly all rat tissues, including the kidney. Interestingly, the GBM proved to be one of the major exceptions. However this CSPG was detected in Bowman's capsule, in tubular and capillary basement membranes, and in the mesangial matrix (McCarthy and Couchman, 1990). The latter is in keeping with the earlier enzyme digestion studies. These findings

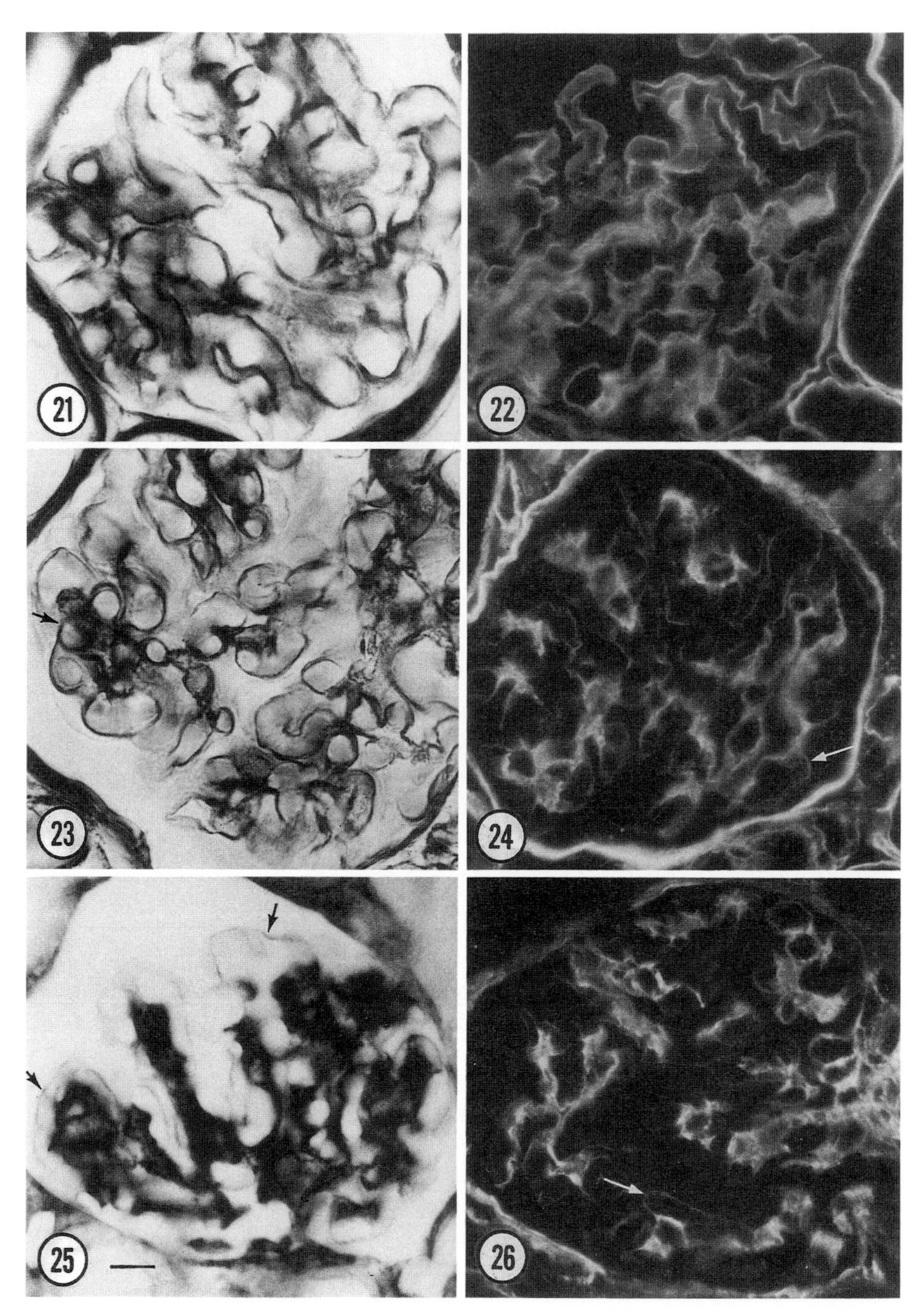

Figures 11-21 to 11-26. Distribution of laminin, type IV collagen, and fibronectin detected by immunofluorescence in unfixed cryostat sections (Figs. 11-22, 11-24, and 11-26) or by immunoperoxidase in sections of aldehyde-fixed kidneys (Figs. 11-21, 11-23, and 11-25). Staining for all three proteins is seen in Bowman's capsule, in the GBM, and in the mesangial matrix. The relative indicate that (1) the glomerulus contains a distinctive population of CSPG, and (2) in addition to differences in collagen IV chains, the mesangial matrix and the GBM differ in their proteoglycan composition.

4.5. Laminin

Laminin is a large (800–900 kDa) noncollagenous sialoglycoprotein composed of one α and two β subunits that was originally purified by Timpl and co-workers (Timpl *et al.*, 1979) from the EHS sarcoma matrix (see Chapter 4). It has been demonstrated by immunocytochemical staining in virtually every basement membrane examined, including the GBM. Laminin is a multidomain protein which has a multiplicity of binding sites—i.e., for collagen, heparin, and cells, and which functions in cell attachment (especially of epithelial cells), neurite outgrowth, and heparin binding.

Laminin was localized to the GBM, mesangium, and Bowman's capsule by immunofluorescence in the original studies of Timpl and co-workers (Rohde *et al.*, 1979) and by Scheinman *et al.* (Scheinman *et al.*, 1980b) (see Figs. 11-21 and 11-22). It has been localized repeatedly at the ultrastructural level within the GBM and mesangial matrix with variable results. In some of these studies, it was noted that the distribution of laminin within the basement membrane layers was not uniform; it was concentrated in the laminae rarae and absent or present at low concentration in the lamina densa (see review by Kerjaschki *et al.*, 1985). However, in a number of other studies, laminin has been detected throughout the GBM (Abrahamson, 1986; Kerjaschki *et al.*, 1985; Laurie *et al.*, 1984), and there is now general agreement that laminin is located throughout all layers of the GBM. The explanation for the different results obtained by different workers probably lies in the different methods and antibodies used. In one recent study, it was clearly shown that different results are obtained with monoclonal antibodies that recognize different regions of the laminin molecule (Abrahamson *et al.*, 1989). Therefore, the simplest explanation for these discrepant localization results is that the antibodies used recognize different epitopes which vary in their accessibility to antibody binding (see Fig. 11-28).

Laminin binds to itself, to HSPG, and to collagen IV at distinctive sites on the molecule, and it also binds to cell surfaces. In the presence of divalent cations, the purified protein can self-assemble *in vitro* into polymeric complexes (see Fig. 11-43). It has been suggested that the polymeric structure might serve to form a mesh in basement membranes thereby contributing to its sieve properties (Yurchenco and Schittny, 1990).

---

distribution of the staining for the three proteins varies: staining for laminin is more intense in the GBM at the periphery (Figs. 11-21 and 11-22); staining for collagen type IV is prominent in both the mesangium and the GBM (arrows, Figs. 11-23 and 11-24); staining for fibronectin is more intense in the mesangium and only faint staining is seen in the GBM (Figs. 11-25 and 11-26). Bar = 10 μm. (From Kerjaschki *et al.*, 1986.) (Reproduced from the *Journal of Cell Biology*, 1980, **87**:691–696, by copyright permission of the Rockefeller University Press.)

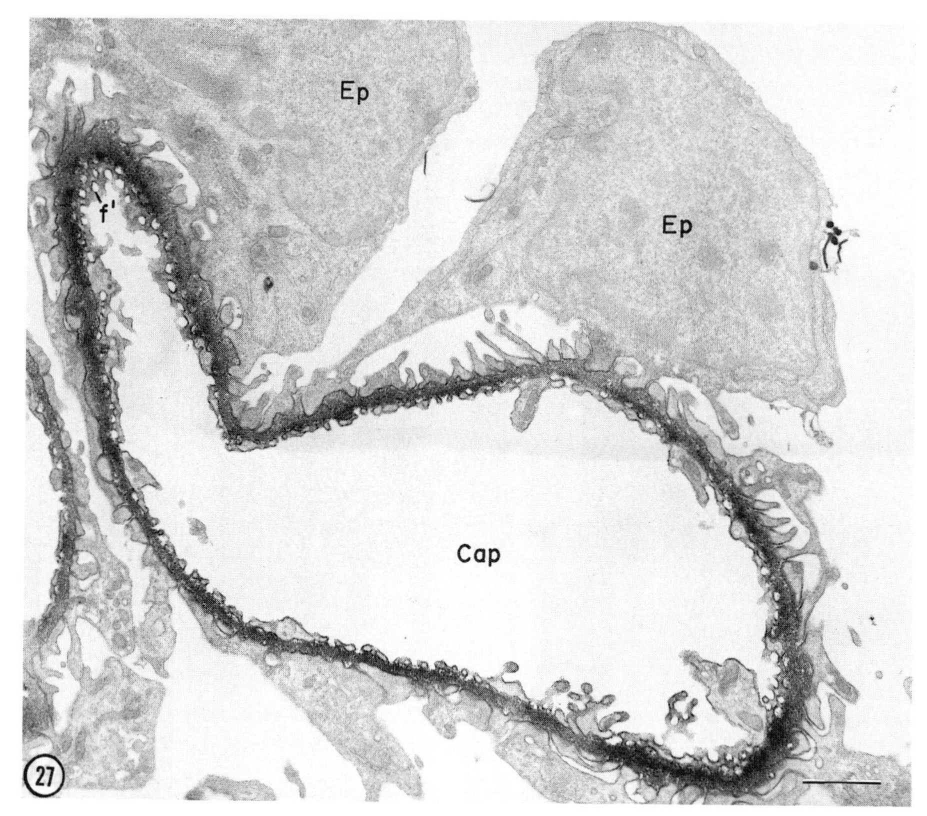

Figure 11-27. Immunoperoxidase localization of type IV collagen in the glomerulus. Peroxidase reaction product is seen throughout all layers of the GBM but is most concentrated in the lamina densa. This figure is from an ~20-μm cryostat section taken from a kidney that was lightly fixed by aldehyde perfusion and incubated with affinity-purified, anti-collagen IV antibodies (prepared by R. Timpl from antiserum of rabbits immunized with laminin purified from the EHS sarcoma matrix) followed by incubation in Fab fragments of sheep anti-rabbit IgG conjugated to peroxidase. Bar = 0.01 μm. (From Courtoy *et al.*, 1982.)

4.6. Fibronectin

Fibronectin is a widespread matrix glycoprotein that is sometimes found associated with basement membranes, especially embryonic basement membranes (see Chapter 4), but is not detectable in all basement membranes.

In the case of the glomerulus, there is general agreement that fibronectin is present at high concentration in the mesangial matrix (Figs. 11-25, 11-26, 11-29, and 11-30); however, there is disagreement concerning whether it is present in the GBM itself (Courtoy *et al.*, 1982; Oberley *et al.*, 1979; Pettersson and Colvin, 1978; Scheinman *et al.*, 1980b) and whether the fibronectin present is derived from the blood or produced locally. The fact that fibronectin can be detected in the Golgi complex of mesangial cells (Fig. 11-30) suggests that it is

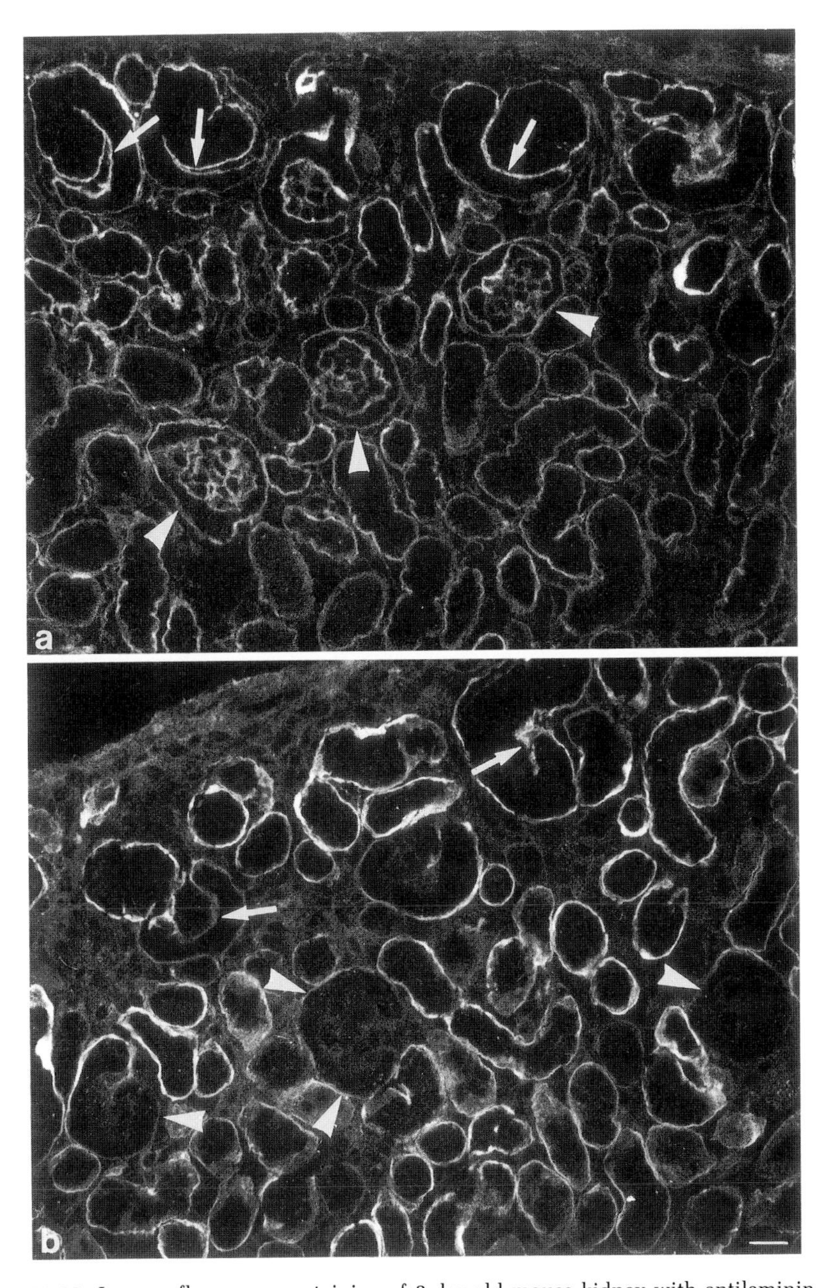

Figure 11-28. Immunofluorescence staining of 2-day-old mouse kidney with antilaminin monoclonal antibodies showing that the two different patterns are seen with two different monoclonal (MAB) antibodies. (a) MAB 5D3 (which recognizes an epitope located near the end of the long arm of laminin) labels basement membranes of the tubules and GBM in all stages of development including S-shaped (arrows) and early capillary loop stages (arrowheads). (b) With MAB 8B3 (which recognizes an epitope near the center of the laminin cross), only the tubular basement membranes and those in vascular clefts of S-shaped nephrons (arrows) are labeled, whereas GBMs of more advanced stage nephrons (arrowheads) are negative. These results suggest that there is a temporal change in the presence of some laminin epitopes during the assembly of the GBM. Bar = 25 μm. (From Abrahamson *et al.*, 1990.)

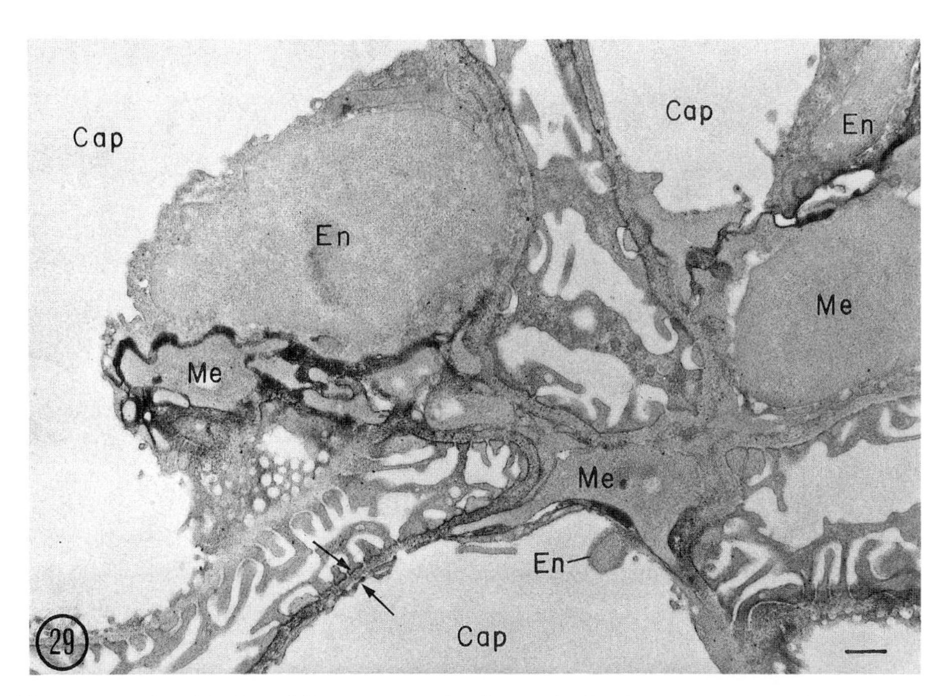

Figure 11-29. Localization of fibronectin in the glomerulus by the immunoperoxidase technique *in situ*. Peroxidase reaction product is concentrated in the mesangial matrix between the endothelial (En) and the mesangial (Me) cells. It is also present in the laminae rarae in some places (arrows). Specimen processed as in Fig. 11-27 except that it was incubated with affinity-purified antibodies to fibronectin (prepared by R. Hynes from monospecific antisera of rabbits immunized against fibronectin isolated from NIL8 hamster cells). Bar = 0.5 μm. (From Courtoy *et al.*, 1980.)

synthesized by these cells (Courtoy *et al.*, 1982). Within the mesangial matrix, fibronectin is especially concentrated between mesangial cells and endothelial cells (Fig. 11-29), suggesting that it may play a role in the attachment of these cells to one another or to the mesangial matrix (Courtoy *et al.*, 1980).

4.7. Entactin/Nidogen

Entactin is a 150-kDa, sulfated glycoprotein originally isolated from the matrix laid down by PYS-2 cells, a tertatocarcinoma-derived cell line (Carlin *et al.*, 1981). It appears to be identical or closely related to nidogen which was isolated from the EHS sarcoma matrix (Martin *et al.*, 1988; Timpl, 1989). This protein has been demonstrated by immunohistochemistry in virtually all basement membranes, including the GBM (Laurie *et al.*, 1984) where it has been detected in all three layers. The main property of this protein is that it forms strong complexes with laminin (to the extent that specific antibodies to either

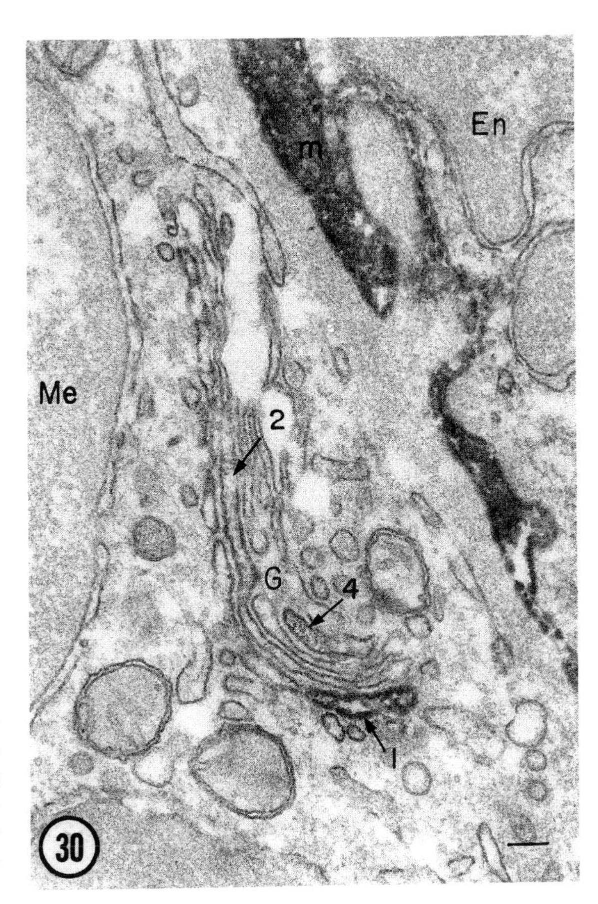

Figure 11-30. Immunoperoxidase localization of fibronectin as in Fig. 11-29. Fibronectin can be detected intracellularly within several of the stacked Golgi (G) cisternae (1, 2, and 4) of a mesangial cell (Me), as well as extracellularly in the mesangial matrix (m) at the mesangial–endothelial (En) interface. The presence of fibronectin within a biosynthetic compartment suggests that this protein is produced locally by mesangial cells. Bar = 0.1 μm. (From Courtoy *et al.*, 1980.)

laminin or nidogen precipitate both proteins), and it may function to modulate the interactions of laminin (Yurchenco and Schittny, 1990).

4.8. Other Components

Apart from collagen type IV, proteoglycans, laminin, entactin, and fibronectin, attempts to characterize other components of the GBM have made little headway. A number of other proteins have been reported to show cross-reactivity (by immunofluorescence) to the GBM. The list of such proteins that have been purified (to a variable extent), used for the production of specific antibodies, and shown to immunostain the GBM includes the microfibrillar protein of elastic tissue (Kelley *et al.*, 1977; Risteli *et al.*, 1980) and type V collagen (Roll *et al.*, 1980). There is also a long history of immunohistochemical localization of glomerular antigens, especially nephritogenic antigens, carried out on poorly or incompletely characterized antigens or antibodies, consideration of which is beyond the scope of this chapter.

4.9. Cell Surface Receptors for Basement Membrane Components in Glomeruli

In the glomerulus, the endothelium and epithelium are firmly attached to the GBM, and to disturb that attachment requires the removal of the cells (by detergent treatment) or removal of the GBM by enzyme digestion. In analogy to information on other systems, it is logical to assume that the attachment is effected through specific cell surface receptors which bind and interact with

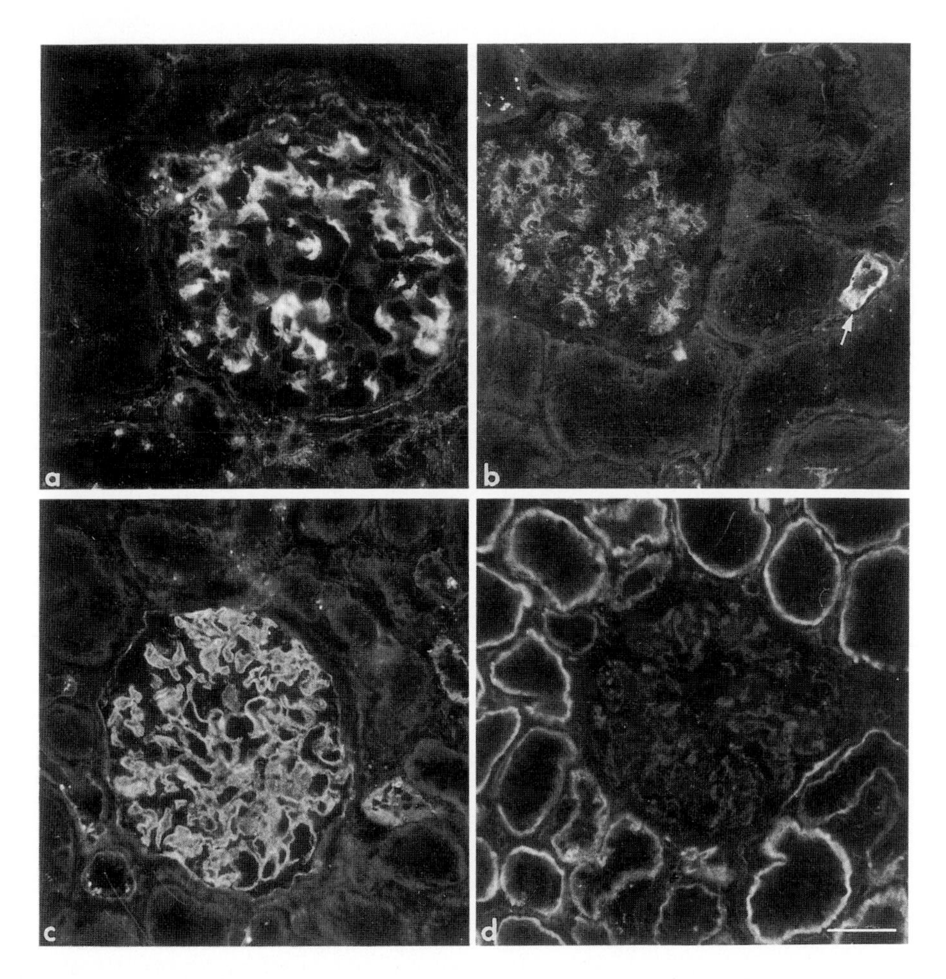

Figure 11.31. Immunofluorescence localization of α-integrin receptors in mature glomeruli of the developing human kidney. α1 is expressed in mesangial regions (a), α2 in glomerular and peritubular (arrow) capillary endothelia (b), α3 at the base of the glomerular epithelium (c), and α6 in all tubules (d). Bar = 50 μm. (From Korhonen *et al.*, 1990.)

individual components of the GBM and mesangial matrix. It has become clear that binding of cells to extracellular matrix components is receptor-mediated, and that extracellular matrix components interact with, and exert their effects on cells through binding of these fixed, multivalent ligands to specific cell surface receptors in a manner comparable to other types of receptor–ligand interaction. By now, receptors for most of the major basement membrane components have been described. Many of these receptors belong to the integrin family which bind to RGD sequences in matrix proteins. To date, at least two integrins which bind to collagen IV, five to fibronectin, and four to laminin have been described (see Chapter 10) and more are being discovered every day. In addition, several nonintegrin matrix receptors have also been described—e.g., a laminin receptor that recognizes a unique YIGSR sequence (Graf *et al.*, 1987) and 38-kDa protein that recognizes the core protein of basement membrane HSPG (Clément *et al.*, 1989).

Characterization of extracellular matrix receptors in the kidney has been rather limited to date. In one study, the presence of a β-1 integrin receptor for fibronectin was demonstrated by immunoelectron microscopy in human glomeruli where it was localized on those surfaces of endothelial, mesangial, and epithelial cells facing the GBM of mesangial matrix (Kerjaschki *et al.*, 1989). The distribution of α1–6 subunits has recently been surveyed by immunofluorescence in the developing human glomerulus (Korhonen *et al.*, 1990). It was found that the α1 subunit was characteristically expressed in mesangial regions, α2 in the glomerular endothelium, α3 in the epithelial cells, and α6 in all tubules (see Fig. 11-31). These results suggest that distinct integrins play different roles during nephron development, with the α1β1 integrin complexes functioning as the main basement membrane receptor for mesangial cells, the α2β1 receptor for endothelial cells, and the α3β1 receptor for podocytes. However, it can be anticipated that multiple receptors may be expressed on each glomerular cell type, as cultured mammalian cell lines express from two to ten different integrins alone (see Chapter 10).

5. What Do We Know about the Function of the GBM as a Filter?

5.1. Clearance Studies Demonstrate Glomerular Size and Charge Selectivity

As mentioned at the outset, the glomerulus serves to filter the blood plasma. Clearance studies done by physiologists in the 1940s and 1950s using various probe molecules (proteins and dextrans) established that the glomerular capillary wall as a whole behaves like a sieve in that it has size-selective properties. That is, the passage of macromolecules is increasingly restricted as their molecular mass and effective diameter increase up to the size of albumin (70 kDa). Albumin is the major plasma protein that must be retained in the blood, and normally only traces are filtered. Based on such studies, it was

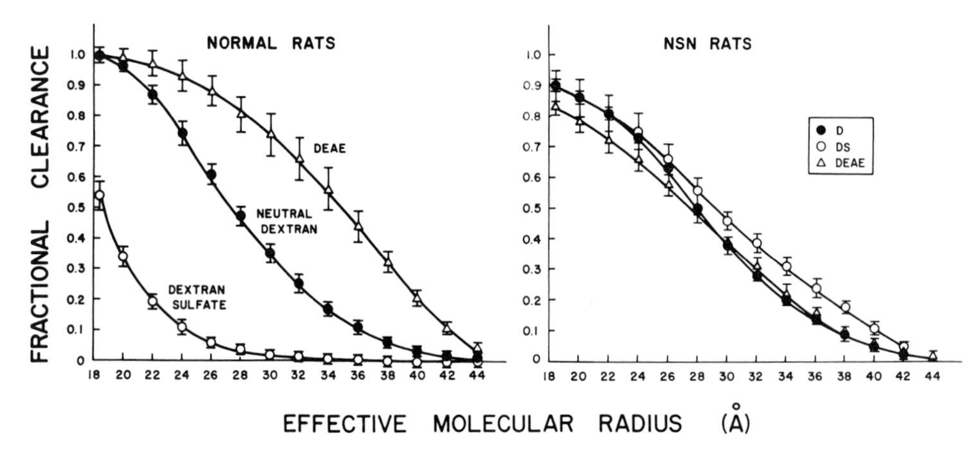

Figure 11-32. Clearance curves showing the size- and charge-selective properties of the normal glomerulus (left) and the loss of charge discrimination that occurs in some glomerular diseases associated with proteinuria (right). Size discrimination is manifest by the fact that there is increasing restriction to glomerular filtration with increasing effective molecular radius of the molecules, regardless of their charge. Charge discrimination is manifest by the fact that in normal rats, the clearance of negatively charged dextran sulfate is retarded, and that of positively charged DEAE dextran is increased over that of neutral dextran of a comparable size. In rats with nephrotoxic serum nephritis (NSN), an experimentally induced glomerular disease in which there is proteinuria, the fractional clearance of all three forms of dextran are similar, indicating a loss of charge discrimination in this disease. (From Bohrer *et al.*, 1978.)

postulated that the walls of glomerular capillaries, as well as other capillaries, contain pores, ~ 9 nm in diameter. Work in the late 1970s further established the importance of charge as a factor in glomerular filtration (Bohrer *et al.*, 1978; Brenner *et al.*, 1978; Venkatachalam and Rennke, 1978, 1980). Negatively charged molecules are filtered in much smaller amounts than neutral molecules of the same size (Fig. 11-32). Thus, the physiologic findings have established that the glomerulus has both size-selective and charge-selective properties, and they imply that the glomerular capillary wall is negatively charged.

5.2. Use of Electron-Dense Tracers

Early work by cell biologists and electron microscopists in this field was preoccupied with efforts to determine where in the glomerular capillary wall the hypothetical pores postulated by the physiologists were located, i.e., to determine which glomerular layer represents the critical barrier serving to retain albumin in the circulation. Many studies were done in which electron-dense tracers were used to identify the barrier, and conflicting results were obtained concerning whether the GBM or the epithelial slits were the main barrier that serves to retain plasma proteins (reviewed by Farquhar, 1975, 1980). At present, however, there is general agreement (1) that the GBM is the main filter serving to retain albumin and other plasma proteins in the circulation, and (2) that the epithelial slits do not serve in this capacity (Farquhar, 1975, 1978, 1980; Farquhar *et al.*, 1982; Farquhar and Kanwar, 1980; Laliberté *et al.*, 1978; Rennke *et al.*, 1975; Ryan *et al.*, 1976; Ryan and Karnovsky, 1976; Venkatachalam and Rennke, 1978, 1980). This conclusion was reached based on results obtained with a variety of anionic (native ferritin, catalase, endogenous albumin, or IgG) and neutral (dextrans) tracers (see Figs. 11-33 and 11-34), all of which indicate that the anionic and neutral macromolecules do not penetrate beyond the lamina rara interna of the GBM, and thereby identify the GBM as the main size and charge barrier since it prevents the filtration of anionic or neutral macromolecules > 70 kDa. The epithelial slits do serve to retain large macromolecular complexes such as protamine sulfate–heparan complexes (Seiler *et al.*, 1977) and antigen–antibody complexes (Kerjaschki *et al.*, 1987) that form on the epithelial side of the GBM.

That the GBM is a negatively charged structure was first demonstrated by the studies of Rennke *et al.* (Rennke *et al.*, 1975) with differently charged ferritins. Their studies showed that, whereas native anionic (pI 4.8) ferritin fails to penetrate beyond the lamina rara interna (Farquhar *et al.*, 1961), cationized ferritins with increasing isoelectric points penetrate the GBM to a greater and greater extent, implying that the GBM as a whole is a negatively charged structure.

If the GBM is the main filter, one can ask: What is the function of the adjoining endothelial, epithelial, and mesangial cells? The cells function to synthesize and maintain the GBM and to modulate its activities by controlling access to it and exit from it (Farquhar, 1978). The endothelium varies access to it by varying the number and size of its fenestrae, and the epithelium varies exit from it by virtue of variation in the number and width of the filtration slits. This is particularly evident in certain glomerular diseases. For example, in preclampsia, a hypertensive disease of pregnant women, the endothelium is swollen and the number of fenestral openings is drastically reduced which serves to reduce glomerular filtration. In so-called minimal change nephrosis, the collective slit area is reduced to 10% of normal and many of the filtration slits are converted into tight junctions (Farquhar and Palade, 1961).

5.3. Nature and Location of Anionic Sites in Glomeruli

With the question of the location of the main filter resolved, attention in this field shifted to attempting to determine the molecular components of the GBM responsible for establishing the size and especially the charge-barrier function of the glomerulus. The latter is a matter of considerable importance because loss of the charge-barrier function has been found in several experimental glomerular diseases (Fig. 11-32) used as models for human diseases that are characterized by leakage of plasma proteins (primarily albumin) into the urine (Brenner *et al.*, 1978).

The approach taken was to use various cationic probes (lysozyme, cationized ferritin, ruthenium red, polyethyleneimine) as stains to determine the

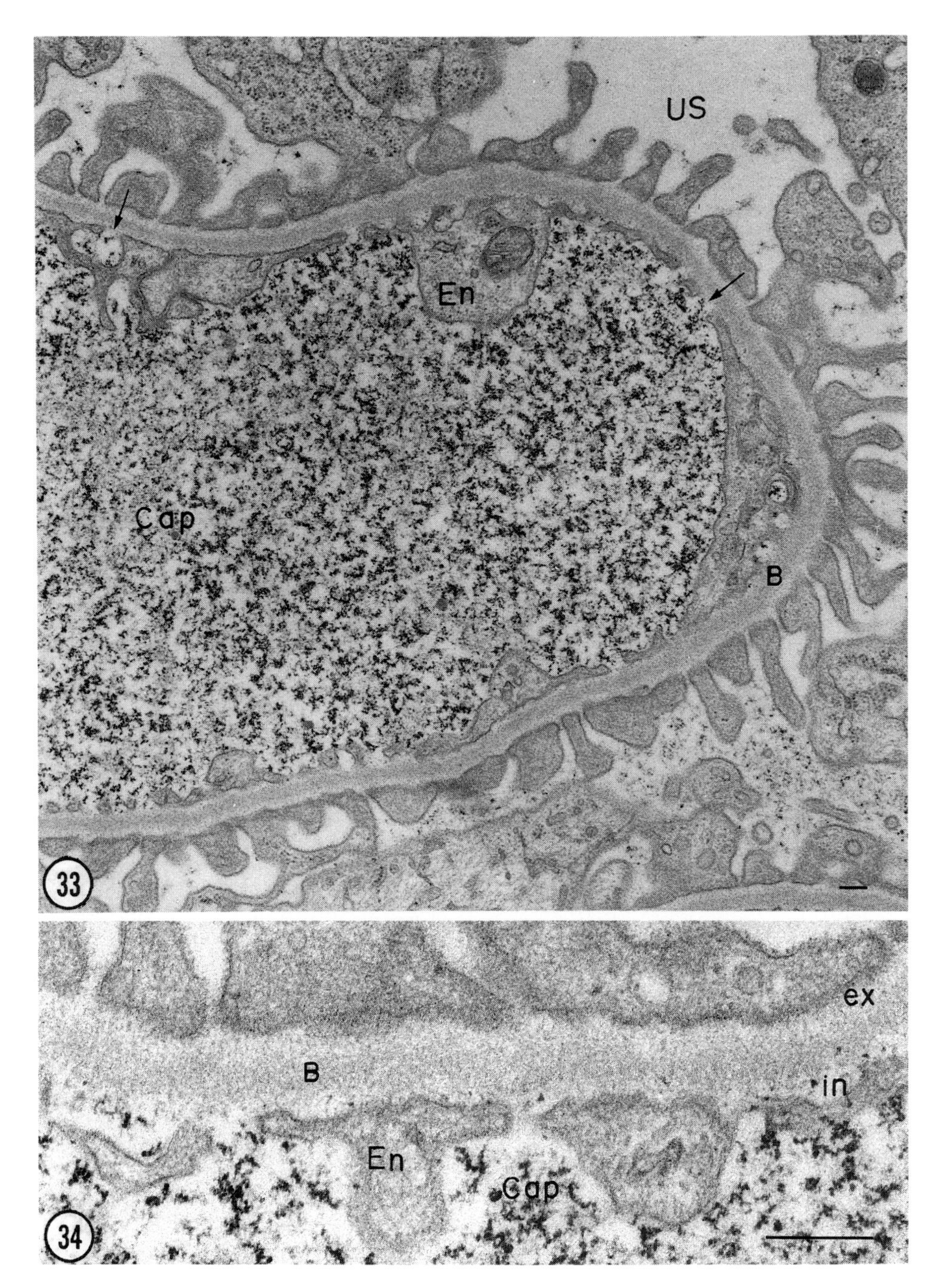

Figures 11-33 and 11-34. Portions of glomeruli from a normal rat sacrificed 3½ hr after the injection of dextran (125 kDa). The capillary lumina (Cap) are filled with dense dextran particles. The latter penetrate the fenestrae of the endothelium (En) (arrows) to reach the lamina rara interna (in) of the GBM. No dextran is seen in the lamina densa (B), lamina rara externa (ex), or epithelial slits. Bars = 0.2 μm. (Figure 11-33 from Farquhar, 1980; Fig. 11-34 from Caulfield J. P., and Farquhar, M. G.,1974, *J. Cell Biol.* **63:**883–903.) (Reproduced from the *Journal of Cell Biology*, 1974, **63:** 883–903, by copyright permission of the Rockefeller University Press.)

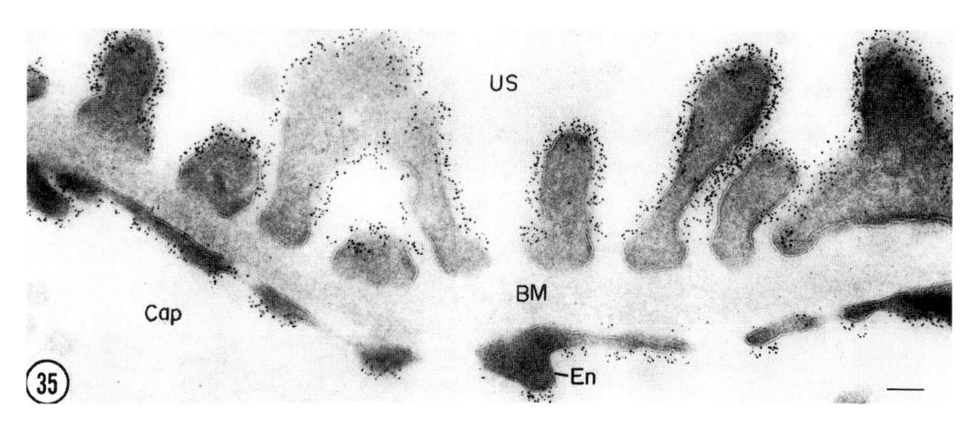

Figure 11-35. Immunogold labeling of podocalyxin in an ultrathin frozen section. Gold labeling for podocalyxin is only found on the apical and lateral surfaces of the podocyte foot processes above the slit diaphragms; no gold particles can be found on the lateral cell surfaces below the slit diaphragms or at the base of the foot processes facing the GBM. A weaker but distinct gold labeling is also present on the luminal surface of the endothelium (En). Ultrathin frozen sections were prepared as in Fig. 11-19, incubated with mouse monoclonal antipodocalyxin IgG followed by rabbit anti-mouse IgG and goat anti-rabbit IgG coupled to 5-nm gold. (From Schnabel *et al.*, 1989.)

nature and location of anionic groups in the GBM (Caulfield and Farquhar, 1976; Reeves *et al.*, 1980; Kanwar and Farquhar, 1979a). In the course of these studies, the network of proteoglycan granules concentrated in the laminae rarae was discovered (Figs. 11-7 to 11-9). Cationized ferritin with an isoelectric point of 7.3–7.5 proved to be a very useful probe for this purpose, as it binds to the anionic sites in the laminae rarae where the GAG chains of the HSPG are located but not to the lamina densa or cell surfaces (Figs. 11-13 and 11-14), thereby demonstrating that the sites with the highest charge density or concentration of negatively charged groups in the glomerulus correspond to the sites where the HSPG are located in the laminae rarae.

Negatively charged groups are also present on the surfaces of the endothelium and epithelium. As mentioned already, it was established some time ago that the epithelial cells possess an unusually thick, sialic acid-rich, cell surface coat—the so-called epithelial polyanion (see Fig. 11-5)—which is made up in large part of podocalyxin. Podocalyxin is the main sialoprotein of the glomerulus. It is found not only on the glomerular epithelium, but also in smaller amounts on the endothelium (Kerjaschki *et al.*, 1984) (see Fig. 11-35). It seems unlikely that podocalyxin contributes in a major way to the glomerular charge-barrier function, because it is restricted in its distribution to the luminal domain of the endothelial cell membrane and the apical domain of the epithelial cell membrane facing the urinary spaces. it is not detected on the surfaces of the cells facing the GBM (Fig. 11-35). However, epithelial polyanion (podocalyxin) does play an important role in maintaining the epithelial slits open by charge repulsion and in maintaining the normal foot process and slit arrangement, as neutralization of the coat by perfusion with polycations causes

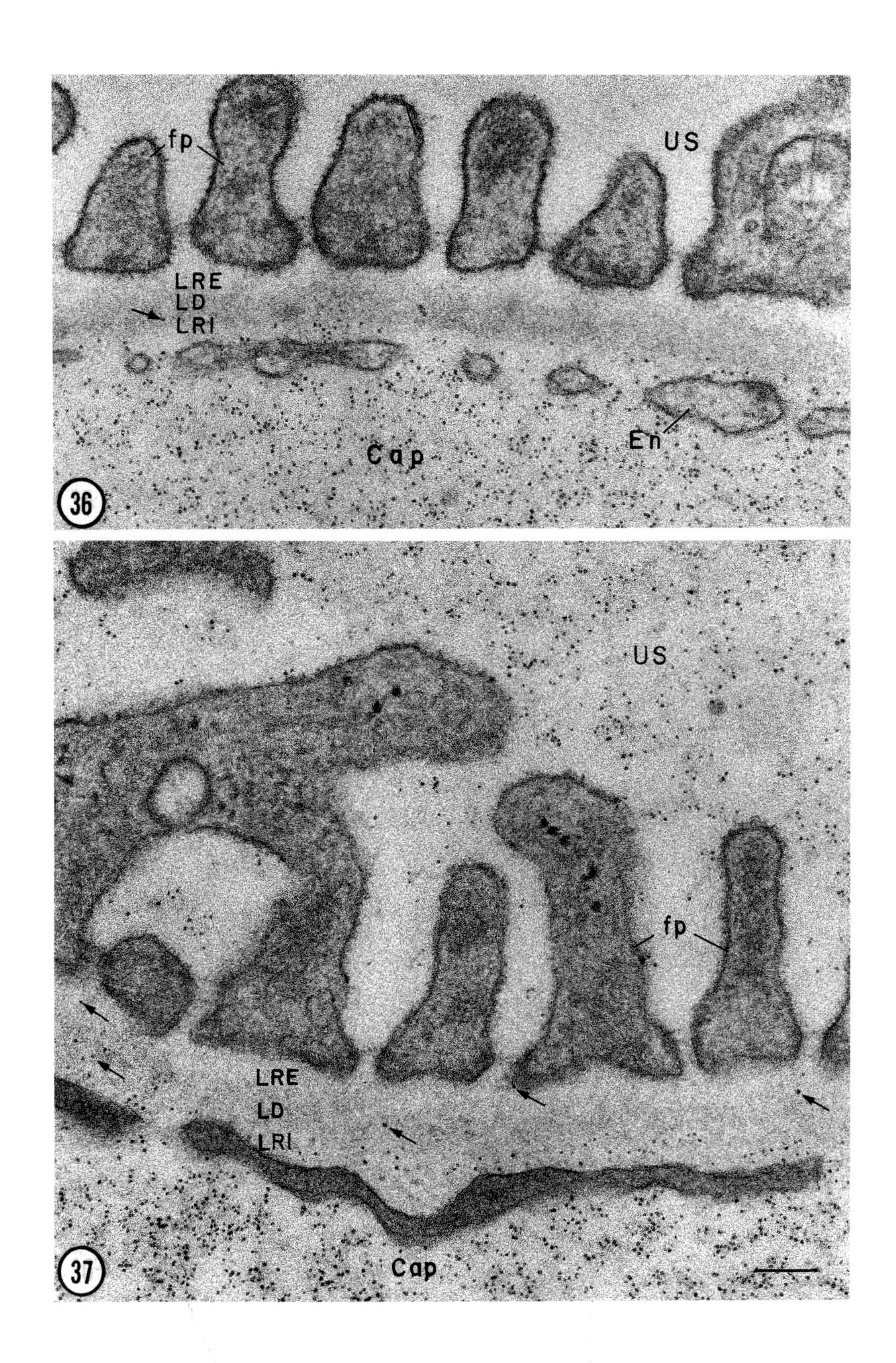
fp
US
LRE
LD
LRI
En
Cap
36
US
fp
LRE
LD
LRI
Cap
37

collapse of the filtration slits and disrupts the foot processes, whereas reperfusion with polyanions restores them to their normal state (Seiler *et al.*, 1977). Moreover, in experimental nephrotic syndrome where the foot process and slit arrangement is distorted, the sialic acid content of podocalyxin has been shown to be reduced to one-third of normal (Kerjaschki *et al.*, 1985).

5.4. Role of HSPG in GBM Permeability to Macromolecules

The GAG moieties of proteoglycans are among the most highly negatively charged molecules known; hence, the presence of HSPG in the GBM makes them automatic candidates for components that might play a role in establishing or in maintaining the charge-barrier properties of the GBM. GAG have also been shown to exert a steric exclusion effect on proteins in solution (Comper and Laurent, 1978; Lindahl and Höök, 1978). Therefore, it could logically be assumed that HSPG might be one of the GBM components responsible for endowing the GBM with its properties as a selective filter. Direct evidence that this is the case was provided by the demonstration that there is increased leakage of anionic ferritin (Kanwar and Farquhar, 1980) and [^{125}I]-BSA (Rosenzweig and Kanwar, 1982) into the urinary spaces after removal of GAG from the GBM by perfusion of kidneys with GAG-degrading enzymes *in situ*. Although native anionic ferritin does not normally penetrate the GBM beyond the lamina rara interna, if heparan sulfate is removed by enzyme digestion there is increased leakage of both ferritin (480 kDa) and BSA (70 kDa) into the urinary spaces (Figs. 11-36 to 11-39). Thus, the presence of heparan sulfate is essential for maintenance of the selective permeability properties of the GBM. The fact that cationized ferritin (pI 7.5–9.0) can penetrate the lamina densa to reach and bind to the lamina rara externa (Kanwar and Farquhar, 1979a; Rennke *et al.*, 1975) indicates that the porosity of the lamina rara interna and lamina densa is equal to or greater than the size of ferritin (~ 12 nm). Normally, ferritin and other anionic proteins are prevented from entering the GBM mesh by charge repulsion due to the presence of the GAG moieties of the HSPG. If the net negative charge of the ferritin is reduced, however, it can penetrate the lamina densa. The same undoubtedly applies to other large molecules, of which the most important form the pathologic standpoint are circulating autoantibodies (see below).

Figures 11-36 and 11-37. Figure 11-36 shows a portion of a glomerular capillary from a control kidney perfused with native ferritin (10 mg/ml). Most of the ferritin molecules are restricted to the capillary lumen (Cap) or to the lamina rara interna (LRI) of the GBM. Occasionally, molecules are seen in the lamina densa (LD) (arrow), but they are rarely seen in the lamina rara externa (LRE) or in the urinary spaces (US). Figure 11-37 shows a similar field from a kidney perfused with heparinase prior to perfusion with native ferritin. Numerous ferritin molecules have penetrated the lamina densa (arrows) and are present in the urinary spaces in large numbers. Bar = 0.1 μm. (From Kanwar *et al.*, 1980.) (Reproduced from the *Journal of Cell Biology*, 1980, **86:**688–693, by copyright permission of the Rockefeller University Press.)

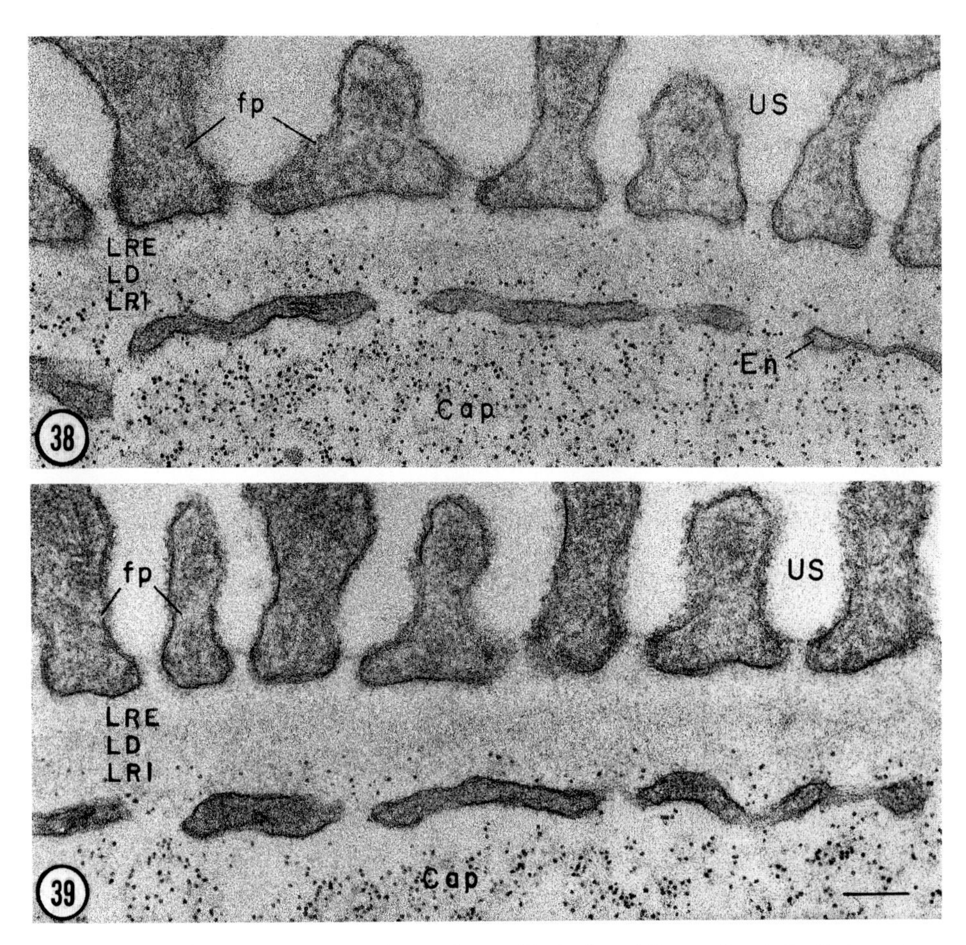

Figures 11-38 and 11-39. Portions of glomerular capillaries from kidneys digested with heparinase (Fig. 11-38) or chondroitinase ABC (Fig. 11-39), followed by perfusion with native ferritin. After heparinase treatment, increased amounts of ferritin have penetrated the GBM and are seen in the lamina densa (LD) and lamina rara externa (LRE), whereas after chondroitinase ABC, as in controls (Fig. 11-36), little or no ferritin is seen beyond the lamina rara interna (LRI). Bar = 0.1 μm. (From Kanwar *et al.*, 1980.) (Reproduced from the *Journal of Cell Biology*, 1980, **86:**688–693, by copyright permission of the Rockefeller University Press.)

It should be added that the role of GAG in modifying basement membrane permeability is probably not limited to the GBM because similar proteoglycan networks are widely, probably universally, distributed in all renal basement membranes (including those of the renal tubule, Bowman's capsule, and the arteriolar or peritubular capillary endothelium (see Figs. 11-40 and 11-41) and in basement membranes in many other locations as well (Hay, 1981; Trelstad *et al.*, 1974).

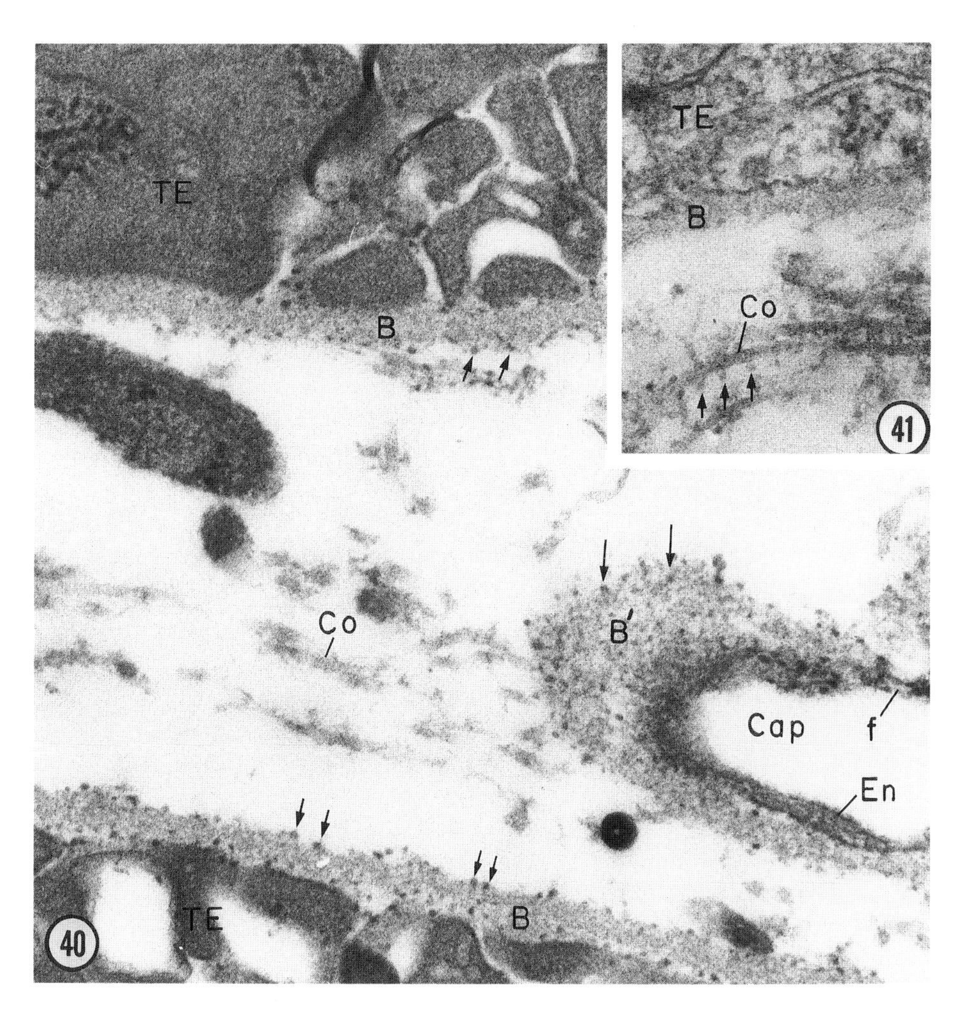

Figures 11-40 and 11-41. Peritubular region of the cortex from a kidney stained with ruthenium red, demonstrating the wide distribution of HSPG granules in renal basement membranes. Figure 11-40 shows the basement membranes (B) at the base of two tubule epithelial cells (TE) and the basement membrane (B′) of a peritubular capillary (Cap), which is cut in grazing section. All of these basement membranes contain proteoglycan granules that are stained with ruthenium red (arrows). Figure 11-41 is a similar field from a kidney perfused with heparitinase prior to staining, demonstrating that the particles in the tubular basement membranes (B) as well as those in the endothelial basement membranes (not shown) are removed by this treatment. The proteoglycans associated with the collagen fibrils (Co) in the interstitia (arrows) are unaffected by this enzyme but are removed by chondroitinase ABC. En, endothelium; f, endothelial fenestra with its diaphragm. Both ×63,000. (From Farquhar and Kanwar, 1982.)

6. What Do We Know about the Biosynthesis and Assembly of GBM and Mesangial Matrix Components?

6.1. Which Glomerular Cells Make GBM Components?

Studies on the biosynthesis of the GBM have had as their main aim determination of the glomerular cell types that manufacture the GBM, the mesangial matrix, or their individual components. In principle, all three cell types, endothelial, epithelial, and mesangial cells, should be capable of making basement membrane components, as it has been shown that endothelial, epithelial, and smooth muscle cells derived from other sources are capable of making basement membrane components in culture (see review by Hay, 1981). Also, as already noted, during glomerular development both the epithelial cells and the presumptive endothelial cells produce a basement membrane, and the two basement membranes eventually merge to form the mature GBM.

The epithelial cell was pinpointed as the main site of GBM synthesis many years ago when basement membrane-like sheets resembling the lamina densa were found inside the rough endoplasmic reticulum of glomerular epithelial cells and demonstrated to react with antibodies prepared against whole GBM (Andres *et al.*, 1962). Moreover, when the GBM was labeled with silver (administered in the drinking water), metallic silver was preferentially deposited along the epithelial side and gradually displaced toward the luminal side, suggesting that the GBM was synthesized by the epithelial cells and removed by the endothelial and/or mesangial cells (Walker, 1973). Autoradiographic studies also suggested that the epithelial cells were the primary source of the GBM (Farquhar *et al.*, 1985; Romen *et al.*, 1976).

More recently, many workers have succeeded in culturing cells of glomerular origin and in establishing apparently homogeneous epithelial and mesangial cell lines. Glomerular endothelial cells have proved to be difficult to propagate in culture. The information obtained from biosynthetic studies on cultures of these homogeneous epithelial and mesangial cell lines is as follows: (1) Both epithelial and mesangial cells have the capacity to synthesize type IV collagen (140–170 kDa) (Foidart *et al.*, 1980; Killen and Striker, 1979; Scheinman and Fish, 1978) and fibronectin (Foidart *et al.*, 1980; Killen and Striker, 1979) in culture. (2) Mesangial cells, like their smooth muscle cell cousins, also produce interstitial collagens, types I, III, and V (Striker *et al.*, 1980), a finding that explains the occasional presence of collagen fibrils in the mesangium of the renal glomerulus *in situ* and their increase in the same location as pathologic conditions. (3) Both epithelial and mesangial cells produce proteoglycans, but the type of proteoglycans they produce differs: Epithelial cells synthesize predominantly HSPG, and mesangial cells produce predominantly CSPG (Farin *et al.*, 1980; Yaoita *et al.*, 1990). (4) HSPG made by epithelial cells is predominantly of the basement membrane type (Stow *et al.*, 1989).

The study of epithelial and mesangial cell lines in culture has been very informative with respect to indicating some of the biosynthetic capabilities of these two cell types. However, information on the specific capabilities of these cell types *in situ* is more limited. Fibronectin has been localized to the Golgi complex (see Fig. 11-30) of mesangial cells *in situ* (Courtoy *et al.*, 1980), indicating that these cells are capable of producing fibronectin. Based on autoradiographic (Farquhar *et al.*, 1985) and immunocytochemical (Stow *et al.*, 1985) data, it appears that epithelial cells are the main producers of HSPG destined for the GBM. The latter was localized to the RER and Golgi of epithelial cells. Mesangial cells are apparently the main source of CSPG which is deposited mainly in the mesangial matrix (Kanwar *et al.*, 1984; McCarthy *et al.*, 1989).

6.2. Synthesis of GBM Components

The biosynthesis and turnover of basement membrane components has been investigated by radiolabeling *in vivo*, by kidney perfusion, and by incubation if isolated glomeruli *in vitro*, or glomerular cells in culture. Previously, synthesis of collagen IV (Heathcote and Grant, 1981; Kefalides, 1979; Spiro, 1978) and of GAG (Lemkin and Farquhar, 1981) or proteoglycans (Kanwar *et al.*, 1981; Klein *et al.*, 1986) has been studied. The latter studies have demonstrated that the glomerulus synthesizes predominantly HSPG (85%) with smaller amounts of chondroitin/dermatan sulfate proteoglycans (15%).

Turnover (degradation) of glomerular components has not received as much attention as biosynthesis, but the data available are already interesting. They show that whereas turnover of the collagenous proteins in the GBM is very slow (Price and Spiro, 1977) (weeks or months rather than hours), the turnover of proteoglycans is very rapid (see Chapter 5). The half-life of glomerular GAG is 1–7 days (Beavan *et al.*, 1989; Cohen and Surma, 1982), i.e., the same range as that of GAG derived from other sources. The rapid rate of turnover of proteoglycans demonstrates that they are the most labile elements in the GBM which is in keeping with the key role they play in GBM permeability and pathology.

The synthesis of fibronectin, laminin, and entactin has been studied in numerous cell types (Chapter 4), but it has not yet been investigated in the glomerulus. Therefore, little or nothing is known of their rate of turnover.

6.3. Basement Membrane Components Self-Assemble into Basement Membranelike Sheets

A characteristic feature of extracellular matrix proteins in general, and of basement membrane components in particular, is that they are all large, multidomain proteins that are capable of multiple interactions both with cells and with other extracellular matrix components (Kleinman *et al.*, 1986). For example, laminin and fibronectin contain binding sites for heparin (or heparan) and collagen IV, and entactin forms complexes with laminin. Monomers of individual basement membrane proteins are also capable of self-assembly into homopolymers. Type IV collagen can assemble into a stable three-dimensional network, laminin can aggregate into large polymers in a calcium-dependent fashion, and the HSPG core protein can self-assemble into dimers and oligomers (Yurchenco and Schittny, 1990) (see Fig. 11-43).

More strikingly, when laminin, collagen IV, and HSPG (in a ratio of 1:1:0.1) are mixed together at a pH of 7.4 at 35°C *in vitro*, they assemble into orderly sheets (Fig. 11-42) that are remarkably similar in appearance and thickness to the lamina densa of basement membranes (Grant *et al.*, 1989). It appears that heparin and heparan can modulate laminin and collagen IV polymerization and thereby modify the basement membrane meshwork (Yurchenco *et al.*, 1987). Their binding and polymer modifying characteristics are favored by a high degree of sulfation. Thus, in principle, any condition that leads to changes in the content and/or the sulfation of HSPG could lead to changes in the molecular architecture of basement membranes. In the case of the GBM, control over the assembly process is crucial for regulating basement membrane porosity.

Clarification of the structure and self-assembly properties of the major basement membrane components has led to the generation of structural models of basement membrane organization (LeBlond and Inoue, 1989; Yurchenco and Schittny, 1990). One current model (Timpl and Dziadek, 1986; Yurchenco and Schittny, 1990; Yurchenco *et al.*, 1986) illustrated in Fig. 11-44 is the following: Type IV collagen forms a skeletal latticelike, polymeric array by C-terminal and side-by-side associations which become stabilized by covalent cross-links. Laminin polymerizes forming sheetlike nets or patches. Collagen IV and laminin are held together through entactin/nidogen bridges and weak laminin–collagen interactions. HSPG cores bind together to form oligomers which are firmly anchored to the basement membrane in unknown sites. The heparan sulfate chains which are peripherally oriented bind weakly to laminin and collagen. HSPG is proposed to act as a space-filling, reversible cross-linking agent binding laminin to laminin, laminin to collagen, and perhaps collagen to collagen. A three-dimensional network of irregular fuzzy strands referred to as cords has been described in thin sections of basement membranes (LeBlond and Inoue, 1989), but it is not clear how these cords relate to the three-dimensional networks assembled *in vitro* (Yurchenco and Schittny, 1990).

6.4. How Is Diversity in Basement Membrane Structure and Function Generated?

It has become evident that there are considerable differences in the organization and functions of basement membranes in different locations. If basement membrane components are capable of self-assembly *in vitro*, the question that arises is: How are these differences generated? The answer is that variations in basement membrane organization must be created by differences in the relative amounts of the individual basement membrane components present, by differences in the receptor populations present on cell surfaces which affect their interaction with basement membrane components, and by variations in the

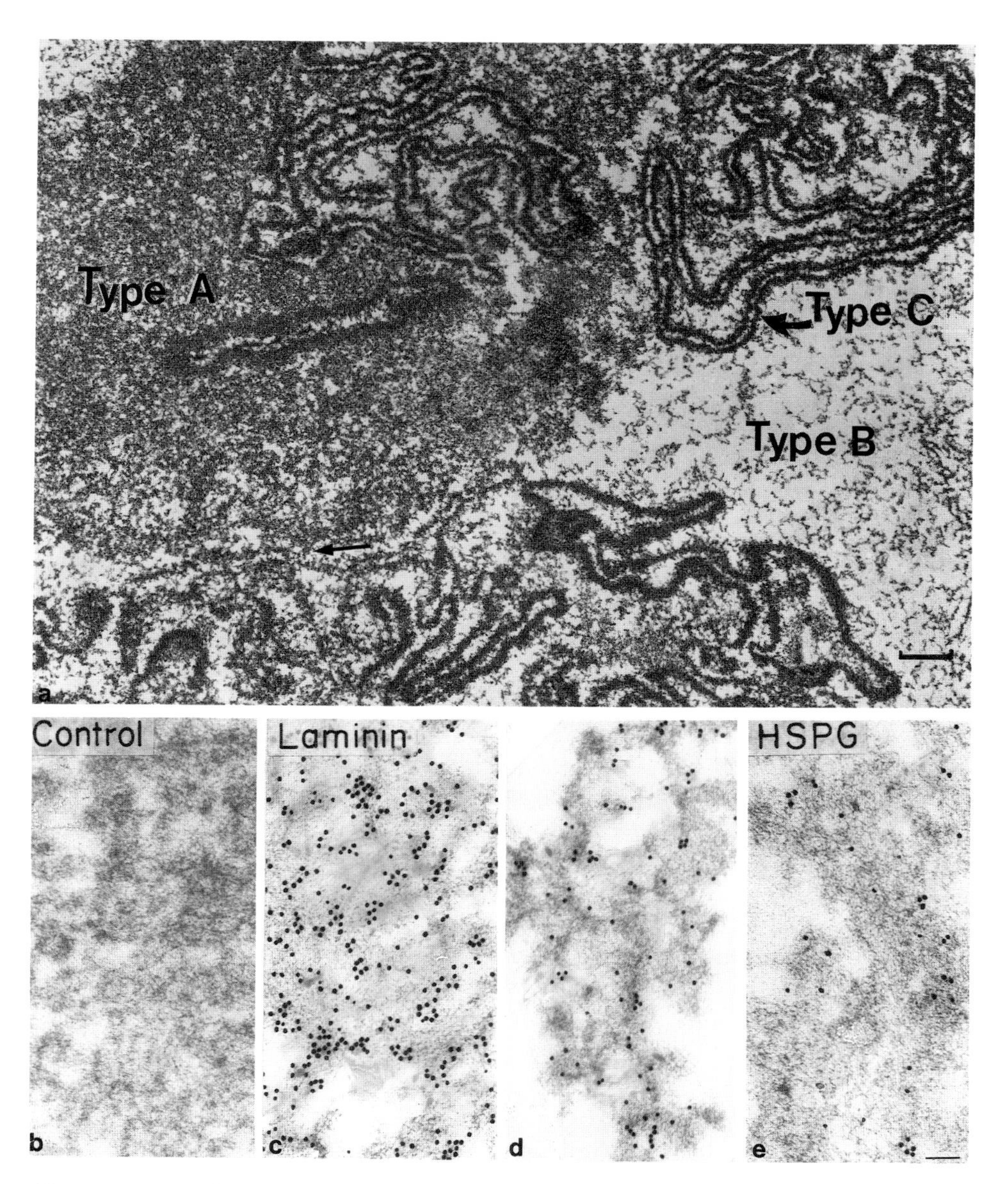

Figure 11-42. (a) Electron micrograph of the precipitate produced by incubation of preparations of laminin, collagen IV, and HSPG at 35°C for 1 hr. Fixation was in permanganate and embedding in Epon. Three structures, referred to as types A, B, and C, may be seen. A mass of type A material is formed of irregular deposits of anastomosing strands separated by narrow spaces. Type B is seen as strips scattered between the other two types, as shown at lower right. The type C structures are the densely stained, thick, wavy lines at upper right and base; they are cross sections of sheets and may be arranged in pairs. The arrow points to the material which may be transitional between types A and C. (b–e) Immunogold labeling of type C aggregates. Plastic (Lowacryl) sections were exposed to either nonimmune control serum (b), laminin (c), collagen IV (d), or HSPG (e) followed by amplification with goat anti-rabbit IgG and protein A gold. a, bar = 1 μm. b–e, bar = 0.1 μm. (From Grant *et al.*, 1989.) (Reproduced from the *Journal of Cell Biology*, 1989, **108:**1567–1574, by copyright permission of the Rockefeller University Press.)

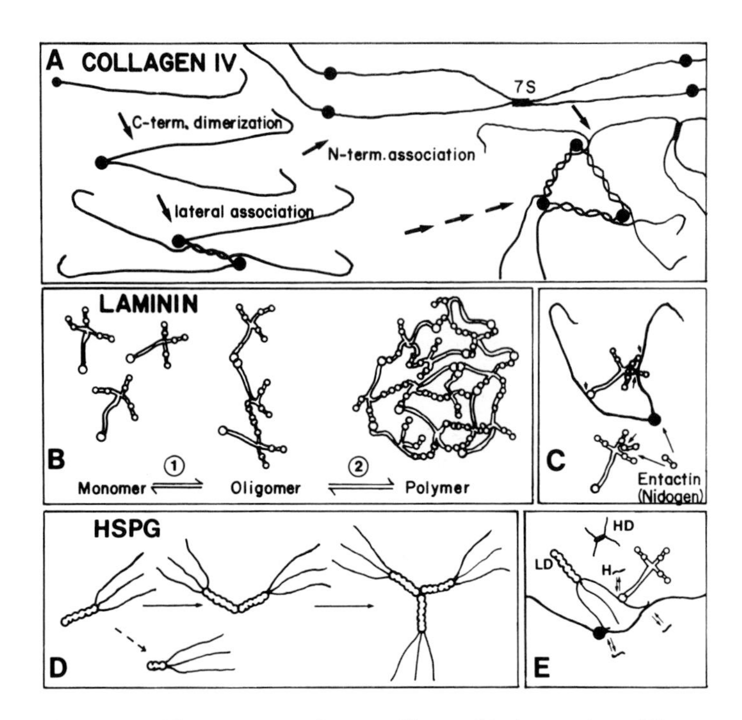

Figure 11-43. Summary of basement membrane self-assembly interactions. Diagrammatic representation of binding interactions between protomers proposed to contribute to basement membrane architecture based on *in vitro* and tissue studies. (A) Type IV collagen forms polymers through COOH-terminal dimerization, lateral associations, and NH_2-terminal (7 S) bond formation. (B) Purified laminin will form polymers in a two-step self-assembly mediated through terminal domains. (C) The COOH-terminal domain of entactin/nidogen can bind both laminin and the triple-helical region of type IV collagen, acting as a bridge. There is also evidence that the ends of the arms of laminin can directly interact with collagen. (D) The low-density heparan sulfate proteoglycan self-interacts through its core to form dimers and trimers. If the end of the core is cleaved, this interaction is lost. (E) Interactions of the heparan sulfate chains (shown with low-density proteoglycan) can occur with high-density form (shown at top of panel) with the main binding site in laminin and with type IV collagen. Competition reaction for these sites with heparin (shown with double arrows) may alter polyanionic charge distribution and the molecular packing of laminin and collagen polymers. These interactions may provide a mechanism to regulate basement membrane architecture. (From Yurchenco and Schittny, 1990.)

binding domains of constituent molecules (e.g., the NC1 region of the collagen chains). It has been demonstrated both by immunocytochemistry (Desjardins *et al.*, 1990; Grant and LeBlond, 1988; Laurie *et al.*, 1984) and by biochemical analysis that basement membranes in different sites within the kidney and elsewhere vary in the relative amounts of individual basement membrane components present. For example, the GBM is much richer in fibronectin than the lens capsule which contains relatively greater amounts of entactin that exceed laminin in both cases (Mohan and Spiro, 1986). It also appears that collagen IV, laminin, and HSPG belong to families of several members with structural differences. Four collagen chains [α1, α2, α3, and α4(IV)], three lamininlike molecules (laminin, S-laminin, and merosin), and a number of HSPG have been

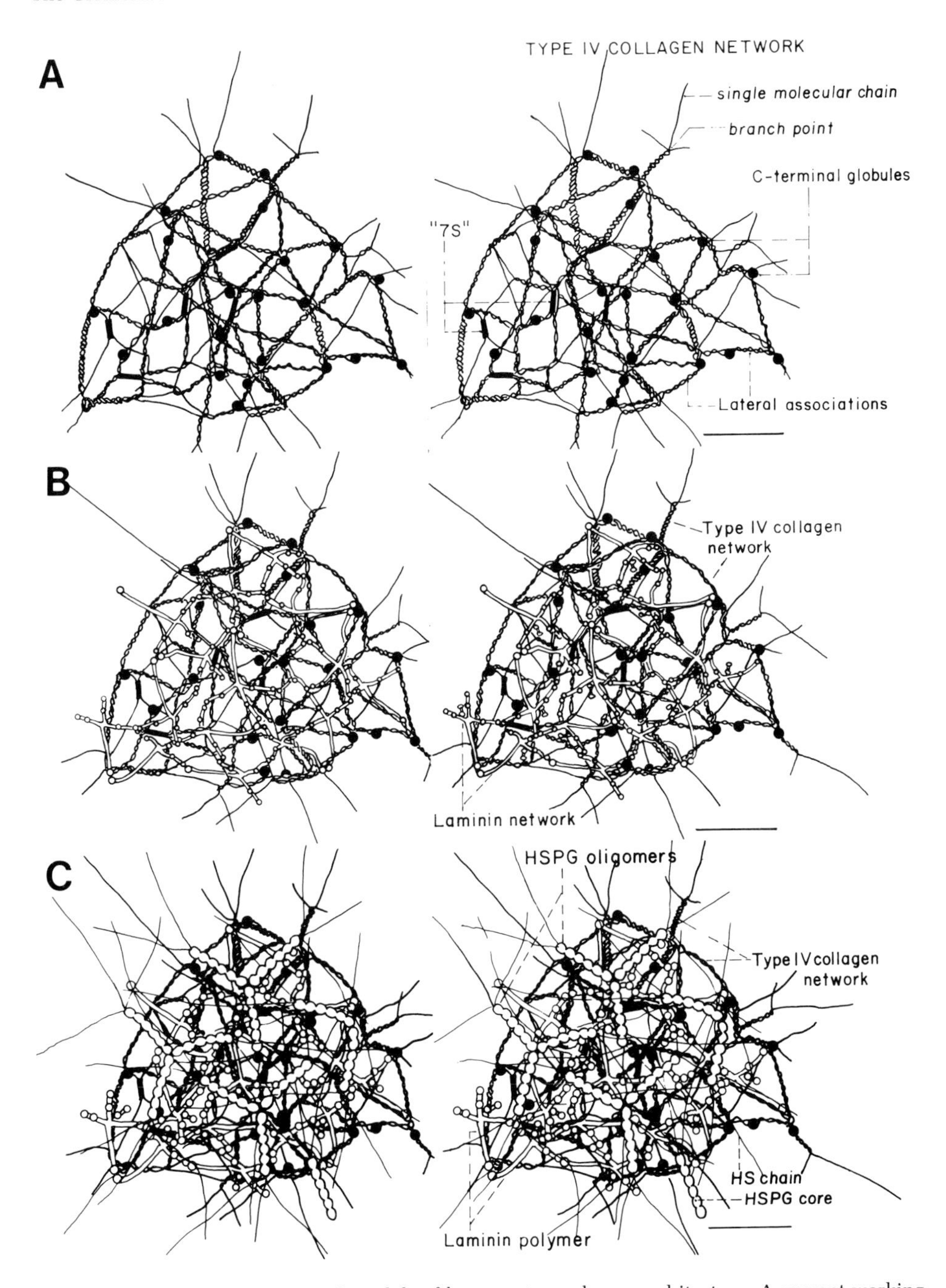

Figure 11-44. Three-dimensional models of basement membrane architecture. A current working hypothesis is shown here in three stereo pairs drawn from photographs of three-dimensional wire-and-bead models of a type IV collagen network patch (A), collagen plus laminin heteropolymer (B), and heteropolymer with associated proteoglycan (C). Type IV collagen and laminin form double polymer networks. Entactin/nidogen bridges collagen and laminin with additional laminin-to-collagen interactions. HSPG complexes, anchored in the network, interact with laminin and collagen through their polyanionic chains. (From Yurchenco and Schittny, 1990.)

isolated from various basement membrane matrices. Of particular interest is the diversity that exists among basement membrane HSPG isolated from various basement membrane sources since these molecules appear to play a modulatory role in basement membrane assembly. They have been shown to vary in the size of their core proteins and in the size, charge, and structure of their GAG chains. In the case of the GBM the heparan sulfate chains are clustered at the periphery of the HSPG molecule, and their structure is diverse, including novel 3-*O*-sulfated glucosamine (Edge and Spiro, 1990). Since proteoglycans are known to play a role in determining the size of interstitial collagen fibrils, it is reasonable to assume that they also play a key role in regulating the assembly and porosity of the basement membrane meshwork.

7. Pathology of GBM and Mesangial Matrix

Pathologic changes in the GBM are found in a variety of diseases, some of which are primary (e.g., Goodpasture's syndrome, an anti-GBM disease) whereas others are secondary (diabetes mellitus) (see Timpl and Dziadek, 1986, for a review of this topic). The GBM is also quite vulnerable to immune and toxic injury, due to the fact that it is directly exposed to the blood plasma via its open fenestrae. The latter does not apply to basement membranes in any other location. Such changes often drastically affect glomerular filtration. The glomerulus normally serves as a highly selective, macromolecular filter with a "porosity" or meshwork fixed at about 9 nm designed to retain albumin, the main plasma protein. If this porosity is increased, proteinuria ensues, whereas if glomeruli and basement membranes are thickened and scarred, as happens in many diseases affecting the kidney, filtration can be decreased or eliminated and renal insufficiency ensues, requiring renal dialysis to maintain the lives of patients. Indeed, renal insufficiency leading to renal dialysis constitutes a major economic problem in Western countries. A complete discussion of this problem is beyond the scope of this chapter; however, a brief description will be given of several examples of renal diseases in which some insight has been gained regarding molecular mechanisms of renal injury that may facilitate our understanding of normal glomerular function.

One of the common denominators that is emerging from studies of pathologically altered GBMs is that changes in the glomerular content of HSPG and/or in the anionic sites in the GBM are often associated with glomerular diseases associated with leakage of albumin into the urine (albuminura). The list of conditions in which this association has been made (see Farquhar *et al.*, 1985) includes diabetic nephropathy, congenital nephrotic syndrome, the NZB mouse (an animal model of lupus), and Heymann nephritis (an animal model of immune complex nephritis). A situation in which there is some knowledge of the molecular pathology involved is in diabetes mellitus. Approximately 50% of long-term, poorly controlled diabetics develop diabetic nephropathy, a condition characterized by thickening of the GBM (and other basement membranes) and expansion of the mesangial matrix (Farquhar *et al.*, 1959) which is part of a generalized microangiopathy. Paradoxically, these changes are associated with a reduction in glomerular filtration coupled with increased proteinuria. Thus, it is clear that in spite of the basement membrane thickening early in the disease, the basement membrane is leakier than normal, indicating a change in its porosity. Eventually, the glomeruli become totally scarred and renal filtration ceases. Biochemical analysis of basement membrane fractions from diabetics demonstrates that their content of HSPG and laminin are reduced whereas their collagen IV content is increased (Shimomura and Spiro, 1987). Similarly, in animals with experimentally induced diabetes, synthesis of HSPG is decreased (Brown *et al.*, 1982; Cohen and Surma, 1981; Rohrbach *et al.*, 1983). This leads one to speculate that the decrease in the HSPG content leads to defects in the assembly and molecular architecture of the GBM. A similar situation may pertain in the other diseases in which there is either a decrease in the synthesis of HSPG or a defect in its incorporation into the GBM. The basis for the decreased synthesis of HSPG in diabetics remains unknown.

A dramatic example of a situation in which changes in HSPG perturb GBM synthesis and removal is after *in vivo* injection of antibodies raised against basement membrane HSPG (Miettinen *et al.*, 1986). When anti-HSPG IgG was given to rats, it rapidly (within 2–5 min) bound to the GBM (Fig. 11-18B) where it was concentrated in the laminae rara (interna and externa) of the GBM (Fig. 11-45A) and remained bound there for up to 60 days. The binding led to a distinct thickening of the GBM up to 3–5 times normal with thickening occurring on the epithelial side of the GBM (Fig. 11-45B,C). Thus, it appears that binding of antibody to HSPG in the laminae rarae somehow perturbs the orderly laying down and removal of the GBM.

Another example of a situation in which there is GBM thickening is in Goodpasture's disease which is an anti-GBM glomerulonephritis (reviewed by Timpl and Dziadek, 1986). This disease is caused by circulating autoantibodies that recognize and bind to the GBM. It has recently been shown that the pathogenic epitopes responsible for binding to the GBM are present on the NC1 domain of α3(IV) collagen chains (Kalluri *et al.*, 1990). They are not present on α1, α2, or α4(IV) chains.

In general, antibodies that recognize GBM components such as laminin (Abrahamson and Caulfield, 1982) and HSPG (Miettinen *et al.*, 1986) or those that recognize membrane proteins on the base of the epithelium facing the GBM (Dekan *et al.*, 1990) can readily bind to the GBM (due to its aforementioned exposed nature). Other diseases in which circulating autoantibodies commonly bind to the GBM are streptococcal glomerulonephritis and mercuric chloride-induced nephritis. In Heymann nephritis, which is a model of membranous nephropathy in humans, the autoantigen is a 330-kDa epithelial membrane glycoprotein (Kerjaschki and Farquhar, 1983). When circulating antibodies bind to the latter, the immune complexes are shed from the cell surface and become cross-linked to the GBM (Kerjaschki *et al.*, 1987), gradually increasing in size. In all these cases, deposition of autoantibodies (together with complement factors) disturbs the porosity, biosynthesis, and turnover of the GBM,

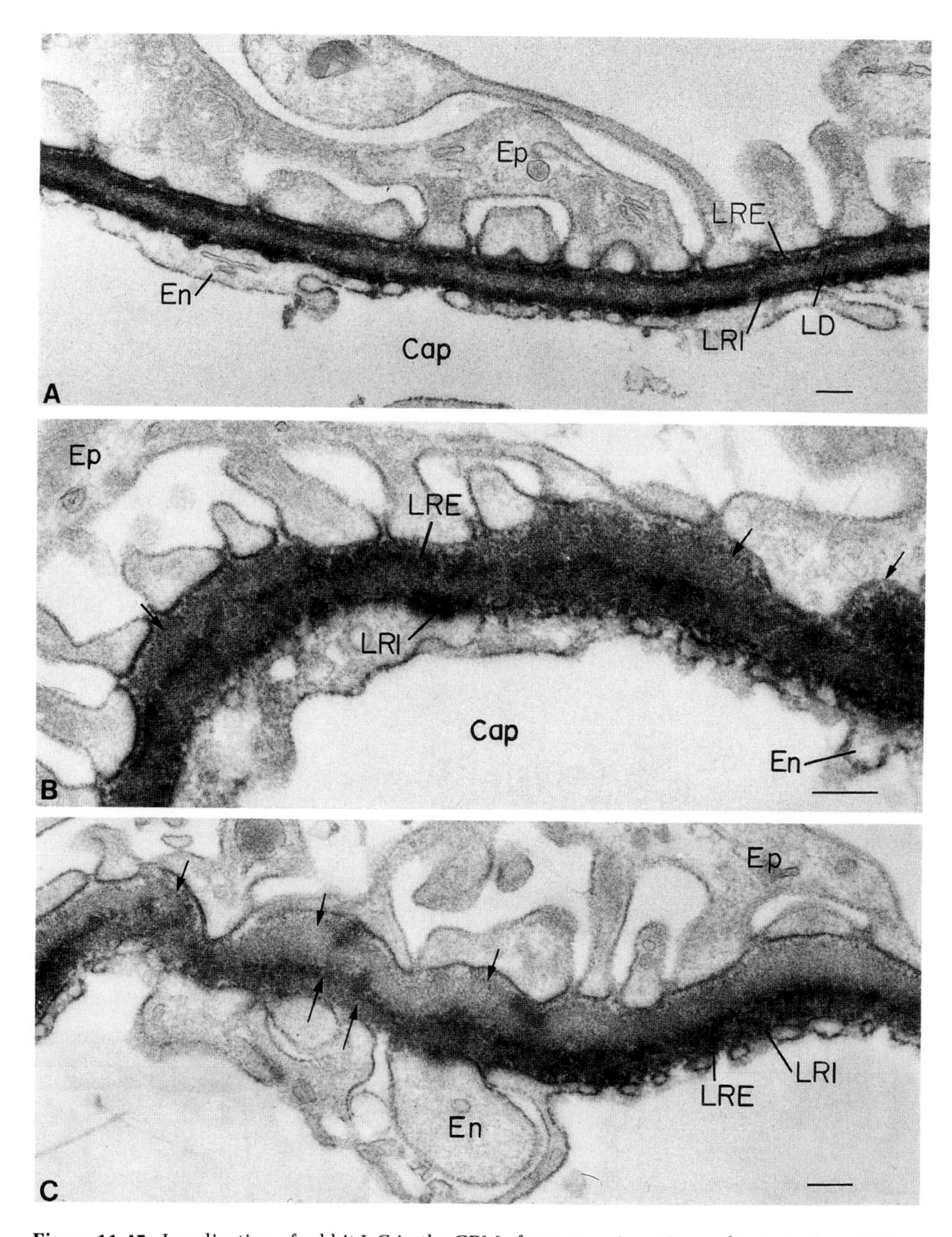

Figure 11-45. Localization of rabbit IgG in the GBM of rats at various times after injection of 20 mg of anti-HSPG IgG. (A) Two hours postinjection. Rabbit IgG is concentrated in the lamina rara interna (LRI) and externa (LRE) of the GBM, with little or no staining of the lamina densa (LD). (B, C) Two months postinjection. Rabbit IgG is still present in the GBM, but newly laid down deposits of basement membrane are see on the epithelial side of the GBM, giving it a scalloped appearance. The new layer of basement membrane is only weakly reactive (B) or unreactive (C) for bound IgG (short arrows). It appears that antibody binding perturbs the orderly biosynthesis and removal of GBM components. In some areas, the original laminae rarae (LRI, LRE) can be detected (B), but in other areas (C) the stained LRI has disappeared, leaving only the former LRE that can be recognized because it stains for rabbit IgG (long arrows, C). Bars = 0.2 μm. (From Miettinen *et al.*, 1986.)

leading eventually to thickening of the GBM, fibrosis of glomeruli, and destruction of normal glomerular function.

8. Concluding Remarks

Knowledge of the structure, derivation, composition, and function of the GBM has been reviewed. It can be asked: To what extent can we relate the information available on the structure and function of the GBM to other basement membranes? As stated at the outset, basement membranes are the natural substrates on which all cells except connective tissue cells rest. Their basic layered organization appears to be similar from one location to another. The functions they perform include (1) to delimit connective tissue–non-connective tissue boundaries, thereby maintaining the orderly organization of organs and tissues; (2) to attach cell layers to extracellular matrix and promote their differentiation; (3) to provide a scaffolding for tissue regeneration and repair (Vracko, 1978); (4) to bind and store growth factors (Folkman *et al.*, 1988); and (5) filtration of macromolecules, as stressed for the GBM. These can be considered general properties applicable to all basement membranes. For example, filtration of macromolecules must be a general function of all basement membranes regardless of their location. All epithelia and muscle cells are nourished by the interstitial fluid that percolates out from the capillaries. To reach these cells, circulating molecules must penetrate in sequence the endothelium, its basement membrane, the ground substance of the connective tissue, and the basement membrane of the cell layer in question. It follows that the ability of a circulating molecule to reach the cells depends on the permeability properties of all of the layers between them and the circulation, which typically includes two basement membranes.

In some cases, the anatomical relationships of basement membranes are modified to facilitate a given function. A prime example is the glomerulus, where the filtration pathway has been dramatically shortened by opening of the endothelial fenestrae and by fusion of the endothelial and epithelial basement membranes. In other sites such as the skin, where the epithelial layers are subjected to considerable mechanical stress, an unusually tight epithelial connective tissue attachment is required; additional elements known as anchoring filaments (Palade and Farquhar, 1965) composed of type VII collagen (Keene *et al.*, 1987) are often added to reinforce the construction along the dermal face of the basement membrane.

Are all basement membranes alike? Morphologically, there is overall similarity among basement membranes, but this apparent organizational similarity at the electron microscopic level hides considerable chemical diversity, as manifest, for example, by the fact that there are qualitative and quantitative differences in the staining of basement membranes for different matrix components, e.g., for fibronectin, for CSPG, and for different epitopes of laminin and HSPG and by the fact that biochemical analyses have revealed differences in the relative amounts of collagenous and noncollagenous components in basement membranes from one location to another. In addition, some proteins, such as the bullous pemphigoid antigen of the epidermal basement membrane, are found only in certain basement membranes (Martin *et al.*, 1988). It can be anticipated that considerable molecular heterogeneity—perhaps to the point of ultimate specificity— may be detected in the future as cell biologists and biochemists probe more deeply into the molecular structure and chemical composition of basement membranes.

With the realization that HSPGs are a universal component of basement membranes, the definition of an important role for them in basement membrane biology and pathology has been documented. Because of their many important properties, i.e., their ability to interact with all known components of basement membranes (collagens, fibronectin, laminin) as well as cell membrane components, their ability to influence the physical form of collagen IV, their high rate of turnover in basement membranes in comparison to the collagenous proteins, and their established role in morphogenesis (Chapter 9), it appears that proteoglycans play a key role in the molecular organization and specificities of basement membranes, as well as in their pathological derangements in disease.

What problems remain for the future? In the case of the glomerulus, considerable attention is focused at present on understanding how the laying down of the GBM and the mesangial matrix is controlled. In analogy to other situations, involvement of growth factors can be expected; in fact, it was recently demonstrated that TGF-β plays a role in regulating production of CSPG by cultured mesangial cells it is responsible for increased synthesis of mesangial matrix in experimental glomerulonephritis (Border *et al.*, 1990). Another important problem to be tackled is how the assembly of basement membrane and its fixed porosity is controlled. It is clear that basement membrane components are capable of self-assembly into homopolymers and heteropolymers *in vitro*, and when placed together *in vitro* they assemble into basement membrane-like sheets with a heteropolymeric meshlike structure. It will be fruitful to explore further the role of HSPG in modulating the assembly of the fine and carefully controlled GBM meshwork.

ACKNOWLEDGMENT. The original research summarized herein was supported by research grant DK 17724 from the National Institutes of Health.

References

Abrahamson, D. R., 1985, Origin of the glomerular basement membrane visualized after in vivo labeling of laminin in newborn rat kidneys, *J. Cell Biol.* **100:**1988–2000.

Abrahamson, D. R., 1986, Post-embedding colloidal gold immunolocalization of laminin to the lamina rara interna, lamina densa and lamina rara externa of renal glomerular basement membranes, *J. Histochem. Cytochem.* **34:**847–853.

Abrahamson, D. R., and Caulfield, J. P., 1982, Proteinuria and structural alterations in rat glomerular basement membranes induced by intravenously injected antilaminin immunoglobin G, *J. Exp. Med.* **156:**128.

Abrahamson, D. R., and St. John, P. L., 1990, Temporal changes in laminin distribution during glomerular development, *J. Am. Soc. Nephrol.* **1:**49P.

Abrahamson, D. R., Irwin, M. H., St. John, P. L., Perry, E. W., Accavitti, M. A., Heck, L. W., and Couchman, J. R., 1989, Selective immunoreactivities of kidney basement membranes to monoclonal antibodies against laminin: Localization of the end of the long arm and the short arms to discrete microdomains, *J. Cell Biol.* **109:**3477–3491.

Andres, G. A., Jorgan, K. C., Hsu, K. C., Rifkind, R. A., and Seegal, B. C., 1962, Electron microscopic studies of experimental nephritis with ferritin-conjugated antibody: The basement membranes and cisternae of visceral epithelial cells in nephritic rat glomeruli, *J. Exp. Med.* **115:**929–936.

Beavan, L. A., Davies, M., Couchman, J. R., Williams, M. A., and Mason, R. M., 1989, In vivo turnover of the basement membrane and other heparan sulfate proteoglycans of rat glomerulus, *Arch. Biochem. Biophys.* **269:**576–585.

Bohrer, M. P., Baylis, C., Humes, H. D., Glassock, R. J., Robertson, C. R., and Brenner, B. M., 1978, Permselectivity of the glomerular capillary wall: Facilitated filtration of circulating polycations, *J. Clin. Invest.* **61:**72–78.

Border, W. A., Okuda, S., Languino, L. R., Sporn, M. B., and Ruoslahti, E., 1990, Suppression of experimental glomerulonephritis by antiserum against transforming growth factors β_1, *Nature* **346:**371–374.

Brenner, B. M., Hostetter, T. H., and Humes, H. D., 1978, Molecular basis of proteinuria of glomerular origin, *N. Engl. J. Med.* **298:**826–833.

Brown, D. M., Klein, D. J., Michael, A. F., and Oegema, T. R., 1982, ^{35}S-glycosaminoglycan and ^{35}S-glycopeptide metabolism by diabetic glomeruli and aorta, *Diabetes* **31:**418–425.

Butkowski, R. J., Wieslander, J., Kleppel, M., Michael, A. F., and Fish, A. J., 1989, Basement membrane collagen in the kidney: Regional localization of novel chains related to collagen IV, *Kidney Int.* **35:**1195–1202.

Carlin, B., Jaffe, R., Bender, B., and Chung, A. E., 1981, Entactin, a novel basal lamina-associated sulfated glycoprotein, *J. Biol. Chem.* **256:**5209–5214.

Caulfield, J. P., and Farquhar, M. G., 1976, Distribution of anionic sites in glomerular basement membranes. Their possible role in filtration and attachment, *Proc. Natl. Acad. Sci. USA* **73:**1646–1650.

Clément, B., Segui-Real, B., Hassell, J. R., Martin, G. R., and Yamada, Y., 1989, Identification of a cell surface-binding protein for the core protein of the basement membrane proteoglycan, *J. Biol. Chem.* **264:**12467–12471.

Cohen, M. P., and Surma, M. L., 1981, [^{35}S] sulfate incorporation into glomerular basement membrane glycosaminoglycans is decreased in experimental diabetes, *J. Lab. Clin. Med.* **98:**715–722.

Cohen, M. P., and Surma, M. L., 1982, In vivo biosynthesis and turnover of ^{35}S-labeled glomerular basement membrane, *Biochim. Biophys. Acta* **716:**337–340.

Comper, W. D., and Laurent, T. C., 1978, Physiological function of connective tissue polysaccharides, *Physiol. Rev.* **58:**255–315.

Couchman, J. R., 1987, Heterogeneous distribution of a basement membrane heparan sulfate proteoglycan in rat tissues, *J. Cell Biol.* **105:**1901–1916.

Courtoy, P. J., Kanwar, Y. S., Hynes, R. O., and Farquhar, M. G., 1980, Fibronectin localization in the rat glomerulus, *J. Cell Biol.* **87:**691–696.

Courtoy, P. J., Timpl, R., and Farquhar, M. G., 1982, Comparative distribution of laminin, type IV collagen, and fibronectin in the rat kidney cortex, *J. Histochem. Cytochem.* **30:**874–886.

Dekan, G., Miettinen, A., Schnabel, E., and Farquhar, M. G., 1990, Binding of monoclonal antibodies to glomerular endothelium, slit membranes and epithelium after in vivo injection: Localization of antigen and bound IgG by immunoelectron microscopy, *Am. J. Pathol.* **137:**913–927.

Dekan, G., Gabel, C. A., and Farquhar, M. G., 1991, Sulfate contributes to the negative charge of podocalyxin—The major sialoglycoprotein of the filtration slits, *Proc. Natl. Acad. Sci. USA* **88:**5398–5402.

Desjardins, M., Gros, F., Wieslander, J., Gubler, M.-C, and Bendayan, M., 1990, Heterogeneous distribution of monomeric elements from the globular domain (NC1) of type IV collagen in renal basement membranes as revealed by high resolution quantitative immunocytochemistry, *Lab. Invest.* **63:**637–646.

Edge, A. S. B., and Spiro, R. G., 1990, Characterization of novel sequences containing 3-O-sulfated glucosamine in glomerular basement membrane heparan sulfate and localization of sulfated disaccharides to a peripheral domain, *J. Biol. Chem.* **265**:15874–15881.

Farin, F., Killen, P., and Striker, G., 1980, Biosynthesis of heparan sulfate (glycosaminoglycans) by glomerular cells in vitro, *Fed. Proc.* **39(2)**:873.

Farquhar, M. G., 1975, The primary glomerular filtration barrier—Basement membrane or epithelial slits? *Kidney Int.* **8**:197–211.

Farquhar, M. G., 1978, Structure and function in glomerular capillaries. Role of the basement membrane in glomerular filtration, in: *Biology and Chemistry of Basement Membranes* (N. A. Kefalides, ed.), pp. 43–80, Academic Press, New York.

Farquhar, M. G., 1980, Role of the basement membrane in glomerular filtration: Results obtained with electron dense tracers, in: *Functional Ultrastructure of the Kidney* (A. B. Maunsbach, T. S. Olsen, and E. I. Christensen, eds.), pp. 31–51, Academic Press, New York.

Farquhar, M. G., 1982, The glomerular basement membrane—a selective macromolecular filter, in: *The Cell Biology of Extracellular Matrix*, (E. D. Hay, ed.), pp. 375–378, Plenum Press, New York.

Farquhar, M. G., and Kanwar, Y. S., 1980, Characterization of anionic sites in the glomerular basement membranes of normal and nephrotic rats, *Renal Pathophysiology* (A. Leaf, G. Giebisch, L. Bolis, and S. Gorini, eds.), pp. 57–74, Raven Press, New York.

Farquhar, M. G., and Palade, G. E., 1961, Glomerular permeability. II. Ferritin transfer across the glomerular capillary wall in nephrotic rats, J. Exp. Med. **114**:699–716.

Farquhar, M. G., Hopper, Jr., J., and Moon, H. D., 1959, Diabetic glomerulosclerosis: Electron and light microscopic studies, *Am. J. Path.* **35**:721–755.

Farquhar, M. G., Wissig, S. L., and Palade, G. E., 1961, Glomerular permeability: I. Ferritin transfer across the normal glomerular capillary wall, *J. Exp. Med.* **113**:47–66.

Farquhar, M. G., Courtoy, P. J., Lemkin, M. C., and Kanwar, J. S., 1982, Current knowledge of the functional architecture of the glomerular basement membrane, in: *New Trends in Basement Membrane Research* (K. Kuhn, R. Timpl, and H. Schone, eds.), pp. 9–29, Raven Press, New York.

Farquhar, M. G., Lemkin, M. C., and Stow, J. L., 1985, Role of proteoglycans in glomerular function and pathology, in: *Nephrology*, Volume 1 (R. R. Robinson, ed.), pp. 580–600, Springer-Verlag, New York.

Foidart, J.-M., Foidart, J. B., and Mahieu, P. R., 1980, Synthesis of collagen and fibronectin by glomerular cells in culture, *Renal Physiol.* **3**:183–192.

Folkman, J., Klagsbrun, M., Sasse, J., Wadzinski, M., Ingber, D., and Vlodavsky, I., 1988, A heparin-binding angiogenic protein—basic fibroblast factor—is stored within basement membrane, *Am. J. Path.* **130**:393–400.

Graf, J., Iwamoto, Y., Sasaki, M., Martin, G. R., Kleinman, H. K., Robey, F. A., and Yamada, Y., 1987, Identification of an amino acid sequence in laminin mediating cell attachment, chemotaxis, and receptor binding, *Cell* **48**:989–996.

Grant, D. S., and LeBlond, C. P., 1988, Immunogold quantitation of laminin, type IV collagen, and heparan sulfate proteoglycan in a variety of basement membranes, *J. Histochem. Cytochem.* **36**:271–283.

Grant, D. S., LeBlond, C. P., Kleinman, H. K., Inoue, S., and Hassell, J. R., 1989, The incubation of laminin, collagen IV, and heparan sulfate proteoglycan at 35°C yields basement-like structures, *J. Cell Biol.* **108**:1567–1574.

Hay, E. D., 1981, Extracellular matrix, *J. Cell Biol.* **91**:205–226S.

Heathcote, J. G., and Grant, M. E., 1981, The molecular organization of basement membranes, *Int. Rev. Connect. Tissue Res.* **9**:191–264.

Hudson, B. G., Wieslander, J., Wisdom, B. J., and Noelken, M., 1989, Goodpasture syndrome: Molecular architecture and function of basement membrane antigen, *Lab. Invest.* **61**:256–269.

Kallunki, P., and Tryggvason, K., 1990, Cloning and partial sequence analysis of cDNA for the human basement membrane heparan sulfate proteoglycan, *J. Cell Biol. Abstr.* **111**:2220.

Kalluri, R., Gunwar, S., Ballester, F., Timoneda, J., Chonko, A., Edwards, S., Noelken, M. E., and Hudson, B. G., 1990, The molecular identity of dimeric subunits of the noncollagenous domain of glomerular basement membrane collagen that react with Goodpasture antibodies, *J. Am. Soc. Nephrol.* **1**:527.

Kanwar, Y. S., and Farquhar, M. G., 1979a, Anionic sites in the glomerular basement membrane. In vivo and in vitro localization to the laminae rarae by cationic probes, *J. Cell Biol.* **81**:137–153.

Kanwar, Y. S., and Farquhar, M. G., 1979b, Presence of heparan sulfate in the glomerular basement membrane, *Proc. Natl. Acad. Sci. USA* **76**:1303–1307.

Kanwar, Y. S., and Farquhar, M. G., 1979c, Isolation of glycosaminoglycans (heparan sulfate) from the glomerular basement membrane, *Proc. Natl. Acad. Sci. USA* **76**:4493–4497.

Kanwar, Y. S., Linker, A., and Farquhar, M. G., 1980, Increased permeability of the glomerular basement membrane to ferritin after removal of glycosaminoglycans (heparan sulfate) by enzyme digestion, *J. Cell Biol.* **86**:688–693.

Kanwar, Y. S., Hascall, V. C., and Farquhar, M. G., 1981, Partial characterization of newly synthesized proteoglycans isolated from the glomerular basement membrane, *J. Cell Biol.* **90**:527–532.

Kanwar, Y. S., Rosenzweig, L. J., and Jakubowski, M. L., 1983, Distribution of *de novo* synthesized sulfated glycosaminoglycans in the glomerular basement membrane and mesangial matrix, *Lab. Invest.* **49**:216–225.

Kanwar, Y. S., Veis, A., Kimura, J. H., and Jakubowski, M. L., 1984, Characterization of heparan sulfate-proteoglycan of glomerular basement membranes, *Proc. Natl. Acad. Sci. USA* **81**:762–766.

Karnovsky, M. J., 1979, The structural basis for glomerular filtration, in: *Kidney Disease*, (J. Churg, B. H. Spargo, F. K. Mostofi, and M. R. Abell, eds.), pp. 1–41, Williams & Wilkins, Baltimore.

Keene, D. R., Sakai, L. Y., Lunstrum, G. P., Morris, N. P., and Burgeson, R. E., 1987, Type VII collagen forms an extended network of anchoring fibrils, *J. Cell Biol.* **104**:611–621.

Kefalides, N. A., 1973, Structure and biosynthesis of basement membranes, *Int. Rev. Connect. Tissue Res.* **6**:63–104.

Kefalides, N. A., 1978, Current status of chemistry and structure of basement membranes, in: *Biology and Chemistry of Basement Membranes* (N. A. Kefalides, ed.), pp. 215–228, Academic Press, New York.

Kefalides, N. A., 1979, Biochemistry and metabolism of basement membranes, *Int. Rev. Cytol.* **61**:167–228.

Kelley, V. E., Kewley, M. A., Steven, F. S., and Williams, G., 1977, Immunofluorescence studies with a specific antiserum to the microfibrillar protein of elastic fibres. Location in elastic and nonelastic connective tissues, *Immunology* **33**:381–386.

Kerjaschki, D., and Farquhar, M. G., 1983, Immunocytochemical localization of the Heymann nephritis antigen (gp330) in glomerular epithelial cells of normal Lewis rats, *J. Exp. Med.* **157**:667–686.

Kerjaschki, D., Sharkley, D. J., and Farquhar, M. G., 1984, Identification and characterization of podocalyxin—the major sialoglycoprotein of the renal glomerular epithelial cell, *J. Cell Biol.* **98**:1591–1596.

Kerjaschki, D., Vernillo, A. T., and Farquhar, M. G., 1985, Reduced sialylation of podocalyxin—the major sialoprotein of the rat kidney glomerulus—in aminonucleoside nephrosis, *Am. J. Pathol.* **118**:343–349.

Kerjaschki, D., Sawada, H., and Farquhar, M. G., 1986, Immunoelectron microscopy in kidney research: Some contributions and limitations, *Kidney Int.* **30**:229–245.

Kerjaschki, D., Miettinen, A., and Farquhar, M. G., 1987, Initial events in the formation of immune deposits in passive Heymann nephritis: Gp330–antigp330 immune complexes form in epithelial coated pits and rapidly become attached to the GBM, *J. Exp. Med.* **166**:109–128.

Kerjaschki, D., Ojha, P. P., Susani, M., Horvat, R., Binder, S., Hovorka, A., Hillemanns, P., and Pytela, R., 1989, A β1-integrin receptor for fibronectin in human kidney glomeruli, *Am. J. Pathol.* **134**:481–489.

Killen, P. D., and Striker, G. F., 1979, Human glomerular visceral epithelium cells synthesize a basal lamina collagen in vitro, *Proc. Natl. Acad. Sci. USA* **76**:3518–3522.

Klein, D. J., Brown, D. M., and Oegema, T. R., 1986, Partial characterization of heparan and dermatan sulfate proteoglycans synthesized by normal rat glomeruli, *J. Biol. Chem.* **261:**16636–16652.

Klein, D. J., Brown, D. M., Oegema, T. R., Brenchley, P. E., Anderson, J. C., Dickinson, M. A. J., Horigan, E. A., and Hassell, J. R., 1988, Glomerular basement membrane proteoglycans are derived from a large precursor, *J. Cell Biol.* **106:**963–970.

Kleinman, H. K., McGarvey, M. L., Hassell, J. R., Star, V. L., Cannon, F. B., Laurie, G. W., and Martin, G. R., 1986, Basement membrane complexes with biological activity, *Biochemistry* **25:**312–318.

Korhonen, M., Ylanne, J., Laitinen, L., and Virtanen, I., 1990, The α1–α6 subunits of integrins are characteristically expressed in distinct segments of developing and adult human nephron, *J. Cell Biol.* **111:**1245–1254.

Krakower, C. A., and Greenspon, S. A., 1951, Localization of nephrotoxic antigen within the isolated renal glomerulus, *Arch. Pathol.* **51:**629–639.

Laliberté, F., Sapin, C., Belair, M. F., Druet, P., and Bariety, J., 1978, The localization of the filtration barrier in normal rat glomeruli by ultrastructural immunoperoxidase techniques, *Biol. Cell.* **31:**15–26.

Laurie, G. W., LeBlond, C. P., Inoue, S., Martin, G. R., and Chung, A., 1984, Fine structure of the glomerular basement membrane and immunolocalization of five basement membrane components to the lamina densa (basal lamina) and its extensions in both glomeruli and tubules of the rat kidney, *Am. J. Anat.* **169:**463–481.

LeBlond, C. P., and Inoue, S., 1989, Structure, composition, and assembly of basement membrane, *Am. J. Anat.* **185:**367–390.

Ledbetter, S. R., Tyree, B., Hassell, J. R., and Horigan, E. A., 1985, Identification of the precursor protein to basement membrane heparan sulfate proteoglycans, *J. Biol. Chem.* **260:**8106–8113.

Lemkin, M. L., and Farquhar, M. G., 1981, Sulfated and nonsulfated glycosaminoglycans and glycopeptides are synthesized by kidney in vivo and incorporated into glomerular basement membranes, *Proc. Natl. Acad. Sci. USA* **78:**1726–1730.

Lindahl, U., and Höök, M., 1978, Glycosaminoglycans and their binding to biological macromolecules, *Ann. Rev. Biochem.* **47:**385–417.

Martin, G. R., Timpl, R., and Kühn, K., 1988, Basement membrane proteins: Molecular structure and function, *Adv. Protein Chem.* **39:**1–50.

McCarthy, K. J., and Couchman, J. R., 1990, Basement membrane chondroitin sulfate proteoglycans: Localization in adult rat tissues, *J. Histochem. Cytochem.* **38:**1479–1486.

McCarthy, K. J., Accavitti, M. A., and Couchman, J. R., 1989, Immunological characterization of a basement membrane-specific chondroitin sulfate proteoglycan, *J. Cell Biol.* **109:**3187–3198.

Meezan, E., Brendel, K., Hjelle, J. T., and Carlson, E. C., 1978, A versatile method for the isolation of ultrastructurally and chemically pure basement membranes without sonication, in: *Biology and Chemistry of Basement Membranes* (N. A. Kefalides, ed.), pp. 31–42, Academic Press, New York.

Miettinen, A., Stow, J. L., Mentone, S., and Farquhar, M. G., 1986, Antibodies to basement membrane heparan sulfate proteoglycans bind to the laminae rarae of the glomerular basement membrane (GBM) and induce subepithelial GBM thickening, *J. Exp. Med.* **163:**1064–1084.

Mohan, P. S., and Spiro, R. G., 1986, Macromolecular organization of basement membranes, *J. Biol. Chem.* **261:**4328–4336.

Mynderse, L. A., Hassel, J., Kleinman, H. K., Martin, G. R., and Martinez-Hernandez, A., 1983, Loss of heparan sulfate proteoglycan from glomerular basement membrane of nephrotic rats, *Lab. Invest.* **48:**292.

Noonan, D. M., Horigan, E. A., Ledbetter, S. R., Vogeli, G., Sasaki, M., Yamada, Y., and Hassell, J. R., 1988, Identification of cDNA clones encoding different domains of the basement membrane heparan sulfate proteoglycan, *J. Biol. Chem.* **263:**16379–16387.

Oberley, T. D., Mosher, D. F., and Mills, M. D., 1979, Localization of fibronectin within the renal glomerulus and its production by cultured glomerular cells, *Am. J. Path.* **96:**651–667.

Palade, G. E., and Farquhar, M. G., 1965, A special fibril of the dermis, *J. Cell Biol.* **27:**215–222.

Parthasarathy, N., and Spiro, R. G., 1981, Characterization of the glycosaminoglycan component of the renal glomerular basement membrane and its relationship to the peptide portion, *J. Biol. Chem.* **256**:507–513.

Parthasarathy, N., and Spiro, R. G., 1984, Isolation and characterization of the heparan sulfate proteoglycan of the bovine glomerular basement membrane, *J. Biol. Chem.* **259**:12749–12755.

Paulsson, M., 1987, Noncollagenous proteins of basement membranes, *Collagen Relat. Res.* **7**:443–461.

Pettersson, E. E., and Colvin, R. B., 1978, Cold-insoluble globulin (fibronectin, LETS protein) in normal and diseased human glomeruli: Papain-sensitive attachment to normal glomerular and deformation in crescents, *Clin. Immunol. Immunopathol.* **11**:425–536.

Pietromonaco, S. F., and Farquhar, M. G., 1988, Identification and characterization of a cDNA encoding the core protein of heparan sulfate proteoglycans from the rat glomerular basement membrane, *J. Cell Biol.* **107**:909.

Price, R. G., and Spiro, R. G., 1977, Studies on the metabolism of the renal glomerular basement membrane: Turnover measurements in the rat with the use of radiolabeled amino acids, *J. Biol. Chem.* **252**:8597–8602.

Reeves, W., Kanwar, Y. S., and Farquhar, M. G., 1980, Assembly of the glomerular filtration surface. Differentiation of anionic sites in glomerular capillaries of newborn rat kidney, *J. Cell Biol.* **85**:735–753.

Rennke, H. G., Cotran, R. S., and Venkatachalam, M. A., 1975, Role of molecular charge in glomerular permeability. Tracer studies with cationized ferritin, *J. Cell Biol.* **67**:638–646.

Risteli, J., Bachinger, H. P., Engle, J., Furthmayr, H., and Timpl, R., 1980, 7-S collagen: characterization of an unusual basement membrane structure, *Eur. J. Biochem.* **108**:239–250.

Rohde, H. G., Wick, G., and Timpl, R., 1979, Immuochemical characterization of the basement membrane glycoprotein laminin, *Eur. J. Biochem.* **102**:195–201.

Rohrbach, D. H., Wagner, C. W., Star, V. L., Martin, G. R., Brown, K. S., and Yoon, J. W., 1983, Reduced synthesis of basement membrane heparan sulfate proteoglycan in streptozotocin-induced diabetic mice, *J. Biol. Chem.* **258**:11672–11677.

Roll, F. J., Madri, J. A., Albert, J., and Furthmayr, H., 1980, Codistribution of collagen types IV and AB in the basement membrane and mesangium of the kidney, *J. Cell Biol.* **85**:597–616.

Romen, W., Schultze, H., and Hempel, K., 1976, Synthesis of the glomerular basement membrane in the rat kidney. Autoradiographic studies with the light and electron microscope, *Virchows Arch. B* **20**:125–137.

Rosenzweig, L. J., and Kanwar, Y. S., 1982, Removal of sulfated (heparan sulfate) or nonsulfated (hyaluronic acid) glycosaminoglycans results in increased permeability of the glomerular basement membrane to ^{125}I-bovine serum albumin, *Lab. Invest.* **47**:177–184.

Ryan, G. B., and Karnovsky, M. J., 1976, Distribution of endogenous albumin in the rat glomerulus: Role of hemodynamic factors in glomerular barrier function, *Kidney Int.* **9**:36–45.

Ryan, G. B., Hein, S. J., and Karnovsky, M. J., 1976, Glomerular permeability in proteins. Effects of hemodynamic factors on the distribution of endogenous immunoglobin G and exogenous catalase in the rat glomerulus, *Lab. Invest.* **34**:415–427.

Sariola, H., Timpl, R., Von der Mark, K., Mayne, R., Fitch, J. M., Linsenmayer, T. F., and Ekblom, P., 1984, Dual origin of glomerular basement membrane, *Dev. Biol.* **101**:86–96.

Scheinman, J. I., and Fish, A. J., 1978, Human glomerular cells in culture. The subcultured cell types bearing glomerular antigens, *Am. J. Pathol.* **92**:125–139.

Scheinman, J. I., Foidart, J.-M., and Michael, A. F., 1980a, The immunohistology of glomerular antigens. V. The collagenous antigens of the glomerulus, *Lab. Invest.* **43**:373–381.

Scheinman, J. I., Foidart, J.-M., Gehron-Robey, P., Fish, A. J., and Michael, A. F., 1980b, The immunohistology of glomerular antigens. IV. Laminin a defined noncollagen basement membrane glycoprotein, *Clin. Immunol. Immunopathol.* **15**:141–154.

Schnabel, E., Dekan, G., Miettinen, A., and Farquhar, M. G., 1989, Biogenesis of podocalyxin—the major glomerular sialoglycoprotein in rat kidney, *Eur. J. Cell Biol.* **48**:313–326.

Schnabel, E., Anderson, J. M., and Farquhar, M. G., 1990, The tight junction protein Z0-1 is concentrated along slit diaphragms of the glomerular epithelium *J. Cell Biol.* **111**:1255–1263.

Seiler, M. W., Rennke, H. G., Venkatachalam, M. A., and Cotran, R. S., 1977, Pathogenesis of polycation-induced alterations ("fusion") of glomerular epithelium, *Lab. Invest.* **36**:48–61.

Shimomura, H., and Spiro, R., 1987, Studies on macromolecular components of human glomerular basement membrane and alterations in diabetes, *Diabetes* **36:**374–381.

Spiro, R. G., 1972, Basement membranes and collagens, in: *Glycoproteins* (A. Gottschalk, ed.), 2nd ed., Part B, pp. 964–999, Elsevier, Amsterdam.

Spiro, R. G., 1978, Nature of the glycoprotein components of the basement membranes, *Ann. N.Y. Acad. Sci.* **312:**106–121.

Stow, J. L., Sawada, H., and Farquhar, M. G., 1985, Basement membrane heparan sulfate proteoglycans are concentrated in the laminae rarae and in podocytes of the rat renal glomerulus, *Proc. Natl. Acad. Sci. USA* **82:**3296–3300.

Stow, J. L., Soroka, C. J., MacKay, K., Striker, L., Striker, G., and Farquhar, M. G., 1989, Basement membrane heparan sulfate proteoglycan is the main proteoglycan synthesized by glomerular epithelial cells in culture, *Am. J. Pathol.* **135:**637–646.

Striker, G. E., Killen, P. D., and Farin, F. M., 1980, Human glomerular cells *in vitro:* Isolation and characterization, *Transplant. Proc.* **12:**88–99.

Timpl, R., 1989, Structure and biological activity of basement membrane proteins, *Eur. J. Biochem.* **180:**487–502.

Timpl, R., and Dziadek, M., 1986, Structure, development, and molecular pathology of basement membranes, *Int. Rev. Exp. Pathol.* **29:**1–112.

Timpl, R., Rohde, H., Robey, P. C., Rennard, S. I., Foidart, J.-M., and Martin, G. R., 1979, Laminin—A glycoprotein from basement membranes, *J. Biol. Chem.* **254:**9933–9937.

Trelstad, R. L., Hayashi, K., and Toole, B. P., 1974, Epithelial collagens and glycosaminoglycans in the embryonic cornea. Macromolecular order and morphogenesis in the basement membrane, *J. Cell Biol.* **62:**815–830.

Van Den Heuvel, L. P. W. J., Van Den Born, J., Van De Velden, T. J. A. M., Veerkamp, J. H., Monnens, L. A. H., Schroder, C. H., and Berden, J. H. M., 1989, Isolation and partial characterization of heparan sulphate proteoglycan from the human glomerular basement membrane, *Biochem. J.* **264:**457–465.

Venkatachalam, M. A., and Rennke, H. G., 1978, The structural and molecular basis of glomerular filtration, *Circ. Res.* **43:**337–347.

Venkatachalam, M. A., and Rennke, H. G., 1980, Glomerular filtration of macromolecules: Structural, molecular and functional determinants, in: *Renal Pathophysiology* (A. Leaf, G. Giebisch, L. Bolis, and S. Gorini, eds.), pp. 43–56, Raven Press, New York.

Vracko, R., 1978, Anatomy of basal lamina scaffold and its role in maintenance of tissue structure, in: *Biology and Chemistry of Basement Membranes* (N. A. Kefalides, ed.), pp. 165–176, Academic Press, New York.

Walker, F., 1973, The origin, turnover and removal of glomerular basement membrane, *J. Path.* **110:**233–244.

Yaoita, E., Oguri, K., Okayama, E., Kawasaki, K., Kobayashi, S., Kihara, I., and Okayama, M., 1990, Isolation and characterization of proteoglycan synthesized by cultured mesangial cells, *J. Biol. Chem.* **265:**522–531.

Yurchenco, P. D., and Schittny, J. C., 1990, Molecular architecture of basement membranes, *FASEB J.* **4:**1577–1590.

Yurchenco, P. D., Tsilibary, E. C., Charonis, A. S., and Furthmayr, H., 1986, Models for the self-assembly of basement membrane, *J. Histochem. Cytochem.* **34:**93–102.

Yurchenco, P. D., Cheng, Y.-S., and Ruben, G. C., 1987, Self-assembly of a high molecular weight basement membrane heparan sulfate proteoglycan into dimers and oligomers, *J. Biol. Chem.* **262:**17668–17676.

Chapter 12

Collagen and Other Matrix Glycoproteins in Embryogenesis

ELIZABETH D. HAY

1. Introduction

This chapter is devoted to the study of embryogenesis and we shall concentrate our attention on what is known about the roles of the major glycoproteins, collagen, fibronectin, laminin, and tenascin, in the development of vertebrate embryos. The role of interactions of cells with the proteoglycan/glycoprotein class of ECM is covered in depth by Toole in Chapter 9, but will not be completely ignored in this chapter, because none of these molecules is an isle unto itself. Indeed, a given cell may possess receptors for and interact with a variety of ECM molecules simultaneously (Chapter 10). Ultimately, we must also understand the interactions of matrix molecules with each other (Chapter 7), as well as with the cell, if we are to understand completely their roles in morphogenesis.

The effects of collagen, fibronectin, and laminin on cells have been the subject of extensive research in the decade since the first edition of this book appeared. A common approach to the study of the interaction of a cell with a matrix molecule is to analyze the ability of the cell to adhere to a planar substratum coated with the molecule in question. This approach is reviewed in Chapters 10 and 12 of the first edition of this book and more recent data derived from studies of cells plated onto ECM-coated substrata *in vitro* are discussed in other chapters in the present edition. Other subjects discussed in the original

Abbreviations used in this chapter: bFGF, basic fibroblast growth factor; BM, basement membrane (basal lamina); CAM, cell adhesion molecule; ECM, extracellular matrix; GAG, glycosaminoglycan; G protein, most commonly is a plasmalemma component involved in signal transduction, but the term is also used for a different protein in the VSV coat; HA, hyaluronic acid, hyaluronan; HSPG, heparan sulfate proteoglycan; IF, intermediate filament(s); MDCK, Madin–Darby canine kidney cell; PG, proteoglycan; TEM, transmission electron microscopy; TGF, transforming growth factor; VSV, vesicular stomatitis virus.

ELIZABETH D. HAY • Department of Anatomy and Cellular Biology, Harvard Medical School, Boston, Massachusetts 02115.

Cell Biology of Extracellular Matrix, Second Edition, edited by Elizabeth D. Hay, Plenum Press, New York, 1991.

edition of this chapter that will not be developed in the present chapter include collagen type transitions in limb regeneration and the induction of cartilage in minced muscle by bone morphogenetic protein. Known to be attached to the collagen in bone (Urist *et al.*, 1979), this factor(s) has been shown more recently to belong to the TGF-β family. Other examples of cooperative effects of growth factors and ECM will be discussed here. It is becoming increasingly clear that ECM can affect cell growth and differentiation not only by the direct attachment of its component molecules to receptors on the cell's surface, but also by binding and/or activating other molecules, like TGF-β and bFGF (Taub *et al.*, 1990; Runyan *et al.*, 1990; Vukicevic *et al.*, 1991), that in turn also react with the cell. While we will emphasize in this chapter what we think are the direct effects of matrix glycoproteins on embryogenesis, the student must bear in mind the complexity introduced by the known ability of ECM to bind non-ECM molecules (Chapter 8, Section 4.3).

2. Distribution and General Functions in the Early Embryo

In the avian (Fig. 12-1) and mammalian embryo, basal laminae first appear under the ectoderm and endoderm when the primitive streak is forming, and the mesenchymal cells migrating from the primitive streak between ectoderm and endoderm move along these basal laminae (Hay, 1968) to fill the space between the two layers, forming a third germ layer, the mesoderm. The formation of the germ layers is called gastrulation. As the primitive streak regresses the notochord forms in front of it from mesodermal mesenchyme. The notochord is an ECM-covered epithelial rod that establishes the primary body axis of the vertebrate embryo and is present at some time in all members of the phylum Chordata (of which vertebrates are a subphylum).

The somites consist of approximately 40 paired epithelial balls that condense from mesodermal mesenchyme lateral to the notochord to delineate the segmental plan of the avian and mammalian embryonic bodies. A lumen forms in the middle of each condensing somitic mass (inset, Fig. 12-1) and a basal lamina is deposited around it, following development of characteristic epithelial apical–basal polarity (Section 3) by the cells. Subsequently, intermediate (nephrogenic) and lateral (coelomic) mesodermal epithelia condense from mesenchyme peripheral to the somites (lower right, Fig. 12-2). Later, the somite gives rise to the myotome (upper right Fig. 12-2) and various mesenchymal cells (Section 5) at about the time that crest mesenchyme begins to migrate away from the neural tube on newly forming ECM. Subsequently, definitive organs begin to appear in the embryo as a result of tissue interactions between epithelial cells and mesenchymal cells that often involve ECM (Section 4).

2.1. Collagen

The first collagen to appear in the early embryo is in the basal laminae of the ectoderm and endoderm and then, of the notochord and the condensing

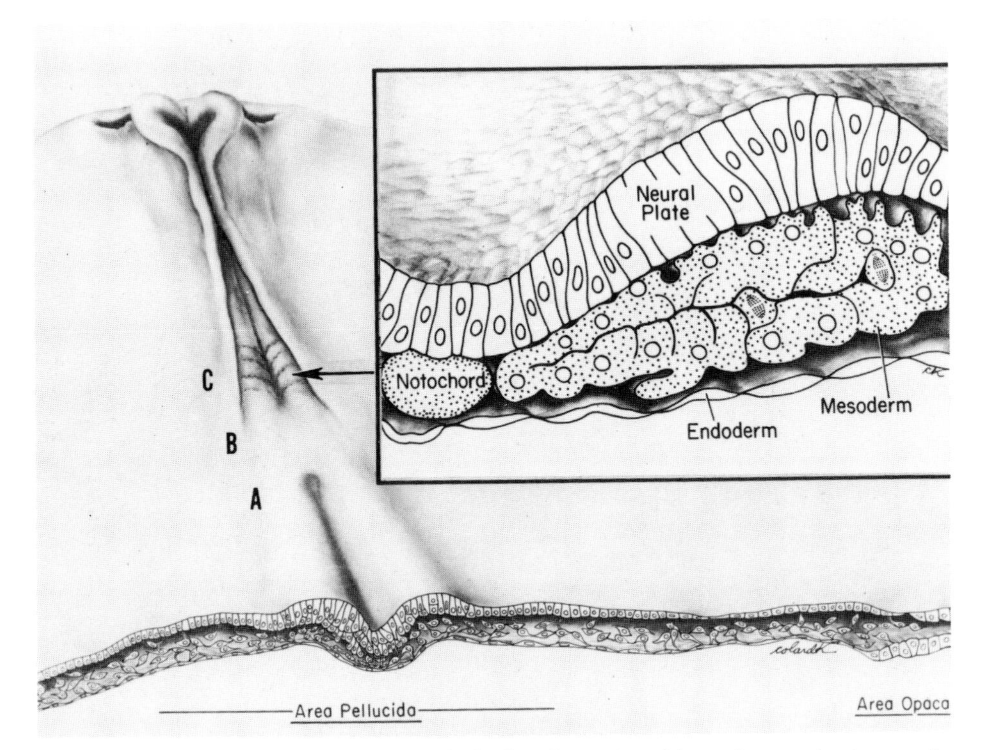

Figure 12-1. Diagram of a stage 8 (1 day old) chick embryo viewed from above and cut across the middle of the primitive streak. In the region of the streak, primary mesenchymal cells are detaching from the overlying ectoderm to migrate between ectoderm and endoderm. Some fuse with endoderm. In the area between A (Hensen's node) and B, the notochord has formed in the midline, and mesoderm is organized into somitomeres that will give rise to somites (Meier, 1979). In the region between B and C, the neural plate is starting to develop. At level C, three somites have formed, one of which is shown in cross section in the inset. Between levels B and C, the first definitive connective tissue fibrils are appearing, the basal laminae are thickening, and the matrix between neural plate and mesoderm is becoming rich in sulfated GAG. (From Hay, 1968.)

mesodermal epithelia, and of the neural tube. The earliest basal laminae probably contain type IV collagen, which, along with other components, provides a structure that both epithelial cells and mesenchymal cells can adhere to. At some point, these embryonic basal laminae develop filtering and the various structural roles of definitive basement membranes (Chapter 11).

The first fibrillar collagen appears around the notochord and is probably type II collagen, a very widely distributed collagen in the early avian embryo (Kosher and Solursh, 1989; Fitch *et al.*, 1989). In addition to the structural strength it gives to the primitive skeletal rod, type II collagen may help mediate the inductive effects of notochord (Section 4) on growth and differentiation. Type I collagen appears around the notochord and neural tube (upper right, Fig. 12-2) at the time of formation of the neural crest (McCarthy and Hay, 1991). The collagen fibrils are too small (10 nm in diameter) to demonstrate overt striations, but immunohistochemistry (Fig. 12-3) shows them to be widespread in

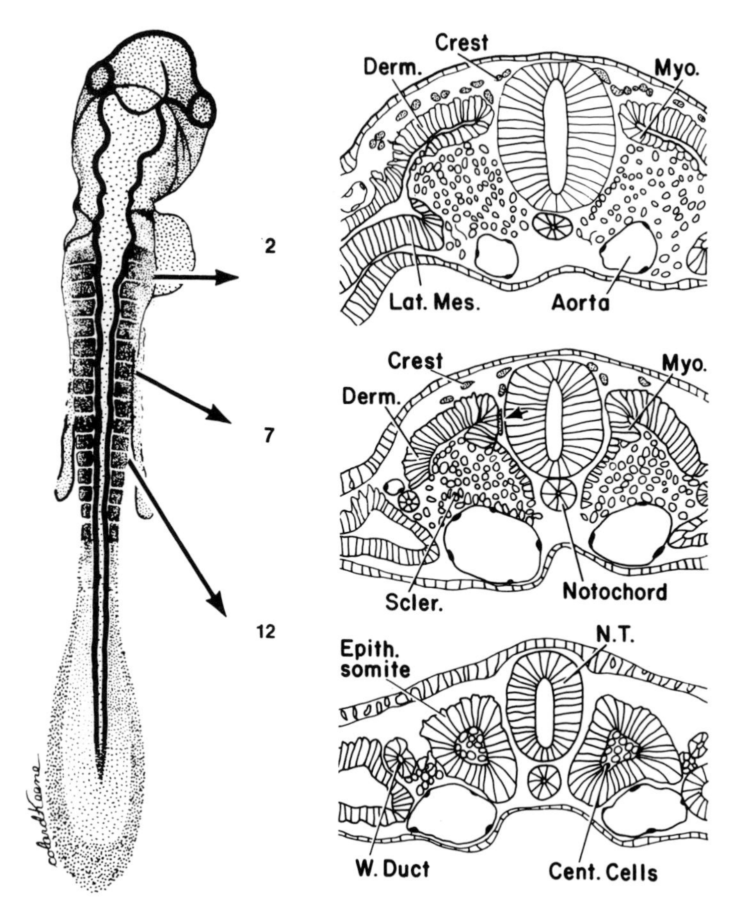

Figure 12-2. Camera lucida drawing of sections of the 16-somite (stage 12) chick embryo at 2 days. A whole mount is shown on the left. These levels are of interest because they show the epithelial (Epith.) somite (level of somite 12, lower right), beginning of neural crest (Crest) and sclerotome (Scler.) formation (level of somite 7, middle right), and a more advanced stage of mesenchyme and myotome (Myo.) formation (level of somite 2, upper right). Derm., dermatome; Lat. Mes., lateral mesoderm; N.T., neural tube; W., Wolffian (nephric); Cent., central; arrow, level 7, ventral crest cell. (From McCarthy and Hay, 1991.)

Figure 12-3. Peroxidase–antiperoxidase localization of anticollagen I (top) and antitenascin (bottom) at level 2 in the 2-day-old chick embryo shown in Fig. 12-2. The stain in the plasmalemma (arrow) is an artifact of aldehyde fixation. Collagen I (Coll I) is not found in the basement membrane (bm, top) of the neural tube (nt) or in BM elsewhere at any stage of development, whereas tenascin (Ten) is present early on (bm, bottom) and then disappears from BM. Collagen I is present in striated fibrils and also in 10-nm-wide nonstriated fibrils. Tenascin is also present in the 10-nm-wide fibrils, which may aggregate into larger groups (fibril, bottom). Tenascin occurs in interstitial bodies (isb) along with fibronectin (not shown). cp, cell process (probably neural crest); myo, myotome; nc, neural crest. Bar = 1 μm. (From McCarthy and Hay, 1991.)

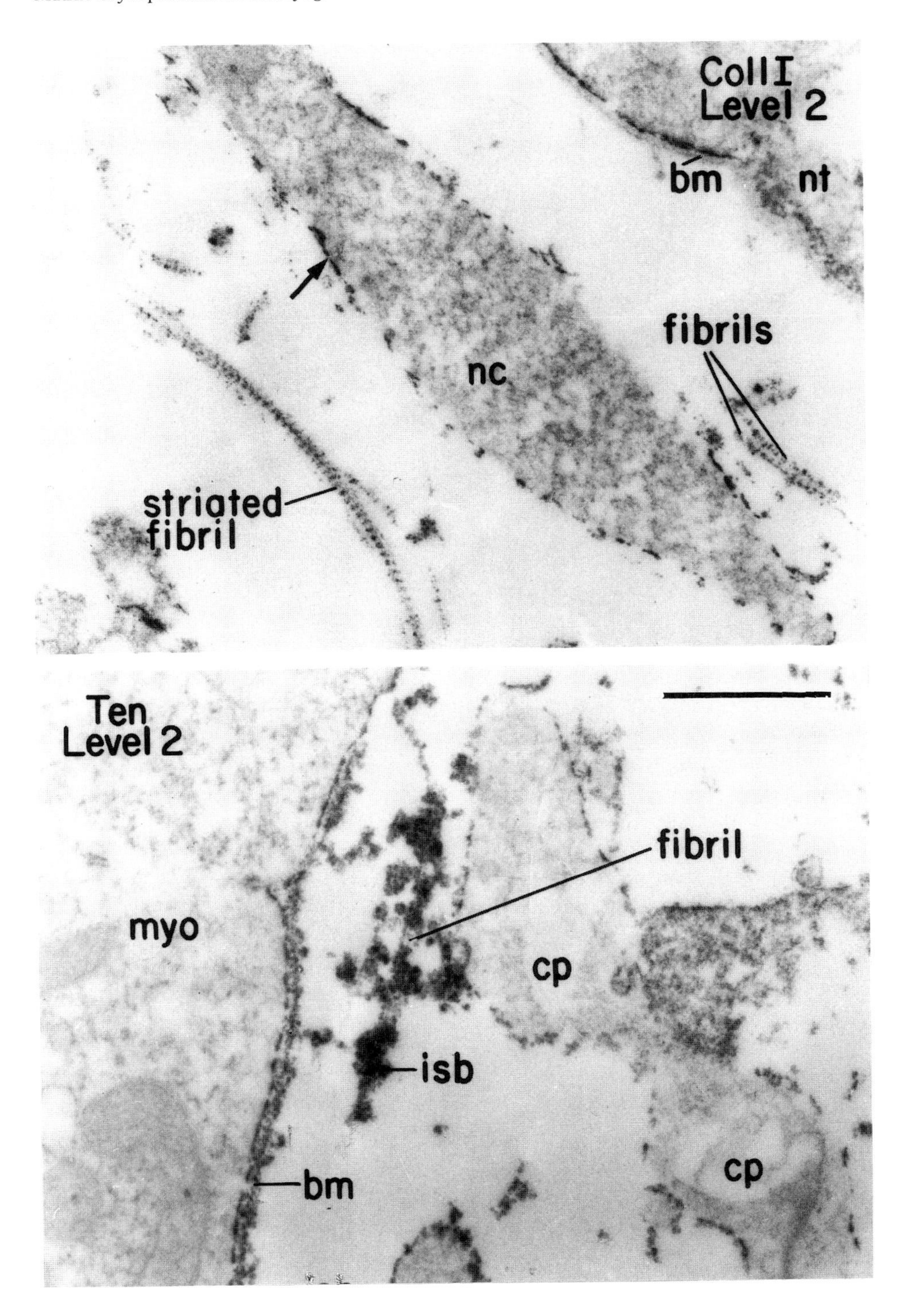

CollI Level 2
bm
nt
fibrils
nc
striated fibril
Ten Level 2
fibril
myo
cp
isb
bm
cp

the embryo at this stage between the ectoderm and somite and in other pathways that mesenchymal cells will use to migrate (Fig. 12-3). *In vitro* studies show that type I collagen gels promote the migration of crest and other mesenchymal cells quite well (Bilozur and Hay, 1988, Section 2.2). In addition to providing tensile strength and supporting adhesion and migration, collagens I and II have been shown to promote differentiation of embryonic cells (Section 4).

Possible functions for some of the numerous recently described collagens are mentioned in Chapters 1 and 6, but little is known of their role in embryogenesis at this time. It seems likely that other fibrillar types of collagen, such as III and V, can substitute for collagen type I in part, for Mov 13 mutant mice develop and survive to age 15 days (Jaenisch *et al.*, 1983). However, some tissues of the Mov 13 mouse embryo express type I collagen (Schwarz *et al.*, 1990).

2.2. Fibronectin

Fibronectin is a ubiquitous component of the ECM of the early embryo. In the avian embryo (Mayer *et al.*, 1981), it is present in the earliest formed basal laminae and in the 10-nm fibrils described above. It also occurs in clumps of nonfibrillar matrix called interstitial bodies (Low, 1970; Mayer *et al.*, 1981). Antibodies to fibronectin and its integrin receptor disrupt neural crest migration *in vivo* (Bronner-Fraser, 1986, 1990), but it is likely that the other ECM glycoproteins also present in the early embryo (e.g., collagen, laminin, and tenascin) are equally important substrata for mesenchymal cell migration (Sections 2.1, 2.3, 2.4). Fibronectin is essential for differentiation of red blood cells (Patel and Lodish, 1987) and it promotes spreading and growth of endothelial cells (see Chapter 8 for discussion of angiogenesis and vasculogenesis). Fibronectin interacts with many ECM molecules and is likely to contribute importantly to the structural organization of embryonic matrix, in addition to its role as a cell adhesion molecule (see Chapter 4; Hynes, 1981, 1990).

Studies of neural crest migration *in vitro* concluded that the cells prefer to migrate (1) on fibronectin tracks on culture dishes (Rovasio *et al.*, 1983) or (2) on either fibronectin or laminin tracks on the dishes (Newgreen, 1984). Studies of crest cell migration in 3D collagen or BM gels, however, consistently demonstrate that these mesenchymal cells can migrate in collagen or laminin with or without fibronectin or RGD-dependent receptors (Tucker and Erickson, 1984; Bilozur and Hay, 1988). Indeed, in BM gels the YIGSR peptide of laminin, but not RGD, inhibits the migration (Fig. 12-4). The behavior of cells in 3D matrix *in vitro* is quite likely to resemble their behavior in 3D matrix *in vivo* and suggests that collagen, laminin, and fibronectin are each suitable substrata. It is possible that mesenchymal cells can adapt their interaction with ECM to take advantage of whichever of these matrix proteins is present in the 3D gel (for further review see Duband and Thiery, 1987; Bilozur and Hay, 1988).

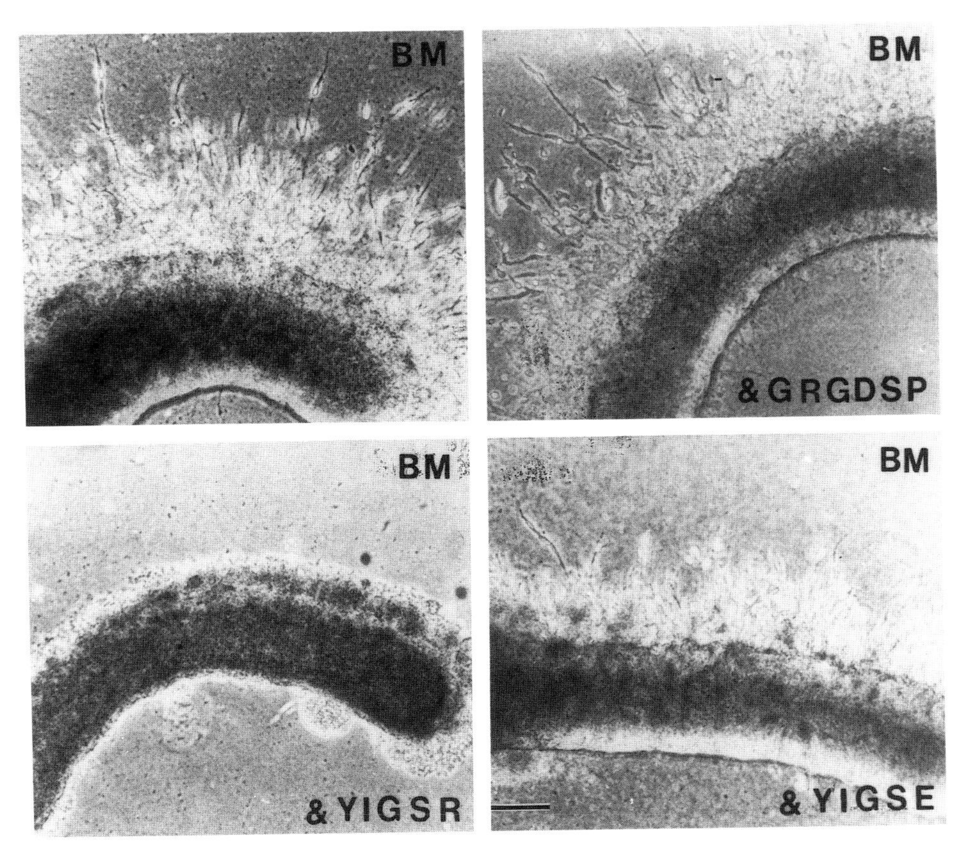

Figure 12-4. Migration of neural crest cells in 3D basement membrane (BM) gels after 22 hr in culture. Neural tubes prior to crest emigration were dissected from 2-day-old chick embryos and placed in the gels to allow crest to emigrate and then migrate. Upper left, control culture. Upper right, 500 μg/ml GRGDSP added at the beginning of culture did not affect migration. Lower left, 400 μg/ml YIGSR inhibited migration by interfering with laminin interaction. Lower right, 400 μg/ml of a control peptide YIGSE did not inhibit migration. Bar = 10 μm. (From Bilozur and Hay, 1988.)

2.3. Laminin

Laminin is present in the earliest of embryonic basal laminae and is a significant substratum for mesenchymal cell migration (Section 2.2; Duband and Thiery, 1987). As a component of the BM, it serves an important adhesion function for epithelia in general. It has also been reported to occur within embryonic interstices by light microscopy (Krotoski *et al.*, 1986), but was not observed in interstitial fibrils or interstitial bodies by electron microscopic immunohistochemistry (see McCarthy and Hay, 1991). Laminin, collagen, and fibronectin may promote epithelial differentiation (specific protein synthesis) by mechanisms that involve indirect effects on the cytoskeleton (Sections 4, 7) or direct effects on genes (Section 8). Laminin also plays a role in epithelial polarization of developing kidney tubules (Section 4.3), and it promotes the differentiation of embryonal carcinoma cells into neurons (Sweeney *et al.*, 1990).

These different effects of laminin may involve different cell binding sites on this large, complex molecule, which can take several different forms (see Graf *et al.*, 1987; Mercurio, 1990; and Chapter 4). Neural crest migration on basal laminae (Erickson, 1987) may involve interaction not only with the YIGSR binding site in laminin (Bilozur and Hay, 1988), but also with the laminin–HSPG complex (Bronner-Fraser and Lallier, 1988). Migration of neuron cell processes is greatly promoted by laminin through interaction with a site near the junction of the large globular domain with the rigid domain of the long arm (Chapter 4). A peptide from the A chain of laminin promotes the metastatic potential of melanoma cells (Kanemoto *et al.*, 1990).

The development of the BM gel, now marketed by Collaborative Research as "Matrigel," was a major advance in technology available to investigators for study of the role of laminin and other BM components in morphogenesis (Kleinman *et al.*, 1986). Use of BM gels in studying mesenchymal cell migration has been discussed (Fig. 12-4; Section 2.2). When Sertoli cells are cultured in 3D BM gels, they reorganize into cords that resemble testicular seminiferous cords and support spermatogenesis. RGD and YIGSR inhibit cord formation in BM gels (Hadley *et al.*, 1990). BM gels greatly promote mammary gland differentiation (Section 8). Growth factors contained in Matrigel collaborate with ECM to promote kidney tubulogenesis (Taub *et al.*, 1990) and osteogenesis (Vukicevic *et al.*, 1991). Matrigel is a product of tumor cells growing in mice (Kleinman *et al.*, 1986). Now that we are aware of multiple forms of laminin (Mercurio, 1990), it will be important in the future to compare their effects on developing cells.

2.4. Tenascin

Tenascin (also called cytotactin, see Chapter 3) is a recently described ECM molecule that occurs in embryonic fibrils (Fig. 12-3), in some basement membranes, and in other configurations in adult matrices (see McCarthy and Hay, 1991). Its functions in cell adhesion and migration are currently the subject of some controversy. It was initially assigned a more important role than fibronectin or laminin in neural crest migration, because its distribution in the early embryo seemed more restricted to crest cell pathways (Crossin *et al.*, 1986; Tan *et al.*, 1987; Mackie *et al.*, 1988). However, fibronectin, laminin, and collagen I have been shown to occur in crest pathways and to play a role in crest migration (Section 2.2). The guilt-by-association argument is strongest for the rostral half of the sclerotome, which is rich in tenascin and is a specific neural crest ventral route (see Rickman *et al.*, 1985; McCarthy and Hay, 1991, for review). Basal lamina is also present on the myotome surface upon which the crest cells are migrating (Loring and Erickson, 1987), and these basal laminae contain tenascin (Fig. 12-3), as well as laminin.

Experimental studies on the role of tenascin in cell attachment and migration include an *in vivo* approach by Bronner-Fraser (1988), who injected anti-tenascin into avian embryos and observed abnormal crest cells in the cranial region, but not in the trunk. *In vitro,* crest cells on tenascin may be round, suggesting weak attachment (Mackie *et al.*, 1988; Halfter *et al.*, 1989), or they may seem to attach to tenascin better than to fibronectin (Tan *et al.*, 1987). Tenascin inhibits spreading of embryonic amphibian cells on fibronectin (Riou *et al.*, 1990), but avian crest cells move faster *in vitro* on tenascin than on fibronectin (Halfter *et al.*, 1989). Tenascin has been reported to promote chondrogenesis on planar substrata *in vitro* (Mackie *et al.*, 1987), perhaps by rounding the cells. It may mediate intercellular interaction between neurons and glia *in vitro* (Grumet *et al.*, 1985). Little reference to tenascin is made in the sections that follow, because its role has yet to be clearly defined in the growth and differentiation of the embryo (see Chapter 4 and Erickson and Bourdon, 1989, for further discussion of *in vitro* studies).

3. Cell Polarity and Tissue Phenotype

Two tissue types comprise the early embryo under discussion here: epithelium and mesenchyme. The basic tissue organization of the primitive chordate is epithelial and so the early vertebrate embryo would be expected to recreate this plan and it does (Figs. 12-1, 12-2). Mesenchymal cells then derive from these primitive epithelia by tissue transitions (Section 5). Before considering in more detail the roles of collagen, fibronectin, and laminin in tissue transitions and tissue interactions in the embryo, we need to describe the features that characterize the several tissue phenotypes.

3.1. Epithelium

An epithelium is a tissue composed of contiguous cells, often cuboidal (Fig. 12-5), closely linked by cell junctions, that reside on top of BM. If the tissue is isolated from BM by enzyme digestion and placed on top of a type I collagen stroma or gel, the basal epithelial surface flattens and the cells remain on top of the ECM (Overton, 1977; Sugrue and Hay, 1981). Mature epithelial cells do not invade ECM when they are placed on top of it. Mesenchymal cells, on the other hand, are characterized by an outstanding ability to invade the interstices of almost any noncompact ECM.

Because epithelial cells sit on top of ECM, they have a surface (basal) that faces ECM and another surface (apical) that faces the "free" world. The free or apical surface may face the lumen of a duct or the outside of the body. The Golgi apparatus is often in the apical cytoplasm of cells secreting products into lumens. The apical surface may contain microvilli and it possesses a sialic acid-

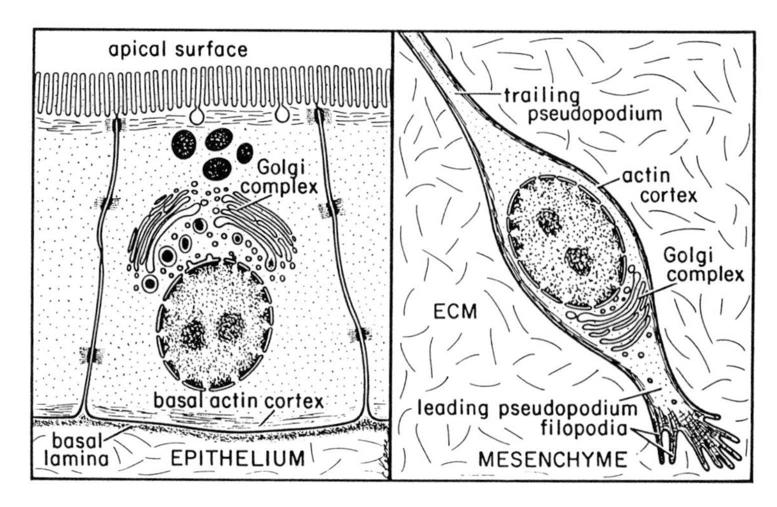

Figure 12-5. Diagram showing the major differences between an epithelial cell (left) and a mesenchymal cell (right). The epithelial cell sits on top of a basal lamina, under which connective tissue fibrils polymerize. The mesenchymal cell resides in the ECM and if it is migrating, it has a leading edge containing the Golgi complex. In the epithelial cell, the Golgi complex is usually in the apical cytoplasm, the free surface is specialized for dealing with the outside world, and the basal surface interacts with underlying ECM; the lateral surfaces form numerous junctions with adjacent cells. Mesenchymal cells interact with ECM all around the cell and do not have specialized cell junctions.

rich glycoprotein surface coat and enzymes that are quite unlike those on the basolateral surface (see Gumbiner, 1990, for a recent review). Studies of epithelial polarity and sorting have been greatly advanced by the availability of the MDCK epithelial cell line and of viruses that bud preferentially from the apical (e.g., influenza) or basolateral (e.g., VSV) epithelial surface (see Zuk *et al.*, 1989b).

Although the basolateral surface behaves as if basal and lateral were homogeneous for many functions, such as budding of VSV, and expression of integrins and syndecan, subtle (and not so subtle) differences do exist between basal and lateral surfaces. Cell junctions and adhesion molecules occur on lateral surfaces. Receptors that interact that basal lamina components would be expected to be best developed on the basal surface (e.g., $\alpha 6\beta 4$, Chapter 10) where they interact with the cytoskeleton, on the one side, and ECM, on the other (Section 7). Indeed, when keratinocytes, such as occur in stratified epithelia, mature, they lose the fibronectin receptor, $\alpha 5\beta 1$ (Adams and Watt, 1990). The integrin subunits $\beta 1$, $\alpha 2$, and $\alpha 3$ are localized in areas of cell contact in keratinocytes and antibodies to $\beta 1$ disrupt cell–cell adhesion under conditions that do not disrupt cell–substratum adhesion (Larjava *et al.*, 1990).

The lateral epithelial surface contains calcium-dependent cell adhesion molecules. The major CAMs that do not require calcium, N-CAM (brain, muscle) and Ng-CAM (neuron–glia), are not well developed in surface epithelia. Of the CAMs that require calcium (e.g., N-cadherin, brain; P-cadherin, placenta; and L-CAM, liver), L-CAM, the chicken equivalent of E-cadherin and uvomorulin, is the one that is widely distributed in epithelia. Transfection with L-CAM induces firoblastic cells to aggregate and polarize like epithelia (Takeichi *et al.*, 1989; McNeil *et al.*, 1990).

3.2. Mesenchyme

Mesenchymal cells in general do not express CAMs, desmosomes, or occluding junctions. They migrate as individuals, but gap junctions can be detected between them, probably in regions of collision (Hay, 1980). When migrating in ECM, they have a highly elongated, bipolar shape and they exhibit front end–back end polarity. That is to say, the front end extends filopodia (or ruffled membranes if the cells are moving on planar substrata), while the rear end is attenuated and recoils from time to time. The Golgi apparatus and actin mRNA are in the front end, which is the site of synthesis of new actin and plasmalemma (Section 6.2). Mesenchymal cell membranes contain integrins, such as $\alpha 5\beta 1$, for interacting with the surrounding matrix (Chapter 10) and, on occasion, syndecan (Chapter 9). A typical mesenchymal cell has vimentin IF and secretes fibrillar collagens and fibronectin (Section 5).

3.3. Muscle and Nerve

Muscular and nervous tissue can derive from either epithelium or mesenchyme. The neurons of the central nervous system are derived from neuroepithelium, but the neurons of peripheral ganglia and the Schwann cells enclosing peripheral neurite processes derive from neural crest mesenchyme. Although there is close contact between cells and cell processes in the central and peripheral nervous system, cell junctions are not well developed. N-CAMs appear early in the development of the neural tube and later are involved in neuron–neuron and neuron–glia interactions (see Grumet *et al.*, 1985; Cunningham and Edelman, 1990). Nerve growth cones migrating through ECM develop integrins (for review, see Lander, 1990). Neuron outgrowth on muscle involves ECM receptors and calcium-dependent and -independent CAMs (Bixby *et al.*, 1987).

Striated muscle is of two types, cardiac and skeletal. Cardiac muscle derives from coelomic epithelium and myotome from somite epithelium. During evolution of the vertebrate, myogenic mesenchymal cells also began to derive from the somites, probably at the myotome–dermatome junction. They form the striated muscle of the face, limbs, and ventral body wall (see Hay, 1990, for review). As is the case for neurons, there is little structural detail in the mature skeletal muscle fiber to belie its dual origin from either epithelium or mesenchyme. Mesenchyme-derived muscle cells are used almost exclusively for myogenic experiments *in vitro*. CAMs and ECM are involved in fusion of myoblasts and subsequent development of myotubes (Sanes *et al.*, 1986; Pouliot *et al.*, 1990) and neuromuscular junctions (Bayne *et al.*, 1984).

4. Tissue Interactions in the Embryo

Embryonic inductions occurring between two tissue types, such as epithelium and mesenchyme, are often referred to as tissue interactions to specify that the induction is taking place after the formation of the tissues from the germ layers, i.e., after gastrulation is complete. *Induction* is a term that can be extremely vague in usage, often referring to any exchange of information between cells. Moreover, there are early interactions between cells in the embryo that do not involve tissues. For example, determination of mesoderm at the time of formation of the amphibian blastula is the result of the action of growth factors derived from vegetal pole cells. These diffusable molecules act before gastrulation to induce neighboring animal pole cells to enter the mesodermal pathway (Altaba and Melton, 1990). The embryologist studying epithelial–epithelial and epithelial–mesenchymal inductions uses the definition of Grobstein (1955, p. 234), that embryonic induction is "developmentally significant interaction between closely associated but dissimilarly derived tissue masses."

The tissue interactions that Grobstein studied involve ECM in many, if not all, cases. As we saw above, the first ECM appears at the time the ectodermal and endodermal epithelia differentiate, and the potential molecules for involvement in tissue interaction include tenascin (Chiquet-Ehrismann *et al.*, 1986), fibronectin, laminin, and collagen. The primitive streak mesenchyme moves into the space between epiblast and hypoblast to form the notochord and mesoderm proper (Fig. 12-1). The first recognized embryonic induction between tissues, the induction by the notochord of the neural folds and their rounding up into a tube, can be classified as epithelial–epithelial.

4.1. Epithelial–Epithelial Interactions

The primary effect of the notochord on the neuroepithelium is cytoskeletal: The neural cells develop extensive microtubules, elongate, develop actin to constrict their apices, then roll up into a tube. Because the notochord has a thick ECM sheath, it seems likely that this ECM plays a role in neurulation. Another effect of notochord, to induce the somites to aggregate, is probably a physical effect of the presence of a notochord sheath, because epithelial somites can be induced merely by physically dividing the embryo along the midline (Lipton and Jacobson, 1974). The neural tube, however, determined the somite pattern.

Two epithelial–epithelial interactions occur early in eye development. The optic cup induces the lens to invaginate (upper left, Fig. 12-6) and then the lens induces the overlying ectoderm to secrete the primary corneal stroma (upper right and lower left, Fig. 12-6). ECM is present between retina and lens and between lens and corneal epithelium. Evidence that the ECM (a capsule of thick BM) produced by the lens plays a role in induction of the primary corneal stroma was obtained by Dodson and Hay in 1971 who cultured isolated corneal epithelium on the cell-free lens capsule (reviewed by Hay, 1981). On top of the

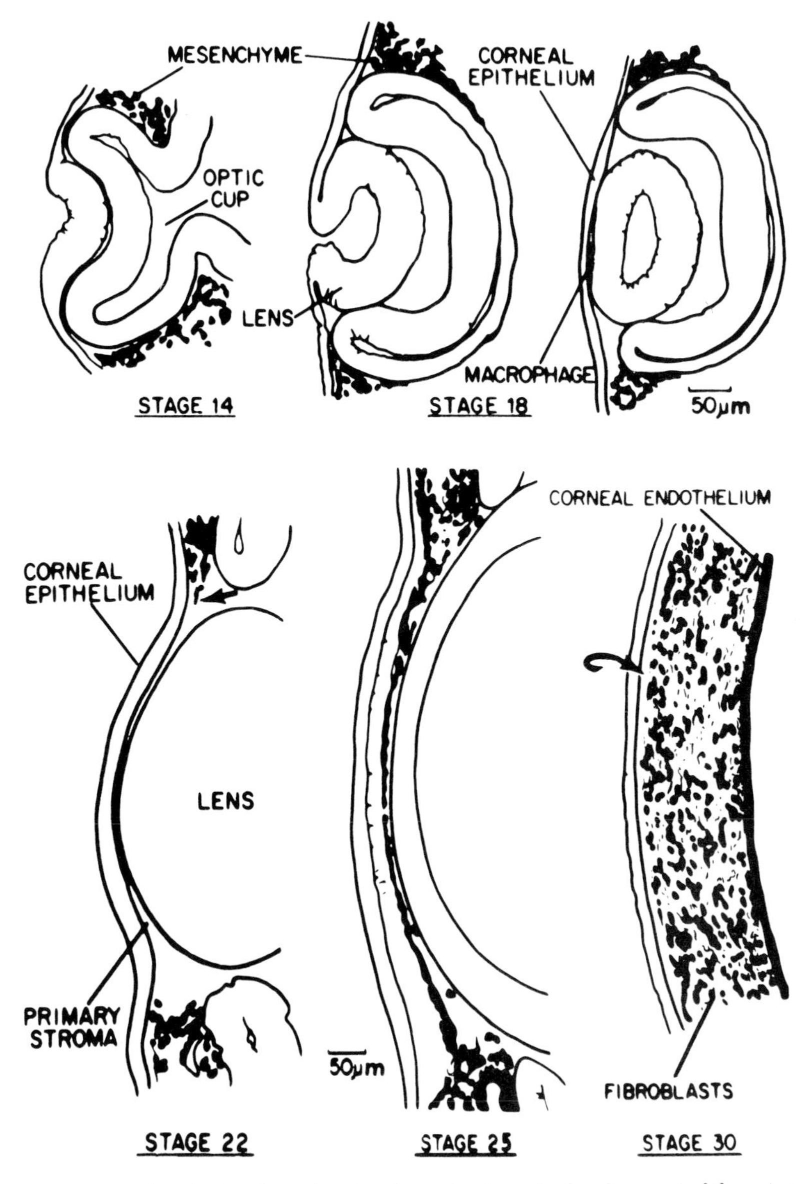

Figure 12-6. Camera lucida drawings showing the early stages in development of the avian cornea. The lens placode invaginates at stage 14 and induces formation of the optic cup. At 3 days of incubation, the lens vesicle pinches off, the optic cup begins to form retina, and the overlying ectoderm becomes the corneal epithelium (stage 18). Macrophages clean up debris associated with lens vesicle formation. By stage 22 (4 days), the corneal epithelium has secreted the primary corneal stroma, and the mesenchymal cells destined to become the corneal endothelium have started to invade the area (straight arrow). Endothelial cell migration is almost complete at stage 25 (4½–5 days). During stage 27 (5–5½ days), junctions between the endothelial cells are established and the primary stroma swells (not shown). It is then immediately invaded by the corneal fibroblasts from the sides of the eye. By stage 30 (6½–7 days), the fibroblasts occupy all layers of the stroma except for a narrow juxtaepithelial zone (curved arrow). (Reproduced with permission of Karger from Hay and Revel, 1969.)

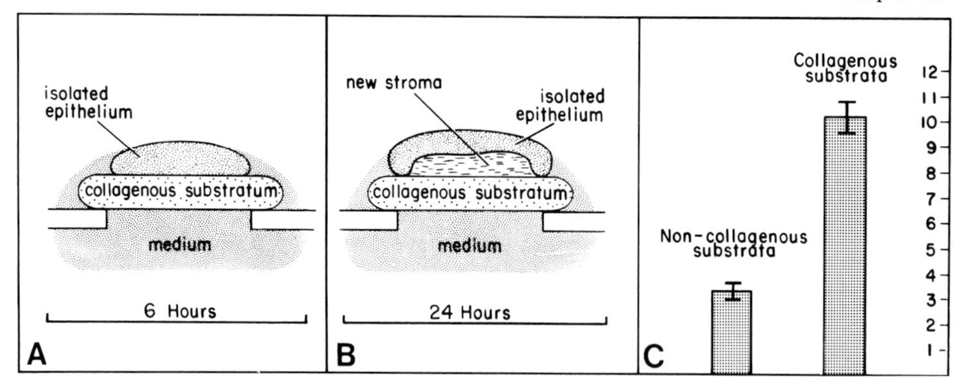

Figure 12-7. Diagrammatic summary (Hay, 1981) of the experiments of Dodson and Hay and Meier and Hay on the production of primary corneal stroma by embryonic avian corneal epithelium. The corneal epithelium is isolated from the embryo and placed on a collagenous substratum (A). Within 6 hr, the basal cell surface is completely smooth, and by 24 hr, the epithelium has produced a new corneal stroma (B). The graph on the right expresses stromal synthesis as cpm [^{3}H]proline in collagen produced by epithelia in 24 hr. The effects of lens capsule containing type IV collagen, rat tail tendon gels containing type I collagen, and purified chondrosarcoma type II collagen are the same (C). Epithelia grown on noncollagenous substrata (Millipore filter, plastic, glass, agar, albumin) produced a small "baseline" amount of collagen. Morphologically recognizable corneal stroma could not be detected in these latter cultures, and the basal surface of the epithelium continued to bleb.

ECM, the corneal epithelium flattened its basal surface and produced collagenous corneal stroma *in vitro* (Fig. 12-7) whereas on plastic it did not. Meier and Hay in 1974 compared the effect of a variety of collagenous substrata on corneal epithelial synthesis as judged by incorporation of [^{3}H]proline into new stroma and found that types I, II, and IV were equally effective in stimulating stroma production (Fig. 12-7C; see Hay, 1981).

Sugrue and Hay (1981) studied the effect of solubilized ECM molecules on corneal epithelial differentiation. When the epithelium is isolated from the embryonic ECM by enzymes, its basal surface blebs and continues to do so if it is placed on a Millipore filter *in vitro* (Fig. 12-8, left). When purified collagens, fibronectin, or laminin are added to the dish, they pass through the filter to contact the basal epithelial surface, which flattens and reorganizes its basal actin cortex (Fig. 12-8, right). The ECM-stimulated epithelia (Fig. 12-8, right) produce over twice as much collagen as epithelia that continue to bleb (Fig. 12-8, left), as judged by incorporation of [^{3}H]proline into OH-proline in new collagen (Sugrue and Hay, 1986). It does not polymerize on the filters (Fig. 12-8); presumably, a polymerized ECM substratum is necessary to prime the assembly of stroma (Fig. 12-7). The interaction with purified collagen, laminin, and fibronectin molecules occurs in the absence of serum and is not likely to involve ECM-bound growth factors. The data are consistent with the hypothesis (see Section 7) that ECM interaction with the epithelial actin cytoskeleton steps up the synthesis of specific differentiated products (collagen in this case); the interaction is mediated by ECM receptors (Sugrue, 1987).

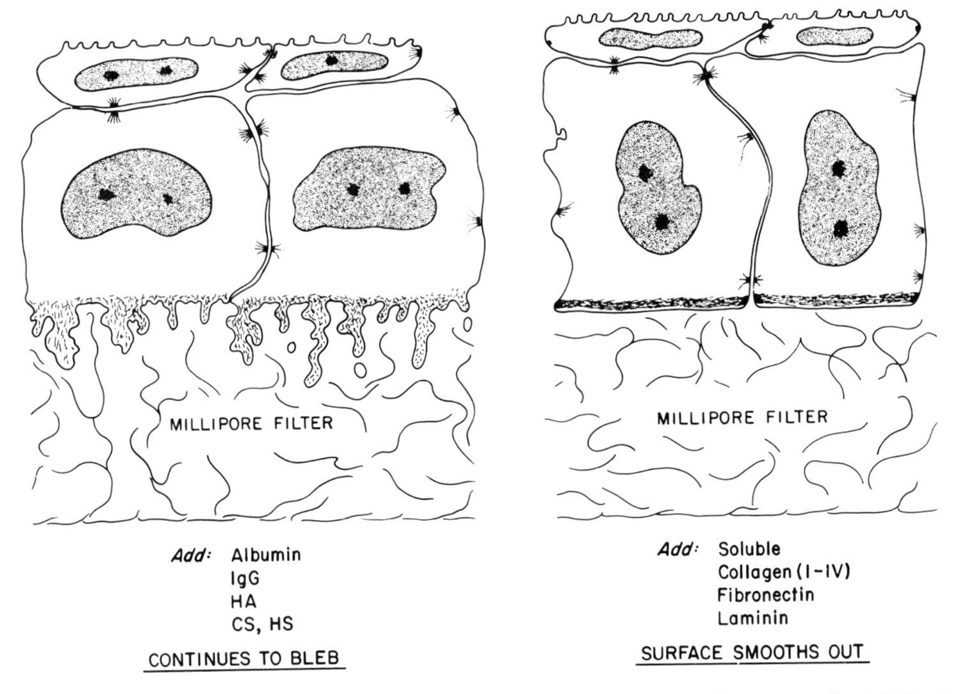

Figure 12-8. Diagrammatic summary of experiments testing the effect of a variety of solubilized ECM and other molecules on the basal corneal epithelial cell surface. The isolated epithelium is placed on a Millipore filter. The basal surface continues to bleb after addition of soluble albumin, IgG, HA, chondroitin and heparan sulfates in various concentrations. The basal surface smooths out and the basal actin cytoskeleton reorganizes in 4–6 hr after adding 10–100 μg/ml of solubilized collagens, fibronectin, or laminin. The ECM molecules are interacting via receptors in the plasmalemma with actin-binding proteins and actin in the cell cortex. (From Sugrue and Hay, 1981.)

4.2. Mesenchymal–Epithelial Interactions

The best studied example of a mesenchymal –epithelial interaction involving ECM is the effect of mesenchyme on branching of gland epithelium such as that of the salivary gland (Grobstein, 1955; Bernfield and Banerjee, 1978; Spooner *et al.*, 1986). Evidence from *in vitro* studies indicates that mesenchyme promotes branching by removing GAG from the basal surface of the epithelium at the tips of developing lobes (Fig. 12-9). The GAG-poor areas are able to grow out and then new clefts develop. Clefts subsequently seem to be immobilized by collagen, which also stabilizes GAG (see Hay, 1981, and Toole, 1981, in the first edition of this book for more detailed review). The theory requires that considerable information be built into the epithelial surface. New clefts form on the epithelial outgrowths even in regions that are exposed to mesenchyme-derived hyaluronidases (Chapter 8). The fact that collagen derived from mesenchyme mainly accumulates in the clefts (see Wessells, 1977; Spooner *et al.*, 1986) also suggests local control of collagen polymerization.

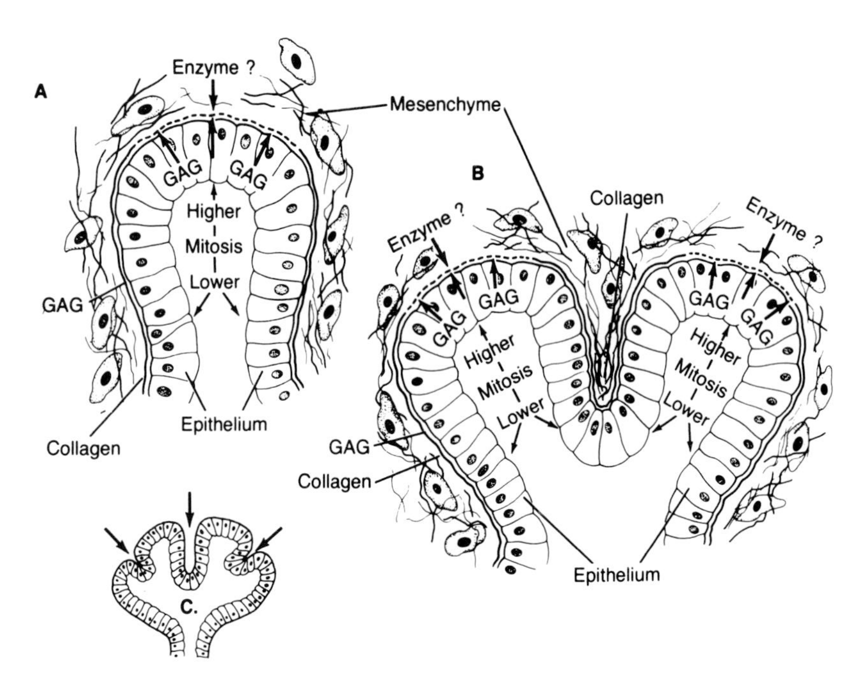

Figure 12-9. The effect of mesenchyme on branching of an epithelial gland is diagrammed here. A depicts an early stage. At this time, the mesenchyme produces hyaluronidase, removing GAG; these "bare" areas (dashed lines) of epithelium grow out. In B, a cleft has formed; in clefts, GAG and collagen are turned over slowly and seem to stabilize the cleft. Mitosis is higher at the outgrowing tips. The gland shown in B then takes on the appearance shown in C. (From Wessells, 1977.)

Thus, it can be said that the gland epithelium is "determined" to react with the mesenchyme products in a particular way and that this induction is "permissive," i.e., that the tissue already contains differentiated products and knows what to do; the correct ECM environment merely promoted the program. Indeed, all of the tissue interactions promoted by ECM in the embryo are probably permissive, if by an "instructive" interaction one means that a totally new program is turned on in a completely "undetermined" cell. The requirement of a particular gland epithelium for a specific mesenchyme during embryonic induction (see Hay, 1981) may mean merely that additional growth factors or other molecules bound to ECM are sometimes involved. Not all fibroblasts make exactly the same products (Hay, 1980).

The interaction of mesenchyme with developing muscle cells is not strictly a mesenchymal–epithelial interaction, but it needs to be mentioned because it was the classic work of Konigsberg and Hauschka (1965) that first drew attention to the possibility that the effect of mesenchyme in tissue interaction could be mediated by as mundane a molecule as collagen was thought to be. The effect of collagen produced by fibroblasts is to promote attachment and fusion of myoblasts, which then develop striated myofibrils. The collagen required is not specific, and fibronectin is needed (Turner, 1986).

Lipton (see Kühl *et al.*, 1984) observed that even if the exogenous ECM requirement for differentiation is met, the developing myotubes do not form basal laminae *in vitro* unless living fibroblasts are present. Von der Mark and co-workers demonstrated in cultures of mixed quail, chick, or mouse myoblasts and muscle fibroblasts, that the basal laminae contain fibroblast-derived type IV collagen and laminin (see Kühl *et al.*, 1984). They were able to do this because they had produced species-specific antibodies that recognized either the avian or the rodent product. TEM confirmed that muscle surfaces on which fibroblast products accumulate contain basal laminae. Sanderson *et al.* (1986) report, on the contrary, that muscle fibroblasts do not make BM. However, they had no species-specific antibodies to study mixed myoblast/fibroblast cultures. If production of type IV collagen and laminin by fibroblasts requires interaction with muscle, then synthesis of these products might not be detected in the isolated fibroblasts they studied.

Another example of mesenchymal ECM as a promotor of differentiation is the interaction of Schwann cells (neural crest mesenchyme) with neurons. Peripheral nerve cells cultured alone lack basal lamina and extracellular fibrils. If Schwann cells are added, they ensheath the nerve cell processes and both basal laminae and fibrils appear (Bunge *et al.*, 1982). In the absence of fibroblasts, a collagenous substratum is required for the Schwann cell–nerve fiber interaction to occur.

4.3. Epithelial–Mesenchymal Interactions

A well-studied example of the effect of epithelial-derived ECM on mesenchymal cell differentiation is the induction of cartilage in the sclerotome mesenchyme by the notochord and/or neural tube (Kosher and Church, 1975; Lash and Vasan, 1978). In culture on agar, sclerotome mesenchyme eventually differentiates into cartilage, but addition of notochord or neural tube, both of which produce collagen, promotes chondrogenesis in 2–3 days *in vitro*. In the absence of inducing tissue, type II collagen is slightly better in promoting chondrogenesis than is type I.

Certain other epithelia also promote chondrogenesis in mesenchyme. Pigmented neural retina induces cranial neural crest to enter a chondrogenic pathway by a matrix-mediated mechanism that may involve type II collagen (Thorogood *et al.*, 1986). Limb bud epithelium normally promotes outgrowth rather than chondrogenesis in limb mesenchyme, but added to tooth mesenchyme *in vitro*, it stimulates chondrogenesis by presumptive odontoblasts that would have formed bone (see von der Mark, 1980, and Hay, 1981, for review). Interestingly, presumptive odontoblasts *in vivo* express syndecan transiently during interactions with the normal oral epithelium (Chapter 9).

ECM has been particularly well characterized recently in an epithelial–mesenchymal interaction that is also an example of mesenchymal–epithelial transition (Section 5). In the embryo, the nephric duct gives rise to a ureteric epithelial bud that grows into nearby metanephric mesenchyme, where it in-

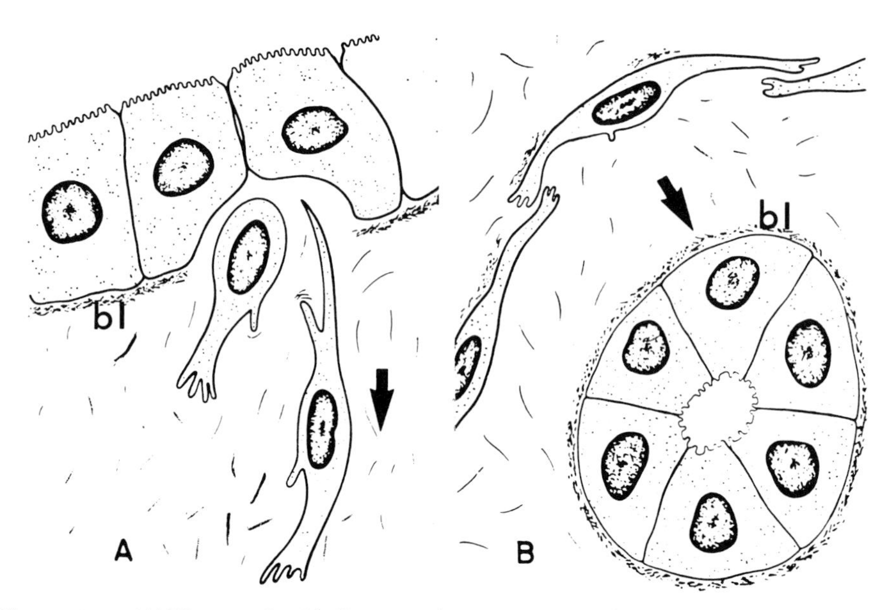

Figure 12-10. (A) Diagram of epithelium transforming to mesenchyme in the embryo. There is no basal lamina (bl) in the region where the new mesenchymal cells are emigrating. The cells extend new front ends containing filopodia into the ECM. (B) Diagram of primary mesenchyme (such as nephrogenic mesenchyme) transforming into epithelium. The cells begin to make laminin, which condenses into basal lamina (bl) that helps polarize the new epithelial cells. (From Hay, 1984.)

duces the formation of kidney tubules. The nature of the inducing molecule(s) is unknown, but it may be ECM in a form capable of moving through Nucleopore filters (Saxen, 1987). Apparently, certain other epithelia also secrete the same or similar inducing molecule(s), for embryonic spinal cord can substitute for ureteric bud as the inducer. Thus, the system is typically studied by setting up embryonic spinal cord transfilter to metanephric mesenchyme (Ekblom *et al.*, 1986; Saxen, 1987). One of the first changes detectable is expression of syndecan by the mesenchyme as it prepares to condense under the influence of the inductor (Vainio *et al.*, 1989). Later, syndecan is lost during maturation of the nephron. Collagen types I and III, and fibronectin disappear, and laminin and L-CAM appear, during mesenchymal condensation (Ekblom *et al.*, 1986; Ekblom, 1989). Antibodies to laminin A, but not to L-CAM, block the epithelial polarization (Fig. 12-10, right) that would now take place (Ekblom *et al.*, 1986; Gumbiner, 1990). Laminin A chain mRNA appears transiently at the time of polarization, whereas β chain mRNAs are expressed earlier (Ekblom *et al.*, 1990). Recognition of the laminin E8 site by an α integrin is involved in the polarization (Sorokin *et al.*, 1990).

Thus, epithelial ECM, syndecan, CAMs, and apical–basal polarity turn on during mesenchymal–epithelial transformation. Any one of these could play a "causative" role. However, it seems likely, in light of the discussion of epithelial–mesenchymal transformation to follow, that the inducer activates a master regulatory gene for the epithelial phenotype in the mesenchyme that turns on appropriate CAMs and ECMs.

5. Epithelial–Mesenchymal Transitions

In the preceding section, we discussed an example of a mesenchymal–epithelial transition, that of nephrogenic mesenchyme to kidney epithelium, and saw that it can be induced by a neighboring epithelium, the ureter. Most, if not all, transitions of this type (mesenchyme to epithelium) involve the primary mesenchyme. This mesenchyme, which is derived from the primitive streak epithelium, forms all of the meseodermal epithelia of the higher vertebrate, including notochord, somite, nephrogenic, and lateral mesoderm (Fig. 12-2).

Some of these mesodermal epithelia give rise to the definitive or secondary (Hay, 1968) mesenchyme that forms connective tissue and muscle. Ectoderm can also give rise to mesenchyme: the neural crest of the early embryo and, as we shall see, part of the palate mesenchyme of the older embryo. In this section, we will discuss a general mechanism for the transformation of epithelium to mesenchyme, introduce some specific examples, and examine the role of ECM in normal and ECM-induced epithelial–mesenchymal transformation. Mesenchymal–epithelial transformation (Section 4.2) is not discussed further here, but it is easy to imagine that its mechanism will be the reverse of that described in the section to follow.

5.1. The Mechanism of Epithelial–Mesenchymal Transition

For an epithelium to give rise to mesenchyme (Fig. 12-10, left), some of its component cells must lose apical–basal polarity and acquire the front end–back end polarity of the migrating mesenchymal cell. In simple epithelia, such as compose the early embryo, the elongating epithelial cell is still attached to adjacent epithelial cells when it begins to extend filopodia into the underlying matrix (Fig. 12-11A). BM if present (as around the somite) disappears in the area of the transformation, implying that the presumptive mesenchymal cells might secrete proteases (Chapter 8). However, a basal lamina is not deposited under many early epithelia destined to form mesenchyme, such as the dorsal surface of the neural tube (Martins-Green and Erickson, 1987) and epiblast of the primitive streak (Hay, 1968).

The filopodia-rich new front end undoubtedly contains newly synthesized actin. This actin is converted to actin cortex along the newly forming pseudopodium of the transforming epithelial cell (Fig. 12-11B). It seems likely that actin cortex is associated with new plasmalemma containing, among other molecules, newly synthesized ECM receptors. McClay and co-workers have shown that the epithelial cells of the echinoderm blastula that are destined to form mesenchyme change their cell surface proteins, lose affinity for them-

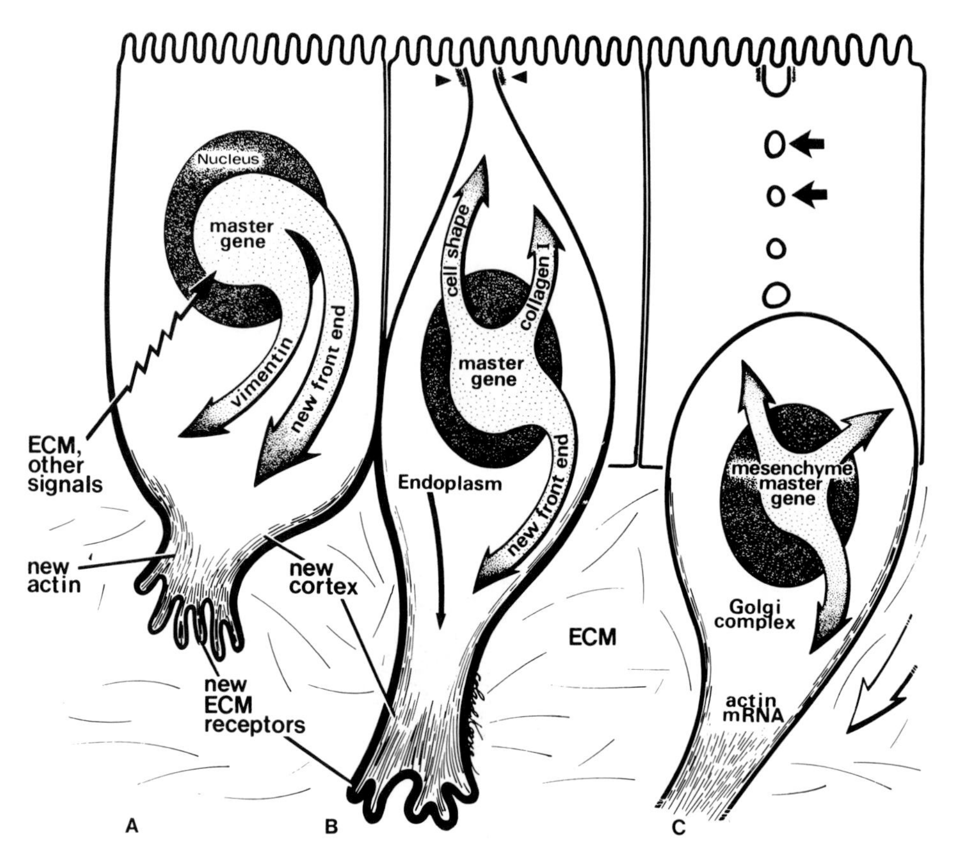

Figure 12-11. Diagram of a theory for transformation of epithelium to mesenchyme. (A) At the beginning of the transformation, the epithelial cell produces a new front end that is covered with filopodia as it moves out into the matrix. We suggest that a master gene turns on the mesenchymal program at this time. (B) The leading pseudopodium becomes longer due to the pushing of myosin-rich endoplasm that is sliding along the new actin cortex. (C) Finally, the mesenchymal cell breaks off its rear end in blobs (short arrows) that remain in the epithelium and then completes its emigration. (From Hay, 1989b.)

selves and for the vitelline membrane (hyaline), and gain affinity for fibronectin at the beginning of gastrulation (McClay and Ettensohn, 1987). Epithelial cells transforming to mesenchymal cells would also be expected to turn off epithelial CAMs during the process of losing contacts and junctions with their neighbors. Indeed, N-CAM and L-CAM, which are present in the embryonic epiblast, disappear from the primitive streak (Cunningham and Edelman, 1990).

The regulatory processes by which the transforming epithelial cell turns on the new front end mechanism is not understood, but must involve intracellular transport mechanisms for localizing the new plasmalemma, actin, and associated proteins in the former base of the cell. A dramatic change occurs in the IF cytoskeleton. Keratin, if present, is lost and vimentin becomes the principal IF protein (Franke *et al.*, 1982; Greenburg and Hay, 1988; Fitchett and Hay, 1989).

We have proposed that endoplasm attached to the vimentin IF, and possibly to the microtubules, slides along the actin cortex (Fig. 12-11) via a myosin-based motor into the new front end (Hay, 1985, 1989a,b; Section 6). Thus, all of the contents of the endoplasm (nucleus, cytoplasmic organelles) would progressively move out of the epithelium, leaving old plasmalemma behind, possibly in the form of blobs (arrows, Fig. 12-11C) still attached to old epithelial junctions and membranes. This mesenchymal motility theory will be developed further in Section 6.

The changes in cell shape and cytoskeleton during epithelial–mesenchymal transformation are so well coordinated that it is tempting to speculate that a master gene is turned on (Hay, 1989a,b; Fig. 12-11) for the mesenchymal phenotype that activates the new front end mechanism and the cell shape transition, as well as synthesis of vimentin, new ECM receptors, and mesenchymal ECM. At the same time, the epithelial program for CAMs, cell junctions, and apical–basal polarity should turn off. A group of candidates for master genes regulating muscle-specific synthetic events have now been isolated (Olsen, 1990), and, while it will be difficult to search for such genes in embryonic epithelial–mesenchymal transformations, the idea is worth further study.

A role for loss of CAMs and the epithelial cell–matrix receptor, syndecan, in epithelial–mesenchymal transformation is suggested by two recent lines of work. (1) It is possible to induce mesenchyme-like L cells to form epithelial-like aggregates by transfection with cDNA for L-CAM (Takeichi *et al.*, 1989). In the presence of calcium, L-CAM-induces L cells to develop basolateral (epithelial) polarity of fodrin and Na^{+}/K^{+}-ATPase (McNeil *et al.*, 1990). (2) Bernfield and co-workers transfected mammary epithelial cells with antisense syndecan DNA and found that the cells become elongated, losing epithelial characters and gaining mesenchymal cell shape. Thus, loss of syndecan, the HSPG cell surface molecule believed to stabilize epithelial morphology (Bernfield and Sanderson, 1990), results in at least partial assumption of mesenchymal morphology. While CAMs and syndecans may influence tissue phenotype, it is tempting to think that the master genes for tissue phenotype are upstream to the genes regulating cell surface molecules.

Evidence that ECM (Fig. 12-11) and associated molecules, such as growth factors, act as signals in epithelial–mesenchymal transformation has accumulated from studies injecting the embryo with anti-ECM molecules (Section 2), from studies of ECM-induced transitions in definitive epithelia (Section 5.3), and from studies of cardiac and neural crest mesenchyme to be described in the next section.

5.2. Examples of Epithelial–Mesenchymal Transformations in the Embryo

A well-studied example of epithelial–mesenchymal transformation occurs during cardiogenesis (Markwald *et al.*, 1984). The heart develops from an endothelial tube, the endocardium, that is surrounded by an ECM called cardiac jelly and a layer of coelomic epithelium called myocardium. During formation of the endocardial cushion mesenchyme, elongating cells detach from the endothelium and take on the characteristics of mesenchymal cells as they move into the cardiac jelly. By placing the endocardium on top of 3D collagen gel before the presumptive mesenchyme cells have emigrated, it has been possible to measure the rate of emigration and migration of the mesenchyme cells and to show that exogenous HA and the myocardium promote their emigration and migration (see Markwald *et al.*, 1984, for review).

The role of myocardium in inducing cardiac epithelial–mesenchymal transformation has been the subject of considerable recent work. Both extracts of ECM and conditioned medium from cultures of myocardium (see Sinning *et al.*, 1988) stimulate the formation of mesenchyme. TGF-β is one of the molecules found in the ECM that contributes to the regional specificity of the interaction (Potts and Runyan, 1989). Antisense TGF-β3 RNA inhibits cardiac epithelial–mesenchymal transformation (Potts *et al.*, 1991), although TGF-β is unable by itself to induce the transformation (Potts and Runyan, 1989). Protein kinase C stimulates, and inhibitors of it block, the transformation in response to myocardium (Runyan *et al.*, 1990). In response to myocardial conditioned medium, the endothelium of the atrioventricular canal (but not of the transition-incompetent ventricle) increases intracellular calcium. The transformation is inhibited by pertussis toxin, suggesting that this tissue interaction is mediated by G protein and one or more kinases (Runyan *et al.*, 1990). The cells are migrating on fibronectin and laminin; the YIGSR peptide of laminin, but not RGD, significantly inhibits movement of the cardiac mesenchymal cells on laminin (Davis *et al.*, 1989; Loeber and Runyan, 1990). They can also use an unidentified ECM substratum for the cell surface receptor, galactosyltransferase (Loeber and Runyan, 1990).

The abundant ECM surrounding the neural tube (Tosney, 1982) has been implicated in neural crest emigration and, as we have seen (Section 2), 3D ECM gels have been used to study crest emigration and migration *in vitro*, demonstrating that laminin and collagen, as well as fibronectin, can play important roles in the process. Without ECM such as fibronectin, crest does not emigrate on planar substrata (Rovasio *et al.*, 1983). It was believed that HA might promote crest cell migration (Toole, 1981). However, Anderson and Meier (1982) were unable to perturb crest migration in the embryo by injections of hyaluronidase, implying that HA plays a secondary, if any, role in neural crest movements.

Another variation of the relative roles of fibrillar ECM versus HA in epithelial–mesenchymal transformation is provided by the model of sclerotome formation. Here, the ventromedial wall of the somite swells as if it were accumulating HA and breaks up into mesenchymal cells without detectable fibrillar ECM being present (Solursh *et al.*, 1979; McCarthy and Hay, 1991). Later, the sclerotome mesenchyme migrates through nearby fibrillar ECM containing collagen and fibronectin, to surround the neural tube and notochord, where it gives rise to the vertebral column. Other mesenchymal cells arise in a more conventional style from the outer somite wall, the dermatome, and form connective tissue. Striated muscle develops from somite epithelium (myotome, Fig. 12-2) and also from myogenic mesenchymal cells that we suggested emigrate from the dermatome–myotome junction (Section 3.3).

A rather interesting epithelial–mesenchymal transformation has recently been documented in the palate (Fitchett and Hay, 1988), which may be an example of widespread use of this process to remove unwanted epithelia during embryonic remodeling (Trelstad *et al.*, 1982). The secondary palate of the rodent forms from two epithelium-lined palatal shelves that approach each other and fuse along the midline. Before fusion takes place, the outer layer (periderm) of the two-cell-thick epithelium sloughs at the midline and the basal epithelial layer on the medial half of each shelf contacts that of the opposite shelf, where fusion is accomplished by formation of desmosomes joining the two basal epithelial layers (Fitchett and Hay, 1989). Then, the cells extend filopodia through the disintegrating basal laminae, acquire vimentin IF, elongate, and move out into adjacent ECM as mesenchyme. Keratin, which contributes to the cohesion of the fused shelves by attaching to the epithelial desmosomes, disappears late in the transformation. It would seem that the gradual transformation of the fused midline epithelium to mesenchyme is less disruptive to palate continuity than would be removal of epithelial cells by massive cell death. The palate epithelium destined to transform into mesenchyme turns off syndecan and L-CAM (Fitchett *et al.*, 1990), a process that may or may not play a causative role in the transformation (see Section 5.1).

5.3. Activation of the Mesenchymal Program by Definitive Epithelia in 3D Gels

It came as a surprise to discover that definitive epithelia, such as that of the lens, cornea, thyroid, and notochord, could be induced to transform into mesenchyme by submerging them in 3D collagen gels (Greenburg and Hay, 1982). The fibrillar nature of the type I collagen gel seems to play the important role, for submerging the epithelia in BM gels does not induce the transformation (Greenburg and Hay, 1986, 1988). Moreover, the effect cannot be a simple chemical interaction, because epithelia placed on top of collagen gels or surrounded by solubilized collagen in the medium (Sugrue and Hay, 1981) do not give rise to mesenchymal cells. Whatever the mechanism will turn out to be, the phenomenon gives credence to the idea that it is relatively easy to switch on the mesenchymal gene regulatory program in epithelial cells. However, once transformed into mesenchyme, the cells seem to be unable to revert to epithelium, even when grown on BM (Zuk *et al.*, 1989a). Similarly, the secondary mesenchyme that forms from mesodermal epithelia in the embryo rarely if ever, reverts to epithelium. Only the primary mesenchyme derived from the primitive streak seems primed to turn on the epithelial master program during normal embryogenesis.

Anterior lens epithelium on lens capsule was dissected from either embry-

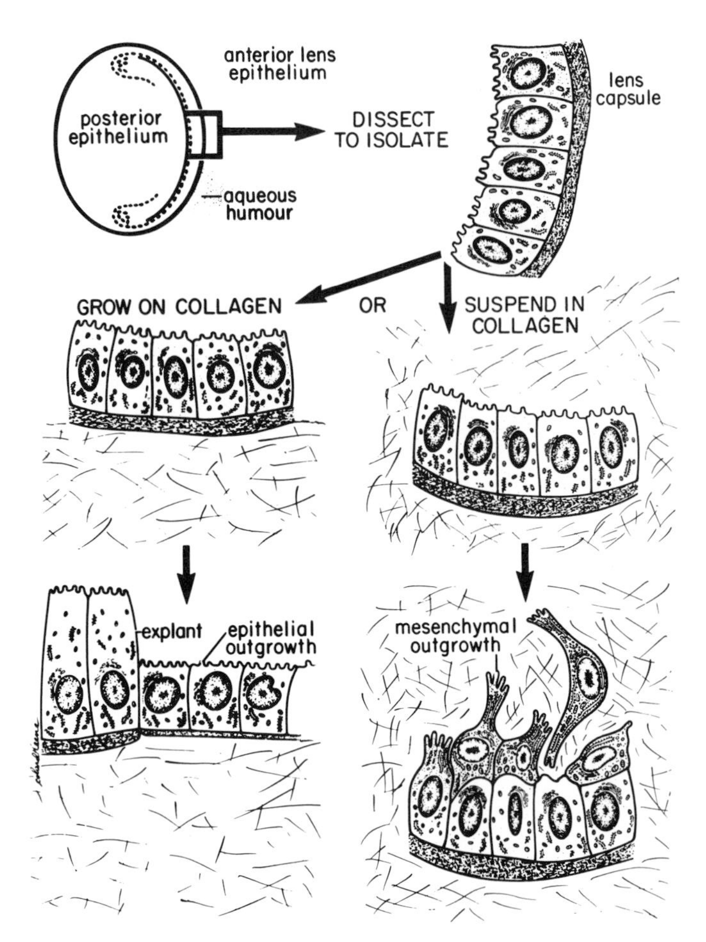

Figure 12-12. Diagram illustrating the fate of cells derived from adult or embryonic lens explants placed either on the surface (left) or suspended within (right) type I collagen gels. Explants of anterior lens epithelium with the associated basal lamina (lens capsule) are placed in two different environments. On the left, the explant is placed on the gel surface and a monolayer of epithelial cells spreads out over the gel surface and lays down a new basal lamina. In contrast, on the right, explants suspended within the gel form multiple cell layers. The apical cell surface acquires motility and the cells elongate, forming bipolar mesenchymelike cells that break free of the explant and migrate within the gel. (From Greenburg and Hay, 1986.)

onic or adult lens and inserted into gelling type I collagen (Fig. 12-12). No mesenchyme is normally present around the lens and none appears if the explant is placed on top of the collagen gel (Fig. 12-12, lower left). The transformation of epithelial cells to mesenchyme begins after about 3 days in culture and occurs from the former free surface in direct contact with the collagen gel (Fig. 12-12, lower right). Filopodia extend into the ECM, and the cells elongate as they emigrate from the epithelium, lose lens crystallins, and begin expressing type I collagen (Greenburg and Hay, 1986). They appear to downregulate $\alpha3$ and $\alpha6$ integrin subunits, which with $\beta1$ mediate epithelial contact (Section 3),

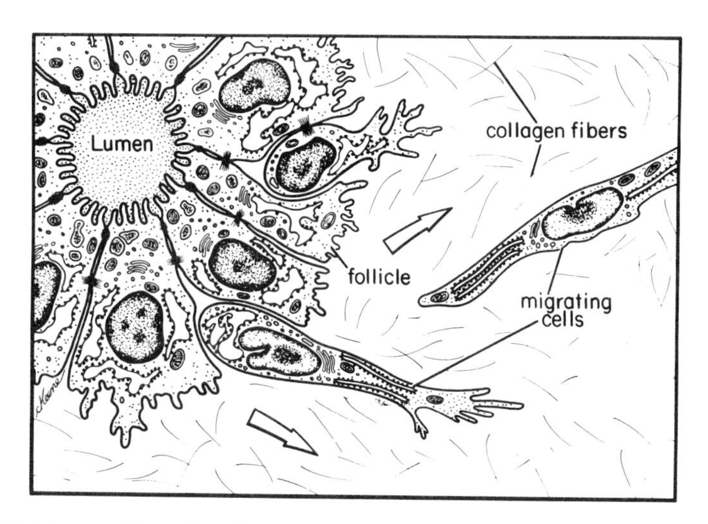

Figure 12-13. Diagram illustrating the events occurring during epithelial–mesenchymal transformation of adult thyroid follicles suspended in type I collagen gels. The basal surface is in contact with ECM and the mesenchymelike cells emerge from this surface. They have thin fibroblast-type RER, and they lose dilated RER and other thyroid characteristics. (From Greenburg and Hay, 1988.)

and to upregulate α5β1, the major fibronectin receptor (Zuk and Hay, 1990). They acquire mesenchymal RER and lose laminin and type IV collagen. Since the lens epithelium, unlike most epithelia, has a vimentin cytoskeleton, a keratin-rich epithelium, that of the thyroid, was studied to analyze IF cytoskeletal changes.

Thyroid follicles were isolated by enzyme treatment and suspended in type I collagen gels. Within a few days, the basal epithelial cells multilayer, lose thyroglobulin, and elongate into the matrix (Fig. 12-13). The newly formed mesenchymal cells contain keratin for a week after emigrating from the basal side of the follicle, a fact that proves their origin from the epithelium rather than from a mesenchymal contaminant (Fig. 12-14, upper right). Vimentin IF are synthesized by the transforming cells (Fig. 12-14, lower left) and coexist with keratin while it slowly turns over (Greenburg and Hay, 1988). The behavior of an MDCK epithelial cell line was also examined. Only by complex manipulations on the surface of the gel was the transformation to mesenchymelike cells induced, and the resulting fusiform cells retain both keratin and vimentin, as is characteristic of the MDCK cell. They do not turn on type I collagen (Zuk *et al.*, 1989b). Thus, the cultured cell line, unlike primary epithelial cultures, seems to have lost the ability to activate the complete mesenchymal genetic program, but retains the ability to give rise to cells that express front end–back end polarity and mesenchymal motility.

Cancer cells may also express front end–back end polarity and mesenchymal motility when emigrating out of an epithelium into ECM without turning on the entire mesenchymal program (see Hay, 1989a,b, 1990). Tumors, such as the spindle cell carcinoma (Battifora, 1976), may contain mixtures of

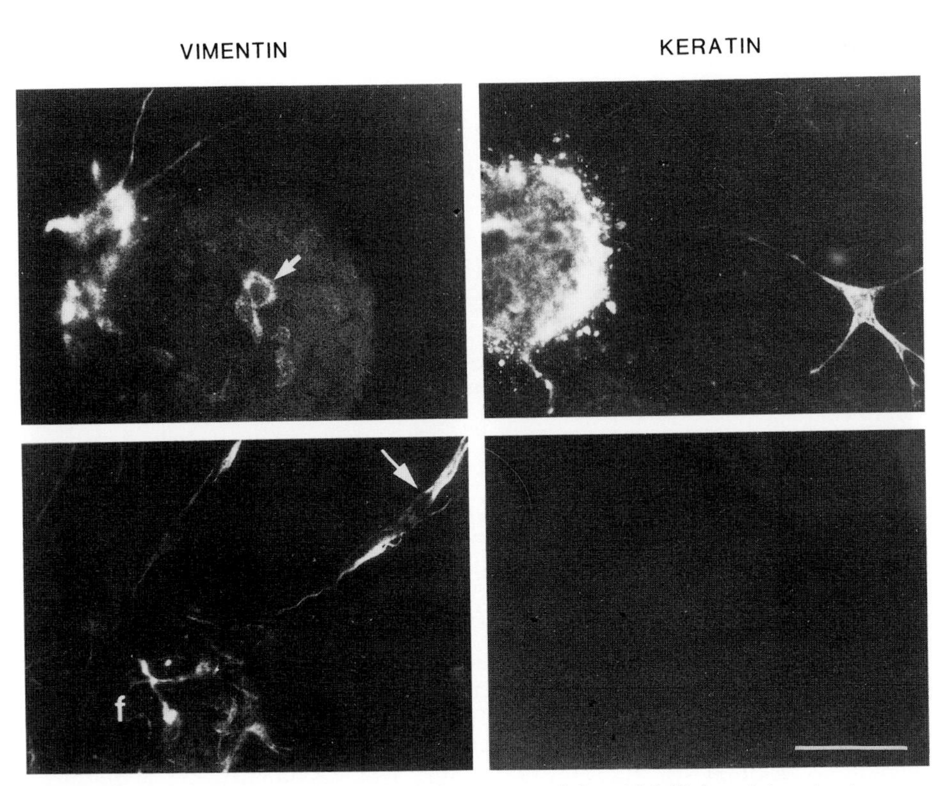

Figure 12-14. Immunofluorescence of whole mounts of thyroid follicles giving rise to mesenchymelike cells in collagen gels. Upper left, a follicle after 3 days in culture shows vimentin-rich, elongating cells preparing to emigrate into the ECM. One vimentin-rich cell (arrow) has not yet elongated. Lower left, a highly elongated mesenchymal cell is vimentin-rich. The epithelial follicle does not contain vimentin. Upper right, a mesenchyme-like cell migrating from a keratin-rich follicle (left) still contains keratin at 3 days. Lower right, at 22 days the mesenchymal cells contain no vimentin. Bar = 50 μm. (From Greenburg and Hay, 1988.)

malignant epithelioid and mesenchymelike cells. Some mammary tumor cell lines spontaneously transform into cells with fibroblast cell shape and motility that do not turn on type I collagen (Dulbecco *et al.*, 1981). TGF-α induces a rat bladder carcinoma cell line to lose desmosomes, acquire vimentin, and become fibroblastlike and very motile, but it is not known whether type I collagen synthesis is activated (Gavrilović *et al.*, 1990). In general, metastasis is associated with loss of E-cadherin and gain of fibroblast morphology by carcinoma cells (Frixen *et al.*, 1991).

Other pathologies in which epithelial cells turn on a mesenchymal program have been reported. In subcapsular cataracts, fibroblastlike cells that arise from lens epithelium produce striated collagen fibrils (Font and Brownstein, 1974). Polymorphonuclear leukocytes induce corneal endothelial cells to transform into fibroblasts that produce collagen types I, III, and V (Kay, 1986). These cells may cause a corneal pathology called posterior collagenous layer.

6. Mechanism of Mesenchymal Cell Migration

The mesenchymal cell moves slowly (1 μm/min) through the ECM, a rate that probably reflects the time it takes to synthesize new actin-rich front end for the myosin-rich endoplasm to slide into (Section 6.2). Nerve growth cones also move slowly, probably using a variant of the mesenchymal motility mechanism, while blood cells and macrophages move so fast that it has been said that synthesis could not be part of their motility mechanism (Zigmond, 1989). Other cell migrations that occur in the embryo include two dramatic examples of homing to a specific target: primordial germ cells to the gonad and nephric duct epithelium to the cloaca. Little is known about the role that ECM might play in their motility, except that fibronectin is present in the germ cell pathway (see Hynes, 1981, 1990).

6.1. General Rules for Mesenchymal Cell Migration

Mesenchymal cells make strong attachments to ECM when moving on ECM-coated dishes or within collagen gels (Fukamizu and Grinnell, 1990; Hay, 1985). It has been proposed that ECM forms specific pathways (Section 2) that route the cells by *contact guidance* to their specific targets. While it is true, for example, that neural crest will follow a track of fibronectin or laminin (Newgreen, 1984), there seems to be nothing specific about the process: if crest cells are moved about in the embryo, they migrate on whatever ECM is available (Erickson, 1985). By making ECM that cells can migrate on available in the right places at the time of epithelia–mesenchymal transformations, the embryo assures that mesenchymal cells can move to relevant destinations. Interestingly, some ECM (perinotochordal) cannot be invaded by crest cells (Newgreen *et al.*, 1986).

Along with the concept of contact guidance, the concepts of *chemotaxis and haptotaxis* have evolved, the one invoking a soluble chemical gradient toward a target and the other a fixed structural gradient. Clearly, white blood cells use chemotaxis to find wounds, and mesenchymal cells in culture move across a filter to the side richest in fibronectin or laminin (Graf *et al.*, 1987), but there are no known examples of soluble or insoluble chemical gradients in the embryo. *Target recognition molecules* alone could explain the homing of a neural crest cell to the adrenal medulla or the germ cell to the gonad.

Contact inhibition plays an important role both *in vivo* and *in vitro* in guiding cells away from crowded areas toward distant goals. As observed in 3D collagen gels or in the cornea itself (Bard and Hay, 1975), invading corneal fibroblasts can be seen to freeze the filopodia on the leading pseudopodium when they contact another cell (Fig. 12-15, interval 10–15). About 15 min later, a new leading pseudopodium appears elsewhere on the cell, and the cell moves off in another direction (Fig. 12-15, arrow, interval 30). The net result of the collisions is that the fibroblasts invading the cornea will move toward the central, less crowded regions of the stroma. As we shall see now, the leading

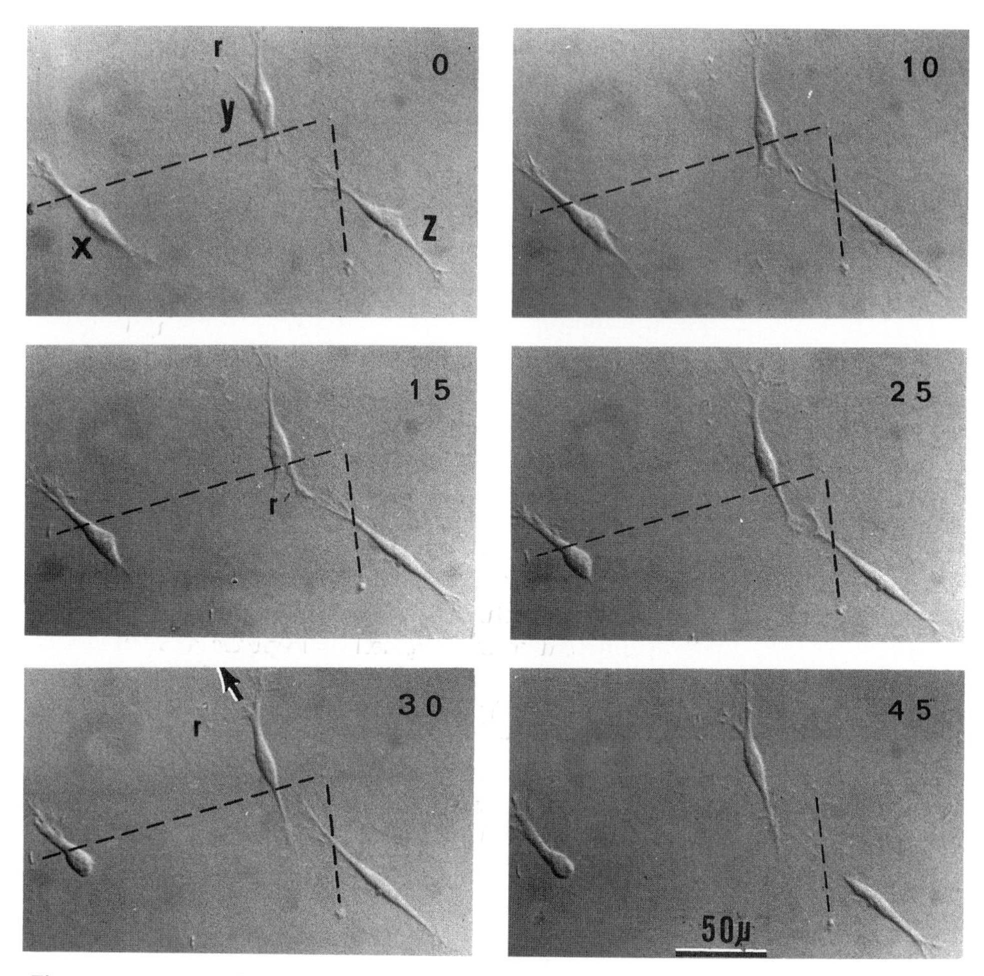

Figure 12-15. Nomarski light micrographs of corneal fibroblasts from a 6-day-old corneal stroma moving in a type I collagen gel. They were photographed at 5-min intervals (for the whole set, see Bard and Hay, 1975). The sequence shows the movement of a single cell on the left and contact inhibition between two cells on the right. The single cell (x) is bipolar at the start (time 0), but between time 10 and time 15, it begins to retract its trailing process and by time 25, it has become pear-shaped. About an hour later, this cell was once again elongated. The pair of cells y and z (time 0) initially moved toward each other with active filopodia on the leading ends. They contacted each other at time 15 and stopped moving (time 25). Cell y produces a new cell process (arrow, time 30) on what is now its leading edge. The dotted lines are in the same location in each frame. r', a cell process that retracts between time 15 and 25, r, location of a bleb left behind by retraction of cell process r (time 0). Bar = 50 μm. (From Bard and Hay, 1975.)

end of the mesenchymal cell is the center for insertion of new membrane and actin. The fact that contact with another cell results in translocation of this center elsewhere implies that efficient intracytoplasmic mechanisms exist to move actin mRNA and the membrane-generating organelle around. However, the mechanism of contact inhibition is not understood at this time.

6.2. The Fixed Cortex Theory of Mesenchymal Cell Motility

The evidence that the front end of the moving fibroblast is new comes mainly from recent experiments by S. J. Singer and his collaborators using virus budding to detect points of new membrane insertion. However, the likelihood that the front end of the migrating mesenchymal cell is new was appreciated some time ago (Abercrombie *et al.*, 1970). Bergmann *et al.* (1983) and Kupfer *et al.* (1987) infected fibroblasts with temperature-sensitive VSV (whole virus or just the DNA template for G protein), moved to the permissive temperature for budding, and fixed the cells at short intervals thereafter. G protein, the major component of the virus coat, was localized by immunohistochemistry to first appear on the leading edge of the cell. Moreover, the Golgi apparatus, the membrane-generating organelle, was also shown to be localized in the front end.

Concomitant with the above studies, Robert Singer and colleagues localized actin mRNA to the front end of moving mesenchymal cells using *in situ* hybridization. Antisense RNA or cDNA hybridized to mRNA is identified in the cytoplasm of fixed, permeabilized cells by its radioactivity (autoradiograms) or by histochemical criteria (Lawrence and Singer, 1986). The intracellular factory (mRNA) producing actin is located just proximal to the ruffles and/or filopodia on the front end of moving cells. Filopodia and ruffles are likely to be the center for polymerization of G-actin into the F-actin that will compose the filamentous cell cortex (Fig. 12-16). Filopodia contain only F-actin (and associated proteins such as talin) as judged by immunohistochemistry, phalloidin staining, and TEM (Fig. 12-16C); they do not stain for myosin, and the filopodial cell surface is rich in β1 integrins (Tomasek *et al.*, 1982; Tomasek and Hay, 1984; Daniels and Hay, 1990). The same is probably true for ruffles, the planar substratum equivalent of filopodia.

Now the task is to reconcile these two sets of data, the fact that moving cells (1) make very firm ECM attachments and (2) are constantly creating a new front end. No existing motility theory comes to grips with the problem (see review by Bilozur and Hay, 1988). Cells on ECM-coated dishes are imagined to attach and reattach to (walk along) the substratum and the new plasmalemmal proteins appearing on the front end are envisioned to move to the rear end for possible recirculation (Singer and Kupfer, 1986).

We have proposed a new theory (Hay, 1985, 1989a,b) of mesenchymal cell movement, which states (1) that the cell membrane and actin cortex (ectoplasm) are stably fixed to ECM by receptors and (2) that the myosin-rich endoplasm slides along the fixed actin into the newly forming front end, breaking off ectoplasm behind. Thus, the points labeled 1, 2 in Fig. 12-17 would appear to move backwards when, in fact, they do not move. The rate of flow of the endoplasm forward would be limited by the rate of synthesis and polymerization of new actin in the front end. Myosin in the endoplasm would need to attach to IF and/or microtubules encompassing the cell organelles in order to transpose the organelles forward by moving along the actin cortex.

What happens to the rear end? It is well known that fibroblasts thin their rear ends and then recoil their tails rapidly from time to time, becoming pear-

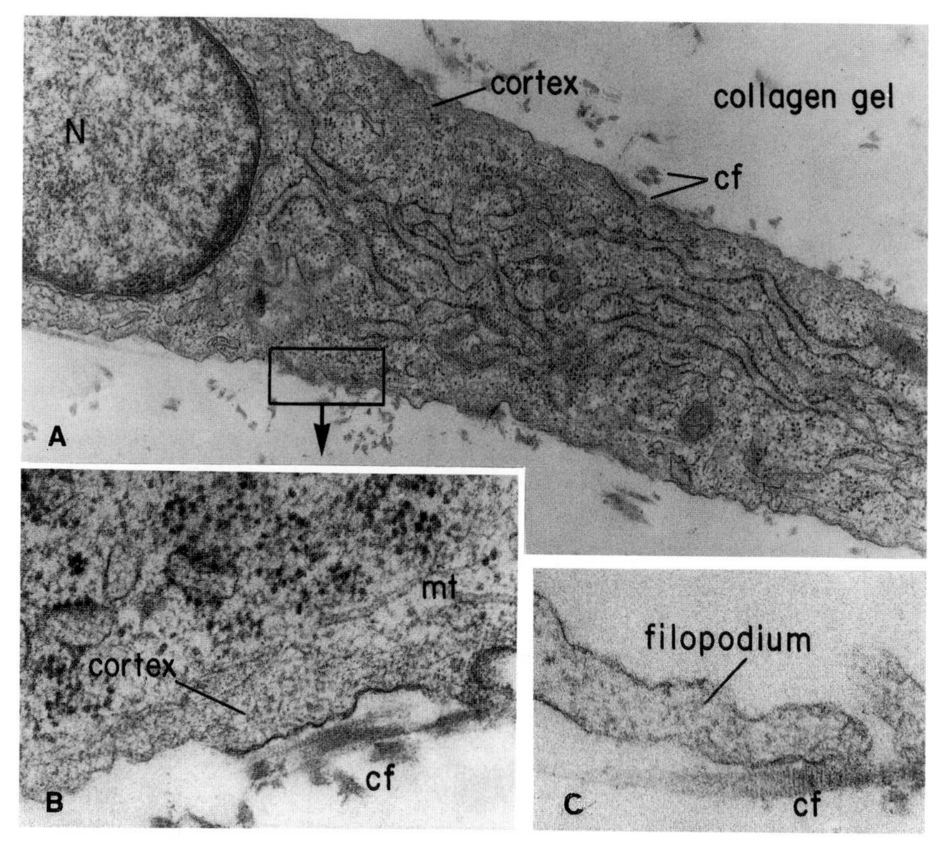

Figure 12-16. Electron micrographs of fibroblasts in 3D collagen gel containing collagen fibrils (cf). (A) At low power, the actin cortex can be seen all around the cell. (B) Higher-power view of the cortex revealing its filamentous nature and association with microtubules (mt). (C) In filopodia, the only visible component is a fine filamentous meshwork of actin. Bar = 1 μm (A), 200 nm (B), and 100 nm (C). (From Tomasek *et al.*, 1982.)

shaped (see cell x, Fig. 12-15, 0 and 30). Pieces (blobs) of cytoplasm bound to ECM break off during the recoil and are left behind in the matrix (points 1, 2, Fig. 12-17C). The blobs contain actin, talin, and integrin, but no myosin (Daniels and Hay, 1990). Moreover, the force of the pull against the matrix not only breaks off parts of the rear end, but also reorganizes the ECM and "contracts" the gel (Fukamizu and Grinnell, 1990; Harris *et al.*, 1981).

This theory could also explain mesenchymal cell emigration from an epithelium, as we have suggested (Section 5.1). It also has the advantage of using a well-established sliding filament mechanism of actin–myosin movement; the actin filaments of the cortex need only be oriented with barbed ends forward. Indeed, Huxley (1973) speculated that in nonmuscle cell motility, myosin does slide past actin attached to the cell membrane. However, he did not fully realize the plasmalemma would have to be fixed to the exterior and myosin to the cell

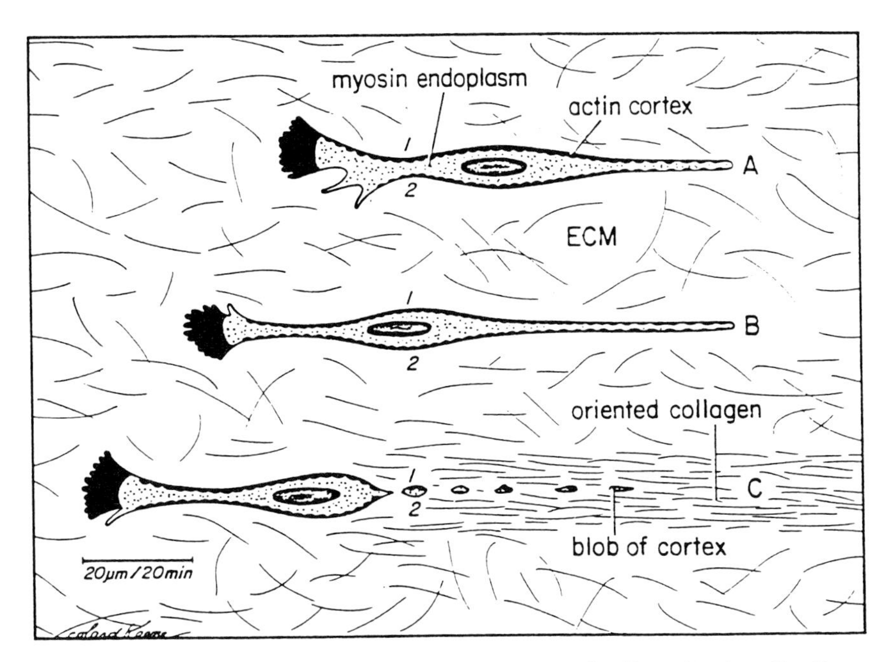

Figure 12-17. Diagram of the fixed cortex theory of mesenchymal cell motility in a 3D collagenous matrix. (A) Two actin–integrin–ECM complexes are labeled 1 and 2. (B) These complexes do not move and are left behind as myosin-rich endoplasm slides into the new front end on the left. (C) At periodic intervals, the rear end retracts (recoils) and pieces of actin cortex, such as 1, 2, are left behind in blobs in the matrix. The pull on the matrix orients (contracts) the collagen. (From Hay, 1989a.)

organelles for this mechanism to move the whole endoplasm forward. Against the theory is the observation that photobleached spots in the actin cortex of fibroblasts *in vitro* move toward the nucleus (Wang, 1985). However, these *in vitro* studies viewed fibroblasts on planar substrata, whose artifactual upper surface is not fixed and is known to move in directions irrelevant to migration (Abercrombie *et al.*, 1970). The work needs to be repeated bleaching only the part of the cell attached to ECM.

7. Interaction of the Actin Cytoskeleton with ECM

The fixed cortex theory of mesenchymal cell movement requires that actin interact directly, or indirectly via actin-binding proteins, with ECM receptors that fix actin to the ECM. Two lines of evidence for actin–ECM interaction were obtained in 1977–1978 by Mautner and Hynes, who showed that destruction of stress fibers by cytochalasin releases fibronectin from the cell surface, and by Hynes and Destree, who showed by immunohistochemistry that stress fibers and newly synthesized fibronectin are colocalized in cells cultured on planar substrata (see Hynes, 1981). While the stress fibers and focal contacts that

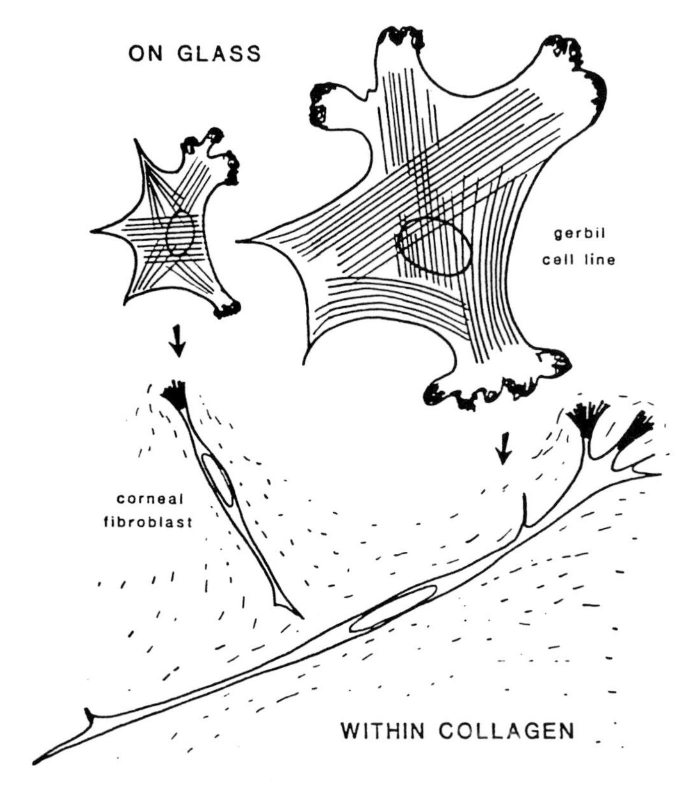

Figure 12-18. When fibroblasts are grown on planar substrata, they flatten, ruffle, and form bundles of actin, myosin, and associated proteins called "stress fibers." These attach to focal contacts (not shown) with the substratum. Within 3D matrix, the same fibroblasts elongate and develop filopodia on the leading edge. (From Hay, 1985.)

develop in fibroblasts over time on planar substrata (Fig. 12-18) are artifacts that rarely occur *in vivo* (Tomasek *et al.*, 1987), they are useful for showing that actin and actin-binding proteins can colocalize with both integrins and ECM (Burridge *et al.*, 1988). In mesenchymal cells that do not develop stress fibers on planar substrata (Duband *et al.*, 1986) or in 3D gels (Daniels and Hay, 1990), the actin cortex goes all around the cell and integrin cell surface staining is diffuse. In mammary epithelial cells, actin and syndecan colocalize (Bernfield and Sanderson, 1990).

A third line of evidence uses cross-linking agents to preserve actin–ECM complexes for extraction and characterization. By this method, direct data have been gathered supporting the idea that $\alpha5\beta1$ integrin interacts with cytoplasmic actin via talin, vinculin, and α actinin in living fibroblasts (Takenaga *et al.*, 1990).

A fourth line of evidence can be classified as behavioral and involves observations on cells after removal and subsequent restoration of ECM. Embryonic corneal epithelium removed by enzyme treatment from BM and stroma, begins to bleb and loses its organized basal actin cortex (Fig. 12-19). Four hours

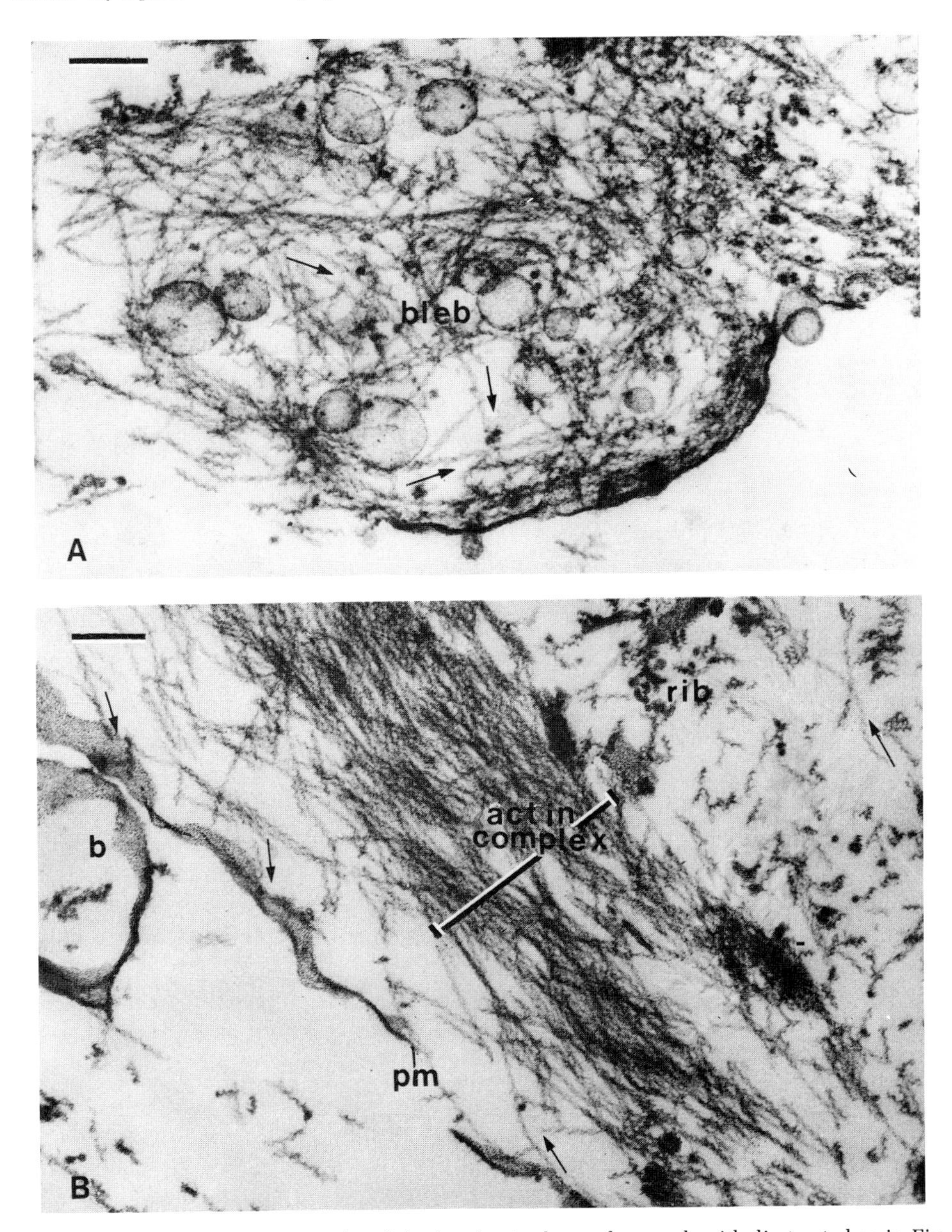

Figure 12-19. Electron micrographs of the basal cytoplasm of corneal epithelia treated as in Fig. 12-8 and labeled with S-1 fragments of heavy meromyosin to decorate actin filaments. The direction of the S-1 arrowheads is indicated for some of the filaments by small arrows. In a bleb (A), the actin filaments are disorganized. After addition of solubilized collagen, the basal surface smooths out and the actin complex that composes the basal cytoskeleton re-forms (B). The preparations were treated with Triton X-100 so the basal plasmalemma (pm) is disrupted (permeabilized) by the treatment, making it difficult to evaluate the exact relation of the actin filaments to the plasmalemma in this kind of a preparation. Some S-1 debris is present in the basal extracellular space (lower left, B); no visible ECM has polymerized. b, blister (remnant) of Triton-solubilized membrane; rib, ribosomes associated with actin filaments in the basal cytoplasm. Bars = 200 nm. (From Sugrue and Hay, 1981.)

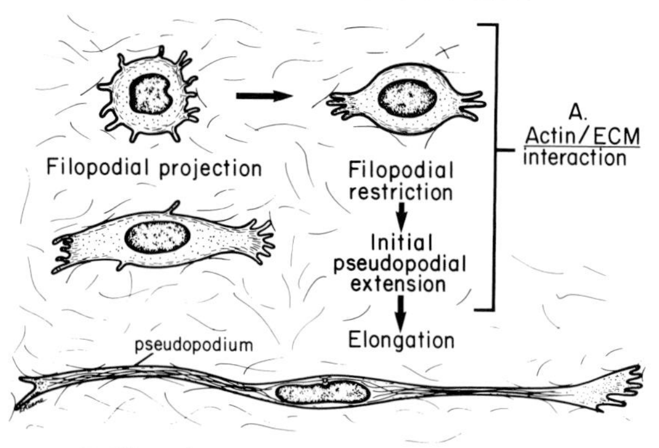

Figure 12-20. Diagram showing the stages in elongation of fibroblasts in 3D matrix. The isolated cell is round. In the ECM, it extends filopodia that become restricted to opposite ends of the cell, which is now bipolar in shape. This takes about 4 hr (A). Final elongation (B) requires both actin cortex and microtubules, and takes about 12 hr. (From Tomasek and Hay, 1984.)

after being placed on a collagenous substratum, it flattens its basal surface and reorganizes its basal actin cortex, and within 18 hr it produces a new stroma (Meier and Hay, 1974). Solubilized collagen, fibronectin, or laminin in the media will also reorganize the basal actin cortex (Fig. 12-19) and stimulate collagen synthesis (Sugrue and Hay, 1981, 1986), presumably via ECM receptors (Sugrue, 1987). When mesenchymal cells are removed from ECM by enzyme treatment, they become rounded and may bleb. On resuspension in collagen, they immediately extend filopodia and by 4 hr have reorganized the actin skeleton and become bipolar (Fig. 12-20; Tomasek *et al.*, 1982; Tomasek and Hay, 1984). This sequence does not take place in the presence of the actin-disrupting drug, cytochalasin D. Cytochalasin D also demolishes the ability of embryonic corneal epithelium to flatten its basal surface and step up collagen synthesis in response to ECM (Svoboda and Hay, 1987).

In Section 4, we put forth the idea that the stimulatory effect of ECM on collagen synthesis by embryonic corneal epithelium might be due to its effect on the cytoskeleton. Forming an organized actin cortex might well affect the organization of other cytoskeletal structures on which the mRNAs (and RER) of the cell need to reside to be active (Penman and co-workers, reviewed in Hay and Svoboda, 1988). The ECM-induced step-up of collagen synthesis by corneal epithelium does not involve a significant increase in collagen mRNA transcription (K. K. Svoboda, unpublished), which supports involvement of a post-transcriptional event such as attachment of mRNA to the cytoskeleton. In the next section, we consider the question of whether ECM interactions with cells can activate gene transcription.

8. ECM and Gene Transcription

Evidence for involvement of transcriptional regulation in certain ECM-mediated events comes from the work of Bissell and colleagues (see Lee *et al.*,

1984; Schmidhauser *et al.*, 1990). Mammary gland cells freshly isolated from mice by enzyme digestion are able to maintain casein synthesis *in vitro* when grown on floating collagen gels, whereas they produce little or no casein on plastic substrata coated (or not coated) with type I collagen. Emerman and co-workers (see Hay, 1981) showed that the cells producing casein on floating gels had regained epithelial morphology (cuboidal or columnar shape, as opposed to flattened on planar substrata). It was believed that the reason the floating collagen gel worked better than collagen attached to plastic was that the cells could contract the gel and thereby maintain normal epithelial shape (see Lee *et al.*, 1984, for review).

Subsequently, Bissell and co-workers reported a large increase in casein mRNA in the cultures on floating collagen gels, and an even larger increase in casein mRNA in mammary epithelium grown on BM gels, suggesting that regulation of casein production by ECM was at the transcriptional level. While some investigators argue for a posttranscriptional control mechanism, recent studies of a mouse mammary epithelial cell line transfected with β casein–CAT fusion genes show that BM and prolactin bring about maximal casein expression by causing DNA to transcribe high levels of casein mRNA (see Schmidhauser *et al.*, 1990). However, BM seems to stimulate whey acidic protein expression by a posttranscriptional mechanism (Chen and Bissell, 1989).

Since BM stimulates mammary cells to produce casein, Streuli and Bissell (1990) investigated the question of whether or not the cells on floating type I collagen gel produce basal lamina. They do. The BM-free cells on plastic, however, produce much larger amounts of mRNA for laminin, type IV collagen, and fibronectin than do cells on floating gels, suggesting that they are trying to compensate for their inability on plastic to polymerize a basal lamina. Thus, conditions favoring BM deposition enable the cells to create a microenvironment favoring their continued differentiation and function as a mammary epithelium (Streuli and Bissell, 1990). Such a microenvironment might contain growth factors bound to ECM whose effect is modulated by ECM (Stoker *et al.*, 1990; Taub *et al.*, 1990).

9. Summary and Concluding Remarks

In this chapter, we have discussed the morphogenetic roles that have been proposed for collagens (types I–IV), fibronectins, laminins, and, in preliminary fashion, for tenascin, drawing our major examples from vertebrate embryogenesis. In addition to performing vital skeletal functions that hold the body together, these and other ECM molecules described in this book communicate to embryonic and adult cells, information that promotes or demotes their differentiation. The molecules may interact directly on cells, or indirectly by harboring factors such as TGF-β. One of the major lines of direct communication is via effects on the organization of the actin cytoskeleton.

The mesenchymal cell migrating within ECM has an actin cortex all around its periphery, organized by the surrounding ECM via ECM receptors, such as integrins, that interact with actin and/or actin-binding proteins (Fig.

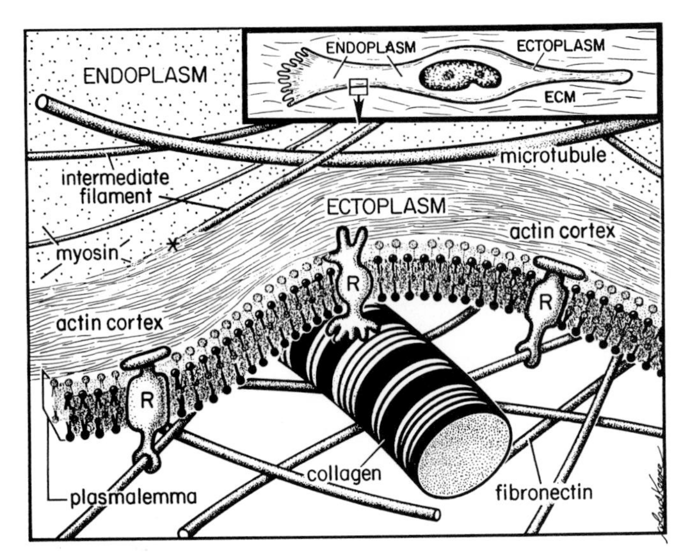

Figure 12-21. Diagram of the interaction of the ectoplasm (actin cortex, plasmalemma) of a fibroblast with ECM. The endoplasm is myosin-rich and contains intermediate filaments and microtubules to which the cell organelles are probably attached. Both microtubules and intermediate filaments contact the actin cortex and may end there (asterisk). R, plasmalemmal receptor for ECM that probably interacts with actin via actin-binding proteins.

12-21). It is possible that microtubules and IF (Fig. 12-21) receive information about ECM via the actin cortex. We propose that the interaction of the actin cortex with ECM via plasmalemma receptors is as stable in the case of the fibroblast as it is in epithelium. Epithelia placed on top of an ECM gel or on a filter containing purified collagen, fibronectin, or laminin, flatten the basal surface by organizing actin into a basal cortex. In order to move through matrix, epithelial cells transform into mesenchymal cells by developing new front ends into which the myosin-rich endoplasm moves. We propose that the migrating mesenchymal cell leaves much of its fixed actin cortex behind in blobs attached to the ECM.

The embryo of the higher vertebrate begins its development by forming two epithelial germ layers, the epiblast and hypoblast. Epithelial–mesenchymal transformation from the primitive streak creates the mesoderm which, together with the epiblast, produces the ECM through which the late-forming mesenchymal cells, including the neural crest, migrate. Both the epithelial cells and the new mesenchymal cells can interact with collagen, fibronectin, tenascin, and laminin of the embryonic matrix. Their cell surface must be viewed as in a dynamic equilibrium, containing a variety of actin-interactive ECM receptors, whose composition can change continuously as the environment changes.

The corneal epithelial cell responds to culture with purified ECM molecules by organizing its actin cortex and doubling its production of collagen. Such posttranscriptional effects on protein synthesis could be due to reorganization of mRNA on the cytoskeleton. Thus, attachment to ECM may promote the differentiation of epithelia by effects on organization of the synthetic machinery. In addition, ECM affects cell shape, cell elongation, and epithelia invagination in the early embryo by effects on the cytoskeleton.

The inductive effects between the two major tissue types that are clearly mediated by ECM are of the permissive type, i.e., they promote rather than initiate a line of differentiation. Remaining to be explored in detail are the complex matrix interactions with cells that involve non-ECM molecules bound to ECM. The myocardium-induced transformation of cardiac endothelium to mesenchyme requires TGF-β3, in addition to ECM. TGF-β bound to collagen can initiate bone morphogenesis and growth factors are required for promotion of kidney tubulogenesis and osteogenesis by BM. Signal transduction using G protein and kinases could be involved in these cases.

An example of an ECM-mediated effect activating transcription of genes is the effect of 3D collagen gels on definitive epithelia suspended within them. The cells turn on a mesenchymal "master gene" program and switch off epithelial programs. The mechanism is entirely unknown. Nor is the mechanism understood by which floating collagen gels or BM enhance the production of casein and casein mRNA by differentiated mammary cells that lose this function on plastic.

The challenge today to the cell biologist attempting to understand how molecules function during embryogenesis is the apparent complexity of the juxtacellular ECM environment. One current approach of great popularity is isolation of a small part of the system, such as a single gene, followed by disruption or enhancement of that single gene in transgenic embryos. Yet, we saw in this chapter an example of an inactivated collagen gene, the Mov 13, whose expression can be activated in certain tissues by an unknown factor(s) residing in cells or their environment. What emerges from this book is the picture of a cell creating and interacting with ECM using a variety of organelles to synthesize, sort, and secrete proteins. The organization and function of these organelles are controlled by the cytoskeleton, whose organization in turn is controlled by the ECM and other products that these cells and/or their neighbors have secreted. The action takes place at the cell surface, which is enriched not only in ECM receptors, but also in receptors involved in hormone and growth factor interactions. And the embryonic cells so busy producing and interacting with ECM may at the same time be preparing enzymes to eliminate ECM. How all of these things can take place simultaneously in the embryo is hard to comprehend.

The reason for the existence of this book, however, is not to overwhelm the student with complexity. It is the conviction of the authors that an understanding of ECM and cell–matrix interaction will clarify many aspects of cell and developmental biology, leading to a more accurate, and eventually more simplified and exciting, concept of a cell's relation to its environment in the embryo, in the adult, and in disease.

ACKNOWLEDGMENTS. The original research reported in this chapter was supported by National Institutes of Health Grant HD-00143.

References

Abercrombie, M., Heaysman, J. E. M., and Pegrum, S. M., 1970, The locomotion of fibroblasts in culture. III. Movement of particles on the dorsal surface of the lamella, *Exp. Cell Res.* **62:**389–398.

Adams, J. C., and Watt, F. M., 1990, Changes in keratinocyte adhesion during terminal differentiation: Reduction in fibronectin binding precedes α5β1 integrin loss from the cell surface, *Cell* **63:**425–435.

Altaba, A. R. I., and Melton, D. A., 1990, Axial patterning and the establishment of polarity in the frog embryo, *Trends Genet.* **6:**57–64.

Anderson, C. B., and Meier, S., 1982, The effect of hyaluronidase treatment on the distribution of cranial neural crest cells in the chick embryo, *J. Exp. Zool.* **221:**329–335.

Bard, J. B. L., and Hay, E. D., 1975, The behavior of fibroblasts from the developing avian cornea: Morphology and movement *in situ* and *in vitro, J. Cell Biol.* **67:**400–418.

Battifora, H., 1976, Spindle cell carcinoma: Ultrastructural evidence of squamous origin and collagen production by the tumor cells, *Cancer* **37:**2275–2282.

Bayne, E. K., Anderson, M. J., and Fambrough, D. M., 1984, Extracellular matrix organization in developing muscle: Correlation with acetycholine receptor aggregates, *J. Cell Biol.* **99:**1486–1501.

Bergmann, J. E., Kupfer, A., and Singer, S. J., 1983, Membrane insertion at the leading edge of motile fibroblasts, *Proc. Natl. Acad. Sci. USA* **80:**1367–1371.

Bernfield, M. R., and Banerjee, S. D., 1978, The basal lamina in epithelial–mesenchymal morphogenetic interactions, in: *Biology and Biochemistry of Basement Membrane* (N. A. Kefalides, ed.), pp. 137–148, Academic Press, New York.

Bernfield, M., and Sanderson, R. D., 1990, Syndecan, a morphogenetically regulated cell surface proteoglycan that binds extracellular matrix and growth factors, *Philos. Trans. R. Soc. London Ser. B* **327:**171–186.

Bilozur, M., and Hay, E. D., 1988, Neural crest migration in 3D extracellular matrix utilizes laminin fibronectin and collagen, *Dev. Biol.* **125:**19–33.

Bilozur, M., and Hay, E. D., 1989, Cell Migration into Neural Tube lumen provides evidence for the "fixed cortex" theory of cell motility, *Cell Motil. Cytoskeleton* **14:**469–484.

Bixby, J. L., Pratt, R. S., Lilien, J., and Reichardt, T. F., 1987, Neurite outgrowth on muscle cell surfaces involves extracellular matrix receptors as well as Ca^{2+}-dependent and -independent cell adhesion molecules, *Proc. natl. Acad. Sci. USA* **84:**2555–2559.

Bronner-Fraser, M., 1986, An antibody to a receptor for fibronectin and laminin perturbs cranial neural crest development *in vivo, Dev. Biol.* **117:**528–536.

Bronner-Fraser, M., 1988, Distribution and function of tenascin during cranial neural crest development in the chick, *J. Neural Sci. Res.* **21:**135–147.

Bronner-Fraser, M. (ed.), 1990, *The Role of the Cell Surface and Extracellular Matrix in Development*, W. B. Saunders Co., New York.

Bronner-Fraser, M., and Lallier, T., 1988, A monoclonal antibody against a laminin–heparan sulfate proteoglycan complex perturbs cranial neural crest migration *in vivo, J. Cell Biol.* **106:**1321–1329.

Bunge, M. B., Williams, A. K., Wood, P. M., Vitto, J., and Jeffrey, J. J., 1982, Comparison of nerve cell and nerve cell plus Schwann cell cultures, with particular emphasis on basal lamina and collagen formation, *J. Cell Biol.* **84:**184–202.

Burridge, K., Fath, K., Kelly, T., Nuckolls, G., and Turner, C., 1988, Focal adhesions: Transmembrane junctions between the extracellular matrix and the cytoskeleton, *Annu. Rev. Cell Biol.* **4:**487–525.

Chen, L.-H., and Bissell, M. J., 1989, A novel regulatory mechanism for whey acidic protein gene expression, *Cell Regul.* **1:**45–54.

Chiquet-Ehrismann, R., Mackie, E. J., Pearson, C. A., and Sakakura, T., 1986, Tenascin: An extracellular matrix molecule involved in tissue interactions during fetal development and oncogenesis, *Cell. Mol. Neurobiol.* **47:**131–139.

Crossin, K. L., Hoffman, S., Grumet, M., Thiery, J.-P., and Edelman, G., 1986, Site restricted expression of cytotactin during development of the chicken embryo, *J. Cell Biol.* **102:**1917–1930.
Cunningham, B. A., and Edelman, G. M., 1990, Structure, expression, and cell surface modulation of cell adhesion molecules, in: *Morphoregulatory Molecules* (G. M. Edelman, B. A. Cunningham, and J.-P. Thiery, eds., pp. 9–40, Wiley, New York.
Daniels, K. J., and Hay, E. D., 1990, Distribution of β1 integrins, matrix, actin, and actin-associated proteins during migration of elongated fibroblasts in 3D collagen gels, *J. Cell Biol.* **111:**140a.
Davis, L. A., Ogle, R. C., and Little, C. D., 1989, Embryonic heart mesenchymal cell migration in laminin, *Dev. Biol.* **133:**37–43.
Duband, J.-L., and Thiery, J.-P., 1987, Distribution of laminin and collagens during avian neural crest development, *Development* **101:**461–478.
Duband, J.-L. Rocker, S., Chen, W.-T., Yamada, K. M., and Thiery, J.-P., 1986, Cell adhesion and migration in the early vertebrate embryo: Location and possible role of the putative fibronectin receptor complex, *J. Cell Biol.* **102:**160–178.
Dulbecco, R., Henahan, M., Bowman, M., Okada, S., Battifora, H., and Unger, M., 1981, Generation of fibroblast-like cells from cloned epithelial mammary cells *in vitro:* A possible new cell type, *Proc. Natl. Acad. Sci. USA* **78:**2345–2349.
Ekblom, M., Klein, G., Mugrauer, G., Fecker, L., Deutzmann, R., Timpl, R., and Ekblom, P., 1990, Transient and locally restricted expression of laminin A chain mRNA by developing epithelial cells during kidney organogenesis, *Cell* **60:**337–346.
Ekblom, P., 1989, Developmentally regulated conversion of mesenchyme to epithelium, *FASEB J.* **3:**2141–2150.
Ekblom, P., Vestweber, D., and Kemler, R., 1986, Cell–matrix interactions and cell adhesion during development, *Annu. Rev. Cell Biol.* **2:**27–47.
Erickson, C. A., 1985, Control of neural crest cell dispersion in the trunk of the avian embryo, *Dev. Biol.* **111:**138–157.
Erickson, C. A., 1987, Behavior of neural crest cells on embryonic basal lamina, *Dev. Biol.* **120:**38–49.
Erickson, H. P., and Bourdon, M. A., 1989, Tenascin: An extracellular matrix protein prominent in specialized embryonic tissues and tumors, *Annu. Rev. Cell Biol.* **5:**71–92.
Fitch, J. M., Mentzer, A., Mayne, R., and Linsenmayer, T. F., 1989, Independent deposition of collagen types II and IX at epithelial-mesenchynal interfaces, *Development* **105:**85–95.
Fitchett, J., and Hay, E. D., 1989, Medial edge epithelium transforms to mesenchyme after embryonic palatal shelves fuse, *Dev. Biol.* **131:**455–474.
Fitchett, J. E., McAlmon, K. R., Hay, E. D., and Bernfield, M., 1990, Epithelial cells lose syndecan prior to epithelial–mesenchymal transformation in the developing rat palate, *J. Cell Biol.* **111:**145a.
Font, R. L., and Brownstein, S., 1974, A light and electron microscopic study of anterior subcapsular cataracts, *Am. J. Ophthalmol.* **78:**972–984.
Franke, W. W., Grund, C., Kuhn, C., Jackson, B. W., and Illmensee, K., 1982, Formation of cytoskeletal elements during mouse embryogenesis. III. Primary mesenchymal cells and the first appearance of vimentin filaments, *Differentiation* **23:**43–59.
Frixen, U. H., Behrens, J., Sachs, M., Eberle, G., Voss, B., Warda, A., Löchner, D., and Birchmeier, W., 1991, E-cadherin-mediated cell–cell adhesion prevents invasiveness of human carcinoma cells, *J. Cell. Biol.* **113:**173–185.
Fukamizu, H., and Grinnel, F., 1990, Spatial organization of extracellular matrix and fibroblast activity: effects of serum, transforming growth factor B, and fibronectin, *Exp. Cell Res.* **190:**276–282.
Gavrilović, J., Moens, G., Thiery, J.-P., and Jouanneau, J., 1990, Expression of transfected transforming growth factor α induces a motile fibroblast-like phenotype with extracellular matrixdegrading potential in a rat bladder carcinomal cell line, *Cell Regul.* **1:**1003–1004.
Graf, J., Iwamoto, Y., Sasaki, M., Martin, G. R., Kleinman, H. K., Robey, F. A., and Yamada, Y., 1987, Identification of an amino acid sequence in laminin mediating cell attachment, chemotaxis, and receptor binding, *Cell* **48:**989–996.
Greenburg, G., and Hay, E. D., 1982, Epithelia suspended in collagen gels can lose polarity and express characteristics of migrating mesenchymal cells, *J. Cell Biol.* **95:**333–339.

Greenburg, G., and Hay, E. D., 1986, Cytodifferentiation and tissue phenotype change during transformation of embryonic lens epithelium to mesenchyme-like cells, *Dev. Biol.* **115**:363–379.

Greenburg, G., and Hay, E. D., 1988, Cytoskeleton and thyroglobulin expression change during transformation of thyroid epithelium to mesenchyme-like cells, *Development* **102**:605–622.

Grobstein, C., 1955, Tissue interaction in morphogenesis of mouse embryonic rudiments *in vitro*, in: *Aspects of Synthesis and Order in Growth* (D. Rudnick, ed.), pp. 233–256, Princeton University Press, Princeton, N.J.

Grumet, M., Hoffman, S., Crossin, K. L., and Edelman, G. M., 1985, Cytotactin, an extracellular matrix protein of neural and nonneural tissues that mediates glia–neuron interaction, *Proc. Natl. Acad. Sci. USA* **82**:8075–8079.

Gumbiner, B., 1990, Generation and maintenance of epithelial cell polarity, *Curr. Opin. Cell Biol.* **2**:881–889.

Hadley, M. A., Weeks, B. S., Kleinman, H. K., and Dym, M., 1990, Laminin promotes formation of cord-like structures by Sertoli cells, *Dev. Biol.* **140**:318–327.

Halfter, W., Chiquet-Ehrismann, R., and Tucker, R. P., 1989, The effect of tenascin and embryonic basal lamina on the behavior and morphology of neural crest cells *in vitro*, *Dev. Biol.* **132**:14–25.

Harris, A. K., Stopak, D., and Wild, P., 1981, Fibroblast traction as a mechanism for collagen morphogenesis, *Nature* **290**:249–251.

Hay, E. D., 1968, Organization and fine structure of epithelium and mesenchyme in the developing chick embryo, in: *Epithelial–Mesenchymal Interactions* (R. Fleischmajer and R. E. Billingham, eds.), pp. 31–55, Williams & Wilkins, Baltimore.

Hay, E. D., 1980, Development of the vertebrate cornea, *Int. Rev. Cytol.* **63**:263–322.

Hay, E. D., 1981, Collagen and embryonic development, in: *Cell Biology of Extracellular Matrix*, pp. 379–409, Plenum Press, New York.

Hay, E. D., 1984, Cell–matrix interaction in the embryo: Cell shape, cell surface, cell skeletons, and their role in differentiation, in: *The Role of Extracellular Matrix in Development* (R. Trelstad, ed.), pp. 1–31, Liss, New York.

Hay, E. D., 1985, Interaction of migrating embryonic cells with extracellular matrix, *Exp. Biol. Med.* **10**:174–193.

Hay, E. D., 1989a, Extracellular matrix, cell skeletons, and embryonic development, *Am. J. Med. Genet.* **34**:14–29.

Hay, E. D., 1989b, Theory for epithelial–mesenchymal transformation based on the "fixed cortex" cell motility model, *Cell Motil. Cytoskeleton* **14**:455–457.

Hay, E. D., 1990, Epithelial–mesenchymal transitions, *Semin. Dev. Biol.* **1**:347–356.

Hay, E. D., and Revel, J.-P., 1969, *Fine Structure of the Developing Cornea*, Karger, Basel.

Hay, E. D., and Svoboda, K. K., 1988, Extracellular matrix interaction with the cytoskeleton, in: *Cell Shape: Determinants, Regulation and Regulatory Role* (W. D. Stein and F. Bronner, eds.), pp. 147–171, Academic Press, New York.

Huxley, H. E., 1973, Muscular contraction and cell motility, *Nature* **243**:445–449.

Hynes, R. O., 1981, Fibronectin and its relation to cellular structure and behavior, in: *Cell Biology of Extracellular Matrix* (E. D. Hay, ed.), pp. 295–334, Plenum Press, New York.

Hynes, R. O., 1990, *Fibronectins*, Springer-Verlag, Berlin.

Jaenisch, R., Harbors, K., Schnieke, A., Lohler, J., Chumakov, I., Jahner, D., Grotkopp, D., and Hoffman, E., 1983, Germline integration of Moloney murine leukemia virus at the Mov13 locus leads to recessive lethal mutation and early embryonic death, *Cell* **32**:209–216.

Kanemoto, T., Reich, R., Royce, L., Greatorex, D., Adler, S. H., Martin, R., Yamada, Y., and Kleinman, H. K., 1990, The identification of an amino acid sequence from the laminin A chain that stimulates metastasis and collagenase IV production, *Proc. Natl. Acad. Sci. USA* **87**:2279–2283.

Kay, E. P., 1986, Rabbit corneal endothelial cells modulated by polymorphonuclear leucocytes are fibroblasts, *Invest. Ophthalmol. Vis. Sci.* **27**:891–897.

Kleinman, H. K., McCarvey, M. L., Hassell, J. R., Star, V. L., Cannon, F. B., Laurie, G. W., and Martin, G. R., 1986, Basement membrane complexes with biological activity, *Biochemistry* **25**:312–318.

Konigsberg, I. R., and Hauschka, S. D., 1965, Cell and tissue interactions in the reproduction of cell type, in: *Reproduction: Molecular, Subcellular, and Cellular* (M. Locke ed.), pp. 243–290, Academic Press, New York.

Kosher, R. A., and Church, R. L., 1975, Stimulation of *in vitro* chondrogenesis by procollagen and collagen, *Nature (London)* **258**:327–330.

Kosher, R. A., and Solursh, M., 1989, Widespread distribution of type II collagen during embryonic chick development, *Dev. Biol.* **131**:558–566.

Krotoski, H. K., Domingo, C., and Bronner-Fraser, M., 1986, Distribution of a putative cell surface receptor for fibronectin and laminin in the avian embryo, *J. Cell Biol.* **103**:1061–1072.

Kühl, U., Öcalan, M., Timpl, R., Mayne, R., Hay, E., and von der Mark, K., 1984, Role of muscle fibroblasts in the deposition of type IV collagen in the basal lamina of myotubes, *Differentiation* **28**:164–172.

Kupfer, A., Kronebusch, P. J., Rose, J. K., and Singer, J., 1987, A critical role for the polarization of membrane recycling in cell motility, *Cell Motil. Cytoskeleton* **8**:182–189.

Lander, A. D., 1990, Mechanisms by which molecules guide axons, *Curr. Opin. Cell Biol.* **2**:907–913.

Larjava, H., Peltonen, J., Akiyama, S. K., Yamada, S. S., Gralnick, H. R., Vitto, J., and Yamada K. M., 1990, Novel function for β1 integrins in keratinocyte cell–cell interactions, *J. Cell Biol.* **110**:803–815.

Lash, J. W., and Vasan, N. S., 1978, Somite chondrogenesis *in vitro*. Stimulation by exogenous matrix components, *Dev. Biol.* **66**:151–171.

Lawrence, J. B., and Singer, R., 1986, Intracellular localization of messenger RNAs for cytoskeletal proteins, *Cell* **45**:407–415.

Lee, Y.-H., Parry, G., and Bissell, M. J., 1984, Modulation of secreted proteins of mouse mammary epithelial cells by the collagenous substrata, *J. Cell Biol.* **98**:146–155.

Lipton, B. H., and Jacobson, A. G., 1974, Experimental analysis of the mechanisms of somite morphogenesis, *Dev. Biol.* **38**:91–103.

Loeber, C. P., and Runyan, R. B., 1990, A comparison of fibronectin, laminin, and galactosyltransferase adhesion mechanisms during embryonic cardiac mesenchymal cell migration *in vitro*, *Dev. Biol.* **140**:401–412.

Loring, J. F., and Erickson, C. A., 1987, Neural crest cell migratory pathways in the trunk of the chick embryo, *Dev. Biol.* **121**:220–236.

Low, F. N., 1970, Interstitial bodies in the early chick embryo, *Am. J. Anat.* **128**:45–56.

Mackie, E. J., Thesleff, I., and Chiquet-Ehrismann, R., 1987, Tenascin is associated with chondrogenic and osteogenic differentiation *in vivo* and promotes chondrogenesis *in vitro*, J. Cell Biol. **105**:2569–2579.

Mackie, E. J., Tucker, R. P., Halfter, W., Chiquet-Ehrismann, R., and Epperlein, H. H., 1988, The distribution of tenascin coincides with pathways of neural crest cell migration, *Development* **102**:237–250.

Markwald, R. R., Runyan, R. B., Kitten, G. T., Funderburg, F. M., Bernanke, D. H., and Brauer, P. R., 1984, Use of collagen gel culture to study heart development: Proteoglycan and glycoprotein interactions during the formation of endocardial cushion tissue, in: *The Role of Extracellular Matrix in Development* (R. L. Trelstad, ed.), pp. 323–350, Liss, New York.

Martins-Green, M., and Erickson, C. A., 1987, Basal lamina is not a barrier to neural crest cell emigration: Documentation by TEM and by immunofluorescent and immunogold labelling, *Development* **101**:517–533.

Mayer, B. W., Hay, E. D., and Hynes, R. O., 1981, Immunocytochemical localization of fibronectin in embryonic chick trunk and area vasculosa, *Dev. Biol.* **82**:267–286.

McCarthy, R. A., and Hay, E. D., 1991, Collagen I, laminin, and tenascin in early avian embryo: ultrastructure and correlation with neural crest, *Int. J. Devel. Biol.* in press.

McClay, D. R., and Ettensohn, C. A., 1987, Cell adhesion in morphogenesis, *Annu. Rev. Cell Biol.* **3**:319–345.

McNeil, H., Ozzawa, M., Kemler, R., and Nelson, W. J., 1990, Novel function of the cell adhesion molecule uvomorulin as an inducer of cell surface polarity, *Cell* **62**:309–316.

Meier, S., 1979, Development of the chick embryo mesoblast. Formation of the embryonic axis and establishment of the metameric pattern, *Dev. Biol.* **73**:24–45.
Meier, S., and Hay, E. D., 1974, Control of corneal differentiation by extracellular materials. Collagen as a promoter and stabilizer of epithelial stroma production, *Dev. Biol.* **38**:249–270.
Mercurio, A. M., 1990, Laminin: Multiple forms, multiple receptors, *Curr. Opin. Cell Biol.* **2**:845–849.
Newgreen, D. F., 1984, Spreading of explants of embryonic chick mesenchymes and epithelia on fibronectin and laminin, *Cell Tissue Res.* **236**:265–277.
Newgreen, D. F., Scheel, M., and Kastner, V., 1986, Morphogenesis of sclerotome and neural crest in avian embryos. *In vivo* and *in vitro* studies on the role of notochordal extracellular material, *Cell Tissue Res.* **244**:299–313.
Olsen, E. N., 1990, MyoD family: A paradigm for development? *Genes Dev.* **4**:1454–1461.
Overton, J., 1977, Response of epithelial and mesenchymal cells to culture on basement lamella observed by scanning microscopy, *Exp. Cell Res.* **105**:313–323.
Patel, V. P., and Lodish, H. F., 1987, A fibronectin matrix is required for differentiation of murine erythroleukemia cells into reticulocytes, *J. Cell Biol.* **105**:3105–3118.
Potts, J. D., and Runyan, R. B., 1989, Epithelial–mesenchymal cell transformation in the embryonic heart can be mediated, in part, by transforming growth factor β, *Dev. Biol.* **134**:392–401.
Potts, J. D., Dagle, J. M., Walder J. A., Weeks, D. L., and Runyan, R. B., 1991, Epithelial–mesenchymal transformation of cardiac endothelial cells is inhibited by a modified antisense oligodeoxynucleotide TGFβ3, *Proc. Natl. Acad. Sci. USA* **88**:1516–1520.
Pouliot, Y. Holland, P. C., and Blaschuk, O. W., 1990, Developmental regulation of a cadherin during the differentiation of skeletal myoblasts, *Dev. Biol.* **141**:292–298.
Rickman, M., Fawcett, J. W., and Keynes, R. J., 1985, The migration of neural crest cells and the growth of motor axons through the rostral half of the chick somite, *J. Embryol. Exp. Morphol.* **90**:437–455.
Riou, J.-F., Shi, D.-L., Chiquet, M., and Boucaut, J.-C., 1990, Exogenous tenascin inhibits mesodermal cell migration during amphibian gastrulation, *Dev. Biol.* **137**:305–317.
Rovasio, R. A., DeLouvee, A., Yamada, K. M., Timpl, R., and Thiery, J.-P., 1983, Neural crest migration: Requirements for exogenous fibronectin and high cell density, *J. Cell Biol.* **96**:462–473.
Runyan, R. B., Potts, J. D., Sharma, R. V., Loeder, C. P., Chiang, J. J., and Bhalla, R. C., 1990, Signal transduction of a tissue interaction during embryonic heart development, *Cell Regul.* **1**:301–313.
Sanderson, R. D., Fitch, J. M., Linsenmayer, T. R., and Mayne, R., 1986, Fibroblasts promote the formation of a continuous basal lamina during myogenesis *in vitro*, *J. Cell Biol.* **102**:740–747.
Sanes, J. R., Schachner, M., and Covault, J., 1986, Expression of several adhesive macromolecules (N-CAM, L1, J1, NILE, uvomorulin, laminin, fibronectin, and a heparin sulfate proteoglycan) in embryonic, adult, and denervated adult skeletal muscle, *J. Cell Biol.* **102**:420–431.
Saxen, L., 1987, *Organogenesis of the Kidney*, Cambridge University Press, London.
Schmidhauser, C., Bissell, M. J., Myers, C. A., and Casperson, G. F., 1990, Extracellular matrix and hormones transcriptionally regulate bovine β-casein 5′ sequences in stably transfected mouse mammary cells, *Proc. Natl. Acad. Sci. USA* **87**:9118–9122.
Schwarz, M., Harbers, K., and Kratochwil, K., 1990, Transcription of a mutant collagen I gene is a cell type and stage-specific marker for odontoblast and osteoblast differentiation, *Development* **108**:717–726.
Singer, S. J., and Kupfer, A., 1986, The directed migration of enkaryotic cells, *Annu. Rev. Cell Biol.* **2**:337–365.
Sinning, A. R., Lepera, R. C., and Markwald, R. R., 1988, Initial expression of type I collagen in chick cardiac mesenchyme is dependent on myocardial stimulation, *Dev. Biol.* **130**:167–174.
Solursh, M., Fisher, M., Meier, S., and Singley, C. T., 1979, The role of extracellular matrix in the formation of the sclerotome, *J. Embryol. Exp. Morphol.* **54**:75–98.
Sorokin, L., Sonnenberg, A., Aumailly, M., Timpl, R., and Ekblom, P., 1990, Recognition of the

laminin E8 cell-binding site by an integrin possessing the α6 subunit is essential for epithelial polarization in developing kidney tubules, *J. Cell Biol.* **111**:1265–1273.

Spooner, B. S., Thompson-Pletscher, H. A., Stokes, B., and Bassett, K., 1986, Extracellular matrix involvement in gland branching, in: *The Cell Surface in Development and Cancer* (M. S. Steinberg, ed.), pp. 225–260, Plenum Press, New York.

Stoker, A. W., Streuli, C. H., Martins-Green, M., and Bissell, M. J., 1990, Designer microenvironments for the analysis of cell and tissue function, *Curr. Opin. Cell Biol.* **2**:864–874.

Streuli, C. H., and Bissell, M. J., 1990, Expression of extracellular matrix components is regulated by substratum, *J. Cell Biol.* **110**:1405–1415.

Sugrue, S. P., 1987, Identification of collagen binding proteins from embryonic chick corneal epithelial cells, *J. Biol. Chem.* **262**:3338–3343.

Sugrue, S. P., and Hay, E. D., 1981, Response of basal epithelial cell surface and cytoskeleton to solubilized extracellular matrix molecules, *J. Cell Biol.* **91**:45–54.

Sugrue, S. P., and Hay, E. D., 1986, The identification of extracellular matrix (ECM) binding sites on the basal surface of embryonic corneal epithelium and the effect of ECM binding on collagen production, *J. Cell Biol.* **102**:1907–1916.

Svoboda, K. K. H., and Hay, E. D., 1987, Embryonic corneal epithelial interaction with exogenous laminin and basal lamina is F-actin dependent, *Dev. Biol.* **123**:455–469.

Sweeney, T. M., Ogle, R. C., and Little, C. D., 1990, Laminin potentiates differentiation of PCC4 uva embryonal carcinoma into neurons, *J. Cell Sci.* **97**:23–31.

Takeichi, M., Hatta, K., Nose, A., Nagafuchi, A., and Matsunga, M., 1989, Cadherin-mediated cell sorting and axonal guidance, in: *The Assembly of the Nervous System* (L. T. Landmesser, ed.), pp. 129–136, Liss, New York.

Takenaga, K., Olden, K., and Yamada, K. M., 1990, Analysis of the interaction between the fibronectin receptor and cytoskeletal proteins in human fibroblasts. A chemical crosslinking study, *J. Cell Biol.* **111**:299a.

Tan, S.-S., Crossin, K. L., Hoffman, S., and Edelman, G. M., 1987, Asymmetric expression in somites of cytotactin and its proteoglycan ligand is correlated with neural crest cell distribution, *Proc. Natl. Acad. Sci. USA* **84**:7977–7981.

Taub, M., Wag, Y., Szczesny, T. M., and Kleinman, H. K., 1990, Epidermal growth factor or transforming growth factor α is required for kidney tubulogenesis in matrigel cultures in serum-free medium, *Proc. Natl. Acad. Sci. USA* **87**:4002–4006.

Thorogood, P., Bee, J., and von der Mark, K., 1986, Transient expression of collagen type II at epitheliomesenchymal interfaces during morphogenesis of the cartilaginous neurocranium, *Dev. Biol.* **116**:497–509.

Tomasek, J. J., and Hay, E. D., 1984, Analysis of the role of microfilaments and microtubules in the acquisition of bipolarity and the subsequent elongation of fibroblasts in hydrated collagen gels, *J. Cell Biol.* **99**:536–549.

Tomasek, J. J., Hay, E. D., and Fujiwara, K., 1982, Collagen modulates cell shape and cytoskeleton of embryonic corneal fibroblasts: Distribution of actin, α-actinin and myosin, *Dev. Biol.* **92**:107–122.

Tomasek, J. J., Schultz, R. J., and Haaksma, C. J., 1987, Extracellular matrix–cytoskeletal connections at the surface of the specialized contractile fibroblast (myofibroblast) in Dupuytren's disease, *J. Bone Jt. Surg.* **69A**:1400–1407.

Toole, B. P., 1981, Glycosaminoglycans in morphogenesis, in: *Cell Biology of Extracellular Matrix* (E. D. Hay, ed.), pp. 259–294, Plenum Press, New York.

Tosney, K. W., 1982, The segregation and early migration of cranial neural crest in the avian embryo, *Dev. Biol.* **89**:13–24.

Trelstad, R. L., Hayashi, A., Hayashi, K., and Donahoe, P. K., 1982, The epithelial–mesenchymal interface of the male rat Mullerian duct: Loss of basement membrane integrity and ductal regression, *Dev. Biol.* **92**:27–40.

Tucker, R. P., and Erickson, C. A., 1984, Morphology and behavior of quail neural crest cells in artificial three-dimensional extracellular matrices, *Dev. Biol.* **104**:390–405.

Turner, D. C., 1986, Cell–cell and cell–matrix interactions in the morphogenesis of muscle, in: *The*

Cell Surface in Development and Cancer (M. S. Steinberg, ed.), pp. 206–224, Plenum Press, New York.
Urist, M. R., Mikulski, A., and Leitz, A., 1979, Solubilized and insolubilized bone morphogenetic protein, *Proc. Natl. Acad. Sci. USA* **76**:1828–1832.
Vainio, S., Lehtonen, E., Jalkanen, M., Bernfield, M., and Saxen, L., 1989, Epithelial–mesenchymal interactions regulate the stage-specific expression of a cell surface proteoglycan, syndecan, in the developing kidney, *Dev. Biol.* **134**:382–391.
von der Mark, K., 1980, Immunological Studies on collagen type transition in chondrogenesis, in: *Immunological Approaches to Embryonic Development and Differentiation* (M. Friedlander, ed.), pp. 199–225, Academic Press, New York.
Vukicevic, S., Kleinman, H. K., Luyten, F. P., Roberts, A. B., Roche, N. S., and Reddi, A. H., 1991, Growth Factors in reconstituted basement membrane (matrigel) regulate the network formation of MC3T3-E1 osteoblastic cells, *J. Clin. Invest.* in press.
Wang, Y.-L., 1985, Exchange of actin subunits at the leading edge of living fibroblasts: Possible role of treadmilling, *J. Cell Biol.* **101**:597–602.
Wessells, N. K., 1977, *Tissue Interactions and Development*, Benjamin, New York.
Zigmond, S. H., 1989, Cell locomotion and chemotaxis, *Curr. Opin. Cell Biol.* **1**:80–86.
Zuk, A., and Hay, E. D., 1990, Changes in expression of the β1 integrin family occur during transformation of lens epithelium to mesenchyme in 3D collagen gels, *J. Cell Biol.* **111**:265a.
Zuk, A., Kleinman, H. K., and Hay, E. D., 1989a, Culture on basement membrane does not reverse the phenotype of lens derived mesenchymelike cells, *Int. J. Dev. Biol.* **33**:487–490.
Zuk, A., Matlin, K., and Hay, E. D., 1989b, Type I collagen gel induces Madin–Darby canine kidney cells to become fusiform in shape and lose apical–basal polarity, *J. Cell Biol.* **108**:903–920.

Index